26th aug, 2004
W. G Y.

Springer
Berlin
Heidelberg
New York
Barcelona
Budapest
Hong Kong
London
Milan
Paris
Santa Clara
Singapore
Tokyo

Alfio Quarteroni   Alberto Valli

Numerical Approximation of Partial Differential Equations

With 59 Figures and 17 Tables

Springer

Alfio Quarteroni
Dipartimento di Matematica
Politecnico di Milano
Piazza Leonardo da Vinci, 32
I-20133 Milano
Italy

Alberto Valli
Dipartimento di Matematica
Università di Trento
I-38050 Povo (Trento)
Italy

Cataloging-in-Publication Data applied for

Die Deutsche Bibliothek - CIP-Einheitsaufnahme

Quarteroni, Alfio:
Numerical approximation of partial differential equations :
with 17 tables / Alfio Quarteroni ; Alberto Valli. - 2., corr.
printing. - Berlin ; Heidelberg ; New York ; Barcelona ;
Budapest ; Hong Kong ; London ; Milan ; Paris ; Santa Clara ;
Singapore ; Tokyo : Springer, 1997
  (Springer series in computational mathematics ; 23)

  ISBN 3-540-57111-6
NE: Valli, Alberto:; GT

Second, Corrected Printing 1997

Mathematics Subject Classification (1991): 65Mxx, 65Nxx, 65Dxx, 65Fxx, 35Jxx, 35Kxx, 35Lxx, 35Q30, 76Mxx

ISSN 0179-3632
ISBN 3-540-57111-6 Springer-Verlag Berlin Heidelberg New York

Springer-Verlag Berlin Heidelberg New York
a part of Springer Science+Business Media

© Springer-Verlag Berlin Heidelberg 1994
Printed in Germany

Typesetting: Camera-ready copy produced from the authors using a Springer TEXmacro package
SPIN 11017424   46/3111 - 5 - Printed on acid free paper

A Fulvia e Tiziana

Preface

Everything is more simple than one thinks
but at the same time more complex than one can understand
Johann Wolfgang von Goethe

To reach the point that is unknown to you,
you must take the road that is unknown to you
St. John of the Cross

This is a book on the numerical approximation of partial differential equations (PDEs). Its scope is to provide a thorough illustration of numerical methods (especially those stemming from the variational formulation of PDEs), carry out their stability and convergence analysis, derive error bounds, and discuss the algorithmic aspects relative to their implementation.

A sound balancing of theoretical analysis, description of algorithms and discussion of applications is our primary concern.

Many kinds of problems are addressed: linear and nonlinear, steady and time-dependent, having either smooth or non-smooth solutions. Besides model equations, we consider a number of (initial-) boundary value problems of interest in several fields of applications.

Part I is devoted to the description and analysis of general numerical methods for the discretization of partial differential equations.

A comprehensive theory of Galerkin methods and its variants (Petrov-Galerkin and generalized Galerkin), as well as of collocation methods, is developed for the spatial discretization. This theory is then specified to two numerical subspace realizations of remarkable interest: the finite element method (conforming, non-conforming, mixed, hybrid) and the spectral method (Legendre and Chebyshev expansion).

For unsteady problems we will illustrate finite difference and fractional-step schemes for marching in time. Finite differences will also be extensively considered in Parts II and III in the framework of convection-diffusion problems and hyperbolic equations. For the latter we will also address, briefly, the schemes based on finite volumes.

For the solution of algebraic systems, which are typically very large and sparse, we revise classical and modern techniques, either direct and iterative with preconditioning, for both symmetric and non-symmetric matrices. A short account will be given also to multi-grid and domain decomposition methods.

Parts II and III are respectively devoted to steady and unsteady problems. For each (initial-) boundary value problem we consider, we illustrate the main theoretical results about well-posedness, i.e., concerning existence, uniqueness and a-priori estimates. Afterwards, we reconsider and analyze the previously mentioned numerical methods for the problem at hand, we derive the corresponding algebraic formulation, and we comment on the solution algorithms.

To begin with, we consider all classical equations of mathematical physics: elliptic equations for potential problems, parabolic equations for heat diffusion, hyperbolic equations for wave propagation phenomena. Furthermore, we discuss extensively advection-diffusion equations for passive scalars and the Navier-Stokes equations (together with their linearized version, the Stokes problem) for viscous incompressible flows. We also derive the equations of fluid dynamics in their general form.

Unfortunately, the limitation of space and our own experience have resulted in the omission of many important topics that we would have liked to include (for example, the Saint-Venant model for shallow water equations, the system of linear elasticity and the biharmonic equation for membrane displacement and thin plate bending, the drift-diffusion and hydrodynamic models for semiconductor devices, the Navier-Stokes and Euler equations for compressible flows).

This book is addressed to graduate students as well as to researchers and specialists in the field of numerical simulation of partial differential equations.

As a graduate text for Ph.D. courses it may be used in its entirety. Part I may be regarded as a one quarter introductory course on variational numerical methods for PDEs. Part II and III deal with its application to the numerical approximation of time-independent and time-dependent problems, respectively, and could be taught through the two remaining quarters. However, other solutions may work well. For instance, supplementing Part I with Chapters 6, 11 and most part of 14 may be suitable for a one semester course. The rest of the book could be covered in the second semester. Following a different key, Part I plus Chapters 8, 9, 10, 12, 13 and 14 can be regarded as an introduction to numerical fluid dynamics. Other combinations are also envisageable.

The authors are grateful to Drs. C. Byrne and J. Heinze of Springer-Verlag for their encouragement throughout this project. The assistence of the technical staff of Springer-Verlag has contributed to the final shaping of the manuscript.

This book benefits from our experience in teaching these subjects over the past years in different academical institutions (the University of Minnesota at Minneapolis, the Catholic University of Brescia and the Polythecnic of Milan for the first author, the University of Trento for the second author), and from students' reactions. Help was given to us by several friends and collaborators who read parts of the manuscript or provided figures or tables. In this connection we are happy to thank V.I. Agoshkov, Yu.A. Kuznetsov, D. Ambrosi, L. Bergamaschi, S. Delladio, M. Manzini, M. Paolini, F. Pasquarelli, L. Stolcis, E. Zampieri, A. Zaretti and in particular C. Bernini, P. Gervasio and F. Saleri.

We would also wish to thank Ms. R. Holliday for having edited the language of the entire manuscript. Finally, the expert and incredibly adept typing of the TEX-files by Ms. C. Foglia has been invaluable.

Milan and Trento   Alfio Quarteroni
May, 1994   Alberto Valli

In the second printing of this book we have corrected several misprints, and introduced some modifications to the original text.

More precisely, we have sligthly changed Sections 2.3.4, 3.4.1, 8.4 and 12.3, and we have added some further comments to Remark 8.2.1.

We have also completed the references of those papers appeared after 1994.

Milan and Trento   Alfio Quarteroni
December, 1996   Alberto Valli

Table of Contents

Part II. Approximation of Boundary Value Problems

Part III. Approximation of Initial-Boundary Value Problems

1. Introduction

Numerical approximation of partial differential equations is an important branch of Numerical Analysis. Often, it demands a knowledge of many aspects of the problem.

First of all, the physical background of the problem is required in order to understand the behaviour of expected solutions. This may often lead to the choice of convenient numerical methods.

Secondly, modern formulation of the problem based on the variational (weak) form ought to be considered, as it allows the search for generalized solutions in Hilbert (or Banach) functional spaces. Variational techniques yield a-priori estimates for the solution, which in turn indicate in which kind of norms any virtual numerical solution can be proven to be stable. Furthermore, results about smoothness of the mathematical solutions may suggest the numerical methodology to be used, and consequently, determine the kind of accuracy that can be achieved. The latter is pointed out from the error analysis.

Clearly, specific attention should be paid to the algorithmic aspects concerned with the choice of any numerical method.

This book aims at providing general ideas on numerical approximation of partial differential equations, although (obviously) not all possible existing methods will be considered. In this respect, we mainly focus on variational numerical methods for the discretization of space derivatives, and on finite difference and fractional-step methods for advancing, in time, unsteady problems.

Whenever possible, we present the unifying approach behind a-priori different numerical strategies, provide general theory for analysis and illustrate a variety of algorithms that can be used to compute the effective numerical solution of the problem at hand, taking into consideration its algebraic structure. Consequently, we try to avoid using technicalities (or tricks, or algorithms) that work only in very specific situations, or that are not sustained from a sound theoretical background. Some problems (and methods) are discussed on a case-to-case basis, but very often they are included in a single logical unit (say Chapter, or Section).

1.1 The Conceptual Path Behind the Approximation

We consider a great number of mathematical problems, and numerical methods for their solution. For the approximation of any given boundary value problem, we schematically illustrate in Fig. 1.1.1 the decision path that needs to be followed.

Level [1] is the boundary value problem at hand under its weak formulation accounting for the prescribed boundary conditions.

Level [2] provides the kind of discretization (or numerical method) that can be pursued in order to reduce the given problem to one having finite dimension. Of course, the strategy adopted will determine the structure of the numerical problem.

Throughout this book we mainly consider two kinds of discretization. The former is the Galerkin method, together with its remarkable variant, the Petrov-Galerkin method, which is based on an integral formulation of the differential problem. The second discretization we consider, is the collocation method, which is, instead, based on the fulfillment of the differential equations at some selected points of the computational domain. We then reformulate the collocation method under a generalized Galerkin mode, precisely combining the Galerkin approach with numerical evaluation of integrals using Gaussian formulae.

At a lower extent, we will address finite difference schemes for space discretization, especially for nonlinear convection-diffusion equations and for problems of wave propagation. For the latter we will also present the approach based on the finite volume method, which is very popular in computational fluid dynamics.

Finally, we will illustrate shortly the elementary principles of the domain decomposition method, an approach which offers the best promise for the parallel solution of large problems in the field of scientific computing.

Other approaches are often encountered in the literature as well, but they will only be addressed incidentally in this book.

Level [3] specifies the nature of the subspaces used in the approximation. Typically, we have piecewise-polynomial functions of low degree when using finite elements, and global algebraic polynomials of high degree for spectral methods. These two remarkable cases will be discussed and analyzed in some of their variants (mixed finite elements, Legendre and Chebyshev spectral collocation methods). The choice operated at this level determines the functional structure of the numerical solution, the kind of accuracy that can be achieved, besides affecting the topological form of the resulting algebraic system.

At level [4] the selection of convenient algorithms needs to be accomplished to solve the algebraic problem, exploiting, at most, the topological structure and the properties of the associated matrices. We illustrate all the important methods available nowadays for solving large scale symmetric and

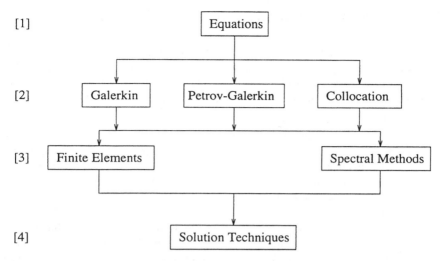

Fig. 1.1.1. The conceptual path behind the approximation

non-symmetric systems, including a short presentation of preconditioning and multi-grid techniques.

For initial-boundary value problems, between levels [1] and [2] the time-discretization needs to be carried out. Quite often, this is performed using finite difference divided quotients to approximate the time derivatives. However, other approaches can be pursued, especially for differential operators of complex structure. An instance is provided by strategies based on fractional-steps, that allow the splitting of the operator into subpieces that are advanced in time in an independent fashion. Both finite difference and fractional-step approaches will be extensively investigated in this book. We will briefly mention other strategies as well, e.g., discontinuous finite elements, characteristic Galerkin and Taylor-Galerkin schemes.

Let us finally underline that, in Part I of the book, the flow reported in Fig. 1.1.1 will be followed in the reverse order (i.e., bottom up). Proceeding backward from level [4] to level [2], we start from algorithms for solving linear algebraic systems (Chapter 2), then discuss the basic concepts of finite element (Chapter 3) and polynomial (Chapter 4) approximations. In Chapter 5 we introduce the discretization methods of level [2], as well as those for the time differencing. At this stage we also present three examples of boundary value problems in order to provide the reader with some concrete guidelines.

All the numerical results presented in this book have been obtained in a double precision mode on a IBM RISC 560 (having 50 MHZ) with 128 MB RAM. Peak performance is 100 MFlops (31 MFlops on LINPACK, 84 MFlops on TPP). Compiler is IBM AIX XL FORTRAN/6000 Version 2.3.

1.2 Preliminary Notation and Function Spaces

In this Section we introduce some definitions and notations which will often be used in the sequel. For a complete presentation we refer the interested reader to, e.g., Yosida (1974) or Brezis (1983) (see also Brezzi and Gilardi (1987) for a comprehensive and easy-to-read proofless presentation).

(i) *Hilbert and Banach spaces*
    Let V be a (real) linear space. A *scalar product* on V is a bilinear map $(\cdot, \cdot) : V \times V \to \mathbb{R}$ that $(w, v) = (v, w)$ for each $w, v \in V$ (symmetry), $(v, v) \geq 0$ for each $v \in V$ (positivity), and $(v, v) = 0$ if and only if $v = 0$.
    A *seminorm* is a map $|| \cdot || : V \to \mathbb{R}$ such that $||v|| \geq 0$ for each $v \in V$, $||cv|| = |c| \, ||v||$ for each $c \in \mathbb{R}$ and $v \in V$, and $||w + v|| \leq ||w|| + ||v||$ for each $w, v \in V$ (triangular inequality).
    A *norm* on V is a seminorm satisfying the additional property that $||v|| = 0$ if, and only if, $v = 0$. Two norms $|| \cdot ||$ and $||| \cdot |||$ on V are equivalent if there exist two positive constants M_1 and M_2 such that

$$M_1 ||v|| \leq |||v||| \leq M_2 ||v||$$

for each $v \in V$.
    It is readily verified that at any scalar product it is associated a norm through the following definition: $||v|| := (v, v)^{1/2}$. Moreover, at any norm we can associate a distance: $d(w, v) := ||w - v||$.
    A linear space V endowed with a scalar product (respectively, a norm) is called *pre-hilbertian* (respectively, *normed*) space. A sequence v_n is a Cauchy sequence in a normed space V if it is a Cauchy sequence with respect to the distance $d(w, v) = ||w - v||$. If any Cauchy sequence in a pre-hilbertian (normed) space V is convergent, the space V is called a *Hilbert* (respectively, *Banach*) space.
    In a Hilbert space the *Schwarz inequality* holds:

(1.2.1) $$|(w, v)| \leq ||w|| \, ||v|| \quad \text{for each } w, v \in V \ .$$

(ii) *Dual spaces*
    If $(V, || \cdot ||_V)$ and $(W, || \cdot ||_W)$ are normed spaces, we denote by $\mathcal{L}(V; W)$ the set of linear continuous functionals from V into W, and for $L \in \mathcal{L}(V; W)$ we define the norm

(1.2.2) $$||L||_{\mathcal{L}(V;W)} := \sup_{\substack{v \in V \\ v \neq 0}} \frac{||Lv||_W}{||v||_V} \ .$$

Thus $\mathcal{L}(V; W)$ is a normed space; if W is a Banach space, then $\mathcal{L}(V; W)$ is a Banach space, too. If $W = \mathbb{R}$, the space $\mathcal{L}(V; \mathbb{R})$ is called the *dual space* of V and is denoted by V'.

The bilinear form $\langle \cdot, \cdot \rangle$ from $V' \times V$ into $\mathbb{R}$ defined by $\langle L, v \rangle := L(v)$ is called the *duality pairing* between V' and V. As a consequence of the Riesz representation theorem (see, e.g., Yosida (1974), p. 90), if V is a Hilbert space, the dual V' is a Hilbert space which can be canonically identified to V.

(iii) *Weak and weak* convergence*

In a normed space V it is possible to introduce another type of convergence, which is called *weak convergence*. It is defined as follows: a sequence v_n is called weakly convergent to $v \in V$ if $L(v_n)$ converges to $L(v)$ for each $L \in V'$. It can be proven that the weak limit v, if it exists, is unique. Clearly, if the sequence v_n converges to v in V, it is also weakly convergent. The converse is not true unless V is finite dimensional.

In a dual space V' a third type of convergence can be introduced, the *weak* convergence*. It is defined as follows: a sequence of functionals $L_n \in V'$ is called weakly* convergent to $L \in V'$ if $L_n(v)$ converges to $L(v)$ for each $v \in V$. Also the weak* limit L, if it exists, is unique. Moreover, it can be shown that the weak convergence in V' implies the weak* convergence.

(iv) L^p *spaces*

We now introduce some spaces of functions which are the basis for the modern theory of partial differential equations. Let Ω be an open set contained in $\mathbb{R}^d$, $d \geq 1$, and consider in Ω the Lebesgue measure. A very important family of Banach spaces is the following one. Let $1 \leq p \leq \infty$, and consider the set of measurable functions v such that

$$(1.2.3) \qquad \int_\Omega |v(\mathbf{x})|^p d\mathbf{x} < \infty \ , \ 1 \leq p < \infty \ ,$$

or, when $p = \infty$,

$$(1.2.4) \qquad \sup\{|v(\mathbf{x})| \,|\, \mathbf{x} \in \Omega\} < \infty \ .$$

These spaces are usually denoted by $L^p(\Omega)$ and the associated norm is

$$(1.2.5) \qquad ||v||_{L^p(\Omega)} := \left(\int_\Omega |v(\mathbf{x})|^p d\mathbf{x} \right)^{1/p} \ , \ 1 \leq p < \infty \ ,$$

or, when $p = \infty$,

$$(1.2.6) \qquad ||v||_{L^\infty(\Omega)} := \sup\{|v(\mathbf{x})| \,|\, \mathbf{x} \in \Omega\} \ .$$

More precisely, $L^p(\Omega)$ is indeed the space of classes of equivalence of measurable functions, satisfying (1.2.3) or (1.2.4) with respect to the equivalence relation: $w \equiv v$ if w and v are different on a subset having zero measure. In other words, in the space $L^p(\Omega)$ two functions, different on a subset which has zero measure, are identified each other. Thus the definition of the space $L^\infty(\Omega)$ in (1.2.4) and of its norm in (1.2.6) should be modified in the following way: $v \in L^\infty(\Omega)$ if

$$\inf\{M \geq 0 \,|\, |v(\mathbf{x})| \leq M \text{ almost everywhere in } \Omega\} < \infty \ ,$$

and

(1.2.7) $\|v\|_{L^{\infty}(\Omega)} := \inf\{M \geq 0 \,|\, |v(\mathbf{x})| \leq M \text{ almost everywhere in } \Omega\}$,

where "almost everywhere in Ω" means "except on a subset of Ω having zero measure".

The space $L^2(\Omega)$ is indeed a Hilbert space, endowed with the scalar product

$$(w, v)_{L^2(\Omega)} := \int_{\Omega} w(\mathbf{x})\, v(\mathbf{x})\, d\mathbf{x} \ .$$

For reasons which will be clear in the sequel, the norm in $L^2(\Omega)$ is denoted by $\|\cdot\|_{0,\Omega}$, or simply $\|\cdot\|_0$ when no confusion about the domain Ω is possible. Moreover, the scalar product $(\cdot,\cdot)_{L^2(\Omega)}$ is often indicated by $(\cdot,\cdot)_{0,\Omega}$ or simply by $(\cdot,\cdot)$.

If $1 \leq p < \infty$, the dual space of $L^p(\Omega)$ is given by $L^{p'}(\Omega)$, where $(1/p) + (1/p') = 1$ (and $p' = \infty$ if $p = 1$). Moreover, the Hölder inequality holds:

(1.2.8) $\left| \int_{\Omega} w(\mathbf{x})\, v(\mathbf{x})\, d\mathbf{x} \right| \leq \|w\|_{L^p(\Omega)} \|v\|_{L^{p'}(\Omega)}$.

Notice that for $p = 2$ the Hölder inequality is the Schwarz inequality (1.2.1) for the Hilbert space $L^2(\Omega)$.

Moreover, from (1.2.8) it easily follows that $L^q(\Omega) \subset L^p(\Omega)$ if $p \leq q$ and Ω has finite measure.

(v) *Distributions*

Let us recall that $C_0^{\infty}(\Omega)$ (or $\mathcal{D}(\Omega)$) denotes the space of infinitely differentiable functions having compact support, i.e., vanishing outside a bounded open set $\Omega' \subset \Omega$ which has a positive distance from the boundary $\partial\Omega$ of Ω).

It is useful to define the concept of convergence for sequences of $\mathcal{D}(\Omega)$. We say that $v_n \in \mathcal{D}(\Omega)$ converges to $v \in \mathcal{D}(\Omega)$ if exists a closed bounded subset $K \in \Omega$ such that v_n vanishes outside K for each n, and for every non-negative multi-index $\boldsymbol{\alpha}$ the derivative $D^{\boldsymbol{\alpha}} v_n$ converges to $D^{\boldsymbol{\alpha}} v$ uniformly in Ω. We recall that if $\boldsymbol{\alpha} = (\alpha_1, ..., \alpha_d)$, α_i non-negative integers, then

$$D^{\boldsymbol{\alpha}} v := \frac{\partial^{|\boldsymbol{\alpha}|} v}{\partial x_1^{\alpha_1} ... \partial x_d^{\alpha_d}} \ ,$$

where $|\boldsymbol{\alpha}| := \alpha_1 + ... + \alpha_d$ is the length of $\boldsymbol{\alpha}$.

The space of linear functionals on $\mathcal{D}(\Omega)$ which are continuous with respect to the convergence introduced above is denoted by $\mathcal{D}'(\Omega)$ and its elements are called *distributions*. If $L \in \mathcal{D}'(\Omega)$ and $v \in \mathcal{D}(\Omega)$, we usually denote $L(v)$ by the duality pairing $\langle L, v \rangle$.

It is easily seen that each function $w \in L^p(\Omega)$ can be associated to the following distribution:

$$v \longrightarrow \int_{\Omega} w(\mathbf{x})\, v(\mathbf{x})\, d\mathbf{x} \quad, \quad v \in \mathcal{D}(\Omega) \quad .$$

However, the Dirac functional

$$v \longrightarrow \delta(v) := v(0) \quad, \quad v \in \mathcal{D}(\mathbb{R}) \quad ,$$

is a distribution which cannot be represented through any function belonging to $L^p(\Omega)$.

We are now in a position for introducing the *derivative* of a distribution. Let α be a non-negative multi-index and L a distribution. Then $D^\alpha L$ is the distribution defined as follows

(1.2.9) $$\langle D^\alpha L, v \rangle := (-1)^{|\alpha|} \langle L, D^\alpha v \rangle \quad \forall\, v \in \mathcal{D}(\Omega) \quad .$$

Notice that, from this definition, a distribution turns out to be infinitely differentiable. On the other hand, when L is a smooth function by integrating by parts it is easily verified that the derivative in the sense of distributions coincides with the usual derivative.

Let us also recall that the Dirac distribution δ is the distributional derivative of the Heaviside function:

$$H(x) := \begin{cases} 1 & , \quad x \geq 0 \\ 0 & , \quad x < 0 \end{cases} \quad .$$

Finally, we say that the α-derivative of a distribution L is a function belonging to $L^p(\Omega)$ if there exists a function g_α in $L^p(\Omega)$ such that

(1.2.10) $$\langle D^\alpha L, v \rangle = \int_{\Omega} g_\alpha(\mathbf{x})\, v(\mathbf{x})\, d\mathbf{x} \quad \forall\, v \in \mathcal{D}(\Omega) \quad .$$

(vi) *Sobolev spaces*

We finally introduce another class of functions, which will be most often used in the sequel, since they furnish the natural environment for the variational theory of partial differential equations. A comprehensive presentation of these spaces can be found in Adams (1975).

The *Sobolev space* $W^{k,p}(\Omega)$, k a non-negative integer and $1 \leq p \leq \infty$, is the space of functions $v \in L^p(\Omega)$ such that all the distributional derivatives of v of order up to k are a function of $L^p(\Omega)$. In short
(1.2.11)
$$W^{k,p}(\Omega) := \{ v \in L^p(\Omega) \,|\, D^\alpha v \in L^p(\Omega) \text{ for each non-negative}$$
$$\text{multi-index } \alpha \text{ such that } |\alpha| \leq k \} \quad .$$

Clearly, for each p, $1 \leq p \leq \infty$, $W^{0,p}(\Omega) = L^p(\Omega)$ and $W^{k_2,p}(\Omega) \subset W^{k_1,p}(\Omega)$ when $k_1 \leq k_2$. For $1 \leq p < \infty$, $W^{k,p}(\Omega)$ is a Banach space with respect to the norm

$$(1.2.12) \qquad ||v||_{k,p,\Omega} := \left(\sum_{|\alpha| \leq k} ||D^{\alpha} v||_{L^p(\Omega)}^p \right)^{1/p} .$$

Further, its seminorm is defined as follows:

$$(1.2.13) \qquad |v|_{k,p,\Omega} := \left(\sum_{|\alpha| = k} ||D^{\alpha} v||_{L^p(\Omega)}^p \right)^{1/p} .$$

On the other hand, $W^{k,\infty}(\Omega)$ is a Banach space with respect to the norm

$$(1.2.14) \qquad ||v||_{k,\infty,\Omega} := \max_{|\alpha| \leq k} ||D^{\alpha} v||_{L^\infty(\Omega)} ,$$

while the corresponding seminorm is denoted by

$$(1.2.15) \qquad |v|_{k,\infty,\Omega} := \max_{|\alpha| = k} ||D^{\alpha} v||_{L^\infty(\Omega)} .$$

In particular, when $p = 2$ we will write $H^k(\Omega)$ instead of $W^{k,2}(\Omega)$, $|| \cdot ||_{k,\Omega}$ and $| \cdot |_{k,\Omega}$ (or even $|| \cdot ||_k$ and $| \cdot |_k$ if no confusion is possible) instead of $|| \cdot ||_{k,2,\Omega}$ and $| \cdot |_{k,2,\Omega}$.

Notice that $H^k(\Omega)$ is indeed a Hilbert space with respect to the scalar product

$$(1.2.16) \qquad (w, v)_{k,\Omega} := \sum_{|\alpha| \leq k} (D^{\alpha} w, D^{\alpha} v)_{0,\Omega} .$$

Finally, we denote by $W_0^{k,p}(\Omega)$ the closure of $C_0^\infty(\Omega)$ with respect to the norm $|| \cdot ||_{k,p,\Omega}$, and with $W^{-k,p'}(\Omega)$ the dual space of $W_0^{k,p}(\Omega)$. As before, when $p = 2$ we will write $H_0^k(\Omega)$ and $H^{-k}(\Omega)$ instead of $W_0^{k,2}(\Omega)$ and $W^{-k,2}(\Omega)$, respectively.

It can also be proven that $W_0^{0,p}(\Omega) = L^p(\Omega)$ and that, if Ω has a Lipschitz continuous boundary, $W^{k,p}(\Omega)$ is indeed the closure of $C^\infty(\overline{\Omega})$ with respect to the norm $|| \cdot ||_{k,p,\Omega}$. In other words, $C^\infty(\overline{\Omega})$ is *dense* in $W^{k,p}(\Omega)$.

It is sometimes useful to consider the Sobolev space $W^{s,p}(\Omega)$, where $s \in \mathbb{R}$ and $1 \leq p \leq \infty$. We don't dwell on this here, but we just recall that, if $\Omega = \mathbb{R}^d$ and $p = 2$, $W^{s,2}(\mathbb{R}^d) = H^s(\mathbb{R}^d)$ can be characterized as follows by means of the Fourier transform $\hat{v}(\boldsymbol{\xi})$:

$$(1.2.17) \qquad H^s(\mathbb{R}^d) = \left\{ v \in L^2(\mathbb{R}^d) \, | \, (1 + |\boldsymbol{\xi}|^2)^{s/2} \hat{v}(\boldsymbol{\xi}) \in L^2(\mathbb{R}^d) \right\} .$$

When considering vector-valued functions $\mathbf{v} : \Omega \to \mathbb{R}^d$, the space

$$(1.2.18) \qquad H(\mathrm{div}; \Omega) := \left\{ \mathbf{v} \in (L^2(\Omega))^d \, | \, \mathrm{div}\, \mathbf{v} \in L^2(\Omega) \right\}$$

is also often used. It is endowed with the graph norm, i.e.,

(1.2.19) $$\|\mathbf{v}\|_{H(\mathrm{div};\Omega)} := (\|\mathbf{v}\|_{0,\Omega}^2 + \|\operatorname{div}\mathbf{v}\|_{0,\Omega}^2)^{1/2} \ .$$

Similarly to the preceding cases, if Ω has a Lipschitz continuous boundary it can be proven that $H(\mathrm{div};\Omega)$ is the closure of $(C^\infty(\overline{\Omega}))^d$ with respect to the norm $\|\cdot\|_{H(\mathrm{div};\Omega)}$.

Another important class of Sobolev spaces is given by $W^{s,p}(\Sigma)$, where $s \geq 0$, $1 \leq p \leq \infty$ and Σ is a suitable subset of the boundary $\partial\Omega$. Their definition requires the introduction of some technical tools, especially if Σ is a non-smooth hypersurface (for instance, the boundary of a polygonal domain). For this we refer to Adams (1975) or Brezzi and Gilardi (1987); however, we return on a characterization of these spaces in the following Section. When $\Sigma = \partial\Omega$, the dual space of $W^{s,2}(\partial\Omega) = H^s(\partial\Omega)$ is denoted by $H^{-s}(\partial\Omega)$.

When considering space-time functions $v(t,\mathbf{x})$, $(t,\mathbf{x}) \in Q_T := (0,T) \times \Omega$, it is natural to introduce the space
(1.2.20)
$$L^q(0,T;W^{s,p}(\Omega)) := \Big\{ v : (0,T) \to W^{s,p}(\Omega) \,|\, v \text{ is measurable}$$
$$\text{and } \int_0^T \|v(t)\|_{s,p,\Omega}^q \, dt < \infty \Big\} \ ,$$

$1 \leq q < \infty$, endowed with the norm

(1.2.21) $$\|v\|_{L^q(0,T;W^{s,p}(\Omega))} := \left(\int_0^T \|v(t)\|_{s,p,\Omega}^q \, dt \right)^{1/q} \ .$$

In a similar way one can define $L^\infty(0,T;W^{s,p}(\Omega))$ and $C^0([0,T];W^{s,p}(\Omega))$.

Some technical results are needed to define the Sobolev spaces $W^{r,q}(0,T; W^{s,p}(\Omega))$, $r \geq 0$; we refer to Lions and Magenes (1968a) or Brezzi and Gilardi (1987) for that. To give an idea, for a Banach space V the space $H^1(0,T;V)$ can be defined as follows:

(1.2.22) $$H^1(0,T;V) := \Big\{ v \in L^2(0,T;V) \,|\, \frac{\partial v}{\partial t} \in L^2(0,T;V) \Big\} \ ,$$

where $\frac{\partial v}{\partial t}$ has to be intended "in the sense of distributions with value in V". As we have only introduced real-valued distributions, the technical point is to give a precise meaning to the latter sentence. However, in several interesting cases the definition can be given in a different way. For instance, when $V = L^2(\Omega)$, $H^1(0,T;L^2(\Omega))$ is the space of functions of $L^2(Q_T)$ having the distributional time-derivative in $L^2(Q_T)$.

1.3 Some Results About Sobolev Spaces

In this Section we present without proofs some relevant properties enjoyed by functions belonging to Sobolev spaces. We will mainly limit ourselves to the Hilbert spaces $H^s(\Omega)$, referring the reader to Lions and Magenes (1968a, 1968b) or Adams (1975) for the general case and all the proofs.

Let us start with the so-called *trace* theorems. The trace on the boundary $\partial\Omega$ of a function $v \in H^s(\Omega)$ is, in a sense to make precise, the value of v restricted to $\partial\Omega$. Indeed, the latter statement has not even a meaning, as a function in $H^s(\Omega)$ is not univocally defined on subsets having measure equal to zero. If we denote by $C^0(\overline{\Omega})$ the space of continuous functions on $\overline{\Omega}$, the precise result reads as follows:

Theorem 1.3.1 (Trace theorem). *Let Ω be a bounded open set of $\mathbb{R}^d$ with Lipschitz continuous boundary $\partial\Omega$ and let $s > 1/2$.*
a. *There exists a unique linear continuous map $\gamma_0 : H^s(\Omega) \to H^{s-1/2}(\partial\Omega)$ such that $\gamma_0 v = v_{|\partial\Omega}$ for each $v \in H^s(\Omega) \cap C^0(\overline{\Omega})$.*
b. *There exists a linear continuous map $\mathcal{R}_0 : H^{s-1/2}(\partial\Omega) \to H^s(\Omega)$ such that $\gamma_0 \mathcal{R}_0 \varphi = \varphi$ for each $\varphi \in H^{s-1/2}(\partial\Omega)$.*
Analogous results also hold true if we consider the trace γ_Σ over a Lipschitz continuous subset Σ of the boundary $\partial\Omega$.

We have thus seen that any function belonging to $H^{s-1/2}(\Sigma)$, $s > 1/2$, is the trace on Σ of a function in $H^s(\Omega)$. This provides a useful characterization of the space $H^{s-1/2}(\Sigma)$.

Let us also notice that in particular the preceding theorem yields the existence of a constant $C^* > 0$ such that

$$(1.3.1) \qquad \int_{\partial\Omega} (\gamma_0 v)^2 \, d\gamma \leq C^* \int_\Omega (v^2 + |\nabla v|^2) \, d\mathbf{x} \qquad \forall\, v \in H^1(\Omega) \ ,$$

where $d\gamma$ denotes the surface measure on $\partial\Omega$.

For vector functions belonging to $H(\text{div}; \Omega)$ the following trace result can be proven:

Theorem 1.3.2 (Trace theorem for vector functions). *Let Ω be a bounded open set of $\mathbb{R}^d$ with Lipschitz continuous boundary $\partial\Omega$.*
a. *There exists a unique linear continuous map $\gamma^* : H(\text{div}; \Omega) \to H^{-1/2}(\partial\Omega)$ such that $\gamma^* \mathbf{v} = (\mathbf{v} \cdot \mathbf{n})_{|\partial\Omega}$ for each $\mathbf{v} \in H(\text{div}; \Omega) \cap (C^0(\overline{\Omega}))^d$.*
b. *There exists a linear continuous map $\mathcal{R}^* : H^{-1/2}(\partial\Omega) \to H(\text{div}; \Omega)$ such that $\gamma^* \mathcal{R}^* \varphi = \varphi$ for each $\varphi \in H^{-1/2}(\partial\Omega)$.*

Here we have denoted by $\mathbf{n}$ the unit outward normal vector on $\partial\Omega$. Let us notice moreover that the normal trace of a vector function $\mathbf{v} \in H(\text{div}; \Omega)$ over a subset Σ different from the boundary $\partial\Omega$ does not belong in general

which is usually denoted by $H_{00}^{-1/2}(\Sigma)$
.)).

s it is possible to characterize the spaces
))^d (here the closure has to be intended
;Ω)). As a matter of fact, if the boundary
e

, $H_0(\text{div}; \Omega) = \left\{ \mathbf{v} \in H(\text{div}; \Omega) \,|\, \gamma^* \mathbf{v} = 0 \right\}$.

for the space

$\left\{ v \in H^1(\Omega) \,|\, \gamma_\Sigma v = 0 \right\}$.

will find widespread application in the sequel,
uality:

inequality). *Assume that Ω is a bounded con-
that Σ is a (non-empty) Lipschitz continuous
. Then there exists a constant $C_\Omega > 0$ such that*

$$^2(\mathbf{x}) \, d\mathbf{x} \le C_\Omega \int_\Omega |\nabla v(\mathbf{x})|^2 \, d\mathbf{x}$$

of the density of $C^\infty(\overline{\Omega})$ in $H^1(\Omega)$ (under the as-
Lipschitz continuous), it is easily proven that for each
owing *Green formula* holds:

(1.3.3) $\int_\Omega$ (D$_j$ $\mathbf{x} = - \int_\Omega w \, D_j v \, d\mathbf{x} + \int_{\partial\Omega} \gamma_0 w \, \gamma_0 v \, n_j \, d\gamma$, $j = 1, ..., d$,

where we have denoted by D_j the partial derivative $\frac{\partial}{\partial x_j}$.

Similarly, if $\mathbf{w} \in H(\text{div}; \Omega)$ and $v \in H^1(\Omega)$, we find

(1.3.4) $\int_\Omega (\text{div}\, \mathbf{w}) \, v \, d\mathbf{x} = - \int_\Omega \mathbf{w} \cdot \nabla v \, d\mathbf{x} + \int_{\partial\Omega} \gamma^* \mathbf{w} \, \gamma_0 v \, d\gamma$.

As we already noticed, functions belonging to Sobolev spaces $W^{s,p}(\Omega)$ are
not univocally defined over subsets having measure equal to zero. However,
if some restrictions on the indices s and p is assumed, these functions indeed
turn out to be regular functions. This is made clear by the following theorem

Theorem 1.3.4 (Sobolev embedding theorem). *Assume that Ω is a (bounded
or unbounded) open set of $\mathbb{R}^d$ with a Lipschitz continuous boundary, and that
$1 \le p < \infty$. Then the following continuous embeddings hold:*
a. If $0 \le sp < d$, then $W^{s,p}(\Omega) \subset L^{p^*}(\Omega)$ for $p^* = dp/(d - sp)$;
b. If $sp = d$, then $W^{s,p}(\Omega) \subset L^q(\Omega)$ for any q such that $p \le q < \infty$;

c. *If* $sp > d$, *then* $W^{s,p}(\Omega) \subset C^0(\overline{\Omega})$.

Let us remind that a map $\Psi : X \to Y$ between two Banach spaces is said *compact* if it is continuous and moreover, given any bounded sequence $\{x_n\} \in X$, it is possible to select a subsequence $\{x_{n_k}\}$ such that $\Psi(x_{n_k})$ is convergent in Y. We can now state another important result, which gives further information on the Sobolev embeddings above.

Theorem 1.3.5 (Rellich-Kondrachov compactness theorem). *Assume that Ω is a bounded open set of $\mathbb{R}^d$ with a Lipschitz continuous boundary, and that $1 \leq p < \infty$. Then the following embeddings are compact:*
a. *If* $0 < sp < d$, *then* $W^{s,p}(\Omega) \subset L^q(\Omega)$ *for any q such that $1 \leq q < p^* = dp/(d-sp)$;*
b. *If* $sp = d$, *then* $W^{s,p}(\Omega) \subset L^q(\Omega)$ *for any q such that $1 \leq q < \infty$;*
c. *If* $sp > d$, *then* $W^{s,p}(\Omega) \subset C^0(\overline{\Omega})$;*
d. *If* $p > 2d/(d+2)$, *then* $L^p(\Omega) \subset H^{-1}(\Omega)$.*
In particular, $H^k(\Omega)$ is compactly embedded into $H^{k-1}(\Omega)$, k a non-negative integer.

Notice that *d.* follows from *a.* (where we have chosen $s = 1$, $p = 2$, so that $p^* = 2d/(d-2)$) via the Schauder duality theorem (see Yosida (1974), p. 282).

In the one-dimensional case, from the Sobolev theorem we have in particular that $H^1(\Omega) \subset C^0(\overline{\Omega})$. A more precise estimate of the C^0-norm is provided by the following interpolation inequality:

Theorem 1.3.6 (Gagliardo-Nirenberg interpolation inequality). *Let (a, b) be a bounded interval. Then the following inequality holds*

$$\max_{a \leq x \leq b} |v(x)| \leq \left(\frac{1}{b-a} + 2 \right)^{1/2} ||v||_0^{1/2} ||v||_1^{1/2} \ \forall\, v \in H^1(a,b) \ .$$

More general interpolation results are stated below.

Theorem 1.3.7 (Interpolation theorem). *Assume that Ω is an open set of $\mathbb{R}^d$ with a Lipschitz continuous boundary. Let $s_1 < s_2$ be two real numbers, and set $r = (1-\theta)s_1 + \theta s_2$, $0 \leq \theta \leq 1$. Then there exists a constant $C > 0$ such that*

$$||v||_r \leq C ||v||_{s_1}^{1-\theta} ||v||_{s_2}^\theta \ \forall\, v \in H^{s_2}(\Omega) \ .$$

We close this Section by considering some properties which are valid for space-time functions.

Theorem 1.3.8 *Assume that Ω is an open set of $\mathbb{R}^d$ with a Lipschitz continuous boundary, and let $s \geq 0$ and $r > 1/2$ be two real numbers. Then for each θ such that $0 \leq \theta \leq 1$ the space*

(1.3.5) $$L^2(0,T;H^s(\Omega)) \cap H^r(0,T;L^2(\Omega))$$

is continuously embedded into the space

$$H^{\theta r}(0,T;H^{(1-\theta)s}(\Omega)) \cap C^0([0,T];H^{\sigma_0}(\Omega)) \ ,$$

where $\sigma_0 := \frac{(2r-1)s}{2r}$.

If, furthermore, the set Ω is bounded and $s > 0$, then the space defined in (1.3.5) is compactly embedded into the space

$$H^{r_1}(0,T;H^{s_1}(\Omega)) \cap C^0([0,T];H^{\sigma_1}(\Omega)) \ ,$$

for each $s_1 \geq 0$, $0 \leq r_1 < r(1 - \frac{s_1}{s})$ and $0 \leq \sigma_1 < \sigma_0$.

For additional references on interpolation spaces we refer, e.g., to Bergh and Löfström (1976).

1.4 Comparison Results

In this Section we present two comparison results, which will be useful in the stability and convergence analysis of initial-boundary value problems.

Lemma 1.4.1 (Gronwall lemma). Let $f \in L^1(t_0,T)$ be a non-negative function, g and φ be continuous functions on $[t_0,T]$. If φ satisfies

(1.4.1) $$\varphi(t) \leq g(t) + \int_{t_0}^t f(\tau)\varphi(\tau) \qquad \forall\, t \in [t_0,T] \ ,$$

then

(1.4.2) $$\varphi(t) \leq g(t) + \int_{t_0}^t f(s)g(s)\exp\left(\int_s^t f(\tau)d\tau\right) \qquad \forall\, t \in [t_0,T] \ .$$

If moreover g is non-decreasing, then

(1.4.3) $$\varphi(t) \leq g(t)\exp\left(\int_{t_0}^t f(\tau)\right) \qquad \forall\, t \in [t_0,T] \ .$$

Proof. Set $R(t) := \int_{t_0}^t f(\tau)\varphi(\tau)$. From (1.4.1) it satisfies

$$\frac{dR}{dt}(t) = f(t)\varphi(t) \leq f(t)[g(t) + R(t)] \ ,$$

hence

$$\frac{d}{dt}\left[R(t)\exp\left(-\int_{t_0}^t f(\tau)d\tau\right)\right]$$

$$(1.4.4) \qquad = \left[\frac{dR}{dt}(t) - R(t)f(t)\right]\exp\left(-\int_{t_0}^t f(\tau)d\tau\right)$$

$$\le f(t)g(t)\exp\left(-\int_{t_0}^t f(\tau)d\tau\right) \quad .$$

Integrating (1.4.4) over (t_0, t) we find

$$R(t)\exp\left(-\int_{t_0}^t f(\tau)d\tau\right) \le \int_{t_0}^t f(s)g(s)\exp\left(-\int_{t_0}^s f(\tau)d\tau\right) \quad ,$$

and (1.4.2) follows.

If g is non-decreasing we find

$$\varphi(t) \le g(t)\left[1 + \int_{t_0}^t f(s)\exp\left(\int_s^t f(\tau)d\tau\right)\right] \quad ,$$

which yields (1.4.3). $\qquad\qquad\qquad\qquad\qquad\qquad\qquad\qquad\qquad \square$

Often, the Gronwall lemma will be used in the special case in which

$$(1.4.5) \qquad g(t) = \varphi(0) + \int_0^t \psi(s) \quad , \quad \psi(s) \ge 0 \quad .$$

Another comparison result, which is the discrete counterpart of Gronwall lemma, is the following one.

Lemma 1.4.2 (Discrete Gronwall lemma). *Assume that k_n is a non-negative sequence, and that the sequence ϕ_n satisfies*

$$(1.4.6) \qquad \begin{cases} \phi_0 \le g_0 \\ \phi_n \le g_0 + \displaystyle\sum_{s=0}^{n-1} p_s + \sum_{s=0}^{n-1} k_s\phi_s \quad , \quad n \ge 1 \quad . \end{cases}$$

Then ϕ_n satisfies

$$(1.4.7) \qquad \begin{cases} \phi_1 \le g_0(1 + k_0) + p_0 \\ \phi_n \le g_0 \displaystyle\prod_{s=0}^{n-1}(1 + k_s) + \sum_{s=0}^{n-2} p_s \prod_{\tau=s+1}^{n-1}(1 + k_\tau) + p_{n-1} \quad , \quad n \ge 2 \quad . \end{cases}$$

Moreover, if $g_0 \ge 0$ and $p_n \ge 0$ for $n \ge 0$, it follows

$$(1.4.8) \qquad \phi_n \le \left(g_0 + \sum_{s=0}^{n-1} p_s\right)\exp\left(\sum_{s=0}^{n-1} k_s\right) \quad , \quad n \ge 1 \quad .$$

Proof. First of all, let us remark that (1.4.8) is an easy consequence of (1.4.7), as $(1 + k_s) \leq \exp(k_s)$ for each $s \geq 0$. Moreover, (1.4.7)$_1$ is trivially obtained by the assumptions.

Before coming to the proof of (1.4.7)$_2$, we also recall that by an induction argument one can readily show that

$$
(1.4.9) \qquad 1 + k_0 + \sum_{s=1}^{n} k_s \prod_{r=0}^{s-1}(1 + k_r) = \prod_{s=0}^{n}(1 + k_s) \quad , \quad n \geq 1 ,
$$

and

$$
(1.4.10) \qquad \sum_{s=1}^{n}(1 + k_s)p_{s-1} + \sum_{s=2}^{n} k_s \sum_{r=0}^{s-2} p_r \prod_{\tau=r+1}^{s-1}(1 + k_\tau)
$$
$$
= \sum_{s=0}^{n-1} p_s \prod_{\tau=s+1}^{n}(1 + k_\tau) \quad , \quad n \geq 2 .
$$

The proof of (1.4.7)$_2$ is also obtained by induction. In fact, first of all we have

$$
\phi_2 \leq g_0 + p_0 + p_1 + k_0\phi_0 + k_1\phi_1
$$
$$
\leq (1 + k_0)g_0 + p_0 + p_1 + k_1[g_0(1 + k_0) + p_0]
$$
$$
\leq g_0 \prod_{s=0}^{1}(1 + k_s) + p_0(1 + k_1) + p_1 \quad ,
$$

i.e., (1.4.7)$_2$ holds for $n = 2$. Let us now assume that (1.4.7)$_2$ is satisfied for any index $2 \leq s \leq n$ and prove that it holds for the index $n + 1$ as well. We have

$$
\phi_{n+1} \leq g_0 + \sum_{s=0}^{n} p_s + \sum_{s=0}^{n} k_s\phi_s
$$
$$
\leq g_0 + p_n + \sum_{s=1}^{n} p_{s-1} + k_0 g_0 + k_1[g_0(1 + k_0) + p_0]
$$
$$
+ \sum_{s=2}^{n} k_s \left[g_0 \prod_{r=0}^{s-1}(1 + k_r) + \sum_{r=0}^{s-2} p_r \prod_{\tau=r+1}^{s-1}(1 + k_\tau) + p_{s-1} \right]
$$
$$
\leq g_0 \left[(1 + k_0) + \sum_{s=1}^{n} k_s \prod_{r=0}^{s-1}(1 + k_r) \right]
$$
$$
+ p_n + \sum_{s=1}^{n}(1 + k_s)p_{s-1} + \sum_{s=2}^{n} k_s \sum_{r=0}^{s-2} p_r \prod_{\tau=r+1}^{s-1}(1 + k_\tau) \quad .
$$

The thesis thus follows from (1.4.9) and (1.4.10).   □

More general inequalities of the form (1.4.7) can be found, e.g., in Lakshmikantham and Trigiante (1988).

2. Numerical Solution of Linear Systems

The solution of linear algebraic systems lies at the heart of most calculations in scientific computing. Here we describe some of the most popular methods that are applied to systems of general form. For a more complete presentation we refer the reader to the literature that we will quote throughout this Chapter. Special techniques for systems arising from the discretization of partial differential equations are discussed in this book for each specific situation.

2.1 Direct Methods

We consider the real algebraic system

$$(2.1.1) \qquad\qquad A\mathbf{x} = \mathbf{b} \ ,$$

where A is a non-singular matrix of dimension n.

Direct methods produce the exact solution of (2.1.1) in a finite number of steps (in the absence of round-off errors). The most classical direct method is the *Gaussian Elimination method* (GEM), which consists of decomposing A into the product LU where L and U are, respectively, a lower triangular and an upper triangular matrix. The decomposition is achieved through the n-step process

$$(2.1.2) \qquad A = A^{(1)} \to A^{(2)} \to ... \to A^{(n)} = U = L^{-1}A \ .$$

The key idea is that going from $A^{(k)}$ to $A^{(k+1)}$ amounts to transform the matrix $A^{(k)}$ in such a way that the elements of the k-th column under the diagonal of the new matrix $A^{(k+1)}$ are set to zero. Fig. 2.1.1 provides a schematic representation of the process: the shaded region is the one which changes in the k-th step.

The corresponding algorithm can be described in compact form as follows:

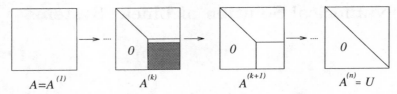

$$A=A^{(1)} \qquad A^{(k)} \qquad A^{(k+1)} \qquad A^{(n)} = U$$

Fig. 2.1.1. The Gaussian elimination process

$$(2.1.3) \quad \begin{cases} \text{For } k = 1, ..., n-1 \\ \quad \text{For } i = k+1, ..., n \\ \qquad m_{ik} = a_{ik}^{(k)} / a_{kk}^{(k)} \\ \qquad \text{For } j = k+1, ..., n \\ \qquad\quad a_{ij}^{(k+1)} = a_{ij}^{(k)} - m_{ik}a_{kj}^{(k)} \\ \qquad \text{end } j \\ \quad \text{end } i \\ \text{end } k \ . \end{cases}$$

Here $a_{ij}^{(1)}$ are the entries of the matrix $A = A^{(1)}$, and the elements m_{ik}, which are called *multipliers*, are the non-diagonal entries of the lower factor L. Moreover, one finds that $l_{ii} = 1$ for $i = 1, ..., n$. This algorithm requires $O(n^3)$ operations. Unfortunately, it can terminate prematurely if A has a singular leading principal submatrix, or, equivalently, when at least one of the pivots $a_{kk}^{(k)}$ vanishes.

Even if this is never the case, the GEM may be unstable because of the possibility of arbitrarily small pivots. This problem is generally alleviated by resorting to a pivoting strategy. The latter consists of interchanging at each step k of the elimination process the last $n - k$ columns (partial columnwise pivoting), or the last $n - k$ rows (partial rowise pivoting), or else the last $n - k$ rows and columns (complete pivoting) of $A^{(k)}$ in order to maximize the absolute value of the new k-th diagonal element, say $\tilde{a}_{kk}^{(k)}$. A schematic representation is given in Fig. 2.1.2.

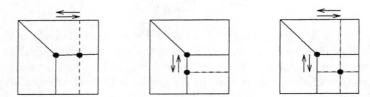

Fig. 2.1.2. Partial columnwise (left), rowise (center) and complete (right) pivoting

This in turn amounts to multiply each $A^{(k)}$ by two permutation matrices, obtaining

$$\tilde{A}^{(k)} = P_k A^{(k)} \Pi_k \ ,$$

before going on with the elimination procedure directly on $\tilde{A}^{(k)}$. The global complete pivoting process leads to the decomposition $PA\varPi = LU$ for some permutation matrices P and $\varPi$. On the other hand, the partial pivoting produces $QA = LU$, where Q is another permutation matrix.

The partial pivoting requires n^2 comparisons between elements, while $2n^3/3$ comparisons are needed for the complete pivoting. In practice, the partial pivoting strategy is often preferred to the complete one.

Here, we recall that, if we don't care about roundoff errors, the GEM can be carried out without any pivoting provided A is *strictly diagonally dominant*, i.e.,

$$(2.1.4) \qquad |a_{ii}| > \sum_{\substack{j=1 \\ j \neq i}}^{n} |a_{ij}| \quad \text{for } i = 1, ..., n \ ,$$

or else A is an *M-matrix*, i.e.,

$$(2.1.5) \quad a_{ij} \leq 0 \ \text{ for } i,j = 1,...,n, i \neq j, \text{ and } (A^{-1})_{ij} \geq 0 \ \text{ for } i,j = 1,...,n$$

(see, e.g., Varga (1962), p. 85).

It is, worthwhile to remember the following result (see Varga (1962), Ortega (1988)):

Lemma 2.1.1 *If A is strictly diagonally dominant, and further $a_{ij} \leq 0$ for all i,j with $i \neq j$ and $a_{ii} > 0$ for $i = 1,...,n$, then A is an M-matrix.*

Proof. Split A as $A = D - K$, where $D := \text{diag}(a_{ii})$ is the diagonal matrix whose entries are the a_{ii}'s. Clearly, D is non-singular and $K_{ij} \geq 0$.

Owing to the Gerschgorin circle theorem, (e.g., Varga (1962), p. 16) it follows that all eigenvalues of $D^{-1}K$ are inside the unit circle in the complex plane. Then we have

$$A^{-1} = (I - D^{-1}K)^{-1}D^{-1} = \sum_{j=0}^{\infty} (D^{-1}K)^j D^{-1} \ ,$$

where I denotes the identity matrix. Therefore, $(A^{-1})_{ij} \geq 0$ and the conclusion holds. $\qquad \square$

As seen in (2.1.2), the Gaussian elimination process amounts to decompose the original matrix as $A = LU$. Therefore, the solution to (2.1.1) leads to the solution of the triangular systems $L\mathbf{y} = \mathbf{b}$ and $U\mathbf{x} = \mathbf{y}$, which in turns can be accomplished by forward and backward substitution, respectively, using $O(n^2)$ operations.

Let us notice that the factors L and U can be computed directly by the following "compact" algorithm. Bearing in mind that $l_{kk} = 1$ for $k = 1,...,n$, it reads:

$$
(2.1.6) \quad
\begin{cases}
\text{For } k = 1, ..., n \\
\quad \text{For } j = k, ..., n \\
\qquad u_{kj} = a_{kj} - \sum_{p=1}^{k-1} l_{kp} u_{pj} \\
\quad \text{end } j \\
\quad \text{For } i = k+1, ..., n \text{ and } k \neq n \\
\qquad l_{ik} = \dfrac{1}{u_{kk}} \left(a_{ik} - \sum_{p=1}^{k-1} l_{ip} u_{pk} \right) \\
\quad \text{end } i \\
\text{end } k \ .
\end{cases}
$$

(Here, and in the sequel, it is understood that the sums don't apply if the lower index is less than the upper index, i.e., in this case, for $k = 1$.) The decomposition $A = LU$ in which the diagonal elements u_{kk} are equal to 1 can be achieved by a similar algorithm.

When all leading principal submatrices of A are non-singular, there exist two lower triangular matrices L, M, whose diagonal entries are all equal to 1, and a diagonal matrix D such that

$$(2.1.7) \qquad A = LDM^T \ ,$$

where M^T denotes the transpose of the matrix M. Once this decomposition is available, it is necessary to solve the three special systems

$$L\mathbf{y} = \mathbf{b} \ , \quad D\mathbf{z} = \mathbf{y} \ , \quad M^T\mathbf{x} = \mathbf{z} \ .$$

If A is symmetric then $M = L$, so that (2.1.7) becomes $A = LDL^T$.

A square matrix is said to be *positive definite* (briefly, $A > 0$) if

$$(2.1.8) \qquad (A\mathbf{w}, \mathbf{w}) > 0 \quad \text{for all } \mathbf{w} \in \mathbb{R}^n \ , \ \mathbf{w} \neq \mathbf{0} \ ,$$

where $(\cdot, \cdot)$ is the euclidean scalar product in $\mathbb{R}^n$. We warn the reader that, in the literature, this property is sometimes referred to as *positive real* (see, e.g., Young (1971)). The notation positive definite is in these cases used only for symmetric matrices.

If A is positive definite and symmetric, the Sylvester theorem ensures that all its leading principal matrices are non-singular, and A admits the following *Cholesky decomposition*

$$(2.1.9) \qquad A = LL^T \ ,$$

where L is a lower triangular matrix with positive diagonal entries. This decomposition is unique. The elements of L can be obtained by the following algorithm, which can be carried out without pivoting:

$$(2.1.10) \quad \begin{cases} \text{For } k = 1, ..., n \\ \qquad l_{kk} = \left(a_{kk} - \sum_{p=1}^{k-1} l_{kp}^2 \right)^{1/2} \\ \qquad \text{For } i = k+1, ..., n \text{ and } k \neq n \\ \qquad\qquad l_{ik} = \frac{1}{l_{kk}} \left(a_{ik} - \sum_{p=1}^{k-1} l_{ip} l_{kp} \right) \\ \qquad \text{end } i \\ \text{end } k \quad . \end{cases}$$

This algorithm requires $O(n^3)$ operations (but about one half of those necessary for the complete Gauss elimination process), plus n square roots.

In the case of a symmetric indefinite matrix, A can still admit an LDL^T decomposition. However, the entries in the factors can have arbitrary magnitude. To avoid this, some form of pivoting is mandatory, and this can be accomplished efficiently (without altering the $O(n^3)$ operations needed to compute LDL^T) by the algorithms of Bunch and Kaufman and that of Aasen. An account is given in Golub and Van Loan (1989), Sect. 4.4; see also Barwell and George (1976).

A different kind of decomposition for a general matrix A is the so-called *QR decomposition*, which is less cheap but more stable than the LU decomposition. For any real matrix $A \in \mathbb{R}^{m \times n}$ (i.e., with m rows and n columns), with $m \geq n$, having rank n, there exist a unique matrix $Q \in \mathbb{R}^{m \times n}$ satisfying

$$(2.1.11) \quad Q^T Q = D \quad , \quad D = \text{diag}(d_1, ..., d_n) \quad , \quad d_k > 0 \text{ for } k = 1, ..., n \quad ,$$

and an upper triangular matrix $R \in \mathbb{R}^{n \times n}$ with $r_{kk} = 1$, $k = 1, ..., n$, such that

$$(2.1.12) \qquad\qquad A = QR \quad .$$

This decomposition is particularly successful for the solution to the *least squares* problem. Indeed, in view of (2.1.12) the *normal equations*

$$(2.1.13) \qquad\qquad A^T A \mathbf{x} = A^T \mathbf{b}$$

lead to

$$0 = A^T(\mathbf{b} - A\mathbf{x}) = R^T Q^T(\mathbf{b} - A\mathbf{x}) = R^T(Q^T\mathbf{b} - Q^T QR\mathbf{x}) \quad ,$$

whence

$$(2.1.14) \qquad\qquad D\mathbf{y} = Q^T\mathbf{b} \quad , \quad R\mathbf{x} = \mathbf{y} \quad .$$

Thus, the computation of the solution of (2.1.13) simply requires the solution of two systems, one diagonal and the other upper triangular.

In turn, the calculation of the factors Q and R can be performed by several algorithms. One is based on a modified Gram-Schmidt orthogonalization method, another (the Householder method) on the use of the reflection matrices, and a third one (the Givens method) on the use of the rotation matrices, which require twice as many operations than Householder method (e.g., Isaacson and Keller (1966) and Golub and Van Loan (1989)).

2.1.1 Banded Systems

In many practical situations, when the system (2.1.1) arises from the discretization of a differential problem, the matrix A is banded, i.e., there is an integer q (the bandwidth), $1 \leq q < n$, such that

$$(2.1.15) \qquad a_{ij} = 0 \text{ if } i, j \text{ are such that } |i - j| > q \ .$$

In particular, A has at most $2q + 1$ non-zero elements per row.

When solving banded systems, the triangular factors of A are also banded, allowing for remarkable saving, in terms of storage and operations. Indeed, if A has a band of width q and can be decomposed as LU, then both triangular factors L and U are banded with bandwidth q. The modification of the Gaussian elimination algorithm (2.1.3) is straightforward, and requires $O(nq^2)$ operations. The corresponding solution of a triangular system by forward or backward elimination requires $O(nq)$ operations.

In general, when the GEM is implemented with a pivoting strategy, the original band structure of A is not perserved for the factors. More precisely, the bandwidth of U can be bigger than that of A's upper triangle ($= 2q$ in the case of partial pivoting). Even worse, nothing can be said in general about L's bandwidth, although it can be shown that L has at most $q + 1$ non-zeros per column.

If A is symmetric, positive definite and satisfies (2.1.15), its Cholesky decomposition (2.1.9) reads

$$(2.1.16) \quad \left\{ \begin{array}{l} \text{For } k = 1, ..., n \\ \quad s_k = \max(1, k - q) \ , \ r_k = \min(n, k + q) \\ \quad l_{kk} = \left(a_{kk} - \displaystyle\sum_{p=s_k}^{k-1} l_{kp}^2 \right)^{1/2} \\ \quad \text{For } i = k + 1, ..., r_k \text{ and } k \neq n \\ \qquad l_{ik} = \dfrac{1}{l_{kk}} \left(a_{ik} - \displaystyle\sum_{p=s_i}^{k-1} l_{ip} l_{kp} \right) \\ \quad \text{end } i \\ \text{end } k \ , \end{array} \right.$$

and requires $O(nq^2)$ operations (but about one half of those needed by the complete GEM) plus n square roots.

Remark 2.1.1 *(Sparse systems).* If A has a bandwidth q which is much smaller than its dimension n, then A is sparse. More generally, A is a sparse matrix if the number of its non-zero elements is $O(n)$.

Unlike the case of banded matrices, quite often the topological structure of sparse matrices is not well defined. This is for instance the case of systems associated with partial differential equations over unstructured meshes, and/or with a careless ordering of grid-points.

A direct method for a sparse system should be capable of exploiting the sparsity of A. As a matter of fact, a decomposition method applied to a sparse matrix without caring about its structure can lead to a substantial *fill-in*, i.e., generate a considerable number of new non-zero elements.

The fill-in can be reduced rearranging rows and columns (which, for partial differential equations, amounts to renumber the nodes) according to special criteria. Amongst these, we mention the method of nested dissection and its variants.

Another approach may involve an intelligent use of data structures. For an analysis of special direct methods for the solution of sparse systems we refer to Bunch and Rose (1976), Björk, Plemmons and Schneider (1981), George and Liu (1981), Duff, Erisman and Reid (1986).   □

Remark 2.1.2 *(Special purpose techniques)*. Special implementations of the GEM are carried out in specific situations. We discuss some of them in Section 6.3.2. Here we will mention three remarkable instances:

(i) the *frontal method* that consists of carrying out the Gaussian elimination process for the stiffness finite element matrix in a suitable order (see the review paper by Liu (1992));

(ii) the *fast Poisson solver* for the finite element (or finite difference) system associated with the discretization of the Poisson equation on structured mesh (see Buzbee, Golub and Nielson (1970), Dorr (1970); see also Vajteršic (1993));

(iii) the Haidvogel-Zang diagonalization of the matrix associated with the spectral Galerkin approximation of the Poisson problem (see Haidvogel and Zang (1979)) or with the spectral collocation approximation of the Helmholtz equation (see Haldenwang, Labrosse, Abboudi and Deville (1984)).   □

2.1.2 Error Analysis

To begin with, we will recall a few basic notations about matrices. A *matrix norm* is a function $f : \mathbb{R}^{n \times n} \to \mathbb{R}$ that satisfies for all matrices $A, B \in \mathbb{R}^{n \times n}$:

$$f(A) \geq 0 \quad (f(A) = 0 \text{ if and only if } A = 0)$$
$$f(A + B) \leq f(A) + f(B)$$
$$f(\alpha A) = |\alpha| f(A) , \quad \alpha \in \mathbb{R}$$
$$f(AB) \leq f(A) \, f(B).$$

A double bar notation with subscripts usually designates matrix norms, i.e., $\|A\| = f(A)$.

The Frobenius norm

$$(2.1.17) \qquad ||A||_F := \left(\sum_{i,j=1}^{n} |a_{ij}|^2 \right)^{1/2}$$

and the p-norms

$$(2.1.18) \qquad ||A||_p := \max_{|\mathbf{w}|_p=1} |A\mathbf{w}|_p \ , \quad 1 \le p \le \infty \ ,$$

are the most frequently used matrix norms in numerical linear algebra. We recall that the p-norms for vectors of $\mathbb{R}^n$ are defined as

$$(2.1.19) \qquad |\mathbf{w}|_p := \left(\sum_{j=1}^{n} |w_j|^p \right)^{1/p} \ , \quad 1 \le p < \infty \ ,$$

and

$$(2.1.20) \qquad |\mathbf{w}|_\infty := \max_{1 \le j \le n} |w_j| \ .$$

More generally, for any vector norm $| \cdot |$ in $\mathbb{R}^n$ it is possible to define its associated *natural* matrix norm as

$$||A|| := \max_{|\mathbf{w}|=1} |A\mathbf{w}| \ .$$

Denoting by $\lambda_i(A)$, $i = 1, ..., n$, the eigenvalues of A, the *spectral radius* of A is defined as

$$(2.1.21) \qquad \rho(A) := \max\{|\lambda_i(A)| \, | \, i = 1, ..., n\} \ .$$

It is possible to prove that $\rho(A)$ is the infimum of $||A||$ taken over all natural matrix norms (see, e.g., Isaacson and Keller (1966), p. 14).

We now notice that

$$(2.1.22) \qquad ||A||_1 = \max_{1 \le j \le n} \sum_{i=1}^{n} |a_{ij}| \ , \quad ||A||_\infty = \max_{1 \le i \le n} \sum_{j=1}^{n} |a_{ij}| \ .$$

Moreover, the following characterization holds (e.g., Varga (1962), p. 11)

$$(2.1.23) \qquad ||A||_2 = \sqrt{\rho(A^T A)} = \sqrt{\lambda_{\max}(A^T A)} \ ,$$

and therefore

$$(2.1.24) \qquad ||A||_2 = \rho(A) \quad \text{if} \ \ A \text{ is symmetric} \ .$$

For any matrix norm $|| \cdot ||$, the *condition number* of a non-singular matrix A with respect to that norm is the real number

$$(2.1.25) \qquad \chi(A) := ||A|| \, ||A^{-1}|| \ .$$

If $|| \cdot ||$ is a natural matrix norm, it follows that $\chi(A) \geq 1$. In particular, we denote by $\chi_p(A)$ the condition number of A computed with respect to the norm (2.1.18). Due to (2.1.23), we have

$$(2.1.26) \qquad \chi_2(A) = \sqrt{\frac{\lambda_{\max}(A^T A)}{\lambda_{\min}(A^T A)}} \ .$$

We further define the following quantity

$$(2.1.27) \qquad \chi_{sp}(A) := \frac{\max\{|\lambda_i(A)| \mid i = 1, ..., n\}}{\min\{|\lambda_i(A)| \mid i = 1, ..., n\}} \ .$$

Clearly, $\chi_2(A) = \chi_{sp}(A)$ if A is symmetric. For convenience, we will call $\chi_{sp}(A)$ the *spectral condition number* of the matrix A, even if A is not symmetric. We also refer to Lemma 2.4.1 for a useful characterization of the spectral condition number for matrices of the form $P^{-1}A$.

A matrix A is said to be "well conditioned" if its condition number is not "large". Otherwise, the matrix A is "ill-conditioned". In the latter case, the solution of the linear system (2.1.1) may not be a well posed problem, as small changes in the data can cause large changes in the solution.

The condition number provides a measure of the conditioning of a matrix. To make this statement more precise, suppose that the data A and $\mathbf{b}$ in (2.1.1) are perturbed by the quantities E (a matrix) and $\mathbf{e}$ (a vector). Then, if $\overline{\mathbf{x}}$ denotes the solution to the perturbed system

$$(2.1.28) \qquad (A + E)\overline{\mathbf{x}} = \mathbf{b} + \mathbf{e}$$

the following result provides an estimate for $\overline{\mathbf{x}} - \mathbf{x}$.

Proposition 2.1.1 *Let A be non-singular and suppose that*

$$(2.1.29) \qquad ||E|| < \frac{1}{||A^{-1}||}$$

for a given natural matrix norm $|| \cdot ||$ induced by a vector norm $| \cdot |$. Then if $\mathbf{x}$ and $\overline{\mathbf{x}}$ satisfy (2.1.1) and (2.1.28) we have

$$(2.1.30) \qquad \frac{|\overline{\mathbf{x}} - \mathbf{x}|}{|\mathbf{x}|} \leq \frac{\chi(A)}{1 - ||A^{-1}|| \, ||E||} \left(\frac{|\mathbf{e}|}{|\mathbf{b}|} + \frac{||E||}{||A||} \right) \ .$$

Proof. If B is any matrix such as $||B|| < 1$, then $I - B$ is non-singular and

$$||(I - B)^{-1}|| \leq (1 - ||B||)^{-1}$$

(e.g., Isaacson and Keller (1966), p. 16). Since $||A^{-1}E|| \leq ||A^{-1}|| \, ||E|| < 1$, choosing $B = -A^{-1}E$ we deduce

(2.1.31) $$\|(I + A^{-1}E)^{-1}\| \le \frac{1}{1 - \|A^{-1}\|\,\|E\|} \ .$$

Subtracting (2.1.1) from (2.1.28) we obtain $(A + E)(\overline{\mathbf{x}} - \mathbf{x}) = \mathbf{e} - E\mathbf{x}$ and therefore multiplying this last equation by A^{-1} we find

$$\overline{\mathbf{x}} - \mathbf{x} = (I + A^{-1}E)^{-1}A^{-1}(\mathbf{e} - E\mathbf{x}) \ .$$

Taking the norm of both sides, using (2.1.31) and dividing by $|\mathbf{x}|$ it follows

$$\frac{|\overline{\mathbf{x}} - \mathbf{x}|}{|\mathbf{x}|} \le \frac{\|A^{-1}\|}{1 - \|A^{-1}\|\,\|E\|} \left(\frac{|\mathbf{e}|}{|\mathbf{x}|} + \|E\| \right) \ .$$

From (2.1.1) we may replace $|\mathbf{x}|$ on the right, since

$$|\mathbf{x}| \ge \frac{|\mathbf{b}|}{\|A\|} \ .$$

The conclusion (2.1.30) now follows using the definition (2.1.25). □

2.2 Generalities on Iterative Methods

Iterative methods provide the solution to the system (2.1.1) as the limit of a sequence $\{\mathbf{x}^k\}$, and usually involve the matrix A only through multiplications by given vectors. Typical situations in which iterative methods are involved are large sparse problems, but also in band matrix problems, when the band itself is sparse, making decomposition algorithms difficult to implement.

Generally, any iterative method is based on a suitable splitting of the matrix A

(2.2.1) $$A = P - N \quad \text{with } P \text{ non-singular} \ ,$$

then the sequence $\{\mathbf{x}_k\}$ is generated as follows

(2.2.2) $$P\mathbf{x}^{k+1} = N\mathbf{x}^k + \mathbf{b} \ , \quad k \ge 0 \ ,$$

where $\mathbf{x}^0$ is given.

The evaluation of an iterative method focuses on two issues: how quickly the iterates $\mathbf{x}^k$ converge, and how cheap is the solution to the linear system (2.2.2) that needs to be faced at each step. Concerning the first point, setting

(2.2.3) $$\mathbf{e}^k := \mathbf{x}^k - \mathbf{x} \ , \quad B := P^{-1}N \ ,$$

we deduce from (2.2.2) the error equation

(2.2.4) $$\mathbf{e}^k = B\mathbf{e}^{k-1} = B^k\mathbf{e}^0 \ ,$$

where $\mathbf{e}^0$ is the initial error.

The iterates $\mathbf{x}^k$ converge to the solution $\mathbf{x}$ of (2.1.1) for any starting vector $\mathbf{x}^0$ if, and only if, the spectral radius of the *iteration matrix* B satisfies

$$(2.2.5) \qquad\qquad \rho(B) < 1 \ .$$

In fact, for each $\varepsilon > 0$ we can find a natural matrix norm $|| \cdot ||_*$ such that $||B||_* \leq \rho(B) + \varepsilon$. From (2.2.4) we obtain

$$|\mathbf{e}^k|_* \leq [\rho(B) + \varepsilon]^k |\mathbf{e}^0|_* \ ,$$

which yields convergence if ε is small enough. Conversely, if $\rho(B) \geq 1$ one can choose $\mathbf{x}^0$ such that $\mathbf{e}^0 = \boldsymbol{\omega}^*$, the eigenvector corresponding to the eigenvalue λ_* of maximum absolute value. For this choice (2.2.4) gives

$$|\mathbf{e}^k| = |B^k \mathbf{e}^0| = |\lambda_*|^k |\mathbf{e}^0| = [\rho(B)]^k |\mathbf{e}^0| \ ,$$

and convergence cannot hold, whatever is the vector norm $| \cdot |$ there considered. Clearly, the smaller is $\rho(B)$, the quicker is the convergence.

In the particular case in which B is diagonalizable, there exists a non-singular matrix T such that

$$B = T \Lambda T^{-1} \ ,$$

with $\Lambda := \mathrm{diag}(\lambda_1, ..., \lambda_n)$, λ_j being the eigenvalues of B. From the recurrence relation (2.2.4) it follows

$$\mathbf{e}^k = T \Lambda^k T^{-1} \mathbf{e}^0 \ .$$

Defining $\boldsymbol{\epsilon}^k := T^{-1} \mathbf{e}^k$, we have immediately

$$\boldsymbol{\epsilon}^k = \Lambda^k \boldsymbol{\epsilon}^0 \ ,$$

i.e., $\epsilon_j^k = (\lambda_j)^k \epsilon_j^0$ for $j = 1, ..., n$, $k \geq 0$. Therefore, up to a changement of basis, the effect of a single iteration amounts to damping each component ϵ_j^0 of the initial error by a factor that is given by the corresponding eigenvalue λ_j.

Since Λ is a diagonal matrix, we have $||\Lambda||_p = \rho(\Lambda) = \rho(B)$ for each $1 \leq p \leq \infty$. Therefore we have

$$|\mathbf{e}^k|_p \leq \chi_p(T)[\rho(B)]^k |\mathbf{e}^0|_p \ , \quad |\boldsymbol{\epsilon}^k|_p \leq [\rho(B)]^k |\boldsymbol{\epsilon}^0|_p \ ,$$

where $\chi_p(T) := ||T||_p ||T^{-1}||_p$ is the p-condition number of the diagonalization matrix T.

The number $\sigma := -\log \rho(B)$ is called the *asymptotic rate of convergence* of the iterative procedure (2.2.2). Its inverse $1/\sigma$ represents a measure of the average value of iterations that are needed in order to reduce the norm of the initial error by a factor $1/e$.

Now let us present some sufficient conditions for convergence. A splitting (2.2.1) is said to be *regular* if $(P^{-1})_{ij} \geq 0$ and $N_{ij} \geq 0$ for $i, j = 1, ..., n$.

As a consequence of the Perron-Frobenius theory on non-negative matrices, it can be proven that if (2.2.1) is a regular splitting and $(A^{-1})_{ij} \geq 0$ for $i, j = 1, ..., n$, then $\rho(P^{-1}N) < 1$ (Varga (1962), p. 89). The following result allows for a comparison between regular splittings (Varga (1962), p. 90).

Theorem 2.2.1 *Let A be such that $(A^{-1})_{ij} > 0$ for $i, j = 1, ..., n$. Consider two regular splittings of A: $A = P_1 - N_1$ and $A = P_2 - N_2$. Assume moreover that*

$$(2.2.6) \qquad (N_2)_{ij} \geq (N_1)_{ij} \quad for \ i, j = 1, ..., n$$

and $N_2 - N_1$ is not the null matrix. Then

$$(2.2.7) \qquad \rho(P_1^{-1}N_1) < \rho(P_2^{-1}N_2) < 1 \ .$$

The inequality (2.2.7) states that the first splitting is virtually superior to the latter, provided (2.2.6) holds (notice that this condition is quite easy to check). Another interesting result is the following:

Theorem 2.2.2 (Householder-John). *Let A be symmetric, positive definite and split as in (2.2.1). Suppose that the matrix $P + P^T - A$ is positive definite. Then $\rho(P^{-1}N) < 1$.*

Proof. Let us write $P^{-1}N = I - P^{-1}A$, and let $\lambda \in \mathbb{C}$, $\mathbf{z} \in \mathbb{C}^n$, $\mathbf{z} \neq 0$, be an eigenvalue and the corresponding eigenvector of $I - P^{-1}A$. Thus we have

$$(2.2.8) \qquad (1 - \lambda)P\mathbf{z} = A\mathbf{z} \ ,$$

and, in particular, it follows that $\lambda \neq 1$. Take the scalar product in $\mathbb{C}^n$ of (2.2.8) by $\mathbf{z} = \mathbf{w} + i\mathbf{y}$, $\mathbf{w}, \mathbf{y} \in \mathbb{R}^n$, and consider also the complex conjugate of the result. As $A = A^T$, we have $(A\mathbf{z}, \mathbf{z})_{\mathbb{C}^n} \in \mathbb{R}$ and

$$(A\mathbf{z}, \mathbf{z})_{\mathbb{C}^n} = (1 - \lambda)(P\mathbf{z}, \mathbf{z})_{\mathbb{C}^n} = (1 - \overline{\lambda})(P^T\mathbf{z}, \mathbf{z})_{\mathbb{C}^n} \ .$$

Thus

$$(2.2.9) \qquad \left(\frac{1}{1 - \lambda} + \frac{1}{1 - \overline{\lambda}} - 1 \right) (A\mathbf{z}, \mathbf{z})_{\mathbb{C}^n} = ((P + P^T - A)\mathbf{z}, \mathbf{z})_{\mathbb{C}^n} \ .$$

Taking the real part of (2.2.9) one easily obtains

$$\frac{1 - |\lambda|^2}{|1 - \lambda|^2}[(A\mathbf{w}, \mathbf{w}) + (A\mathbf{y}, \mathbf{y})] = ((P + P^T - A)\mathbf{w}, \mathbf{w}) + ((P + P^T - A)\mathbf{y}, \mathbf{y}) \ ,$$

where $(\cdot, \cdot)$ is the scalar product in $\mathbb{R}^n$. Since both A and $(P + P^T - A)$ are positive definite, we can conclude that $|\lambda| < 1$, i.e., $\rho(P^{-1}N) < 1$. $\qquad \square$

We conclude this Section with another convergence theorem. First, we recall that a matrix is said to be *N-stable* (or *negative-stable*) if all of its eigenvalues have positive real parts (see Young (1971)).

Theorem 2.2.3 *The convergence condition* (2.2.5) *holds if, and only if, $I - B$ is non-singular, and moreover*

$$(2.2.10) \qquad H := (I - B)^{-1}(I + B) \text{ is } N\text{-stable} .$$

Proof. Under the assumption that $I - B$ is non-singular, from (2.2.10) we have $H + I = 2(I - B)^{-1}$, $H - I = 2(I - B)^{-1}B$, then

$$B = (H + I)^{-1}(H - I) .$$

If λ is an eigenvalue of B, then $\mu = \frac{1+\lambda}{1-\lambda}$ is an eigenvalue of H. Moreover, we can write

$$\lambda = \frac{\mu - 1}{\mu + 1} = \frac{(\operatorname{Re}\mu - 1) + i \operatorname{Im}\mu}{(\operatorname{Re}\mu + 1) + i \operatorname{Im}\mu} .$$

Thus $|\lambda| < 1$ if and ony if $\operatorname{Re}\mu > 0$.

On the other hand, if (2.2.5) holds, $I - B$ is clearly non-singular and, if μ is an eigenvalue of H, $\lambda = \frac{\mu-1}{\mu+1}$ is an eigenvalue of B. The thesis thus follows by proceeding as before. $\qquad\qquad\square$

2.3 Classical Iterative Methods

We review here the classical Jacobi, Gauss-Seidel and other relaxation iterative procedures.

2.3.1 Jacobi Method

It is perhaps the simplest iterative scheme, and it can be defined for matrices having non-zero diagonal elements. If the matrix A is represented as

$$(2.3.1) \qquad A = \begin{pmatrix} & & U_A \\ & D_A & \\ L_A & & \end{pmatrix} ,$$

where D_A is the diagonal of A and L_A, U_A are its lower and upper triangular parts, respectively, and we set

$$D := \begin{pmatrix} & & 0 \\ & D_A & \\ 0 & & \end{pmatrix} , \quad E := \begin{pmatrix} & & 0 \\ & 0 & \\ L_A & & \end{pmatrix} , \quad F := \begin{pmatrix} & & U_A \\ & 0 & \\ 0 & & \end{pmatrix} ,$$

the Jacobi method is based on the splitting

(2.3.2) $P := D \ , \quad N := -(E + F) \ .$

The Jacobi iteration matrix is therefore $B_J := -D^{-1}(E + F)$.
If we define the *residual vector* as

(2.3.3) $\mathbf{r}^k := \mathbf{b} - A\mathbf{x}^k \ ,$

the Jacobi iteration can be written as

$$\mathbf{x}^{k+1} - \mathbf{x}^k = D^{-1}\mathbf{r}^k \ .$$

Componentwise, this algorithm reads

(2.3.4) $x_i^{k+1} = \dfrac{1}{a_{ii}} \left(b_i - \displaystyle\sum_{\substack{j=1 \\ j \neq i}}^n a_{ij} x_j^k \right) \ , \quad i = 1, ..., n \ .$

The condition $\rho(B_J) < 1$ is satisfied if A is strictly diagonally dominant (see (2.1.4)). In fact,

$$\rho(B_J) \leq ||D^{-1}(E + F)||_\infty = \max_{1 \leq i \leq n} \sum_{\substack{j=1 \\ j \neq i}}^n \left| \frac{a_{ij}}{a_{ii}} \right| < 1$$

(recall that the spectral radius is less than or equal to any natural matrix norm). Further, it happens that the "more dominant" the diagonal of A, the more rapid the convergence will be (see, however, Golub and Van Loan (1989), pp. 509 and 514).

Condition (2.1.4) may be rather restrictive. In particular, it is satisfied neither from the matrix associated to the five-point finite difference discretization of the Laplacian (e.g., Isaacson and Keller (1966)), nor from the bilinear finite element approximation to the same problem on a regular rectangular mesh (see Chapter 3). For these cases, however, the following result can be successfully applied (Varga (1962); Young (1971), p. 107).

Theorem 2.3.1 *Assume that A is irreducible and weakly diagonally dominant. Then the Jacobi method converges.*

We recall that a matrix A is said to be *irreducible* if there isn't any permutation matrix Π such that $\Pi^T A \Pi$ is a two-by-two block matrix whose block of position (2,1) is the null matrix. Further, A is *weakly diagonally dominant* if

(2.3.5) $\forall\, i = 1, ..., n \ , \ |a_{ii}| \geq \displaystyle\sum_{\substack{j=1 \\ j \neq i}}^n |a_{ij}| \ \text{ and } \ \exists\, k : |a_{kk}| > \displaystyle\sum_{\substack{j=1 \\ j \neq k}}^n |a_{kj}| \ .$

Finally, let us remember that the Jacobi method can fail to converge even if A is a symmetric and positive definite matrix (see, e.g., Young (1971), p. 111). However, it does converge if we further require that also $2D - A$ is positive definite, as it follows from Theorem 2.2.2.

2.3.2 Gauss-Seidel Method

The *Gauss-Seidel method* is based on the splitting

$$(2.3.6) \qquad P := D + E \ , \quad N := -F \ ,$$

hence, as for the Jacobi method, it is defined for matrices having non-zero diagonal elements. The corresponding iteration matrix is

$$B_{GS} := -(D + E)^{-1}F \ .$$

The iteration (2.2.2) can be written as

$$(2.3.7) \quad x_i^{k+1} = \frac{1}{a_{ii}} \left(b_i - \sum_{j=1}^{i-1} a_{ij} x_j^{k+1} - \sum_{j=i+1}^{n} a_{ij} x_j^k \right) \ , \quad i = 1, ..., n \ .$$

(It is understood that the first sum does not apply for $i = 1$, as well as the second one for $i = n$.)

If A is symmetric and positive definite, the Gauss-Seidel method converges. This result follows from Theorem 2.2.2. Indeed in the current case $E^T = F$ and

$$(2.3.8) \qquad P + P^T - A = (D + E) + (D + E)^T - (D + E + F) = D \ ,$$

and D is positive definite.

Another situation in which the Gauss-Seidel iteration converges is that of a matrix A which is strictly diagonally dominant (e.g., Isaacson and Keller (1966), p. 67). The Gauss-Seidel iteration also converges under the assumptions of Theorem 2.3.1 (Young (1971), p. 108).

A comparison between the spectral radii of B_J and B_{GS} can be obtained in several circumstances. In order to give some examples, let us recall the following Theorems (for the proofs we refer, e.g., to Varga (1962) or Young (1971), p. 120).

Theorem 2.3.2 (Stein-Rosenberg). *If A is such that $(B_J)_{ij} \geq 0$ for $i, j = 1, ..., n$, then one and only one of the following mutually exclusive relations is valid:*

$$(2.3.9) \qquad \begin{aligned} \rho(B_{GS}) = \rho(B_J) = 0 \ , \ \ \rho(B_{GS}) < \rho(B_J) < 1 \ , \\ \rho(B_{GS}) = \rho(B_J) = 1 \ , \ \ \rho(B_{GS}) > \rho(B_J) > 1 \ . \end{aligned}$$

Thus the Jacobi matrix and the Gauss-Seidel matrix are either both convergent, or both divergent.

Theorem 2.3.3 *Let A be a tridiagonal matrix. If λ is an eigenvalue of B_J then λ^2 is an eigenvalue of B_{GS}. Similarly, if μ is an eigenvalue of B_{GS} greater than zero, then the square roots of μ are eigenvalues of B_J. In particular, this yields that the Gauss-Seidel method converges if, and only if, the Jacobi iteration converges, moreover*

$$(2.3.10) \qquad\qquad \rho(B_{GS}) = [\rho(B_J)]^2 \ .$$

2.3.3 Relaxation Methods (S.O.R. and S.S.O.R.)

The *Successive Over-Relaxation (S.O.R.) method* is an iteration procedure that accelerates the Gauss-Seidel method by the help of a positive parameter ω. Still assuming that $a_{ii} \neq 0$ for $i = 1, ..., n$, in order to get the new iterate $\mathbf{x}^{k+1}$ from $\mathbf{x}^k$ we set:

$$(2.3.11) \qquad x_i^{k+1} = (1 - \omega)x_i^k + \frac{\omega}{a_{ii}}\left(b_i - \sum_{j=1}^{i-1} a_{ij}x_j^{k+1} - \sum_{j=i+1}^{n} a_{ij}x_j^k \right)$$

for $k \geq 0$, $i = 1, ..., n$. This amounts to using these splitting matrices

$$(2.3.12) \qquad P = \frac{D}{\omega} + E \ , \quad N = \left(\frac{1}{\omega} - 1\right)D - F \ .$$

The corresponding iteration matrix is

$$(2.3.13) \qquad B_\omega := (D + \omega E)^{-1}[(1 - \omega)D - \omega F] \ ,$$

and it coincides with B_{GS} if $\omega = 1$. Notice that in general neither P nor B_ω are symmetric matrices, even if A satisfies this property.

As far as convergence, we have the following necessary condition for ω.

Theorem 2.3.4 (Kahan). *The S.O.R. iteration matrix verifies*

$$(2.3.14) \qquad\qquad \rho(B_\omega) \geq |\omega - 1| \ ,$$

therefore a necessary condition for convergence is

$$(2.3.15) \qquad\qquad 0 < \omega < 2 \ .$$

Proof. If $\{\lambda_i \,|\, i = 1, ..., n\}$ denote the eigenvalues of B_ω, from (2.3.13) we easily get

$$\prod_{i=1}^{n} |\lambda_i| = |\det B_\omega| = |\det D|^{-1} |\omega - 1|^n |\det D| \ .$$

Thus $[\rho(B_\omega)]^n \geq |\omega - 1|^n$, hence (2.3.15) follows. □

Under the assumptions of Theorem 2.3.1, the S.O.R. method converges for all $0 < \omega \leq 1$ (Young (1971), p. 107). Furthermore, for a special class of matrices, condition (2.3.15) is also sufficient for convergence, as stated by the following

Theorem 2.3.5 (Ostrowski-Reich). *If A is symmetric and positive definite, the S.O.R. method converges if, and only if, (2.3.15) is satisfied.*

Proof. In view of applying Theorem 2.2.2 to the S.O.R. splitting (2.3.12) we notice that

$$P + P^T - A = \left(\frac{D}{\omega} + E\right)^T + \left(\frac{1}{\omega} - 1\right) D - F = \left(\frac{2}{\omega} - 1\right) D \ ,$$

since $F = E^T$ due to the symmetry of A. The assumption that A is positive definite implies that both D and $(2/\omega - 1)D$ enjoy the same property provided (2.3.15) holds. □

A symmetric version of the S.O.R. method is defined as follows. For the transition from $\mathbf{x}^k$ to $\mathbf{x}^{k+1}$ we compute firstly the S.O.R. iterate (2.3.11) that we denote by $\mathbf{y}^{k+1}$. Then, we apply a *backward* S.O.R. iteration, obtaining

$$(2.3.16) \quad x_i^{k+1} = (1 - \omega)y_i^{k+1} + \frac{\omega}{a_{ii}} \left(b_i - \sum_{j=1}^{i-1} a_{ij}y_j^{k+1} - \sum_{j=i+1}^{n} a_{ij}x_j^{k+1} \right)$$

for $i = n, ..., 1$. This algorithm, which is called *Symmetric S.O.R. (S.S.O.R.) method*, is based on the splitting

$$(2.3.17) \quad \begin{aligned} P &= \frac{1}{\omega(2 - \omega)}(D + \omega E)D^{-1}(D + \omega F) \\ N &= \frac{1}{\omega(2 - \omega)}[(1 - \omega)D - \omega E]D^{-1}[(1 - \omega)D - \omega F] \ , \end{aligned}$$

and has the following iteration matrix

$$(2.3.18) \quad \begin{aligned} B_\omega^s &:= (D + \omega F)^{-1}D(D + \omega E)^{-1} \\ &\quad \times [(1 - \omega)D - \omega E]D^{-1}[(1 - \omega)D - \omega F] \ . \end{aligned}$$

Notice that, if A is symmetric, then P and N are symmetric, too. Moreover, if A is symmetric and D is positive definite, B_ω^s is similar to a symmetric non-negative definite matrix, and thus it has real non-negative eigenvalues. In fact, under this assumption we have $F = E^T$ and the matrix $(D + \omega F)^{-1}D(D + \omega E)^{-1}$ is symmetric and positive definite. By employing the Cholesky method to decompose it, we find

$$(D + \omega E^T)^{-1}D(D + \omega E)^{-1} = LL^T ,$$

and therefore

$$L^{-1}B_\omega^s L = L^T[(1 - \omega)D - \omega E]D^{-1}[(1 - \omega)D - \omega E^T]L ,$$

which is a symmetric and non-negative definite matrix.

The Ostrowski-Reich theorem applies to the S.S.O.R. iteration, too. In fact, if A is symmetric and positive definite, applying Theorem 2.2.2 one finds

$$P + P^T - A = \frac{2}{\omega(2 - \omega)}(D + \omega E)D^{-1}(D + \omega E^T) - A$$

$$= \frac{2 - \omega}{2\omega}D + \frac{2\omega}{(2 - \omega)} \left(\frac{1}{2}D^{1/2} + ED^{-1/2}\right) \left(\frac{1}{2}D^{1/2} + ED^{-1/2}\right)^T ,$$

where $D^{1/2}D^{1/2} = D$ and $D^{-1/2} = (D^{1/2})^{-1}$. (Recall that D is a diagonal matrix with positive entries.) Thus $P + P^T - A$ is positive definite if and only if $0 < \omega < 2$. Moreover, one easily obtains that $\rho(B_\omega^s) \geq (\omega - 1)^2$, therefore a necessary condition for convergence is $0 < \omega < 2$.

For a few structured problems the optimal value of the relaxation parameter ω, i.e., the one that minimizes $\rho(B_\omega)$, is known. An instance is provided by block tridiagonal systems for which the diagonal blocks of the matrix A are non-singular. This is, e.g., the case of the five-point finite difference discretization to the Poisson equation in a rectangle. In such cases, if all eigenvalues of B_J (the Jacobi iteration matrix) are real and $\rho(B_J) < 1$, the optimal relaxation parameter is given by

$$(2.3.19) \qquad \omega^* = \frac{2}{1 + \sqrt{1 - [\rho(B_J)]^2}} ,$$

and correspondingly $\rho(B_{\omega^*}) = \omega^* - 1$.

In more complicated problems, the determination of an appropriate parameter may require a fairly sophisticated eigenvalue analysis. For a complete survey we refer to Wachpress (1966), Young (1971), O'Leary (1976).

2.3.4 Chebyshev Acceleration Method

Another way to accelerate the convergence of the iterative method (2.2.2) makes use of Chebyshev polynomials. The idea is to modify the sequence $\{\mathbf{x}^k\}$ into

$$(2.3.20) \qquad \mathbf{y}^m := \sum_{k=0}^{m} \gamma_k^m \mathbf{x}^k \ ,$$

where the coefficients γ_k^m need to be determined according to an optimality criterium. The first requirement is that $\sum_{k=0}^{m} \gamma_k^m = 1$, as we wish that $\mathbf{y}^m = \mathbf{x}$ (the exact solution of (2.1.1)) in the case in which $\mathbf{x}^k = \mathbf{x}$ for all $k = 0, ..., m$. Thus, in particular, $\mathbf{y}^0 = \mathbf{x}^0$.

If we set

$$\boldsymbol{\varepsilon}^k := \mathbf{y}^k - \mathbf{x} \ , \quad P_m(t) := \sum_{k=0}^{m} \gamma_k^m t^k \ ,$$

we obtain the recursion formula

$$(2.3.21) \qquad \boldsymbol{\varepsilon}^m = P_m(B)\boldsymbol{\varepsilon}^0 \ ,$$

where B is the iteration matrix introduced in (2.2.3). The problem is led to the search of that algebraic polynomial P_m of degree less than or equal to m satisfying $P_m(1) = 1$ which minimizes the spectral radius of $P_m(B)$.

If B is symmetric and its real eigenvalues λ_i satisfy

$$-1 < \alpha \le \lambda_1 \le ... \le \lambda_n \le \beta < 1 \ , \quad \alpha < \beta \ ,$$

it follows that $P_m(B)$ is also symmetric with eigenvalues given by $P_m(\lambda_i)$, therefore

$$(2.3.22) \qquad \rho(P_m(B)) = \max_{1 \le i \le n} |P_m(\lambda_i)| \le \max_{\alpha \le \lambda \le \beta} |P_m(\lambda)| \ .$$

Minimizing, among all polynomials of degree less than or equal to m such that $P_m(1) = 1$, the quantity at the right hand side of the previous inequality yields

$$P_m(t) = \frac{T_m\left(\dfrac{\beta + \alpha - 2t}{\beta - \alpha}\right)}{T_m\left(\dfrac{\beta + \alpha - 2}{\beta - \alpha}\right)} \ ,$$

where $T_m(x)$ are the Chebyshev polynomials (see, e.g., Rivlin (1974) and also Section 4.3.1).

With this choice we obtain from (2.3.21) and (2.3.22)

$$(2.3.23) \qquad |\boldsymbol{\varepsilon}^m|_2 \le ||P_m(B)||_2 \, |\boldsymbol{\varepsilon}^0|_2 \le \left| T_m\left(\frac{\beta + \alpha - 2}{\beta - \alpha}\right) \right|^{-1} |\boldsymbol{\varepsilon}^0|_2 \ ,$$

where

$$|\mathbf{v}| := |\mathbf{v}|_2 = (\mathbf{v}, \mathbf{v})^{1/2} = \left(\sum_{i=1}^{n} v_i^2\right)^{1/2} \ , \quad \mathbf{v} \in \mathbb{R}^n \ ,$$

denotes the length (i.e., the euclidean norm) of the vector $\mathbf{v}$. Here we have used the property (2.1.24).

More generally, let us assume that the iteration matrix B is not necessarily symmetric, but similar to a symmetric matrix, i.e., there exists a non-singular matrix W such that $W^{-1}BW$ is symmetric, and that its eigenvalues satisfy the same relations as before. In this case, setting $\mathcal{P} := W^{-T}W^{-1}$ (here W^{-T} denotes $(W^{-1})^T = (W^T)^{-1}$), which is a symmetric and positive definite matrix, we can introduce the scalar product

$$(\mathbf{u}, \mathbf{v})_{\mathcal{P}} := (\mathcal{P}\mathbf{u}, \mathbf{v}) \ .$$

The matrix B turns out to be symmetric with respect to this scalar product, as

$$(B\mathbf{u}, \mathbf{v})_{\mathcal{P}} = (\mathcal{P}B\mathbf{u}, \mathbf{v}) = (W^{-1}B\mathbf{u}, W^{-1}\mathbf{v}) = (W^{-1}BWW^{-1}\mathbf{u}, W^{-1}\mathbf{v})$$
$$= (W^{-1}\mathbf{u}, W^{-1}BWW^{-1}\mathbf{v}) = (\mathcal{P}\mathbf{u}, B\mathbf{v}) = (\mathbf{u}, B\mathbf{v})_{\mathcal{P}} \ .$$

The same is true for the matrix $P_m(B)$, therefore, using now the relation (2.1.24) but with the euclidean scalar product substituted by the scalar product $(\cdot, \cdot)_{\mathcal{P}}$, the matrix norm of $P_m(B)$ associated to the vector norm $|\mathbf{v}|_{\mathcal{P}} := (\mathbf{v}, \mathbf{v})_{\mathcal{P}}^{1/2}$ is equal to the spectral radius $\rho(P_m(B))$. Hence we can conclude

$$(2.3.24) \qquad |\varepsilon^m|_{\mathcal{P}} \leq \left| T_m\left(\frac{\beta + \alpha - 2}{\beta - \alpha}\right)\right|^{-1} |\varepsilon^0|_{\mathcal{P}} \ ,$$

and this inequality generalizes (2.3.23).

Since the Chebyshev polynomials satisfy a three-term recursion formula (see (4.3.4)), it can be seen that the sequence (2.3.20) can be generated avoiding the storage of $\mathbf{x}^0, ..., \mathbf{x}^m$. As a matter of fact, the Chebyshev acceleration procedure can be carried out as follows (e.g., Hageman and Young (1981), p. 48 and also Golub and Van Loan (1989), pp. 511–513):

$$(2.3.25) \qquad \mathbf{y}^{m+1} = \omega_{m+1}(\mathbf{y}^m - \mathbf{y}^{m-1} + \gamma\mathbf{z}^m) + \mathbf{y}^{m-1} \ , \quad m \geq 1 \ ,$$

where

$$P\mathbf{z}^m = \mathbf{b} - A\mathbf{y}^m \ , \quad \omega_{m+1} = 2\frac{2 - \beta - \alpha}{\beta - \alpha}\frac{T_m(\mu)}{T_{m+1}(\mu)} \ ,$$

$$\mu = 1 + 2\frac{1 - \beta}{\beta - \alpha} \ , \quad \gamma = \frac{2}{2 - \alpha - \beta} \ ,$$

with $\mathbf{y}^0 = \mathbf{x}^0$ and $\mathbf{y}^1 = \mathbf{x}^1$.

We refer to (2.3.25) as to the *Chebyshev acceleration method* associated with (2.2.2).

The estimate of α and β may be difficult to ascertain in all but a few structured problems. On the other hand, a sharp determination of these parameters is mandatory for the acceleration to be effective.

Remark 2.3.1 If A is symmetric and D is positive definite, the iteration matrix B_J of the Jacobi method is similar to a symmetric matrix (just take $W = D^{-1/2}$), and its eigenvalues are real and symmetrically distributed about the origin. Thus the Chebyshev acceleration procedure can be applied. This is also true for the S.S.O.R. method, but not for the Gauss-Seidel and the S.O.R. iterations.   □

2.3.5 The Alternating Direction Iterative Method

Other classical methods have been introduced by Douglas (1955), Peaceman and Rachford (1955), and Douglas and Rachford (1956) for solving linear systems obtained by the finite difference discretization of elliptic and parabolic problems. We will review several of these methods in Section 5.7; here we want to present the general ideas in an abstract case, giving sufficient conditions for the convergence of the *Peaceman-Rachford method* (also called *alternating direction iterative method*).

Let A be a positive definite matrix, i.e., $(A\mathbf{w}, \mathbf{w}) > 0$ for each $\mathbf{w} \in \mathbb{R}^n$, $\mathbf{w} \neq \mathbf{0}$, and suppose that A is split as

$$(2.3.26) \qquad\qquad A = A_1 + A_2 \ ,$$

where A_k are non-negative definite matrices, i.e., $(A_k\mathbf{w}, \mathbf{w}) \geq 0$ for each $\mathbf{w} \in \mathbb{R}^n$, $k = 1, 2$, and either A_1 or A_2 is positive definite.

Let us consider the following scheme (see, e.g., Il'in (1966)):

$$(2.3.27) \qquad \begin{cases} \gamma(\mathbf{x}^{k+1/2} - \mathbf{x}^k) + A_1\mathbf{x}^{k+1/2} = \mathbf{b} - A_2\mathbf{x}^k \\[2mm] \gamma(\mathbf{x}^{k+1} - \mathbf{x}^{k+1/2}) + A_2\mathbf{x}^{k+1} = A_2\mathbf{x}^k + \rho\gamma(\mathbf{x}^{k+1/2} - \mathbf{x}^k) \ , \end{cases}$$

where $\gamma > 0$, $(1 + \rho) \neq 0$. For $\rho = 1$ we have the *Peaceman-Rachford scheme*, and for $\rho = 0$ the *Douglas-Rachford scheme*. It can be easily seen that the matrices $(\gamma I + A_1)$ and $(\gamma I + A_2)$ are positive definite, hence non-singular.

From a continuity argument it follows that if $\mathbf{x}^k$ converges, also $\mathbf{x}^{k+1/2}$ is convergent. Further, the limit is the same; as a matter of fact, if we define

$$\mathbf{y} := \lim_k \mathbf{x}^k \ , \quad \mathbf{z} := \lim_k \mathbf{x}^{k+1/2} \ ,$$

then from (2.3.27) we find

$$\begin{cases} \gamma(\mathbf{z} - \mathbf{y}) + A_1\mathbf{z} = \mathbf{b} - A_2\mathbf{y} \\[2mm] \gamma(1 + \rho)(\mathbf{y} - \mathbf{z}) = 0 \ , \end{cases}$$

hence $\mathbf{y} = \mathbf{z} = A^{-1}\mathbf{b}$.

To prove the convergence of $\mathbf{x}^k$ the following result due to Kellogg (1963) will be useful.

Proposition 2.3.1 (Kellogg lemma). *If G is a non-negative definite matrix and σ is a real non-negative parameter, then*

$$(2.3.28) \qquad ||(I - \sigma G)(I + \sigma G)^{-1}||_2 \leq 1 \ .$$

If G is positive definite and σ is positive, then

$$(2.3.29) \qquad ||(I - \sigma G)(I + \sigma G)^{-1}||_2 < 1 \ .$$

Here $||\cdot||_2$ denotes the matrix norm subordinated to the euclidean vector norm (see (2.1.18) and (2.1.19)).

Proof. We have

$$||(I - \sigma G)(I + \sigma G)^{-1}||_2^2$$

$$= \sup_{\substack{\mathbf{w} \in \mathbb{R}^n \\ \mathbf{w} \neq 0}} \frac{((I - \sigma G)(I + \sigma G)^{-1}\mathbf{w}, (I - \sigma G)(I + \sigma G)^{-1}\mathbf{w})}{(\mathbf{w}, \mathbf{w})}$$

$$= \sup_{\substack{\mathbf{y} \in \mathbb{R}^n \\ \mathbf{y} \neq 0}} \frac{((I - \sigma G)\mathbf{y}, (I - \sigma G)\mathbf{y})}{((I + \sigma G)\mathbf{y}, (I + \sigma G)\mathbf{y})}$$

$$= \sup_{\substack{\mathbf{y} \in \mathbb{R}^n \\ \mathbf{y} \neq 0}} \frac{(\mathbf{y}, \mathbf{y}) - 2\sigma(G\mathbf{y}, \mathbf{y}) + \sigma^2(G\mathbf{y}, G\mathbf{y})}{(\mathbf{y}, \mathbf{y}) + 2\sigma(G\mathbf{y}, \mathbf{y}) + \sigma^2(G\mathbf{y}, G\mathbf{y})} \ .$$

Since $\sigma(G\mathbf{y}, \mathbf{y}) \geq 0$, (2.3.28) follows at once. On the other hand, (2.3.29) holds if $\sigma(G\mathbf{y}, \mathbf{y}) > 0$ for each $\mathbf{y} \in \mathbb{R}^n$, $\mathbf{y} \neq \mathbf{0}$. $\qquad\square$

Let us go back to the iterations (2.3.27). Eliminating $\mathbf{x}^{k+1/2}$ we find that the iteration operator is given by

$$(2.3.30) \qquad B = (\gamma I + A_2)^{-1}(\gamma I + A_1)^{-1}(\gamma^2 I - \rho \gamma A + A_1 A_2) \ .$$

The spectral radius of B is thus the same of

$$(2.3.31) \qquad B_* = (\gamma I + A_1)^{-1}(\gamma^2 I - \rho \gamma A + A_1 A_2)(\gamma I + A_2)^{-1} \ .$$

Let us take now $\rho = 1$, i.e., consider the Peaceman-Rachford scheme. It can be easily checked that

$$(\gamma^2 I - \gamma A + A_1 A_2) = (\gamma I - A_1)(\gamma I - A_2) \ .$$

Hence, by Proposition 2.3.1,

$$||B_*||_2 \leq ||(I - \gamma^{-1}A_1)(I + \gamma^{-1}A_1)^{-1}||_2 \, ||(I - \gamma^{-1}A_2) \cdot (I + \gamma^{-1}A_2)^{-1}||_2 < 1 \ ,$$

and this ensures the convergence of $\mathbf{x}^k$.

For an analysis of the Il'in and Douglas-Rachford schemes we refer, e.g., to Yanenko (1971), pp. 59–66. Moreover, several methods are known to accelerate the convergence of $\mathbf{x}^k$, especially exploiting information on the spectrum of A and choosing a different parameter γ_k at each step and using them cyclically. A thorough discussion is given in Young (1971), Chap. 17.

2.4 Modern Iterative Methods

In this Section we present some families of iterative methods whose acceleration parameters need not be determined on the basis of an eigenvalue analysis,
but rather they can be computed by a close formula depending upon available
iterates. They include gradient and conjugate gradient methods with preconditioners. A thorough presentation of these and more general methods can be
found, e.g., in Marchuk and Kuznetsov (1974), Hestenes (1980), Golub and
Van Loan (1989).

2.4.1 Preconditioned Richardson Method

Iteration (2.2.2) can be restated equivalently as

$$P(\mathbf{x}^{k+1} - \mathbf{x}^k) = \mathbf{r}^k \ , \quad k \in \mathbb{N} \ ,$$

where $\mathbf{r}^k$ is the residual at the step k (see (2.3.3)). According to the terminology that will be used later, the matrix P in the splitting (2.2.1) can therefore
be regarded as a *preconditioner* for the matrix A. Let us recall that P needs
to be non-singular and "easy" to invert.

This process can be generalized as follows

$$(2.4.1) \qquad P(\mathbf{x}^{k+1} - \mathbf{x}^k) = \alpha \mathbf{r}^k \ , \quad k \in \mathbb{N} \ ,$$

where, for each k, $\alpha \neq 0$ is a real acceleration parameter. The iteration (2.4.1)
is called the *preconditioned Richardson method*: *stationary* if α is constant for
all k, *dynamical* if α changes along the iterations.

The stationary case can be formulated recursively as follows. Let $\mathbf{x}^0$ be
given, and $\mathbf{r}^0 := \mathbf{b} - A\mathbf{x}^0$. Subsequent iterations are made according to

$$(2.4.2) \qquad \begin{cases} P\mathbf{z}^k := \mathbf{r}^k \\[2mm] \mathbf{x}^{k+1} := \mathbf{x}^k + \alpha \mathbf{z}^k \\[2mm] \mathbf{r}^{k+1} := \mathbf{r}^k - \alpha A\mathbf{z}^k \end{cases}$$

for $k \geq 0$. At each step, (2.4.2) requires the solution of a linear system with
matrix P, along with a matrix-vector multiplication involving the original
matrix A. The preconditioned Richardson iteration matrix is

$$(2.4.3) \qquad R_\alpha := I - \alpha P^{-1} A \ .$$

From now on we denote with η_i the eigenvalues of $P^{-1}A$.

Theorem 2.4.1 *For any non-singular matrix P, the stationary preconditioned Richardson method converges if, and only if,*

(2.4.4) $$|\eta_i|^2 < \frac{2}{\alpha}\mathrm{Re}\,\eta_i \quad \forall\, i = 1, ..., n \ .$$

Proof. Let us apply Theorem 2.2.3 to the current situation. Owing to (2.4.3) we have

$$(I - R_\alpha)^{-1}(I + R_\alpha) = \frac{2}{\alpha}A^{-1}P - I \ .$$

Denoting by μ_i the eigenvalues of $A^{-1}P$ it follows from Theorem 2.2.3 that $\rho(R_\alpha) < 1$ if, and only if, $(2/\alpha)\mathrm{Re}\,\mu_i > 1$ for all $i = 1, ..., n$. Condition (2.4.4) now follows, noting that $\eta_i = 1/\mu_i$ for all $i = 1, ..., n$. □

Let us notice that, if the sign of the real parts of the eigenvalues of $P^{-1}A$ is not constant, then the stationary preconditioned Richardson method cannot converge.

Corollary 2.4.1 *If $P^{-1}A$ has real eigenvalues, then the stationary preconditioned Richardson method converges if, and only if,*

(2.4.5) $$1 < \frac{2}{\alpha\eta_i} \quad \forall\, i = 1, ..., n \ .$$

If we assume from now on that $P^{-1}A$ is symmetric and positive definite, then condition (2.4.5) reduces to

$$0 < \alpha < \frac{2}{\eta_{\max}} \ ,$$

where $\eta_{\max}$ is the maximum eigenvalue of $P^{-1}A$. Moreover, denoting with $\{\mathbf{v}_i \,|\, i = 1, ..., n\}$ the orthonormal basis given by the eigenvectors of $P^{-1}A$, and expanding the error $\mathbf{e}^k$ (see (2.2.3)) as

$$\mathbf{e}^k = \sum_{i=1}^{n} e_i^k \mathbf{v}_i \ , \quad k \in \mathbb{N} \ ,$$

we deduce from (2.2.4) that

(2.4.6) $$e_i^{k+1} = (1 - \alpha\eta_i)e_i^k \ , \quad i = 1, ..., n \ , \quad k \in \mathbb{N} \ .$$

The real numbers $\rho_i := |1 - \alpha\eta_i|$ are called the *reduction factors* (see Fig. 2.4.1). Clearly, convergence is achieved if, and only if, $\rho_i < 1$ for each $i = 1, ..., n$, as the spectral radius of R_α is given by $\max_{1 \leq i \leq n} \rho_i$.

The optimal value of α is the one α^* for which

$$|1 - \alpha^*\eta_{\min}| = |1 - \alpha^*\eta_{\max}| \ ,$$

where $\eta_{\min}$ and $\eta_{\max}$ denote the minimum and maximum eigenvalues of $P^{-1}A$, respectively. Therefore

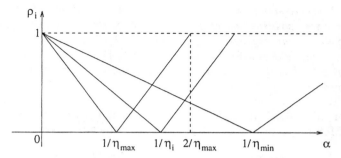

Fig. 2.4.1. Reduction factors of the stationary preconditioned Richardson method

$$(2.4.7) \qquad \alpha^* = \frac{2}{\eta_{\min} + \eta_{\max}} \ .$$

Correspondingly we have

$$\rho^* := \rho(R_{\alpha^*}) = 1 - \alpha^* \eta_{\min} = \frac{\eta_{\max} - \eta_{\min}}{\eta_{\max} + \eta_{\min}} \ .$$

If $P^{-1}A$ is symmetric and positive definite, we have from (2.1.27)

$$(2.4.8) \qquad \chi_{sp}(P^{-1}A) = \frac{\eta_{\max}}{\eta_{\min}} \ .$$

Therefore we conclude that

$$(2.4.9) \qquad |e^{k+1}| \le \rho^* |e^k| \qquad \forall\, k \in \mathbb{N} \ ,$$

with

$$(2.4.10) \qquad \rho^* = \frac{\chi_{sp}(P^{-1}A) - 1}{\chi_{sp}(P^{-1}A) + 1} \ .$$

The term ρ^* is called reduction factor of the stationary preconditioned Richardson method with optimal parameter α^*. The smaller $\chi_{sp}(P^{-1}A)$, the faster the convergence will be.

We must point out that if the assumption that $P^{-1}A$ is symmetric and positive definite doesn't hold (which is very often the case), the previous argument still applies, provided both P and A be symmetric and positive definite. Indeed, $P^{-1}A$ still has real and positive eigenvalues, as $P^{-1}A$ turns out to be symmetric and positive definite with respect to the scalar product $(\cdot, \cdot)_P$ defined as

$$(2.4.11) \qquad (\mathbf{u}, \mathbf{v})_P := (P\mathbf{u}, \mathbf{v}) \ , \quad |\mathbf{u}|_P := \sqrt{(\mathbf{u}, \mathbf{u})_P} \quad \forall\, \mathbf{u}, \mathbf{v} \in \mathbb{R}^n \ ,$$

where $(\cdot, \cdot)$ denotes the euclidean scalar product. Clearly, $(\cdot, \cdot)_P$ is a scalar product in $\mathbb{R}^n$, and $|\cdot|_P$ is the associated vector norm. As a matter of fact the following result holds.

Lemma 2.4.1 *Assume that P is symmetric and positive definite. Then, if A is symmetric (positive definite) $P^{-1}A$ is symmetric (positive definite) with respect to the scalar product $(\cdot, \cdot)_P$ defined in (2.4.11). Further, if A is symmetric the spectral condition number $\chi_{sp}(P^{-1}A)$ is equal to the condition number of $P^{-1}A$ with respect to the natural matrix norm $\|\cdot\|_P$ induced by the vector norm $|\cdot|_P$.*

Proof. For all $\mathbf{u} \in \mathbb{R}^n$ it is $(P^{-1}A\mathbf{u}, \mathbf{u})_P = (A\mathbf{u}, \mathbf{u})$, thus $P^{-1}A$ is positive definite. Moreover, for any $\mathbf{u}, \mathbf{v} \in \mathbb{R}^n$, owing to the symmetry of A and P we have

$$(P^{-1}A\mathbf{u}, \mathbf{v})_P = (A\mathbf{u}, \mathbf{v}) = (\mathbf{u}, A\mathbf{v}) = (P^{-1}P\mathbf{u}, A\mathbf{v})$$
$$= (P\mathbf{u}, P^{-1}A\mathbf{v}) = (\mathbf{u}, P^{-1}A\mathbf{v})_P \ .$$

Finally, we notice that

$$\|P^{-1}A\|_P = \rho(P^{-1}A) \ .$$

This equality can be obtained by repeating the proof of (2.1.23), provided that we replace the euclidean scalar product with the scalar product $(\cdot, \cdot)_P$, and denote by A^T the matrix satisfying $(A\mathbf{u}, \mathbf{v})_P = (\mathbf{u}, A^T\mathbf{v})_P$ for each $\mathbf{u}, \mathbf{v} \in \mathbb{R}^n$. Therefore

$$\chi_{sp}(P^{-1}A) = \frac{\max\{|\eta_i| \mid i = 1, ..., n\}}{\min\{|\eta_i| \mid i = 1, ..., n\}}$$
$$= \rho(P^{-1}A)\,\rho(A^{-1}P) = \|P^{-1}A\|_P\,\|A^{-1}P\|_P \ .$$

This concludes the proof. $\qquad\square$

Let us now assume that A is positive definite and that P is both symmetric and positive definite. From (2.4.2) it follows

$$\mathbf{z}^{k+1} = \mathbf{z}^{k+1}(\alpha) = P^{-1}(\mathbf{r}^k - \alpha A\mathbf{z}^k) \ ,$$

therefore

$$|\mathbf{z}^{k+1}|_P^2 = |P^{-1}\mathbf{r}^k|_P^2 + \alpha^2|P^{-1}A\mathbf{z}^k|_P^2 - 2\alpha(P^{-1}\mathbf{r}^k, P^{-1}A\mathbf{z}^k)_P \ .$$

Now we choose that value of α, say α_k, which minimizes $|\mathbf{z}^{k+1}(\alpha)|_P^2$, i.e.,

(2.4.12) α_k is such that $\left(\dfrac{d}{d\alpha}|\mathbf{z}^{k+1}(\alpha)|_P^2\right)_{|\alpha = \alpha_k} = 0 \ .$

A simple calculation yields

(2.4.13) $$\alpha_k = \frac{(\mathbf{z}^k, A\mathbf{z}^k)}{(A\mathbf{z}^k, P^{-1}A\mathbf{z}^k)}$$

The numerator is always positive unless $\mathbf{z}^k = \mathbf{0}$ in which case $\mathbf{r}^k = \mathbf{0}$, i.e., $\mathbf{x}^k = \mathbf{x}$ (the exact solution).

The iteration scheme (2.4.2) in which α is replaced by α_k, given in (2.4.13), is called *preconditioned Richardson-Minimum Residual (PR-MR) method*. Indeed, if P was the identity then $\mathbf{z}^{k+1} = \mathbf{r}^{k+1}$ and α_k would minimize the euclidean norm $|\mathbf{r}^{k+1}|$ of the new residual vector $\mathbf{r}^{k+1}$. This method (for $P = I$) has been introduced and analyzed by Krasnosel'skiĭ and Kreĭn (1952).

It is easily proven that

$$\frac{|\mathbf{z}^k|_P^2 - |\mathbf{z}^{k+1}|_P^2}{|\mathbf{z}^k|_P^2} = \frac{(A\mathbf{z}^k, \mathbf{z}^k)^2}{(P^{-1}A\mathbf{z}^k, A\mathbf{z}^k)(P\mathbf{z}^k, \mathbf{z}^k)} ,$$

thus $|\mathbf{z}^k|_P$ is non-increasing and convergent. As a consequence,

$$\frac{(A\mathbf{z}^k, \mathbf{z}^k)^2}{(P^{-1}A\mathbf{z}^k, A\mathbf{z}^k)} \to 0 .$$

Since A is positive definite, there exists $\sigma > 0$ such that $(A\mathbf{w}, \mathbf{w}) \geq \sigma|\mathbf{w}|^2$ for each $\mathbf{w} \in \mathbb{R}^n$. This implies

$$\frac{(A\mathbf{z}^k, \mathbf{z}^k)^2}{(P^{-1}A\mathbf{z}^k, A\mathbf{z}^k)} \geq \frac{\sigma^2|\mathbf{z}^k|^4}{||P^{-1}A||_2||A||_2|\mathbf{z}^k|^2} = \frac{\sigma^2|\mathbf{z}^k|^2}{||P^{-1}A||_2||A||_2} ,$$

i.e., $\mathbf{z}^k \to \mathbf{0}$ and $\mathbf{x}^k \to \mathbf{x} = A^{-1}\mathbf{b}$. (Here $|| \cdot ||_2$ is the matrix norm associated to the euclidean scalar product.) For an alternative proof of this convergence result, see Kuznetsov (1969).

Assuming further that A is symmetric, it follows that $P^{-1}A$ is symmetric and positive definite with respect to the scalar product $(\cdot, \cdot)_P$ (see Lemma 2.4.1). Thus, there exists a matrix B, the square root of $P^{-1}A$, symmetric and positive definite with respect to $(\cdot, \cdot)_P$ and such that $B^2 = P^{-1}A$ and $P^{-1}AB = BP^{-1}A$. Then we can write

$$\frac{|\mathbf{z}^k|_P^2 - |\mathbf{z}^{k+1}|_P^2}{|\mathbf{z}^k|_P^2} = \frac{(B\mathbf{z}^k, B\mathbf{z}^k)_P^2}{(P^{-1}AB\mathbf{z}^k, B\mathbf{z}^k)_P(A^{-1}PB\mathbf{z}^k, B\mathbf{z}^k)_P} .$$

As a consequence of the following Kantorovich inequality (see, e.g., Luenberger (1984), p. 217; Todd (1977), p. 185)

$$\frac{(P^{-1}A\mathbf{w}, \mathbf{w})_P (A^{-1}P\mathbf{w}, \mathbf{w})_P}{|\mathbf{w}|_P^4} \leq \frac{(\eta_{\max} + \eta_{\min})^2}{4\eta_{\max}\eta_{\min}} ,$$

the following estimate holds

(2.4.14) $$|\mathbf{z}^{k+1}|_P \leq \rho^*|\mathbf{z}^k|_P , \quad k \in \mathbb{N} ,$$

with ρ^* still given by (2.4.10).

Assuming that A is symmetric and positive definite, it follows that, we can define the *preconditioned Richardson-Steepest Descent (PR-SD) method*, which has been introduced and analyzed by Kantorovich (1948). It reads as follows. From (2.4.2) we deduce

$$\mathbf{r}^k = \mathbf{b} - A\mathbf{x}^k = A(\mathbf{x} - \mathbf{x}^k) = -A\mathbf{e}^k \quad ,$$

where $\mathbf{e}^k$ is the error term defined in (2.2.3). From (2.2.4) and (2.4.3) it follows

$$\mathbf{e}^{k+1} = \mathbf{e}^{k+1}(\alpha) = (I - \alpha P^{-1}A)\mathbf{e}^k \quad ,$$

hence (defining by $| \cdot |_A$ the vector norm induced by the scalar product $(\mathbf{u}, \mathbf{v})_A := (A\mathbf{u}, \mathbf{v})$)

$$|\mathbf{e}^{k+1}|_A^2 = (A\mathbf{e}^{k+1}, \mathbf{e}^{k+1}) = -(\mathbf{r}^{k+1}, \mathbf{e}^{k+1})$$
$$= -(\mathbf{r}^k, \mathbf{e}^k) + \alpha[(\mathbf{r}^k, P^{-1}A\mathbf{e}^k) + (A\mathbf{z}^k, \mathbf{e}^k)] - \alpha^2(A\mathbf{z}^k, P^{-1}A\mathbf{e}^k) \quad .$$

The PR-SD iteration is characterized by taking in (2.4.2) $\alpha = \alpha_k$, where

$$(2.4.15) \qquad \alpha_k \text{ is such that } \left(\frac{d}{d\alpha} |\mathbf{e}^{k+1}(\alpha)|_A^2 \right)_{|\alpha=\alpha_k} = 0 \quad .$$

Writing $\mathbf{x}^{k+1}(\alpha) = \mathbf{x}^k + \alpha\mathbf{z}^k$, this is equivalent to the minimization of $\phi(\mathbf{x}^{k+1}(\alpha))$, where $\phi(\mathbf{w})$ is the quadratic functional

$$\phi(\mathbf{w}) := \frac{1}{2}(A\mathbf{w}, \mathbf{w}) - (\mathbf{b}, \mathbf{w})$$

(see also the following Section 2.4.2). An elementary calculation yields

$$\alpha_k = \frac{1}{2} \frac{(\mathbf{r}^k, P^{-1}A\mathbf{e}^k) + (A\mathbf{z}^k, \mathbf{e}^k)}{(A\mathbf{z}^k, P^{-1}A\mathbf{e}^k)} = \frac{1}{2} \frac{(\mathbf{r}^k, \mathbf{z}^k) + (A\mathbf{z}^k, A^{-1}\mathbf{r}^k)}{(A\mathbf{z}^k, \mathbf{z}^k)} \quad ,$$

therefore

$$(2.4.16) \qquad \alpha_k = \frac{(\mathbf{z}^k, \mathbf{r}^k)}{(\mathbf{z}^k, A\mathbf{z}^k)} \quad .$$

Again the numerator is different from 0 unless $\mathbf{r}^k = \mathbf{0}$.

Noticing that

$$\frac{|\mathbf{e}^k|_A^2 - |\mathbf{e}^{k+1}|_A^2}{|\mathbf{e}^k|_A^2} = \frac{(\mathbf{z}^k, \mathbf{z}^k)_P^2}{(P^{-1}A\mathbf{z}^k, \mathbf{z}^k)_P \, (A^{-1}P\mathbf{z}^k, \mathbf{z}^k)_P} \quad ,$$

using again the Kantorovich inequality, the following error estimate holds

$$(2.4.17) \qquad |\mathbf{e}^{k+1}|_A \leq \rho^* |\mathbf{e}^k|_A \quad , \quad k \in \mathbb{N} \quad ,$$

with the reduction factor ρ^* given by (2.4.10).

A possible definition of the *asymptotic rate of convergence* of the above iteration is

$$(2.4.18) \qquad \sigma := - \log \rho^* \quad .$$

(Sometimes, an alternative concept is used to express the convergence property of a method, based on averaging over k steps the reduction factor of the residuals.)

Its reciprocal, say $\mathcal{I}$, measures the average number of iterations required to reduce the initial error by a factor $1/e$. The larger the convergence rate, the fewer iterations are required to obtain a solution to a given accuracy. For the preconditioned Richardson method described so far, in all cases (2.4.7), (2.4.13) and (2.4.16), the necessary number of iterations increases as

$$(2.4.19) \qquad \mathcal{I} \simeq \frac{1}{2}\chi_{sp}(P^{-1}A) \ .$$

The choice of matrix P in (2.4.1) can therefore be regarded as a way to reduce the condition number of A, and, consequently, the number of iterations $\mathcal{I}$. In this respect, P can be considered a *preconditioner* of A, and the methods introduced in this Section are preconditioned Richardson methods. The case in which P is the identity matrix is referred to as the Richardson method.

The issue of preconditioning a linear system is developed in some detail in Section 2.5.

Remark 2.4.1 If P is the identity matrix then $\mathbf{z}^k = \mathbf{r}^k$. In such case, (2.4.16) becomes

$$\alpha_k = \frac{|\mathbf{r}^k|^2}{(\mathbf{r}^k, A\mathbf{r}^k)} \ ,$$

and (2.4.2) yields

$$(2.4.20) \qquad \mathbf{x}^{k+1} = \mathbf{x}^k + \frac{|\mathbf{r}^k|^2}{(\mathbf{r}^k, A\mathbf{r}^k)}\mathbf{r}^k \ .$$

This is the *Gradient method* or the *Richardson-Steepest Descent method*, already considered by Cauchy (1847). When P is different from the identity, the iterative scheme (2.4.2), with α_k chosen as in (2.4.16), can also be called the *preconditioned Gradient method*. $\square$

Remark 2.4.2 As the iteration (2.2.2) is equivalent to

$$P(\mathbf{x}^{k+1} - \mathbf{x}^k) = \mathbf{r}^k \ , \quad k \in \mathbb{N} \ ,$$

we can interpret Jacobi, Gauss-Seidel, S.O.R. and S.S.O.R. methods as stationary preconditioned Richardson methods, with the acceleration parameter $\alpha = 1$ and the preconditioner P given by (2.3.2), (2.3.6), (2.3.12) and (2.3.17), respectively. $\square$

2.4.2 Conjugate Gradient Method

Assume that A is symmetric and positive definite. Then it is easily verified that solving (2.1.1) is equivalent to minimizing the quadratic functional

$$(2.4.21) \qquad \phi(\mathbf{w}) := \frac{1}{2}(A\mathbf{w}, \mathbf{w}) - (\mathbf{b}, \mathbf{w}) \ .$$

Indeed, the minimum value of ϕ is $-(1/2)(A^{-1}\mathbf{b}, \mathbf{b})$, and it is attained for $\mathbf{x} = A^{-1}\mathbf{b}$. We also notice that

$$(2.4.22) \qquad \mathbf{r}^k = -\nabla\phi(\mathbf{x}^k) \ ,$$

i.e., the residual defined in (2.3.3) is the negative gradient of ϕ at $\mathbf{x}^k$.

When $P = I$, the Richardson methods invariably update the iterate by incrementing it in the direction of the negative gradient of ϕ (see (2.4.2)):

$$(2.4.23) \qquad \mathbf{x}^{k+1} := \mathbf{x}^k + \alpha_k \mathbf{r}^k \ .$$

In *conjugate direction methods* the update of the iterate is obtained from (2.4.23) by replacing $\mathbf{r}^k$ with a new direction $\mathbf{p}^k$ that is not parallel to that of the gradient. These methods were originally proposed by Hestenes and Stiefel (1952) (see also Hestenes (1980)) as direct methods for solving symmetric and positive definite linear systems, as (in the absence of round-off errors) they produce the exact solution in a finite number of steps. Nowadays, they are implemented as iterative schemes which are capable of producing very accurate results in a small number of iterations.

In the *Conjugate Gradient (CG) method*, the directions $\{\mathbf{p}^k\}$ are A-*conjugate*, i.e., they satisfy the orthogonality property $(\mathbf{p}^j, A\mathbf{p}^m) = 0$ for $m \neq j$. In particular

$$(2.4.24) \qquad (\mathbf{p}^{k+1}, A\mathbf{p}^k) = 0 \qquad \forall\, k \in \mathbb{N} \ .$$

The idea of the method is based on the following remark. Let $\mathbf{p}^0, ..., \mathbf{p}^m$ be linearly independent vectors, $\mathbf{x}^0$ being an initial guess, and construct the sequence

$$\mathbf{x}^{k+1} = \mathbf{x}^k + \alpha_k \mathbf{p}^k \quad , \quad 0 \leq k \leq m \ ,$$

where α_k are non-zero real numbers. Then $\mathbf{x}^{k+1}$ minimizes the functional ϕ on the $(k+1)$-hyperplane

$$\mathbf{w} = \mathbf{x}^0 + \sum_{j=0}^{k} \gamma_j \mathbf{p}^j \quad , \quad \gamma_j \in \mathbb{R} \ ,$$

if, and only if, $\mathbf{p}^j$ are A-conjugate and $\alpha_k = \frac{(\mathbf{r}^k, \mathbf{p}^k)}{(\mathbf{p}^k, A\mathbf{p}^k)}$ (see Hestenes (1980), pp. 102–103).

Therefore, the CG method can be formulated as follows. Let $\mathbf{x}^0$ be an initial guess, $\mathbf{r}^0 := \mathbf{b} - A\mathbf{x}^0$, and set $\mathbf{p}^0 := \mathbf{r}^0$. For each $k \in \mathbb{N}$, the k-th iteration is made according to the formulae:

$$(2.4.25) \qquad \alpha_k := \frac{(\mathbf{r}^k, \mathbf{p}^k)}{(\mathbf{p}^k, A\mathbf{p}^k)} = \frac{|\mathbf{r}^k|^2}{(\mathbf{p}^k, A\mathbf{p}^k)}$$

$$(2.4.26) \qquad \mathbf{x}^{k+1} := \mathbf{x}^k + \alpha_k \mathbf{p}^k$$

$$(2.4.27) \qquad \mathbf{r}^{k+1} := \mathbf{r}^k - \alpha_k A\mathbf{p}^k$$

$$(2.4.28) \qquad \beta_{k+1} := -\frac{(\mathbf{r}^{k+1}, A\mathbf{p}^k)}{(\mathbf{p}^k, A\mathbf{p}^k)} = \frac{|\mathbf{r}^{k+1}|^2}{|\mathbf{r}^k|^2}$$

$$(2.4.29) \qquad \mathbf{p}^{k+1} := \mathbf{r}^{k+1} + \beta_{k+1}\mathbf{p}^k \quad .$$

To prove that the coefficients α_k and β_{k+1} can also be written in the second form indicated above, one verifies at first that

$$(2.4.30) \qquad (\mathbf{p}^k, \mathbf{r}^{k+1}) = 0 \ , \ (\mathbf{p}^k, A\mathbf{p}^k) = (\mathbf{r}^k, A\mathbf{p}^k) \ , \ (\mathbf{r}^{k+1}, \mathbf{r}^k) = 0 \ .$$

(To show these last two relations, the symmetry of A plays a crucial role.)

Let us notice that procedure (2.4.25)-(2.4.29) is well defined unless $\mathbf{p}^k = 0$. In this case, however, it follows

$$\mathbf{r}^k = -\beta_k \mathbf{p}^{k-1} \quad , \quad k \geq 1 \ ,$$

and moreover

$$|\mathbf{r}^k|^2 = -\beta_k(\mathbf{p}^{k-1}, \mathbf{r}^k) = 0 \ .$$

Consequently $\mathbf{x}^k = A^{-1}\mathbf{b}$ is the solution we are looking for.

On the other hand, the iterate $\mathbf{x}^{k+1}$ is equal to the previous one $\mathbf{x}^k$ if, and only if, $\mathbf{p}^k = 0$ or $\alpha_k = 0$, which again means $\mathbf{x}^k = A^{-1}\mathbf{b}$.

For any fixed direction $\mathbf{p}^k$, let us set

$$\mathbf{e}^{k+1}(\alpha) := \mathbf{x}^{k+1}(\alpha) - \mathbf{x} = (\mathbf{x}^k + \alpha\mathbf{p}^k) - \mathbf{x} \ .$$

The value α_k given in (2.4.25) is the one for which

$$(2.4.31) \qquad \left(\frac{d}{d\alpha}|\mathbf{e}^{k+1}(\alpha)|_A^2\right)_{|\alpha=\alpha_k} = 0 \ .$$

Therefore, α_k is the optimal parameter which minimizes the value of the functional $\phi(\mathbf{x}^k + \alpha\mathbf{p}^k)$ among all possible $\alpha \in \mathbb{R}$. On the other hand, the value β_{k+1} given in (2.4.28) is the one ensuring that (2.4.24) holds.

Let us introduce

$$(2.4.32) \qquad K_k(\mathbf{r}^0) := \mathrm{span}\{\mathbf{r}^0, A\mathbf{r}^0, ..., A^{k-1}\mathbf{r}^0\} \ ,$$

the *Krylov space* of order k generated from $\mathbf{r}^0$. We will see below that $K_k(\mathbf{r}^0) = \mathrm{span}\{\mathbf{p}^0, \mathbf{p}^1, ..., \mathbf{p}^{k-1}\}$, therefore, for each $k \geq 1$, $\mathbf{x}^k$ also satisfies

$$\phi(\mathbf{x}^k) \leq \phi(\mathbf{w}) \qquad \forall \, \mathbf{w} \in \mathbf{x}^0 + K_k(\mathbf{r}^0) \ .$$

For a complete derivation of the CG method we refer, e.g., to Golub and Van Loan (1989). Hereafter we recall some of the main properties of this method. We claim that

$$(2.4.33) \qquad (\mathbf{r}^j, \mathbf{r}^m) = 0 \ \text{ and } \ (\mathbf{p}^j, A\mathbf{p}^m) = 0 \ \text{ for } j \neq m \ .$$

This result is proven by induction on m. First of all, we have already noticed that $(\mathbf{r}^1, \mathbf{r}^0) = 0$ and $(A\mathbf{p}^1, \mathbf{p}^0) = 0$. Assume now that

$$(\mathbf{r}^l, \mathbf{r}^j) = (A\mathbf{p}^l, \mathbf{p}^j) = 0 \quad \text{ for } \ 0 \leq j < l \leq m \ .$$

We want to show that the same relation holds true for $0 \leq j < l \leq m + 1$. It has been already proven for $j = m$ and $l = m + 1$ (see (2.4.30) and (2.4.24)). Taking $1 \leq j < m$ and $l = m + 1$, it follows

$$(\mathbf{r}^{m+1}, \mathbf{r}^j) = (\mathbf{r}^m, \mathbf{r}^j) - \alpha_m(A\mathbf{p}^m, \mathbf{r}^j) = -\alpha_m(A\mathbf{p}^m, \mathbf{p}^j - \beta_{j-1}\mathbf{p}^{j-1}) = 0$$

by the induction hypothesis. As $\mathbf{r}^0 = \mathbf{p}^0$, the same is also true for $j = 0$. Moreover, using the symmetry of A we find for $0 \leq j < m$

$$(\mathbf{p}^{m+1}, A\mathbf{p}^j) = (\mathbf{r}^{m+1}, A\mathbf{p}^j) + \beta_{m+1}(A\mathbf{p}^m, \mathbf{p}^j) = (\mathbf{r}^{m+1}, \mathbf{r}^j - \mathbf{r}^{j+1})\alpha_j^{-1} = 0 \ ,$$

when $\alpha_j \neq 0$. However, if $\alpha_j = 0$ it follows at once that $\mathbf{r}^j = \mathbf{p}^j = \mathbf{0}$, and the proof is complete.

The first relation in (2.4.33) implies that $\mathbf{r}^k = \mathbf{0}$ for some $k \leq n$ (the order of the matrix A). Thus, convergence to the exact solution is achieved (in principle) in a number of steps not greater than n. Furthermore, if A has m distinct eigenvalues, with $m < n$, then convergence occurs in m iterations. However, round-off errors contaminate residuals and conjugate directions, making the fulfillment of (2.4.33) unrealistic. On the other hand, often, the order n of the matrix is so large that performing all of the n steps of the CG procedure would be inconvenient.

The good convergence properties of the CG iteration is reflected by the estimate

$$(2.4.34) \qquad |\mathbf{e}^k|_A \leq 2 \left(\frac{\sqrt{\chi_{sp}(A)} - 1}{\sqrt{\chi_{sp}(A)} + 1} \right)^k |\mathbf{e}^0|_A \ , \qquad k \in \mathbb{N} \ .$$

This can be proven in the following way. First of all, from (2.4.26) one has

$$\mathbf{x}^{k+1} - \mathbf{x}^0 = \sum_{s=0}^{k} \alpha_s \mathbf{p}^s \ ,$$

and consequently, taking $0 \leq j \leq k$ and setting $\mathbf{x} = A^{-1}\mathbf{b}$

$$(A\mathbf{p}^j, \mathbf{x}^{k+1} - \mathbf{x}^0) = \alpha_j(A\mathbf{p}^j, \mathbf{p}^j) = (\mathbf{p}^j, \mathbf{r}^j) = (A\mathbf{p}^j, \mathbf{x} - \mathbf{x}^j) = (A\mathbf{p}^j, \mathbf{x} - \mathbf{x}^0) \ ,$$

by (2.4.33). This shows that $\mathbf{x}^{k+1} - \mathbf{x}^0$ is the orthogonal projection of $\mathbf{x} - \mathbf{x}^0$ onto the space spanned by $\mathbf{p}^0, \mathbf{p}^1, ..., \mathbf{p}^k$, with respect to the scalar product $(\mathbf{u}, \mathbf{v})_A = (A\mathbf{u}, \mathbf{v})$.

On the other hand, we claim that for $k \geq 0$

$$\text{span}\{\mathbf{p}^0, \mathbf{p}^1, ..., \mathbf{p}^k\} = \text{span}\{\mathbf{r}^0, A\mathbf{r}^0, ..., A^k\mathbf{r}^0\} =: K_{k+1}(\mathbf{r}^0) .$$

In fact, from (2.4.29) and $\mathbf{p}^0 = \mathbf{r}^0$ it follows at once

$$\text{span}\{\mathbf{p}^0, \mathbf{p}^1, ..., \mathbf{p}^k\} = \text{span}\{\mathbf{r}^0, \mathbf{r}^1, ..., \mathbf{r}^k\} , \quad k \geq 0 .$$

Employing now an induction argument, assume that $\text{span}\{\mathbf{p}^0, \mathbf{p}^1, ..., \mathbf{p}^k\} = K_{k+1}(\mathbf{r}^0)$. Then, $A\mathbf{p}^k \in K_{k+2}(\mathbf{r}^0)$ and from (2.4.27) $\mathbf{r}^{k+1} \in K_{k+2}(\mathbf{r}^0)$, i.e., $\text{span}\{\mathbf{p}^0, \mathbf{p}^1, ..., \mathbf{p}^{k+1}\} \subset K_{k+2}(\mathbf{r}^0)$.

On the other hand, $A^k\mathbf{r}^0 \in \text{span}\{\mathbf{p}^0, \mathbf{p}^1, ..., \mathbf{p}^k\}$ by the induction hypothesis and consequently $A^{k+1}\mathbf{r}^0 \in \text{span}\{A\mathbf{p}^0, A\mathbf{p}^1, ..., A\mathbf{p}^k\}$. Thus, from (2.4.27) we have $A^{k+1}\mathbf{r}^0 \in \text{span}\{\mathbf{r}^0, \mathbf{r}^1, ..., \mathbf{r}^{k+1}\}$ and the thesis follows.

Thus $\mathbf{x}^{k+1} - \mathbf{x}^0$ is the projection of $\mathbf{x} - \mathbf{x}^0$ onto $K_{k+1}(\mathbf{r}^0)$, and

$$|\mathbf{x} - \mathbf{x}^{k+1}|_A = |\mathbf{x} - \mathbf{x}^0 - (\mathbf{x}^{k+1} - \mathbf{x}_0)|_A = \min_{\mathbf{w} \in K_{k+1}(\mathbf{r}^0)} |\mathbf{x} - \mathbf{x}^0 - \mathbf{w}|_A .$$

Since $\mathbf{r}^0 = A(\mathbf{x} - \mathbf{x}^0)$, any $\mathbf{w} \in K_{k+1}(\mathbf{r}^0)$ can be written as

$$\mathbf{w} = \sum_{j=1}^{k+1} \gamma_j A^j (\mathbf{x} - \mathbf{x}^0) ,$$

and consequently

$$|\mathbf{x} - \mathbf{x}^{k+1}|_A = \min_{p \in \mathbb{P}_{k+1}^*} |p(A)(\mathbf{x} - \mathbf{x}^0)|_A ,$$

where $\mathbb{P}_{k+1}^*$ is the set of polynomials $p : \mathbb{R} \to \mathbb{R}$ of degree less than or equal to $k + 1$ and such that $p(0) = 1$.

From the assumption $A = A^T > 0$, we know that there exists an orthonormal basis given by eigenvectors of A, and that the eigenvalues λ_j of A are strictly positive. By expanding $\mathbf{x} - \mathbf{x}^0$ with respect to these eigenvectors, some simple calculations yield

$$|p(A)(\mathbf{x} - \mathbf{x}^0)|_A \leq \max_{1 \leq j \leq n} |p(\lambda_j)| \, |\mathbf{x} - \mathbf{x}^0|_A .$$

We choose

$$p(x) = \frac{T_{k+1}\left(\dfrac{\lambda_{\max} + \lambda_{\min} - 2x}{\lambda_{\max} - \lambda_{\min}}\right)}{T_{k+1}\left(\dfrac{\lambda_{\max} + \lambda_{\min}}{\lambda_{\max} - \lambda_{\min}}\right)} ,$$

where $T_{k+1}(x)$ is the Chebyshev polynomial of degree $k + 1$ introduced in Section 4.3.1. Since $|T_{k+1}(y)| \leq 1$ for $|y| \leq 1$ it follows that

$$\max_{x\in[\lambda_{\min},\lambda_{\max}]}|p(x)| \leq \left[T_{k+1}\left(\frac{\lambda_{\max}+\lambda_{\min}}{\lambda_{\max}-\lambda_{\min}}\right)\right]^{-1} .$$

Evaluating this expression, one finds that

$$\max_{x\in[\lambda_{\min},\lambda_{\max}]}|p(x)|$$

$$\leq 2\left[1+\left(\frac{\sqrt{\lambda_{\max}}-\sqrt{\lambda_{\min}}}{\sqrt{\lambda_{\max}}+\sqrt{\lambda_{\min}}}\right)^{2k+2}\right]^{-1}\left(\frac{\sqrt{\lambda_{\max}}-\sqrt{\lambda_{\min}}}{\sqrt{\lambda_{\max}}+\sqrt{\lambda_{\min}}}\right)^{k+1},$$

i.e., (2.4.34) holds.

An interesting consequence is that the reciprocal of the rate of convergence σ behaves now as

$$(2.4.35) \qquad\qquad \mathcal{I} \simeq \frac{1}{2}\sqrt{\chi_{sp}(A)} .$$

This is more favorable than (2.4.19) when P is the identity matrix.

However, if $\chi_{sp}(A)$ is large, a preconditioning procedure is advisable. If we still denote the preconditioner by P, a symmetric and positive definite matrix, the *preconditioned Conjugate Gradient (PCG) method* reads:

$$(2.4.36) \qquad\qquad \alpha_k := \frac{(\mathbf{z}^k, \mathbf{r}^k)}{(\mathbf{p}^k, A\mathbf{p}^k)}$$

$$(2.4.37) \qquad\qquad \mathbf{x}^{k+1} := \mathbf{x}^k + \alpha_k \mathbf{p}^k$$

$$(2.4.38) \qquad\qquad \mathbf{r}^{k+1} := \mathbf{r}^k - \alpha_k A\mathbf{p}^k$$

$$(2.4.39) \qquad\qquad P\mathbf{z}^{k+1} := \mathbf{r}^{k+1}$$

$$(2.4.40) \qquad\qquad \beta_{k+1} := \frac{(\mathbf{z}^{k+1}, \mathbf{r}^{k+1})}{(\mathbf{z}^k, \mathbf{r}^k)}$$

$$(2.4.41) \qquad\qquad \mathbf{p}^{k+1} := \mathbf{z}^{k+1} + \beta_{k+1}\mathbf{p}^k ,$$

for $k \geq 0$. In this case, initialization is accomplished as in the CG method, taking an initial guess $\mathbf{x}^0$ and setting $\mathbf{r}^0 := \mathbf{b} - A\mathbf{x}^0$, $\mathbf{p}^0 := \mathbf{z}^0 := P^{-1}\mathbf{r}^0$.

The error estimate (2.4.34) becomes

$$(2.4.42) \qquad |e^k|_A \leq 2\left(\frac{\sqrt{\chi_{sp}(P^{-1}A)}-1}{\sqrt{\chi_{sp}(P^{-1}A)}+1}\right)^k |e^0|_A , \quad k \in \mathbb{N} .$$

The proof is performed as in the non-preconditioned case, upon operating the following changes: the euclidean scalar product $(\mathbf{u}, \mathbf{v})$ is replaced by the scalar product $(\mathbf{u}, \mathbf{v})_P$ induced by P (see (2.4.11)), the matrix A by $P^{-1}A$ and the residual $\mathbf{r}^k$ by $\mathbf{z}^k = P^{-1}\mathbf{r}^k$.

The reduction factor may be considerably smaller than in the CG iteration if P is suitably chosen. On the other hand, we stress that at each step of the PCG algorithm the linear system (2.4.39) needs to be solved.

We conclude by noticing that a frequently used stopping criterium for conjugate gradient iterations is

(2.4.43)
$$|\mathbf{r}^k| \leq \varepsilon |\mathbf{b}|$$

where $\varepsilon > 0$ is a prescribed tolerance.

Conjugate gradient iterations can also be used to accelerate the convergence of any other method that is based on the procedure (2.2.2), so long as P (the preconditioner) is symmetric and positive definite. This includes, in particular, Jacobi and S.S.O.R. iterations. The idea, which is due to Concus, Golub and O'Leary (1976), consists of finding suitable formulae for parameters ω_k, γ_k that accelerate the convergence of $\mathbf{x}^k$ to $\mathbf{x}$. This algorithm reads as follows. Let us set $\mathbf{x}^{-1} := \mathbf{x}^0 := \mathbf{0}$, $\mathbf{r}^0 := \mathbf{b}$; then for $k \geq 0$

(2.4.44)
$$\begin{cases} P\mathbf{z}^k := \mathbf{r}^k \\[2mm] \gamma_k := \dfrac{(\mathbf{z}^k, P\mathbf{z}^k)}{(\mathbf{z}^k, A\mathbf{z}^k)} \\[2mm] \text{if } k = 0 \text{ set } \omega_1 = 1 \text{ ; if } k \geq 1 \text{ set} \\[2mm] \omega_{k+1} := \left[1 - \dfrac{\gamma_k (\mathbf{z}^k, P\mathbf{z}^k)}{\gamma_{k-1}\omega_k(\mathbf{z}^{k-1}, P\mathbf{z}^{k-1})} \right]^{-1} \\[2mm] \mathbf{x}^{k+1} := \mathbf{x}^{k-1} + \omega_{k+1}(\gamma_k \mathbf{z}^k + \mathbf{x}^k - \mathbf{x}^{k-1}) \\[2mm] \mathbf{r}^{k+1} := \mathbf{b} - A\mathbf{x}^{k+1} \ . \end{cases}$$

For an interesting historical survey on conjugate gradient methods see Golub and O'Leary (1989).

We summarize the basic features of each method in Table 2.4.1. In the first column we enter the various methods, in the second column the assumptions on the matrix A (the symbol > 0 denotes that the matrix is positive definite). Column 3 shows the optimal value of the parameter α_k and column 4 the quantity which is minimized at the k-th iteration. Finally, the last column shows the error estimate after k iterations (with $c = \chi_{sp}(P^{-1}A)$). The preconditioner P is always assumed to be symmetric and positive definite.

2.5 Preconditioning

From the error estimates shown throughout the previous Sections, it is clear that the performance of a direct or an iterative procedure may strongly benefit from a low condition number of the system matrix. For ill-conditioned systems, the idea is to resort to a preconditioner, i.e., to a non-singular matrix P, and consider the equivalent system

Table 2.4.1. Summary on the iterative methods presented in Section 2.4.
(*) Let us remind that the stationary Richardson method converges under weaker assumptions on the matrices A and P, as stated in Theorem 2.4.1

	A	α_k	Minimization of	Error estimate
Stationary Richardson (*)	$A = A^T > 0$	$\dfrac{2}{\eta_{\min} + \eta_{\max}}$	$\rho(I - \alpha P^{-1}A)$	$|e^k|_P \leq \left(\dfrac{c-1}{c+1}\right)^k |e^0|_P$
Richardson- Minimum Residual	$A > 0$	$\dfrac{(z^k, Az^k)}{(Az^k, P^{-1}Az^k)}$	$|z^{k+1}|_P$ $(= |e^{k+1}|_{A^T P^{-1} A})$	$|z^k|_P \leq \left(\dfrac{c-1}{c+1}\right)^k |z^0|_P$ (if $A = A^T$)
Richardson- Steepest Descent (Gradient)	$A = A^T > 0$	$\dfrac{(z^k, r^k)}{(z^k, Az^k)}$	$|e^{k+1}|_A$ or $\phi(x^{k+1})$	$|e^k|_A \leq \left(\dfrac{c-1}{c+1}\right)^k |e^0|_A$
Conjugate Gradient	$A = A^T > 0$	$\dfrac{(z^k, r^k)}{(p^k, Ap^k)}$	$|e^{k+1}|_A$ or $\phi(x^{k+1})$	$|e^k|_A \leq 2\left(\dfrac{\sqrt{c}-1}{\sqrt{c}+1}\right)^k |e^0|_A$

(2.5.1)
$$P^{-1}A\mathbf{x} = P^{-1}\mathbf{b} \ .$$

The basic requirements in order for P to be a good preconditioner of A are that P is "easy" to invert. Furthermore, the condition number of $P^{-1}A$ should be definitely smaller than the one of A. P is said an *optimal* preconditioner if the condition number of $P^{-1}A$ is bounded uniformly with respect to n (the order of A).

The drawback with (2.5.1) is that $P^{-1}A$ could fail to be symmetric or positive definite even if both factors are so. However, if both P and A are symmetric and positive definite, owing to Lemma 2.4.1 we can apply the conjugate gradient iteration directly to (2.5.1), by replacing A with $P^{-1}A$, $\mathbf{b}$ with $P^{-1}\mathbf{b}$, $\mathbf{r}^k$ with $\mathbf{z}^k = P^{-1}\mathbf{r}^k$, $(\cdot, \cdot)$ with $(\cdot, \cdot)_P$ and $|\cdot|$ with $|\cdot|_P$, through (2.4.25)-(2.4.29). If is not difficult to see that what we find is precisely the preconditioned conjugate gradient iteration (2.4.36)-(2.4.41).

An alternative way for obtaining a linear system associated to a symmetric and positive definite matrix is as follows. Let us write the symmetric and positive definite preconditioner P as

(2.5.2)
$$P = C^2 \ ,$$

where C is a symmetric and non-singular matrix, for instance, the square root of P. (We recall that the square root $\sqrt{P}$ is defined as follows. Since P can be written as $P = T \Lambda T^T$, where $\Lambda := \mathrm{diag}(\lambda_1, ..., \lambda_n)$ is the diagonal matrix of eigenvalues and $T := (\boldsymbol{\omega}^1|...|\boldsymbol{\omega}^n)$ is the matrix of right eigenvectors, satisfying $T^{-1} = T^T$, the square root is $\sqrt{P} := T\Lambda^{1/2}T^T$, with $\Lambda^{1/2} := \mathrm{diag}(\sqrt{\lambda_1}, ..., \sqrt{\lambda_n})$.) Then instead of (2.1.1) we consider the equivalent system

(2.5.3) $A_*\mathbf{x}_* = \mathbf{b}_*$ where $A_* := C^{-1}AC^{-1}$, $\mathbf{b}_* := C^{-1}\mathbf{b}$, $\mathbf{x}_* := C\mathbf{x}$.

The matrix A_* is symmetric and positive definite with respect to the usual euclidean scalar product, therefore the CG iteration (2.4.25)-(2.4.29) can be applied directly to (2.5.3) and produces a sequence of approximants $\{\mathbf{x}_*^k\}$. It can easily be shown that the resulting algorithm, when expressed in term of the iterate $\mathbf{x}^k := C^{-1}\mathbf{x}_*^k$, fits again under the form (2.4.36)-(2.4.41). In particular, the preconditioner P solely is needed to define the iterative procedure, and the matrix C has only resulted to be an auxiliary tool for the theoretical justification of the scheme.

A third possible procedure is to write the preconditioner P as

(2.5.4)
$$P = HH^T \ ,$$

where H is a non-singular matrix, for instance, the lower triangular matrix obtained by Cholesky decomposition. Then one considers the equivalent system

$$A^*\mathbf{x}^* = \mathbf{b}^* \ \text{where} \ A^* := H^{-1}AH^{-T} \ , \ \mathbf{b}^* : H^{-1}\mathbf{b} \ , \ \mathbf{x}^* := H^T\mathbf{x}$$

(here, H^{-T} denotes $(H^{-1})^T = (H^T)^{-1}$). Also the matrix A^* turns out to be symmetric and positive definite with respect to the usual euclidean scalar product, and the CG iteration (2.4.25)-(2.4.29), rewritten in term of $\mathbf{x}^k$, takes the form (2.4.36)-(2.4.41).

Concerning the spectral properties of $P^{-1}A$, A_* and A^*, we have the following result:

(2.5.5) $\begin{cases} \text{when } P \text{ is given by (2.5.2),} \\ \quad \text{the matrices } P^{-1}A \text{ and } A_* \text{ have the same eigenvalues ;} \\ \text{when } P \text{ is given by (2.5.4),} \\ \quad \text{the matrices } P^{-1}A \text{ and } A^* \text{ have the same eigenvalues .} \end{cases}$

This follows from the similarity transformations

$$C(P^{-1}A)C^{-1} = C^{-1}AC^{-1} = A_* \; , \; H^T(P^{-1}A)H^{-T} = H^{-1}AH^{-T} = A^* \; .$$

We conclude that the preconditioned matrix $P^{-1}A$ has the same spectral condition number as A_* (when $P = C^2$) and A^* (when $P = HH^T$).

To estimate $\chi_{sp}(P^{-1}A)$ the following result is often quite useful.

Theorem 2.5.1 *Let A and M be two symmetric and positive definite matrices such that*

(2.5.6) $\qquad \exists \, K_1, K_2 > 0 : \quad K_1(M\mathbf{w}, \mathbf{w}) \leq (A\mathbf{w}, \mathbf{w}) \leq K_2(M\mathbf{w}, \mathbf{w})$

for each $\mathbf{w} \in \mathbb{R}^n$. Then

(2.5.7) $\qquad\qquad\qquad\qquad \chi_{sp}(M^{-1}A) \leq \dfrac{K_2}{K_1} \; .$

Proof. Since $M^{-1}A$ is symmetric and positive definite with respect to $(\cdot, \cdot)_M$, it has positive eigenvalues η_i and real eigenvectors $\boldsymbol{\omega}_i$, $i = 1, ..., n$. Moreover, from the spectral relation

$$M^{-1}A\boldsymbol{\omega}_i = \eta_i\boldsymbol{\omega}_i$$

it follows that

$$(M^{-1}A\boldsymbol{\omega}_i, \boldsymbol{\omega}_i)_M = \eta_i(\boldsymbol{\omega}_i, \boldsymbol{\omega}_i)_M \; ,$$

thus η_i is given by the Rayleigh quotient

(2.5.8) $\qquad\qquad\qquad\qquad \eta_i = \dfrac{(M^{-1}A\boldsymbol{\omega}_i, \boldsymbol{\omega}_i)_M}{(\boldsymbol{\omega}_i, \boldsymbol{\omega}_i)_M} \; .$

From (2.5.6) we deduce

$$K_1 \leq \frac{(M^{-1}A\mathbf{w}, \mathbf{w})_M}{(\mathbf{w}, \mathbf{w})_M} \leq K_2 \quad \forall \, \mathbf{w} \in \mathbb{R}^n \; , \; \mathbf{w} \neq \mathbf{0} \; ,$$

and therefore

$$K_1 \leq \eta_i \leq K_2 \text{ for any eigenvalue } \eta_i \text{ of } M^{-1}A \ ,$$

from which (2.5.7) follows.   □

Now the question is how to choose the preconditioner P so that $P^{-1}A$ (or, equivalently, $C^{-1}AC^{-1}$, $H^{-1}AH^{-T}$) has an improved condition number. Clearly, for a preconditioned iteration method to be an effective procedure, we must be able to solve easily linear systems associated to P. Hereafter, we always assume that A is symmetric and positive definite.

Keeping in mind the decomposition (2.3.2), a simple preconditioner is provided by the diagonal matrix

$$(2.5.9) \qquad\qquad P := D = \mathrm{diag}(a_{11}, ..., a_{nn}) \ ,$$

i.e., the Jacobi preconditioner, or else by

$$(2.5.10) \qquad P := \mathrm{diag}(c_1, ..., c_n) \ , \text{ with } c_i := \left(\sum_{j=1}^{n} a_{ij}^2 \right)^{1/2} .$$

Another way is to use the S.S.O.R. preconditioner

$$(2.5.11) \qquad\qquad P := (D + \omega E)D^{-1}(D + \omega E)^T \ ,$$

for some $\omega \in (0, 2)$. In this case a matrix H that satisfies $P = HH^T$ is given by

$$(2.5.12) \quad H := (D + \omega E)(D^{1/2})^{-1} \ , \text{ with } D^{1/2} := \mathrm{diag}(\sqrt{a_{11}}, ..., \sqrt{a_{nn}}) \ .$$

With the choice of the S.S.O.R. preconditioner, if A is the matrix associated with the five-point finite difference discretization of the Poisson equation on a square based on line ordering and $\omega = \omega^*$ (see (2.3.19)), then $\chi_{sp}(P^{-1}A) = O(h^{-1})$, whereas $\chi_{sp}(A) = O(h^{-2})$, where h is the grid-spacing.

Another strategy involves computing an *incomplete Cholesky decomposition* of A, say H, and choosing (if H is non-singular)

$$(2.5.13) \qquad\qquad P := HH^T \ .$$

The simplest way to compute H is to modify (2.1.10) as follows: we set $h_{ij} = l_{ij}$ if $a_{ij} \neq 0$, otherwise $h_{ij} = 0$. The corresponding algorithm reads:

(2.5.14)
$$\begin{cases} \text{For } k = 1, ..., n \\ \quad h_{kk} = \left(a_{kk} - \sum_{p=1}^{k-1} h_{kp}^2 \right)^{1/2} \\ \quad \text{For } i = k+1, ..., n \text{ and } k \neq n \\ \qquad \text{if } a_{ik} = 0 \\ \qquad\quad \text{then } h_{ik} = 0 \\ \qquad \text{else} \\ \qquad\qquad h_{ik} = \frac{1}{h_{kk}} \left(a_{ik} - \sum_{p=1}^{k-1} h_{ip} h_{kp} \right) \\ \quad \text{end } i \\ \text{end } k \ . \end{cases}$$

(Here, it is understood that the sums don't apply for $k = 1$.) With this approach, P preserves the same sparsity pattern of A. Unfortunately, algorithm (2.5.14) is not always *stable*. Classes of matrices whose incomplete Cholesky decomposition is stable are identified in Manteuffel (1979).

For a non-symmetric matrix A, especially when A is banded, another preconditioning strategy can be based on the so-called *incomplete LU (ILU) decomposition*. We compute the LU decomposition of A but drop any fill-in in L or U outside of the original structure of A.

Higher accuracy incomplete decomposition are also used: $ILU(p)$ allows for p additional diagonals in L and U. We refer to Meijerink and van der Vorst (1981), Evans (1983) and van der Vorst (1989) for a survey.

Other examples are provided by block preconditioners for block matrices (see Golub and Van Loan (1989) for a short presentation) or else polynomial preconditioners. Polynomial preconditioning consists of choosing a polynomial π and replacing the original system (2.1.1) by

$$\pi(A)A\mathbf{x} = \pi(A)\mathbf{b} \ ,$$

where $\pi(A)$ is chosen according to some optimality criterion. The new system is then solved by an iterative procedure. A survey can be found in Saad (1989), where the issue of implementing such preconditioners on supercomputers is also addressed.

We would also like to point out that in specific situations, other kinds of preconditioners can be successfully used. One instance is provided by preconditioners based on different numerical realizations, such as those given by finite element or finite difference discretizations of a differential operator that is being approximated by spectral methods (see Section 6.3.3).

Other examples occur for those preconditioners keeping track of the structure of a differential operator the system matrix refers to. Among other cases, we mention the preconditioners of the pressure matrix arising from the approximation of the Stokes problem (see Section 9.6.1).

2.6 Conjugate Gradient and Lanczos like Methods for Non-Symmetric Problems

This subject is currently receiving much attention, though a totally satisfactory iterative scheme is not yet available. The interested reader should keep abreast of the literature in this field.

If A is a non-singular matrix of order n, throughout this Section we set

$$(2.6.1) \qquad A_S := \frac{1}{2}(A + A^T) \ , \quad A_{SS} := \frac{1}{2}(A - A^T) \ .$$

A_S is the symmetric part of A while A_{SS} is its skew-symmetric part. If A is positive definite so is A_S. When using a preconditioner P, in the definition (2.6.1) A should be replaced by $P^{-1}A$.

In view of the good performance of CG iteration for symmetric systems, a natural approach could be to convert any non-singular system (2.1.1) into the symmetric and positive definite one (*normal equations*, such as those of least squares method, see (2.1.13))

$$(2.6.2) \qquad A^T A \mathbf{x} = A^T \mathbf{b} \ ,$$

and then apply the CG method to (2.6.2). This would lead to the following algorithm (known as *Conjugate Gradient Normal Residual (CGNR) method*.

Let $\mathbf{x}^0$ be given, $\mathbf{r}^0 := \mathbf{b} - A\mathbf{x}^0$, $\mathbf{p}^0 := A^T \mathbf{r}^0$. For each $k \geq 0$ the k-th iteration becomes:

$$(2.6.3) \qquad \alpha_k := \frac{|A^T \mathbf{r}^k|^2}{|A\mathbf{p}^k|^2}$$

$$(2.6.4) \qquad \mathbf{x}^{k+1} := \mathbf{x}^k + \alpha_k \mathbf{p}^k$$

$$(2.6.5) \qquad \mathbf{r}^{k+1} := \mathbf{r}^k - \alpha_k A\mathbf{p}^k$$

$$(2.6.6) \qquad \beta_{k+1} := \frac{|A^T \mathbf{r}^{k+1}|^2}{|A^T \mathbf{r}^k|^2}$$

$$(2.6.7) \qquad \mathbf{p}^{k+1} := A^T \mathbf{r}^{k+1} + \beta_{k+1} \mathbf{p}^k \ .$$

This procedure still converges (theoretically) in no more than n iterations. However, compared to the CG iterations for the symmetric case, it requires extra matrix-vector multiplications. Moreover, the condition number of the transformed system (2.6.2) is higher than that of the original system (2.1.1), as $\lambda_i(A^T A) = \sigma_i^2(A)$, with $\sigma_i(A)$ denoting the singular values of a matrix A (e.g., Golub and Van Loan (1989), p. 427).

2.6.1 GCR, Orthomin and Orthodir Iterations

We recall that if A is symmetric and positive definite the CG method (2.4.25)-(2.4.29) applied to (2.1.1) minimizes at each step the energy norm $|\mathbf{e}^k|_A$

(see (2.4.31)), by selecting direction vectors that are A-conjugate (i.e., A-orthogonal).

The *Conjugate Residual (CR) method* is a variant of CG that minimizes the residual $|\mathbf{r}^k|$ at each iteration, choosing direction vectors that are $A^T A$-orthogonal. Both methods have the finite termination property (i.e., they converge in n steps to the solution in exact arithmetic).

All generalizations of the CG and CR methods for non-symmetric systems form a set of directions which are based on the space

$$(2.6.8) \qquad K_m(\mathbf{r}^0) := \text{span}\{\mathbf{r}^0, A\mathbf{r}^0, A^2\mathbf{r}^0, ..., A^{m-1}\mathbf{r}^0\} \ ,$$

where m is a positive integer. We recall that $K_m(\mathbf{r}^0)$ is the Krylov space of order m generated from $\mathbf{r}^0$.

For a positive definite matrix A, Concus and Golub (1976) and Widlund (1978) derived the *Generalized Conjugate Gradient (GCG) method*, which requires at each iteration the solution of a symmetric system. Eisenstat, Elman and Schultz (1983) proposed the *Generalized Conjugate Residual (GCR) method*, which forms a new direction vector from the current residual by forcing the $A^T A$-orthogonality to all preceding direction vectors. The initialization is: $\mathbf{x}^0$ is a given vector, $\mathbf{r}^0 := \mathbf{b} - A\mathbf{x}^0$, $\mathbf{p}^0 := \mathbf{r}^0$. The k-th iterate ($k \geq 0$) is:

$$(2.6.9) \qquad \alpha_k := \frac{(\mathbf{r}^k, A\mathbf{p}^k)}{|A\mathbf{p}^k|^2}$$

$$(2.6.10) \qquad \mathbf{x}^{k+1} := \mathbf{x}^k + \alpha_k \mathbf{p}^k$$

$$(2.6.11) \qquad \mathbf{r}^{k+1} := \mathbf{r}^k - \alpha_k A\mathbf{p}^k$$

$$(2.6.12) \qquad \beta_j^k := -\frac{(A\mathbf{r}^{k+1}, A\mathbf{p}^j)}{|A\mathbf{p}^j|^2} \ , \quad k - k_0 + 1 \leq j \leq k$$

$$(2.6.13) \qquad \mathbf{p}^{k+1} := \mathbf{r}^{k+1} + \sum_{j=k-k_0+1}^{k} \beta_j^k \mathbf{p}^j$$

$$(2.6.14) \qquad A\mathbf{p}^{k+1} = A\mathbf{r}^{k+1} + \sum_{j=k-k_0+1}^{k} \beta_j^k A\mathbf{p}^j$$

for $k_0 = k + 1$.

If A is positive definite, then the GCR method converges and

$$(2.6.15) \qquad |\mathbf{r}^k| \leq \left[1 - \frac{(\lambda_{\min}(A_S))^2}{\lambda_{\max}(A^T A)}\right]^{k/2} |\mathbf{r}^0| \ .$$

Otherwise, the GCR method can break down.

If $k_0 < k + 1$ this procedure is known as *Orthomin(k_0) method* (see Vinsome (1976)). Therefore, Orthomin(k_0) is the truncated GCR method (storing k_0 directions), and it forms a new direction from the current residual by forcing $A^T A$-orthogonality to the last k_0 direction vectors.

The GCR and Orthomin(k_0) methods minimize the residual error over all or k_0 preceding direction vectors, respectively. A related procedure is the *Orthodir method*, which was proposed by Young and Jea (1980), and it forms a new direction by multiplying the current direction by A and forcing $A^T A$-orthogonality to all preceding directions. For positive definite matrices it minimizes the residual norm $|\mathbf{r}^k|$ at each iteration. This method is similar to the GCR algorithm (2.6.9)-(2.6.14) except that (2.6.12), (2.6.13) and (2.6.14) need to be replaced respectively by

$$(2.6.16) \qquad \beta_j^k := -\frac{(A^2 \mathbf{p}^k, A\mathbf{p}^j)}{|A\mathbf{p}^j|^2} \quad , \quad k - k_0 + 1 \le j \le k$$

$$(2.6.17) \qquad \mathbf{p}^{k+1} := A\mathbf{p}^k + \sum_{j=k-k_0+1}^{k} \beta_j^k \mathbf{p}^j$$

$$(2.6.18) \qquad A\mathbf{p}^{k+1} = A^2 \mathbf{p}^k + \sum_{j=k-k_0+1}^{k} \beta_j^k A\mathbf{p}^j \ .$$

Orthodir corresponds to the choice $k_0 = k + 1$, and it forms a sequence of $A^T A$-orthogonal residuals. Its truncated version *Orthodir*(k_0) is obtained for some $0 \le k_0 < k + 1$. If A is symmetric or skew-symmetric, then Orthodir(2) is equivalent to Orthodir.

For non-symmetric indefinite matrices Orthodir is guaranteed to converge if n iterations are carried out. However convergence is not warranted for Orthodir(k_0).

2.6.2 Arnoldi and GMRES Iterations

One of the most interesting methods for non-symmetric matrices is the *Generalized Minimal Residual (GMRES) method* that was derived by Saad and Schultz (1986). Its main feature is that it *cannot* break down even for problems with indefinite symmetric parts unless it has already converged. GMRES iterations are based on the *Arnoldi method* for computing eigenvalues of non-symmetric matrices. Given an arbitrary vector $\mathbf{v}^1$ such that $|\mathbf{v}^1| = 1$, the Arnoldi method is a Galerkin procedure on the Krylov subspace $K_m(\mathbf{v}^1)$ (see (2.6.8)) for approximating the eigenvalues of A. More precisely, it finds a set of eigenvalue estimates $\{\lambda_1, ..., \lambda_m\}$ so that there exist non-zero $\mathbf{w}^i \in K_m(\mathbf{v}^1)$, $i = 1, ..., m$, for which

$$(A\mathbf{w}^i - \lambda_i \mathbf{w}^i, \mathbf{v}) = 0 \quad , \quad i = 1, ..., m \quad , \quad \forall \, \mathbf{v} \in K_m(\mathbf{v}^1) \ .$$

This is accomplished by constructing an orthonormal matrix $V_m = (\mathbf{v}^1|...|\mathbf{v}^m)$ whose columns $\{\mathbf{v}^j \, | \, j = 1, ..., m\}$ span $K_m(\mathbf{v}^1)$, and then computing the eigenvalues of $V_m^T A V_m$. The Arnoldi algorithm can be described as follows (see Arnoldi (1951)).

Choose an initial vector $\mathbf{v}^1$ such that $|\mathbf{v}^1| = 1$, and a step number m. Then

(2.6.19)

$$\begin{cases} \text{For } j = 1, ..., m \\ \quad \text{For } i = 1, ..., j \\ \qquad h_{i,j} = (A\mathbf{v}^j, \mathbf{v}^i) \\ \qquad \hat{\mathbf{v}}^{j+1} = A\mathbf{v}^j - \sum_{i=1}^{j} h_{i,j}\mathbf{v}^i \\ \quad \text{end } i \\ \quad h_{j+1,j} = |\hat{\mathbf{v}}^{j+1}| \\ \quad \mathbf{v}^{j+1} = \hat{\mathbf{v}}^{j+1}/h_{j+1,j} \\ \text{end } j \ . \end{cases}$$

This method is essentially a Gram-Schmidt process for orthonormalizing the Krylov sequence $\{A^k\mathbf{v}^1\}$, $k = 0, ..., m-1$. The orthonormal matrix V_m is such that $V_m^T A V_m = H_m$, where H_m is the $m \times m$ upper Hessenberg matrix whose (i,j) entry is the scalar $h_{i,j}$. Let us recall that H is an upper (respectively, lower) Hessenberg matrix if $H_{ij} = 0$ for $j > i + 1$ ($i > j + 1$, respectively).

If A is symmetric, H_m is symmetric and tridiagonal. Otherwise, the Arnoldi method generalizes the symmetric Lanczos algorithm to non-symmetric matrices (see, e.g., Golub and Van Loan (1989)).

The Arnoldi method is used to approximate the solution of (2.1.1) in the *Full Orthogonalization method (FOM)* (Saad (1981)). The approximate solution is $\mathbf{x}^j = V_j\mathbf{y}^j + \mathbf{x}^0$, where $\mathbf{y}^j := H_j^{-1}\beta\mathbf{e}^1, \beta := |\mathbf{r}^0|$ and $\mathbf{e}^1 = (1, 0, ..., 0)$ is the first unit vector in $\mathbb{R}^m$. FOM is theoretically equivalent to the *Orthores method* developed by Young and Jea (1980).

The GMRES method uses the Arnoldi basis to compute a point $\mathbf{x}^m \in \mathbf{x}^0 + K_m(\mathbf{r}^0)$ whose residual norm $|\mathbf{r}^m|$ is minimum, m being an increasing index that will be upgraded until convergence is achieved. Its description is as follows. We start setting $m = 1$.

(2.6.20)

$$\begin{cases} \text{(i) Choose } \mathbf{x}^0 \text{ and compute } \mathbf{r}^0 := \mathbf{b} - A\mathbf{x}^0, \beta := |\mathbf{r}^0| \\ \quad \text{and } \mathbf{v}^1 := \mathbf{r}^0/|\mathbf{r}^0|; \\ \\ \text{(ii) perform the Arnoldi process (2.6.19), then define} \\ \quad H_m \text{ as the } (m+1) \times m \text{ upper Hessenberg matrix} \\ \quad \text{whose non-zero entries are the coefficients } h_{i,j}; \\ \\ \text{(iii) form the approximate solution } \mathbf{x}^m := \mathbf{x}^0 + V_m\mathbf{y}^m, \\ \quad \text{where } \mathbf{y}^m \text{ minimizes } J(\mathbf{y}) := |\beta\mathbf{e}^1 - H_m\mathbf{y}| \text{ among} \\ \quad \text{all vectors } \mathbf{y} \in \mathbb{R}^m; \\ \\ \text{(iv) restart: if satisfied then stop, or else set } m + 1 \leftarrow m \\ \quad \text{and go to (i).} \end{cases}$$

GMRES is mathematically equivalent to Orthodir for arbitrary non-singular matrices A. Thus, in exact arithmetic it cannot break down although it may be very slow or even stagnate in cases when the matrix is not positive definite. For positive definite matrices it is equivalent to GCR. However, for large step numbers, it requires one third of the multiplications and one half the storage of these methods.

It has been shown that minimizing $J(\mathbf{y})$ is equivalent to obtaining

$$\min_{\mathbf{z} \in K_m(\mathbf{v}^1)} |\mathbf{b} - A(\mathbf{x}^0 + \mathbf{z})| \ .$$

The practical implementation of this step can be accomplished by decomposing H_m into $Q_m R_m$ using plane rotations (see Section 2.1 for QR decompositions). This decomposition can be updated progressively as a new column appears, i.e., at every step of the Arnoldi process. This is important as it enables one to obtain the residual norm of the approximate solution without computing $\mathbf{x}^m$. In turns this allows the user to decide when to stop the process without wasting needless operations.

These implementation aspects are thoroughly discussed in the original paper by Saad and Schultz. Moreover, from the theoretical viewpoint they prove that GMRES cannot break down, regardless of the positiveness of A. In particular, it converges even if A is not positive definite, while CGR doesn't. Concerning its rate of convergence, we report a result that was proven by Elman, Saad and Saylor (1986). Assume that A is diagonalizable,

(2.6.21) $$A = T \Lambda T^{-1}$$

where Λ is the diagonal matrix of eigenvalues $\{\lambda_j\}_{j=1,\ldots,n}$ and $T = (\boldsymbol{\omega}^1|\ldots|\boldsymbol{\omega}^n)$ is the matrix of right eigenvectors of A. Both T and Λ may be complex. Suppose that the initial residual is dominated by m eigenvectors, i.e.,

$$\mathbf{r}^0 = \sum_{j=1}^m \alpha_j \boldsymbol{\omega}^j + \mathbf{e} \ ,$$

where $|\mathbf{e}|$ is small in comparison to $|\sum_{j=1}^m \alpha_j \boldsymbol{\omega}^j|$. Moreover, let us assume that if some complex $\boldsymbol{\omega}^j$ appears in the previous sum, then its conjugate $\overline{\boldsymbol{\omega}}^j$ appears as well. Under these assumptions the residual norm after k steps of GMRES satisfies

(2.6.22) $$|\mathbf{r}^k| \leq ||T||_2 ||T^{-1}||_2 c_k |\mathbf{e}| = \chi_2(T) c_k |\mathbf{e}| \ ,$$

where

$$c_k = \max_{p > k} \prod_{j=1}^k \left| \frac{\lambda_p - \lambda_j}{\lambda_j} \right|$$

and $|| \cdot ||_2$ is the matrix norm associated with the euclidean vector norm $| \cdot |$, i.e.,

$$||T||_2 := \max\{|T\mathbf{v}| \mid \mathbf{v} \in \mathbb{C}^n \ , \ |\mathbf{v}| = 1\} \ .$$

Very often, c_k is of order one, hence the k steps of GMRES reduce the residual norm to the order of $|e|$ provided that the condition number of T is not too large.

From a practical point of view, when m (the dimension of the Krylov space) increases, the number of vectors requiring storage increases the same as m and the number of multiplications the same as $(1/2)m^2 n$. To remedy this difficulty, the GMRES method can be used iteratively, i.e., the algorithm can be restarted every k_0 steps, where k_0 is some fixed integer parameter. The corresponding algorithm, called $GMRES(k_0)$, can be described as follows (as usual P denotes the preconditioner):

(2.6.23)

$\quad$ (i) Choose $\mathbf{x}^0$ and compute $\mathbf{z}^0 := P^{-1}(\mathbf{b} - A\mathbf{x}^0)$ and
$\quad\quad$ $\mathbf{v}^1 := \mathbf{z}^0/|\mathbf{z}^0|$;

$\quad$ (ii) for $j = 1, ..., k_0$
$\quad\quad$ $h_{i,j} := (P^{-1}A\mathbf{v}^j, \mathbf{v}^i)$ for $i = 1, ..., j$
$\quad\quad$ $\hat{\mathbf{v}}^{j+1} := P^{-1}A\mathbf{v}^j - \sum_{i=1}^{j} h_{i,j}\mathbf{v}^i$
$\quad\quad$ $h_{j+1,j} := |\hat{\mathbf{v}}^{j+1}|$
$\quad\quad$ $\mathbf{v}^{j+1} := \hat{\mathbf{v}}^{j+1}/h_{j+1,j}$.
$\quad\quad$ Define H_{k_0} as the $(k_0 + 1) \times k_0$ upper Hessenberg matrix
$\quad\quad$ whose non-zero entries are the coefficients $h_{i,j}$;

$\quad$ (iii) form the approximate solution $\mathbf{x}^{k_0} := \mathbf{x}^0 + V_{k_0}\mathbf{y}^{k_0}$,
$\quad\quad$ where $\mathbf{y}^{k_0}$ minimizes $|\beta\mathbf{e}^1 - H_{k_0}\mathbf{y}|$ among all vectors
$\quad\quad$ $\mathbf{y} \in \mathbb{R}^{k_0}$ and $\beta := |\mathbf{z}^0|$;

$\quad$ (iv) restart: compute $\mathbf{r}^{k_0} := \mathbf{b} - A\mathbf{x}^{k_0}$. If satisfied then stop,
$\quad\quad$ or else set $\mathbf{x}^0 \leftarrow \mathbf{x}^{k_0}$ and go to (i).

This method is mathematically equivalent to Orthodir(k_0), thus its convergence is not guaranteed for non-symmetric and indefinite matrices. In some cases, it might be convenient to use a *variable* preconditioner along the iterations. This amounts to replace P by P_j in (ii).

2.6.3 Bi-CG, CGS and Bi-CGSTAB Iterations

Now we present three variants of the CG method for non-symmetric matrices.

The residual vectors $\mathbf{r}^k$ generated by the CG method satisfy a three-term recurrence. This property is lost when A is non-symmetric. Also, when A is not positive definite, it does not make sense to minimize $|\mathbf{e}^k|_A$ at each step as $|\cdot|_A$ does not define a norm.

The *Bi-Conjugate Gradient (Bi-CG) method*, which has been introduced by Fletcher (1976), constructs a residual $\mathbf{r}^k$ orthogonal with respect to another row of vectors $\tilde{\mathbf{r}}^0$, $\tilde{\mathbf{r}}^1$,...,$\tilde{\mathbf{r}}^{k-1}$, and, viceversa, $\tilde{\mathbf{r}}^k$ is orthogonal with respect to $\mathbf{r}^0$, $\mathbf{r}^1$,...,$\mathbf{r}^{k-1}$. This method also terminates within n steps at most, but there is no minimization property as in CG (or in GMRES) for the intermediate steps.

A description of the algorithm is as follows. Choose $\mathbf{x}^0$ and set $\mathbf{p}^0 := \mathbf{r}^0 := \mathbf{b} - A\mathbf{x}^0$, $\tilde{\mathbf{p}}^0 := \tilde{\mathbf{r}}^0$, where $\tilde{\mathbf{r}}^0$ is a vector such that $\rho_0 := (\tilde{\mathbf{r}}^0, \mathbf{r}^0) \neq 0$. (e.g., $\tilde{\mathbf{r}}^0 = \mathbf{r}^0$). Then for each $k \geq 0$ do:

$$(2.6.24) \qquad \mathbf{v}^k := A\mathbf{p}^k$$

$$(2.6.25) \qquad \alpha_k := \frac{\rho_k}{(\mathbf{v}^k, \tilde{\mathbf{p}}^k)}$$

$$(2.6.26) \qquad \mathbf{x}^{k+1} := \mathbf{x}^k + \alpha_k \mathbf{p}^k$$

$$(2.6.27) \qquad \mathbf{r}^{k+1} := \mathbf{r}^k - \alpha_k \mathbf{v}^k \;;\; \tilde{\mathbf{r}}^{k+1} := \tilde{\mathbf{r}}^k - \alpha_k A^T \tilde{\mathbf{p}}^k$$

$$(2.6.28) \qquad \rho_{k+1} := (\mathbf{r}^{k+1}, \tilde{\mathbf{r}}^{k+1})$$

$$(2.6.29) \qquad \beta_{k+1} := \frac{\rho_{k+1}}{\rho_k}$$

$$(2.6.30) \qquad \mathbf{p}^{k+1} := \mathbf{r}^{k+1} + \beta_{k+1}\mathbf{p}^k \;;\; \tilde{\mathbf{p}}^{k+1} := \tilde{\mathbf{r}}^{k+1} + \beta_{k+1}\tilde{\mathbf{p}}^k \;.$$

If A is symmetric, then Bi-CG coincides exactly with CG, but it does twice the work.

In case of convergence, both $\{\mathbf{r}^k\}$ and $\{\tilde{\mathbf{r}}^k\}$ converge toward zero, but only the convergence of the $\{\mathbf{r}^k\}$ is exploited. Based on this observation, Sonneveld (1989) proposed a modification that concentrates every effort on the $\mathbf{r}^k$ vectors. The algorithm, which is called *Conjugate Gradient-Squared (CGS) method*, can be represented by the following scheme. (We provide directly a version for the preconditioned system (2.5.1).) Choose $\mathbf{x}^0$ and set $\mathbf{w}^0 := \mathbf{p}^0 := \mathbf{r}^0 := \mathbf{b} - A\mathbf{x}^0$. Choose moreover $\tilde{\mathbf{r}}^0$, a vector such that $\rho_0 := (\tilde{\mathbf{r}}^0, \mathbf{r}^0) \neq 0$. Then for each $k \geq 0$ do:

$$(2.6.31) \qquad P\hat{\mathbf{p}}^k := \mathbf{p}^k$$

$$(2.6.32) \qquad \mathbf{v}^k := A\hat{\mathbf{p}}^k$$

$$(2.6.33) \qquad \alpha_k := \frac{\rho_k}{(\mathbf{v}^k, \tilde{\mathbf{r}}^0)}$$

$$(2.6.34) \qquad \mathbf{q}^k := \mathbf{w}^k - \alpha_k \mathbf{v}^k$$

$$(2.6.35) \qquad P\hat{\mathbf{u}}^k := \mathbf{w}^k + \mathbf{q}^k$$

$$(2.6.36) \qquad \mathbf{x}^{k+1} := \mathbf{x}^k + \alpha_k \hat{\mathbf{u}}^k$$

$$(2.6.37) \qquad \mathbf{r}^{k+1} := \mathbf{r}^k - \alpha_k A\hat{\mathbf{u}}^k$$

$$(2.6.38) \qquad \rho_{k+1} := (\mathbf{r}^{k+1}, \tilde{\mathbf{r}}^0)$$

$$(2.6.39) \qquad \beta_{k+1} := \frac{\rho_{k+1}}{\rho_k}$$

$$(2.6.40) \qquad \mathbf{w}^{k+1} := \mathbf{r}^{k+1} + \beta_{k+1}\mathbf{q}^k$$

$$(2.6.41) \qquad \mathbf{p}^{k+1} := \mathbf{w}^{k+1} + \beta_{k+1}(\mathbf{q}^k + \beta_{k+1}\mathbf{p}^k) \;.$$

It can be proven that, in absence of roundoff errors, both Bi-CG and CGS terminate after at most n steps by zero division, since ρ_k will be 0 for some $k \leq n$. An essential requirement for convergence in considerably less than n steps is that A is an N-stable matrix (i.e., its eigenvalues have positive real parts). In this case a suitable vector $\tilde{\mathbf{r}}^0$ exists (though not easy to find in practice) such that both Bi-CG and CGS converge at a rate comparable to that of CG.

It can be shown that CGS generates residual vectors $\mathbf{r}^k$ given by

$$\mathbf{r}^k = P_k^2(A)\mathbf{r}^0 \ ,$$

where $P_k(A)$ is the k-th degree polynomial in A for which $P_k(A)\mathbf{r}^0$ is equal to the residual at the k-th iteration step obtained by means of Bi-CG. The residual reduction operator in CGS is therefore the square of that arising in Bi-CG. In particular this explains why A^T is no longer needed in CGS.

Neither Bi-CG nor CGS minimizes the residual, hence in practical calculations we may observe many local peaks in the convergence curve for both Bi-CG and CGS.

Instead of simply squaring the Bi-CG polynomial, as in CGS, one could use a product of $P_k(A)$ times a suitable polynomial of degree k. In the *Bi-CGSTAB method*, introduced by van der Vorst (1992), the residual vector turns out to be

$$\mathbf{r}^k = Q_k(A)P_k(A)\mathbf{r}^0 \ ,$$

where $Q_k(x) = \prod_{i=1}^{k}(1 - \alpha_i x)$ and α_i are suitable constants.

The Bi-CGSTAB algorithm can be formulated as follows. Choose $\mathbf{x}^0$ and set $\mathbf{p}^0 := \mathbf{r}^0 := \mathbf{b} - A\mathbf{x}^0$. Choose moreover $\tilde{\mathbf{r}}^0$, a vector such that $\rho_0 := (\tilde{\mathbf{r}}^0, \mathbf{r}^0) \neq 0$. Then for each $k \geq 0$ do:

$$(2.6.42) \qquad P\hat{\mathbf{p}}^k := \mathbf{p}^k$$

$$(2.6.43) \qquad \mathbf{v}^k := A\hat{\mathbf{p}}^k$$

$$(2.6.44) \qquad \alpha_k := \frac{\rho_k}{(\mathbf{v}^k, \tilde{\mathbf{r}}^0)}$$

$$(2.6.45) \qquad \mathbf{s}^k := \mathbf{r}^k - \alpha_k \mathbf{v}^k$$

$$(2.6.46) \qquad P\hat{\mathbf{s}}^k := \mathbf{s}^k$$

$$(2.6.47) \qquad \mathbf{t}^k := A\hat{\mathbf{s}}^k$$

$$(2.6.48) \qquad P\hat{\mathbf{t}}^k := \mathbf{t}^k$$

$$(2.6.49) \qquad \omega_k := \frac{(\hat{\mathbf{t}}^k, \hat{\mathbf{s}}^k)}{(\hat{\mathbf{t}}^k, \hat{\mathbf{t}}^k)}$$

$$(2.6.50) \qquad \mathbf{x}^{k+1} := \mathbf{x}^k + \alpha_k \hat{\mathbf{p}}^k + \omega_k \hat{\mathbf{s}}^k$$

$$(2.6.51) \qquad \mathbf{r}^{k+1} := \mathbf{s}^k - \omega_k \mathbf{t}^k$$

$$(2.6.52) \qquad \rho_{k+1} := (\mathbf{r}^{k+1}, \tilde{\mathbf{r}}^0)$$

$$(2.6.53) \qquad \mathbf{p}^{k+1} := \mathbf{r}^{k+1} + \frac{\rho_{k+1}\alpha_k}{\rho_k \omega_k}(\mathbf{p}^k - \omega_k \mathbf{v}^k) \ .$$

For an unfavourable choice of $\bar{\mathbf{r}}^0$, ρ_k or $(\mathbf{v}^k, \bar{\mathbf{r}}^0)$ can be 0 or very small. In this case, one has to restart, e.g., with $\bar{\mathbf{r}}^0$ and $\mathbf{x}^0$ given by the last available values of $\mathbf{r}^k$ and $\mathbf{x}^k$.

In exact arithmetic Bi-CGSTAB is also a finite termination method. The theoretical properties of Bi-CGSTAB are very much the same as those of CGS. The essential difference is that often Bi-CGSTAB is more smoothly converging than CGS, i.e., its oscillations are less pronounced. The computational cost per iteration of the two methods, as well as the memory requirement, are comparable.

2.7 The Multi-Grid Method

The classical iterative methods fail to be effective whenever the spectral radius of the iteration matrix B is close to one. A Fourier analysis shows that large eigenvalues are associated with errors of low frequency (we will discuss this issue on a simple example below). Therefore the smooth component (low frequencies) of the error holds back convergence, whereas high frequency errors are rapidly damped.

The basic idea of the method is to change the grid, whenever the linear system at hand arises from a discretization of differential equations over a certain mesh. As a matter of fact, errors that are smooth on a grid of width h can be attacked on a coarser grid, while high frequency errors that are not visible on the coarse grid of width $2h$ can be resolved on the fine grid.

2.7.1 The Multi-Grid Cycles

The basic multi-grid algorithm can be presented in an abstract fashion in the following form. We are concerned with a family of linear problems

$$(2.7.1) \qquad A_j \mathbf{x}_{(j)} = \mathbf{b}_{(j)} \ , \quad j = 0, ..., m \ ,$$

where A_j is an invertible linear operator on a finite dimensional space V^j, with $\dim V^j < \dim V^{j+1}$, $j = 0, ..., m-1$, and $\mathbf{b}_{(j)}$ is a given right hand side. The goal is to solve (2.7.1) for $j = m$. For this reason we occasionally omit the superscript m for the highest level, thus writing

$$V = V^m \ , \ A = A_m \ , \ \mathbf{b} = \mathbf{b}_{(m)} \ , \ \mathbf{x} = \mathbf{x}_{(m)} \ .$$

The operators A_j, $j < m$, are auxiliary and will be used in the multi-grid algorithm.

For each j, problem (2.7.1) should be regarded as the discretization of a differential equation on a grid whose characteristic mesh spacing is h_j. If we assume that $h_{j-1} = 2h_j$ and set $h = h_m$, then (2.7.1) provides a family of discretizations with a refinement factor 2.

To solve each of the problems (2.7.1) we consider a basic iterative method

(2.7.2) $$\mathbf{x}_{(j)}^{k+1} = \mathbf{x}_{(j)}^{k} + P_j(\mathbf{b}_{(j)} - A_j\mathbf{x}_{(j)}^{k}) \ ,$$

where P_j is a preconditioner for A_j. The associated error $\mathbf{e}_{(j)}^{k}$ is transformed according to

(2.7.3) $$\mathbf{e}_{(j)}^{k+1} = B_j\mathbf{e}_{(j)}^{k} \ , \quad B_j := I - P_jA_j \ .$$

The multi-grid algorithm can be regarded as an acceleration of (2.7.2) for $j = m$, and can be formulated as follows.

1. *Pre-smoothing on the fine grid*
Being known $\mathbf{x}_{(j)}^{0}$, do n_1 times

(2.7.4) $$\mathbf{x}_{(j)}^{k} = \mathbf{x}_{(j)}^{k-1} + P_j(\mathbf{b}_{(j)} - A_j\mathbf{x}_{(j)}^{k-1}) \ , \quad k = 1, ..., n_1 \ .$$

2. *Coarse grid correction*
(i) Form the residual (or defect) $\mathbf{r}_{(j)} = \mathbf{b}_{(j)} - A_j\mathbf{x}_{(j)}^{n_1}$;
(ii) restrict the residual on the coarse grid setting $\mathbf{r}_{(j-1)} := R_j^{j-1}\mathbf{r}_{(j)}$, where

$$R_j^{j-1} : V^j \rightarrow V^{j-1}$$

is a *restriction operator*;
(iii) solve the defect problem

(2.7.5) $$A_{j-1}\mathbf{x}_{(j-1)} = \mathbf{r}_{(j-1)} \ ;$$

(iv) correct the solution in V^j by setting

$$\overline{\mathbf{x}}_{(j)} = \mathbf{x}_{(j)}^{n_1} + \Pi_{j-1}^{j}\mathbf{x}_{(j-1)} \ ,$$

where

$$\Pi_{j-1}^{j} : V^{j-1} \rightarrow V^j$$

is a *prolongation operator*.
3. *Post-smoothing on the fine grid*
Set $\hat{\mathbf{x}}_{(j)}^{0} := \overline{\mathbf{x}}_{(j)}$ and do n_2 times

(2.7.6) $$\hat{\mathbf{x}}_{(j)}^{k} = \hat{\mathbf{x}}_{(j)}^{k-1} + P_j(\mathbf{b}_{(j)} - A_j\hat{\mathbf{x}}_{(j)}^{k-1}) \ , \quad k = 1, ..., n_2 \ .$$

Both prolongation and restriction operators are full-rank linear mappings. One of the numbers of pre- or post-smoothing steps n_1 and n_2 may be 0. In the most general case, both n_1 and n_2 may depend on j and may be determined adaptively.

Concerning the coarse grid problem (2.7.5), if the dimension of V^{j-1} is low it can be solved exactly by a direct method. In this case we have a *two-grid algorithm*. The basic multi-grid cycles are obtained when the coarse grid problem (2.7.5) is not solved exactly but recursively by γ applications of the multi-grid algorithm itself. The cases $\gamma = 1$ and $\gamma = 2$ are referred to as *V-cycles* and *W-cycles*, respectively.

The structure of one iteration step (*cycle*) of a multi-grid method is illustrated by a few pictures which are given in Fig. 2.7.1. The symbols $\circ$, $\square$, $\backslash$ and $/$ mean smoothing, solving exactly, fine-to-coarse and coarse-to-fine transfer, respectively.

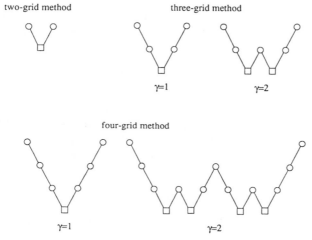

Fig. 2.7.1. Structure of one multi-grid cycle for different number of grids and values of γ

2.7.2 A Simple Example

Consider the two-point boundary value problem

(2.7.7)
$$\begin{cases} -u''(x) = f(x) \ , \quad 0 < x < 1 \ , \\ u(0) = u(1) = 0 \ . \end{cases}$$

For each $j = 1, 2$ divide the interval $\Omega = (0, 1)$ in a grid $\mathcal{G}^j$ having $l_j = 2^j$ intervals of equal length $h_j = 2^{-j}$, whose endpoints are given by $\{x_l = lh_j \mid l = 0, ..., l_j\}$.

The finite difference approximation to (2.7.1) based on the second order centered formula at the nodes x_l yields the linear system (see, e.g., Isaacson and Keller (1966)):

(2.7.8)
$$A_j \boldsymbol{\xi}_{(j)} = \mathbf{f}_{(j)} \ ,$$

where

$$\boldsymbol{\xi}_{(j)} := (u_1, ..., u_{l_j-1}) \ , \quad \mathbf{f}_{(j)} = (f(x_1), ..., f(x_{l_j-1})) \ ,$$

u_l is the approximation of $u(x_l)$ we are looking for, and A_j is the finite difference matrix

$$A_j := h_j^{-2} \operatorname{tridiag}(-1, 2, -1) .$$

For the set up of the basic multi-grid cycle, the restriction operator R_j^{j-1} is given by the stencil

$$R_j^{j-1} = \frac{1}{2} \operatorname{tridiag}\left(\frac{1}{2}, 1, \frac{1}{2}\right)$$

and is barely obtained by a three-point averaging. The prolongation operator Π_{j-1}^j is the adjoint of R_j^{j-1} with respect to the discrete scalar product

$$\langle \boldsymbol{\xi}, \boldsymbol{\eta} \rangle_j := \frac{1}{h_j} \sum_{i=1}^{l_j-1} \xi_i \eta_i .$$

For a fixed j, the spectral radius of the iteration matrix related to the usual Jacobi, Gauss-Seidel and Richardson methods on the grid $\mathcal{G}^j$ behaves like $1 - O(h_j^2)$, while S.O.R. with optimal relaxation parameter has the spectral radius equal to $1 - O(h_j)$ (see, e.g., Varga (1962)). Consider for instance the Richardson iteration

$$(2.7.9) \qquad \boldsymbol{\xi}_{(j)}^{k+1} = \boldsymbol{\xi}_{(j)}^k + h_j^2 \theta_j (\mathbf{f}_{(j)} - A_j \boldsymbol{\xi}_{(j)}^k) , \quad k \geq 0 ,$$

whose acceleration parameter is $\alpha_j = h_j^2 \theta_j$, $\theta_j > 0$. The iteration matrix is

$$B_j = I - h_j^2 \theta_j A_j .$$

Its eigenvalues are

$$(2.7.10) \qquad \lambda_{(j),i} = 1 - 4\theta_j \sin^2\left(ih_j \frac{\pi}{2}\right) , \quad i = 1, ..., l_j - 1 .$$

Their behaviour, when $\theta_j = 1/4$, is reported in Fig. 2.7.2. The abscissa refers to the value of ih_j. The corresponding eigenvector $\boldsymbol{\omega}_{(j),i}$ read (componentwise)

$$(2.7.11) \qquad (\boldsymbol{\omega}_{(j),i})_s = \sqrt{2h_j} \sin(ih_j s\pi) , \quad i, s = 1, ..., l_j - 1 .$$

If we define the error at the k-th step by

$$(2.7.12) \qquad \mathbf{e}_{(j)}^k := \boldsymbol{\xi}_{(j)}^k - \boldsymbol{\xi}_{(j)} = \sum_{i=1}^{l_j-1} \beta_{i,k} \boldsymbol{\omega}_{(j),i} , \quad k \geq 0 ,$$

by recursion we obtain that $\beta_{i,k} = \beta_{i,0}(\lambda_{(j),i})^k$, $i = 1, ..., l_j - 1$.

We can therefore deduce from (2.7.10) that a relaxation method like (2.7.9) is very efficient in *smoothing* the errors, i.e., in reducing the high frequency error components (see Fig. 2.7.3), whereas for low frequencies the amplitude reduction is very slight. High frequencies on the grid $\mathcal{G}^j$ are those

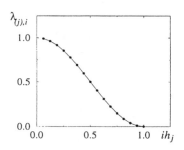

Fig. 2.7.2. Behaviour of the eigenvalues for $\theta_j = 1/4$

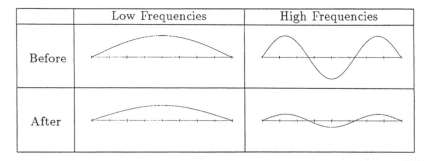

Fig. 2.7.3. Typical error smoothing behaviour of relaxation methods. "Before" and "after" stand for "before relaxation" and "after relaxation", respectively

$\beta_{i,0}$ with $l_j/2 < i \leq l_j$, while low frequencies are those corresponding to $1 \leq i \leq l_j/2$.

The smoothing properties of an iterative method is measured by a smoothing factor, the worst (largest) factor by which high frequency error components are reduced per relaxation step. It tipically happens that the smoothing factor can be smaller than $1/2$ independently of h_j.

When passing to the coarser grid, whose subintervals have length $h_{j-1} = 2^{-(j-1)} = 2h_j$, the high frequency components on the fine grid are not visible on the coarse one, due to aliasing. Therefore, they cannot be modified by use of this grid. We illustrate this in Fig. 2.7.4 for $h_j = 1/8$ (corresponding to $l_j = 8$) and $h_{j-1} = 1/4$. Instead, defect corrections smooth the error components that are associated with low frequencies on the fine grid.

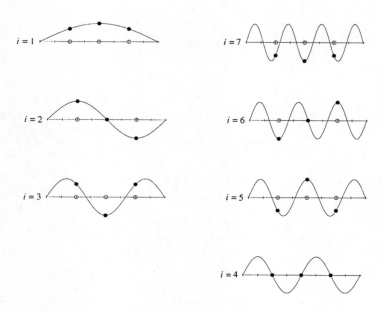

Fig. 2.7.4. $\omega_{(j),i}(x) = \sqrt{2h_j}\sin(i\pi x)$: low ($i = 1, 2, 3$) and high ($i = 4, 5, 6, 7$) frequency components for $h_j = 1/8$ and $h_{j-1} = 1/4$. Low components are visible on the coarse grid $\mathcal{G}^{j-1}$, whereas high frequencies are not

2.7.3 Convergence

In the case of the two-grid cycle, the error is transformed according to

$$(2.7.13) \qquad \mathbf{e}_{(j)}^{\text{new}} = M_j \mathbf{e}_{(j)}^{\text{old}} \,,$$

where M_j is the two-grid iteration operator

$$(2.7.14) \qquad M_j := B_j^{n_2} T_j B_j^{n_1}$$

and T_j is the coarse-grid correction operator

$$(2.7.15) \qquad T_j := I - \Pi_{j-1}^j A_{j-1}^{-1} R_j^{j-1} A_j \,.$$

Convergence holds provided that $\rho_j := \|M_j\| \le C < 1$ for each $j = 0, \dots, m$ for a suitable natural matrix norm. Once an estimate of the two-grid convergence factor ρ_j has been obtained, one can also find an estimate of the convergence factor for the multi-grid V-cycle and W-cycle by a perturbation argument and recursion.

Define the convergence factor as

$$(2.7.16) \qquad\qquad \varepsilon_j := \max_{\mathbf{e}^p_{(j)} \neq 0} \frac{|\mathbf{e}^{p+1}_{(j)}|}{|\mathbf{e}^p_{(j)}|} \quad ,$$

where $\mathbf{e}^p_{(j)}$ is the error before and $\mathbf{e}^{p+1}_{(j)}$ after one step of the multi-grid iteration. It can be shown that for the W-cycle the following recursion inequality holds:

$$\varepsilon_j \leq \rho_j + C\,\varepsilon^2_{j-1} \quad , \quad j = 1, ..., m \quad ,$$

provided that the restriction and prolongation operators are bounded in suitable norms.

If ρ_j is bounded from above, independently of j, by a small enough constant (this can be achieved by using sufficiently many smoothing steps), it follows that $\varepsilon_m \leq C < 1$. This implies that the multi-grid convergence factor is bounded away from 1 independently of the number of levels m. If we assume that the cost of smoothing on V_j is proportional to N_j (the dimension of V_j), then the cost of both the V- and the W-cycle are proportional to N_j. Therefore the multi-grid cycle has optimal asymptotical computational complexity.

A historical remark on multi-grid method as well as a description of the basic algorithms are presented in Stüben and Trottenberg (1981). For a theoretical analysis see, among others, Brandt (1977), Hackbusch (1981a, 1985), Bank and Dupont (1981), Mandel, McCormick and Bank (1987), Bramble (1993), Douglas and Douglas (1993) and Yserentant (1993).

2.8 Complements

The availability of advanced-architecture computers with vector and parallel facilities is having a significant impact on algorithm development in numerical linear algebra. It is beyond the scope of this Chapter to address this issue. However, we mention the monographs by Golub and Van Loan (1989), Chap. 6, Dongarra, Duff, Sorensen and van der Vorst (1991) and Golub and Ortega (1993), and the review article by Demmel, Heath and van der Vorst (1993) as pioneering attempts towards a systematic approach to this very promising field.

3. Finite Element Approximation

In this Chapter we present the properties of the classical finite element approximation. We underline the three basic aspects of this method: the existence of a triangulation of Ω, the construction of a finite dimensional subspace consisting of piecewise-polynomials, and the existence of a basis of functions having small support. Then, we introduce the interpolation operator and we estimate the interpolation error. Some final remarks will be devoted to several projection operators upon finite element subspaces and their approximation properties.

These results provide the basis for deriving error estimates for finite element approximation of partial differential equations that will be considered in the forthcoming Chapters.

3.1 Triangulation

Let the set $\Omega \subset \mathbb{R}^d$, $d = 2, 3$, be a polygonal domain, i.e., Ω is an open bounded connected subset such that $\overline{\Omega}$ is the union of a finite number of polyhedra.

Here, we consider a finite decomposition

$$(3.1.1) \qquad \overline{\Omega} = \bigcup_{K \in \mathcal{T}_h} K \ ,$$

where:

$(3.1.2)$ each K is a polyhedron with $\overset{\circ}{K} \neq \emptyset$;

$(3.1.3)$ $\overset{\circ}{K}_1 \cap \overset{\circ}{K}_2 = \emptyset$ for each distinct $K_1, K_2 \in \mathcal{T}_h$;

$(3.1.4)$ if $F = K_1 \cap K_2 \neq \emptyset$ (K_1 and K_2 distinct elements of $\mathcal{T}_h$) then F is a common face, side, or vertex of K_1 and K_2;

$(3.1.5)$ $\mathrm{diam}(K) \leq h$ for each $K \in \mathcal{T}_h$.

$\mathcal{T}_h$ is called a *triangulation* of $\overline{\Omega}$ (see Fig. 3.1.1. for a description of admissibile and forbidden configurations).

For simplicity, in the sequel we assume further that each element K of $\mathcal{T}_h$ can be obtained as $K = T_K(\hat{K})$, where $\hat{K}$ is a reference polyhedron and

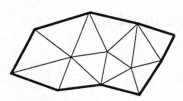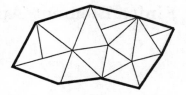

Fig. 3.1.1. Admissibile (left) and non-admissible (right) triangulations

T_K is a suitable invertible affine map, i.e., $T_K(\hat{\mathbf{x}}) = B_K\hat{\mathbf{x}} + \mathbf{b}_K$, B_K being a non-singular matrix.

We will confine ourselves to consider two different cases:

(3.1.6)   the reference polyhedron $\hat{K}$ is the unit d-simplex, i.e., the triangle of vertices $(0,0)$, $(1,0)$, $(0,1)$ (when $d = 2$), or the tetrahedron of vertices $(0,0,0)$, $(1,0,0)$, $(0,1,0)$, $(0,0,1)$ (when $d = 3$). As a consequence, each $K = T_K(\hat{K})$ is a triangle or a tetrahedron.

(3.1.7)   the reference polyhedron $\hat{K}$ is the unit d-cube $[0, 1]^d$. As a consequence, each $K = T_K(\hat{K})$ is a parallelogram (when $d = 2$) or a parallelepiped (when $d = 3$).

In the latter case, the triangulation is made by d-rectangles if for each $K \in T_h$ the matrix B_K defining the affine transformation T_K is diagonal.

Notice that dealing with general quadrilaterals or hexahedrons would require admitting that each component of the (invertible) transformation T_K is no longer an affine map but a linear polynomial with respect to each single variable $x_1, ..., x_d$. We will not consider this case, and we refer the interested reader to Ciarlet (1978, 1991), where a general approximation theory for finite elements is presented.

3.2 Piecewise-Polynomial Subspaces

A second basic aspect of the finite element method consists of determining a finite dimensional space X_h, which should result in a suitable approximation of the infinite dimensional space X from case to case under consideration.

Here the point is that the functions $v_h \in X_h$ are piecewise-polynomials, i.e., for each $K \in T_h$ the space

$$P_K := \{v_{h|K} \,|\, v_h \in X_h\}$$

consists of algebraic polynomials.

To be more precise, let us denote by $\mathbb{P}_k$, $k \geq 0$, the space of polynomials of degree less than or equal to k in the variables $x_1, ..., x_d$, and by $\mathbb{Q}_k$ the space of polynomials that are of degree less than or equal to k with respect to each variable $x_1, ..., x_d$. An easy computation shows that

(3.2.1) $\dim \mathbb{P}_k = \begin{pmatrix} d+k \\ k \end{pmatrix}$, $\dim \mathbb{Q}_k = (k+1)^d$.

Moreover the following inclusions hold true

$$\mathbb{P}_k \subset \mathbb{Q}_k \subset \mathbb{P}_{dk} .$$

We also introduce the following space of vector polynomials

(3.2.2) $\mathbb{D}_k := (\mathbb{P}_{k-1})^d \oplus \mathbf{x}\, \mathbb{P}_{k-1}$, $k \geq 1$,

where $\mathbf{x} \in \mathbb{R}^d$ is the independent variable. One can verify that

(3.2.3) $\dim \mathbb{D}_k = (d+k)\dfrac{(d+k-2)!}{(d-1)!(k-1)!}$

and that $(\mathbb{P}_{k-1})^d \subset \mathbb{D}_k \subset (\mathbb{P}_k)^d$ (see, e.g., Roberts and Thomas (1991)).

3.2.1 The Scalar Case

We are now in a position to define the most commonly used spaces X_h. In case (3.1.6) we set

(3.2.4) $X_h = X_h^k := \{v_h \in C^0(\overline{\Omega}) \mid v_{h|K} \in \mathbb{P}_k \ \forall\, K \in \mathcal{T}_h\}$, $k \geq 1$,

which will be called the space of *triangular finite elements*.
    In case (3.1.7) we define

(3.2.5) $X_h = X_h^k := \{v_h \in C^0(\overline{\Omega}) \mid v_{h|K} \circ T_K \in \mathbb{Q}_k \ \forall\, K \in \mathcal{T}_h\}$, $k \geq 1$,

which is called the space of *parallelepipedal finite elements*.
    Let us remark that for case (3.2.5) the finite dimensional space $P_K = \{v_{h|K} \mid v_h \in X_h^k\}$ is different from $\mathbb{Q}_k$, except when K is a rectangle with sides parallel to the coordinate axes. In fact, only for this case T_K is represented by a diagonal matrix plus a translation, and $\mathbb{Q}_k$ is invariant under this type of transformations. In general, one has $\mathbb{P}_k \subset P_K \subset \mathbb{P}_{dk}$.
    In both cases, (3.2.4) and (3.2.5), it is worthwhile to notice that

$$X_h^k \subset H^1(\Omega) \qquad \forall\, k \geq 1 .$$

This is in fact a consequence of the following result (notice that for the sake of simplicity, here, and in the sequel, we write $H^s(K)$ instead of $H^s(\overset{\circ}{K})$):

Proposition 3.2.1 *A function $v : \Omega \to \mathbb{R}$ belongs to $H^1(\Omega)$ if and only if*
a. $v_{|K} \in H^1(K)$ for each $K \in \mathcal{T}_h$;
b. for each common face $F = K_1 \cap K_2$, $K_1, K_2 \in \mathcal{T}_h$, the trace on F of $v_{|K_1}$ and $v_{|K_2}$ is the same.

Proof. Using *a.*, define the functions $w_j \in L^2(\Omega), j = 1, ..., d$, through

$$w_{j|K} := D_j(v_{|K}) \quad \forall \, K \in \mathcal{T}_h \ .$$

In order to show that $v \in H^1(\Omega)$, we simply have to prove that $w_j = D_j v$.

By using the Green formula (see (1.3.3)), we can write for each $\varphi \in \mathcal{D}(\Omega)$

$$\int_\Omega w_j \varphi = \sum_K \int_K w_j \varphi = - \sum_K \int_K (v_{|K}) D_j \varphi + \sum_K \int_{\partial K} v_{|K} \, \varphi \, n_{K,j} \ ,$$

where $\mathbf{n}_K$ is the unit normal vector on ∂K. Since φ is vanishing on $\partial \Omega$, and $\mathbf{n}_{K_1} = -\mathbf{n}_{K_2} =: \mathbf{n}$ on a common face $F = K_1 \cap K_2$, we have by b.

$$\int_\Omega w_j \varphi = - \int_\Omega v D_j \varphi + \sum_F \int_F (v_{|K_1} - v_{|K_2}) \, \varphi \, n_j = - \int_\Omega v D_j \varphi \ ,$$

i.e., $w_j = D_j v$.

On the other hand, if we assume that $v \in H^1(\Omega)$ it follows at once that a. holds. Moreover, we have $w_j = D_j v$, hence by proceeding as before one finds

$$\sum_F \int_F (v_{|K_1} - v_{|K_2}) \, \varphi \, n_j = 0 \quad \forall \, \varphi \in \mathcal{D}(\Omega) \ , \ j = 1, ..., d \ ,$$

i.e., b. is satisfied. □

3.2.2 The Vector Case

In the vector case, when $\hat{K}$ satisfies (3.1.6), we define

$$(3.2.6) \qquad W_h^k := \{ \mathbf{v}_h \in H(\mathrm{div}; \Omega) \mid \mathbf{v}_{h|K} \in \mathbb{D}_k \ \forall \, K \in \mathcal{T}_h \} \ , \quad k \geq 1 \ .$$

An analogous definition can be given for case (3.1.7). We refer the interested reader to Thomas (1977), Raviart and Thomas (1977), Nédélec (1980); however, we do not consider that case here. The following result gives a clear indication as to which conditions must be satisfied on the boundary of the elements K to ensure that $\mathbf{v}_h \in W_h^k$.

Proposition 3.2.2 *Let* $\mathbf{v} : \Omega \to \mathbb{R}^d$ *be such that*
a. $\mathbf{v}_{|K} \in (H^1(K))^d$ for each $K \in \mathcal{T}_h$;
b. for each common face $F = K_1 \cap K_2$, $K_1, K_2 \in \mathcal{T}_h$, the trace on F of the normal component $\mathbf{n} \cdot \mathbf{v}_{|K_1}$ and $\mathbf{n} \cdot \mathbf{v}_{|K_2}$ is the same.
Then $\mathbf{v} \in H(\mathrm{div}; \Omega)$. Conversely, if $\mathbf{v} \in H(\mathrm{div}; \Omega)$ and a. is satisfied, then b. holds.

Proof. Define $w \in L^2(\Omega)$ through

$$w_{|K} := \mathrm{div}(\mathbf{v}_{|K}) \quad \forall \, K \in \mathcal{T}_h \ .$$

By using the Green formula (see (1.3.4)), we have for each $\varphi \in \mathcal{D}(\Omega)$

$$\langle \operatorname{div} \mathbf{v}, \varphi \rangle = -\int_{\Omega} \mathbf{v} \cdot \nabla \varphi = -\sum_K \int_K (\mathbf{v}_{|K}) \cdot \nabla \varphi$$

$$= \sum_K \int_K \operatorname{div}(\mathbf{v}_{|K}) \varphi - \sum_F \int_F (\mathbf{n} \cdot \mathbf{v}_{|K_1} - \mathbf{n} \cdot \mathbf{v}_{|K_2}) \varphi = \int_{\Omega} w \varphi \ ,$$

i.e., $\operatorname{div} \mathbf{v} = w$.

On the other hand, if $\mathbf{v} \in H(\operatorname{div}; \Omega)$ we have $w = \operatorname{div} \mathbf{v}$. Since $\mathbf{v}_{|K} \in (H^1(K))^d$, the trace on F is well defined, and we obtain

$$\sum_F \int_F (\mathbf{n} \cdot \mathbf{v}_{|K_1} - \mathbf{n} \cdot \mathbf{v}_{|K_2}) \varphi = 0 \quad \forall \, \varphi \in \mathcal{D}(\Omega) \ ,$$

hence $b.$ holds. $\qquad\qquad\qquad\qquad\qquad\qquad\qquad\qquad\qquad\qquad\qquad$ $\square$

3.3 Degrees of Freedom and Shape Functions

It is now necessary to construct a basis for the space X_h in such a way that the basis functions can be easily described.

An important point is concerned with the choice of a set of *degrees of freedom* on each element K (i.e., the parameters which permit to uniquely identify a function in $\mathbb{P}_k$, $\mathbb{Q}_k$ or $\mathbb{D}_k$). We recall that their number is given by (3.2.1) or (3.2.3).

3.3.1 The Scalar Case: Triangular Finite Elements

Let us start with the case (3.2.4), $d = 2$. To identify $v_{h|K}$, when $k = 1$ we have to choose three degrees of freedom on each element K, with the additional constraint that $v_h \in C^0(\overline{\Omega})$. The simplest choice is that of the values at the vertices of each K.

It can be noticed that, if we consider

$$Y_h^1 := \{v_h \in L^2(\Omega) \mid v_{h|K} \in \mathbb{P}_1 \ \forall \, K \in \mathcal{T}_h\}$$

instead of X_h^1 defined in (3.2.4), we are free to choose the degrees of freedom on K as the values at three arbitrary points (not necessarily coincident with the vertices). For instance, one can take as nodes three internal points, or else the midpoints of each side. The latter choice clearly implies continuity only at the midpoints; with the former one the functions belonging to Y_h^1 generally have two different definitions on the sides common to adjacent triangles. However, let us go back to the case (3.2.4).

When $k = 2$, we assume that the element degrees of freedom are given by the value at the vertices and in the middle point of each side. To give insight of the proof in more general cases, let us show this assertion. Denote the vertices of the triangle K by $\mathbf{a}^i$, $i = 1, 2, 3$, and the midpoints by $\mathbf{a}^{ij}$, $i < j$, $i, j = 1, 2, 3$.

Proposition 3.3.1 *A function $p \in \mathbb{P}_2$ is uniquely determined by the six values $p(\mathbf{a}^i)$, $1 \le i \le 3$, and $p(\mathbf{a}^{ij})$, $1 \le i < j \le 3$.*

Proof. Since the number of degrees of freedom is equal to the dimension of $\mathbb{P}_2$ $(= 6)$, we have only to prove that if $p(\mathbf{a}^i) = p(\mathbf{a}^{ij}) = 0$ then $p \equiv 0$.

With this aim, let us notice that the restriction of p over each side is a quadratic function of one variable vanishing in three distinct points, hence p is vanishing over each side. Thus we can write

$$p(\mathbf{x}) = c\,p_1(\mathbf{x})\,p_2(\mathbf{x})\,p_3(\mathbf{x}) \ ,$$

where $p_i(\mathbf{x})$ are linear functions, each one vanishing on one side of K. Since $p \in \mathbb{P}_2$, it follows $c = 0$.   □

Furthermore, let us remark that this choice of degrees of freedom guarantees that $v_h \in C^0(\overline{\Omega})$, since the degrees of freedom on each side uniquely identify the restriction of v_h on that side.

In a similar way, one can prove that the degrees of freedom for a "cubic" triangle $(k = 3)$ are given by ten values at the following nodes: the three vertices; two other nodes on each side, dividing it into three subintervals of equal length; the center of gravity.

The situation for the three cases $k = 1, 2, 3$ is illustrated on the upper part of Fig. 3.3.1.

We are not going to describe in detail the general case $k > 3$, but refer the interested reader, e.g., to Ciarlet (1978, 1991).

When $d = 3$, it is not difficult to see that the degrees of freedom are the values at the nodes indicated in Fig. 3.3.2 (there, only the nodes on the visible faces are pictured).

It is worthwhile to note that this choice of degrees of freedom reduces the problem of the identification of a polynomial $p \in \mathbb{P}_k$ from the three-dimensional case to the two-dimensional one. In fact, on each face the degrees of freedom are the same as that for the corresponding two-dimensional triangle; hence, if a polynomial vanishes at the nodes indicated in Fig. 3.3.2, it vanishes on each face, too. Consequently, one is left with a polynomial p of degrees 1, 2 or 3, vanishing over four distinct planes, which implies $p \equiv 0$.

A basis for X_h^k is now easily constructed. In particular, by denoting $\mathbf{a}_j$, $j = 1, ..., N_h$, the global set of nodes in $\overline{\Omega}$, is sufficient to choose those functions $\varphi_i \in X_h^k$ such that

(3.3.1) $\varphi_i(\mathbf{a}_j) = \delta_{ij}$, $i, j = 1, ..., N_h$ (δ_{ij} is the Kronecker symbol) .

These basis functions are called *shape functions*. They are illustrated on the lower part of Fig. 3.3.1 for the cases $k = 1, 2, 3$, below the corresponding

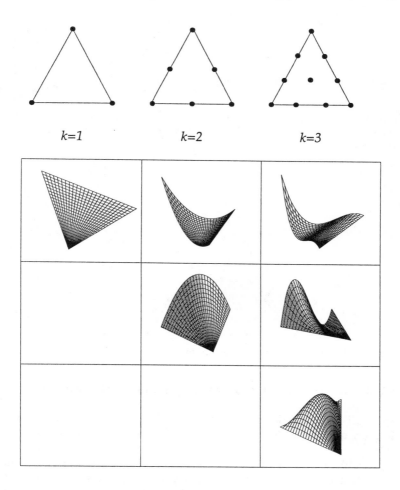

Fig. 3.3.1. Degrees of freedom (upper) and shape functions (lower) for triangular elements

value of k. The upper row refers to those associated with vertices, the mid one to nodes lying on sides (excluding the vertices), while the lower to nodes internal to Ω.

It is important to notice that the support of each shape function is "small", i.e., it is given by a few elements of the triangulation. In Fig. 3.3.3, we show the possible support of a linear shape function, depending on the position of the associated node.

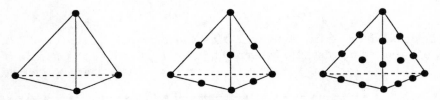

Fig. 3.3.2. $d = 3$ and, from left to right, $k = 1$, $k = 2$, $k = 3$. The number of degrees of freedom is 4, 10, 20, respectively. Only the nodes on the visible faces are pictured

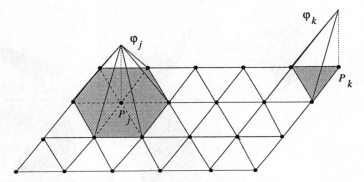

Fig. 3.3.3. The support of linear shape functions

3.3.2 The Scalar Case: Parallelepipedal Finite Elements

We shall now consider the case (3.2.5). We first describe the degrees of freedom on the reference square $\hat{K} = [0,1]^2$, then we will consider the general case.

It will be shown in Proposition 3.3.2 that when $k = 1$ the degrees of freedom are the values at the vertices of the square; when $k = 2$ one has to add the values at the midpoint of each side and at the center of gravity. When $k = 3$ one has to consider the values at the vertices and at the points of coordinates $1/3$ and $2/3$, as drawn in Fig. 3.3.4, which is the counterpart of Fig. 3.3.1 for parallepipedal elements.

Let us prove that a function in $\mathbb{Q}_k$ is uniquely determined by its values at these nodes.

Proposition 3.3.2 *If* $q \in \mathbb{Q}_k$ $(k = 1, 2, 3)$ *vanishes at the nodes described in Fig. 3.3.4, then* $q \equiv 0$.

Proof. We start from the case $k = 1$. The restriction of q to each side is a linear polynomial of one variable. Hence q vanishes over each side and therefore it can be written as

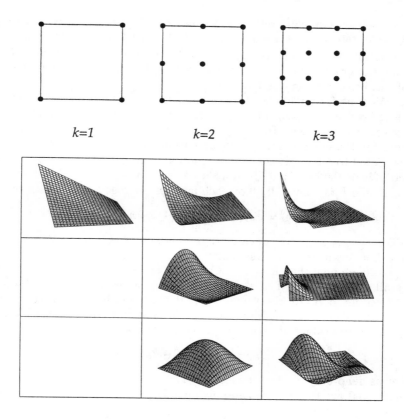

Fig. 3.3.4. Degrees of freedom (upper) and shape functions (lower) for parallelepipedal elements

$$q(\mathbf{x}) = c_1 x_1(1 - x_1)x_2(1 - x_2) \;,$$

which implies $c_1 = 0$.

A similar argument applied to the cases $k = 2$ and $k = 3$ implies that q has the form

$$q(\mathbf{x}) = c_2 x_1 \left(\frac{1}{2} - x_1\right)(1 - x_1)x_2(1 - x_2) \qquad (k = 2)$$

or

$$q(\mathbf{x}) = c_3 x_1 \left(\frac{1}{3} - x_1\right)\left(\frac{2}{3} - x_1\right)(1 - x_1)x_2(1 - x_2) \qquad (k = 3) \;.$$

Since $x_1^3 x_2^2 \notin \mathbb{Q}_2$ and $x_1^4 x_2^2 \notin \mathbb{Q}_3$, it follows that $c_2 = c_3 = 0$. $\qquad \square$

The three-dimensional case can be treated quite similarly. Starting from $\hat{K} = [0,1]^3$, the nodes are given as in Fig. 3.3.5 (where only the nodes on the visible faces are pictured; for $k = 2$ there is also an internal node in the center of gravity; for $k = 3$ there are also eight internal nodes at the vertices of a cube of coordinates $1/3$ and $2/3$).

As done for triangular finite elements, we can reduce the problem of the identification of a polynomial $p \in \mathbb{Q}_k$ from the three-dimensional case to the two-dimensional one. In fact, on each face the nodes are the same as the ones for the corresponding two-dimensional square, and a function $p \in \mathbb{Q}_k$ restricted to a face gives rise to a function in $\mathbb{Q}_k$ (which depends only on two variables). Hence, if $p \in \mathbb{Q}_k$ vanishes at each node indicated in Fig. 3.3.5, in particular it is identically 0 on each face. In the case $k = 1$ we can then write

$$p(\mathbf{x}) = cx_1(1 - x_1)x_2(1 - x_2)x_3(1 - x_3) \ .$$

Since $x_1^2 x_2^2 x_3^2 \notin \mathbb{Q}_1$ we conclude that $c = 0$, i.e., $p \equiv 0$. The other cases $k = 2$ and $k = 3$ can be treated similarly.

Assume now that $K = T_K(\hat{K})$, where T_K is an affine invertible map. We recall that $v_h \in X_h^k$ if $v_h \in C^0(\overline{\Omega})$ and $v_{h|K} \circ T_K \in \mathbb{Q}_k$. Hence the degrees of freedom in K are the values of v_h at the nodes $\mathbf{a}_{j,K} = T_K(\hat{\mathbf{a}}_j)$, where $\hat{\mathbf{a}}_j$ are the nodes in $[0,1]^d$. Moreover, the total number of degrees of freedom is given by the values of v_h at the global set of nodes

$$\Sigma_h := \{ \mathbf{a}_{j,K} \mid K \in T_h \} \subset \overline{\Omega} \ .$$

Let us denote by $\mathbf{a}_j$ these nodes, $j = 1, ..., N_h$. The basis functions (or shape functions) are those piecewise-polynomials $\varphi_j \in X_h^k$ such that

$$\varphi_j(\mathbf{a}_i) = \delta_{ij} \ , \qquad i, j = 1, ..., N_h \ ,$$

and are reported on the lower part of Fig. 3.3.4.

3.3.3 The Vector Case

Let us first consider now the vector case (3.2.6) for two-dimensional domains. To begin with, on each K we have to find the degrees of freedom, whose number must be $(k + 2)k$ (see (3.2.3)). Moreover, we have to choose them in such a way that $\mathbf{v}_h \in H(\text{div}; \Omega)$, and, from Proposition 3.2.2, it is necessary and sufficient that $\mathbf{n} \cdot \mathbf{v}_{h|K_1} = \mathbf{n} \cdot \mathbf{v}_{h|K_2}$ on each common side $F = K_1 \cap K_2$.

The fact that $(\mathbf{n} \cdot \mathbf{q})_{|F} \in \mathbb{P}_{k-1}$ and $\text{div}\,\mathbf{q} \in \mathbb{P}_{k-1}$ for each $\mathbf{q} \in \mathbb{D}_k$ suggests that k degrees of freedom can be given by the values of $\mathbf{n} \cdot \mathbf{q}$ at k distinct points of each side. This is sufficient for the case $k = 1$, since we need to fix three degrees of freedom. If $k = 2$ we add the values of $\int_K q_1$ and $\int_K q_2$; if $k = 3$ we also take $\int_K q_1 x_1, \int_K q_1 x_2, \int_K q_2 x_1$ and $\int_K q_2 x_2$. For the general case $k > 3$, we refer the interested reader to Roberts and Thomas (1991).

Let us show that this choice of degrees of freedom permits us to identify a polynomial $\mathbf{q} \in \mathbb{D}_k$ in a unique way.

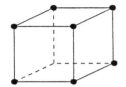 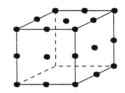 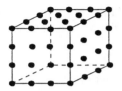

Fig. 3.3.5. $d = 3$ and, from left to right, $k = 1$, $k = 2$ and $k = 3$. The number of degrees of freedom is 8, 27 and 64 respectively. Only the nodes on the visible faces are pictured

Proposition 3.3.3 *Let* $k = 1, 2$ *or* 3. *Assume that* $\mathbf{q} \in \mathbb{D}_k$ *is such that* $\mathbf{n} \cdot \mathbf{q}$ *vanishes at* k *distinct points on each side of* K. *Assume moreover that*

$$(3.3.2) \qquad \int_K q_1 = \int_K q_2 = 0 \qquad (if\ k \geq 2)$$

and

$$(3.3.3) \quad \int_K x_1 q_1 = \int_K x_2 q_1 = \int_K x_1 q_2 = \int_K x_2 q_2 = 0 \quad (only\ if\ k = 3)\ .$$

Then $\mathbf{q} \equiv 0$.

Proof. First of all, $(\mathbf{n} \cdot \mathbf{q})_{|F} \in \mathbb{P}_{k-1}$, hence it vanishes on each side F of K.

Therefore, by the Green formula (1.3.4) and (3.3.2), (3.3.3), for each $\varPsi \in \mathbb{P}_{k-1}$ we have

$$\int_K \varPsi \operatorname{div} \mathbf{q} = -\int_K \nabla\varPsi \cdot \mathbf{q} + \int_{\partial K} \varPsi\, \mathbf{n} \cdot \mathbf{q} = 0\ ,$$

since $\nabla\varPsi \in (\mathbb{P}_{k-2})^2$. As $\operatorname{div} \mathbf{q} \in \mathbb{P}_{k-1}$, it follows that $\operatorname{div} \mathbf{q} = 0$ in K.

Remember now that we can write

$$\mathbf{q}(\mathbf{x}) = \mathbf{p}_{k-1}(\mathbf{x}) + \mathbf{x} p_{k-1}^*(\mathbf{x})\ ,$$

where $\mathbf{p}_{k-1} \in (\mathbb{P}_{k-1})^2$ and the polynomial p_{k-1}^* is a homogeneous function of degree $k - 1$. It follows that

$$0 = \operatorname{div} \mathbf{q} = \operatorname{div} \mathbf{p}_{k-1} + 2 p_{k-1}^* + \mathbf{x} \cdot \nabla p_{k-1}^*$$
$$= \operatorname{div} \mathbf{p}_{k-1} + (2 + k - 1) p_{k-1}^*\ ,$$

where the Euler identity for homogeneous functions has been used. Thus $p_{k-1}^* \in \mathbb{P}_{k-2}$ and consequently $\mathbf{q} \in (\mathbb{P}_{k-1})^2$.

Since $\operatorname{div} \mathbf{q} = 0$, we can find a polynomial $w \in \mathbb{P}_k$ (unique up to an additive constant) such that

$$\mathbf{q} = (D_2 w, -D_1 w)\ .$$

Moreover, since $(\mathbf{n} \cdot \mathbf{q})_{|F} = 0$, we can assume that w is vanishing on each side F, and consequently

$$w(\mathbf{x}) = c_0 p_1(\mathbf{x}) p_2(\mathbf{x}) p_3(\mathbf{x}) \ ,$$

where $p_i(\mathbf{x})$ are linear functions, each one vanishing on one side of K.

This completes proof if $k = 1$ or $k = 2$, since $w \in \mathbb{P}_k$. When $k = 3$, using again (3.3.2) and (3.3.3) we obtain for each $\mathbf{r} \in (\mathbb{P}_1)^2$

$$0 = \int_K \mathbf{q} \cdot \mathbf{r} = \int_K [(D_2 w) r_1 - (D_1 w) r_2] = -\int_K w(D_2 r_1 - D_1 r_2) \ .$$

Choosing $\mathbf{r}$ such that $D_2 r_1 - D_1 r_2 = c_0$, it follows

$$c_0^2 \int_K p_1 p_2 p_3 = 0 \ ,$$

i.e., $c_0 = 0$ and $\mathbf{q} \equiv 0$.   □

In particular we have proved that, if $\mathbf{q} \in \mathbb{D}_k$ is such that $\mathbf{n} \cdot \mathbf{q}$ is vanishing at k distinct points of a side of K, then $\mathbf{n} \cdot \mathbf{q}$ is vanishing on that side. This implies that a piecewise-polynomial $\mathbf{v}_h$, identified in each K by the degrees of freedom described above, indeed belongs to $H(\mathrm{div}; \Omega)$ in view of Proposition 3.2.2 .

The three-dimensional case can be treated in a similar way. The degrees of freedom are given by the values of $\mathbf{n} \cdot \mathbf{q}$ at $k(k+1)/2$ distinct points of each face, plus (if $k \geq 2$) the integrals $\int_K \mathbf{w} \cdot \mathbf{q}$ for each $\mathbf{w} \in (\mathbb{P}_{k-2})^3$. The total number of degrees of freedom on each triangle K is given by (3.2.3), i.e., is equal to $k(k+1)(k+3)/2$. We are not going to enter into further details, but will refer rather to Nédélec (1980), Roberts and Thomas (1991).

The construction of a basis of W_h^k is somehow less evident than for X_h^k, since it is not true now that all degrees of freedom are pointvalues of $\mathbf{v}_h$. This remains true for $\mathbf{n} \cdot \mathbf{v}_h$ at the nodes chosen on the interelement boundaries; in addition, we also have to consider the value of the integrals over K described before (the so-called K-moments). Let us denote by $m_j(\mathbf{v})$, $j = 1, ..., N_{1,h}$, the values $(\mathbf{n} \cdot \mathbf{v})(\mathbf{a}_j)$, where $\mathbf{a}_j$ is the set of all nodes in $\bar{\Omega}$, and by $m_l(\mathbf{v})$, $l = N_{1,h} + 1, ..., N_h$, the set of all K-moments of the function $\mathbf{v}$. A basis of W_h^k is constructed by requiring that

(3.3.4) $m_s(\varphi_i) = \delta_{is} \ , \qquad i, s = 1, ..., N_h \ .$

3.4 The Interpolation Operator

The identification of the degrees of freedom and of the shape functions easily leads to the definition of an *interpolation operator*, i.e., an operator defined on the space of continuous functions and valued in the finite element spaces X_h^k or W_h^k defined in (3.2.4)-(3.2.6).

The simplest case is provided by X_h^k. In fact, for each $v \in C^0(\overline{\Omega})$ we can set

$$(3.4.1) \qquad \pi_h^k(v) := \sum_{i=1}^{N_h} v(\mathbf{a}_i)\varphi_i \ ,$$

where $\mathbf{a}_i$ are the nodes on $\overline{\Omega}$, and φ_i are the corresponding shape functions. The interpolant $\pi_h^k(v)$ is the unique function in X_h^k which takes the same values of the given function v at all nodes $\mathbf{a}_i$.

In a similar way, we can introduce a *local* interpolation operator, i.e., an operator acting on functions defined in K, and giving a polynomial in $\mathbb{P}_k$ or $\mathbb{Q}_k$. If $\mathbf{a}_{i,K}$, $i = 1, ..., M_K$, are the nodes in K, we set

$$(3.4.2) \qquad \pi_K^k(v) := \sum_{i=1}^{M_K} v(\mathbf{a}_{i,K})\varphi_{i|K} \qquad \forall \, v \in C^0(K) \ ,$$

where φ_i are the shape functions.

It can be verified at once that

$$(3.4.3) \qquad \pi_h^k(v)_{|K} = \pi_K^k(v_{|K}) \qquad \forall \, K \in \mathcal{T}_h \ , \ v \in C^0(\overline{\Omega}) \ .$$

3.4.1 Interpolation Error: the Scalar Case

We provide now some estimates for the interpolation errors $v - \pi_h^k(v)$. Due to the localization property (3.4.3), the problem can be reduced to the estimate of $v - \pi_h^k(v)$ in each element $K \in \mathcal{T}_h$. The key idea is to obtain suitable estimates over the reference element $\hat{K}$, and then to convert them over any given element K, using the properties of the affine map T_K.

The choice of the norm for measuring the interpolation error is another important issue. At first we are interested to estimate the error $v - \pi_K^k(v)$ with respect to the seminorms $|\cdot|_{m,K}$ of the Sobolev spaces $H^m(K)$, $m \geq 0$ (see Section 1.2).

In the sequel, we assume that $v \in C^0(K)$, so that $\pi_K^k(v)$ is well defined (as a consequence of Theorem 1.3.4, it is sufficient to assume $v \in H^2(K)$). The estimate of the interpolation error will be done in four steps. The first one is to show that any Sobolev seminorm in K is bounded from above and below by the corresponding seminorm in $\hat{K}$, times factors depending on the map T_K. To be more explicit, let us write

$$T_K(\hat{\mathbf{x}}) = B_K \hat{\mathbf{x}} + \mathbf{b}_K \ , \qquad \hat{\mathbf{x}} \in \hat{K} \ ,$$

where B_K is a $(d \times d)$-matrix. We have:

Proposition 3.4.1 *For any $v \in H^m(K)$, $m \geq 0$, define $\hat{v} := v \circ T_K$. Then $\hat{v} \in H^m(\hat{K})$, and there exists a constant $C = C(m, d)$ such that*

$$(3.4.4) \qquad |\hat{v}|_{m,\hat{K}} \leq C \, \|B_K\|^m \, |\det B_K|^{-1/2} \, |v|_{m,K} \qquad \forall \, v \in H^m(K)$$

and

$$(3.4.5) \qquad |v|_{m,K} \leq C \, \|B_K^{-1}\|^m \, |\det B_K|^{1/2} \, |\hat{v}|_{m,\hat{K}} \qquad \forall \, \hat{v} \in H^m(\hat{K}) \ ,$$

where $\| \cdot \|$ is the matrix norm associated to the euclidean norm in $\mathbb{R}^d$.

Proof. Since $C^\infty(K)$ is dense in $H^m(K)$, it is sufficient to prove (3.4.4) for a smooth v. As

$$|\hat{v}|^2_{m,\hat{K}} = \sum_{|\alpha|=m} \int_{\hat{K}} |D^\alpha \hat{v}|^2 \ ,$$

using the chain rule to compute $D^\alpha \hat{v}$ one easily finds

$$\|D^\alpha \hat{v}\|_{0,\hat{K}} \leq C \, \|B_K\|^m \sum_{|\beta|=m} \|(D^\beta v) \circ T_K\|_{0,\hat{K}} \ .$$

Mapping back to K, one obtains

$$\|D^\alpha \hat{v}\|_{0,\hat{K}} \leq C \, \|B_K\|^m \, |\det B_K|^{-1/2} \, |v|_{m,K} \ ,$$

hence (3.4.4) by summing up for $|\alpha| = m$. The proof of (3.4.5) is carried out in a similar way. $\qquad\qquad\square$

It is now necessary to evaluate $\|B_K\|$ and $\|B_K^{-1}\|$ in terms of some geometric quantities related to K. Define

$$h_K := \operatorname{diam}(K) \ , \quad \rho_K := \sup\{\operatorname{diam}(S) \,|\, S \text{ is a ball contained in } K\} \ .$$

The same quantities will be denoted by $\hat{h}$ and $\hat{\rho}$ when they are referred to the reference domain $\hat{K}$.

Proposition 3.4.2 *The following estimates hold*

$$(3.4.6) \qquad \|B_K\| \leq \frac{h_K}{\hat{\rho}} \ , \quad \|B_K^{-1}\| \leq \frac{\hat{h}}{\rho_K} \ .$$

Proof. We can write

$$\|B_K\| = \frac{1}{\hat{\rho}} \, \sup\{|B_K \, \boldsymbol{\xi}| \,|\, |\boldsymbol{\xi}| = \hat{\rho}\} \ .$$

For each $\boldsymbol{\xi}$ satisfying $|\boldsymbol{\xi}| = \hat{\rho}$, we find two points $\hat{\mathbf{x}}, \hat{\mathbf{y}} \in \hat{K}$ such that $\hat{\mathbf{x}} - \hat{\mathbf{y}} = \boldsymbol{\xi}$. Since $B_K \boldsymbol{\xi} = T_K \hat{\mathbf{x}} - T_K \hat{\mathbf{y}}$, we deduce $|B_K \boldsymbol{\xi}| \leq h_K$, hence the first inequality in (3.4.6). The other is proved in a similar way. $\quad\square$

We thus need an estimate for the seminorm of $[v - \pi_K^k(v)] \circ T_K$ in $H^m(\hat{K})$. We will denote $\pi_K^k(v) \circ T_K$ by $[\pi_K^k(v)]^\wedge$. Since the nodes $\mathbf{a}_{i,K}$ in K coincide with $T_K(\hat{\mathbf{a}}_i)$, where $\hat{\mathbf{a}}_i$ are the nodes in $\hat{K}$, and, similarly, the shape functions in $\hat{K}$ are given by $\hat{\varphi}_i = \varphi_i \circ T_K$, we obtain

$$[\pi_K^k(v)]^\wedge = \pi_K^k(v) \circ T_K = \sum_{i=1}^{M_K} v(\mathbf{a}_{i,K})(\varphi_i \circ T_K)$$

(3.4.7)

$$= \sum_{i=1}^{M_K} v(T_K(\hat{\mathbf{a}}_i))\hat{\varphi}_i = \pi_{\hat{K}}^k(\hat{v}) \ .$$

Hence we have to estimate $\hat{v} - \pi_{\hat{K}}^k(\hat{v})$ in $H^m(\hat{K})$. To begin with, let us prove the following result:

Proposition 3.4.3 (Bramble-Hilbert lemma). *Let* $\hat{\lambda} : H^s(\hat{K}) \to H^m(\hat{K})$, $m \geq 0$, $s \geq 0$, *be a linear and continuous mapping such that*

(3.4.8)
$$\hat{\lambda}(\hat{p}) = 0 \qquad \forall \, \hat{p} \in \mathbb{P}_l \ ,$$

where $l \geq 0$. *Then for each* $\hat{v} \in H^s(\hat{K})$

(3.4.9)
$$|\hat{\lambda}(\hat{v})|_{m,\hat{K}} \leq \|\hat{\lambda}\|_{\mathcal{L}(H^s(\hat{K});H^m(\hat{K}))} \inf_{\hat{p} \in \mathbb{P}_l} \|\hat{v} + \hat{p}\|_{s,\hat{K}} \ .$$

In particular, for each $\hat{v} \in H^s(\hat{K})$, $s \geq 2$, *it holds*

$$|\hat{v} - \pi_{\hat{K}}^k(\hat{v})|_{m,\hat{K}} \leq \|I - \pi_{\hat{K}}^k\|_{\mathcal{L}(H^s(\hat{K});H^m(\hat{K}))} \inf_{\hat{p} \in \mathbb{P}_l} \|\hat{v} + \hat{p}\|_{s,\hat{K}}$$

for $0 \leq m \leq s$, $0 \leq l \leq k$.

Proof. Let $\hat{v} \in H^s(\hat{K})$. For any $\hat{p} \in \mathbb{P}_l$ we have from (3.4.8)

$$|\hat{\lambda}(\hat{v})|_{m,\hat{K}} = |\hat{\lambda}(\hat{v} + \hat{p})|_{m,\hat{K}} \leq \|\hat{\lambda}\|_{\mathcal{L}(H^s(\hat{K});H^m(\hat{K}))} \|\hat{v} + \hat{p}\|_{s,\hat{K}} \ ,$$

hence (3.4.9) follows, since $\hat{p}$ is arbitrary.

Notice that the interpolation operator is defined in $H^s(\hat{K})$, as a consequence of the Sobolev embedding theorem $H^s(\hat{K}) \subset C^0(\hat{K})$ for $s \geq 2$ (see Theorem 1.3.4). Therefore the operator $I - \pi_{\hat{K}}^k$ satisfies the assumptions of the Proposition. $\quad\square$

The following Proposition provides the last tool we need to prove the desired estimate for the interpolation error.

Proposition 3.4.4 (Deny-Lions lemma). *For each $l \geq 0$ there exists a constant $C = C(l, d, \hat{K})$ such that*

$$(3.4.10) \qquad \inf_{\hat{p} \in \mathbb{P}_l} \|\hat{v} + \hat{p}\|_{l+1, \hat{K}} \leq C|\hat{v}|_{l+1, \hat{K}} \qquad \forall \, \hat{v} \in H^{l+1}(\hat{K}) \ .$$

Proof. First of all, we prove that there exists a constant $C = C(l, d, \hat{K})$ such that

$$(3.4.11) \qquad \|\hat{v}\|_{l+1, \hat{K}} \leq C \left\{ |\hat{v}|^2_{l+1, \hat{K}} + \sum_{|\alpha| \leq l} \left(\int_{\hat{K}} D^\alpha \hat{v} \right)^2 \right\}^{1/2}$$

for each $\hat{v} \in H^{l+1}(\hat{K})$. We proceed by contradiction. If (3.4.11) didn't hold, then we could find a sequence $\hat{v}_j \in H^{l+1}(\hat{K})$ such that

$$(3.4.12) \qquad \|\hat{v}_j\|_{l+1, \hat{K}} = 1$$

and

$$(3.4.13) \qquad |\hat{v}_j|^2_{l+1, \hat{K}} + \sum_{|\alpha| \leq l} \left(\int_{\hat{K}} D^\alpha \hat{v}_j \right)^2 < \frac{1}{j^2} \ .$$

Since the immersion $H^{l+1}(\hat{K}) \hookrightarrow H^l(\hat{K})$ is compact (see Theorem 1.3.5), we can select a subsequence, still denoted by $\hat{v}_j$, which is strongly convergent in $H^l(\hat{K})$. As a consequence of (3.4.13) $\hat{v}_j$ is indeed a Cauchy sequence in $H^{l+1}(\hat{K})$, therefore a function $\hat{w}$ exists such that $\hat{v}_j$ converge to $\hat{w}$ in $H^{l+1}(\hat{K})$ and $\|\hat{w}\|_{l+1, \hat{K}} = 1$. Moreover, $\int_{\hat{K}} D^\alpha \hat{w} = 0$ for $|\alpha| \leq l$ and $D^\alpha \hat{w} = 0$ for $|\alpha| = l + 1$. This last relation implies that $\hat{w} \in \mathbb{P}_l$, while the former produces $\hat{w} = 0$, which is a contradiction. Hence (3.4.11) is proven.

Now, for each $\hat{v} \in H^{l+1}(\hat{K})$ we can construct a unique $\hat{q} \in \mathbb{P}_l$ such that

$$\int_{\hat{K}} D^\alpha \hat{q} = - \int_{\hat{K}} D^\alpha \hat{v} \ , \qquad \forall \, |\alpha| \leq l \ .$$

Hence applying (3.4.11) to $\hat{v} + \hat{q}$ we obtain

$$\inf_{\hat{p} \in \mathbb{P}_l} \|\hat{v} + \hat{p}\|_{l+1, \hat{K}} \leq \|\hat{v} + \hat{q}\|_{l+1, \hat{K}} \leq C|\hat{v} + \hat{q}|_{l+1, \hat{K}} = C|\hat{v}|_{l+1, \hat{K}}$$

i.e., (3.4.10). □

We are now able to prove the main result of this Section.

Theorem 3.4.1 *Let $0 \leq m \leq l + 1$, where $l = \min(k, s - 1) \geq 1$. Then there exists a constant $C = C(\hat{K}, \pi_{\hat{K}}^k, k, m, s, d)$ such that*

(3.4.14) $$|v - \pi_K^k(v)|_{m,K} \leq C \frac{h_K^{l+1}}{\rho_K^m} |v|_{l+1,K} \qquad \forall\, v \in H^s(K) \ .$$

Proof. Using (3.4.5)-(3.4.7) we have

(3.4.15) $$|v - \pi_K^k(v)|_{m,K} \leq C \frac{1}{\rho_K^m} |\det B_K|^{1/2} |\hat{v} - \pi_{\hat{K}}^k(\hat{v})|_{m,\hat{K}} \ .$$

Since $(I - \pi_{\hat{K}}^k)(\hat{p}) = 0$ for each $\hat{p} \in \mathbb{P}_l$ (let us recall that, if $\hat{K} = [0,1]^d$, then $\pi_{\hat{K}}^k$ is indeed invariant on $\mathbb{Q}_l$), we obtain

$$|\hat{v} - \pi_{\hat{K}}^k(\hat{v})|_{m,\hat{K}} \leq C \inf_{\hat{p} \in \mathbb{P}_l} \|\hat{v} + \hat{p}\|_{s,\hat{K}} \ .$$

Let us firstly assume that $s - 1 < k$, so that $l = s - 1$. From (3.4.10) we have

$$|\hat{v} - \pi_{\hat{K}}^k(\hat{v})|_{m,\hat{K}} \leq C|\hat{v}|_{l+1,\hat{K}} \ .$$

Applying now (3.4.4) and again (3.4.6) we finally find

(3.4.16) $$|\hat{v} - \pi_{\hat{K}}^k(\hat{v})|_{m,\hat{K}} \leq C\, h_K^{l+1} |\det B_K|^{-1/2} |v|_{l+1,K} \ .$$

The result (3.4.14) now follows from (3.4.15) and (3.4.16).

If $s - 1 \geq k$, we have $l = k$ and we can apply the Bramble-Hilbert lemma to the map $I - \pi_{\hat{K}}^k$, which is linear and continuous from $H^{k+1}(\hat{K})$ to $H^m(\hat{K})$. Therefore we obtain

$$|\hat{v} - \pi_{\hat{K}}^k(\hat{v})|_{m,\hat{K}} \leq C \inf_{\hat{p} \in \mathbb{P}_k} \|\hat{v} + \hat{p}\|_{k+1,\hat{K}} \leq C|\hat{v}|_{k+1,\hat{K}} \ ,$$

having used (3.4.10), and the thesis follows as before.   □

Remark 3.4.1 It is worth noting that, if $1 \leq s - 1 < k$, $0 \leq m \leq s$, then one obtains

(3.4.17) $$|v - \pi_K^k(v)|_{m,K} \leq C \frac{h_K^s}{\rho_K^m} |v|_{s,K} \qquad \forall\, v \in H^s(K) \ ,$$

i.e., the same error estimate proven for the interpolation operator $\pi_K^{s-1}(v)$. This shows that high order interpolations on v do not give, in principle, better error estimates if v is not regular enough.   □

We also underline that similar results hold for interpolation in the Sobolev spaces $W^{s,p}(\Omega)$, $p \in [1, \infty]$. A precise result can be found, e.g., in Ciarlet (1991). To give an example, the following Remark is devoted to the case $p = \infty$.

Remark 3.4.2 *(L^∞-interpolation error)*. It is possible to estimate the interpolation error with respect to different norms. For instance, by proceeding as before it is easy to show that

$$|v - \pi_K^k(v)|_{m,\infty,K} \leq C \, [\text{meas}(K)]^{-1/2} \frac{h_K^{l+1}}{\rho_K^m} |v|_{l+1,K} \qquad \forall \, v \in H^s(K)$$

for $l = \min(k, s-1) \geq 1$, $0 \leq m < l+1 - d/2$, $d = 2, 3$ (so that $H^{l+1}(K) \subset W^{m,\infty}(K)$: see Theorem 1.3.4), and

$$|v - \pi_K^k(v)|_{m,\infty,K} \leq C \frac{h_K^{l+1}}{\rho_K^m} |v|_{l+1,\infty,K} \qquad \forall \, v \in W^{s,\infty}(K)$$

for $0 \leq m \leq l+1$. Here $|\cdot|_{m,\infty,K}$ denotes the seminorm in $W^{m,\infty}(K)$ (see Section 1.2).

The latter estimate is in fact trivial, as Bramble-Hilbert and Deny-Lions lemmas still hold if one substitutes $H^s(\hat{K})$ and $H^m(\hat{K})$ with $W^{s,\infty}(\hat{K})$ and $W^{m,\infty}(\hat{K})$, respectively. To prove the former one, let us notice that for a multi-index $\boldsymbol{\alpha}$ such that $|\boldsymbol{\alpha}| = m$ one has

$$\|D^{\boldsymbol{\alpha}} v\|_{\infty,K} \leq C \, \|B_K^{-1}\|^m \, |\hat{v}|_{m,\infty,\hat{K}} \ .$$

Proceeding as in Theorem 3.4.1 and recalling that $H^{l+1}(\hat{K}) \subset W^{m,\infty}(\hat{K})$ for $0 \leq m < l+1 - d/2$, we have

$$|v - \pi_K^k(v)|_{m,\infty,K} \leq \frac{C}{\rho_K^m} |\hat{v}|_{l+1,\hat{K}} \ .$$

Since

$$|\det B_K| = \frac{\text{meas}(K)}{\text{meas}(\hat{K})} \ ,$$

the result follows at once from (3.4.4). □

Before considering the global interpolation error over Ω, let us introduce a definition concerning the family of triangulations $\mathcal{T}_h$.

Definition 3.4.1 A family of triangulations $\mathcal{T}_h$, $h > 0$ is called *regular* if there exists a constant $\sigma \geq 1$ such that

$$(3.4.18) \qquad \qquad \max_{K \in \mathcal{T}_h} \frac{h_K}{\rho_K} \leq \sigma \qquad \forall \, h > 0 \ .$$

We can finally prove an estimate for the global interpolation error.

Theorem 3.4.2 *Let $\mathcal{T}_h$ be a regular family of triangulations and assume that $m = 0, 1$, $l = \min(k, s-1) \geq 1$. Then there exists a constant C, independent of h, such that*

(3.4.19) $\qquad |v - \pi_h^k(v)|_{m,\Omega} \le Ch^{l+1-m}|v|_{l+1,\Omega} \quad \forall \, v \in H^s(\Omega) \ .$

Proof. We localize the estimate over K:

$$|v - \pi_h^k(v)|_{m,\Omega}^2 = \sum_{K \in \mathcal{T}_h} |v - \pi_h^k(v)|_{m,K}^2 \ .$$

From (3.4.14), (3.4.17), (3.4.18) we can write

$$|v - \pi_h^k(v)|_{m,K} \le Ch^{l+1-m}|v|_{l+1,K} \ ,$$

hence the result follows by summing up over K. $\qquad\qquad\qquad\square$

Note that the restriction on the index m is due to the fact that the inclusion $X_h^k \subset H^m(\Omega)$ holds only if $m \le 1$. The construction of a finite dimensional space contained in $H^2(\Omega)$ would require higher order continuity across the interelement boundaries.

3.4.2 Interpolation Error: the Vector Case

If we are considering W_h^k, to define the interpolation operator we must give a meaning to the point value $\mathbf{n} \cdot \mathbf{v}$ at all nodes $\mathbf{a}_i \in \overline{\Omega}$ and to all the K-moments $m_l(\mathbf{v})$ ($l = N_{1,h}+1, ..., N_h$). If $\mathbf{v} \in (C^0(\overline{\Omega}))^d$, this is, of course, easily doable. But it will be useful to define the interpolation operator even in spaces of functions that are not necessarily continuous. To do this, let us first remark that for $\mathbf{q} \in \mathbb{D}_k$, instead of the point values of $\mathbf{n} \cdot \mathbf{q}$ on a face F_K of K, one can consider the following degrees of freedom

$$\int_{F_K} \mathbf{n} \cdot \mathbf{q}\,\psi \ , \qquad \psi \in \mathbb{P}_{k-1}$$

(see Roberts and Thomas (1991)), which are called F_K-moments. Let us denote the global set of these F_K-moments relative to a function $\mathbf{v}$ in the same way as the degrees of freedom associated to pointvalues of $\mathbf{n} \cdot \mathbf{v}$ have been indicated in Section 3.3.3, i.e., $m_i(\mathbf{v})$, $i = 1, ..., N_{1,h}$. Moreover, still denote the set of K-moments by $m_l(\mathbf{v})$, $l = N_{1,h}+1, ..., N_h$ and by φ_i the shape functions defined as in (3.3.4).

We can see now that the interpolation operator $\boldsymbol{\omega}_h^k$ is defined in $(H^1(\Omega))^d$ by setting

(3.4.20) $\qquad\qquad \boldsymbol{\omega}_h^k(\mathbf{v}) := \sum_{i=1}^{N_h} m_i(\mathbf{v})\varphi_i \ .$

We again recall that this is the only one function in W_h^k satisfying

$$m_i(\boldsymbol{\omega}_h^k(\mathbf{v})) = m_i(\mathbf{v}) \qquad \forall \, i = 1, ..., N_h \ .$$

To introduce a local interpolation operator, let us denote by $m_{i,K}(\mathbf{v})$, $i = 1, ..., M_K$, the set of K-moments and F_K-moments relative to K, and define

$$(3.4.21) \qquad \boldsymbol{\omega}_K^k(\mathbf{v}) := \sum_{i=1}^{M_K} m_{i,K}(\mathbf{v}) \varphi_{i|K} \quad \forall\, \mathbf{v} \in (H^1(K))^d \ .$$

One verifies at once that

$$(3.4.22) \qquad \boldsymbol{\omega}_h^k(\mathbf{v})_{|K} = \boldsymbol{\omega}_K^k(\mathbf{v}_{|K}) \quad \forall\, K \in \mathcal{T}_h \ , \ \forall\, \mathbf{v} \in (H^1(\Omega))^d \ .$$

A further important point that oughts to be underlined: the local interpolation operator $\boldsymbol{\omega}_K^k$ satisfies

$$\operatorname{div}(\boldsymbol{\omega}_K^k(\mathbf{v})) = P_K^{k-1}(\operatorname{div} \mathbf{v}) \quad \forall\, \mathbf{v} \in (H^1(K))^d \ ,$$

where P_K^{k-1} is the orthogonal projection in $L^2(K)$ onto $\mathbb{P}_{k-1}$. In fact, the divergence of $\boldsymbol{\omega}_K^k(\mathbf{v})$ is a $\mathbb{P}_{k-1}$ polynomial, hence for each $\psi \in \mathbb{P}_{k-1}$

$$\int_K \psi \operatorname{div}(\boldsymbol{\omega}_K^k(\mathbf{v})) = -\int_K \nabla\psi \cdot \boldsymbol{\omega}_K^k(\mathbf{v}) + \int_{\partial K} \psi\, \mathbf{n} \cdot \boldsymbol{\omega}_K^k(\mathbf{v})$$
$$= -\int_K \nabla\psi \cdot \mathbf{v} + \int_{\partial K} \psi\, \mathbf{n} \cdot \mathbf{v} = \int_K \psi \operatorname{div} \mathbf{v} \ ,$$

owing to the fact that the moments of $\mathbf{v}$ and $\boldsymbol{\omega}_K^k(\mathbf{v})$ are the same, i.e., $m_{i,K}(\mathbf{v}) = m_{i,K}(\boldsymbol{\omega}_K^k(\mathbf{v}))$, $i = 1, ..., M_K$. An analogous property holds for the global interpolation operator $\boldsymbol{\omega}_h^k$, i.e.,

$$(3.4.23) \qquad \operatorname{div}(\boldsymbol{\omega}_h^k(\mathbf{v})) = p_h^{k-1}(\operatorname{div} \mathbf{v}) \quad \forall\, \mathbf{v} \in (H^1(\Omega))^d \ ,$$

where p_h^{k-1} is the $L^2(\Omega)$-orthogonal projection onto

$$Y_h^{k-1} := \{ v_h \in L^2(\Omega) \mid v_{h|K} \in \mathbb{P}_{k-1} \ \forall\, K \in \mathcal{T}_h \} \ .$$

This *commutativity property* is very important for the approximation theory in $H(\operatorname{div}; \Omega)$, and it turns out to be useful also when considering optimal error estimates for boundary value problems (see, e.g., Remark 7.2.4). The introduction of the polynomial spaces $\mathbb{D}_k$ is in fact motivated by the validity of (3.4.23).

We can summarize this result by means of the following diagram:

$$
\begin{array}{ccc}
(H^1(\Omega))^d & \xrightarrow{\ \operatorname{div}\ } & L^2(\Omega) \\[2pt]
\Big\downarrow{\scriptstyle \boldsymbol{\omega}_h^k} & & \Big\downarrow{\scriptstyle p_h^{k-1}} \\[2pt]
W_h^k & \xrightarrow{\ \operatorname{div}\ } & Y_h^{k-1}
\end{array}
$$

We are now going to prove an estimate for the interpolation error $\mathbf{v} - \omega_h^k(\mathbf{v})$, $\mathbf{v} \in (H^1(\Omega))^d$. As in the scalar case, we start from $\mathbf{v} - \omega_K^k(\mathbf{v})$, and we map back and forth on the reference domain $\hat{K}$. First of all, we have to verify how to express the interpolation operator in $\hat{K}$ with respect to ω_K^h. Let us define

$$(3.4.24) \qquad \hat{\mathbf{v}} := |\det B_K| B_K^{-1} \mathbf{v} \circ T_K \ .$$

It can be noticed that $\hat{\mathbf{v}} \in (H^1(\hat{K}))^d$ if and only if $\mathbf{v} \in (H^1(K))^d$, and, most important, $\hat{\mathbf{v}} \in \mathbb{D}_k$ if and only if $\mathbf{v} \in \mathbb{D}_k$. This last invariance property would not be true if we choose to transform a vector field operating on each component as in Proposition 3.4.1, i.e., $\hat{v}_i = v_i \circ T_K$, $i = 1, ..., d$.

Transformation (3.4.24) has some further interesting properties: for any given $\mathbf{v} \in (H^1(K))^d$, $\psi \in H^1(K)$ one obtains

$$\int_{\hat{K}} \hat{\mathbf{v}} \cdot \nabla \hat{\psi} = \int_{\hat{K}} |\det B_K| (B_K^{-1} \mathbf{v} \circ T_K) \cdot (B_K^T \nabla \psi \circ T_K)$$

$$= \int_K (B_K^{-1} \mathbf{v}) \cdot (B_K^T \nabla \psi) = \int_K \mathbf{v} \cdot \nabla \psi \ ,$$

and analogously

$$\int_{\hat{K}} \hat{\psi} \, \mathrm{div} \, \hat{\mathbf{v}} = \int_K \psi \, \mathrm{div} \, \mathbf{v} \ .$$

(Here, the scalar function $\hat{\psi}$ has been defined as in Proposition 3.4.1, i.e., $\hat{\psi} = \psi \circ T_K$.) Hence by Green formula (1.3.4)

$$(3.4.25) \qquad \int_{\partial \hat{K}} \hat{\psi} \hat{\mathbf{v}} \cdot \hat{\mathbf{n}} = \int_{\partial K} \psi \mathbf{v} \cdot \mathbf{n} \ .$$

By applying a density argument, this last equality can be shown to hold even for $\psi \in L^2(\partial K)$.

We claim now that for each $\mathbf{v} \in (H^1(K))^d$

$$(3.4.26) \qquad \omega_{\hat{K}}^k(\hat{\mathbf{v}}) = |\det B_K| B_K^{-1} \omega_K^k(\mathbf{v}) \circ T_K =: [\omega_K^k(\mathbf{v})]^\wedge \ .$$

In fact, by mapping $\hat{K}$ back and forth on K it can be easily seen that the K-moments of $[\omega_K^k(\mathbf{v})]^\wedge$ and $\hat{\mathbf{v}}$ do coincide:

$$\int_{\hat{K}} [\omega_K^k(\mathbf{v})]^\wedge \cdot \hat{\mathbf{w}} = \int_{\hat{K}} \hat{\mathbf{v}} \cdot \hat{\mathbf{w}} \qquad \forall \, \hat{\mathbf{w}} \in (\mathbb{P}_{k-2})^d \ .$$

Moreover, the $F_{\hat{K}}$-moments satisfy

$$\int_{F_{\hat{K}}} \hat{\mathbf{n}} \cdot [\omega_K^k(\mathbf{v})]^\wedge \hat{\psi} = \int_{F_{\hat{K}}} \hat{\mathbf{n}} \cdot \hat{\mathbf{v}} \hat{\psi} \qquad \forall \, \hat{\psi} \in \mathbb{P}_{k-1} \ ,$$

for each face $F_{\hat{K}}$, since in (3.4.25) we can take $\hat{\psi}$ vanishing on the other faces of $\hat{K}$.

Before estimating the interpolation error in $H(\text{div}; K)$, we have to prove a result like Proposition 3.4.1 for the vector fields $\mathbf{v}$ and $\hat{\mathbf{v}}$.

Proposition 3.4.5 *For any* $\mathbf{v} \in H^m(K)$, $m \geq 0$, *define* $\hat{\mathbf{v}}$ *as* (3.4.24). *Then* $\hat{\mathbf{v}} \in (H^m(\hat{K}))^d$, *and there exists a constant* $C = C(m,d)$ *such that*

$$(3.4.27) \quad |\hat{\mathbf{v}}|_{m,\hat{K}} \leq C \, \|B_K^{-1}\| \, \|B_K\|^m \, |\det B_K|^{1/2} \, |\mathbf{v}|_{m,K} \quad \forall \, \mathbf{v} \in (H^m(K))^d$$

$$(3.4.28) \quad |\mathbf{v}|_{m,K} \leq C \, \|B_K\| \, \|B_K^{-1}\|^m \, |\det B_K|^{-1/2} \, |\hat{\mathbf{v}}|_{m,\hat{K}} \quad \forall \, \hat{\mathbf{v}} \in (H^m(K))^d \, .$$

Proof. It is sufficient to prove (3.4.27) ((3.4.28) is similar) for $\mathbf{v} \in (C^\infty(K))^d$. Computing $D^\alpha \hat{\mathbf{v}}$, $|\alpha| = m$, we have

$$\|D^\alpha \hat{\mathbf{v}}\|_{0,\hat{K}} \leq C \, |\det B_K| \, \|B_K^{-1}\| \, \|B_K\|^m \sum_{|\beta|=m} \|(D^\beta \mathbf{v}) \circ T_K\|_{0,\hat{K}} \; .$$

By changing variable this produces

$$\|D^\alpha \hat{\mathbf{v}}\| \leq C \, |\det B_K|^{1/2} \, \|B_K^{-1}\| \, \|B_K\|^m \, |\mathbf{v}|_{m,K} \quad ,$$

and summing on α (3.4.27) holds. $\qquad\square$

We are thus in a position to prove an estimate for the interpolation error in $H(\text{div}; K)$.

Theorem 3.4.3 *Let* $0 \leq m \leq k$, $k \geq 1$. *There exists a constant* $C = C(\hat{K}, \omega_{\hat{K}}^k, k, m, d)$ *such that for each* $\mathbf{v} \in (H^k(K))^d$

$$(3.4.29) \quad |\mathbf{v} - \omega_K^k(\mathbf{v})|_{m,K} \leq C \, \frac{h_K^{k+1}}{\rho_K^{m+1}} \, |\mathbf{v}|_{k,K}$$

and for each $\mathbf{v} \in (H^1(K))^d$ *with* $\text{div} \, \mathbf{v} \in H^k(K)$

$$(3.4.30) \quad |\, \text{div} \, \mathbf{v} - \text{div} \, \omega_K^k(\mathbf{v})|_{m,K} \leq C \, \frac{h_K^k}{\rho_K^m} \, |\, \text{div} \, \mathbf{v}|_{k,K} \quad .$$

Proof. As in the proof of Theorem 3.4.1, we pass at first on the reference set $\hat{K}$. From (3.4.28) and (3.4.26) we have

$$|\mathbf{v} - \omega_K^k(\mathbf{v})|_{m,K} \leq C \, \|B_K\| \, \|B_K^{-1}\|^m \, |\det B_K|^{-1/2} \, |\hat{\mathbf{v}} - \omega_{\hat{K}}^k(\hat{\mathbf{v}})|_{m,\hat{K}} \quad .$$

Since $\omega_{\hat{K}}^k(\hat{\mathbf{v}})$ is invariant on $\mathbb{D}_k$, and $(\mathbb{P}_{k-1})^d \subset \mathbb{D}_k$, by Proposition 3.4.3 and 3.4.4 we obtain

$$|\hat{\mathbf{v}} - \omega_{\hat{K}}^k(\hat{\mathbf{v}})|_{m,\hat{K}} \leq C|\hat{\mathbf{v}}|_{k,\hat{K}} \quad .$$

Applying now (3.4.27) we obtain

$$|\hat{\mathbf{v}} - \omega_{\hat{K}}^k(\mathbf{v})|_{m,\hat{K}} \le C \, \|B_K^{-1}\| \, \|B_K\|^k \, |\det B_K|^{1/2} \, |\mathbf{v}|_{k,K} \ .$$

Hence (3.4.29) follows from (3.4.6).

The proof of (3.4.30) is similar. At first, one remarks that

$$(3.4.31) \qquad [\operatorname{div}\mathbf{v} - \operatorname{div}\omega_K^k(\mathbf{v})]^\wedge = |\det B_K|^{-1} \ \operatorname{div}[\hat{\mathbf{v}} - \omega_{\hat{K}}^k(\hat{\mathbf{v}})] \ .$$

As already observed, $\operatorname{div}\omega_{\hat{K}}^k(\hat{\mathbf{v}}) = P_{\hat{K}}^{k-1}(\operatorname{div}\hat{\mathbf{v}})$, where $P_{\hat{K}}^{k-1}$ is the $L^2(\hat{K})$-orthogonal projection onto $\mathbb{P}_{k-1}$; hence, owing to (3.4.5) we obtain

$$|\operatorname{div}\mathbf{v} - \operatorname{div}\omega_K^k(\mathbf{v})|_{m,K} \le C \, \|B_K^{-1}\|^m \, |\det B_K|^{-1/2} \, |\operatorname{div}\hat{\mathbf{v}} - P_{\hat{K}}^{k-1}(\operatorname{div}\hat{\mathbf{v}})|_{m,\hat{K}} \ .$$

The invariance of $P_{\hat{K}}^{k-1}$ on $\mathbb{P}_{k-1}$ produces

$$|\operatorname{div}\hat{\mathbf{v}} - P_{\hat{K}}^{k-1}(\operatorname{div}\hat{\mathbf{v}})|_{m,\hat{K}} \le C |\operatorname{div}\hat{\mathbf{v}}|_{k,\hat{K}} \ .$$

Using again (3.4.31) (for $\operatorname{div}\mathbf{v}$) and (3.4.4) one obtains at once (3.4.30). $\quad\Box$

Remark 3.4.3 Similarly to what we have noticed in Remark 3.4.1, if $1 \le l < k$ and $0 \le m \le l$ we have for each $\mathbf{v} \in (H^l(K))^d$

$$(3.4.32) \qquad |\mathbf{v} - \omega_K^k(\mathbf{v})|_{m,K} \le C \, \frac{h_K^{l+1}}{\rho_K^{m+1}} \, |\mathbf{v}|_{l,K}$$

and for each $\mathbf{v} \in (H^1(K))^d$ with $\operatorname{div}\mathbf{v} \in H^l(K)$

$$(3.4.33) \qquad |\operatorname{div}\mathbf{v} - \operatorname{div}\omega_K^k(\mathbf{v})|_{m,K} \le C \, \frac{h_K^l}{\rho_K^m} \, |\operatorname{div}\mathbf{v}|_{l,K} \ .$$

In (3.4.33) the case $l = 0$ is also admitted. $\quad\Box$

It is now straigthforward to prove the error estimate for the global interpolation error. Proceeding as in Theorem 3.4.2, we have

Theorem 3.4.4 *Let $\mathcal{T}_h$ be a regular family of triangulations and assume that $k \ge 1$. Then there exists a constant C, independent of h, such that*

$$(3.4.34) \qquad \|\mathbf{v} - \omega_h^k(\mathbf{v})\|_{H(\operatorname{div};\Omega)} \le C h^l (|\mathbf{v}|_{l,\Omega} + |\operatorname{div}\mathbf{v}|_{l,\Omega})$$

for each $\mathbf{v} \in (H^l(\Omega))^d$ with $\operatorname{div}\mathbf{v} \in H^l(\Omega)$, $1 \le l \le k$.

3.5 Projection Operators

The interpolation operator gives optimal error estimates in Sobolev norms whenever the function to be interpolated enjoys the minimal requirement to be continuous. In view of finite element analysis, it is useful to introduce other approximation operators, remarkably the $L^2(\Omega)$- and $H^1(\Omega)$-orthogonal projection operators, which make sense on functions which need not to be continuous.

If H is a Hilbert space and S is a closed subspace of H, let us remind now that we can define in H the orthogonal projection operator over S (say, P_S) in the following way:

$$(3.5.1) \qquad P_S(v) \in S : (P_S(v), \varphi)_H = (v, \varphi)_H \qquad \forall \, \varphi \in S \ .$$

It is characterized by the property

$$(3.5.2) \qquad ||v - P_S(v)||_H = \min_{\varphi \in S} ||v - \varphi||_H$$

(see, e.g., Yosida (1974), p. 82).

We are particularly interested in the following projection operators

$$(3.5.3) \qquad P_h^k : L^2(\Omega) \to X_h^k$$
$$(3.5.4) \qquad P_{1,h}^k : H^1(\Omega) \to X_h^k$$
$$(3.5.5) \qquad p_h^k : L^2(\Omega) \to Y_h^k$$
$$(3.5.6) \qquad \mathbf{Q}_h^k : H(\mathrm{div}; \Omega) \to W_h^k \ ,$$

where X_h^k and W_h^k are defined in (3.2.4)-(3.2.6), $k \geq 1$, and

$$(3.5.7) \qquad Y_h^k := \{v_h \in L^2(\Omega) \mid v_{h|K} \in \mathbb{P}_k \quad \forall \, K \in \mathcal{T}_h \} \, , \ k \geq 0 \ ,$$

in case (3.1.6), or

$$(3.5.8) \qquad Y_h^k := \{v_h \in L^2(\Omega) \mid v_{h|K} \circ T_K \in \mathbb{Q}_k \quad \forall \, K \in \mathcal{T}_h \} \, , \ k \geq 0 \ ,$$

in case (3.1.7).

If $\mathcal{T}_h$ is a regular family of triangulations, as a consequence of (3.5.2), (3.4.19) and (3.4.34) we obtain the following error estimates:

$$(3.5.9) \qquad ||v - P_h^k(v)||_{0,\Omega} \leq Ch^{l+1}|v|_{l+1,\Omega} \ , \ 1 \leq l \leq k \ ,$$
$$(3.5.10) \qquad ||v - P_{1,h}^k(v)||_{1,\Omega} \leq Ch^l|v|_{l+1,\Omega} \ , \ 1 \leq l \leq k \ ,$$

for each $v \in H^{l+1}(\Omega)$ and

$$(3.5.11) \qquad ||\mathbf{v} - \mathbf{Q}_h^k(\mathbf{v})||_{H(\mathrm{div};\Omega)} \leq Ch^l(|\mathbf{v}|_{l,\Omega} + |\,\mathrm{div}\,\mathbf{v}|_{l,\Omega}) \ , \ 1 \leq l \leq k \ ,$$

for each $\mathbf{v} \in (H^l(\Omega))^d$ with $\mathrm{div}\,\mathbf{v} \in H^l(\Omega)$. Moreover

$$(3.5.12) \ ||v - P_{1,h}^k(v)||_{1,\Omega} \leq C|v|_{1,\Omega} \ , \quad ||\mathbf{v} - \mathbf{Q}_h^k(\mathbf{v})||_{H(\mathrm{div};\Omega)} \leq ||\mathbf{v}||_{H(\mathrm{div};\Omega)}$$

as $P_{1,h}^k$ and $\mathbf{Q}_h^k$ are orthogonal projections.

It can be noticed that if $v \in H^{k+1}(\Omega)$, then, under suitable assumptions, the L^2-norm of $v - P_{1,h}^k(v)$ is in fact $O(h^{k+1})$. This can be proved by a duality argument (*Aubin-Nitsche trick*). In fact, *let us assume* that, given an arbitrary $r \in L^2(\Omega)$, the function $\varphi(r) \in H^1(\Omega)$ such that

$$(3.5.13) \qquad (\varphi(r), \psi)_{H^1(\Omega)} = (r, \psi)_{L^2(\Omega)} \qquad \forall\, \psi \in H^1(\Omega) \ ,$$

satisfies $\varphi(r) \in H^2(\Omega)$. As a consequence of the Closed Graph theorem (see, e.g., Yosida (1974), p. 79) there exists a constant $C = C(\Omega)$ such that

$$(3.5.14) \qquad |\varphi(r)|_{2,\Omega} \le C\|r\|_{0,\Omega} \qquad \forall\, r \in L^2(\Omega) \ .$$

(The existence of $\varphi(r)$ verifying (3.5.13) is due to the Riesz representation theorem.) Then we can prove

$$(3.5.15)\ \ \|v - P_{1,h}^k(v)\|_{0,\Omega} \le Ch^{l+1}|v|_{l+1,\Omega} \qquad \forall\, v \in H^{l+1}(\Omega)\,,\ 0 \le l \le k \ .$$

In fact, we can write (using (3.5.13))

$$\|v - P_{1,h}^k(v)\|_{0,\Omega} = \sup_{\substack{r \in L^2(\Omega) \\ r \ne 0}} \frac{(v - P_{1,h}^k(v), r)_{L^2(\Omega)}}{\|r\|_{0,\Omega}}$$

$$= \sup_{\substack{r \in L^2(\Omega) \\ r \ne 0}} \frac{(v - P_{1,h}^k(v), \varphi(r))_{H^1(\Omega)}}{\|r\|_{0,\Omega}} \ .$$

Now take any $v_h \in X_h^k$: from (3.5.1) it follows

$$\|v - P_{1,h}^k(v)\|_{0,\Omega} = \sup_{\substack{r \in L^2(\Omega) \\ r \ne 0}} \frac{(v - P_{1,h}^k(v), \varphi(r) - v_h)_{H^1(\Omega)}}{\|r\|_{0,\Omega}}$$

$$\le \|v - P_{1,h}^k(v)\|_{1,\Omega} \sup_{\substack{r \in L^2(\Omega) \\ r \ne 0}} \frac{\|\varphi(r) - v_h\|_{1,\Omega}}{\|r\|_{0,\Omega}} \ .$$

By choosing $v_h = \pi_h^k(\varphi(r))$, from (3.4.19) we have

$$\|\varphi(r) - \pi_h^k(\varphi(r))\|_{1,\Omega} \le Ch|\varphi(r)|_{2,\Omega} \ ;$$

hence using (3.5.14)

$$(3.5.16) \qquad \|v - P_{1,h}^k(v)\|_{0,\Omega} \le Ch\|v - P_{1,h}^k(v)\|_{1,\Omega} \ .$$

From (3.5.10) and (3.5.12)$_1$ we obtain at once (3.5.15).

We will come back to this remark in Section 6.2, and we will analyze a sufficient condition on Ω ensuring (3.5.14) (see Proposition 6.2.2 and Remark 6.2.1).

Let us consider now the projection operator P_h^k defined in (3.5.3). We can obtain at once

$(3.5.17)$ $\qquad ||v - P_h^k(v)||_{0,\Omega} = \min_{v_h \in X_h^k} ||v - v_h||_{0,\Omega} \leq ||v - P_{1,h}^k(v)||_{0,\Omega}$.

(In contrast, it is worthy to notice that in general the estimate

$$||v - P_{1,h}^k(v)||_{0,\Omega} \leq C \min_{v_h \in X_h^k} ||v - v_h||_{0,\Omega} = C||v - P_h^k(v)||_{0,\Omega}$$

is not true. See Babuška and Osborn (1980) for a counterexample.) Hence from (3.5.15) (for $l = 0$) it follows

$(3.5.18)$ $\qquad ||v - P_h^k(v)||_{0,\Omega} \leq Ch|v|_{1,\Omega} \qquad \forall \, v \in H^1(\Omega)$.

Finally, let us assume that the family of triangulations T_h is *quasi-uniform*, i.e., it is regular and moreover there exists a constant $\tau > 0$ such that

$(3.5.19)$ $\qquad \min_{K \in T_h} h_K \geq \tau h \qquad \forall \, h > 0$.

This yields the so-called *inverse inequality* (whose proof is furnished in Proposition 6.3.2), i.e., there exists a positive constant C_1 such that

$(3.5.20)$ $\qquad ||\nabla v_h||_0 \leq C_1 h^{-1} ||v_h||_0 \qquad \forall \, v_h \in X_h^k$.

Under this assumption we can prove that there exists a positive constant C_2 such that

$(3.5.21)$ $\qquad ||v - P_h^k(v)||_{1,\Omega} \leq C_2 ||v - P_{1,h}^k(v)||_{1,\Omega} \qquad \forall \, v \in H^1(\Omega)$.

In fact, since $P_{1,h}^k(v) \in X_h^k$,

$$||v - P_h^k(v)||_{1,\Omega} \leq ||v - P_{1,h}^k(v)||_{1,\Omega} + ||P_{1,h}^k(v) - P_h^k(v)||_{1,\Omega}$$
$$= ||v - P_{1,h}^k(v)||_{1,\Omega} + ||P_h^k[v - P_{1,h}^k(v)]||_{1,\Omega} \ .$$

Using the inverse inequality (3.5.20) we have

$$||P_h^k[v - P_{1,h}^k(v)]||_{1,\Omega} \leq \sqrt{1 + C_1^2 h^{-2}} \, ||v - P_{1,h}^k(v)||_{0,\Omega}$$

as P_h^k is the $L^2(\Omega)$-orthogonal projection onto X_h^k. Inequality (3.5.21) now follows from (3.5.16).

In the conclusion, we can summarize the properties of these projection operators P_h^k and $P_{1,h}^k$ by the following inequality:

$(3.5.22)$
$$||v - P_h^k(v)||_{0,\Omega} + ||v - P_{1,h}^k(v)||_{0,\Omega}$$
$$+ h(||v - P_h^k(v)||_{1,\Omega} + ||v - P_{1,h}^k(v)||_{1,\Omega})$$
$$\leq Ch^{l+1}|v|_{l+1,\Omega} \qquad \forall \, v \in H^{l+1}(\Omega) \ , \quad 0 \leq l \leq k \ .$$

Finally, let us consider the projection operator p_h^k defined in (3.5.5). It is easily verified that, if P_K^k is the $L^2(K)$-orthogonal projection onto $\mathbb{P}_k$, then

(3.5.23) $$p_h^k(v)_{|K} = P_K^k(v_{|K}) \qquad \forall \, v \in L^2(\Omega) \; ,$$

i.e., p_h^k is a local operator like the interpolation operator π_h^k. Thus, proceding as in the proof of Theorem 3.4.2, if $\mathcal{T}_h$ is a regular family of triangulations we obtain for each $k \geq 0$

(3.5.24) $$\|v - p_h^k(v)\|_{0,\Omega} \leq Ch^{l+1}|v|_{l+1,\Omega} \qquad \forall \, v \in H^{l+1}(\Omega) \; , \; 0 \leq l \leq k \; .$$

3.6 Complements

In this presentation we have not considered domains with curved boundary, nor examples of *non-conforming* approximations (i.e., finite element spaces X_h^k or W_h^k such that $X_h^k \not\subset H^1(\Omega)$ or $W_h^k \not\subset H(\mathrm{div};\Omega)$). The interested reader can refer to Strang and Fix (1973) and Ciarlet (1978, 1991).

The approximation theory we have presented here is restricted to affine-equivalent families of finite elements. The same results hold in more general situations as well. A thorough analysis is provided in Ciarlet (1978, 1991), where several type of finite elements of triangular or parallelepipedal type are considered.

Estimates of the approximation error in fractional order Sobolev spaces have been proven by Dupont and Scott (1980).

An approximation theory based on averaging rather than interpolation is also possible. Clément (1975) has shown how to construct an operator $r_h : H^l(\Omega) \rightarrow X_h^k$, $l \geq 0$, enjoying the same approximation properties as the interpolation operator π_h^k, and furthermore

$$\lim_{h \to 0} |v - r_h(v)|_0 = 0 \quad \forall \, v \in L^2(\Omega)$$

$$\lim_{h \to 0} |v - r_h(v)|_1 = 0 \; , \quad |v - r_h(v)|_0 \leq Ch|v|_1 \quad \forall \, v \in H^1(\Omega) \; .$$

For an historical review on finite elements see Zienkiewicz (1973) and Oden (1991). Classical books on the theory and implementation of the finite element method are the ones by Zienkiewicz (1977) and Oden (1972). Early monographs on the mathematical foundations of finite elements are the books by Strang and Fix (1973) and Oden and Reddy (1976).

In this Chapter we have limited our analysis to the so-called *h-version* of finite element method, which is based on the assumption that the polynomial degree is fixed and the accuracy of the approximation is achieved by refining the mesh-size h. An alternative approach is obtained by fixing the mesh, and increasing the degree of the elements. This is called the *p-version* of the finite element method (p denoting the polynomial degree). It is also possible to refine the mesh and increase the degree of the polynomials at the same time. This leads to the *h-p version* of the finite element method. Early theoretical papers concerning the p- and h-p versions are the ones by Babuška, Szabó and

Katz (1981) and Babuška and Dorr (1981), respectively. We will not address these methods in this book, referring the interested reader to Babuška and Suri (1990), Szabó and Babuška (1991).

4. Polynomial Approximation

This Chapter is devoted to the introduction of basic notions and working tools concerning orthogonal algebraic polynomials. More specifically, we will present some properties of both Chebyshev and Legendre polynomials, concerning projection and interpolation processes. These will provide the background of spectral methods for the approximation of partial differential equations that are considered throughout Part II and III of this book.

4.1 Orthogonal Polynomials

Let $\mathbb{P}_N$ denote the space of all algebraic polynomials of degree less than or equal to N, and let $w(x)$ denote a non-negative, integrable function (i.e., a weight function) over the interval $I = (-1, 1)$.

Define

(4.1.1) $L_w^2(I) := \{v : I \to \mathbb{R} \mid v$ is measurable and $||v||_{0,w} < \infty\}$,

where

(4.1.2) $$||v||_{0,w} := \left(\int_{-1}^{1} |v(x)|^2 \, w(x) \, dx \right)^{1/2}$$

is the norm induced by the scalar product

(4.1.3) $$(u, v)_w := \int_{-1}^{1} u(x) \, v(x) \, w(x) \, dx \quad .$$

Let $\{p_n\}_{n \geq 0}$ denote a system of algebraic polynomials, with the degree of p_n equal to n, which are mutually orthogonal under (4.1.3), i.e.,

(4.1.4) $(p_n, p_m)_w = 0$ whenever $m \neq n$.

The Weierstrass approximation theorem (see, e.g., Yosida (1974), p. 8) implies that this system is complete in $L_w^2(-1, 1)$, i.e., for any function $u \in L_w^2(-1, 1)$ the following expansion holds

$$(4.1.5) \qquad u(x) = \sum_{k=0}^{\infty} \hat{u}_k \, p_k(x) \quad \text{with} \quad \hat{u}_k := \frac{(u, p_k)_w}{||p_k||_{0,w}^2} \quad .$$

The $\hat{u}_k$'s are the expansion coefficients associated with the family $\{p_k\}$. The series converges in the sense of $L_w^2(I)$. Precisely, denoting by

$$(4.1.6) \qquad P_N u(x) := \sum_{k=0}^{N} \hat{u}_k \, p_k(x)$$

the truncation of order N, then

$$(4.1.7) \qquad ||u - P_N u||_{0,w} \to 0 \quad \text{as} \quad N \to \infty \quad .$$

The polynomial $P_N u$ can also be regarded as the orthogonal projection of u upon $\mathbb{P}_N$ with respect to the scalar product (4.1.3), i.e.,

$$(4.1.8) \qquad (P_N u, v_N)_w = (u, v_N)_w \quad \forall \, v_N \in \mathbb{P}_N \quad .$$

Finally, we remember the following Parseval identity that follows from (4.1.5)

$$(4.1.9) \qquad ||u||_{0,w} = \left(\sum_{k=0}^{\infty} |\hat{u}_k|^2 \, ||p_k||_{0,w}^2 \right)^{1/2} \quad .$$

Let us now set

$$(4.1.10) \qquad w^{(\alpha,\beta)}(x) := (1 - x)^{\alpha}(1 + x)^{\beta} \quad ,$$

for $-1 < \alpha, \beta < 1$, $-1 \leq x \leq 1$. The family of polynomials orthogonal under this weight function are the *Jacobi polynomials* and are denoted by $\{J_n^{(\alpha,\beta)}\}_{n \geq 0}$. We report, hereafter, some properties that will be used in the sequel (see also Szegö (1959), Courant and Hilbert (1953)).

Under the normalization

$$J_n^{(\alpha,\beta)}(1) = \binom{n + \alpha}{n} := \frac{\Gamma(n + \alpha + 1)}{n \, ! \, \Gamma(\alpha + 1)} \quad ,$$

$\Gamma(\cdot)$ being the gamma function, an explicit representation is

$$(4.1.11) \quad J_n^{(\alpha,\beta)}(x) = 2^{-n} \sum_{k=0}^{n} \binom{n + \alpha}{k} \binom{n + \beta}{n - k} (x - 1)^{n-k}(x + 1)^k \quad .$$

The following maximum principle holds for all α, β:

$$(4.1.12) \qquad \begin{aligned} \max_{-1 \leq x \leq 1} |J_n^{(\alpha,\beta)}(x)| &= \max\{|J_n^{(\alpha,\beta)}(\pm 1)|\} \\ &= \max\left\{ \binom{n + \alpha}{n}, \binom{n + \beta}{n} \right\} \quad . \end{aligned}$$

The Jacobi polynomials satisfy the Sturm-Liouville eigenvalue problem

(4.1.13) $$L^{(\alpha,\beta)} J_n^{(\alpha,\beta)}(x) = \lambda_n^{(\alpha,\beta)} w^{(\alpha,\beta)}(x) J_n^{(\alpha,\beta)}(x) \ ,$$

where

$$L^{(\alpha,\beta)} := -\frac{d}{dx}\left((1-x^2)\,w^{(\alpha,\beta)}(x)\frac{d}{dx}\right)$$

and $\lambda_n^{(\alpha,\beta)} := n(n+\alpha+\beta+1)$ (see, e.g., Szegö (1959), Abramowitz and Stegun (1966)).

The Jacobi expansion coefficients of a function u decay with a rate that depends solely on the smoothness degree of u. More precisely, if we consider the expansion (4.1.5) with $p_k = J_k^{(\alpha,\beta)}$, we have (see, e.g., Canuto, Hussaini, Quarteroni and Zang (1988), Sect. 9.2.2)

(4.1.14) $$|\hat{u}_k| \le C_m \frac{1}{k^{2m}} \ , \quad k \ge 1 \ .$$

This is valid under the assumption that, for some $m \ge 1$, $u_{(m)} \in L^2_{w^{(\alpha,\beta)}}(I)$ and $\lim_{x\to\pm1}(1-x^2)\,w^{(\alpha,\beta)}(x)\,\frac{d}{dx}u_{(j)}(x) = 0$ for all $j = 1,...,m$, where we have set

$$u_{(0)} := u \ , \quad u_{(j)} := \frac{1}{w^{(\alpha,\beta)}}\,L^{(\alpha,\beta)}u_{(j-1)}$$

The constant C_m depends on u and m.

Special cases of Jacobi polynomials are the *Legendre polynomials*, corresponding to the choice $\alpha = \beta = 0$, and the *Chebyshev polynomials of first kind*, that are obtained when $\alpha = \beta = -1/2$. Both families have great relevance in the approximation of boundary value problems by spectral methods, and will therefore be reconsidered in detail in Sections 4.3 ad 4.4.

4.2 Gaussian Quadrature and Interpolation

For any fixed integer $N \ge 1$, we denote by $\{x_j\}_{j=0,...,N}$ the zeroes of the polynomial $(1-x^2)(J_N^{(\alpha,\beta)})'(x)$ (the prime means derivative). Correspondingly, we denote by $\{w_j\}_{j=0,...,N}$ the real numbers that solve the linear system:

$$\sum_{j=0}^{N}(x_j)^k\,w_j = \int_{-1}^{1} x^k\,w^{(\alpha,\beta)}(x)\,dx \ , \quad 0 \le k \le N \ .$$

The quadrature formula

(4.2.1) $$\sum_{j=0}^{N} f(x_j)\,w_j \simeq \int_{-1}^{1} f(x)\,w^{(\alpha,\beta)}(x)\,dx$$

is the Jacobi Gauss-Lobatto formula; $\{x_j\}$ and $\{w_j\}$ are called, respectively, nodes and weights. A close expression for them will be provided in the next

Sections for both Legendre and Chebyshev cases. Formula (4.2.1) is precise up to the degree $2N - 1$, i.e.,

$$(4.2.2) \qquad \sum_{j=0}^{N} p(x_j)\, w_j = \int_{-1}^{1} p(x)\, w^{(\alpha,\beta)}(x)\, dx \quad \forall\, p \in \mathbb{P}_{2N-1} \ .$$

(see, e.g., Davis and Rabinowitz (1984)).

If u is any continuous function in $\overline{I}$, we denote by $I_N u \in \mathbb{P}_N$ its Lagrangian interpolant at the nodes $\{x_j\}_{j=0,...,N}$, i.e.,

$$(4.2.3) \qquad I_N u(x_j) = u(x_j) \quad \forall\, j = 0, ..., N \ .$$

For any continuous functions u and v in $\overline{I}$, we define the *discrete* scalar product and norm respectively as

$$(4.2.4) \qquad (u,v)_N := \sum_{j=0}^{N} u(x_j)\, v(x_j)\, w_j \quad , \quad ||v||_N := (v,v)_N^{1/2} \ .$$

Let us note that

$$(4.2.5) \qquad (I_N u, v)_N = (u,v)_N \quad \forall\, u, v \in C^0(\overline{I})$$

and also that, owing to (4.2.2),

$$(4.2.6) \qquad (u,v)_N = (u,v)_{w^{(\alpha,\beta)}} \quad \forall\, u, v \text{ such that } uv \in \mathbb{P}_{2N-1} \ .$$

From (4.2.5) we obtain easily

$$(4.2.7) \qquad I_N u(x) = \sum_{k=0}^{N} u_k^* \, J_k^{(\alpha,\beta)}(x) \quad \text{with} \quad u_k^* := \frac{(u, J_k^{(\alpha,\beta)})_N}{\gamma_k^{(\alpha,\beta)}}$$

and $\gamma_k^{(\alpha,\beta)} := ||J_k^{(\alpha,\beta)}||_N^2$. The $\{u_k^*\}$ are the *discrete* expansion coefficients of the function u. Note the formal analogy between (4.2.7) and (4.1.6).

The previous formula produces

$$(4.2.8) \qquad u_k^* = \sum_{j=0}^{N} \left[\frac{1}{\gamma_k^{(\alpha,\beta)}} J_k^{(\alpha,\beta)}(x_j)\, w_j \right] u(x_j) \ , \quad 0 \le k \le N \ ,$$

which is a linear transformation, called the *discrete transform*, between the values of a function u at the nodes $\{x_j\}$ and its discrete expansion coefficients. The *discrete anti-transform*

$$(4.2.9) \qquad u(x_j) = \sum_{k=0}^{N} J_k^{(\alpha,\beta)}(x_j)\, u_k^* \ , \quad 0 \le j \le N \ ,$$

follows easily from (4.2.3) and (4.2.7).

4.3 Chebyshev Expansion

Let us consider the special case of Chebyshev polynomials of the first kind.
We briefly review their basic properties, then we define the interpolation and
projection operators and illustrate their convergence properties.

4.3.1 Chebyshev Polynomials

They are defined as

$$T_n(x) := \frac{J_n^{(-1/2,-1/2)}(x)}{J_n^{(-1/2,-1/2)}(1)} \quad , \quad n \in \mathrm{N} \ ,$$

and, in view of (4.1.13), they satisfy the Sturm-Liouville differential equation

$$(4.3.1) \qquad -(\sqrt{1-x^2}\, T_n'(x))' = n^2\, \frac{1}{\sqrt{1-x^2}}\, T_n(x) \ .$$

For $x \in [-1,1]$ a remarkable characterization is given by

$$(4.3.2) \qquad T_n(x) = \cos n\theta \quad , \quad \theta = \arccos x \ ,$$

while for $|x| \geq 1$ $T_n(x) = (\mathrm{sign}\, x)^n \cosh n\theta$, $\theta = \cosh^{-1}|x|$ (see, e.g., Rivlin
(1974)).

We deduce that, according to (4.1.12),

$$(4.3.3) \qquad |T_n(x)| \leq 1 \ , \quad |x| \leq 1 \quad , \quad T_n(1) = 1 \quad , \quad T_n(-1) = (-1)^n \ ,$$

and that T_n has the same parity of n. Using the trigonometric relation

$$\cos(n+1)\theta + \cos(n-1)\theta = 2\cos\theta\,\cos n\theta$$

we deduce the recursion formula

$$(4.3.4) \qquad T_{n+1}(x) = 2x\, T_n(x) - T_{n-1}(x) \quad , \quad n \geq 1 \ ,$$

with $T_0(x) = 1$ and $T_1(x) = x$.

Let us set

$$(4.3.5) \qquad
\begin{aligned}
(u,v)_w &:= \int_{-1}^{1} u(x)\,v(x)\,\frac{1}{\sqrt{1-x^2}}\,dx \\[2mm]
\|u\|_{0,w} &:= \left(\int_{-1}^{1} u^2(x)\,\frac{1}{\sqrt{1-x^2}}\,dx \right)^{1/2}
\end{aligned}$$

($w(x) := (1-x^2)^{-1/2}$ is called Chebyshev weight function). Operating the
change of variable $x = \cos\theta$, we obtain from elementary trigonometric properties the orthogonality property

$$(4.3.6) \qquad (T_n, T_m)_w = \frac{\pi}{2} c_n \, \delta_{nm} \quad , \quad c_n = \begin{cases} 2 & \text{if } n = 0 \\ 1 & \text{if } n \geq 1 \end{cases} \; .$$

Therefore

$$\|T_n\|_{0,w} = \sqrt{\frac{\pi}{2} c_n} \; .$$

If $u \in L_w^2(I)$, its Chebyshev series is

$$(4.3.7) \qquad u(x) = \sum_{k=0}^{\infty} \hat{u}_k \, T_k(x) \quad , \quad \hat{u}_k = \frac{2}{\pi \, c_k} \, (u, T_k)_w \; .$$

If we assume further that $u' \in L_w^2(I)$, then

$$(4.3.8) \qquad u'(x) = \sum_{k=0}^{\infty} \hat{u}_k^{(1)} T_k(x) \; ,$$

with (see, e.g., Gottlieb and Orszag (1977), Canuto, Hussaini, Quarteroni and Zang (1988), Sect. 2.4.2)

$$(4.3.9) \qquad \hat{u}_k^{(1)} = \frac{2}{c_k} \sum_{\substack{p=k+1 \\ p+k \text{ odd}}}^{\infty} p \, \hat{u}_p \; .$$

If u is a polynomial of degree less than or equal to N, then the expansion coefficients of its first derivative can be computed recursively with $O(N)$ operations through the three-term backward relations

$$(4.3.10) \qquad \begin{cases} \hat{u}_{N+1}^{(1)} = \hat{u}_N^{(1)} = 0 \\[2mm] c_k \, \hat{u}_k^{(1)} = \hat{u}_{k+2}^{(1)} + 2(k+1) \, \hat{u}_{k+1} \quad , \quad k = N-1, ..., 0 \; . \end{cases}$$

Going further, assuming that $u'' \in L_w^2(I)$ one can also infer that

$$(4.3.11) \qquad u''(x) = \sum_{k=0}^{\infty} \hat{u}_k^{(2)} T_k(x) \quad , \quad \hat{u}_k^{(2)} = \frac{1}{c_k} \sum_{\substack{p=k+2 \\ p+k \text{ even}}}^{\infty} p(p^2 - k^2) \, \hat{u}_p \; .$$

If u is a polynomial of degree less than or equal to N, the coefficients $\hat{u}_k^{(2)}$ can still be computed recursively by iterating on the formula (4.3.10).

4.3.2 Chebyshev Interpolation

For any positive integer N, the $N + 1$ Chebyshev Gauss-Lobatto nodes x_j are the zeroes of the polynomial $(1 - x^2) T'_N(x)$. By differentiating (4.3.2) we obtain

$$(4.3.12) \qquad x_j = \cos \frac{\pi j}{N} \ , \quad j = 0, ..., N \ .$$

These nodes are symmetrically distributed about $x = 0$. As far as N increases, they cluster towards the endpoints of the interval. In particular $|x_N - x_{N-1}| = |x_1 - x_0| = O(N^{-2})$. This prevents the formation of Runge's instability appearing in those interpolation processes using equispaced nodes (e.g., Isaacson and Keller (1966), Atkinson (1978)).

The weights of the Chebyshev Gauss-Lobatto integration formula are (see, e.g., Rivlin (1974))

$$(4.3.13) \qquad w_j = \frac{\pi}{d_j N} \ \text{ with } \ d_j = \begin{cases} 2 \text{ for } j = 0, N, \\ 1 \text{ for } j = 1, ..., N - 1 \end{cases} \ .$$

The discrete scalar product and the discrete norm take respectively the form:

$$(4.3.14) \qquad (u, v)_N = \frac{\pi}{N} \sum_{j=0}^{N} \frac{1}{d_j} u(x_j) \, v(x_j)$$

$$\|v\|_N = \left(\frac{\pi}{N} \sum_{j=0}^{N} \frac{1}{d_j} v^2(x_j) \right)^{1/2} \ .$$

Owing to (4.2.2)

$$(4.3.15) \qquad \|T_k\|_N = \|T_k\|_{0,w} \ , \quad k = 0, ..., N - 1 \ ,$$

while, writing $\theta_j = \arccos x_j = \pi j / N$,

$$(4.3.16) \quad \|T_N\|_N^2 = \frac{\pi}{N} \sum_{j=0}^{N} \cos^2(N\theta_j) \frac{1}{d_j} = \frac{\pi}{N} \sum_{j=0}^{N} \frac{1}{d_j} = \pi = 2 \, \|T_N\|_{0,w}^2 \ .$$

Therefore

$$(4.3.17) \qquad \|T_k\|_N = \sqrt{\frac{\pi}{2} d_k} \ , \quad k = 0, ..., N \ .$$

If we denote by $I_N u \in \mathbb{P}_N$ the interpolant of the function u at the Chebyshev nodes (4.3.12), a direct consequence of (4.2.7) and (4.3.14), (4.3.17) is

$$(4.3.18) \quad I_N u(x) = \sum_{k=0}^{N} u_k^* T_k(x) \ , \quad u_k^* = \frac{2}{N d_k} \sum_{j=0}^{N} \frac{1}{d_j} \cos\left(\frac{kj\pi}{N} \right) u(x_j) \ .$$

Due to its trigonometric structure, the *discrete Chebyshev transform*

$$\{u(x_j)\,|\,j = 0, ..., N\} \rightarrow \{u_k^* \,|\, k = 0, ..., N\}$$

can be computed by the Fast Fourier Transform algorithm using $O(N \log_2 N)$ operations, whenever N is a power of 2. The same clearly holds for the anti-transform

$$\{u_k^* \,|\, k = 0, ..., N\} \rightarrow \{u(x_j)\,|\,j = 0, ..., N\} \quad ,$$

since

$$(4.3.19) \qquad u(x_j) = \sum_{k=0}^{N} \cos\left(\frac{kj\pi}{N}\right) u_k^* \quad , \quad j = 0, ..., N \quad .$$

Algorithms for performing Fast Fourier and Chebyshev Transforms are reported in Appendix B of Canuto, Hussaini, Quarteroni and Zang (1988).

A consequence of (4.3.15), (4.3.16) is that the discrete Chebyshev norm is uniformly equivalent to the continuous L_w^2-norm for all polynomials of degree less than or equal to N. Indeed, for all $u_N = \sum_{k=0}^{N} \hat{u}_k T_k$

$$||u_N||_N^2 = \sum_{k=0}^{N} |\hat{u}_k|^2\, ||T_k||_N^2 = \sum_{k=0}^{N-1} |\hat{u}_k|^2\, ||T_k||_{0,w}^2 + 2|\hat{u}_N|^2\, ||T_N||_{0,w}^2 \quad ,$$

and therefore

$$(4.3.20) \qquad ||u_N||_{0,w} \leq ||u_N||_N \leq \sqrt{2}||u_N||_{0,w} \quad \forall\, u_N \in \mathbb{P}_N \quad .$$

In view of the application to the Chebyshev collocation method it is worthwhile to introduce the *Chebyshev pseudo-spectral derivative* $\partial_N u$ of a continuous function u. It is a polynomial in $\mathbb{P}_{N-1}$ defined by

$$(4.3.21) \qquad \partial_N u := \frac{d}{dx}(I_N u) \quad ,$$

i.e., as the exact derivative of the interpolant of u at the nodes (4.3.12). By resorting to the representation

$$(4.3.22) \qquad I_N u(x) = \sum_{j=0}^{N} u(x_j)\,\psi_j(x) \quad ,$$

where the *Lagrange functions* $\psi_j \in \mathbb{P}_N$ are such that $\psi_j(x_k) = \delta_{jk}$ for each $k = 0, ..., N$, we deduce

$$(4.3.23) \qquad (\partial_N u)(x_i) = \sum_{j=0}^{N} \psi_j'(x_i)\,u(x_j) \quad , \quad i = 0, ..., N \quad .$$

The matrix $(D_N)_{ij} := \psi_j'(x_i)$ is named *Chebyshev pseudo-spectral matrix*. Since

$$\psi_j(x) = \frac{(-1)^{j+1} (1 - x^2) T_N'(x)}{d_j N^2 (x - x_j)} \quad ,$$

the entries of D_N can be computed explicitly (see Voigt, Gottlieb and Hussaini (1984) or Canuto, Hussaini, Quarteroni and Zang (1988), Sect. 2.4.2)

$$(4.3.24) \qquad (D_N)_{ij} = \begin{cases} \dfrac{d_i}{d_j} \dfrac{(-1)^{i+j}}{x_i - x_j} & i \neq j \\[2mm] -\dfrac{x_j}{2(1 - x_j^2)} & 1 \leq i = j \leq N - 1 \\[2mm] \dfrac{2N^2 + 1}{6} & i = j = 0 \\[2mm] -\dfrac{2N^2 + 1}{6} & i = j = N \quad . \end{cases}$$

The matrix D_N is not skew-symmetric; its only eigenvalue is 0 with algebraic multiplicity $N + 1$.

We conclude this section by giving several estimates of the Chebyshev interpolation error. Define the weighted Sobolev space

$$(4.3.25) \quad H_w^s(I) := \{v \in L_w^2(I) \mid v^{(k)} \in L_w^2(I) \text{ for } k = 1, ..., s\} \, , \, s \in \mathbb{N} \, ,$$

whose norm is

$$(4.3.26) \qquad ||v||_{s,w} := \left(\sum_{k=0}^{s} ||v^{(k)}||_{0,w}^2 \right)^{1/2} \quad .$$

Here $v^{(k)}$ denotes the k-th order distributional derivative of v (see (1.2.9)).

Assuming that u is a function of $H_w^s(I)$ for some $s \geq 1$, it holds

$$(4.3.27) \qquad ||u - I_N u||_{0,w} \leq C N^{-s} ||u||_{s,w}$$

(see Canuto and Quarteroni (1982)). This result is based on a similar one for the trigonometric interpolation, which is originally due to Kreiss and Oliger (1979). The proof we present here has been given by Pasciak (1980). Denoting by $H_p^s(0, 2\pi)$, $s \geq 1$, the subspace of the Sobolev space $H^s(0, 2\pi)$ consisting of functions whose first $s - 1$ derivatives are periodic, we have:

Lemma 4.3.1 *Set* $\mathcal{S}_N := \text{span}\{e^{ik\theta} : [0, 2\pi] \to \mathbb{C} \mid -N \leq k \leq N-1\}$ *and let* $\mathcal{I}_N : C^0([0, 2\pi]) \to \mathcal{S}_N$ *be the interpolation operator at the nodes* $\theta_j := j\pi/N$, $j = 0, ..., 2N - 1$. *Then for each* $\phi \in H_p^s(0, 2\pi)$ *it holds*

$$(4.3.28) \qquad ||\phi - \mathcal{I}_N \phi||_{0,(0,2\pi)} \leq C N^{-s} |\phi|_{s,(0,2\pi)} \quad .$$

Proof. Let us start by defining the map $\phi \to \hat{\phi}$ in the following way:

$$\hat{\phi}(\theta) := \phi\left(\frac{\theta}{N}\right) \quad ;$$

set moreover $\hat{S}_N := \{\hat{\phi}_N : [0, 2\pi N] \to \mathbb{C} \mid \phi_N \in S_N\}$ and $\hat{\theta}_j := N\theta_j = j\pi$, $j = 0, ..., 2N-1$. Clearly, the interpolation operator $\hat{I}_N : C^0([0, 2\pi N]) \to \hat{S}_N$ at the nodes $\hat{\theta}_j$ satisfies

$$(4.3.29) \qquad (\mathcal{I}_N\phi)^\wedge = \hat{I}_N\hat{\phi} \quad \forall\, \phi \in C^0([0, 2\pi]) \ .$$

In addition, it is at once verified that for each $\phi \in H^s(0, 2\pi)$ it holds

$$(4.3.30) \qquad |\hat{\phi}|_{l,(0,2\pi N)} = N^{-l} N^{1/2} |\phi|_{l,(0,2\pi)} \ , \quad 0 \le l \le s \ .$$

Thus, by proceeding as in Bramble-Hilbert lemma (see Proposition 3.4.3)

$$
\begin{aligned}
(4.3.31) \qquad \|\phi - \mathcal{I}_N\phi\|_{0,(0,2\pi)} &= N^{-1/2} \|\hat{\phi} - \hat{I}_N\hat{\phi}\|_{0,(0,2\pi N)} \\
&\le N^{-1/2} \|I - \hat{I}_N\| \inf_{\hat{\psi}_N \in \hat{S}_N} \|\hat{\phi} - \hat{\psi}_N\|_{s,(0,2\pi N)}
\end{aligned}
$$

(the norm of $I - \hat{I}_N$ is the norm in $\mathcal{L}(H_p^s(0, 2\pi N); L^2(0, 2\pi N))$.) Choosing $\hat{\psi}_N = (\mathcal{P}_N\phi)^\wedge$, $\mathcal{P}_N$ being the L^2-orthogonal projection on S_N, from (4.3.30) we obtain

$$
\begin{aligned}
(4.3.32) \qquad \|\hat{\phi} - (\mathcal{P}_N\phi)^\wedge\|_{s,(0,2\pi N)} &= \left(\sum_{l=0}^{s} |\hat{\phi} - (\mathcal{P}_N\phi)^\wedge|_{l,(0,2\pi N)}^2\right)^{1/2} \\
&= \left(\sum_{l=0}^{s} N^{-2l} N |\phi - \mathcal{P}_N\phi|_{l,(0,2\pi)}^2\right)^{1/2} \ .
\end{aligned}
$$

The space $H_p^s(0, 2\pi)$ consists of functions for which it is permissible to differentiate termwise the Fourier series s times, provided the convergence is in $L^2(0, 2\pi)$. Thus from the Parseval identity

$$(4.3.33) \qquad |\phi|_{l,(0,2\pi)}^2 = \|\phi^{(l)}\|_{0,(0,2\pi)}^2 = 2\pi \sum_{k=-\infty}^{\infty} |k|^{2l} |\hat{\phi}_k|^2 \ ,$$

it follows

$$
\begin{aligned}
(4.3.34) \qquad |\phi - \mathcal{P}_N\phi|_{l,(0,2\pi)}^2 &= 2\pi \sum_{|k| \ge N} |k|^{2l} |\hat{\phi}_k|^2 \\
&\le CN^{2(l-s)} |\phi|_{s,(0,2\pi)}^2 \ , \quad 0 \le l \le s \ ,
\end{aligned}
$$

where the symbol $\sum_{|k| \ge N}$ means that the sum is taken for $k < -N$ and $k \ge N$. Finally, from (4.3.32) we obtain

$$(4.3.35) \qquad \|\hat{\phi} - (\mathcal{P}_N\phi)^\wedge\|_{s,(0,2\pi N)} \le CN^{1/2-s} |\phi|_{s,(0,2\pi)} \ .$$

It remains to show that

$$||I - \hat{\mathcal{I}}_N||_{\mathcal{L}(H_p^s(0,2\pi N);L^2(0,2\pi N))} \leq C \ .$$

Clearly, it is enough to consider $\hat{\mathcal{I}}_N$. From the orthogonality relation

$$\frac{1}{2N} \sum_{j=0}^{2N-1} e^{ip\theta_j} = \begin{cases} 1 \text{ if } p = 0, \pm 2N, \pm 4N, ... \\ 0 \text{ otherwise} \end{cases} \ ,$$

(see, e.g., Kreiss and Oliger (1979)) it follows that for each $\phi_N, \psi_N \in \mathcal{S}_N$ it holds

$$(\phi_N, \psi_N)_{0,(0,2\pi)} = \frac{\pi}{N} \sum_{j=0}^{2N-1} \phi_N(\theta_j)\overline{\psi_N(\theta_j)} \ .$$

As a consequence we have

$$||\hat{\mathcal{I}}_N\hat{\phi}||_{0,(0,2\pi N)} = N^{1/2}||\mathcal{I}_N\phi||_{0,(0,2\pi)} = N^{1/2} \left(\frac{\pi}{N} \sum_{j=0}^{2N-1} |(\mathcal{I}_N\phi)(\theta_j)|^2 \right)^{1/2}$$

$$= \sqrt{\pi} \left(\sum_{j=0}^{2N-1} |\hat{\phi}(\hat{\theta}_j)|^2 \right)^{1/2} \leq C||\hat{\phi}||_{s,(0,2\pi N)} \ ,$$

having used the Gagliardo-Nirenberg inequality (see Theorem 1.3.6) on each interval $(\hat{\theta}_j, \hat{\theta}_{j+1})$. $\qquad\square$

The proof of (4.3.27) is now easily performed. In fact, for each continuous function $u : [-1, 1] \to \mathbb{R}$ define $u^* : [0, 2\pi] \to \mathbb{R}$ as

$$(4.3.36) \qquad\qquad u^*(\theta) := u(\cos\theta) \ .$$

By changing variable, it follows at once

$$(4.3.37) \qquad ||u||_{0,w} = \frac{1}{\sqrt{2}}||u^*||_{0,(0,2\pi)}$$

$$(4.3.38) \qquad ||u^*||_{s,(0,2\pi)} \leq C||u||_{s,w} \ \forall \, u \in H_w^s(-1,1) \ , \ s \geq 0 \ .$$

Moreover, $u^* \in H_p^s(0, 2\pi)$. Define now

$$\mathcal{S}_N^* := \left\{ \phi_N : [0, 2\pi] \to \mathbb{C} \mid \phi_N(\theta) = \sum_{k=-N}^{N} \hat{\phi}_k e^{ik\theta} \ , \ \hat{\phi}_N = \hat{\phi}_{-N} \right\} \ ,$$

and let $\mathcal{I}_N^* : C^0([0, 2\pi]) \to \mathcal{S}_N^*$ be the interpolation operator at the nodes $\theta_j = \pi j/N$, $j = 0, ..., 2N-1$. It is easily verified that Lemma 4.3.1 still applies to $\mathcal{S}_N^*$ and $\mathcal{I}_N^*$, and that, for each $u \in C^0([-1,1])$, $(I_N u)^* \in \mathcal{S}_N^*$. Moreover, u^* and $(I_N u)^*$ match at θ_j, $j = 0, ..., N$. Since they are even functions with respect to the point $\theta = \pi$, and the nodes θ_j are symmetrically distributed about to the same point, they indeed are coincident at all the nodes θ_j, $j = 0, ..., 2N - 1$. Therefore

$$(4.3.39) \qquad (I_N u)^* = \mathcal{I}_N^* u^* .$$

From (4.3.37), (4.3.28) and (4.3.38) we can thus conclude

$$\begin{aligned}
||u - I_N u||_{0,w} &= \frac{1}{\sqrt{2}} ||u^* - \mathcal{I}_N^* u^*||_{0,(0,2\pi)} \\
&\leq CN^{-s} |u^*|_{s,(0,2\pi)} \\
&\leq CN^{-s} ||u||_{s,w} \ \forall \ u \in H_w^s(-1,1) .
\end{aligned}$$

In the $H_w^1(I)$-norm the behaviour of the Chebyshev interpolation error is still optimal; indeed (see Bernardi and Maday (1992), p. 80)

$$(4.3.40) \qquad ||u - I_N u||_{1,w} \leq CN^{1-s} ||u||_{s,w} .$$

In particular, from this inequality we obtain $||I_N u||_{1,w} \leq C ||u||_{1,w}$ and the following error estimate for the pseudo-spectral derivative:

$$(4.3.41) \qquad ||u' - \partial_N u||_{0,w} \leq CN^{1-s} ||u||_{s,w} .$$

In the maximum norm, the interpolation error satisfies

$$(4.3.42) \qquad \max_{-1 \leq x \leq 1} |u(x) - I_N u(x)| \leq CN^{1/2-s} ||u||_{s,w} .$$

This follows easily from (4.3.27), (4.3.40), the Gagliardo-Nirenberg inequality (see Theorem 1.3.6) and the obvious fact that $||v||_k \leq ||v||_{k,w}$, $k = 0, 1$.

We would also like to remind that the Chebyshev Gauss-Lobatto quadrature produces a convergent process for any function u such that uw is Riemann integrable over I. Indeed

$$(4.3.43) \qquad \int_{-1}^1 u(x) \frac{1}{\sqrt{1-x^2}} \, dx = \lim_{N \to \infty} \frac{\pi}{N} \sum_{j=0}^N \frac{1}{d_j} u(x_j)$$

(e.g., Szegö (1959), p. 342). For a smoother function u, an order of convergence can be inferred. In particular, for all $v_N \in \mathbb{P}_N$ and $u \in H_w^s(I)$, $s \geq 1$, we have

$$(4.3.44) \qquad |(u, v_N)_w - (u, v_N)_N| \leq CN^{-s} ||u||_{s,w} ||v_N||_{0,w} .$$

This result is often useful in the analysis of collocation schemes. Its proof reads as follows. Denote by $I_{N-1} u \in \mathbb{P}_{N-1}$ the interpolant of u at the nodes $\cos[\pi j/(N-1)]$, $j = 0, ..., N-1$. From (4.2.5) and (4.2.6) we have

$$\begin{aligned}
|(u, v_N)_w - (u, v_N)_N| &\leq |(u, v_N)_w - (I_{N-1} u, v_N)_w| \\
&\quad + |(I_{N-1} u, v_N)_N - (I_N u, v_N)_N| .
\end{aligned}$$

Then, using (4.3.20) we find

$$|(u, v_N)_w - (u, v_N)_N| \leq C(||u - I_{N-1}u||_{0,w} + ||u - I_N u||_{0,w})||v_N||_{0,w} \quad .$$

The quadrature error estimate (4.3.44) follows now from (4.3.27).

4.3.3 Chebyshev Projections

If u is any function of $L_w^2(I)$, the truncated of order N of its Chebyshev series (4.3.7) is

$$(4.3.45) \qquad P_N u(x) = \sum_{k=0}^{N} \hat{u}_k T_k(x) \quad ,$$

and converges to u in the norm $|| \cdot ||_{0,w}$. We remind here that $P_N u$ is the L_w^2-orthogonal projection of u upon $\mathbb{P}_N$ (see (4.1.8)).

If $u \in H_w^s(I)$ for some $s \geq 0$, then

$$(4.3.46) \qquad ||u - P_N u||_{0,w} \leq CN^{-s}||u||_{s,w}$$

(see Canuto and Quarteroni (1982)). This result could be inferred from (4.1.14). However, for the sake of completeness we report here a proof that is based on a more direct argument. Let us define

$$\tilde{S}_N := \text{span}\{e^{ik\theta} : [0, 2\pi] \to \mathbb{C} \mid -N \leq k \leq N\} \quad ,$$

and let $\tilde{\mathcal{P}}_N$ be the L^2-orthogonal projection on $\tilde{S}_N$. Recalling (4.3.36), at first one easily verifies that

$$(4.3.47) \qquad \tilde{\mathcal{P}}_N u^* = (P_N u)^* \quad .$$

In fact,

$$(P_N u)^*(\theta) = \sum_{k=0}^{N} \hat{u}_k T_k^*(\theta) = \sum_{k=0}^{N} \hat{u}_k \cos k\theta \quad .$$

Writing $\cos k\theta = (e^{ik\theta} + e^{-ik\theta})/2$, (4.3.47) follows at once. Since the projection $\tilde{\mathcal{P}}_N$ clearly satisfies an error estimate like (4.3.34), we finally have

$$||u - P_N u||_{0,w} = \frac{1}{\sqrt{2}}||u^* - \tilde{\mathcal{P}}_N u^*||_{0,(0,2\pi)}$$

$$\leq CN^{-s}|u^*|_{s,(0,2\pi)} \leq CN^{-s}||u||_{s,w} \quad .$$

In the $H_w^1(I)$-norm, the truncation error satisfies the inequality (see Canuto and Quarteroni (1982))

$$(4.3.48) \qquad ||u - P_N u||_{1,w} \leq C\sqrt{N}N^{1-s}||u||_{s,w} \quad , \quad s \geq 1 \quad ,$$

thus it fails to be optimal due to the presence of the factor $\sqrt{N}$. In other words, the truncation error doesn't behave like the best approximation error in higher order Sobolev norms.

This non-optimal result leads to introduce the orthogonal projection $P_{1,N}u$ of u upon $\mathbb{P}_N$ with respect to the scalar product of $H^1_w(I)$:

$$(4.3.49) \qquad (u,v)_{1,w} := \int_{-1}^{1} (u'v' + uv)\frac{1}{\sqrt{1-x^2}}\, dx \quad .$$

Therefore, for any function $u \in H^1_w(I)$ we have

$$(4.3.50) \qquad (P_{1,N}u, v_N)_{1,w} = (u, v_N)_{1,w} \quad \text{for all} \ \ v_N \in \mathbb{P}_N \quad .$$

Then

$$(4.3.51) \qquad ||u - P_{1,N}u||_{k,w} \le CN^{k-s}||u||_{s,w} \quad , \quad k = 0, 1 \ , \ s \ge 1 \ .$$

For $k = 1$ this result follows at once from (4.3.40); for $k = 0$ it can be obtained by a duality argument as in the proof of (3.5.16) (see, e.g., Canuto, Hussaini, Quarteroni and Zang (1988), Sect. 9.5.4).

A similar result holds for those functions vanishing at the endpoints $x = \pm 1$, in which case the orthogonal projection $P^0_{1,N}u$ matches the same boundary values of u. More precisely, let us introduce the space

$$(4.3.52) \qquad H^1_{w,0}(I) := \left\{ v \in H^1_w(I) \mid v(-1) = v(1) = 0 \right\}$$

and denote by $\mathbb{P}^0_N$ the space of algebraic polynomials of degree less than or equal to N vanishing at $x = \pm 1$. Owing to the Poincaré inequality

$$(4.3.53) \qquad ||v||_{0,w} \le C||v'||_{0,w} \quad \forall\, v \in H^1_{w,0}(I)$$

(see, e.g., Canuto, Hussaini, Quarteroni and Zang (1988), Sect. 11.1.2) the scalar product

$$(4.3.54) \qquad (u,v)^0_{1,w} := \int_{-1}^{1} u'v' \frac{1}{\sqrt{1-x^2}}\, dx$$

induces on the functions of $H^1_{w,0}(I)$ a norm equivalent to $||\cdot||_{1,w}$. Then $P^0_{1,N}u \in \mathbb{P}^0_N$ is defined through

$$(4.3.55) \qquad (P^0_{1,N}u, v_N)^0_{1,w} = (u, v_N)^0_{1,w} \quad \text{for all} \ \ v_N \in \mathbb{P}^0_N \quad .$$

For all $s \ge 1$ and any function $u \in H^s_w(I) \cap H^1_{w,0}(I)$ the following estimate holds

$$(4.3.56) \qquad ||u - P^0_{1,N}u||_{k,w} \le CN^{k-s}||u||_{s,w} \quad , \quad k = 0, 1 \ , \ s \ge 1 \ .$$

This result is trivial for $k = 1$, as $P^0_{1,N}$ is the $H^1_{0,w}$-orthogonal projection and (4.3.40) and (4.3.53) hold. On the other hand, the proof for $k = 0$ is based on the usual Aubin-Nitsche duality argument, since the solution $\varphi(g)$ of the problem

$$(\varphi(g), v)^0_{1,w} = (g, v)_{0,w}$$

belongs to $H^2_w(I)$ when $g \in L^2_w(I)$ (see Bernardi and Maday (1989), Theor. 4.2).

Furthermore, for any $u \in H^s_w(I) \cap H^1_{w,0}(I)$, $s \geq 1$, it is possible to construct another polynomial $\tilde{u}_N \in \mathbb{P}^0_N$ (besides $I_N u$ and $P^0_{1,N} u$) whose distance from u decays in an optimal way both in the H^1_w-norm and in the L^2_w-norm, i.e.,

$$(4.3.57) \qquad \|u - \tilde{u}_N\|_{k,w} \leq C N^{k-s} \|u\|_{s,w} \quad , \quad k = 0, 1 \ , \quad s \geq 1 \ .$$

Precisely, $\tilde{u}_N$ is the solution to the problem

$$(4.3.58) \qquad \int_{-1}^{1} (u - \tilde{u}_N)'(v_N w)' dx = 0 \qquad \forall \, v_N \in \mathbb{P}^0_N$$

(see, e.g., Canuto, Hussaini, Quarteroni and Zang (1988), Sect. 9.5.2). Notice that the bilinear form appearing in (4.3.58) is the one associated to the operator $-u''$, using the scalar product $(\cdot, \cdot)_w$.

4.4 Legendre Expansion

Here we are going to replicate for the Legendre polynomials what has been done throughout the previous section for the Chebyshev polynomials.

4.4.1 Legendre Polynomials

The Legendre polynomials are Jacobi polynomials with $\alpha = \beta = 0$, i.e., they are defined as

$$(4.4.1) \qquad L_n(x) := \frac{J_n^{(0,0)}(x)}{J_n^{(0,0)}(1)} \quad , \quad n \in \mathbb{N} \ .$$

According to (4.1.13) they satisfy the Sturm-Liouville equation

$$(4.4.2) \qquad ((1 - x^2) L_n'(x))' + n(n+1) L_n(x) = 0 \ .$$

If L_n is normalized so that $L_n(1) = 1$, then

$$(4.4.3) \qquad |L_n(x)| \leq 1 \ , \quad |x| \leq 1 \ , \quad L_n(1) = 1 \ , \quad L_n(-1) = (-1)^n \ .$$

L_n has the same parity as n, and the three-term recursion formula reads

$$(4.4.4) \qquad L_{n+1}(x) = \frac{2n+1}{n+1} x \, L_n(x) - \frac{n}{n+1} L_{n-1}(x) \ , \quad n \geq 1 \ ,$$

where $L_0(x) = 1$ and $L_1(x) = x$ (see, e.g., Szegö (1959)).

Legendre polynomials are orthogonal under the weight function $w(x) = 1$. More precisely, setting

$$(4.4.5) \qquad (u, v) := \int_{-1}^{1} u(x)\, v(x)\, dx \quad , \quad ||u||_0 := \left(\int_{-1}^{1} u^2(x)\, dx \right)^{1/2}$$

(these are the usual scalar product and norm of $L^2(-1, 1)$, see Section 1.2), we have (e.g., Szegö (1959))

$$(4.4.6) \qquad (L_n, L_m) = \left(n + \frac{1}{2} \right)^{-1} \delta_{nm} \ .$$

For any function $u \in L^2(-1, 1)$ its Legendre series is

$$(4.4.7) \qquad u(x) = \sum_{k=0}^{\infty} \hat{u}_k L_k(x) \quad , \quad \hat{u}_k = \left(k + \frac{1}{2} \right)(u, L_k) \ .$$

About differentiation, let us recall the definition of the Sobolev space

$$(4.4.8) \qquad H^s(I) := \left\{ v \in L^2(I) \mid v^{(k)} \in L^2(I) \text{ for } k = 1, ..., s \right\} \ , \quad s \in \mathbb{N} \ ,$$

whose norm is

$$(4.4.9) \qquad ||v||_s := \left(\sum_{k=0}^{s} ||v^{(k)}||_0^2 \right)^{1/2}$$

(see Section 1.2). If we assume that $u \in H^1(I)$, then $u' \in L^2(I)$ and its Legendre series is (see, e.g., Canuto, Hussaini, Quarteroni and Zang (1988), Sect. 2.3.2)

$$(4.4.10) \qquad u'(x) = \sum_{k=0}^{\infty} \hat{u}_k^{(1)} L_k(x) \quad , \quad \hat{u}_k^{(1)} = (2k + 1) \sum_{\substack{p=k+1 \\ p+k \text{ odd}}}^{\infty} \hat{u}_p \ .$$

Similarly, if $u \in H^2(I)$, then

$$(4.4.11) \qquad u''(x) = \sum_{k=0}^{\infty} \hat{u}_k^{(2)} L_k(x)$$

$$\hat{u}_k^{(2)} = \left(k + \frac{1}{2} \right) \sum_{\substack{p=k+2 \\ p+k \text{ even}}}^{\infty} [p(p + 1) - k(k + 1)]\, \hat{u}_p \ .$$

If u is any polynomial of $\mathbb{P}_N$, the following backward recursion formula can be applied to get the Legendre coefficients $\{\hat{u}_k^{(q)}\}$ of its q-th derivative, $q \geq 1$, as a function of those of its derivative of lower order:

$$(4.4.12) \quad \hat{u}_{k-1}^{(q)} = (2k - 1)\left(\hat{u}_k^{(q-1)} + \frac{1}{2k + 3} \hat{u}_{k+1}^{(q)} \right) \ , \quad k = N, N - 1, ..., 1 \ ,$$

keeping in mind that $\hat{u}_{N+1}^{(q)} = ... = \hat{u}_{N+1-q}^{(q)} = 0$.

4.4.2 Legendre Interpolation

Let N be a positive integer. Here, we review the basic formulae of the polynomial interpolation at the Legendre Gauss-Lobatto nodes, i.e., at the $N+1$ zeroes of the polynomial $(1 - x^2) L'_N(x)$, ordered from right to left. As for the Chebyshev case, these nodes $\{x_j\}_{j=0,...,N}$ are symmetrically distributed around $x = 0$, and they cluster towards the endpoints of the interval (-1,1) as far as N increases.

The weights of the Legendre Gauss-Lobatto integration formula are (see, e.g., Davis and Rabinowitz (1984))

$$(4.4.13) \qquad w_j = \frac{2}{N(N+1)} \, \frac{1}{L_N^2(x_j)} \ , \quad j = 0, ..., N \ .$$

It can be shown that

$$\frac{2}{N(N+1)} \le w_j \le \frac{C}{N} \ , \quad j = 0, ..., N$$

(see, e.g., Bernardi and Maday (1992), p. 76).

The discrete approximations to the scalar product of $L^2(I)$ and its associated norm are

$$(4.4.14) \qquad \begin{aligned} (u,v)_N &= \frac{2}{N(N+1)} \sum_{j=0}^{N} \frac{1}{L_N^2(x_j)} \, u(x_j) \, v(x_j) \ , \\ ||v||_N &= \Big(\frac{2}{N(N+1)} \sum_{j=0}^{N} \frac{v^2(x_j)}{L_N^2(x_j)} \Big)^{1/2} \ , \end{aligned}$$

where the $\{x_j\}$ are the Legendre Gauss-Lobatto nodes. In view of (4.2.2) and (4.4.6) we have

$$(4.4.15) \qquad \begin{aligned} ||L_k||_N &= ||L_k||_0 \ , \quad k = 0, ..., N-1 \ , \\ ||L_N||_N &= \Big(\frac{2}{N}\Big)^{1/2} = \Big(2 + \frac{1}{N}\Big)^{1/2} ||L_N||_0 \ . \end{aligned}$$

In particular, these relations produce

$$(4.4.16) \qquad ||u_N||_0 \le ||u_N||_N \le \Big(2 + \frac{1}{N}\Big)^{1/2} ||u_N||_0 \quad \forall \, u_N \in \mathbb{P}_N \ .$$

It follows that for all polynomials of degree less than or equal to N the discrete norm defined in (4.4.14) is uniformly equivalent (with respect to N) to the L^2-norm.

For any continuous function u we can now introduce its interpolant $I_N u \in \mathbb{P}_N$ that matches u at the $N+1$ Legendre Gauss-Lobatto nodes $\{x_j\}_{j=0,...,N}$. In accordance with (4.2.7), and owing to (4.4.6), (4.4.15) we have

(4.4.17)
$$I_N u(x) = \sum_{k=0}^{N} u_k^* L_k(x)$$

with

(4.4.18)
$$\begin{cases} u_k^* = \dfrac{2k+1}{N(N+1)} \displaystyle\sum_{j=0}^{N} u(x_j)\, L_k(x_j) \dfrac{1}{L_N^2(x_j)} \quad , \quad k = 0, ..., N-1 \\[4mm] u_N^* = \dfrac{1}{N+1} \displaystyle\sum_{j=0}^{N} u(x_j)\, \dfrac{1}{L_N(x_j)} \quad . \end{cases}$$

Formula (4.4.18) is the *discrete Legendre transform*. It follows from (4.4.17) that its inverse is given by

(4.4.19)
$$u(x_j) = \sum_{k=0}^{N} L_k(x_j)\, u_k^* \quad , \quad j = 0, ..., N \quad .$$

Similarly to what we have done for the Chebyshev interpolation, for any continuous function u we can now define a *Legendre pseudo-spectral derivative* $\partial_N u$. It is precisely the polynomial in $\mathbb{P}_{N-1}$ which is formally defined as in (4.3.21), but now $I_N u$ is the Legendre interpolant of u given by (4.4.17). We still have a representation like (4.3.22) with the Lagrangian functions ψ_j now expressed by

$$\psi_j(x) = -\frac{1}{N(N+1)} \frac{(1-x^2)\, L_N'(x)}{(x - x_j)\, L_N(x_j)} \quad .$$

The *Legendre pseudo-spectral matrix* D_N associates to the $N+1$ values $\{u(x_j)\,|\,j = 0, ..., N\}$ the $N+1$ values $\{(\partial_N u)(x_j)\,|\,j = 0, ..., N\}$ of the pseudo-spectral derivative of u at the same Legendre Gauss-Lobatto nodes. Using (4.3.23) and the definition of ψ_j it can be deduced that (see Solomonoff and Turkel (1986) or Canuto, Hussaini, Quarteroni and Zang (1988), Sect. 2.3.2):

(4.4.20)
$$(D_N)_{ij} = \begin{cases} \dfrac{1}{x_i - x_j} \dfrac{L_N(x_i)}{L_N(x_j)} & i \neq j \\[3mm] 0 & 1 \leq i = j \leq N-1 \\[3mm] \dfrac{N(N+1)}{4} & i = j = 0 \\[3mm] -\dfrac{N(N+1)}{4} & i = j = N \quad . \end{cases}$$

The only eigenvalue is 0, with algebraic multiplicity $N+1$.

Concerning the interpolation error estimate, the following result holds (see Bernardi and Maday (1992), pp. 77–78). If $u \in H^s(I)$ for some $s \geq 1$, then

(4.4.21) $$||u - I_N u||_k \leq CN^{k-s}||u||_s \ , \quad k = 0, 1 \ ,$$

in analogy with the Chebyshev interpolation (see (4.3.27), (4.3.40)).

Using the Gagliardo-Nirenberg inequality (see Theorem 1.3.6) we deduce, in particular, the following error behaviour in the maximum norm

(4.4.22) $$\max_{-1 \leq x \leq 1} |u(x) - I_N u(x)| \leq CN^{1/2-s}||u||_s \ .$$

Also, we obtain that the error induced by the pseudo-spectral derivative is

(4.4.23) $$||u' - \partial_N u||_0 \leq CN^{1-s}||u||_s \ .$$

Concerning the Legendre Gauss-Lobatto integration (4.4.14), by proceeding as in the proof of (4.3.44) and using (4.4.21) we can show that for all $u \in H^s(I), s \geq 1$, and $v_N \in \mathbb{P}_N$

(4.4.24) $$|(u, v_N) - (u, v_N)_N| \leq CN^{-s}||u||_s||v_N||_0 \ .$$

We also notice that taking $k = s = 1$ in (4.4.21) we obtain

(4.4.25) $$||I_N u||_1 \leq C||u||_1 \ ,$$

which means that I_N is a continuous operator upon $H^1(I)$.

Remark 4.4.1 An inequality like (4.4.16) or (4.3.20) can also be established for exact and discrete maximum norms. Precisely, let us define

(4.4.26) $$||v||_\infty := \max_{-1 \leq x \leq 1} |v(x)| \ , \quad ||v||_{\infty,N} := \max_{0 \leq j \leq N} |u(x_j)| \ ,$$

where $\{x_j\}$ are the Gauss-Lobatto nodes. Then

(4.4.27) $$||u_N||_\infty \leq ||u_N||_{\infty,N} \leq \delta_N ||u_N||_\infty \quad \forall \ u_N \in \mathbb{P}_N \ .$$

Using the Chebyshev nodes, the function δ_N can grow at most logarithmically with N. Indeed (e.g., Natanson (1965))

$$\delta_N < \frac{2}{\pi} \log N + 1 \ .$$

In the Legendre case, the logarithmic growth is actually a lower bound, as it is possible to find a constant $\varepsilon > 0$ such that (Erdös (1961))

$$\delta_N > \frac{2}{\pi} \log N - \varepsilon \ .$$

Inequalities (4.4.27) are useful whenever discrete maximum norms are adopted to measure the error behaviour of computed solutions. $\qquad \Box$

4.4.3 Legendre Projections

For any function $u \in L^2(I)$, the truncated of order N of its Legendre series (4.4.7) is

$$(4.4.28) \qquad P_N u(x) = \sum_{k=0}^{N} \hat{u}_k L_k(x) \ ,$$

and $||u - P_N u||_0 \to 0$ as $N \to \infty$. Further, it has been proven by Canuto and Quarteroni (1982) that if $u \in H^s(I)$ for some $s \geq 0$ then

$$(4.4.29) \qquad ||u - P_N u||_0 \leq C N^{-s} ||u||_s \ ,$$

which is an optimal rate of convergence, since $P_N : L^2(I) \to \mathbb{P}_N$ is the orthogonal projection operator. Moreover

$$(4.4.30) \qquad ||u - P_N u||_1 \leq C \sqrt{N} N^{1-s} ||u||_s \ , \quad s \geq 1 \ .$$

As for the Chebyshev case, this estimate is non-optimal. We are therefore led to introduce the orthogonal projection $P_{1,N} : H^1(I) \to \mathbb{P}_N$, which is defined as follows

$$(4.4.31) \qquad (P_{1,N} u, v_N)_1 = (u, v_N)_1 \quad \text{for all } v_N \in \mathbb{P}_N \ ,$$

where

$$(4.4.32) \qquad (u, v)_1 := \int_{-1}^{1} (u'v' + uv) \, dx$$

denotes the scalar product of $H^1(I)$. The above projection yields an approximation error which is optimal in both the L^2 and H^1 norms. As a matter of fact

$$(4.4.33) \qquad ||u - P_{1,N} u||_k \leq C N^{k-s} ||u||_s \ , \quad k = 0, 1 \ , \ s \geq 1 \ .$$

This follows from (4.4.21) when $k = 1$, and by an usual duality argument when $k = 0$ (see also Maday and Quarteroni (1981)).

The definition of orthogonal projection can also be extended to account for the boundary behaviour of the function u. In fact, let us introduce the space

$$(4.4.34) \qquad H_0^1(I) := \left\{ v \in H^1(I) \mid v(\pm 1) = 0 \right\}$$

and recall that, due to the Poincaré inequality (1.3.2), the scalar product

$$(4.4.35) \qquad (u, v)_1^0 := \int_{-1}^{1} u'v' \, dx$$

induces upon $H_0^1(I)$ a norm equivalent to $||\cdot||_1$. We now define $P_{1,N}^0 : H_0^1(I) \to \mathbb{P}_N^0$ through

(4.4.36) $(P^0_{1,N}u, v_N)^0_1 = (u, v_N)^0_1$ for all $v_N \in \mathbb{P}^0_N$.

As in the case of the projection operator $P_{1,N}$, for all $s \geq 1$ and all $u \in H^1_0(I) \cap H^s(I)$ it holds

(4.4.37) $||u - P^0_{1,N}u||_k \leq CN^{k-s}||u||_s$, $k = 0,1$.

4.5 Two-Dimensional Extensions

Here we show how the previous definitions and results extend to the two-dimensional domain $\Omega = (-1, 1)^2$. For any $N \in \mathbb{N}$ we denote by $\mathbb{Q}_N$ the space of algebraic polynomials of degree less than or equal to N with respect to each single variable x_i, $i = 1, 2$, and by $\mathbb{Q}^0_N$ the subspace of those polynomials that vanish on $\partial\Omega$. We introduce the space (see Section 1.3)

(4.5.1) $L^2(\Omega) := \left\{ v : \Omega \to \mathbb{R} \mid \int_\Omega v^2(\mathbf{x})\, d\mathbf{x} < \infty \right\}$,

endowed with the following scalar product and norm

(4.5.2) $(u, v) := \int_\Omega u(\mathbf{x})\, v(\mathbf{x})\, d\mathbf{x}$, $||v||_0 := \left(\int_\Omega v^2(\mathbf{x})\, d\mathbf{x} \right)^{1/2}$.

Similarly we define $w(\mathbf{x}) := (1 - x_1^2)^{-1/2}(1 - x_2^2)^{-1/2}$ and

(4.5.3) $L^2_w(\Omega) := \left\{ v : \Omega \to \mathbb{R} \mid \int_\Omega v^2(\mathbf{x})\, w(\mathbf{x})\, d\mathbf{x} < \infty \right\}$.

This space is equipped by the following scalar product and norm

(4.5.4)

$$(u, v)_w := \int_\Omega u(\mathbf{x})\, v(\mathbf{x})\, w(\mathbf{x})\, d\mathbf{x}$$

$$||v||_{0,w} := \left(\int_\Omega v^2(\mathbf{x})\, w(\mathbf{x})\, d\mathbf{x} \right)^{1/2} .$$

4.5.1 The Chebyshev Case

The Chebyshev polynomials in Ω are defined by tensor product as follows

(4.5.5) $T_{km}(\mathbf{x}) := T_k(x_1)\, T_m(x_2)$, $k, m \in \mathbb{N}$.

In view of (4.3.6) we have the orthogonality relationship

(4.5.6) $(T_{km}, T_{ln})_w = \dfrac{\pi^2}{4}\, c_k\, c_m\, \delta_{kl}\, \delta_{mn}$.

The truncated Chebyshev series of order N is

$$(4.5.7) \quad P_N u(\mathbf{x}) := \sum_{k,m=0}^{N} \hat{u}_{km} T_{km}(\mathbf{x}) \quad , \quad \hat{u}_{km} := \frac{4}{\pi^2} \frac{1}{c_k c_m} (u, T_{km})_w \quad .$$

Clearly, P_N is the orthogonal projection operator from $L_w^2(\Omega)$ upon $\mathbb{Q}_N$.

The remainder of the series can be estimated as follows. For any $s \in \mathbb{N}$ we first define the weighted Sobolev space
$(4.5.8)$
$$H_w^s(\Omega) := \{v : \Omega \to \mathbb{R} \,|\, D^{\boldsymbol{\alpha}} v \in L_w^2(\Omega)$$
$$\text{for all } \boldsymbol{\alpha} = (\alpha_1, \alpha_2) \in \mathbb{N}^2 \text{ with } |\boldsymbol{\alpha}| \leq s\} \quad ,$$

where $D^{\boldsymbol{\alpha}} v = \frac{\partial}{\partial x_1^{\alpha_1}} \frac{\partial}{\partial x_2^{\alpha_2}} v$, and $|\boldsymbol{\alpha}| = \alpha_1 + \alpha_2$. This space is endowed with the norm

$$(4.5.9) \qquad \qquad ||v||_{s,w} := \left(\sum_{|\boldsymbol{\alpha}| \leq s} ||D^{\boldsymbol{\alpha}} v||_w^2 \right)^{1/2} \quad .$$

Then for all $u \in H_w^s(\Omega)$ with $s \geq 0$

$$(4.5.10) \qquad \qquad ||u - P_N u||_{0,w} \leq C N^{-s} ||u||_{s,w} \quad ,$$

while, if $s \geq 1$,

$$(4.5.11) \qquad \qquad ||u - P_N u||_{1,w} \leq C\sqrt{N} N^{1-s} ||u||_{s,w}$$

(see Canuto and Quarteroni (1982)).

We now introduce the orthogonal projection $P_{1,N} : H_w^1(\Omega) \to \mathbb{Q}_N$ through

$$(4.5.12) \qquad \qquad (P_{1,N}, v_N)_{1,w} = (u, v_N)_{1,w} \quad \forall \, v_N \in \mathbb{Q}_N \quad ,$$

where

$$(4.5.13) \qquad \qquad (u, v)_{1,w} := (u, v)_w + (\nabla u, \nabla v)_w$$

is the scalar product of $H_w^1(\Omega)$. Then, for all $u \in H_w^s(\Omega), s \geq 1$,

$$(4.5.14) \qquad \qquad ||u - P_{1,N} u||_{k,w} \leq C N^{k-s} ||u||_{s,w} \quad , \quad k = 0, 1 \quad .$$

Similarly, defining

$$(4.5.15) \qquad \qquad H_{w,0}^1(\Omega) := \{v \in H_w^1(\Omega) \,|\, v = 0 \quad \text{on} \quad \partial\Omega\}$$

and remembering that the scalar product

$$(4.5.16) \qquad \qquad (u, v)_{1,w}^0 := (\nabla u, \nabla v)_w$$

induces on $H_{w,0}^1(\Omega)$ a norm equivalent to $|| \cdot ||_{1,w}$ (still due to the Poincaré inequality), we define $P_{1,N}^0 : H_{w,0}^1(\Omega) \to \mathbb{Q}_N^0$ as follows:

$$(4.5.17) \qquad \qquad (P_{1,N}^0 u, v_N)_{1,w}^0 = (u, v_N)_{1,w}^0 \quad \forall \, v_N \in \mathbb{Q}_N^0 \quad .$$

Then, for all $u \in H_w^s(\Omega) \cap H_{w,0}^1(\Omega)$, with $s \geq 1$,

$$(4.5.18) \qquad \|u - P_{1,N}^0 u\|_{k,w} \leq CN^{k-s}\|u\|_{s,w} \ , \quad k = 0,1 \ .$$

For $k = 1$ the above inequalities (4.5.14), (4.5.18) can be found in Canuto, Hussaini, Quarteroni and Zang (1988), Sect. 9.7.3. On the other hand, for $k = 0$ they are proven by the Aubin-Nitsche duality argument (see Bernardi and Maday (1992), pp. 60–61).

We now turn to the two-dimensional Chebyshev interpolation. Fixed $N \geq 1$, this time the Chebyshev Gauss-Lobatto nodes and weights are (see (4.3.13))

$$(4.5.19) \ \mathbf{x}_{ij} := \left(\cos\frac{\pi i}{N}, \cos\frac{\pi j}{N}\right) \ , \quad w_{ij} := \left(\frac{\pi}{N}\right)^2 \frac{1}{d_i d_j} \ , \quad i,j = 0, ..., N \ .$$

The nodes are obtained by cartesian product of the one-dimensional nodes (4.3.12), the weights are the product of the corresponding one-dimensional weights (4.3.13). For any functions $u, v \in C^0(\overline{\Omega})$ we also define the discrete scalar product and norm

$$(4.5.20) \qquad (u,v)_N := \sum_{i,j=0}^{N} u(\mathbf{x}_{ij}) v(\mathbf{x}_{ij}) w_{ij} \ , \quad \|v\|_N^2 := (v,v)_N$$

and note that

$$(4.5.21) \qquad (u,v)_N = (u,v)_w \quad \text{if} \quad uv \in \mathbb{Q}_{2N-1} \ .$$

Moreover, from (4.3.20) it follows that $\|u_N\|_{0,w} \leq \|u_N\|_N \leq 2\|u_N\|_{0,w}$ for all $u_N \in \mathbb{Q}_N$.

For any continuous function u, we still denote by $I_N u \in \mathbb{Q}_N$ its interpolant at the nodes $\{\mathbf{x}_{ij}\}$. Then

$$(4.5.22) \qquad I_N u(\mathbf{x}) = \sum_{k,m=0}^{N} u_{km}^* T_{km}(\mathbf{x}) \ ,$$

where

$$u_{km}^* := \frac{4}{N^2} \frac{1}{d_k d_m} \sum_{i,j=0}^{N} \frac{1}{d_i d_j} u(\mathbf{x}_{ij}) T_{km}(\mathbf{x}_{ij}) \ .$$

The interpolation error estimate is: for any $s \geq 2$ and $u \in H_w^s(\Omega)$

$$(4.5.23) \qquad \|u - I_N u\|_{k,w} \leq CN^{k-s}\|u\|_{s,w} \ , \quad k = 0,1 \ ,$$

(see Canuto and Quarteroni (1982) for $k = 0$ and Bernardi and Maday (1992), p. 84, for $k = 1$).

4.5.2 The Legendre Case

The two-dimensional Legendre polynomials are defined as

$$(4.5.24) \qquad L_{km}(\mathbf{x}) := L_k(x_1) L_m(x_2) \ , \qquad k, m \in \mathbf{N} \ .$$

From (4.4.6) it follows

$$(4.5.25) \qquad (L_{km}, L_{ln}) = \left(k + \frac{1}{2}\right)^{-1} \left(m + \frac{1}{2}\right)^{-1} \delta_{kl} \, \delta_{mn} \ .$$

Let $P_N : L^2(\Omega) \to \mathbb{Q}_N$ denote the orthogonal projection with respect to the $L^2(\Omega)$-scalar product $(\cdot, \cdot)$. Owing to (4.5.25), for any $u \in L^2(\Omega)$, $P_N u$ is the truncated Legendre series

$$(4.5.26) \quad P_N u(\mathbf{x}) := \sum_{k,m=0}^{N} \hat{u}_{km} L_{km}(\mathbf{x}) \ , \quad \hat{u}_{km} := \left(k + \frac{1}{2}\right)\left(m + \frac{1}{2}\right)(u, L_{km}) \ .$$

For any $s \geq 0$ let us recall the definition of the Sobolev space

$$(4.5.27) \qquad \begin{aligned} H^s(\Omega) := \{ v : \Omega \to \mathbb{R} \,|\, D^{\boldsymbol{\alpha}} v \in L^2(\Omega) \\ \text{for all } \boldsymbol{\alpha} \in \mathbf{N}^2 \text{ with } |\boldsymbol{\alpha}| \leq s \} \ , \end{aligned}$$

whose norm is

$$(4.5.28) \qquad \|v\|_s := \left(\sum_{|\boldsymbol{\alpha}| \leq s} \|D^{\boldsymbol{\alpha}} v\|_0^2 \right)^{1/2} \ .$$

Clearly, $H^0(\Omega) = L^2(\Omega)$ (see Section 1.2).

For all $u \in H^s(\Omega)$, with $s \geq 0$,

$$(4.5.29) \qquad \|u - P_N u\|_0 \leq C N^{-s} \|u\|_s \ .$$

Further, if $s \geq 1$

$$(4.5.30) \qquad \|u - P_N u\|_1 \leq C \sqrt{N} N^{1-s} \|u\|_s$$

(see Canuto and Quarteroni (1982)).

We now define $P_{1,N} : H^1(\Omega) \to \mathbb{Q}_N$ the orthogonal projection for the scalar product of $H^1(\Omega)$, i.e.,

$$(4.5.31) \qquad (P_{1,N} u, v_N)_1 = (u, v_N)_1 \ \ \forall \, v_N \in \mathbb{Q}_N \ ,$$

with

$$(4.5.32) \qquad (u, v)_1 := (u, v) + (\nabla u, \nabla v) \ .$$

Similarly, we define $P_{1,N}^0 : H_0^1(\Omega) \to \mathbb{Q}_N^0$ through

$$(4.5.33) \qquad (P_{1,N}^0 u, v_N)_1^0 = (u, v_N)_1^0 \ \ \forall \, v_N \in \mathbb{Q}_N \ ,$$

with

(4.5.34)
$$(u, v)_1^0 := (\nabla u, \nabla v)$$

and

(4.5.35)
$$H_0^1(\Omega) := \{v \in H^1(\Omega) \mid v = 0 \text{ on } \partial\Omega\} .$$

As a consequence of the Poincaré inequality (1.3.2), we notice here that for all functions of $H_0^1(\Omega)$ the scalar product (4.5.34) induces a norm equivalent to $\|\cdot\|_1$.

For all $u \in H^s(\Omega)$, $s \geq 1$,

(4.5.36)
$$\|u - P_{1,N}u\|_k \leq CN^{k-s}\|u\|_s , \quad k = 0, 1 .$$

Similarly, for all $u \in H^s(\Omega) \cap H_0^1(\Omega)$, $s \geq 1$,

(4.5.37)
$$\|u - P_{1,N}^0 u\|_k \leq CN^{k-s}\|u\|_s , \quad k = 0, 1 .$$

Both estimates (4.5.36) and (4.5.37) are optimal; they can be found in Canuto, Hussaini, Quarteroni and Zang (1988), Sect. 9.7.2.

Concerning interpolation, let us define the two-dimensional Legendre Gauss-Lobatto nodes and weights

(4.5.38)
$$\mathbf{x}_{ij} := (x_i, x_j) , \quad w_{ij} := w_i w_j , \quad i, j = 0, ..., N , \quad N \geq 1 ,$$

where $\{x_k\}_{k=0,...,N}$ are the zeroes of $(1 - t^2)L_N'(t)$, while $\{w_k\}_{k=0,...,N}$ are defined in (4.4.13).

In Fig. 4.5.1 we present the Legendre Gauss-Lobatto grid corresponding to three different values of N. The clustering of the nodes towards the boundary is evident. The distribution of the Chebyshev Gauss-Lobatto nodes (4.5.19) is quite similar.

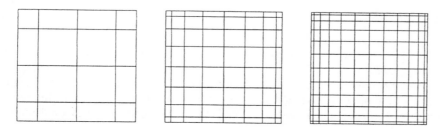

Fig. 4.5.1. Legendre Gauss-Lobatto grid for $N = 4$, $N = 8$ and $N = 12$

The associated discrete scalar product is

(4.5.39)
$$(u, v)_N := \sum_{i,j=0}^{N} u(\mathbf{x}_{ij}) v(\mathbf{x}_{ij}) w_{ij} , \quad u, v \in C^0(\overline{\Omega}) ,$$

and

$$(4.5.40) \qquad (u,v)_N = (u,v) \quad \text{if } uv \in \mathbb{Q}_{2N-1} \ .$$

We can easily deduce from (4.4.15) and (4.4.16) that

$$(4.5.41) \qquad ||u_N||_0 \leq ||u_N||_N \leq \left(2 + \frac{1}{N}\right)||u_N||_0 \quad \text{for all } u_N \in \mathbb{Q}_N \ .$$

For any continuous function u we denote by $I_N u \in \mathbb{Q}_N$ the interpolant of u at the nodes $\{\mathbf{x}_{ij}\}$. Then

$$(4.5.42) \qquad I_N u(\mathbf{x}) = \sum_{k,m=0}^{N} u_{km}^* L_{km}(\mathbf{x}) \ , \quad u_{km}^* := \frac{1}{||L_{km}||_N^2}(u, L_{km})_N \ ,$$

where $||v||_N^2 := (v,v)_N$. It is also possible to write u_{km}^* in a more explicit way like (4.4.18).

The interpolation error estimate is

$$(4.5.43) \qquad ||u - I_N u||_k \leq C N^{k-s} ||u||_s \ , \quad k = 0, 1 \ ,$$

provided $u \in H^s(\Omega)$ for some $s \geq 2$ (see Bernardi and Maday (1992), pp. 82–83).

Remark 4.5.1 *(Bases for polynomial spaces).* A basis for the space $\mathbb{P}_N$ of one-dimensional polynomials of degree less than or equal to N, $N \geq 0$, is provided either by $L_k(\xi)$ or $T_k(\xi)$, $k = 0, ..., N$.

A basis for the space $\mathbb{P}_N^0 := \{p : [-1,1] \to \mathbb{R} \, | \, p \in \mathbb{P}_N \, , \, p(-1) = p(1) = 0\}$, $N \geq 2$, is provided either by

$$\varphi_k(\xi) = L_k(\xi) - \varphi_k^*(\xi) \ , \ 2 \leq k \leq N \ ,$$

or by

$$\varphi_k(\xi) = T_k(\xi) - \varphi_k^*(\xi) \ , \ 2 \leq k \leq N \ ,$$

with $\varphi_k^*(\xi) = 1$ if k is even, or $\varphi_k^*(\xi) = \xi$ if k is odd (see, e.g., Gottlieb and Orszag (1977)).

Alternative bases for $\mathbb{P}_N^0$ are provided by $(1-\xi^2)L_k'(\xi)$ or by $(1-\xi^2)T_k'(\xi)$. Indeed, focussing for instance on the Legendre case, using the Sturm-Liouville equation (4.4.2) the system $(1 - \xi^2)L_k'(\xi)$ is shown to be orthogonal under the scalar product $\int_{-1}^{1} z(\xi)v(\xi)(1 - \xi^2)^{-1}d\xi$.

For spaces of two-dimensional polynomials over the square $[-1,1]^2$ the bases for both $\mathbb{Q}_N$ and $\mathbb{Q}_N^0$ are easily obtained by tensor product of the one-dimensional bases introduced above. $\square$

4.6 Complements

The construction of both Chebyshev and Legendre approximation for the three-dimensional domain $\Omega = (-1, 1)^3$ can be carried out by following the same guidelines of Sections 4.5.1 and 4.5.2. The previously reported error estimates hold true even in the three-dimensional case, and the proof can be performed by the same arguments (see, e.g., Canuto, Hussaini, Quarteroni and Zang (1988), Bernardi and Maday (1992)).

The monographs by Gottlieb and Orszag (1977) and Boyd (1989) contain many useful tools for dealing with polynomial-based expansions. In particular, they provide formulae for coordinate transformation, for mapping infinite into finite domains, for transforming between physical and frequency spaces. See also Canuto, Hussaini, Quarteroni and Zang (1988), where several routines for implementing polynomial transformations and polynomial-based quadratures are reported.

Spectral methods were formerly proposed for the approximation of differential problems with periodic data. For these cases Fourier expansion via trigonometric polynomials takes the place of Chebyshev and Legendre expansions. We will not deal with Fourier spectral methods in this book. The interested reader can refer, e.g., to Gottlieb and Orszag (1977), Canuto, Hussaini, Quarteroni and Zang (1988), Mercier (1989) and Boyd (1989).

For other theoretical results about polynomial approximations in Sobolev spaces see Nikol'skiĭ (1975), Canuto, Hussaini, Quarteroni and Zang (1988), Chap. 9, and also Bernardi and Maday (1992), Funaro (1992), where results on Laguerre and Hermite polynomials are provided, too.

5. Galerkin, Collocation and Other Methods

This Chapter is devoted to a short presentation of some classical techniques for the discretization of (initial-) boundary value problems.

We start with an abstract boundary value problem and present three simple examples. After the reformulation of the problem in a weak (or variational) form, we introduce four different families of space approximations, precisely the Galerkin, Petrov-Galerkin, collocation and generalized Galerkin methods. For each of them we review the main stability and convergence properties.

We then face the problem of time-discretization of initial-boundary value problems. Therefore, we must consider some elementary examples of finite-difference schemes for the time derivative, and also illustrate some classical fractional-step and operator-splitting methods.

5.1 An Abstract Reference Boundary Value Problem

As usual, Ω denotes a bounded domain of $\mathbb{R}^d$, $d = 2, 3$, whose boundary is $\partial\Omega$. We consider a boundary value problem of the form

(5.1.1)
$$\begin{cases} Lu = f & \text{in } \Omega \\ Bu = 0 & \text{on } \partial\Omega^* \end{cases},$$

where f is a given function, u is the unknown, L is a linear differential operator and B is an affine boundary operator. Finally, $\partial\Omega^*$ is a subset of $\partial\Omega$ (possibly the whole boundary). Most often, L is an unbounded operator in a space H that can be either $L^2(\Omega)$ or $L_w^2(\Omega)$. The latter is the weighted Hilbert space introduced in (4.5.3), in the framework of Chebyshev spectral approximation. The solution u is looked for in a space $X \subset H$, such that L and B have a meaning for functions belonging to X.

Problem (5.1.1) can generally be reformulated in a weak (or variational) form. From one hand, this approach allows the search of weak solutions, which don't necessarily satisfy the equations (5.1.1) in a pointwise manner. This considerably enlarges the field of physical applications to account for problems with non-smooth data. Furthermore, the weak formulation is a convenient form in order to design approximation methods such as the Galerkin method and its extensions (Petrov-Galerkin, generalized Galerkin, etc.).

At this stage, we prefer to be concise and heuristic in many points, as we aim to illustrate the basic ideas underlying the different methodologies. The correct mathematical framework will be made precise from case to case in the forthcoming chapters.

Formally speaking, the weak formulation can be derived after multiplication of the differential equation by a suitable set of test functions and performing an integration upon the domain. Most often, the Green formula of integration by parts

$$(5.1.2) \qquad \int_\Omega \frac{\partial u}{\partial x_i} v \, d\mathbf{x} = - \int_\Omega u \frac{\partial v}{\partial x_i} \, d\mathbf{x} + \int_{\partial\Omega} uv \, n_i \, d\gamma \qquad i = 1, ..., d \ ,$$

where $\mathbf{n} = (n_1, ..., n_d)$ is the unit outward normal vector on $\partial\Omega$, is used at this stage with the aim of reducing the order of differentiation for the solution u.

As a result, we obtain a problem that reads

$$(5.1.3) \qquad \text{find } u \in W : \mathcal{A}(u, v) = \mathcal{F}(v) \qquad \forall \, v \in V \ ,$$

where W is the space of admissible solutions and V is the space of test functions. Both W and V can be assumed to be Hilbert spaces. $\mathcal{F}$ is a linear functional on V that accounts for the right hand side f as well as for possible non-homogeneous boundary terms. Finally, $\mathcal{A}(\cdot, \cdot)$ is a bilinear form corresponding to the differential operator L.

The boundary conditions on u can be enforced directly in the definition of W (this is the case of the so-called *essential* boundary conditions). Otherwise, they can be achieved indirectly through a suitable choice of the bilinear form $\mathcal{A}$ as well as the functional $\mathcal{F}$ (*natural* boundary conditions).

In most cases, $W = V$. At any rate, the nature of the spaces W and V is so that all operations involved in the formulation (5.1.3) make sense from a mathematical point of view.

In order to make this presentation slightly less abstract, we hereby present three examples that will be thoroughly reconsidered in the sequel of this book. Again, we wish to make clear that we are not committed to any rigorous formalism.

Example 1. The Poisson problem
The problem reads

$$(5.1.4) \qquad \begin{cases} -\Delta u = f & \text{in } \Omega \\ u = 0 & \text{on } \partial\Omega \ , \end{cases}$$

where f is a given function and

$$\Delta := \sum_{i=1}^{d} \frac{\partial^2}{\partial x_i^2}$$

is the Laplace operator. This problem arises in a wide variety of applications, and will be considered in Chapter 6.

In order to find its variational formulation, we recall the following Green formula for the Laplacian:

$$(5.1.5) \qquad -\int_{\Omega} \Delta u\, v\, dx = \int_{\Omega} \nabla u \cdot \nabla v\, dx - \int_{\partial\Omega} \frac{\partial u}{\partial n} v\, d\gamma \ ,$$

which follows from (5.1.2) noticing that $\Delta u = \operatorname{div} \nabla u$. From (5.1.4) and (5.1.5) we can easily deduce that u satisfies a problem like (5.1.3) where

$$(5.1.6) \qquad \begin{cases} W = V = H_0^1(\Omega) \\ \mathcal{A}(u,v) := \int_{\Omega} \nabla u \cdot \nabla v\, dx \\ \mathcal{F}(v) := \int_{\Omega} fv\, dx \ , \end{cases}$$

with $H_0^1(\Omega)$ defined in Section 1.2.

Example 2. The Stokes problem
This time we look for a solution to the problem

$$(5.1.7) \qquad \begin{cases} -\nu \Delta \mathbf{u} + \nabla p = \mathbf{f} & \text{in } \Omega \\ \operatorname{div} \mathbf{u} = 0 & \text{in } \Omega \\ \mathbf{u} = 0 & \text{on } \partial\Omega \ , \end{cases}$$

where ν is a positive constant and $\mathbf{f}$ is a given vector function. This problem is considered in Chapter 9. Its derivation within the framework of Navier-Stokes equations for viscous incompressible flows is carried out in Chapter 10. In this respect, $\mathbf{u}$ and p denote the velocity and the pressure of the fluid, respectively. The first equation is the (linearized) momentum equation, while the second equation expresses the incompressibility constraint.

Problem (5.1.7) can be formulated as (5.1.1) setting

$$u = \begin{pmatrix} \mathbf{u} \\ p \end{pmatrix} \ , \quad f = \begin{pmatrix} \mathbf{f} \\ 0 \end{pmatrix} \ , \quad L = \begin{pmatrix} -\nu\Delta & \nabla \\ \operatorname{div} & 0 \end{pmatrix} \ , \quad B = \begin{pmatrix} I & 0 \\ 0 & 0 \end{pmatrix} \ ,$$

where I is the identity operator.

Its variational form can be derived as follows. We define

$$(5.1.8) \quad \mathcal{V} = (H_0^1(\Omega))^d \ , \quad \mathcal{Q} = L_0^2(\Omega) := \left\{ q \in L^2(\Omega)\Big|\ \int_{\Omega} q\, dx = 0 \right\} \ .$$

$\mathcal{V}$ is the space of admissible velocities, while $\mathcal{Q}$ is that of admissible pressures. The vanishing on the average is enforced in order to find a unique pressure. Using both (5.1.2) and (5.1.5) for the first equation of (5.1.7) we obtain

$$\nu \int_\Omega \nabla \mathbf{u} \cdot \nabla \mathbf{v} \, dx - \int_\Omega p \operatorname{div} \mathbf{v} \, dx = \int_\Omega \mathbf{f} \cdot \mathbf{v} \, dx \quad \forall \, \mathbf{v} \in \mathcal{V} \ ,$$

while from the second equation we deduce

$$\int_\Omega q \operatorname{div} \mathbf{u} \, dx = 0 \quad \forall \, q \in \mathcal{Q} \ .$$

Therefore, (5.1.7) can be reformulated as (5.1.3) setting

$$(5.1.9) \qquad \begin{cases} W = V = \mathcal{V} \times \mathcal{Q} \\ \mathcal{A}(u,v) := \nu \int_\Omega \nabla \mathbf{u} \cdot \nabla \mathbf{v} \, dx - \int_\Omega p \operatorname{div} \mathbf{v} \, dx + \int_\Omega q \operatorname{div} \mathbf{u} \, dx \\ \mathcal{F}(v) := \int_\Omega \mathbf{f} \cdot \mathbf{v} \, dx \ , \end{cases}$$

for $u = (\mathbf{u}, p)$, $v = (\mathbf{v}, q) \in V$. For further details on this derivation we refer to Chapter 9 (however, notice the change of notations for the functional spaces).

Example 3. A steady advection problem
Let $\mathbf{a}$, a_0 and f be some given functions on Ω, and consider the problem

$$(5.1.10) \qquad \begin{cases} \mathbf{a} \cdot \nabla u + a_0 u = f & \text{in } \Omega \\ u = \varphi & \text{on } \partial \Omega^{in} \ , \end{cases}$$

where

$$\partial \Omega^{in} := \{ \mathbf{x} \in \partial \Omega \mid \mathbf{a}(\mathbf{x}) \cdot \mathbf{n}(\mathbf{x}) < 0 \}$$

is the so-called inflow boundary and φ is given there. According to the values of $\mathbf{a}(\mathbf{x})$, $\partial \Omega^{in}$ can be empty, it can coincide with $\partial \Omega$, or else it can be a proper subset of $\partial \Omega$. The time-dependent version of (5.1.10) is an inflow-outflow hyperbolic problem which is considered in Chapter 14.

A possible weak formulation of (5.1.10) can be obtained as follows. Integrating by parts as in (5.1.2) we have

$$\int_\Omega \mathbf{a} \cdot \nabla u \, v \, dx = - \int_\Omega u \operatorname{div}(\mathbf{a} v) \, dx + \int_{\partial \Omega^{in}} \mathbf{a} \cdot \mathbf{n} \, \varphi v \, d\gamma$$
$$+ \int_{\partial \Omega \setminus \partial \Omega^{in}} \mathbf{a} \cdot \mathbf{n} \, u v \, d\gamma \quad \forall \, v \in H^1(\Omega) \ .$$

We emphasize that the boundary condition has been accounted for in the integration-by-parts process. If we set

$$(5.1.11) \qquad \begin{cases} W = L^2(\Omega) \ , \quad V = H^1(\Omega) \\ \mathcal{A}(u,v) := - \int_\Omega u \operatorname{div}(\mathbf{a} v) \, dx + \int_{\partial \Omega \setminus \partial \Omega^{in}} \mathbf{a} \cdot \mathbf{n} \, u v \, d\gamma \\ \hspace{6cm} + \int_\Omega a_0 u v \, dx \\ \mathcal{F}(v) := \int_\Omega f v \, dx - \int_{\partial \Omega^{in}} \mathbf{a} \cdot \mathbf{n} \, \varphi v \, d\gamma \ , \end{cases}$$

we are able to formulate (5.1.10) as (5.1.3).

5.1.1 Some Results of Functional Analysis

Before considering the numerical approximations to (5.1.3), we present two basic functional theorems about existence and uniqueness of the solution. The first one refers to the special case $W = V$ and therefore to the problem

(5.1.12)   find $u \in V : \mathcal{A}(u,v) = \mathcal{F}(v)$ $\forall \, v \in V$.

Theorem 5.1.1 (Lax-Milgram lemma). *Let V be a (real) Hilbert space, endowed with the norm $||\cdot||$, $\mathcal{A}(u,v) : V \times V \to \mathbb{R}$ a bilinear form and $\mathcal{F}(v) : V \to \mathbb{R}$ a linear continuous functional, i.e., $\mathcal{F} \in V'$, where V' denotes the dual space of V. Assume moreover that $\mathcal{A}(\cdot,\cdot)$ is continuous, i.e.,*

(5.1.13) $\exists \, \gamma > 0 : |\mathcal{A}(w,v)| \leq \gamma ||w|| \, ||v||$ $\forall \, w,v \in V$,

and coercive, i.e.,

(5.1.14) $\exists \, \alpha > 0 : \mathcal{A}(v,v) \geq \alpha ||v||^2$ $\forall \, v \in V$.

Then, there exists a unique $u \in V$ solution to (5.1.12) and

(5.1.15) $||u|| \leq \dfrac{1}{\alpha} ||\mathcal{F}||_{V'}$.

Proof. Due to the Riesz representation theorem (e.g., Yosida (1974), p. 90), we can write

$$\mathcal{F}(v) = (\mathsf{R}\mathcal{F}, v)_V \qquad \forall \, v \in V$$

and for each fixed $w \in V$

$$\mathcal{A}(w,v) = (\mathsf{A}w, v)_V \qquad \forall \, v \in V ,$$

where $(\cdot,\cdot)_V$ is the scalar product in V, and the bijection $\mathsf{R} : V' \to V$ and $\mathsf{A} : V \to V$ are linear continuous operators. More precisely, R is an isometric operator since

$$||\mathsf{R}\mathcal{F}|| = \sup_{\substack{v \in V \\ v \neq 0}} \frac{(\mathsf{R}\mathcal{F}, v)_V}{||v||} = \sup_{\substack{v \in V \\ v \neq 0}} \frac{\mathcal{F}(v)}{||v||} = ||\mathcal{F}||_{V'} \ ;$$

furthermore

$$||\mathsf{A}w|| = \sup_{\substack{v \in V \\ v \neq 0}} \frac{(\mathsf{A}w, v)_V}{||v||} = \sup_{\substack{v \in V \\ v \neq 0}} \frac{\mathcal{A}(w,v)}{||v||} \leq \gamma ||w|| ,$$

by (5.1.13).

Problem (5.1.12) is thus equivalent to the following one: for each $\mathcal{F} \in V'$, find a unique $u \in V$ such that

(5.1.16) $\mathsf{A}u = \mathsf{R}\mathcal{F}$,

i.e., to prove that A is a bijection.

Let us start showing that A is injective. We have

(5.1.17) $$||v||^2 \leq \frac{1}{\alpha}(Av, v)_V \leq \frac{1}{\alpha}||Av|| \, ||v|| \ ,$$

and consequently $||v|| \leq (1/\alpha) \, ||Av||$. The uniqueness is thus proven.

We can now show that the range $\mathcal{R}(A)$ of A is closed and $\mathcal{R}(A)^{\perp} = \{0\}$, which is equivalent to prove that $\mathcal{R}(A) = V$. (If S is a closed subspace of a Hilbert space X with scalar product $(\cdot, \cdot)_X$, its orthogonal subspace $S^{\perp}$ is made by all values $y \in X$ for which $(y, x)_X = 0$ for all $x \in S$.) Suppose that $Av_n \to w$ in V. From (5.1.17)

$$||v_n - v_m|| \leq \frac{1}{\alpha}||Av_n - Av_m|| \ ,$$

hence v_n is a Cauchy sequence. Set $v = \lim v_n$. Since A is continuous, $Av = w$, and $\mathcal{R}(A)$ is closed. Now take $z \in \mathcal{R}(A)^{\perp}$; then

$$0 = (Az, z)_V = \mathcal{A}(z, z) \geq \alpha ||z||^2 \ ,$$

i.e., $z = 0$.

To conclude, we have only to prove (5.1.15). Choosing $v = u$ in (5.1.12), we find

$$||u||^2 \leq \frac{1}{\alpha}\mathcal{A}(u, u) = \frac{1}{\alpha}\mathcal{F}(u) \leq \frac{1}{\alpha}||\mathcal{F}||_{V'}||u|| \ ,$$

which produces (5.1.15). □

Remark 5.1.1 *(The symmetric case).* If the bilinear form is symmetric, i.e.,

(5.1.18) $$\mathcal{A}(w, v) = \mathcal{A}(v, w) \qquad \forall \, w, v \in V \ ,$$

then $\mathcal{A}(\cdot, \cdot)$ defines a scalar product on V, and the Riesz representation theorem suffices to infer existence and uniqueness for the solution of (5.1.12). We also remind now, that in this case the solution of (5.1.12) can be regarded as the unique solution to the *minimization problem*

(5.1.19) $$\text{find } u \in V : J(u) \leq J(v) \qquad \forall \, v \in V \ ,$$

where

(5.1.20) $$J(v) := \frac{1}{2}\mathcal{A}(v, v) - \mathcal{F}(v)$$

is a quadratic functional. □

Remark 5.1.2 *(The complex case).* If V is a complex Hilbert space, we have to modify the assumptions on the form $\mathcal{A}(\cdot, \cdot)$. First of all, $\mathcal{A}$ is required to be sesquilinear, i.e.,

$$\mathcal{A}(w, c_1 v_1 + c_2 v_2) = \overline{c_1}\mathcal{A}(w, v_1) + \overline{c_2}\mathcal{A}(w, v_2)$$

for each $w, v_1, v_2 \in V$, and $c_1, c_2 \in \mathbb{C}$. Here $\overline{c_k}$ is the conjugate of the complex number c_k, $k = 1, 2$. Moreover, the coerciveness assumption now takes the form

$$(5.1.21) \qquad |\mathcal{A}(v,v)| \geq \alpha||v||^2 \quad \forall\, v \in V \ .$$

The proof of the Lax-Milgram lemma holds essentially unchanged.

In the real case, condition (5.1.21) is not less restrictive than (5.1.14). In fact, it can be shown that it reduces to either one of inequalities $\mathcal{A}(v,v) \geq \alpha||v||^2$ or $-\mathcal{A}(v,v) \geq \alpha||v||^2$ for each $v \in V$, i.e., either $\mathcal{A}(\cdot, \cdot)$ or $-\mathcal{A}(\cdot, \cdot)$ has to satisfy (5.1.14). □

In the more general case in which W is actually different than V, problem (5.1.3) can be analyzed by the following result, which is an extension of the Lax-Milgram lemma due to Nečas (1962).

Theorem 5.1.2 *Let W and V be two (real) Hilbert spaces, with norms $|||\cdot|||$ and $||\cdot||$, respectively. Assume that there exist two positive constants α and γ such that the bilinear form $\mathcal{A} : W \times V \to \mathbb{R}$ satisfies*

$$(5.1.22) \qquad |\mathcal{A}(w,v)| \leq \gamma|||w|||\,||v|| \quad \forall\, w \in W,\ v \in V$$

$$(5.1.23) \qquad \sup_{\substack{v \in V \\ v \neq 0}} \frac{\mathcal{A}(w,v)}{||v||} \geq \alpha|||w||| \quad \forall\, w \in W$$

$$(5.1.24) \qquad \sup_{w \in W} \mathcal{A}(w,v) > 0 \quad \forall\, v \in V,\ v \neq 0 \ .$$

Then, for any $\mathcal{F} \in V'$, there exists a unique solution $u \in W$ of (5.1.3) which satisfies

$$(5.1.25) \qquad |||u||| \leq \frac{||\mathcal{F}||_{V'}}{\alpha} \ .$$

Proof. It is very similar to that of Lax-Milgram lemma (Theorem 5.1.1). In fact, due to the Riesz representation theorem we can construct a linear and continuous operator $\mathsf{A} : W \to V$ such that

$$(5.1.26) \qquad \mathcal{A}(w,v) = (\mathsf{A}w, v)_V \quad \forall\, v \in V$$

(where $(\cdot, \cdot)_V$ denotes the scalar product in V). Furthermore

$$||\mathsf{A}w|| \leq \gamma|||w||| \quad \forall\, w \in W \ .$$

The problem thus reduces to find, for any given $\mathcal{F} \in V'$, a unique $u \in W$ such that

$$(5.1.27) \qquad \mathsf{A}u = \mathsf{R}\mathcal{F}$$

(here $\mathsf{R} : V' \to V$ is the isometric operator constructed in Theorem 5.1.1).

From (5.1.23) and (5.1.26) it follows at once that $Aw = 0$ implies $w = 0$, i.e., A is injective.

Moreover, the range $\mathcal{R}(A)$ of A is closed. In fact, if $Aw_n \rightarrow v$ in V, we have

$$|||w_n - w_m||| \leq \frac{1}{\alpha} \sup_{\substack{v \in V \\ v \neq 0}} \frac{(A(w_n - w_m), v)_V}{||v||} \leq \frac{1}{\alpha} ||A(w_n - w_m)|| \ ,$$

hence $w_n \rightarrow w$ in W, and $Aw_n \rightarrow Aw = v$ in V. Finally, if $z \in \mathcal{R}(A)^{\perp}$, i.e.,

$$(Aw, z)_V = \mathcal{A}(w, z) = 0 \quad \forall \, w \in W \ ,$$

from (5.1.24) it follows $z = 0$, hence A is surjective. We have thus shown that for each $\mathcal{F} \in V'$ there exists a unique solution u to (5.1.27).

Moreover

$$\alpha |||u||| \leq \sup_{\substack{v \in V \\ v \neq 0}} \frac{(Au, v)_V}{||v||} = \sup_{\substack{v \in V \\ v \neq 0}} \frac{(R\mathcal{F}, v)_V}{||v||} = ||\mathcal{F}||_{V'} \ ,$$

hence (5.1.25) holds.   □

5.2 Galerkin Method

We now consider the case in which $W = V$, and therefore the problem at hand is (5.1.12).

Let $h > 0$ be a parameter that is going to be "small" in the applications (h is the mesh spacing for finite elements, or else the reciprocal of the polynomial degree of the spectral solution; see Chapters 3 and 4).

Let

(5.2.1) $\{V_h \,|\, h > 0\}$

denote a family of finite dimensional subspaces of V. We assume that

(5.2.2)   for all $v \in V$, $\displaystyle\inf_{v_h \in V_h} ||v - v_h|| \rightarrow 0$ as $h \rightarrow 0$.

The Galerkin approximation to (5.1.12) reads: given $\mathcal{F} \in V'$,

(5.2.3)   find $u_h \in V_h : \mathcal{A}(u_h, v_h) = \mathcal{F}(v_h) \quad \forall \, v_h \in V_h$.

It is therefore an *internal* approximation to (5.1.12). From the algebraic point of view, let $\{\varphi_j \,|\, j = 1, ..., N_h\}$ be a basis for the vector space V_h, so that we can set

(5.2.4) $\displaystyle u_h(\mathbf{x}) = \sum_{j=1}^{N_h} \xi_j \, \varphi_j(\mathbf{x})$.

Then from (5.2.3) we deduce the following linear system of dimension N_h:

(5.2.5)
$$A\boldsymbol{\xi} = \mathbf{F} \; ,$$

with $\boldsymbol{\xi} = (\xi_j)$, $\mathbf{F} := (\mathcal{F}(\varphi_i))$, $A_{ij} := \mathcal{A}(\varphi_j, \varphi_i)$ for $i, j = 1, ..., N_h$. The matrix A is called the *stiffness matrix*. Its properties will be reported in the following Remark 5.2.2.

For the analysis of problem (5.2.3) we have

Theorem 5.2.1 *Under the assumptions of Theorem 5.1.1 there exists a unique solution u_h to (5.2.3), which furthermore is stable since*

(5.2.6)
$$||u_h|| \leq \frac{||\mathcal{F}||_{V'}}{\alpha} \; .$$

Moreover, if u is the solution to (5.1.12), it follows

(5.2.7)
$$||u - u_h|| \leq \frac{\gamma}{\alpha} \inf_{v_h \in V_h} ||u - v_h|| \; ,$$

hence u_h converges to u, owing to (5.2.2).

Proof. Since V_h is a subspace of V, the assumptions (5.1.13) and (5.1.14) allow the application of the Lax-Milgram lemma also to the problem (5.2.3), yielding existence and uniqueness of u_h. Moreover, taking $v_h = u_h$ and using (5.1.14) we obtain

$$\alpha ||u_h||^2 \leq \mathcal{A}(u_h, u_h) = \mathcal{F}(u_h) \leq ||\mathcal{F}||_{V'} ||u_h|| \; .$$

Thus (5.2.6) follows, providing stability as both $||\mathcal{F}||_{V'}$ and α are independent of h. Now, subtracting the equation (5.2.3) from (5.1.12) (the latter being restricted to the test functions of V_h) we obtain

(5.2.8)
$$\mathcal{A}(u - u_h, v_h) = 0 \quad \forall \, v_h \in V_h \; .$$

Using (5.1.13), (5.1.14) and (5.2.8) for $v_h = w_h - u_h$, $w_h \in V_h$, it follows

$$\alpha ||u - u_h||^2 \leq \mathcal{A}(u - u_h, u - u_h) = \mathcal{A}(u - u_h, u - w_h)$$
$$\leq \gamma ||u - u_h|| \, ||u - w_h|| \quad \forall \, w_h \in V_h \; ,$$

from where (5.2.7) holds.   □

Estimate (5.2.7) is usually referred to as *Céa lemma*.

Remark 5.2.1 *(The symmetric case).* When $\mathcal{A}(\cdot, \cdot)$ is symmetric, Galerkin method is referred to as the Ritz method. In this case existence and uniqueness of (5.2.3) still follows from the Riesz representation theorem. Furthermore, we notice that (5.2.8) states that the error $u - u_h$ is orthogonal to V_h under the scalar product induced from $\mathcal{A}$. Thus, u_h turns out to be the orthogonal projection of u upon V_h with respect to the same scalar product, and the inequality (5.2.7) follows at once with a better constant $(\gamma/\alpha)^{1/2}$. $\square$

Remark 5.2.2 *(Properties of the stiffness matrix A).* The matrix A defined in (5.2.5) is positive definite, i.e., for any $\boldsymbol{\eta} \in \mathbb{R}^{N_h}$, $\boldsymbol{\eta} \neq \mathbf{0}$, $(A\boldsymbol{\eta}, \boldsymbol{\eta}) > 0$, where $(\cdot, \cdot)$ denotes the euclidean scalar product. Indeed, let $\eta_h \in V_h$ be the function defined as

$$(5.2.9) \qquad \eta_h(\mathbf{x}) = \sum_{j=1}^{N_h} \eta_j \, \varphi_j(\mathbf{x}) \ .$$

Then

$$(A\boldsymbol{\eta}, \boldsymbol{\eta}) = \sum_{i,j=1}^{N_h} \eta_i \, \mathcal{A}(\varphi_j, \varphi_i) \, \eta_j = \mathcal{A}(\eta_h, \eta_h) \ ,$$

and the conclusion holds owing to (5.1.14). In particular, any eigenvalue of A has positive real part.

Thus, the existence and uniqueness of a solution to (5.2.3) can also be proven by a pure algebraic argument without resorting to the Lax-Milgram lemma.

When the bilinear form $\mathcal{A}$ is symmetric, it follows immediately that A is also symmetric. $\square$

5.3 Petrov-Galerkin Method

This is the case in which the numerical problem reads

$$(5.3.1) \qquad \text{find } u_h \in W_h : \mathcal{A}_h(u_h, v_h) = \mathcal{F}_h(v_h) \quad \forall \, v_h \in V_h \ ,$$

where $\{W_h \,|\, h > 0\}$ and $\{V_h \,|\, h > 0\}$ are two families of finite dimensional spaces such that $W_h \neq V_h$ but $\dim W_h = \dim V_h = N_h$, for all $h > 0$.

Problem (5.3.1) can be regarded as an approximation to (5.1.3) provided $\mathcal{A}_h : W_h \times V_h \to \mathbb{R}$ and $\mathcal{F}_h : V_h \to \mathbb{R}$ are convenient approximations to $\mathcal{A}$ and $\mathcal{F}$, respectively, (possibly coinciding with $\mathcal{A}$ and $\mathcal{F}$), while W_h is a subspace of W and V_h one of V. Spaces W and V need not be necessarily different, though.

The algebraic restatement of (5.3.1) is accomplished in the following way. Let $\{\varphi_j \,|\, j = 1, ..., N_h\}$ be a basis of W_h, and $\{\psi_i \,|\, i = 1, ..., N_h\}$ one of V_h. Expanding again the solution u_h of (5.3.1) as in (5.2.4) we obtain the following system of dimension N_h:

$$(5.3.2) \qquad\qquad A\boldsymbol{\xi} = \mathbf{F} \ ,$$

with $\boldsymbol{\xi} = (\xi_j)$, $\mathbf{F} := (\mathcal{F}_h(\psi_i))$, $A_{ij} := \mathcal{A}_h(\varphi_j, \psi_i)$.

For the analysis of stability and convergence of (5.3.1) we have the following theorem, which is due to Babuška (e.g., Babuška and Aziz (1972)).

Theorem 5.3.1 *Under the assumptions of Theorem 5.1.2, suppose further that $\mathcal{F}_h : V_h \to \mathbb{R}$ is a linear map and that $\mathcal{A}_h : W_h \times V_h \to \mathbb{R}$ is a bilinear form satisfying the same properties (5.1.23), (5.1.24) of $\mathcal{A}$ by replacing W with W_h, V with V_h, and the constant α by α_h. Then, there exists a unique solution u_h to (5.3.1) that satisfies*

$$(5.3.3) \qquad |||u_h||| \leq \frac{1}{\alpha_h} \sup_{\substack{v_h \in V_h \\ v_h \neq 0}} \frac{\mathcal{F}_h(v_h)}{||v_h||} \ .$$

Moreover, if u is the solution of (5.1.3), it follows

$$
\begin{aligned}
(5.3.4) \qquad |||u - u_h||| \leq \inf_{w_h \in W_h} &\left[\left(1 + \frac{\gamma}{\alpha_h}\right) |||u - w_h||| \right. \\
&\left. + \frac{1}{\alpha_h} \sup_{\substack{v_h \in V_h \\ v_h \neq 0}} \frac{|\mathcal{A}(w_h, v_h) - \mathcal{A}_h(w_h, v_h)|}{||v_h||} \right] \\
+ \frac{1}{\alpha_h} \sup_{\substack{v_h \in V_h \\ v_h \neq 0}} &\frac{|\mathcal{F}(v_h) - \mathcal{F}_h(v_h)|}{||v_h||} \ .
\end{aligned}
$$

Proof. For any fixed h, existence and uniqueness follow from Theorem 5.1.2. Moreover, (5.3.3) follows using (5.1.23) for $\mathcal{A}_h$ and (5.3.1). Indeed,

$$|||u_h||| \leq \frac{1}{\alpha_h} \sup_{\substack{v_h \in V_h \\ v_h \neq 0}} \frac{\mathcal{A}_h(u_h, v_h)}{||v_h||} = \frac{1}{\alpha_h} \sup_{\substack{v_h \in V_h \\ v_h \neq 0}} \frac{\mathcal{F}_h(v_h)}{||v_h||} \ .$$

Concerning (5.3.4), for all $w_h \in W_h$ and $v_h \in V_h$ we have

$$\mathcal{A}_h(u_h - w_h, v_h) = \mathcal{A}(u - w_h, v_h) + \mathcal{A}(w_h, v_h) - \mathcal{A}_h(w_h, v_h) + \mathcal{F}_h(v_h) - \mathcal{F}(v_h) \ .$$

Using (5.1.22) along with the discrete counterpart of (5.1.23) produces

$$
\alpha_h \ |||u_h - w_h||| \leq \gamma |||u - w_h||| + \sup_{\substack{v_h \in V_h \\ v_h \neq 0}} \left\{ \frac{|\mathcal{A}(w_h, v_h) - \mathcal{A}_h(w_h, v_h)|}{||v_h||} \right.
$$
$$
\left. + \frac{|\mathcal{F}(v_h) - \mathcal{F}_h(v_h)|}{||v_h||} \right\} \ .
$$

Owing to the triangle inequality

$$|||u - u_h||| \leq |||u - w_h||| + |||u_h - w_h||| \ ,$$

the result (5.3.4) follows easily. $\square$

Examples of Petrov-Galerkin approximations are furnished by the so-called τ-method (see, e.g., Canuto, Hussaini, Quarteroni and Zang (1988), Sect. 10.4) and by the "upwind" treatment of advection-diffusion equations (see Section 8.2.2).

Moreover, the Petrov-Galerkin approximation can sometimes take the form

(5.3.5) find $u_h \in W_h : \mathcal{A}_h(u_h, w_h + L_h w_h) = \mathcal{F}_h(w_h + L_h w_h)$ $\forall\, w_h \in W_h$,

where L_h is a suitable operator related to L (according to some criteria). Problem (5.3.5) can be written in the form (5.3.1) provided we set

$$V_h := \{w_h + L_h w_h \mid w_h \in W_h\} .$$

Examples of this kind can be found in Section 14.3, where the so-called GALS, SUPG and DWG methods are used to approximate pure advection equations (see Remark 14.3.1).

5.4 Collocation Method

Collocation methods are used in the framework of several kind of finite dimensional approximations, remarkably for boundary element, finite element and spectral methods. For the sake of brevity, here we focus only on the latter case, where the collocation approach is by far more successful than any other one.

We can therefore restrict our interest to the domain $\Omega = (-1, 1)^d$, $d = 2, 3$, and suppose that we are looking for a discrete solution which is an algebraic polynomial of degree less than or equal to N. Also, we are given $(N + 1)^d$ collocation points $\mathbf{x}_i \in \overline{\Omega}$ which are the nodes of a Gauss-Lobatto integration formula (see (4.5.19) or (4.5.38)).

The spectral collocation solution is a function $u_N \in \mathbb{Q}_N$ (the space of algebraic polynomials of degree less than or equal to N with respect to each single variable x_i, $i = 1, ..., d$) that satisfies

(5.4.1)
$$\begin{cases} L_N u_N = f & \text{at } \mathbf{x}_i \in \overline{\Omega} \setminus \partial\Omega^* \\[2mm] B_N u_N = 0 & \text{at } \mathbf{x}_i \in \partial\Omega^* , \end{cases}$$

where, typically, L_N is an approximation to L obtained by replacing any derivative by the pseudo-spectral derivative, i.e., the exact derivative of the interpolant at the nodes $\{\mathbf{x}_i\}$ (see Sections 4.3.2 and 4.4.2). The same is true for B_N.

Thus, the boundary conditions are satisfied at all nodes lying on $\partial\Omega^*$, whereas the differential equation is enforced at all remaining Gauss-Lobatto nodes. More precisely, we point out that when the condition $Bu = 0$ expresses a natural boundary condition, at all nodes of $\partial\Omega^*$ it is not imposed exactly, but a suitable linear combination between it and the residual of the differential equation is enforced. This translates, at the discrete level, the fulfillment of the boundary conditions in a natural fashion.

As we will see in Section 6.2.2, when the original problem takes the form (5.1.12), this approach leads to solving

$$(5.4.2) \qquad \text{find } u_N \in V_N : \mathcal{A}_N(u_N, v_N) = \mathcal{F}_N(v_N) \quad \forall\, v_N \in V_N \ ,$$

where $V_N = \mathbb{Q}_N \cap V$ and $\mathcal{A}_N$ is an approximation to the form $\mathcal{A}$ in which all integrals are evaluated through the Gauss-Lobatto quadrature formula. The same kind of approximation is carried out for the right hand side.

In the form (5.4.2), which is referred to as the *weak* form of the spectral collocation method, whereas (5.4.1) will be called the *strong* form, this problem appears indeed as a generalized Galerkin method, and goes therefore under the stability and convergence analysis that is presented in next Section.

From the algebraic point of view, let $\{\psi_j\}$ denote the Lagrangian basis of V_N, i.e.,

$$(5.4.3) \qquad \psi_j \in \mathbb{Q}_N : \psi_j(\mathbf{x}_i) = \delta_{ij} \qquad \forall\, i,j = 1, ..., (N+1)^d \ .$$

Setting

$$u_N(\mathbf{x}) = \sum_j \xi_j \psi_j(\mathbf{x}) \ , \quad \text{where} \quad \xi_j = u_N(\mathbf{x}_j) \ ,$$

it follows from (5.4.1) that

$$(5.4.4) \qquad \begin{cases} \displaystyle\sum_j (L_N \psi_j)(\mathbf{x}_i)\,\xi_j = f(\mathbf{x}_i) & \text{for all } i \text{ such that } \mathbf{x}_i \in \overline{\Omega} \setminus \partial\Omega^* \\[2mm] \displaystyle\sum_j (B_N \psi_j)(\mathbf{x}_i)\,\xi_j = 0 & \text{for all } i \text{ such that } \mathbf{x}_i \in \partial\Omega^* \ . \end{cases}$$

This produces a linear system for $\boldsymbol{\xi} = (\xi_j)$, whose matrix can be expressed in terms of the pseudo-spectral derivative matrices. Therefore, the spectral collocation method is also called the *pseudo-spectral* method.

5.5 Generalized Galerkin Method

In this case, the finite dimensional problem reads

$$(5.5.1) \qquad \text{find } u_h \in V_h : \mathcal{A}_h(u_h, v_h) = \mathcal{F}_h(v_h) \quad \forall\, v_h \in V_h \ ,$$

where $\{V_h \mid h > 0\}$ is a family of finite dimensional subspaces of V. Here $\mathcal{A}_h$ and $\mathcal{F}_h$ denote convenient approximations to $\mathcal{A}$ and $\mathcal{F}$, respectively. This is a special subcase of (5.3.1) and includes the collocation method in its weak form (5.4.2). We present, however, a separate analysis, as some differences in the convergence proof arise in the present situation. Other instances of the generalized Galerkin method are provided by the Galerkin finite element method that makes use of numerical integration (see Section 6.2.3) and by stabilization methods for advection-diffusion problems or for the Stokes problem (see Sections 8.3 and 9.4).

We recall that $\mathcal{F}_h(\cdot)$ is a linear form defined over V_h and $\mathcal{A}_h(\cdot,\cdot)$ is a bilinear form defined over $V_h \times V_h$, and they do not necessarily make sense when applied to elements of V. As a matter of fact, typically $\mathcal{A}_h$ and $\mathcal{F}_h$ involve pointvalues of u_h or/and its derivatives, which do not necessarily exist for functions of V (e.g., this is never the case for the three examples discussed above).

The following result, known as first Strang lemma, holds.

Theorem 5.5.1 *Under the assumptions of Theorem 5.1.1, suppose further that $\mathcal{F}_h(\cdot)$ is a linear map and the bilinear form $\mathcal{A}_h(\cdot,\cdot)$ is uniformly coercive over $V_h \times V_h$. This means that there exists $\alpha^* > 0$ such that for all $h > 0$*

$$(5.5.2) \qquad \mathcal{A}_h(v_h, v_h) \geq \alpha^* ||v_h||^2 \quad \forall\, v_h \in V_h \ .$$

Then there exists a unique solution u_h to (5.5.1), which satisfies

$$(5.5.3) \qquad ||u_h|| \leq \frac{1}{\alpha^*} \sup_{\substack{v_h \in V_h \\ v_h \neq 0}} \frac{\mathcal{F}_h(v_h)}{||v_h||} \ ,$$

and, if u is the solution to (5.1.12),

$$(5.5.4) \qquad \begin{aligned} ||u - u_h|| \leq \inf_{w_h \in V_h} & \left[\left(1 + \frac{\gamma}{\alpha^*}\right) ||u - w_h|| \right. \\ & \left. + \frac{1}{\alpha^*} \sup_{\substack{v_h \in V_h \\ v_h \neq 0}} \frac{|\mathcal{A}(w_h, v_h) - \mathcal{A}_h(w_h, v_h)|}{||v_h||} \right] \\ + \frac{1}{\alpha^*} \sup_{\substack{v_h \in V_h \\ v_h \neq 0}} & \frac{|\mathcal{F}(v_h) - \mathcal{F}_h(v_h)|}{||v_h||} \ . \end{aligned}$$

Proof. As in the proof of Theorem 5.3.1, existence, uniqueness and inequality (5.5.3) follow from the Lax-Milgram lemma, owing to (5.5.2). The proof of (5.5.4), although not too different from that of (5.3.4), is reported for the reader's convenience.

Let w_h be an arbitrary element in the space V_h. Setting $\sigma_h := u_h - w_h$, and using (5.5.2) we obtain

$$\begin{aligned} \alpha^* ||\sigma_h||^2 &\leq \mathcal{A}_h(\sigma_h, \sigma_h) \\ &= \mathcal{A}(u - w_h, \sigma_h) + \mathcal{A}(w_h, \sigma_h) - \mathcal{A}_h(w_h, \sigma_h) + \mathcal{F}_h(\sigma_h) - \mathcal{F}(\sigma_h) \ . \end{aligned}$$

Assuming $\sigma_h \neq 0$, due to (5.1.13) one has

$$\alpha^* ||\sigma_h|| \leq \gamma ||u - w_h|| + \frac{|\mathcal{A}(w_h, \sigma_h) - \mathcal{A}_h(w_h, \sigma_h)|}{||\sigma_h||} + \frac{|\mathcal{F}_h(\sigma_h) - \mathcal{F}(\sigma_h)|}{||\sigma_h||}$$

$$\leq \gamma ||u - w_h|| + \sup_{\substack{v_h \in V_h \\ v_h \neq 0}} \frac{|\mathcal{A}(w_h, v_h) - \mathcal{A}_h(w_h, v_h)|}{||v_h||}$$

$$+ \sup_{\substack{v_h \in V_h \\ v_h \neq 0}} \frac{|\mathcal{F}_h(v_h) - \mathcal{F}(v_h)|}{||v_h||} .$$

The above inequality is clearly true also when $\sigma_h = 0$. Combining it with the triangular inequality $||u - u_h|| \leq ||u - w_h|| + ||\sigma_h||$, and taking the infimum with respect to $w_h \in V_h$ we obtain the desired result. $\qquad \square$

The following result will also be useful:

Proposition 5.5.1 *Under the assumptions of Theorem 5.5.1, suppose further that the bilinear form $\mathcal{A}_h(\cdot, \cdot)$ is defined at (u, v_h), where u is the solution to (5.1.12) and $v_h \in V_h$, and satisfies for a suitable $\gamma^* > 0$*

$$(5.5.5) \qquad |\mathcal{A}_h(u - w_h, v_h)| \leq \gamma^* ||u - w_h|| \, ||v_h|| \qquad \forall \, w_h, v_h \in V_h \ ,$$

uniformly with respect to $h > 0$. Then the following convergence estimate holds:

$$(5.5.6) \qquad \begin{aligned} ||u - u_h|| &\leq \left(1 + \frac{\gamma^*}{\alpha^*} \right) \inf_{w_h \in V_h} ||u - w_h|| \\ &+ \frac{1}{\alpha^*} \sup_{\substack{v_h \in V_h \\ v_h \neq 0}} \frac{|\mathcal{A}_h(u, v_h) - \mathcal{F}_h(v_h)|}{||v_h||} . \end{aligned}$$

Proof. Let w_h be an arbitrary element in V_h. We can write:

$$\begin{aligned} \mathcal{A}_h(u_h - w_h, u_h - w_h) &= \mathcal{A}_h(u - w_h, u_h - w_h) \\ &+ \mathcal{F}_h(u_h - w_h) - \mathcal{A}_h(u, u_h - w_h) . \end{aligned}$$

Therefore, using (5.5.2) and (5.5.5) we find

$$||u_h - w_h|| \leq \frac{\gamma^*}{\alpha^*} ||u - w_h|| + \frac{1}{\alpha^*} \sup_{\substack{v_h \in V_h \\ v_h \neq 0}} \frac{|\mathcal{F}_h(v_h) - \mathcal{A}_h(u, v_h)|}{||v_h||} .$$

The triangular inequality gives at once (5.5.6). $\qquad \square$

From the algebraic point of view, if we expand u_h as in (5.2.4), problem (5.5.1) produces a linear system like (5.2.5) where $\mathcal{F}(\cdot)$ and $\mathcal{A}(\cdot, \cdot)$ are replaced by $\mathcal{F}_h(\cdot)$ and $\mathcal{A}_h(\cdot,\cdot)$, respectively. The coerciveness assumption (5.5.2) guarantees that A is a positive definite matrix; moreover, it is also symmetric if $\mathcal{A}_h(\cdot,\cdot)$ is a symmetric form.

Remark 5.5.1 *(Non-conforming approximation).* An extreme case of (5.5.1) is the one in which V_h is not a subspace of the space V, thus the bilinear form $\mathcal{A}$ is not necessarily defined on $V_h \times V_h$. We assume that a norm $||\cdot||_h$ and the approximate bilinear form $\mathcal{A}_h(\cdot,\cdot)$ are defined in $(V + V_h)$, and that the approximate linear functional $\mathcal{F}_h(\cdot)$ is defined on V_h. We require moreover that there exist constants $\alpha^* > 0$, $\gamma^* > 0$ such that for each $h > 0$

$$\mathcal{A}_h(v_h, v_h) \geq \alpha^* ||v_h||_h^2 \quad \forall\, v_h \in V_h$$
$$|\mathcal{A}_h(w, v_h)| \leq \gamma^* ||w||_h\, ||v_h||_h \quad \forall\, w \in (V + V_h)\,,\; v_h \in V_h\;.$$

Then by the so-called second Strang lemma we find the following error estimate

$$||u - u_h||_h \leq \left(1 + \frac{\gamma^*}{\alpha^*}\right) \inf_{w_h \in V_h} ||u - w_h||_h + \frac{1}{\alpha^*} \sup_{\substack{v_h \in V_h \\ v_h \neq 0}} \frac{|\mathcal{A}_h(u, v_h) - \mathcal{F}_h(v_h)|}{||v_h||_h}\,.$$

The proof is quite similar to that of Proposition 5.5.1. See also, e.g., Ciarlet (1991), pp. 212–213. □

A schematic illustration of the various formulations presented in the preceding Sections is reported in Fig. 5.5.1. There we have denoted with $\overline{\Omega}_h$ and $\partial\Omega_h^*$ the set of collocation nodes in $\overline{\Omega}$ and $\partial\Omega^*$, respectively.

5.6 Time-Advancing Methods for Time-Dependent Problems

In this Section we address the issue of time-discretization for initial-boundary value problems. This presentation is very short, and it only serves the purpose of giving the reader a flavour of the several possible ways to face this problem. Detailed discussions are provided all across Part III of this book for several instances.

Keeping in mind the abstract framework of Section 5.1, we now define the following problem

(5.6.1)
$$\begin{cases} \dfrac{\partial u}{\partial t} + Lu = f & \text{in } Q_T := (0, T) \times \Omega \\[2mm] Bu = 0 & \text{on } \Sigma_T^* := (0, T) \times \partial\Omega^* \\[2mm] u = u_0 & \text{on } \Omega,\ \text{for } t = 0\;, \end{cases}$$

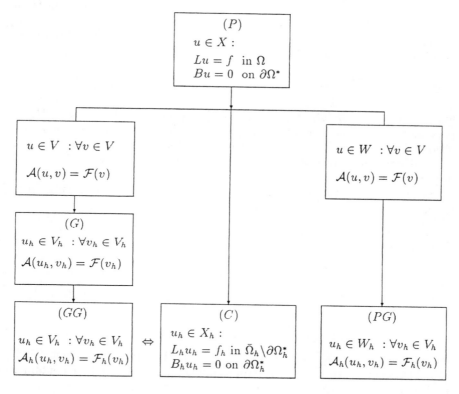

Fig. 5.5.1. Galerkin (G), Petrov-Galerkin (PG), Generalized Galerkin (GG) and Collocation (C) approximations to the boundary value problem (P)

where $T > 0$ is a prescribed time-level, $\frac{\partial}{\partial t}$ denotes time-differentiation, u (the unknown) and f are functions of $t \in (0, T)$ and $\mathbf{x} \in \Omega$, and $u_0 = u_0(\mathbf{x})$ is the assigned initial datum. The differential operator L, as well as the boundary operator B, can now depend on t.

When u is a vector-valued function, the time differentiation needs not necessarily apply to all the components of u. This is, for instance, the case of the Stokes problem that is introduced below. The corresponding initial-boundary value problems are referred to as Differential-Algebraic Equations (see Brenan, Campbell and Petzold (1989)).

We are not going to discuss the general assumptions that ensure existence and uniqueness of a solution to (5.6.1). Nonetheless, we can suppose that a weak formulation of the above problem can be stated. For this, we assume that there exist three Hilbert spaces V, W, H such that W and V are contained into H with dense, continuous inclusion. (If not otherwise specified, $H = L^2(\Omega)$.) The scalar product of H is denoted by $(\cdot, \cdot)$. We assume that

$u_0 \in H$, and $f \in L^2(0, T; H)$. Furthermore, we assume that there exists a bilinear form $\mathcal{A}(\cdot, \cdot)$ continuous on $W \times V$.

The weak formulation of (5.6.1) reads: find $u \in L^2(0, T; W) \cap C^0([0, T]; H)$ (see Section 1.3 for the definition of these spaces) such that

$$(5.6.2) \qquad \frac{d}{dt}(u(t), v) + \mathcal{A}(u(t), v) = \mathcal{F}(t, v) \qquad \forall \, v \in V$$

and $u = u_0$ at $t = 0$. The right hand side is a compact notation for $(f(t), v)$ plus another possible term depending upon non-homogeneous boundary conditions. Equations (5.6.2) are to be intended in the sense of distributions over $(0, T)$. Under the previous assumptions on the data all operations indicated in (5.6.2) make sense.

We now provide three examples which are nothing but the unsteady counterparts of those presented in Section 5.1.

Example 1. The heat equation
The problem we consider reads

$$(5.6.3) \qquad \begin{cases} \dfrac{\partial u}{\partial t} - \Delta u = f & \text{in } Q_T \\[2mm] u = 0 & \text{on } \Sigma_T := (0, T) \times \partial \Omega \\[2mm] u = u_0 & \text{on } \Omega, \text{ for } t = 0 \quad, \end{cases}$$

and will be extensively addressed in Chapter 11. It describes the time evolution of the temperature u of an isotropic and homogeneous medium contained in $\overline{\Omega}$ under an external heat source f; u_0 is the initial temperature.

Its weak formulation is: find $u : Q_T \to \mathbb{R}$ such that $u(t, \cdot) \in H_0^1(\Omega)$ and

$$(5.6.4) \qquad \begin{aligned} \int_\Omega \frac{\partial u}{\partial t}(t, \mathbf{x})\, v(\mathbf{x})\, d\mathbf{x} &+ \int_\Omega \nabla u(t, \mathbf{x}) \cdot \nabla v(\mathbf{x})\, d\mathbf{x} \\ &= \int_\Omega f(t, \mathbf{x})\, v(\mathbf{x})\, d\mathbf{x} \quad \forall \, v \in H_0^1(\Omega) \end{aligned}$$

for almost every $t \in (0, T)$; moreover, $u = u_0$ at $t = 0$. It is therefore a special case of (5.6.2), with symbols' explanation provided in (5.1.6).

Example 2. The unsteady Stokes problem
The time-dependent version of problem (5.1.7) is

$$(5.6.5) \qquad \begin{cases} \dfrac{\partial \mathbf{u}}{\partial t} - \nu \Delta \mathbf{u} + \nabla p = \mathbf{f} & \text{in } Q_T \\[2mm] \operatorname{div} \mathbf{u} = 0 & \text{in } Q_T \\[2mm] \mathbf{u} = \mathbf{0} & \text{on } \Sigma_T \\[2mm] \mathbf{u} = \mathbf{u}_0 & \text{on } \Omega, \text{ for } t = 0 \quad. \end{cases}$$

These equations describe the unsteady motion of a viscous incompressible fluid confined into Ω, subject to a volumic density $\mathbf{f}$ of external forces and having an initial velocity $\mathbf{u}_0$, under the assumption that this motion is slow. Their derivation within the framework of time-dependent Navier-Stokes equations is presented in Chapter 13.

The weak form of (5.6.5) reads: find $\mathbf{u} : Q_T \to \mathbb{R}^d$ and $p : Q_T \to \mathbb{R}$ such that $\mathbf{u}(t, \cdot) \in (H_0^1(\Omega))^d$, $p(t, \cdot) \in L_0^2(\Omega)$ and

(5.6.6)
$$
\begin{cases}
\displaystyle\int_\Omega \frac{\partial \mathbf{u}}{\partial t}(t, \mathbf{x}) \cdot \mathbf{v}(\mathbf{x}) \, dx + \nu \int_\Omega \nabla \mathbf{u}(t, \mathbf{x}) \cdot \nabla \mathbf{v}(\mathbf{x}) \, dx \\[2mm]
\qquad - \int_\Omega p(t, \mathbf{x}) \operatorname{div} \mathbf{v}(\mathbf{x}) \, dx \\[2mm]
\qquad = \int_\Omega \mathbf{f}(t, \mathbf{x}) \cdot \mathbf{v}(\mathbf{x}) \, dx \qquad \forall \, \mathbf{v} \in (H_0^1(\Omega))^d \\[2mm]
\int_\Omega q(\mathbf{x}) \operatorname{div} \mathbf{u}(t, \mathbf{x}) \, dx = 0 \qquad \forall \, q \in L_0^2(\Omega)
\end{cases}
$$

for almost every $t \in (0, T)$; moreover, $\mathbf{u} = \mathbf{u}_0$ at $t = 0$. Again, (5.6.6) can be written in the abstract form (5.6.2) by adopting the notations (5.1.9).

Example 3. A linear transport problem
Under the notation adopted in (5.1.10), the transport initial-boundary value problem is (see also Chapter 14):

(5.6.7)
$$
\begin{cases}
\dfrac{\partial u}{\partial t} + \mathbf{a} \cdot \nabla u = f & \text{in } Q_T \\[2mm]
u = \varphi & \text{on } \Sigma_T^{in} := (0, T) \times \partial\Omega^{in} \\[2mm]
u = u_0 & \text{on } \Omega, \text{ for } t = 0 \quad .
\end{cases}
$$

The above equations describe the transport of a quantity u whose speed of propagation at any point $\mathbf{x}$ is given by the vector function $\mathbf{a}$. For the sake of simplicity, we can assume that $\mathbf{a}$ is smooth and doesn't depend on t so that the inflow boundary $\partial\Omega^{in}$ doesn't vary along the time.

The weak form of (5.6.7) reads: find $u : Q_T \to \mathbb{R}$ such that $u(t, \cdot) \in L^2(\Omega)$ and

(5.6.8)
$$
\begin{aligned}
\int_\Omega \frac{\partial u}{\partial t}(t, \mathbf{x}) \, v(\mathbf{x}) \, dx &- \int_\Omega u(t, \mathbf{x}) \operatorname{div}(\mathbf{a} v)(\mathbf{x}) \, dx \\
&+ \int_{\partial\Omega \setminus \partial\Omega^{in}} (\mathbf{a} \cdot \mathbf{n})(\mathbf{x}) \, u(t, \mathbf{x}) \, v(\mathbf{x}) \, dx \\
= \int_\Omega f(t, \mathbf{x}) \, v(\mathbf{x}) \, dx & \\
&- \int_{\partial\Omega^{in}} (\mathbf{a} \cdot \mathbf{n})(\mathbf{x}) \, \varphi(t, \mathbf{x}) \, v(\mathbf{x}) \, dx \quad \forall \, v \in H^1(\Omega)
\end{aligned}
$$

for almost every $t \in (0, T)$; moreover $u = u_0$ at $t = 0$. This is an instance of (5.6.2), as it follows from (5.1.11).

5.6.1 Semi-Discrete Approximation

The abstract problem (5.6.2) needs to be discretized with respect to both the time and space variables.

The space discretization can be carried out as done in the case of a steady problem. For instance, the Galerkin approximation to (5.6.2) is obtained by generalizing the approach (5.2.3). The resulting problem reads: for each $t \in [0,T]$ find $u_h(t,\cdot) \in V_h \subset V$ such that

$$(5.6.9) \quad \frac{d}{dt}(u_h(t),v_h) + \mathcal{A}(u_h(t),v_h) = \mathcal{F}(t,v_h) \quad \forall \, v_h \in V_h \ , \quad t \in (0,T) \ ,$$

with $u_h(0) = u_{0,h}$, the latter being a suitable element of V_h that approximates the initial value u_0. (Typically, the initial value is provided either from the interpolant of u_0, or its projection with respect to the scalar product of $L^2(\Omega)$ or else that of V.)

Other kind of approximations, based upon Petrov-Galerkin, or generalized Galerkin, or else the collocation method, can be introduced similarly.

Problem (5.6.9) is often called a *semi-discrete* (or *continuous-in-time*) approximation to (5.6.2). Setting

$$(5.6.10) \quad u_h(t,\mathbf{x}) = \sum_{j=1}^{N_h} \xi_j(t)\,\varphi_j(\mathbf{x}) \quad \text{for } t \geq 0 \ , \quad u_{0,h}(\mathbf{x}) = \sum_{j=1}^{N_h} \xi_{0,j}\,\varphi_j(\mathbf{x})$$

where $\{\varphi_j\}$ still denotes a basis of V_h, it is easily seen that (5.6.9) produces a system of ordinary differential equations for the unknown vectors $\mathbf{u}(t)$ that reads

$$(5.6.11) \quad M\frac{d\boldsymbol{\xi}(t)}{dt} + A\boldsymbol{\xi}(t) = \mathbf{F}(t) \quad , \quad \boldsymbol{\xi}(0) = \boldsymbol{\xi}_0 \ .$$

For each $t \geq 0$, $\boldsymbol{\xi}(t) = (\xi_j(t))$, $\mathbf{F}(t) := (\mathcal{F}(t,\varphi_i))$, while $\boldsymbol{\xi}_0 = (\xi_{0,j})$ and

$$(5.6.12) \quad M_{ij} := (\varphi_i,\varphi_j) \ , \quad A_{ij} := \mathcal{A}(\varphi_j,\varphi_i) \ , \quad i,j = 1,...,N_h \ .$$

The stiffness matrix A is the same as that introduced in the steady case (see (5.2.5)). It is independent of t if the operator L is independent, too. This will be our assumption throughout this Section, for the sake of exposition. The symmetric and positive definite matrix M, which is always independent of t, is called the *mass matrix*.

5.6.2 Fully-Discrete Approximation

The discretization in time of (5.6.9) can be accomplished in several possible ways. To start with, we partition the time-interval $[0,T]$ into $\mathcal{N}$ subintervals $[t_n,t_{n+1}]$ of length $\Delta t = T/\mathcal{N}$, with $t_0 = 0$ and $t_\mathcal{N} = T$. We then look for an approximation to $u(t)$ at each time-level t_n, and denote by u_h^n the corresponding finite dimensional function. In particular, u_h^0 is a convenient approximation to u_0. A classical and widespread practice to achieve a full discretization of problem (5.6.9) is to resort to a discretization to the time-derivative by a finite difference scheme. The latter is nothing but a divided quotient that is obtained from the Taylor expansion formula. The basic idea is illustrated on the Cauchy problem for the ordinary differential equation

(5.6.13)
$$\begin{cases} \dfrac{dy}{dt}(t) = \psi(t, y(t)) \ , \quad 0 < t < T \\[2mm] y(0) = y_0 \ , \end{cases}$$

where ψ is a continuous function from $[0, T] \times \mathbb{R}$ into $\mathbb{R}$ which is further Lipschitz continuous with respect to y, uniformly in $t \in [0, T]$.

Among the many possible finite difference schemes, we start illustrating the simple θ-scheme, $0 \le \theta \le 1$, according to which the above problem is replaced by

(5.6.14) $$\frac{1}{\Delta t}(y^{n+1} - y^n) = \theta \, \psi(t_{n+1}, y^{n+1}) + (1 - \theta) \, \psi(t_n, y^n)$$

for $n = 0, 1, ..., \mathcal{N} - 1$, with $y^0 = y_0$ and y^n approximating $y(t_n)$.

The θ-scheme applied to (5.6.9) produces

(5.6.15) $$\frac{1}{\Delta t}(u_h^{n+1} - u_h^n, v_h) + \mathcal{A}(\theta u_h^{n+1} + (1 - \theta) u_h^n, v_h)$$
$$= \theta \, \mathcal{F}(t_{n+1}, v_h) + (1 - \theta) \, \mathcal{F}(t_n, v_h) \quad \forall \, v_h \in V_h$$

for $n = 0, 1, ..., \mathcal{N} - 1$, with $u_h^0 = u_{0,h}$. The extreme cases $\theta = 0$ and $\theta = 1$ define the well-known forward and backward Euler methods, respectively, which are first order accurate with respect to the time-step Δt. The scheme associated to the case $\theta = 1/2$ is known as the Crank-Nicolson one, and it is potentially second order accurate.

Other second order schemes can be obtained by resorting to two-step methods, i.e., involving the three time levels t_{n+1}, t_n and t_{n-1}. A first example is provided by the second order Adams-Bashforth method

(5.6.16) $$\frac{1}{\Delta t}(y^{n+1} - y^n) = \frac{3}{2}\psi(t_n, y^n) - \frac{1}{2}\psi(t_{n-1}, y^{n-1}) \ , \quad n = 1, 2, ..., \mathcal{N} - 1 \ .$$

Another instance is given by the second order backward differentiation method

(5.6.17) $$\frac{1}{2\Delta t}(3y^{n+1} - 4y^n + y^{n-1}) = \psi(t_{n+1}, y^{n+1}) \ , \quad n = 1, 2, ..., \mathcal{N} - 1 \ .$$

Both schemes require an initialization of y^1 which must be also second order accurate.

The applications of (5.6.16) and (5.6.17) to the discretization of (5.6.9) is straightforward.

An order of accuracy equal to p is formally achievable by the Taylor formula, under the assumption that the function u is continuous, together with its derivatives of order $p + 1$.

For the time being, we confine ourselves to notice that the algebraic restatement of (5.6.15) reads

$$(5.6.18) \qquad M\boldsymbol{\xi}^{n+1} + \theta \, \Delta t \, A\boldsymbol{\xi}^{n+1} = \boldsymbol{\eta}^{n+1} \quad , \quad n = 0, 1, ..., \mathcal{N} - 1 \quad ,$$

with

$$\boldsymbol{\eta}^{n+1} := \theta \, \Delta t \, \mathbf{F}(t_{n+1}) + (1 - \theta) \, \Delta t \, \mathbf{F}(t_n) + M\boldsymbol{\xi}^n + (\theta - 1) \, \Delta t \, A\boldsymbol{\xi}^n$$

and $\boldsymbol{\xi}^0 = \boldsymbol{\xi}_0$ (for notation see (5.6.10)-(5.6.12)). The same algebraic system is obtained if the finite difference scheme is applied at first to (5.6.2) and then the resulting steady problem is approximated in space by the Galerkin method.

When $\theta = 0$, the above scheme is explicit (up to the "inversion" of the mass matrix M, which, when not diagonal, is often *lumped* to a diagonal form, see Section 11.4). For any other value of θ, (5.6.18) defines $\boldsymbol{\xi}^{n+1}$ implicitly. Similarly, (5.6.16) is explicit while (5.6.17) is implicit.

From the computational point of view, explicit methods are less challenging than implicit ones. The drawback is that, in general, they are only *conditionally stable*, i.e., they require a time-step Δt sufficiently small with respect to the inverse of the modulus of the largest eigenvalue of A, and therefore small enough compared to h. On the contrary, some implicit methods can be *unconditionally stable*. In such cases the choice of Δt is dictated from accuracy requirements only.

Along Part III of this book the concept of *temporal stability*, that we have advocated here, will be made precise for any kind of time-dependent problem we are going to consider. For the time being, we state that, qualitatively, a method is stable for the Cauchy problem

$$(5.6.19) \qquad \begin{cases} \dfrac{d\boldsymbol{\eta}}{dt}(t) = B\boldsymbol{\eta}(t) \quad , \quad 0 < t < \infty \\[2mm] \boldsymbol{\eta}(0) = \boldsymbol{\eta}_0 \quad , \end{cases}$$

B a given matrix of dimension m, if there exist positive constants C, λ and δ, independent of Δt, such that

$$(5.6.20) \qquad |\boldsymbol{\eta}^n| \leq C \, e^{\lambda t_n} \, |\boldsymbol{\eta}_0| \quad , \quad n \geq 0 \quad ,$$

for all $0 < \Delta t \leq \delta$. Here $\boldsymbol{\eta}^n$ is the computed solution at the time $t_n = n\Delta t$, and $|\cdot|$ is any vector norm in $\mathbb{R}^m$.

The exponential growth allowed by (5.6.20) may be unsuitable for those problems whose solution $\boldsymbol{\eta}(t)$ is bounded for all $t > 0$. In these cases the concept of *absolute stability* is a more appropriate requirement for the time-discretization method. This concept is defined as follows: consider the scalar model problem (5.6.13) where $\psi(t, y(t)) = \mu y(t)$, $\mu \in \mathbb{C}$. A numerical method is said *absolutely stable* for an assigned value $\mu \Delta t$ if the approximate solution y^n generated by that method is vanishing as n goes to ∞. The *region of absolute stability* is the set of all $\mu \Delta t$ for which the method is absolutely stable. Moreover, a method is called *A-stable* if its region of absolute stability includes the region $\{z \in \mathbb{C} \,|\, \mathrm{Re}\ z < 0\}$. It is readily proven that the θ-method is *A*-stable for $\theta \geq 1/2$ (see, e.g., Lambert (1991), p. 244).

If instead of (5.6.19) we consider the semi-discrete problem (5.6.9) (with $\mathcal{F}(\cdot, \cdot) = 0$ and $T = \infty$), a time-advancing method is *stable* with respect to a spatial norm $\| \cdot \|$ in V_h if

$$(5.6.21) \qquad \|u_h^n\| \leq C e^{\lambda t_n} \|u_h^0\| \quad , \quad n \geq 0 \ ,$$

for all $0 < \Delta t \leq \delta_h$, where C, λ and δ_h are independent of Δt and both C and λ are independent of h. If δ_h is bounded from below independently of h, the method is *unconditionally stable*. Otherwise, the functional dependence of δ_h from h is called the *stability limit* of the numerical method. Notice that all norms in V_h are equivalent but (in general) non-uniformly with respect to h. Hence (5.6.21) can be only valid for suitable norms and not for all norms in V_h.

5.7 Fractional-Step and Operator-Splitting Methods

Splitting methods provide another tool for the time-discretization of initial-boundary value problems. They are also often used to achieve the solution of stationary boundary value problems like (5.1.1) as steady state of the corresponding time-dependent problem (5.6.1). In either case, the underlying assumption is that the spatial differential operator can be split into a sum of two (or more) components of simpler structure, which are then successively integrated in time producing less complicated equations.

We consider a problem of the form

$$(5.7.1) \qquad \frac{d\varphi}{dt} + \mathcal{L}\varphi = \psi \quad , \qquad t > 0 \ ,$$

where $\mathcal{L}$ is a matrix that arises from the space discretization of a given differential operator, accounting also for the boundary conditions, ψ is the corresponding right hand side, and φ the unknown solution.

For Galerkin approximations $\mathcal{L} = M^{-1}A$ (see (5.6.11)). However, we can also take $\mathcal{L} = M^{-1/2}AM^{-1/2}$ having operated a change of variable through the square root of the symmetric and positive definite matrix M. In the latter case, if A is symmetric, then $\mathcal{L}$ is symmetric, too.

For spectral collocation approximations the mass matrix M is diagonal or even the identity I (see Section 11.4), therefore in the latter case one can simply consider $\mathcal{L} = A$.

For the sake of simplicity, we assume that $\mathcal{L}$ is the sum of two components only

$$(5.7.2) \qquad\qquad \mathcal{L} = \mathcal{L}_1 + \mathcal{L}_2 \ ,$$

each of them being independent of the time variable. Most existing theory for splitting methods is available only for the case in which

(5.7.3)   both $\mathcal{L}_1$ and $\mathcal{L}_2$ are positive (or non-negative) definite matrices .

Splitting methods, however, find application in a much wider class of problems.

We need also to mention that quite often the splitting is not operated at the algebraic level, as done in (5.7.2), but rather at the differential stage

$$(5.7.4) \qquad\qquad L = L_1 + L_2 \ ,$$

i.e., on the differential operator itself. This approach is even more fascinating than the others, as often it is based on physical considerations: L_1 and L_2 may have specific physical meaning, such as in the case of diffusion and transport processes. The inherent mathematical difficulty however is that (5.7.4) generally requires a splitting between the boundary conditions as well, in order for endowing either L_1 and L_2 with consistent boundary data. We will return on this issue in Chapters 12 and 13, where some fractional-step methods for advection-diffusion equations and for the Navier-Stokes system will be considered.

For the time being, we stay with the situation (5.7.1)-(5.7.2). As usual, we set $t_0 = 0$ and $t_{n+1} = t_n + \Delta t$, for $n \geq 0$. Moreover, we take as φ^0 a convenient approximation of the initial data u_0.

First of all we consider the *Yanenko* splitting:

$$(5.7.5) \qquad \begin{cases} \dfrac{\varphi^{n+1/2} - \varphi^n}{\tau} + \mathcal{L}_1 \varphi^{n+1/2} = 0 \\[2mm] \dfrac{\varphi^{n+1} - \varphi^{n+1/2}}{\tau} + \mathcal{L}_2 \varphi^{n+1} = \psi^n \end{cases}$$

for $n \geq 0$ and $\tau = \Delta t$. This method, that produces a couple of implicit problems, is first order accurate with respect to Δt provided problem's data are sufficiently smooth. In fact, eliminating $\varphi^{n+1/2}$ we have

$$\frac{\varphi^{n+1} - \varphi^n}{\Delta t} + \mathcal{L}\varphi^{n+1} = \psi^n + \Delta t \mathcal{L}_1(\psi^n - \mathcal{L}_2 \varphi^{n+1}) \ .$$

Furthermore, under restriction (5.7.3) the splitting scheme (5.7.5) is unconditionally stable (Marchuk (1990), p. 232).

In the case in which L is the Laplace operator in two dimensions and $\mathcal{L}_i$ is the discrete counterpart of the second derivative $\frac{\partial^2}{\partial x_i^2}$, $i = 1, 2$, the above splitting scheme produces a classical *alternating direction* method. Its extension to a three-stage scheme for a three dimensional Laplace operator is straightforward.

The *predictor-corrector* scheme is a three-step method that reads

$$(5.7.6) \quad \begin{cases} \dfrac{\varphi^{n+1/4} - \varphi^n}{\tau} + \mathcal{L}_1\,\varphi^{n+1/4} = \psi^{n+1/2} \\[2mm] \dfrac{\varphi^{n+1/2} - \varphi^{n+1/4}}{\tau} + \mathcal{L}_2\,\varphi^{n+1/2} = 0 \\[2mm] \dfrac{\varphi^{n+1} - \varphi^n}{2\tau} + \mathcal{L}\,\varphi^{n+1/2} = \psi^{n+1/2} \end{cases}$$

where $\tau = \Delta t/2$ and $\psi^{n+1/2}$ refers to the intermediate time-level $t_{n+1/2} = t_n + \Delta t/2$. The predictor provides a guess $\varphi^{n+1/2}$ which is first order accurate for the solution at $t_{n+1/2}$ through two implicit equations, while the second-order corrector produces the new solution φ^{n+1} explicitly.

The method is formally second order accurate. Indeed, eliminating $\varphi^{n+1/4}$ we obtain

$$\frac{\varphi^{n+1/2} - \varphi^n}{\Delta t} + \frac{1}{2}\mathcal{L}\,\varphi^{n+1/2} + \frac{\Delta t}{4}\mathcal{L}_1\,\mathcal{L}_2\,\varphi^{n+1/2} = \frac{1}{2}\psi^{n+1/2} \quad,$$

and using the third equation in (5.7.6) to express $\mathcal{L}\,\varphi^{n+1/2}$ we find

$$\varphi^{n+1/2} = \frac{1}{2}(\varphi^{n+1} + \varphi^n) - \frac{(\Delta t)^2}{4}\mathcal{L}_1\,\mathcal{L}_2\,\varphi^{n+1/2} \quad.$$

As a consequence, the relation

$$\frac{\varphi^{n+1} - \varphi^n}{\Delta t} + \mathcal{L}\left(\frac{\varphi^{n+1} + \varphi^n}{2}\right) = \psi^{n+1/2} + \frac{(\Delta t)^2}{4}\mathcal{L}\,\mathcal{L}_1\,\mathcal{L}_2\,\varphi^{n+1/2}$$

holds. Since

$$\varphi^{n+1/2} = \left(I + \frac{\Delta t}{2}\mathcal{L}_2\right)^{-1}\left(I + \frac{\Delta t}{2}\mathcal{L}_1\right)^{-1}\left(\varphi^n + \frac{\Delta t}{2}\psi^{n+1/2}\right)$$
$$= \varphi^n + O(\Delta t) \quad,$$

we finally have

$$\frac{\varphi^{n+1} - \varphi^n}{\Delta t} + \mathcal{L}\left(\frac{\varphi^{n+1} + \varphi^n}{2}\right) = \psi^{n+1/2} + \frac{(\Delta t)^2}{4}\mathcal{L}\,\mathcal{L}_1\,\mathcal{L}_2\,\varphi^n + O(\Delta t)^3 \quad,$$

and we recognize the Crank-Nicolson scheme up to $O(\Delta t)^2$. If $\mathcal{L}_1$ and $\mathcal{L}_2$ commute, one indeed finds the expression

$$\frac{\varphi^{n+1} - \varphi^n}{\Delta t} + \mathcal{L}\left(\frac{\varphi^{n+1} + \varphi^n}{2}\right)$$
$$= \psi^{n+1/2} + \frac{(\Delta t)^2}{4}\mathcal{L}_1\,\mathcal{L}_2\left(\psi^{n+1/2} - \frac{\varphi^{n+1} - \varphi^n}{\Delta t}\right) \quad.$$

The predictor-corrector scheme is unconditionally stable provided (5.7.3) holds, even if the corrector step may be unstable when considered separately (see Marchuk (1990), p. 271).

The splitting method of *Peaceman and Rachford* reads

$$(5.7.7) \quad \begin{cases} \dfrac{\varphi^{n+1/2} - \varphi^n}{\tau} + \mathcal{L}_1 \varphi^{n+1/2} = \psi^{n+1/2} - \mathcal{L}_2 \varphi^n \\[2mm] \dfrac{\varphi^{n+1} - \varphi^{n+1/2}}{\tau} + \mathcal{L}_2 \varphi^{n+1} = \psi^{n+1/2} - \mathcal{L}_1 \varphi^{n+1/2} \end{cases}$$

with $\tau = \Delta t/2$. This method is second order accurate (see (5.7.10) below). If $\mathcal{L}_i$ are suitable approximations of the second order derivative $\frac{\partial^2}{\partial x_i^2}$, this scheme is unconditionally stable. However, this is no longer true in three dimensions.

A similar method is that of *Douglas and Rachford*

$$(5.7.8) \quad \begin{cases} \dfrac{\varphi^{n+1/2} - \varphi^n}{\tau} + \mathcal{L}_1 \varphi^{n+1/2} = \psi^n - \mathcal{L}_2 \varphi^n \\[2mm] \dfrac{\varphi^{n+1} - \varphi^{n+1/2}}{\tau} + \mathcal{L}_2 \varphi^{n+1} = \mathcal{L}_2 \varphi^n \end{cases}$$

with $\tau = \Delta t$. This scheme is only first order accurate (see (5.7.10) below), but is unconditionally stable even when generalized to the case $\mathcal{L} = \mathcal{L}_1 + \mathcal{L}_2 + \mathcal{L}_3$, $\mathcal{L}_i$ a suitable approximation of $\frac{\partial^2}{\partial x_i^2}$ (see Marchuk (1990), pp. 277–279).

A splitting scheme that generalizes both (5.7.7) and (5.7.8) is given by the *Il'in* method

$$(5.7.9) \quad \begin{cases} \dfrac{\varphi^{n+1/2} - \varphi^n}{\tau} + \mathcal{L}_1 \varphi^{n+1/2} = \psi^{n+1/2} - \mathcal{L}_2 \varphi^n \\[2mm] \dfrac{\varphi^{n+1} - \varphi^{n+1/2}}{\tau} + \mathcal{L}_2(\varphi^{n+1} - \varphi^n) = \rho \dfrac{\varphi^{n+1/2} - \varphi^n}{\tau} \end{cases}$$

with $\tau = \Delta t/(1 + \rho)$, and $\rho \in (-1, 1]$ is a parameter. When $\rho = 0$ and $\rho = 1$ we recognize in (5.7.9) the schemes (5.7.8) (with $\psi^{n+1/2}$ instead of ψ^n at the right hand side) and (5.7.7), respectively.

Eliminating $\varphi^{n+1/2}$, one obtains

$$(5.7.10) \quad \begin{aligned} &\dfrac{\varphi^{n+1} - \varphi^n}{\Delta t} + \mathcal{L}\left(\dfrac{1}{1 + \rho} \varphi^{n+1} + \dfrac{\rho}{1 + \rho} \varphi^n \right) \\[2mm] &\quad = \psi^{n+1/2} - \dfrac{(\Delta t)^2}{(1 + \rho)^2} \mathcal{L}_1 \mathcal{L}_2 \left(\dfrac{\varphi^{n+1} - \varphi^n}{\Delta t} \right) . \end{aligned}$$

Hence this scheme is second order accurate for $\rho = 1$ (in this case it coincides with the Peaceman-Rachford scheme), and first order accurate in all other cases. Moreover, when $\mathcal{L}_i$ satisfy (5.7.3), it is unconditionally stable for any value of $\rho \in (-1, 1]$ (see Yanenko (1971), p. 65; Marchuk (1990), p. 291).

A three-stage splitting scheme is the so-called ϑ-method (or *Strang* method, see Strang (1968)) that reads

(5.7.11)
$$\begin{cases} \dfrac{\varphi^{n+\vartheta} - \varphi^n}{\vartheta \tau} + \mathcal{L}_1 \varphi^{n+\vartheta} = \boldsymbol{\psi}^n - \mathcal{L}_2 \varphi^n \\[2ex] \dfrac{\varphi^{n+1-\vartheta} - \varphi^{n+\vartheta}}{(1-2\vartheta)\tau} + \mathcal{L}_2 \varphi^{n+1-\vartheta} = \boldsymbol{\psi}^{n+1-\vartheta} - \mathcal{L}_1 \varphi^{n+\vartheta} \\[2ex] \dfrac{\varphi^{n+1} - \varphi^{n+1-\vartheta}}{\vartheta \tau} + \mathcal{L}_1 \varphi^{n+1} = \boldsymbol{\psi}^{n+1-\vartheta} - \mathcal{L}_2 \varphi^{n+1-\vartheta} \end{cases}$$

where $\tau = \Delta t$, while ϑ is a positive real number strictly less than $1/2$. This scheme is second order accurate for $\vartheta = 1 - \frac{\sqrt{2}}{2}$, first order accurate otherwise; on the other hand, a complete analysis of the stability is not still available. For some additional comments, see Glowinski and Le Tallec (1989).

As we have mentioned previously, if $\boldsymbol{\psi}$ does not depend on t the above splitting methods (except the Yanenko and the predictor-corrector schemes) can be used to achieve the solution of the steady problem $\mathcal{L}\boldsymbol{\chi} = \boldsymbol{\psi}$. Indeed, whenever φ^n is convergent, say $\varphi^n \to \varphi^\infty$, it follows that $\varphi^\infty = \boldsymbol{\chi}$. In such kind of application the parameter τ should no longer be regarded as related to the time-step but rather as a parameter to accelerate the convergence process. In this regard, see also Section 2.3.5.

We would like to mention that many other splitting methods are available in the literature; we refer the interested reader to Yanenko (1971), Marchuk (1990). Often, such methods are tailored on specific equations, in order to catch the different nature of the physical processes involved in the whole problem. We will present some interesting examples in Chapters 12 and 13.

Remark 5.7.1 *(About terminology)*. Operator-splitting or fractional-step are often used as synonymous in the literature. For the sake of precision, however, operator-splitting should better refer to the decomposition (5.7.4) of the given differential operator (or the one (5.7.2) of the associated matrix), while fractional-step is anyone of the methods (like, e.g., Douglas-Rachford, Peaceman-Rachford, etc.) that are used to advance the problem from t_n to t_{n+1}, on the basis of the operator-splitting that has been adopted. Hence, on a given operator-splitting several possible fractional-step schemes can be applied. Similarly, the same fractional-step scheme can be used for different operator-splittings (i.e., for both $L = L_1 + L_2$ or $L = L_3 + L_4$, with $L_1 \neq L_3$ and $L_1 \neq L_4$). □

5.8 Complements

Many numerical methods have been neglected in this Chapter, as we do not attempt to be exhaustive.

Finite difference methods perhaps provide the most popular tool for space discretization. We will extensively address this method in Chapters 8 and 14. For a systematic approach to this subject, the reader should refer to the very abundant available literature. In particular, we mention the classical monographs by Forsythe and Wasow (1960), Godunov and Ryabenkiĭ (1987), Richtmyer and Morton (1967) and Samarskiĭ and Andreev (1978), as well as the more recent ones by Heinrich (1987) and Strikwerda (1989), and the references therein.

A very successful approach for the approximation of some integral equations arising from certain elliptic boundary value problems (e.g., the Laplace equation in exterior domains) is the so-called *boundary element* method. Passing through the fundamental solution of the differential operator, the problem is reformulated as an integral one over the domain's boundary, and there it is approximated by seeking the solution in suitable finite dimensional (piecewise-polynomial) subspaces. The interested reader may refer, e.g., to Wendland (1979) and Brebbia, Telles and Wrobel (1984). A more recent monograph on the subject is the one by Chen and Zhou (1992) (see also the references therein).

The discretization by *finite volumes* stems from conservation laws expressed in integral form. Its principal interest is therefore in the field of computational fluid dynamics. We will make a short presentation of this method (and provide many references) in Section 14.6.

Special space-time integration techniques include the so-called *discontinuous Galerkin* and *Taylor-Galerkin* methods. Both methods have been applied to hyperbolic initial-boundary value problems, and will be presented in Sections 14.3.3 and 14.3.4. The discontinuous Galerkin method can also be used in the approximation of parabolic problems (see Sections 11.3.1 and 12.3).

All the convergence results presented so far in Sections 5.2-5.5 are based on the so-called *a priori error analysis*: a numerical method does converge provided it is stable and consistent with the original boundary value problem. This is a generalization of the Lax-Richtmyer theorem of equivalence, which states that, for linear well-posed initial value problems, a consistent difference scheme is convergent if, and only if, it is stable. From case to case, the consistency assumption may take the form (5.2.2) and (5.2.8) for the Galerkin approximation, or (5.2.2) jointly to the fact that as $h \to 0$ the approximate forms $\mathcal{A}_h$ and $\mathcal{F}_h$ tend to $\mathcal{A}$ and $\mathcal{F}$, respectively, in the case of Petrov-Galerkin and generalized Galerkin methods. We will see in the next Chapters that error estimates can be drawn from this analysis, showing that a suitable norm of the error can be bounded by a constant C_{pr} multiplied by a certain power of h.

A complementary approach could be based on the so-called *a-posteriori error analysis*. The latter can only be used after that a computation has been really carried out (i.e., a-posteriori) and aims at providing an upper bound of the error in terms of a constant C_{post} (which can be sharply estimated) multiplied by the residual of the problem (which can be easily computed). As opposed to the a-priori one, the a-posteriori analysis states that if a numerical method is stable and the solution that it provides has a vanishing residual, then the scheme is convergent. The a-posteriori analysis can be of remarkable interest especially for those complex problems (e.g., nonlinear ones) for which some level of adaptivity is required all along the resolution process, in order to keep the error under a desired tolerance.

Developing the a-posteriori error analysis or proposing and analyzing adaptive methods is beyond the scope of this book. However, we aware the reader that this is a very interesting area of investigation, and references can be found, e.g., in the monograph by Eriksson, Estep, Hansbo and Johnson (1995). Let us also mention the earlier papers by Babuška (1976), Babuška and Rheinboldt (1978a, 1978b), Bank and Wieser (1985), Babuška and Miller (1987), Zienkiewicz and Zhu (1987), Ainsworth, Zhu, Craig and Zienkiewicz (1989), Verfürth (1989), Ainsworth and Oden (1993), and the books edited by Babuška, Chandra and Flaherty (1983), Babuška, Zienkiewicz, Gago and de Arantes e Oliveira (1986), Oden (1990) and Demkowicz, Oden and Babuška (1992).

A brief presentation of adaptive methods based on a-priori error estimates can be found in Johnson (1987), Sect. 4.6.

6. Elliptic Problems: Approximation by Galerkin and Collocation Methods

This Chapter deals with second order linear elliptic equations. We present the variational formulation of some classical boundary value problems, accounting for several kind of boundary conditions, and derive existence and uniqueness of the solution. Then we approximate these problems by Galerkin and collocation methods, in the framework of both finite element and spectral methods. For each kind of discretization we analyze its stability and convergence properties, as well as its algorithmic aspects.

6.1 Problem Formulation and Mathematical Properties

Let Ω be a bounded domain of $\mathbb{R}^d$, $d = 2, 3$, with a Lipschitz continuous boundary $\partial\Omega$, and denote by $\mathbf{n}$ the unit outward normal direction on $\partial\Omega$.

We consider the second order linear operator L defined by

$$(6.1.1) \qquad Lw := -\sum_{i,j=1}^{d} D_i(a_{ij}D_jw) + \sum_{i=1}^{d}[D_i(b_iw) + c_iD_iw] + a_0w \ .$$

We have denoted by D_j the partial derivative $\frac{\partial}{\partial x_j}$; $a_{ij} = a_{ij}(\mathbf{x}), b_i = b_i(\mathbf{x}), c_i = c_i(\mathbf{x}), a_0 = a_0(\mathbf{x})$ are given functions. If the coefficients b_i and c_i are regular enough, we can omit either $D_i(b_iw)$ or c_iD_iw in (6.1.1) without loosing generality.

Definition 6.1.1 The differential operator L is said to be elliptic in Ω if there exists a constant $\alpha_0 > 0$ such that

$$\sum_{i,j=1}^{d} a_{ij}(\mathbf{x})\xi_i\xi_j \geq \alpha_0|\boldsymbol{\xi}|^2$$

for each $\boldsymbol{\xi} \in \mathbb{R}^d$ and for almost every $\mathbf{x} \in \Omega$.

To the operator L we may associate the following bilinear form

$$a(w,v) := \int_\Omega \Big[\sum_{i,j=1}^d a_{ij} D_j w D_i v$$

(6.1.2)

$$- \sum_{i=1}^d (b_i w D_i v - c_i v D_i w) + a_0 w v \Big] \;,$$

where w, v are functions defined in Ω. (For the sake of simplicity, here and in the sequel we don't indicate the integration measure dx on Ω or $d\gamma$ on $\partial\Omega$.) Formally speaking, this is obtained by multiplying (6.1.1) by v, integrating over Ω, applying the Green formula (1.3.3) and discarding the boundary terms.

A typical example is the Laplace operator $L = -\Delta$, in which case the associated bilinear form is the Dirichlet form

$$a(w,v) = \int_\Omega \nabla w \cdot \nabla v \;.$$

If we choose a closed subspace V of $H^1(\Omega)$ such that

(6.1.3) $$H_0^1(\Omega) \subset V \subset H^1(\Omega)$$

and we assume

(6.1.4) $$a_{ij}, b_i, c_i, a_0 \in L^\infty(\Omega)$$

(see Section 1.2 for the definition of these functional spaces), then $a(\cdot,\cdot)$ is well defined in $V \times V$. Take note that less stringent assumptions on the coefficients b_i, c_i, a_0 could be made. However, for the sake of simplicity, we assume that (6.1.4) holds.

We are interested in solving the variational problem: given $\mathcal{F}$, which is a continuous linear functional on the Hilbert space V,

(6.1.5)   find $u \in V$: $\mathcal{A}(u,v) = \mathcal{F}(v) \qquad \forall\, v \in V$.

As we will see, the bilinear form $\mathcal{A}(\cdot,\cdot)$ coincides with $a(\cdot,\cdot)$ up to the sum of possible boundary terms.

Depending on the choice of the subspace V, (6.1.5) describes the weak formulation of classical boundary value problems of Dirichlet, Neumann, mixed or Robin type, that we are going to specify below. Before entering into details, let us remind that, when Γ_D is a subset, proper or not, of $\partial\Omega$, then the space $H^1_{\Gamma_D}(\Omega)$ is defined by

(6.1.6) $$H^1_{\Gamma_D}(\Omega) := \{ v \in H^1(\Omega) \,|\, v = 0 \text{ on } \Gamma_D \}$$

($v = 0$ on Γ_D means that the trace of v is vanishing on Γ_D, see Section 1.3). In particular, if $\Gamma_D = \partial\Omega$ we simply write $H_0^1(\Omega)$ instead of $H^1_{\partial\Omega}(\Omega)$.

6.1.1 Variational Form of Boundary Value Problems

Let us specify some remarkable cases of boundary value problems.

(i) *The Dirichlet problem*
Consider the homogeneous Dirichlet problem, i.e., given a function f defined in Ω, find u such that

$$(6.1.7) \qquad \begin{cases} Lu = f & \text{in } \Omega \\ u = 0 & \text{on } \partial\Omega \ . \end{cases}$$

The weak formulation reads as follows: given $f \in L^2(\Omega)$,

$$(6.1.8) \qquad \text{find } u \in H^1_0(\Omega) : a(u,v) = (f,v) \qquad \forall \, v \in H^1_0(\Omega) \ ,$$

where $(\cdot,\cdot)$ is the scalar product in $L^2(\Omega)$ (see Section 1.2).

By the formula of integration by parts (1.3.3) it is verified at once that, if u is a (classical) solution to (6.1.7), then u is also a solution to (6.1.8). It is useful to clarify at which extent problem (6.1.8) is related to the one (6.1.7). Choosing $v \in C_0^\infty(\Omega)$ and integrating by parts, one easily finds that

$$\int_\Omega Lu\,v = \int_\Omega fv \qquad \forall \, v \in C_0^\infty(\Omega)$$

and therefore

$$Lu = f$$

in the sense of $\mathcal{D}'(\Omega)$ (see Section 1.2). Since $f \in L^2(\Omega)$, this equality holds almost everywhere in Ω. Moreover, the Dirichlet boundary condition $u_{|\partial\Omega}$ is achieved in the sense of $H^1_0(\Omega)$, i.e., the trace of u on $\partial\Omega$ vanishes.

(ii) *The Neumann problem*
In this case we want to solve

$$(6.1.9) \qquad \begin{cases} Lu = f & \text{in } \Omega \\ \dfrac{\partial u}{\partial n_L} = g & \text{on } \partial\Omega \ , \end{cases}$$

where

$$(6.1.10) \qquad \frac{\partial u}{\partial n_L} := \sum_{i,j=1}^d a_{ij} D_j u\, n_i - \mathbf{b} \cdot \mathbf{n}u$$

is the *conormal* derivative of u, and g is a given function on $\partial\Omega$. If $\mathbf{b} = 0$ and $a_{ij} = \delta_{ij}$, the Kronecker symbol defined by $\delta_{ii} = 1$, $\delta_{ij} = 0$ for $i \neq j$, then (6.1.10) is nothing but the derivative of u along the unit outward normal direction $\mathbf{n}$. The weak formulation of (6.1.9) reads as follows: given $f \in L^2(\Omega)$, and $g \in L^2(\partial\Omega)$

(6.1.11)   find $u \in H^1(\Omega) : a(u,v) = (f,v) + (g,v)_{\partial\Omega} \quad \forall\, v \in H^1(\Omega)$,

where $(\cdot,\cdot)_{\partial\Omega}$ is the scalar product in $L^2(\partial\Omega)$.

As before, a (classical) solution to (6.1.9) is also a solution to (6.1.11), owing to (1.3.3). Conversely, choosing $v \in C_0^\infty(\Omega)$ one obtains that $Lu = f$ in $\mathcal{D}'(\Omega)$ and then almost everywhere in Ω. Furthermore, taking now $v \in H^1(\Omega)$ and integrating by parts, one is left with

$$\int_\Omega Lu\,v + \int_{\partial\Omega} \frac{\partial u}{\partial n_L} v = \int_\Omega fv + (g,v)_{\partial\Omega} \quad \forall\, v \in H^1(\Omega) ,$$

provided u is regular enough to justify the preceding calculations. Thus

(6.1.12) $$\int_{\partial\Omega} \frac{\partial u}{\partial n_L} v = (g,v)_{\partial\Omega} \quad \forall\, v \in H^1(\Omega) .$$

Due to the arbitrarity of v, we can conclude that $\frac{\partial u}{\partial n_L} = g$ on $\partial\Omega$.

(iii) *The mixed problem*
We now consider the problem

(6.1.13)
$$\begin{cases} Lu = f & \text{in } \Omega \\[2mm] u = 0 & \text{on } \Gamma_D \\[2mm] \dfrac{\partial u}{\partial n_L} = g & \text{on } \Gamma_N , \end{cases}$$

where Γ_D and Γ_N are open disjoint subsets of $\partial\Omega$, and $\overline{\Gamma_D} \cup \overline{\Gamma_N} = \partial\Omega$. It is easily seen that the weak formulation is: given $f \in L^2(\Omega)$ and $g \in L^2(\Gamma_N)$,

(6.1.14) find $u \in H^1_{\Gamma_D}(\Omega) : a(u,v) = (f,v) + (g,v)_{\Gamma_N} \quad \forall\, v \in H^1_{\Gamma_D}(\Omega) .$

As in the preceding cases, it is possible to interpret this weak problem and find that $Lu = f$ almost everywhere in Ω, the trace of u is zero on Γ_D and (at least formally, i.e., assuming that u is smooth enough so that all operations that we are carrying out make sense) $\frac{\partial u}{\partial n_L} = g$ on Γ_N.

(iv) *The Robin (or third type) problem*
In this case the problem reads

(6.1.15)
$$\begin{cases} Lu = f & \text{in } \Omega \\[2mm] \dfrac{\partial u}{\partial n_L} + \kappa u = g & \text{on } \partial\Omega , \end{cases}$$

κ being a given function on $\partial\Omega$. Its weak formulation is as follows:

(6.1.16)
$$\begin{aligned} \text{find } u \in H^1(\Omega) : a(u,v) + (\kappa u, v)_{\partial\Omega} \\ = (f,v) + (g,v)_{\partial\Omega} \quad \forall\, v \in H^1(\Omega) , \end{aligned}$$

where we have assumed that $f \in L^2(\Omega)$, $g \in L^2(\partial\Omega)$, $\kappa \in L^\infty(\partial\Omega)$. By proceeding exactly as before one can deduce from (6.1.16) that $Lu = f$ almost everywhere in Ω and (formally) $\frac{\partial u}{\partial n_L} + \kappa u = g$ on $\partial\Omega$.

Remark 6.1.1 *(The symmetric case)*. If the bilinear form $\mathcal{A}(\cdot, \cdot)$ is coercive and symmetric, i.e.,

$$\mathcal{A}(v, v) \geq \alpha ||v||_1^2 \qquad \forall\, v \in V$$

and

$$\mathcal{A}(w, v) = \mathcal{A}(v, w) \qquad \forall\, w, v \in V \ ,$$

then the problems above are equivalent to the minimization in V of the *energy functional*.

The latter in the Dirichlet case takes the form

$$J(v) := \frac{1}{2}a(v, v) - (f, v) \ .$$

In fact, if u is a solution to (6.1.8), it is easily verified that

$$J(u + v) = J(u) + \frac{1}{2}a(v, v) > J(u) \qquad \forall\, v \neq 0 \ .$$

Conversely, if the functional J attains its minimum at u, it follows that its Gâteaux derivative must be vanishing, i.e.,

$$\left[\frac{d}{d\tau}J(u + \tau v)\right]_{|\tau=0} = 0 \qquad \forall\, v \in V \ .$$

As a consequence

$$\begin{aligned}
0 &= \frac{d}{d\tau}J(u + \tau v)_{|\tau=0} = \lim_{\tau\to 0}\frac{J(u + \tau v) - J(u)}{\tau} \\
&= \lim_{\tau\to 0}\left[a(u, v) + \frac{\tau}{2}a(v, v) - (f, v)\right] \\
&= a(u, v) - (f, v) \qquad \forall\, v \in V \ ,
\end{aligned}$$

hence (6.1.8) holds.

If the problem considered is the equilibrium of an elastic membrane fixed on the boundary and subject to a load of intensity f, the functional $J(v)$ represents the total potential energy associated with the displacement v.

When considering the other boundary value problems, the corresponding definition of $J(v)$ is easily obtained from (6.1.11), (6.1.14) and (6.1.16). Precisely, we find

$$J(v) := \frac{1}{2}a(v, v) - (f, v) - (g, v)_{\partial\Omega}$$

for problem (6.1.11),

$$J(v) := \frac{1}{2}a(v,v) - (f,v) - (g,v)_{\Gamma_N}$$

for problem (6.1.14), and

$$J(v) := \frac{1}{2}[a(v,v) + (\kappa v, v)_{\partial\Omega}] - (f,v) - (g,v)_{\partial\Omega}$$

for problem (6.1.16).   □

6.1.2 Existence, Uniqueness and A-Priori Estimates

Here we review some of the main mathematical properties of the preceding boundary value problems. The basic ingredient for proving the existence of a solution is the Lax-Milgram lemma (see Theorem 5.1.1), which in turn is a generalization of the Riesz representation theorem. With the aim of applying this result, we are going to check that, under suitable assumptions on the data, conditions (5.1.13) and (5.1.14) are satisfied.

We start from the Dirichlet problem (6.1.8). First of all, any $f \in L^2(\Omega)$ defines a continuous linear functional $v \to (f,v)$ on $H_0^1(\Omega)$. Thus we have only to show that $a(\cdot,\cdot)$ is continuous and coercive. Under assumption (6.1.4), the continuity is easily verified. Moreover, the ellipticity assumption gives

$$(6.1.17) \qquad \int_\Omega \sum_{i,j=1}^d a_{ij}D_iv D_jv \geq \alpha_0 \int_\Omega |\nabla v|^2 \qquad \forall\, v \in H_0^1(\Omega) \ ,$$

as choosing $\boldsymbol{\xi} = \nabla v(\mathbf{x})$ in Definition 6.1.1 it follows

$$\sum_{i,j=1} a_{ij}(\mathbf{x})D_iv(\mathbf{x})D_jv(\mathbf{x}) \geq \alpha_0|\nabla v(\mathbf{x})|^2 \quad \text{a.e. in } \Omega \ .$$

Here and in the sequel a.e. means almost everywhere.

Consider now the remaining term in $a(v,v)$. Assuming that $\text{div}(\mathbf{b} - \mathbf{c}) \in L^\infty(\Omega)$, we can write

$$\int_\Omega \left[-\sum_{i=1}^d (b_i - c_i)vD_iv + a_0v^2 \right] = \int_\Omega \left[-\frac{1}{2}\sum_{i=1}^d (b_i - c_i)D_i(v^2) + a_0v^2 \right]$$

$$= \int_\Omega \left[\frac{1}{2}\,\text{div}(\mathbf{b} - \mathbf{c}) + a_0 \right] v^2$$

for each $v \in H_0^1(\Omega)$. If C_Ω is the constant of the Poincaré inequality (see (1.3.2)), i.e.,

$$(6.1.18) \qquad \int_\Omega v^2 \leq C_\Omega \int_\Omega |\nabla v|^2 \qquad \forall\, v \in H_0^1(\Omega) \ ,$$

we can conclude that $a(\cdot,\cdot)$ is coercive provided we assume that for almost every $\mathbf{x} \in \Omega$

(6.1.19) $\dfrac{1}{2}\operatorname{div}[\mathbf{b}(\mathbf{x}) - \mathbf{c}(\mathbf{x})] + a_0(\mathbf{x}) \geq -\eta$, $-\infty < \eta < \dfrac{\alpha_0}{C_\Omega}$.

When considering the Neumann problem (6.1.11) we have $V = H^1(\Omega)$, hence the Poincaré inequality is no longer valid in V. On the other hand, (6.1.17) still holds for each $u \in H^1(\Omega)$. Furthermore,

$$-\int_\Omega \sum_{i=1}^d (b_i - c_i) v D_i v = \int_\Omega \frac{1}{2}\operatorname{div}(\mathbf{b} - \mathbf{c}) v^2 - \int_{\partial\Omega} \frac{1}{2}(\mathbf{b} - \mathbf{c})\cdot\mathbf{n} v^2 \ .$$

If we assume that

(6.1.20) $\dfrac{1}{2}\operatorname{div}[\mathbf{b}(\mathbf{x}) - \mathbf{c}(\mathbf{x})] + a_0(\mathbf{x}) \geq \mu_0 > 0$ a.e. in Ω ,

and further

(6.1.21) $[\mathbf{b}(\mathbf{x}) - \mathbf{c}(\mathbf{x})]\cdot\mathbf{n}(\mathbf{x}) \leq 0$ a.e. on $\partial\Omega$,

we can still conclude that $a(\cdot,\cdot)$ is coercive. In alternative to (6.1.21), a condition on the magnitude of $|\mathbf{b} - \mathbf{c}|$ could be assumed, for instance

(6.1.22) $\|\mathbf{b} - \mathbf{c}\|_{L^\infty(\partial\Omega)} \leq \varepsilon_0$, $0 \leq \varepsilon_0 < \dfrac{2\,\min(\alpha_0, \mu_0)}{C^*}$

where C^* is the constant of the *trace inequality*, i.e., (see (1.3.1))

$$\int_{\partial\Omega} v^2 \leq C^* \int_\Omega (v^2 + |\nabla v|^2) \qquad \forall\, v \in H^1(\Omega) \ .$$

On the other hand, the assumption $g \in L^2(\partial\Omega)$ on the boundary datum entails that $v \to (g, v)_{\partial\Omega}$ is a continuous functional.

In the case of the mixed problem (6.1.14), the Poincaré inequality is still valid in $H^1_{\Gamma_D}(\Omega)$. Thus $a(\cdot,\cdot)$ is coercive if we assume (6.1.19) and, furthermore, (6.1.21) or else (6.1.22) with Γ_N instead of $\partial\Omega$ (the latter case for a suitable ε_0 whose explicit upper bound could be easily computed similarly as we did above). As before, the assumption that $g \in L^2(\Gamma_N)$ is sufficient to guarantee that $v \to (g, v)_{\Gamma_N}$ is a continuous functional.

Concerning the Robin problem (6.1.16), since the Poincaré inequality is no longer valid, coerciveness holds provided we assume (6.1.20) together with

(6.1.23) $\kappa(\mathbf{x}) - \dfrac{1}{2}[\mathbf{b}(\mathbf{x}) - \mathbf{c}(\mathbf{x})]\cdot\mathbf{n}(\mathbf{x}) \geq 0$ a.e. on $\partial\Omega$.

Alternatively to (6.1.23) we could require that $\|\kappa - 1/2(\mathbf{b} - \mathbf{c})\cdot\mathbf{n}\|_{L^\infty(\partial\Omega)}$ is small enough. The continuity of $(\kappa u, v)_{\partial\Omega}$ holds if, for instance, $\kappa \in L^\infty(\partial\Omega)$.

As we already pointed out, alternative assumptions guarantee the coerciveness of the bilinear forms appearing at the left hand side of the previous problems. For instance, another simple condition is given by

(6.1.24)
$$\inf_{\Omega} a_0 - \frac{1}{4\alpha_0}||\mathbf{b} - \mathbf{c}||^2_{L^\infty(\Omega)} > 0 \ .$$

In fact, under this sole assumption one can write

$$a(v,v) \geq \alpha_0 \int_\Omega |\nabla v|^2 - \left|\int_\Omega v(\mathbf{b} - \mathbf{c}) \cdot \nabla v\right| + \int_\Omega a_0 v^2$$

$$\geq (\alpha_0 - \varepsilon) \int_\Omega |\nabla v|^2 + \int_\Omega \left[a_0 - \frac{1}{4\varepsilon}||\mathbf{b} - \mathbf{c}||^2_{L^\infty(\Omega)}\right] v^2$$

for each $\varepsilon > 0$. Choosing ε smaller than but close enough to α_0 and using (6.1.24), the coerciveness follows at once. Let us notice that in this case we are neither requiring extra regularity on the coefficients b_i and c_i, nor conditions like (6.1.21), (6.1.23).

To summarize, in all previous cases we have shown that, under suitable assumptions on the data, the Lax-Milgram lemma can be applied producing existence and uniqueness of a solution $u \in V$ to (6.1.5). An a-priori estimate can also be inferred, namely (see (5.1.15))

(6.1.25)
$$||u||_1 \leq \frac{1}{\alpha}||\mathcal{F}||_{V'} \ ,$$

where α is the coerciveness constant of $\mathcal{A}(\cdot, \cdot)$ and $\mathcal{F}$ is the continuous linear functional appearing at the right hand side of (6.1.8), (6.1.11), (6.1.14) or (6.1.16). For instance, in the Dirichlet case we have

(6.1.26)
$$||\mathcal{F}||_{V'} \leq C||f||_0 \ ,$$

and in the Neumann case

(6.1.27)
$$||\mathcal{F}||_{V'} \leq C(||f||_0 + ||g||_{0,\partial\Omega}) \ .$$

Remark 6.1.2 *(The non-homogeneous Dirichlet problem).* When the boundary datum in the Dirichlet problem is not zero, we have to solve

$$\begin{cases} Lu = f & \text{in } \Omega \\ u = \varphi & \text{on } \partial\Omega \ , \end{cases}$$

where Ω is an open bounded domain of $\mathbb{R}^d$, $d = 2, 3$.

This case can be analyzed as follows. The function φ is extended in the whole Ω (denote by $\tilde{\varphi}$ this extension), then we look for a solution $\tilde{u} = u - \tilde{\varphi}$ of the homogeneous problem

$$\begin{cases} L\tilde{u} = f - L\tilde{\varphi} & \text{in } \Omega \\ \tilde{u} = 0 & \text{on } \partial\Omega \ . \end{cases}$$

In the weak sense, this means: find $\tilde{u} \in H^1_0(\Omega)$, the solution of

$$a(\tilde{u}, v) = (f, v) - a(\tilde{\varphi}, v) \quad \forall \, v \in H_0^1(\Omega) \ .$$

In order to apply the Lax-Milgram lemma we need to show that the mapping $v \rightarrow a(\tilde{\varphi}, v)$ is a continuous functional, i.e., that $\tilde{\varphi} \in H^1(\Omega)$. This is true if $\varphi \in H^{1/2}(\partial\Omega)$ (see Theorem 1.3.1). Moreover, for each $v \in H^1(\Omega)$ it holds

$$|a(\tilde{\varphi}, v)| \leq c||\tilde{\varphi}||_1 ||v||_1 \leq C||\varphi||_{1/2, \partial\Omega} ||v||_1 \ .$$

The a-priori estimate (6.1.25) thus becomes

$$||u||_1 \leq ||\tilde{u}||_1 + ||\tilde{\varphi}||_1 \leq C(||f||_0 + ||\varphi||_{1/2, \partial\Omega}) \ ,$$

stating that the energy norm of the solution is bounded from above in terms of the problem data.

An alternative way to formulate the non-homogeneous Dirichlet problem is presented in Remark 7.1.4 (see also the whole Section 7.1). □

Remark 6.1.3 Another way for proving the existence and uniqueness of a solution to these boundary value problems is based on the Fredholm alternative theorem and the maximum principle. We are not going to present this approach here, but refer the interested reader to Agmon (1965), Grisvard (1985). □

6.1.3 Regularity of Solutions

Assuming additional regularity on the data it is possible to prove that the weak solution of the preceding problems, which is sought in $H^1(\Omega)$, is indeed more regular, i.e., it belongs to $H^s(\Omega)$ for some $s > 1$.

The range of achievable results is extremely broad, depending on the different assumptions on the data; here we just want to present a few of them, without attempting to be exhaustive. For a complete discussion we refer the reader to Grisvard (1985).

It is worthwhile mentioning that the smoothness degree of the solution of a boundary-value problem does affect the order of convergence of a numerical approximation. This statement will be extensively validated all along this Chapter.

Let us start from the non-homogeneous Dirichlet problem (see Remark 6.1.2). The following regularity result holds. Assume that for some $k \geq 0$, $\partial\Omega$ is a C^{k+2} manifold, the coefficients a_{ij}, b_i, belong to $C^{k+1}(\overline{\Omega})$, the coefficients c_i and a_0 belong to $C^k(\overline{\Omega})$, and $f \in H^k(\Omega)$, $\varphi \in H^{k+3/2}(\partial\Omega)$. Then the solution $u \in H^{k+2}(\Omega)$. In particular, if all data are C^∞, then u is C^∞, too. We notice that the assumptions on $\partial\Omega$ and the coefficients could be weakened using the information provided by the Sobolev embedding theorems more carefully.

Another interesting case arises when Ω is a polygonal domain, i.e., $\overline{\Omega}$ is the union of a finite number of polyhedra. In this case, when the dimension d

is equal to 2, the boundary $\partial\Omega$ is only a Lipschitz continuous manifold (while when $d = 3$ this is not always true: see, for instance, Ženíšek (1990), p. 384), and the previous result doesn't apply any longer.

To give an example of the regularity result in this situation, we focus on the homogeneous Dirichlet problem for the Laplace operator:

(6.1.28)
$$\begin{cases} -\Delta u = f & \text{in } \Omega \\ u = 0 & \text{on } \partial\Omega \end{cases}.$$

In this case, assuming that $f \in L^2(\Omega)$ and that Ω is a plane convex polygonal domain, one still has $u \in H^2(\Omega)$.

If Ω is not convex, and $\omega > \pi$ is the measure of the angle of a concave corner of $\partial\Omega$, then it turn out that locally near that corner $(u - \psi)$ is an H^2-function. Here ψ is a suitable function having in the corner a singularity of the type $r^{\pi/\omega}$, r being the distance from the corner itself. In particular, one can always conclude that $u \in H^{3/2}(\Omega)$.

When considering either the Neumann or the Robin problems, the regularity result for a smooth domain Ω is similar to the one for the Dirichlet case. The assumptions on $\partial\Omega$, the coefficients a_{ij}, b_i, c_i and a_0 and the datum f are in fact the same, and requiring $g \in H^{k+1/2}(\partial\Omega)$ and $\kappa \in C^{k+1}(\partial\Omega)$ produces a solution $u \in H^{k+2}(\Omega)$, $k \geq 0$.

Let Ω be a plane polygonal domain and consider, for the sake of simplicity, the homogeneous Neumann problem for the Laplace operator:

(6.1.29)
$$\begin{cases} -\Delta u = f & \text{in } \Omega \\ \dfrac{\partial u}{\partial n} = 0 & \text{on } \partial\Omega \end{cases}.$$

In this case, assuming $f \in L^2(\Omega)$ and the compatibility condition $\int_\Omega f = 0$, one still obtains $u \in H^2(\Omega)$ for a convex domain, and $u \in H^{3/2}(\Omega)$ in the general case.

On the contrary, the solution of the mixed problem in general is not regular. More precisely, there exist examples in which the data and the boundary are smooth, while the solution belongs to $H^s(\Omega)$ for any $s < 3/2$, but not to $H^{3/2}(\Omega)$.

6.1.4 On the Degeneracy of the Constants in Stability and Error Estimates

Let us consider now the issue of the a-priori bounds for the solution of the elliptic problems introduced above. We have already seen from Theorem 5.1.1 that the solution satisfies

$$\|u\|_1 \leq \frac{1}{\alpha}\|\mathcal{F}\|_{V'} \ ,$$

where α is the coerciveness constant related to the bilinear form $a(u,v)$ (or $a(u,v) + (\kappa u, v)_{\partial\Omega}$ in the case of the Robin problem).

To fix the ideas, from now on we will refer to the Dirichlet boundary value problem (6.1.8). If we assume $a_{ij} = \varepsilon\delta_{ij}$ ($\varepsilon > 0$ a constant), div $\mathbf{b} = 0$, $c_i = 0$ and $a_0 \geq 0$, the existence of a unique solution is guaranteed since (6.1.19) is satisfied. Moreover, the coerciveness constant is given by ε, and we have the a-priori estimate

$$\|\nabla u\|_0 \leq \frac{C_\Omega^{1/2}}{\varepsilon}\|f\|_0 \ ,$$

where C_Ω is the constant that appears in the Poincaré inequality (6.1.18). The control on the gradient of u can be very poor if the constant ε is very small. More important, the error estimate for the Galerkin method reads (see (5.2.7))

(6.1.30) $$\|\nabla u - \nabla u_h\|_0 \leq \frac{\gamma}{\varepsilon} \inf_{v_h \in V_h} \|\nabla u - \nabla v_h\|_0 \ .$$

Here γ is the continuity constant related to the bilinear form $a(u,v)$ and can be expressed by

$$\gamma = \varepsilon + C_\Omega^{1/2}\|\mathbf{b}\|_{L^\infty(\Omega)} + C_\Omega\|a_0\|_{L^\infty(\Omega)} \ .$$

Again, γ/ε is a large number if ε is small in comparison with $\|\mathbf{b}\|_{L^\infty(\Omega)}$ and/or $\|a_0\|_{L^\infty(\Omega)}$.

In this situation, the performance of the pure Galerkin method can be quite poor. This is revealed by the onset of oscillations that may appear in the numerical solution near to the boundary of Ω whenever the exact solution u exhibits boundary layers. In such cases, the numerical method needs to be "stabilized" by resorting to a generalized Galerkin approach that damps the numerical oscillations. This issue will be discussed thoroughly in Chapter 8.

6.2 Numerical Methods: Construction and Analysis

In this Section we introduce numerical methods to approximate any elliptic boundary value problem that can be expressed in the form (6.1.5).

For the reasons we have outlined in Section 6.1.4, the effectiveness of the approximation methods we are going to describe depends on the relative magnitude of the ellipticity constant α_0, the advective vector fields $\mathbf{b}$ and $\mathbf{c}$, and the coefficient a_0. The methods we present in this Chapter are well suited for the *diffusion-dominated* case, i.e., when the ellipticity constant α_0 is large enough in comparison with the coefficients b_i, c_i, $i = 1, ..., d$, and a_0. In Chapter 8 we will consider other methods which can be applied to the *advection-dominated* case.

Among the numerical methods presented in Chapter 5, we are going to consider the Galerkin, collocation and generalized Galerkin methods.

6.2.1 Galerkin Method: Finite Element and Spectral Approximations

As pointed out in (5.2.3), the Galerkin approximation to (6.1.5) reads:

(6.2.1)   find $u_h \in V_h : \mathcal{A}(u_h, v_h) = \mathcal{F}(v_h)$ $\forall\, v_h \in V_h$,

where V_h is a suitable finite dimensional subspace of V.

To ensure the existence and uniqueness of the solutions u and u_h to (6.1.5) and (6.2.1), respectively, we are always assuming that the bilinear form $\mathcal{A}(\cdot, \cdot)$ is continuous and coercive, and the linear functional $\mathcal{F}(\cdot)$ is continuous. Moreover, as a consequence of these assumptions, we know that the error estimates (5.2.7) holds. Let us remember that sufficient conditions implying continuity of $\mathcal{A}$ and $\mathcal{F}$ and coerciveness of $\mathcal{A}$ have been made precise in Section 6.1.2 for all boundary value problems considered in Section 6.1.1.

The main point toward proving the convergence of u_h to u is thus to verify that (5.2.2) holds, i.e.,

(6.2.2) $\lim\limits_{h \to 0} \inf\limits_{v_h \in V_h} ||v - v_h|| = 0$ $\forall\, v \in V$,

where we are denoting by $||\cdot||$ the norm of V.

We then prove a general result which requires an apparently weaker version of (6.2.2) (indeed, the following condition (6.2.3) turns out to be equivalent to (6.2.2)).

Proposition 6.2.1 *Assume that the bilinear form $\mathcal{A}(\cdot, \cdot)$ is continuous and coercive in V, and the linear functional $\mathcal{F}(\cdot)$ is continuous in V. Let V_h be a family of finite dimensional subspaces of V. Assume that there exists a subset $\mathcal{V}$ dense in V such that*

(6.2.3) $\inf\limits_{v_h \in V_h} ||v - v_h|| \to 0$ *as* $h \to 0$ $\forall\, v \in \mathcal{V}$.

Then the Galerkin method is convergent, i.e., the solution u_h of (6.2.1) converges in V to the solution u of (6.1.5) with respect to the norm $||\cdot||$.

Proof. Since $\mathcal{V}$ is dense in V, for each $\varepsilon > 0$ we can find $v \in \mathcal{V}$ such that

$$||u - v|| < \varepsilon \ .$$

Moreover, due to (6.2.3) there exist $h_0(\varepsilon) > 0$ and, for any positive $h < h_0(\varepsilon)$, $v_h \in V_h$ such that

$$||v - v_h|| < \varepsilon \ .$$

Hence, using the error estimate (5.2.7)

$$||u - u_h|| \leq \frac{\gamma}{\alpha}||u - v_h|| \leq \frac{\gamma}{\alpha}(||u - v|| + ||v - v_h||) \ ,$$

and the thesis follows.   □

We intend to specify the above results in the framework of both finite element and spectral methods.

We start with the *finite element method,* and assume that $\Omega \subset \mathbb{R}^d$, $d = 2, 3$, is a polygonal domain with Lipschitz boundary and $\mathcal{T}_h$ is a family of triangulations of $\overline{\Omega}$. At first, we want to make precise what is V_h for anyone of the boundary value problems we have considered so far.

(i) *The Dirichlet problem*
In this case, we choose

$$(6.2.4) \qquad V_h = X_h^k \cap H_0^1(\Omega) = \{v_h \in X_h^k \mid v_h = 0 \text{ on } \partial\Omega\} \ , \qquad k \geq 1 \ ,$$

where X_h^k is defined in (3.2.4) if the reference polyhedra $\hat{K}$ is the unit d-simplex. On the other hand, when $\hat{K} = [0,1]^d$ the space X_h^k is defined by (3.2.5).

(ii) *The Neumann problem*
We take

$$(6.2.5) \qquad\qquad\qquad V_h = X_h^k \ , \qquad k \geq 1 \ .$$

(iii) *The mixed problem*
We choose

$$(6.2.6) \qquad V_h = X_h^k \cap H_{\Gamma_D}^1(\Omega) = \{v_h \in X_h^k \mid v_h = 0 \text{ on } \Gamma_D\} \ , \qquad k \geq 1 \ .$$

(Here each triangulation $\mathcal{T}_h$ has been chosen in such a way that no element $K \in \mathcal{T}_h$ intersects both Γ_D and Γ_N.)

(iv) *The Robin problem*
In this case

$$(6.2.7) \qquad\qquad\qquad V_h = X_h^k \ , \qquad k \geq 1 \ .$$

Furthermore, in all cases we assume that the degrees of freedom and the shape functions are those described in Section 3.3. Consequently, for each $v \in C^0(\overline{\Omega})$ the interpolation function $\pi_h^k(v)$ is the one defined in (3.4.1).

We are now in a position to show the convergence of the finite element Galerkin method. To verify that (6.2.2) or (6.2.3) hold, we make use of the approximation result obtained in Chapter 3. The main theorem reads as follows:

Theorem 6.2.1 *Let Ω be a polygonal domain of $\mathbb{R}^d$, $d = 2, 3$, with Lipschitz boundary, and $\mathcal{T}_h$ be a regular family of triangulations of $\overline{\Omega}$ associated to a reference polyhedra $\hat{K}$, which is the unit d-simplex or $[0,1]^d$. Suppose that the* *bilinear form $\mathcal{A}(\cdot, \cdot)$ is continuous and coercive in V and the linear functional $\mathcal{F}(\cdot)$ is continuous in V. Let V_h be defined as stated in (6.2.4)-(6.2.7). Under these assumptions the finite element Galerkin method is convergent. If moreover the exact solution $u \in H^s(\Omega)$ for some $s \geq 2$, the following error estimate holds*

$$(6.2.8) \qquad \|u - u_h\|_1 \leq Ch^l \|u\|_{l+1} \ ,$$

where $l = \min(k, s - 1)$.

Proof. We apply the Proposition 6.2.1. Since $C^\infty(\overline{\Omega})$ is dense in $H^1(\Omega)$, we can choose $V = C^\infty(\overline{\Omega})$ for both Neumann and Robin problem, $V = C^\infty(\overline{\Omega}) \cap H_0^1(\Omega)$ for the Dirichlet problem and $V = C^\infty(\overline{\Omega}) \cap H_{\Gamma_D}^1(\Omega)$ for the mixed problem. Furthermore, for each $v \in V$

$$\inf_{v_h \in V_h} \|v - v_h\|_1 \leq \|v - \pi_h^k(v)\|_1 \ ,$$

hence it converges to zero due to (3.4.19).

We now prove (6.2.8). Under the assumption that $u \in H^s(\Omega)$, $s \geq 2$, from the Sobolev embedding theorem we have $u \in C^0(\overline{\Omega})$, hence the interpolation function $\pi_h^k(u)$ is well-defined. Additionally, $\pi_h^k(u) \in V_h$, since it is easily verified that $\pi_h^k(u) \in H_0^1(\Omega)$ in the Dirichlet case, and $\pi_h^k(u) \in H_{\Gamma_D}^1(\Omega)$ in the mixed case (we can always choose the triangulation $\mathcal{T}_h$ so that its restriction on Γ_D provides a triangulation of Γ_D). From Theorem 3.4.2 we thus have

$$(6.2.9) \qquad \|u - \pi_h^k(u)\|_1 \leq Ch^l \|u\|_{l+1} \ .$$

Moreover, the error estimate (5.2.7) holds, i.e.

$$(6.2.10) \qquad \|u - u_h\|_1 \leq \frac{\gamma}{\alpha} \inf_{v_h \in V_h} \|u - v_h\|_1 \ .$$

Now (6.2.8) follows from (6.2.9) and (6.2.10). $\qquad \square$

The convergence result (6.2.8) is *optimal* in the $H^1(\Omega)$-norm, i.e., it provides the highest possible rate of convergence in the $H^1(\Omega)$-norm allowed by the polynomial degree k. However, looking at the interpolation estimate (3.4.19) for $m = 0$, one could expect that the $L^2(\Omega)$-norm is in fact $O(h^{l+1})$. Indeed, this is true under suitable assumptions. To clarify this assertion we consider the auxiliary problem: given $r \in L^2(\Omega)$,

$$(6.2.11) \qquad \text{find } \varphi(r) \in V : \mathcal{A}(v, \varphi(r)) = (r, v) \quad \forall \, v \in V \ .$$

If $\mathcal{A}$ is continuous and coercive, the existence of $\varphi(r)$ is ensured by the Lax-Milgram lemma. By using a duality argument (also called Aubin-Nitsche trick), we are now in a position to obtain an optimal convergence result in $L^2(\Omega)$.

Proposition 6.2.2 *Let the assumptions of Theorem 6.2.1 be satisfied. Suppose, moreover, that for each $r \in L^2(\Omega)$ the solution $\varphi(r)$ of (6.2.11) belongs to $H^2(\Omega)$, so that, as a consequence of the Closed Graph theorem, there exists a constant $C > 0$ such that*

(6.2.12)
$$||\varphi(r)||_2 \leq C||r||_0 \quad \forall\, r \in L^2(\Omega) \ .$$

Then if $u \in H^s(\Omega)$, $s \geq 2$, the following error estimate holds:

(6.2.13)
$$||u - u_h||_0 \leq Ch^{l+1}||u||_{l+1} \quad , \qquad l = \min(k, s - 1) \ .$$

Proof. By proceeding as in Section 3.5 and using (6.2.11), we can write

$$||u - u_h||_0 = \sup_{\substack{r \in L^2(\Omega) \\ r \neq 0}} \frac{(r, u - u_h)}{||r||_0} = \sup_{\substack{r \in L^2(\Omega) \\ r \neq 0}} \frac{A(u - u_h, \varphi(r))}{||r||_0} \ .$$

For any arbitrary $\varphi_h \in V_h$, one has

$$A(u - u_h, \varphi(r)) = A(u - u_h, \varphi(r) - \varphi_h) \ ,$$

thus

$$(r, u - u_h) \leq \gamma||u - u_h||_1||\varphi(r) - \varphi_h||_1 \ .$$

Since $\varphi(r) \in H^2(\Omega) \subset C^0(\overline{\Omega})$, we can take $\varphi_h = \pi_h^k(\varphi(r))$, and from (3.4.19)

$$(r, u - u_h) \leq C\gamma||u - u_h||_1 h||\varphi(r)||_2 \ .$$

Using now (6.2.12) one obtains

$$||u - u_h||_0 \leq Ch||u - u_h||_1 \ ,$$

thus the thesis follows from (6.2.8). □

Remark 6.2.1 Inequality (6.2.12) is a regularity assumption on the solution of the *adjoint problem* (6.2.11). The solution to (6.2.11) enjoys the same regularity property than the one of the original problem (6.1.5) (see Section 6.1.3). In particular, if Ω is a polygonal domain the solution $\varphi(r)$ belongs to $H^2(\Omega)$ and satisfies (6.2.12), provided that Ω is convex, $a_{ij} \in C^1(\overline{\Omega})$, and $\kappa \in C^1(\partial\Omega)$ (see Grisvard (1976)). This is true for all but the mixed boundary value problem. In fact, we have already noticed in Section 6.1.3 that the solution of the mixed problem belongs to $H^s(\Omega)$ for any $s < 3/2$ but in general not to $H^{3/2}(\Omega)$, even for smooth data. □

Remark 6.2.2 *(The non-homogeneous Dirichlet problem)*. When considering a non-vanishing boundary datum $u = \varphi$ on $\partial\Omega$, as pointed out in Remark 6.1.2 the variational formulation of the Dirichlet problem reads as follows:

$$\text{find } \tilde{u} \in H_0^1(\Omega) : a(\tilde{u}, v) = (f, v) - a(\tilde{\varphi}, v) \quad \forall\, v \in H_0^1(\Omega) \ ,$$

where $\tilde{\varphi} \in H^1(\Omega)$ is a suitable extension of $\varphi \in H^{1/2}(\partial\Omega)$.

Taking V_h as in (6.2.4), a possible finite element approximation is given by:

$$\text{find } \tilde{u}_h \in V_h : a(\tilde{u}_h, v_h) = (f, v_h) - a(\tilde{\varphi}, v_h) \quad \forall\, v_h \in V_h \ .$$

However, this is not a practical way to define an approximate solution, since the construction of the extension operator $\varphi \to \tilde{\varphi}$ is not easily performed.

Assuming that φ belongs not only to $H^{1/2}(\partial\Omega)$ but also to $C^0(\partial\Omega)$, we can alternatively proceed in the following way. Define $\{\mathbf{x}_s \,|\, s = 1, ..., M_h\}$ the nodes on $\partial\Omega$ and $\{\mathbf{a}_i \,|\, i = 1, ..., N_h\}$ the internal nodes, and set

$$V_h^* := \{v_h \in X_h^k \,|\, v_h(\mathbf{x}_s) = \varphi(\mathbf{x}_s) \quad \forall\, s = 1, ..., M_h\} \ .$$

The approximate problem reads:

$$\text{find } u_h \in V_h^* : a(u_h, v_h) = (f, v_h) \quad \forall\, v_h \in V_h \ .$$

We can write any $u_h \in V_h^*$ as

$$u_h(x) = \sum_{i=1}^{N_h} u_h(\mathbf{a}_i)\varphi_i + \sum_{s=1}^{M_h} \varphi(\mathbf{x}_s)\tilde{\varphi}_s =: z_h + \tilde{\varphi}_h \ ,$$

where φ_i and $\tilde{\varphi}_s$ are the basis functions of X_h^k relative to the internal and boundary nodes, respectively. We note that $z_h \in V_h$ and $\tilde{\varphi}_h \in V_h^*$. Assuming that the family of triangulation $\mathcal{T}_h$ is quasi-uniform (see Definition 6.3.1), it can also be noticed that $||\tilde{\varphi}_h||_1 = O(h^{-1/2})$.

Now the problem can be rewritten as

$$\text{find } z_h \in V_h : a(z_h, v_h) = (f, v_h) - a(\tilde{\varphi}_h, v_h) \quad \forall\, v_h \in V_h \ ,$$

therefore it is solvable if $a(\cdot, \cdot)$ is coercive in $V \times V$. Concerning the error estimate, one verifies at once that

$$a(u - u_h, v_h) = 0 \quad \forall\, v_h \in V_h \ ,$$

hence, by proceeding as in Theorem 5.2.1 (Céa lemma),

$$||u - u_h||_1 \leq \frac{\gamma}{\alpha} \inf_{v_h^* \in V_h^*} ||u - v_h^*||_1 \ .$$

If $u \in H^s(\Omega)$, $s \geq 2$, the interpolant $\pi_h^k(u)$ belongs to V_h^*. Hence, proceeding as in the homogeneous Dirichlet case, we conclude that $||u - u_h||_1 = O(h^l)$, $l = \min(k, s - 1)$. Stability follows from convergence.

Alternative approaches to the non-homogeneous Dirichlet problem, based on Lagrangian multipliers and penalty techniques, have been proposed by Babuška (1973a, 1973b) (see also Section 7.1). $\qquad\square$

Remark 6.2.3 *(L$^\infty$-error estimate)*. We are now going to provide an error estimate in $L^\infty(\Omega)$. To begin, we write

$$||u - u_h||_{L^\infty(\Omega)} \leq ||u - \pi_h^k(u)||_{L^\infty(\Omega)} + ||\pi_h^k(u) - u_h||_{L^\infty(\Omega)} \ .$$

In Remark 3.4.2 we have shown that for $1 \leq l \leq k$

$$||u - \pi_K^k(u)||_{L^\infty(K)} \leq C\, h_K^{l+1} \, [\mathrm{meas}(K)]^{-1/2} \, |u|_{l+1,K} \qquad \forall \, u \in H^{l+1}(K) \ .$$

If the family of triangulations is a regular one, we have

$$[\mathrm{meas}(K)]^{-1/2} \leq C h_K^{-d/2} \ ,$$

hence

$$||u - \pi_h^k(u)||_{L^\infty(\Omega)} \leq C h^{l+1-d/2}|u|_{l+1,\Omega} \qquad \forall \, u \in H^{l+1}(\Omega) \ .$$

On the other hand, since in a finite dimensional space all norms are equivalent, we can infer that

$$||\hat{u}_h - \pi_{\hat{K}}^k(\hat{u})||_{L^\infty(\hat{K})} \leq C||\hat{u}_h - \pi_{\hat{K}}^k(\hat{u})||_{0,\hat{K}} \ .$$

Then by (3.4.4) it follows

$$||u_h - \pi_K^k(u)||_{L^\infty(K)} \leq C \, |\det B_K|^{-1/2} \, ||u_h - \pi_K^k(u)||_{0,K}$$
$$\leq C \, [\mathrm{meas}(K)]^{-1/2} \, ||u_h - \pi_K^k(u)||_{0,K} \ .$$

Let us now assume that the family of triangulations $\mathcal{T}_h$ is quasi-uniform, i.e., it is regular and there exists a constant $\tau > 0$ such that

$$\min_{K \in \mathcal{T}_h} h_K \geq \tau h \qquad \forall \, h > 0$$

(see Definition 6.3.1). Then

$$[\mathrm{meas}(K)]^{-1/2} \leq C h^{-d/2} \qquad \forall \, K \in \mathcal{T}_h \ , \quad \forall \, h > 0 \ ,$$

hence

$$||u_h - \pi_h^k(u)||_{L^\infty(\Omega)} \leq C h^{-d/2}||u_h - \pi_h^k(u)||_{0,\Omega} \ .$$

Finally,

$$||u_h - \pi_h^k(u)||_{0,\Omega} \leq ||u - u_h||_{0,\Omega} + ||u - \pi_h^k(u)||_{0,\Omega} \ ,$$

and by Theorem 3.4.2 and Proposition 6.2.2 we have

$$||u_h - \pi_h^k(u)||_{0,\Omega} \leq C h^{l+1}|u|_{l+1,\Omega} \qquad \forall \, u \in H^{l+1}(\Omega) \ .$$

Summing up, we have thus obtained the error estimate

$$||u - u_h||_{L^\infty(\Omega)} \leq C h^{l+1-d/2}|u|_{l+1,\Omega} \qquad \forall \, u \in H^{l+1}(\Omega) \ .$$

A better order of convergence can be obtained by a more technical approach, which make use of weighted norms and seminorms as suggested by Nitsche (1977) (for a different approach, see also Scott (1976)). Under suitable assumptions on the domain $\Omega \subset \mathbb{R}^2$ and the operator L, and taking

$V_h = X_h^1 \cap H_0^1(\Omega)$, where X_h^1 is defined as in (3.2.4), the following error estimate holds

$$||u - u_h||_{L^\infty(\Omega)} + h||\nabla u - \nabla u_h||_{L^\infty(\Omega)}$$
$$\leq Ch^2 |\log h| \, |u|_{W^{2,\infty}(\Omega)} \quad \forall \, u \in W^{2,\infty}(\Omega) \ .$$

Choosing $V_h = X_h^k \cap H_0^1(\Omega)$, $k \geq 2$, the factor $|\log h|$ can be dropped away, even if $\Omega \subset \mathbb{R}^d$, $d \geq 3$. On the contrary, the presence of the $|\log h|$ term cannot be avoided for "linear" triangles (Haverkamp (1984)). Several other results and comments concerning L^∞-error estimates can be found in Ciarlet (1978, 1991). □

Remark 6.2.4 *(Non-coercive bilinear forms)*. We have limited our analysis to coercive bilinear forms, so that existence and uniqueness of the exact solution is a consequence of the Lax-Milgram lemma. Based on a different approach (see Remark 6.1.3), an approximation theory can also be developed for non-coercive forms. An example is provided by Schatz (1974), where the bilinear form is assumed to satisfy the Gårding inequality (see 11.1.6). □

Now we return to the general Galerkin problem (6.2.1). We need to point out that the evaluation of the right hand side of (6.2.1) can be done exactly only in very simple cases. The same occurs in the evaluation of the left hand integrals in (6.2.1), whenever the differential boundary value problem has variable coefficients a_{ij}, b_i, c_i and a_0. In these cases, numerical integration formulae should be used. The resulting scheme is a modified (or generalized) Galerkin method and is analyzed in Section 6.2.3.

We now consider the *spectral method*. We assume from now on that $\Omega = (-1,1)^2$, and that Γ_D is the union (possibly empty) of edges of $\partial\Omega$. As usual, we denote by $\mathbb{Q}_N$ the space of algebraic polynomials of degree less than or equal to N in each variable x_i, $i = 1, 2$. The choice of the family of finite dimensional subspaces is given now by

$$(6.2.14) \qquad V_N = \mathbb{Q}_N^0 := \{v_N \in \mathbb{Q}_N \mid v_N = 0 \text{ on } \partial\Omega\}$$

for the Dirichlet problem, by

$$(6.2.15) \qquad\qquad\qquad V_N = \mathbb{Q}_N$$

for the Neumann and Robin problems, and by

$$(6.2.16) \qquad V_N = \mathbb{Q}_N \cap H_{\Gamma_D}^1(\Omega) = \{v_N \in \mathbb{Q}_N \mid v_N = 0 \text{ on } \Gamma_D\}$$

for the mixed problem.

The interest of the Galerkin method in the spectral framework is much greater for boundary value problems with periodic solutions, in which case trigonometric polynomials are used instead of algebraic ones (e.g., Gottlieb and Orszag (1977)). In the non-periodic case, the spectral collocation method is more often used than the spectral Galerkin method. However, for completeness we report here on the spectral Galerkin case as well.

For simplicity, we focus on the homogeneous Dirichlet problem (6.1.8) and its Galerkin approximation

$$(6.2.17) \qquad \text{find } u_N \in V_N : a(u_N, v_N) = (f, v_N) \quad \forall \, v_N \in V_N \ ,$$

V_N being given by (6.2.14).

Following the same procedure employed for the finite element case, the proof of convergence of the spectral Galerkin method is based on Proposition 6.2.1 and suitable estimates for the interpolation or projection operators. First of all, from (5.2.7) we have

$$||u - u_N||_1 \leq \frac{\gamma}{\alpha} \inf_{v_N \in V_N} ||u - v_N||_1 \ .$$

Assuming that $u \in H^s(\Omega)$, $s \geq 2$, by the results proven in Section 4.5.2 (see in particular (4.5.43) or (4.5.37)), we can choose for instance $v_N = I_N u$, where I_N is the Legendre interpolation operator, and conclude that

$$(6.2.18) \qquad ||u - u_N||_1 \leq \frac{\gamma}{\alpha}||u - I_N u||_1 \leq CN^{1-s}||u||_s \ .$$

We stress the fact that the rate of convergence is virtually bounded only by s, the smoothness degree of the solution, and not by any other parameter as it occurs in finite elements (see (6.2.8)).

A duality argument as in Proposition 6.2.2 yields

$$||u - u_N||_0 \leq CN^{-1}||u - u_N||_1 \ ,$$

having assumed that the solution $\varphi(g)$ of the adjoint problem (6.2.11) satisfies (6.2.12). Hence for $s \geq 2$ we have the optimal error estimate

$$(6.2.19) \qquad ||u - u_N||_0 \leq CN^{-s}||u||_s \ .$$

We are now in a position to prove also the convergence of u_N to u without other assumptions on u except $u \in V$. In fact, $C_0^\infty(\Omega)$ is dense in $H_0^1(\Omega)$, and it is included in $H^s(\Omega)$ for each $s \geq 0$. Hence from (4.5.43) or (4.5.37) it follows that (6.2.3) holds, provided we choose $V = C_0^\infty(\Omega)$.

We conclude this Section by detailing the application of the spectral Galerkin method to the Poisson problem (6.1.28). In this case the approximation space is $\mathbb{Q}_N^0$ (see (6.2.14)). The set of the trial functions

$$(6.2.20) \quad \varphi_{kl}(x_1, x_2) = (1 - x_1^2)(1 - x_2^2)L_k'(x_1)L_l'(x_2) \ , \quad 1 \leq k, l \leq N - 1 \ ,$$

where $\{L_n(\xi)\}$ are the Legendre polynomials, provides a basis for $\mathbb{Q}_N^0$ (see Remark 4.5.1). If the solution to (6.2.17) is sought in the form

$$u_N = \sum_{m,n=1}^{N-1} u_{mn}\varphi_{mn} \quad ,$$

then the unknown coefficients u_{mn} satisfy the linear system

$$\frac{k^2(k+1)^2}{k+1/2} \sum_{n=1}^{N-1} u_{kn}\gamma_{nl} + \frac{l^2(l+1)^2}{l+1/2} \sum_{m=1}^{N-1} u_{ml}\gamma_{km} = F_{kl} \ , \quad 1 \leq k,l \leq N-1 \ ,$$

where

$$\gamma_{ij} = \int_{-1}^{1} L_i'(\xi)L_j'(\xi)(1-\xi^2)^2 d\xi$$

$$F_{kl} = \int_{-1}^{1}\int_{-1}^{1} f(x_1,x_2)\varphi_{kl}(x_1,x_2)dx_1 dx_2 \quad .$$

In deriving these equations we have used the Sturm-Liouville relation (4.4.2) and the orthogonality property (4.4.6).

The exact calculation of the latter quantities may be very complicated if not impossible at all. Moreover, when the differential operator is not as simple as the Laplacian (e.g., in the case of variable coefficients) the entries of system matrix can be quite difficult to obtain in a closed form. We should also point out that the knowledge of the coefficients u_{kl} (the spectral unknowns) allows the determination of the spectral solution u_N at some selected nodes (physical representation) only upon using the Legendre transformation (i.e., evaluating the basis functions φ_{mn} at the same nodes).

This makes the spectral Galerkin method often impracticable, unless one resorts to suitable numerical integration formulae. Its interest is merely theoretical: it is easy to be analyzed, and produces a very accurate approximation to the exact solution whenever the latter is smooth.

An effective implementation of (6.2.17) would require a different set of trial functions (of Lagrangian type), yielding as associated unknowns the values of u_N at some selected Gaussian nodes, and further a Gaussian numerical evaluation of the integrals. (Any other numerical quadrature of lower order would not reproduce the spectral accuracy obtained in (6.2.18).) This produces the spectral collocation method that will be introduced in Section 6.2.2.

Among other different realizations of Galerkin method, we mention here the so-called *p-* and *h-p versions* of the finite element method, as well as the *spectral-element* method.

The first approach, developed by Babuška and coworkers (e.g., Babuška, Szabó and Katz (1981), Dorr (1986)) fits under the general setting of the finite element approximation. However, the finite element space X_h^k is now chosen with respect to a triangulation $\mathcal{T}_h$ into large elements (whose diameter h is no longer destined to go to zero), while the polynomial degree k (here called p) can be quite large. Indeed, convergence is achieved by letting p

increase while keeping the elemental size fixed. This is in contrast with the usual *h-version* of finite elements in which k is low while accuracy is pursued through a mesh refinement strategy (i.e., by letting h go to zero).

A strategy in-between leads to the so-called *h-p version* of finite elements, in which the two parameters are let to vary (h decreases while p increases) according to the local behaviour of the expected solution. There has been an extensive activity in this field in the latest years, both from the point of view of the theoretical analysis (Babuška and Dorr (1981), Guo and Babuška (1986a, 1986b), Babuška and Suri (1990)) and practical applications (Szabó (1990), Szabó and Babuška (1991), Oden and Demkowicz (1991)).

The *spectral-element* method is formally defined as the p-version of the finite element method. The distinguishing feature is that the elements are now quadrilaterals. The choice of trial functions as well as the way of evaluating the integrals differ quite considerably in the two cases though. The original contribution was given in Patera (1984). A review article is the one by Maday and Patera (1989).

A spectral method, alternative to the Galerkin and the collocation ones, is the so-called *τ-method* (or *Lanczos method*). It is sometimes applied to the solution of the Poisson equation (or other linear differential equations with constant coefficients) and consists in looking for $u_N \in \mathbb{Q}_N$ (satisfying the boundary conditions) and projecting the residual upon a space of polynomials of lower degree. In this respect, the $τ$-method is a Petrov-Galerkin technique. A very effective way to diagonalize the $τ$-matrix for the Poisson equation can be found in Haidvogel and Zang (1979).

6.2.2 Spectral Collocation Method

To start with, we consider the Dirichlet problem (6.1.28) in the square $\Omega = (-1,1)^2$. Let N be a given positive integer, and denote by $\mathbf{x}_{ij}$, w_{ij} for $0 \le i,j \le N$ the nodes and weights, respectively, of the Legendre Gauss-Lobatto formula (see (4.5.38)). Then for each $z,v \in C^1(\overline{\Omega})$ we set

$$(6.2.21) \qquad a_N(z,v) := (\nabla z, \nabla v)_N \ ,$$

where $(\cdot, \cdot)_N$ denotes the discrete scalar product that was introduced in (4.5.39). The spectral collocation problem reads:

$$(6.2.22) \qquad \text{find } u_N \in \mathbb{Q}_N^0 : a_N(u_N, v_N) = (f, v_N)_N \qquad \forall \, v_N \in \mathbb{Q}_N^0 \ .$$

Here we are assuming that f is a continuous function in $\overline{\Omega}$.

Since the one-dimensional Legendre Gauss-Lobatto integration formula is exact over $\mathbb{P}_{2N-1}$ (see (4.2.6)), it is possible to prove that for each $z_N, v_N \in \mathbb{Q}_N$

$$(6.2.23) \qquad (\nabla z_N, \nabla v_N)_N = -(\Delta z_N, v_N)_N + \sum_{k=1}^{4} \left(\frac{\partial z_N}{\partial n}, v_N \right)_{N, S_k} \ ,$$

where S_k, $k = 1, ..., 4$, are the sides of $\overline{\Omega}$, and $(\cdot, \cdot)_{N,S_k}$ denotes the Legendre Gauss-Lobatto quadrature formula on S_k. (Notice that in this formula each corner point gives a double contribution.) In fact, for each fixed x_2 the functions $x_1 \rightarrow \frac{\partial z_N}{\partial x_1}$ and $x_1 \rightarrow \frac{\partial v_N}{\partial x_1}$ belong to $\mathbb{P}_{N-1}$, and the same is true, having fixed x_1, for the functions $x_2 \rightarrow \frac{\partial z_N}{\partial x_2}$ and $x_2 \rightarrow \frac{\partial v_N}{\partial x_2}$. Therefore, we can write

$$
\begin{aligned}
(\nabla z_N, \nabla v_N)_N &= \sum_{i,j=0}^{N} \nabla z_N(\mathbf{x}_{ij}) \cdot \nabla v_N(\mathbf{x}_{ij})\, w_{ij} \\
&= \sum_{j=0}^{N} \left(\sum_{i=0}^{N} \frac{\partial z_N}{\partial x_1}(\mathbf{x}_{ij}) \frac{\partial v_N}{\partial x_1}(\mathbf{x}_{ij})\, w_i \right) w_j \\
&\quad + \sum_{i=0}^{N} \left(\sum_{j=0}^{N} \frac{\partial z_N}{\partial x_2}(\mathbf{x}_{ij}) \frac{\partial v_N}{\partial x_2}(\mathbf{x}_{ij})\, w_j \right) w_i \\
&= \sum_{j=0}^{N} w_j \int_{-1}^{1} \frac{\partial z_N}{\partial x_1}(\xi, x_j) \frac{\partial v_N}{\partial x_1}(\xi, x_j)\, d\xi \\
&\quad + \sum_{i=0}^{N} w_i \int_{-1}^{1} \frac{\partial z_N}{\partial x_2}(x_i, \eta) \frac{\partial v_N}{\partial x_2}(x_i, \eta)\, d\eta \ .
\end{aligned}
$$

Integrating by parts and using again (4.2.6) we find at once (6.2.23).

Taking as v_N the Lagrangian function associated to the interior node $\mathbf{x}_{ij}$ for any $1 \le i, j \le N - 1$ and using (6.2.23) we get easily from (6.2.22) that

$$
(6.2.24) \qquad
\begin{cases}
-\Delta u_N(\mathbf{x}_{ij}) = f(\mathbf{x}_{ij}) & \forall\, \mathbf{x}_{ij} \in \Omega \\[2mm]
u_N(\mathbf{x}_{ij}) = 0 & \forall\, \mathbf{x}_{ij} \in \partial\Omega \ .
\end{cases}
$$

Remark 6.2.5 Traditionally, the spectral collocation method was implemented in the pointwise form (6.2.24). The formulation (6.2.22), that stems from a generalized Galerkin approach, is sometimes referred to as the spectral collocation method in *weak* form.

Although the two approaches are fully equivalent to one another for problems with a Dirichlet boundary condition (as the one we are dealing with), they actually differ on the treatment of Neumann data. This will be discussed below. For this reason, the pointwise form of the spectral collocation method could be referred to as *strong* form. $\qquad \square$

Since (6.2.22) fits into the framework of the generalized Galerkin method, its analysis can be carried out on the ground of Theorem 5.5.1 (with the obvious change of notations). For that, we need to check assumption (5.5.2), and this is accomplished in the following

Lemma 6.2.1 *The bilinear form $a_N(\cdot, \cdot)$ satisfies*

(6.2.25) $\qquad a_N(z_N, v_N) \leq 9|z_N|_{1,\Omega}|v_N|_{1,\Omega} \qquad \forall\ z_N, v_N \in \mathbb{Q}_N$

and

(6.2.26) $\qquad a_N(v_N, v_N) \geq |v_N|^2_{1,\Omega} \qquad \forall\ v_N \in \mathbb{Q}_N$.

Proof. If $p, q \in C^0(\overline{\Omega})$, using the Schwarz inequality in its algebraic form gives

$$(p, q)_N \leq \left(\sum_{i,j=0}^N p^2(\mathbf{x}_{ij})w_{ij} \right)^{1/2} \left(\sum_{i,j=0}^N q^2(\mathbf{x}_{ij})w_{ij} \right)^{1/2} ,$$

i.e.,

$$(p, q)_N \leq ||p||_N ||q||_N ,$$

where $|| \cdot ||_N = (\cdot, \cdot)_N^{1/2}$ is the discrete norm. Then

$$a_N(z_N, v_N) \leq ||\nabla z_N||_N ||\nabla v_N||_N$$

and (6.2.25) follows from (4.5.41). Inequality (6.2.26) is proven similarly. $\quad\square$

We notice moreover that

(6.2.27) $\quad ||f||_N^2 = \sum_{i,j=0}^N f^2(\mathbf{x}_{ij})w_{ij} \leq \max_{\mathbf{x} \in \overline{\Omega}} |f^2(\mathbf{x})| \sum_{i,j=0}^N w_{ij} = 4||f||^2_{C^0(\overline{\Omega})}$.

Therefore, in view of the above lemma and using (4.5.41) for estimating $||u_N||_N$ in terms of $||u_N||_0$, the following bound can be easily obtained

$$|u_N|_{1,\Omega} \leq 6C_\Omega^{1/2}||f||_{C^0(\overline{\Omega})} ,$$

producing uniform stability since f is continuous. Here C_Ω is the Poincaré constant (see (6.1.18)).

A straightforward application of (5.5.4) produces:

$$|u - u_N|_{1,\Omega} \leq \inf_{w_N \in \mathbb{Q}_N^0} \left[2|u - w_N|_{1,\Omega} \right.$$

$$+ \sup_{v_N \in \mathbb{Q}_N^0} \frac{|(\nabla w_N, \nabla v_N) - (\nabla w_N, \nabla v_N)_N|}{|v_N|_{1,\Omega}} \Big]$$

$$+ \sup_{v_N \in \mathbb{Q}_N^0} \frac{|(f, v_N) - (f, v_N)_N|}{|v_N|_{1,\Omega}} .$$

If we take $w_N \in \mathbb{Q}_{N-1}^0$, we see that

$$(\nabla w_N, \nabla v_N) = (\nabla w_N, \nabla v_N)_N \qquad \forall\ v_N \in \mathbb{Q}_N^0$$

owing to (4.5.40). On the other hand, applying the two-dimensional extension of (4.4.24) (which is based on (4.5.43)) and the Poincaré inequality (6.1.18) it follows:

$$\frac{|(f, v_N) - (f, v_N)_N|}{|v_N|_{1,\Omega}} \leq C N^{-r} \|f\|_r , \quad r \geq 2 .$$

Using (4.5.37), we deduce

$$\|u - u_N\|_{1,\Omega} \leq C(N^{1-s}\|u\|_{s,\Omega} + N^{-r}\|f\|_{r,\Omega}) ,$$

provided $u \in H^s(\Omega) \cap H_0^1(\Omega)$ for some $s \geq 1$, and $f \in H^r(\Omega)$ for some $r \geq 2$.

A collocation problem like (6.2.24) can be set up using the Chebyshev Gauss-Lobatto nodes (4.5.19) rather than the Legendre Gauss-Lobatto ones. In this case, problem (6.2.24) would still admit the equivalent form (6.2.22) but now $(\cdot, \cdot)_N$ denotes the Chebyshev discrete scalar product (4.5.20), and

$$(6.2.28) \qquad a_N(z, v) := \left(\nabla z, \frac{\nabla(vw)}{w} \right)_N ,$$

where $w(x_1, x_2) = [(1 - x_1^2)(1 - x_2^2)]^{-1/2}$ is the Chebyshev weight function. This form can be proven to satisfy estimates like (6.2.25) and (6.2.26) with constants independent of N, provided the seminorm $|\cdot|_{1,\Omega}$ is replaced by the Chebyshev seminorm

$$|v|_{1,w,\Omega} := \left(\int_\Omega |\nabla v|^2 w \right)^{1/2} .$$

(A proof can be found in Funaro (1981); see also Canuto, Hussaini, Quarteroni and Zang (1988), Theor. 11.2.) Moreover, by proceeding as in (6.2.23) it can be shown that

$$a_N(z_N, v_N) = -(\Delta z_N, v_N)_N \quad \forall z_N, v_N \in \mathbb{Q}_N^0 ,$$

as the function $x_i \to D_i(v_N w)w^{-1}$ belongs to $\mathbb{P}_{N-1}$, $i = 1, 2$. Therefore, using (6.2.28) in (6.2.22) still gives a collocation problem.

By still applying Theorem 5.5.1, it can be shown that the Chebyshev collocation solution is stable and convergent (with respect to the Chebyshev norm) and satisfies an error bound like that obtained in the Legendre case. For a proof we refer to Bernardi and Maday (1992), pp. 90–92.

In the case of different kind of boundary conditions, the spectral collocation scheme (6.2.22) needs to be modified accordingly. Consider for instance the homogeneous Neumann problem:

$$(6.2.29) \qquad \begin{cases} -\Delta u + a_0 u = f & \text{in } \Omega \\ \dfrac{\partial u}{\partial n} = 0 & \text{on } \partial\Omega , \end{cases}$$

with $\inf_\Omega a_0 > 0$ in order to ensure existence and uniqueness of the solution. Its Legendre collocation approximation (in *weak* form) reads

$$(6.2.30) \qquad \text{find } u_N \in \mathbb{Q}_N : (\nabla u_N, \nabla v_N)_N + (a_0 u_N, v_N)_N$$
$$= (f, v_N)_N \qquad \forall\, v_N \in \mathbb{Q}_N \ .$$

The stability and convergence analysis holds substantially unchanged (with obvious modification due to the fact that both test and trial functions don't vanish on $\partial\Omega$ any longer).

Concerning the pointwise interpretation of (6.2.30), using (6.2.23) and proceeding straightforwardly, we obtain

$$(6.2.31) \quad (-\Delta u_N + a_0 u_N)(\mathbf{x}_{ij}) = f(\mathbf{x}_{ij}) \qquad \forall\, \mathbf{x}_{ij} \in \Omega$$

$$(6.2.32) \quad \frac{\partial u_N}{\partial n}(\mathbf{x}_{ij}) = \gamma_{ij}(f + \Delta u_N - a_0 u_N)(\mathbf{x}_{ij}) \qquad \forall\, \mathbf{x}_{ij} \in \partial\Omega \setminus C_P \ ,$$

where C_P denotes the set of the four corner points, $\mathbf{x}_{ij} = (x_i, x_j)$, and

$$\gamma_{ij} = \begin{cases} w_i & \text{if } i = 0 \text{ or } i = N \ , \ \forall\, j = 1, ..., N-1 \\ w_j & \text{if } j = 0 \text{ or } j = N \ , \ \forall\, i = 1, ..., N-1 \ . \end{cases}$$

We recall that w_i (respectively, w_j) is the one-dimensional Legendre Gauss-Lobatto weight associated to the collocation node x_i (respectively, x_j) .

We deduce that the differential equation is still satisfied at each internal collocation point. Instead, at any Gaussian point lying on the boundary, the Neumann condition is no longer satisfied exactly (as it would be in the *strong* form of the collocation method), but only up to a constant (that goes to zero as far as N goes to infinity) multiplied by the residual of the differential equation therein.

If $\mathbf{x}_{ij}$ is a corner point, then (6.2.32) needs to be modified as follows. Suppose, e.g., that $\mathbf{x}_{ij} = \mathbf{x}_{0N} = (1, -1)$ (recall that we have assumed the Legendre nodes ordered from right to left: see Section 4.4.2). Then the homogeneous Neumann condition reads

$$(6.2.33) \quad \left(w_N \frac{\partial u_N}{\partial x_1} - w_0 \frac{\partial u_N}{\partial x_2}\right)(1, -1) = w_0 w_N(f + \Delta u_N - a_0 u_N)(1, -1) \ .$$

The Chebyshev approximation of the Neumann problem deserves a more sophisticated analysis, as the weight function $w(\mathbf{x})$ is singular on the boundary, whereas the solution and the test functions do not vanish on $\partial\Omega$. This problem has been considered by Canuto and Quarteroni (1984) (see also Funaro (1992), pp. 194–196).

Consider now the more general case (6.1.13), where the operator L is the one defined in (6.1.1), its coefficients a_{ij}, b_i, a_0 and the datum f are supposed to be continuous in $\overline{\Omega}$, whereas the boundary datum g is continuous in $\overline{\Gamma_N}$. Moreover, we assume that $\operatorname{div} \mathbf{b}$ is continuous in $\overline{\Omega}$ and, for simplicity, that

$c_i = 0$ for $i = 1, 2$. The first order term in the operator L can be split as follows

$$(6.2.34) \qquad \sum_{i=1}^{2} D_i(b_i z) = \frac{1}{2}(\text{div}\,\mathbf{b})z + \frac{1}{2}\left[\sum_{i=1}^{2} D_i(b_i z) + \sum_{i=1}^{2} b_i D_i z\right].$$

This corresponds to the decomposition of $\text{div}(\mathbf{b}\,z)$, defined in the space $H_0^1(\Omega)$, into its symmetric and skew-symmetric part with respect to the $L^2(\Omega)$ scalar product. Similarly, the bilinear form $\mathcal{A}(z, v)$ can also be written as

$$\mathcal{A}(z, v) = \sum_{i,j=1}^{2} (a_{ij} D_j z, D_i v) - \frac{1}{2}\sum_{i=1}^{2}[(b_i z, D_i v) - (b_i D_i z, v)]$$
$$+ \left(\left(\frac{1}{2}\,\text{div}\,\mathbf{b} + a_0\right)z, v\right) - \frac{1}{2}(\mathbf{b}\cdot\mathbf{n}\,z, v)_{\partial\Omega}$$

for each $z, v \in H^1(\Omega)$.

Therefore we are led to (the weak form of) the Legendre collocation approximation problem:

find $u_N \in V_N$:

$$(6.2.35) \qquad \mathcal{A}_N(u_N, v_N) = (f, v_N)_N + \sum_{k=1}^{4}(g, v_N)_{N, \Gamma_N \cap S_k} \qquad \forall\, v_N \in V_N,$$

where V_N is given in (6.2.16), $(\cdot, \cdot)_{N, \Gamma_N \cap S_k}$ is the Legendre Gauss-Lobatto quadrature formula on $\Gamma_N \cap S_k$ (S_k, $k = 1, ..., 4$, being the sides of $\overline{\Omega}$), and for each $z, v \in C^1(\overline{\Omega})$

$$\mathcal{A}_N(z, v) := \sum_{i,j=1}^{2} (a_{ij} D_j z, D_i v)_N - \frac{1}{2}\sum_{i=1}^{2}[(b_i z, D_i v)_N - (b_i D_i z, v)_N]$$
$$+ \left(\left(\frac{1}{2}\,\text{div}\,\mathbf{b} + a_0\right)z, v\right)_N - \frac{1}{2}\sum_{k=1}^{4}(\mathbf{b}\cdot\mathbf{n}\,z, v)_{N, \Gamma_N \cap S_k}.$$

Notice that $\mathcal{A}_N(\cdot, \cdot)$ is uniformly coercive over V_N if conditions (6.1.19) and (6.1.21) (restricted on Γ_N) are satisfied.

The pointwise interpretation of (6.2.35) is as follows: u_N belongs to $\mathbb{Q}_N$ and satisfies

$$(6.2.36) \quad L_N u_N(\mathbf{x}_{ij}) = f(\mathbf{x}_{ij}) \qquad \forall\, \mathbf{x}_{ij} \in \Omega$$

$$(6.2.37) \quad u_N(\mathbf{x}_{ij}) = 0 \qquad \forall\, \mathbf{x}_{ij} \in \overline{\Gamma_D}$$

$$(6.2.38) \quad \left(\frac{\partial u_N}{\partial n_L} - g\right)(\mathbf{x}_{ij}) = \gamma_{ij}(f - L_N u_N)(\mathbf{x}_{ij}) \qquad \forall\, \mathbf{x}_{ij} \in \Gamma_N \setminus C_P.$$

The operator L_N is a particular *pseudo-spectral* approximation to L. It is precisely obtained from L (where the first order terms have been written in the form (6.2.34)), by replacing any derivative in the first and second order terms by the pseudo-spectral derivative (see Section 4.4.2)). Therefore, setting

(6.2.39) $$D_{N_j} v := D_j(I_N v) \quad \forall \, v \in C^0(\overline{\Omega}) \, , \, j = 1, 2 \, ,$$

we have

$$L_N u_N = - \sum_{i,j=1}^{2} D_{N_i}(a_{ij} D_j u_N) + \frac{1}{2} \sum_{i=1}^{2} [D_{N_i}(b_i u_N) + b_i D_i u_N]$$
(6.2.40)
$$+ \left(\frac{1}{2} \operatorname{div} \mathbf{b} + a_0 \right) u_N \, .$$

Thus (6.2.36) states that at each internal collocation point the pseudo-spectral residual must vanish. On the other hand, the Dirichlet condition is enforced exactly at each node of $\overline{\Gamma_D}$, see (6.2.37), while at each node of Γ_N the Neumann condition is satisfied up to a small constant (that vanishes as far as N goes to ∞) multiplied by the pseudo-spectral residual at the same node. Usual modification arises in (6.2.38) when $\mathbf{x}_{ij}$ is a corner point.

The stability and convergence analysis can be carried out straightforwardly by applying Theorem 5.5.1, and using the interpolation error estimates of Section 4.5.2. The weak formulation of the Neumann condition is especially addressed in Funaro (1986, 1988), Bernardi and Maday (1992), Cividini, Quarteroni and Zampieri (1993). A similar approach can be also performed for the Chebyshev case.

In Tables 6.2.1 and 6.2.2 we report the relative error in the discrete maximum norm (i.e., the maximum is taken over the grid-points solely) generated by either the finite element and spectral collocation methods. The boundary value problem at hand is the Poisson problem (6.1.28). This time however we are dealing with non-homogeneous Dirichlet data $u = \varphi$ on $\partial\Omega$.

The finite element Galerkin method for this non-homogeneous Dirichlet problem is described in Remark 6.2.2. In the current case, we have $a(u, v) = (\nabla u, \nabla v)$. The space V_h is introduced in (6.2.4), with X_h^1 defined in (3.2.4) for $k = 1$. The spectral method is the Legendre collocation one that is formulated in (6.2.24), excepting that now $u_N(\mathbf{x}_{ij}) = \varphi(\mathbf{x}_{ij})$ for each $\mathbf{x}_{ij} \in \partial\Omega$. The parameter N refers to the number of grid-points; precisely, the domain Ω has been subdivided into $2N^2$ triangles of equal size in the case of finite elements, and we are using $(N+1)^2$ Gauss-Lobatto nodes for the collocation problem.

First test case (Table 6.2.1)
The domain is $\Omega = (0,1)^2$, the exact solution is the C^∞-function $u(x_1, x_2) = \cos(4\pi x_1) \sin(4\pi x_2)$. In this situation, the rate of convergence of the finite element error is almost quadratic with respect to the mesh size $h = \sqrt{2}/N$

Table 6.2.1. Behaviour of the error in the discrete maximum norm for the first test case

N	Finite Elements	Spectral Collocation
8	1.948e-1	1.518e-2
12	9.678e-2	6.943e-5
16	5.732e-2	7.700e-8
20	3.823e-2	3.696e-11
24	2.550e-2	1.769e-14
32	1.451e-2	

(as stated in Remark 6.2.3), while the spectral error decays esponentially with respect to N.

Second test case (Table 6.2.2)
The domain is $\Omega = (0, 2) \times (0, 1)$, the exact solution is

$$u(x_1, x_2) = \begin{cases} \sin(\pi x_2) & \text{in } [0, 1] \times [0, 1] \\ \sin(\pi x_2) + (x_1 - 1)^\alpha & \text{in } [1, 2] \times [0, 1] \end{cases}.$$

For $\alpha \geq 1$, this function belongs to the Sobolev space $H^\alpha(\Omega)$. As in the former case, the finite element error decays almost quadratically, no matter how large is $\alpha \geq 2$ (since we are using linear elements). Instead, the rate of decay of the spectral error benefits from the increasing order of regularity of u, as predicted by the convergence theory.

Table 6.2.2. Behaviour of the error in the discrete maximum norm for the second test case

	Finite Elements		Spectral Collocation		
N	$\alpha = 2$	$\alpha = 4$	$\alpha = 2$	$\alpha = 4$	$\alpha = 6$
8	4.102e-4	4.390e-4	2.563e-2	2.878e-4	7.764e-5
12	1.871e-4	2.016e-4	1.892e-2	7.501e-5	8.017e-6
16	1.066e-4	1.139e-4	1.491e-2	2.765e-5	1.598e-6
20	6.847e-5	7.309e-5	1.228e-2	1.245e-5	4.523e-7
24	4.758e-5	5.095e-5	1.043e-2	6.403e-6	1.599e-7

6.2.3 Generalized Galerkin Method

As we already pointed out in Section 5.5, the generalized Galerkin method reads:

(6.2.41)   find $u_h \in V_h : \mathcal{A}_h(u_h, v_h) = \mathcal{F}_h(v_h)$ $\forall \, v_h \in V_h$,

where $\mathcal{A}_h(\cdot, \cdot)$ is a bilinear form defined on $V_h \times V_h$ (and not necessarily on $V \times V$), and $\mathcal{F}_h(\cdot)$ is a linear functional defined on V_h.

Besides the weak formulation for the spectral collocation method we have considered in the preceding Section, we present another example where this method is usefully applied to elliptic boundary value problems. We will focus on the Dirichlet problem for the elliptic operator $Lu = -\Sigma_{ij} D_j(a_{ij} D_i u)$, choosing $V = H_0^1(\Omega)$, $V_h = X_h^k \cap H_0^1(\Omega)$, X_h^k defined in (3.2.4).

We make use of numerical integration to evaluate both the bilinear form and the source term. This means that each integral over the polygonal domain Ω is approximated as

(6.2.42) $$\int_\Omega \varphi \simeq \sum_{K \in \mathcal{T}_h} \sum_{j=1}^M \int_K \omega_{j,K} \varphi(\mathbf{b}_{j,K}) =: Q_h(\varphi) \, ,$$

where the weights $\omega_{j,K}$ and the nodes $\mathbf{b}_{j,K}$ are derived from a quadrature formula defined on the reference element $\hat{K}$. More precisely, these weights and nodes are defined as

$$\omega_{j,K} = (\det B_K)\hat{\omega}_j \, , \quad \mathbf{b}_{j,K} = T_K(\hat{\mathbf{b}}_j) \, ,$$

where $\hat{\omega}_j$ and $\hat{\mathbf{b}}_j$ are the weights and nodes of the quadrature formula chosen on $\hat{K}$, and $T_K(\hat{\mathbf{x}}) = B_K \hat{\mathbf{x}} + \mathbf{b}_K$ is the affine map from $\hat{K}$ onto K. The finite element problem with numerical integration reads as (6.2.41) with $\mathcal{F}_h(v_h) = Q_h(v_h)$ and $\mathcal{A}_h(u_h, v_h) = Q_h(\sum_{i,j} a_{ij} D_j u_h D_i v_h)$.

Due to the presence of pointvalues, both $\mathcal{A}_h(\cdot, \cdot)$ and $\mathcal{F}_h(\cdot)$ should now be defined on V_h and not on V. Furthermore, we need to assume that the coefficients of both the operator L and the right hand side f are continuous functions on $\overline{\Omega}$.

Under the assumption that $\mathcal{A}_h(\cdot, \cdot)$ is uniformly coercive in $V_h \times V_h$ (see (5.5.2)), the error estimate (5.5.4) holds. Besides the approximation error

$$\inf_{w_h \in V_h} \|u - w_h\| \, ,$$

two consistency errors also appear.

We notice at first that, in several important cases, the bilinear form $\mathcal{A}_h(\cdot, \cdot)$ is in fact uniformly coercive in V_h. For instance this is true when $V_h = X_h^k \cap H_0^1(\Omega)$ (X_h^k defined in (3.2.4)), $k = 1, 2, 3$, $Lu = -\Sigma_{ij} D_j(a_{ij} D_i u)$ and either one of the following quadrature formulae in $\hat{K}$ is chosen:

$$(6.2.43) \qquad \int_{\hat{K}} \hat{\varphi} \simeq \text{meas}(\hat{K}) \hat{\varphi}(\hat{\mathbf{b}}_0) \quad , \quad k = 1$$

$$(6.2.44) \qquad \int_{\hat{K}} \hat{\varphi} \simeq \frac{1}{3} \text{meas}(\hat{K}) \sum_{1 \leq i < j \leq 3} \hat{\varphi}(\hat{\mathbf{b}}_{ij}) \quad , \quad k = 2$$

$$(6.2.45) \qquad \int_{\hat{K}} \hat{\varphi} \simeq \frac{1}{60} \text{meas}(\hat{K}) \left[3 \sum_{i=1}^{3} \hat{\varphi}(\hat{\mathbf{b}}_i) \right. \\ \left. + 8 \sum_{1 \leq i < j \leq 3} \hat{\varphi}(\hat{\mathbf{b}}_{ij}) + 27 \hat{\varphi}(\hat{\mathbf{b}}_0) \right] \quad , \quad k = 3 \ .$$

Here, $\hat{\mathbf{b}}_0$ is the center of gravity of $\hat{K}$; $\hat{\mathbf{b}}_i$, $i = 1, 2, 3$, are the vertices; $\hat{\mathbf{b}}_{ij}$, $i, j = 1, 2, 3$, are the midpoints of each side. These formulae are exact on $\mathbb{P}_1, \mathbb{P}_2, \mathbb{P}_3$, respectively. The proof of the uniform coerciveness of $\mathcal{A}_h(\cdot, \cdot)$ is such cases is given, e.g., in Ciarlet (1991), pp. 193–196.

We now look at the consistency errors. Define at first the quadrature error

$$(6.2.46) \qquad E_K(\varphi) := \int_K \varphi - \sum_{j=1}^{L} \omega_{j,K} \varphi(\mathbf{b}_{j,K}) \ .$$

The following results will be useful.

Proposition 6.2.3 *Assume that the quadrature formula on $\hat{K}$ is exact on $\mathbb{P}_{2k-2}$. Then there exists a constant C, independent of h, such that for all $K \in \mathcal{T}_h$*

$$(6.2.47) \qquad |E_K(aD_ipD_jq)| \leq C \, h_K^k \, \|a\|_{W^{k,\infty}(K)} \, \|p\|_{k,K} \, |q|_{1,K}$$

for any $a \in W^{k,\infty}(K)$, $p, q \in \mathbb{P}_k$, $i, j = 1, ..., d$, and

$$(6.2.48) \qquad |E_K(fp)| \leq C \, h_K^k \, [\text{meas}(K)]^{1/2} \, \|f\|_{W^{k,\infty}(K)} \, \|p\|_{1,K}$$

for any $f \in W^{k,\infty}(K)$, $p \in \mathbb{P}_k$.

A proof is given in Ciarlet (1991), pp. 199–203.

We are now in a condition to prove the following error estimate for the approximation (6.2.41) to (6.1.5).

Theorem 6.2.2 *Let $\mathcal{T}_h$ be a regular family of triangulations. Assume that the quadrature formula on $\hat{K}$ is exact on $\mathbb{P}_{2k-2}$ and that its weights $\hat{\omega}_j$ are positive. If the solution $u \in H^{k+1}(\Omega)$, the coefficients $a_{ij} \in W^{k,\infty}(\Omega)$ and the datum $f \in W^{k,\infty}(\Omega)$, then there exists a constant C independent of h such that*

$$(6.2.49) \qquad ||u - u_h||_1 \leq Ch^k \Big[|u|_{k+1} + ||u||_{k+1} \sum_{i,j=1}^{d} ||a_{ij}||_{W^{k,\infty}(\Omega)}$$

$$+ ||f||_{W^{k,\infty}(\Omega)} \Big] \ .$$

Proof. From Theorem 3.4.2 we already know that

$$\inf_{w_h \in V_h} ||u - w_h||_1 \leq \dot{C} h^k |u|_{k+1} \ .$$

Moreover, for any $v_h \in V_h$ we deduce from (6.2.47) that

$$|\mathcal{A}(\pi_h^k(u), v_h) - \mathcal{A}_h(\pi_h^k(u), v_h)| \leq \sum_{K \in \mathcal{T}_h} \sum_{i,j=1}^{d} |E_K(a_{ij} D_j \pi_h^k(u) D_i v_h)|$$

$$\leq Ch^k \left(\sum_{i,j=1}^{d} ||a_{ij}||_{W^{k,\infty}(\Omega)} \right) \left(\sum_{K \in \mathcal{T}_h} ||\pi_h^k(u)||_{k,K} \right) |v_h|_1 \ .$$

In addition, writing $\pi_h^k(u)$ as $\pi_h^k(u) - u + u$, we obtain

$$\sum_{K \in \mathcal{T}_h} ||\pi_h^k(u)||_{k,K} \leq C \left(||u||_k + \sum_{K \in \mathcal{T}_h} ||u - \pi_h^k(u)||_{k,K} \right)$$

$$\leq C(||u||_k + h|u|_{k+1}) \leq C ||u||_{k+1} \ .$$

Proceeding in an analogous way, from (6.2.48) we obtain

$$|\mathcal{F}(v_h) - \mathcal{F}_h(v_h)| \leq \sum_{K \in \mathcal{T}_h} |E_K(f v_h)| \leq C h^k \left[\text{meas}(\Omega) \right]^{1/2} ||f||_{W^{k,\infty}(\Omega)} ||v_h||_1 \ ,$$

thus (6.2.49) follows. $\qquad \qquad \square$

Other examples of the generalized Galerkin method, in the framework of the approximation of advection-diffusion equations or of the Stokes system, will be presented and analyzed in Sections 8.3 and 9.4, respectively.

6.3 Algorithmic Aspects

In this Section we describe and analyze the structure of the algebraic problems arising from the discretization of elliptic boundary value problems via Galerkin or collocation methods. We will mainly focus on the Galerkin finite element problem (6.2.1) and the spectral collocation problem (6.2.22) (or (6.2.35)).

6.3.1 Algebraic Formulation

As we already noticed in Section 5.2, when considering the finite element approximation, the unknowns of the finite dimensional problem (6.2.1) are given by the pointvalues of u_h at the finite element nodes $\mathbf{a}_j$. In fact, each element $u_h \in V_h$ can be represented through the basis functions φ_j:

$$(6.3.1) \qquad u_h(\mathbf{x}) = \sum_{j=1}^{N_h} u_h(\mathbf{a}_j)\varphi_j(\mathbf{x}) \ ,$$

where N_h denotes the total number of degrees of freedom in V_h, i.e., the dimension of V_h. Thus, denoting by

$$(6.3.2) \qquad \boldsymbol{\xi}_{fe} := \{u_h(\mathbf{a}_j)\}_{j=1,\ldots,N_h}$$

and

$$(6.3.3) \qquad \mathbf{F}_{fe} := \{(f,\varphi_i)\}_{i=1,\ldots,N_h} \ ,$$

problem (6.2.1) can be rewritten as

$$(6.3.4) \qquad A_{fe}\boldsymbol{\xi}_{fe} = \mathbf{F}_{fe} \ .$$

The matrix A_{fe} is given by

$$(6.3.5) \qquad (A_{fe})_{ij} := \mathcal{A}(\varphi_j, \varphi_i) \ , \quad i,j = 1,\ldots,N_h \ ,$$

and it is called the finite element *stiffness* matrix.

Also in the spectral collocation case the pointvalues of the approximate solution are chosen as unknown. Consider for example the homogeneous Dirichlet boundary value problem for the Laplace operator (6.1.28), with $\Omega = (-1,1)^2$. Its spectral Legendre collocation approximation (in the weak form) (6.2.22) admits the equivalent formulation (6.2.24). The latter is precisely the one which is more convenient to implement. If we denote by $N_h = (N-1)^2$ the number of internal collocation points $\{\mathbf{x}_i\}$, and by ψ_j, $j = 1,\ldots,N_h$, the Lagrangian basis function associated to $\mathbf{x}_j$, i.e., $\psi_j \in \mathbb{Q}_N^0$ and $\psi_j(\mathbf{x}_i) = \delta_{ij}$, then

$$u_N(\mathbf{x}) = \sum_{j=1}^{N_h} u_N(\mathbf{x}_j)\psi_j(\mathbf{x}) \ .$$

The algebraic reformulation of (6.2.22) reads

$$(6.3.6) \qquad A_{sp}\boldsymbol{\xi}_{sp} = \mathbf{F}_{sp} \ ,$$

where

$$(6.3.7) \qquad \begin{aligned} (A_{sp})_{ij} &:= a_N(\psi_j, \psi_i) \ , \ \boldsymbol{\xi}_{sp} := \{u_N(\mathbf{x}_j)\}_{j=1,\ldots,N_h} \\ \mathbf{F}_{sp} &:= \{(f,\psi_i)_N\}_{i=1,\ldots,N_h} \ . \end{aligned}$$

An analogous derivation can be performed for the Chebyshev case.

The entries of the spectral collocation stiffness matrix A_{sp} can be characterized as follows. Let us limit ourselves to the Legendre case and set $\mathbf{x}_j = (x_{j_1}, x_{j_2})$, $1 \leq j_1, j_2 \leq N - 1$, where $x_j \in [-1, 1]$ denotes the j-th Legendre Gauss-Lobatto node (see Sections 4.4.2), and w_j the corresponding weight (see (4.4.13)). Furthermore, let l_j denote the Lagrangian basis function associated with x_j, i.e., $l_j \in \mathbb{P}_N$ and $l_j(x_i) = \delta_{ij}$ for $i, j = 0, ..., N$. Then

$$a_N(\psi_j, \psi_i) = a_N(l_{j_1} l_{j_2}, l_{i_1} l_{i_2}) = \alpha_{j_1 i_1} \delta_{j_2 i_2} w_{j_2} + \alpha_{j_2 i_2} \delta_{j_1 i_1} w_{j_1} \; ,$$

where (see Bernardi and Maday (1992), p. 102)

$$\alpha_{jr} = \begin{cases} \dfrac{4}{N(N+1)L_N(x_j)L_N(x_r)(x_j - x_r)^2} & \text{if } j \neq r \\ \dfrac{2}{3(1 - x_j^2)L_N^2(x_j)} & \text{if } j = r \end{cases} .$$

Very often, however, the method is implemented in the pure collocation form (6.2.24). For that, we recall that the matrix expressing the pseudo-spectral derivative in one-dimension is given by the matrix D_N in (4.4.20) or (4.3.24). If we enumerate the nodes in $[-1, 1]^2$ rowise from right to left and proceeding downward, the matrix $D_{N,1}$ representing the derivative with respect to x_1 turns out to be a block-matrix. Each block is of dimension $(N + 1) \times (N + 1)$, and the block of position (i, j) is given by

$$(D_{N,1})_{i,j} = \delta_{ij} D_N \quad , \quad i, j = 1, ..., N + 1 \; .$$

On the other hand, the derivative with respect to x_2 is expressed by the block-matrix $D_{N,2}$, where the blocks are given by

$$(D_{N,2})_{ij} = d_{ij} I \quad , \quad i, j = 1, ..., N + 1 \; ,$$

where I is the $(N + 1) \times (N + 1)$ identity matrix, and d_{ij} denote here the entries of D_N. If we define

$$A^* := -(D_{N,1}^2 + D_{N,2}^2) \; ,$$

problem (6.2.24) can be written in the form

$$(6.3.8) \qquad A_{sp}^c \boldsymbol{\xi}_{sp}^c = \mathbf{F}_{sp}^c$$

where the matrix A_{sp}^c is now obtained from A^* by eliminating the unknowns relative to the nodes on the boundary as the solution is required to be 0 there. Moreover

$$(6.3.9) \qquad \boldsymbol{\xi}_{sp}^c := \{u_N(\mathbf{x}_j)\}_{j=1,...,N_h} \; ,$$

as in (6.3.7), while

$$(6.3.10) \qquad\qquad \mathbf{F}^c_{sp} := \{f(\mathbf{x}_i)\}_{i=1,...,N_h} \quad,$$

where $\{\mathbf{x}_i \,|\, i = 1, ..., N_h\}$ are the collocation nodes internal to Ω.

Another way to write A^c_{sp} for problem (6.2.24) is

$$(6.3.11) \qquad\qquad (A^c_{sp})_{ij} = -\Delta\psi_j(\mathbf{x}_i) \quad, \quad i,j = 1, ..., N_h \quad,$$

where ψ_j is the Lagrangian basis function associated to the node $\mathbf{x}_j$. The matrix A^c_{sp} (which is associated with the strong form of the collocation problem) is not symmetric. Indeed we have

$$(6.3.12) \qquad\qquad D A^c_{sp} = A_{sp}$$

where $D = \mathrm{diag}(w_i)$ and $w_i = w_{i_1} w_{i_2}$ is the Gauss-Lobatto weight associated with the i-th node $\mathbf{x}_i = (x_{i_1}, x_{i_2})$ (see (4.5.38)).

A similar approach is pursued for more general boundary value problems such as, e.g., (6.2.35), which is also quite often implemented in the form (6.2.36)-(6.2.38). In this case, ξ^c_{sp} is the set of the values of u_N at all nodes $\mathbf{x}_i$ but those belonging to Γ_D. On the other hand, the matrix A^c_{sp} is obtained from the matrix A^* associated to L_N by suppressing the rows (and columns) corresponding to the nodes on Γ_D, and replacing each row associated to a node of Γ_N by the corresponding equation (6.2.38). The right hand side $\mathbf{F}^c_{sp}$ is defined accordingly. (For details see, e.g., Carlenzoli and Gervasio (1992).)

The extension of the above considerations to the three-dimensional domain $\Omega = (-1,1)^3$ can be carried out easily.

Whenever the unknowns of the finite dimensional problem (stemming from either finite element or spectral discretizations) are the pointvalues of the approximate solution, we say that we are employing the *physical representation* of the approximate solution. An alternative way arises, for instance, in the spectral Galerkin case, when we look for a solution of the form

$$u_N(\mathbf{x}) = \sum_{k=1}^{N_h} u_k p_k(\mathbf{x}) \quad,$$

where p_k can be either the Legendre or the Chebyshev polynomials, and thus the unknowns are the expansion coefficients u_k (see Section 6.2.1). This case is usually referred to as the *frequency representation* of the approximate solution.

6.3.2 The Finite Element Case

We focus now on the stiffness matrix A_{fe} given by (6.3.5). First of all, under the usual assumption of coerciveness for the bilinear form $\mathcal{A}$, it follows that

$$(6.3.13) \qquad\qquad A_{fe} \text{ is positive definite}$$

(see Remark 5.2.2). Moreover, if the bilinear form is symmetric, i.e., it satisfies $\mathcal{A}(z,v) = \mathcal{A}(v,z)$ for all $z, v \in V$, then

(6.3.14) A_{fe} is symmetric .

In this case, several efficient methods to solve the linear system (6.3.4) are available. To this aim, let us refer to Chapter 2, where both families of direct and iterative methods have been analyzed.

Properties (6.3.13), (6.3.14) hold true no matter how the finite dimensional space V_h is chosen. Instead, any finite element space entails the following peculiar property:

(6.3.15) A_{fe} is sparse .

In fact, the support of each individual basis function is small (see Fig. 3.3.4); consequently, the bilinear form satisfies $\mathcal{A}(\varphi_j, \varphi_i) = 0$ if the indices i and j refer to nodes that are "far enough". More precisely, the structure of A_{fe} depends indeed on the way the nodes have been enumerated, as well as on the type of triangulation. In any case, it remains true that $\mathcal{A}(\varphi_j, \varphi_i) = 0$ for all indices i and j such that supp $\varphi_j \cap$ supp $\varphi_i = \emptyset$.

We are now going to provide a precise evaluation of the condition number $\chi(A_{fe})$ of A_{fe} as a function of the grid-size h. We recall here that

$$\chi(A_{fe}) := \|A_{fe}\| \, \|A_{fe}^{-1}\| \ ,$$

where $\| \cdot \|$ is a suitable matrix-norm (see (2.1.25)). When A_{fe} is symmetric we have $\|A_{fe}\|_2 = \rho(A_{fe})$, where $\rho(A_{fe})$ denotes the spectral radius of A_{fe} (see (2.1.24)). Under the additional assumption that A_{fe} is positive definite, the condition number associated to the norm $\| \cdot \|_2$ is given by:

(6.3.16) $$\chi_2(A) = \chi_{sp}(A_{fe}) := \frac{\lambda_{\max}(A_{fe})}{\lambda_{\min}(A_{fe})}$$

(see Section 2.1.2).

We want to show that, under suitable assumptions on the family of triangulations $\mathcal{T}_h$,

(6.3.17) $\chi_{sp}(A_{fe}) = O(h^{-2})$.

The following definition will be useful.

Definition 6.3.1 A family of triangulations $\mathcal{T}_h$, $h > 0$, is said to be *quasi-uniform* if it is regular and moreover there exists a constant $\tau > 0$ such that

(6.3.18) $\min_{K \in \mathcal{T}_h} h_K \geq \tau h \quad \forall \, h > 0$.

We are now in a position to prove the following results, which are the main steps toward the proof of (6.3.17).

Proposition 6.3.1 *Let $\mathcal{T}_h$ be a quasi-uniform family of triangulations of $\overline{\Omega}$. Then there exist positive constants C_1 and C_2 such that for each $v_h \in V_h$, $v_h = \sum_i \eta_i \varphi_i$,*

$$(6.3.19) \qquad C_1 h^d |\boldsymbol{\eta}|^2 \le \|v_h\|_0^2 \le C_2 h^d |\boldsymbol{\eta}|^2 \quad.$$

Proof. Since $\mathcal{T}_h$ is regular, for any given finite element node the number of elements sharing it is bounded uniformly with respect to h. Moreover, quasi-uniformity entails that $h_K \ge \tau h$. Hence it is enough to show that for any element K

$$(6.3.20) \qquad C_1^* h_K^d \sum_{i=1}^{M} \eta_i^2 \le \int_K v_h^2 \le C_2^* h_K^d \sum_{i=1}^{M} \eta_i^2 \quad.$$

Here M is the number of degrees of freedom associated with K.

First we prove that (6.3.20) holds in the reference element $\hat{K}$. Set $\hat{v} = v_h \circ T_K$, where T_K is the affine map from $\hat{K}$ onto K; thus

$$\hat{v} = \sum_{i=1}^{M} \eta_i \hat{\varphi}_i \quad.$$

Define for each $\boldsymbol{\eta} \in \mathbb{R}^M$, $\boldsymbol{\eta} \ne \mathbf{0}$

$$\psi(\boldsymbol{\eta}) := \frac{\int_{\hat{K}} \hat{v}^2}{\sum_{i=1}^{M} \eta_i^2} \quad.$$

This function ψ is clearly positive and continuous, furthermore it is homogeneous of zero degree, i.e., $\psi(t\boldsymbol{\eta}) = \psi(\boldsymbol{\eta})$ for each $t > 0$. Thus ψ has positive minimum and maximum over the compact set $\{\boldsymbol{\eta} \in \mathbb{R}^M \mid \sum_{i=1}^{M} \eta_i^2 = 1\}$, and from this result it follows that

$$C_1^{**} \sum_{i=1}^{M} \eta_i^2 \le \int_{\hat{K}} \hat{v}^2 \le C_2^{**} \sum_{i=1}^{M} \eta_i^2 \quad.$$

Consider now the integral $\int_K v_h^2$. Writing $v_h = \hat{v} \circ T_K^{-1}$, we have

$$\int_K v_h^2 = \int_K (\hat{v} \circ T_K^{-1})^2 = \int_{\hat{K}} \hat{v}^2 \, |\det B_K| \quad.$$

Choosing for instance $v_h = 1$, it follows

$$|\det B_K| = \frac{\text{meas}(K)}{\text{meas}(\hat{K})} \le C h_K^d \quad.$$

On the other hand, since the family $\mathcal{T}_h$ is regular, we have also

$$|\det B_K| \geq C h_K^d \quad ,$$

and thus (6.3.20) holds. $\square$

Proposition 6.3.2 (Inverse inequality for piecewise-polynomials). *Let $\mathcal{T}_h$ be a quasi-uniform family of triangulations of $\overline{\Omega}$. Then there exists a positive constant C_3 such that for each $v_h \in V_h$*

(6.3.21) $$\|\nabla v_h\|_0^2 \leq C_3 h^{-2} \|v_h\|_0^2 \quad .$$

Proof. As before, since $\mathcal{T}_h$ is quasi-uniform it is enough to prove

(6.3.22) $$\int_K |\nabla v_h|^2 \leq C_3^* h_K^{-2} \int_K v_h^2$$

for each $K \in \mathcal{T}_h$. Working upon the reference element $\hat{K}$, as done in the previous proof, and keeping the same notations adopted therein, we can easily see that

$$\psi^*(\boldsymbol{\eta}) := \frac{\int_{\hat{K}} |\nabla \hat{v}|^2}{\int_{\hat{K}} \hat{v}^2} \quad , \quad \boldsymbol{\eta} \neq 0 \quad ,$$

is a homogeneous function of zero degree, bounded on $\mathbb{R}^M \setminus \{\mathbf{0}\}$. Using now (3.4.4)-(3.4.6), it follows

$$\int_K |\nabla v_h|^2 \leq C \|B_K^{-1}\|^2 \int_K v_h^2 \leq \frac{C}{\rho_K^2} \int_K v_h^2 \quad .$$

The thesis now follows owing to the assumption that $\mathcal{T}_h$ is a regular family of triangulations. $\square$

Now we turn to the estimate for $\chi_{sp}(A_{fe})$. Assuming that A_{fe} is symmetric and positive definite, so that its eigenvalues are real and positive and (6.3.16) holds, we need to evaluate the minimum and maximum eigenvalue of A_{fe}. Writing as before $v_h = \sum_i \eta_i \varphi_i$, we have

(6.3.23) $$\frac{(A_{fe}\boldsymbol{\eta}, \boldsymbol{\eta})}{|\boldsymbol{\eta}|^2} = \frac{\mathcal{A}(v_h, v_h)}{|\boldsymbol{\eta}|^2} \quad ,$$

where $(\cdot, \cdot)$ denotes the euclidean scalar product. Since $\mathcal{A}(\cdot, \cdot)$ is continuous and coercive, $\mathcal{A}(\cdot, \cdot)^{1/2}$ is equivalent to the norm of $H^1(\Omega)$. Thus, using (6.3.19) and (6.3.21) it follows

$$\alpha C_1 h^d \leq \frac{(A_{fe}\boldsymbol{\eta}, \boldsymbol{\eta})}{|\boldsymbol{\eta}|^2} \leq \gamma C_2 h^d (1 + C_3 h^{-2}) \quad ,$$

where α and γ are the coerciveness and continuity constants, respectively. Consequently

$$\chi_{sp}(A_{fe}) = \frac{\lambda_{\max}(A_{fe})}{\lambda_{\min}(A_{fe})} \leq \frac{\gamma C_2}{\alpha C_1}(1 + C_3 h^{-2}) = O(h^{-2}) \ .$$

More precisely, we have shown that any eigenvalue λ of A_{fe} satisfies $\alpha C_1 h^d \leq \lambda \leq \gamma C_2(1 + C_3 h^{-2})h^d$.

In Table 6.3.1 we report the spectral condition number $\chi_{sp}(P^{-1}A_{fe})$ of the preconditioned stiffness matrix associated to the Laplace operator in $\Omega = (0,1)^2$ with Dirichlet condition (case (a)) or mixed condition (Dirichlet on horizontal sides, Neumann on the vertical sides; case (b)). The domain is triangulated into N^2 uniform squares, for several values of N, and piecewise-bilinear finite elements are employed (see (3.2.5) for $k = 1$). The preconditioner P is given by the diagonal of A_{fe} and by its incomplete Cholesky decomposition (see (2.5.9) and (2.5.13)).

Table 6.3.1 (a). The spectral condition number $\chi_{sp}(P^{-1}A_{fe})$ for the Dirichlet boundary value problem

N		P	
	I	diag(A_{fe})	$IC(A_{fe})$
4	3.030	3.035	1.097
8	12.228	12.233	1.775
12	27.647	28.651	3.016
16	49.245	49.248	4.764
20	77.017	77.019	7.015
24	110.961	110.963	9.767

Table 6.3.1 (b). The spectral condition number $\chi_{sp}(P^{-1}A_{fe})$ for the mixed boundary value problem

N		P	
	I	diag(A_{fe})	$IC(A_{fe})$
4	7.105	5.960	1.310
8	26.105	23.631	2.711
12	56.973	53.083	5.020
16	99.639	94.316	8.292
20	154.091	147.329	12.494
24	220.322	212.122	17.647

We can compare the spectrum of A_{fe} with the spectrum of the bilinear form $\mathcal{A}(\cdot, \cdot)$ (still supposing that $\mathcal{A}(\cdot, \cdot)$ is symmetric and coercive). The spectrum of $\mathcal{A}(\cdot, \cdot)$ is defined as the set of $\mu \in \mathbb{R}$ such that there exists an eigenfunction $\omega_h \in V_h$, $\omega_h \neq 0$, satisfying

$$\mathcal{A}(\omega_h, v_h) = \mu(\omega_h, v_h) \qquad \forall \, v_h \in V_h \ .$$

Thus, if μ is an eigenvalue and u_h is the corresponding eigenfunction, we have

(6.3.24)
$$\alpha \leq \frac{\mathcal{A}(\omega_h, \omega_h)}{||\omega_h||_0^2} = \mu \leq \gamma \frac{||\omega_h||_1^2}{||\omega_h||_0^2} \leq \gamma(1 + C_3 h^{-2}) \ ,$$

and consequently the eigenvalues μ of $\mathcal{A}(\cdot, \cdot)$ satisfy $\alpha \leq \mu \leq \gamma(1 + C_3 h^{-2})$.

Notice the extra-factor h^d appearing in the spectrum of the stiffness matrix A_{fe}. For this reason, sometimes A_{fe} is scaled by a factor h^{-d}, so that its spectrum is "equivalent" to that of $\mathcal{A}(\cdot, \cdot)$, and turns out to be a correct finite dimensional approximation of the spectrum of an elliptic operator, which has eigenvalues belonging to $[\alpha, \infty)$.

The finite element system (6.3.4) can be solved using the methods of Chapter 2 (for additional remarks we refer, e.g., to the book by Axelsson and Barker (1984)). The sparsity pattern of matrix A, which in turn depends on the way the finite element nodes have been enumerated, can be conveniently exploited. In this respect, besides what is reported in Section 2.1.1, we refer to some special direct techniques that have been devised for the finite element stiffness matrix.

The *frontal method* (together with its *multifrontal* variant (see Liu (1992)), which is especially suited for multiprocessor implementations) is based on the observation that, along a Gaussian elimination procedure, only few nodal unknowns are active during a single elimination step. This is dynamically accounted for by resorting to an active matrix of little dimension that needs to be dealt with in the central computers memory.

Several criteria can be adopted in order to reduce the *fill-in* (i.e., the replacement of zero-entries in the original matrix by non-zero elements in the final *LU*-factors) during the elimination procedures. They include the minimum-degree algorithm, the reversed Cuthill-Mc Kee algorithm, the nested-dissection algorithm and the refined quotient tree algorithm (e.g., George and Liu (1981)).

Another special technique is based on the classical fast Poisson solver in case of structured finite element grids for the Laplace equation in a rectangle (or a cube).

Among iterative methods, the conjugate gradient one is often used when A is symmetric and positive definite. However, when the grid-size h is sufficiently small, the condition number of A is large (see (6.3.17)) and the convergence rate degenerates accordingly (see (2.4.34)). As pointed out in Section 2.5, the cure is to resort to a preconditioner, for instance the incomplete Cholesky decomposition of A (see (2.5.13)).

Another very successful approach is based on the *multi-grid* method, which has been shortly described in an abstract framework in Section 2.7. For a detailed analysis of this method and its applications to partial differential equations we refer to Hackbusch (1985), Bramble (1993).

6.3.3 The Spectral Collocation Case

The Legendre spectral collocation matrix A_{sp} introduced in (6.3.7) is positive definite and symmetric if so is the bilinear form $a_N(\cdot, \cdot)$. For a one-dimensional problem it is full, whereas for the d-dimensional domain $\Omega = (-1, 1)^d$, $d = 2, 3$, the number of its non-zero entries is of the order of N^{d+1} over N^{2d} elements. To give an example, in the case of the approximation of a two-dimensional Neumann boundary value problem its structure is outlined in Fig. 6.3.1. The symbols $\triangle$, $\square$ and $\circ$ correspond to a vertex, a boundary node and an internal node, respectively, whereas the colour white, grey and black indicate a Lagrangian basis function associated to a vertex, a boundary node and an internal node, respectively.

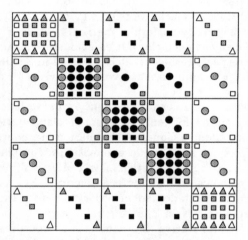

Fig. 6.3.1. The structure of the spectral collocation matrix for $d = 2$ and $N = 4$.

Its conditioning is much more severe than the one of the finite element matrix that refers to a uniform mesh and the same number of grid-points. The same is true for the matrix A_{sp}^c introduced in (6.3.8). Indeed,

$$(6.3.25) \qquad\qquad \chi_{sp}(A_{sp}^c) = O(N^4) \ .$$

This result is true, both for the Legendre and Chebyshev collocation matrices. Hereafter, we present a proof which emulates the one given in the preceding Section for finite elements. We obtain also an estimate like (6.3.25) for the spectral condition number of the matrix A_{sp} in the Chebyshev case. On the contrary, it has been proven that $\chi_{sp}(A_{sp}) = O(N^3)$ in the Legendre case (see Bernardi and Maday (1992), p. 104).

For the sake of simplicity, we confine ourselves to the case of the Laplace operator with homogeneous Dirichlet boundary conditions. We start by considering the Chebyshev case. The matrix A_{sp} is associated with the bilinear form $a_N(\cdot, \cdot)$ defined in (6.2.28), i.e.,

$$a_N(z, v) := \left(\nabla z, \frac{\nabla(vw)}{w} \right)_N \quad ,$$

$(\cdot, \cdot)_N$ being the Chebyshev discrete scalar product defined in (4.5.20). Then

$$(A_{sp})_{ij} := \left(\nabla \psi_j, \frac{\nabla(\psi_i w)}{w} \right)_N = \sum_{k=1}^{N_h} \nabla \psi_j(\mathbf{x}_k) \frac{\nabla(\psi_i w)(\mathbf{x}_k)}{w(\mathbf{x}_k)} w_k \quad ,$$

having denoted by $N_h = (N - 1)^d$ the number of all Gauss-Lobatto nodes internal to $\overline{\Omega}$ induced by a subdivision of any x_1, ..., x_d direction into N subintervals.

Proposition 6.3.3 *There exist two positive constants C_1^* and C_2^* such that for any $v_N \in \mathbb{Q}_N$ with $v_N = \sum_i \eta_i \psi_i$*

(6.3.26) $$C_1^* N^{-d} |\boldsymbol{\eta}|^2 \leq ||v_N||_N^2 \leq C_2^* N^{-d} |\boldsymbol{\eta}|^2 \quad ,$$

where $|| \cdot ||_N$ is the Chebyshev discrete norm defined in (4.5.20).

Proof. We prove it for $d = 1$. Still denoting by w_i the weight associated to the i-th Gauss-Lobatto node we have

$$||v_N||_N^2 = \sum_{i=0}^{N} \eta_i^2 w_i \quad .$$

Since $\pi/2N \leq w_i \leq \pi/N$ for each $i = 0, ..., N$ (see (4.3.13)), the result follows immediately. The generalization to the case $d > 1$ is straightforward. $\square$

The spectral counterpart of (6.3.21) is provided by the following proposition, proven in Canuto and Quarteroni (1982):

Proposition 6.3.4 (Inverse inequality for algebraic polynomials). *There exists a positive constant C_3^* such that for each $v_N \in \mathbb{Q}_N$*

(6.3.27) $$||\nabla v_N||_{0,w}^2 \leq C_3^* N^4 ||v_N||_{0,w}^2 \quad .$$

From (6.3.12) we have $A_{sp} = D A_{sp}^c$, where $D = \text{diag} \{ (\frac{\pi}{N})^d \}$ in the Chebyshev case corresponding to homogeneous Dirichlet data. This has several consequences. At first, since A_{sp}^c has real and positive eigenvalues (see Gottlieb and Lustman (1983), Canuto, Hussaini, Quarteroni and Zang (1988), Sect.

11.4.1), the same occurs to A_{sp}. Furthermore, if $\lambda(A_{sp})$ and $\mu(A_{sp}^c)$ denote eigenvalues of A_{sp} and A_{sp}^c, respectively, we have

$$(6.3.28) \qquad \lambda(A_{sp}) = (\frac{\pi}{N})^d \mu(A_{sp}^c) \ .$$

Now let λ be an eigenvalue of A_{sp} and $\boldsymbol{\xi}$ a corresponding eigenvector. In the current case instead of (6.3.23) we have

$$(6.3.29) \qquad \lambda = \frac{(A_{sp}\boldsymbol{\xi},\boldsymbol{\xi})}{|\boldsymbol{\xi}|^2} = \frac{a_N(v_N,v_N)}{|\boldsymbol{\xi}|^2} \ ,$$

where $v_N = \sum_i \xi_i \psi_i$, and therefore, owing to (6.3.26),

$$(6.3.30) \qquad \frac{a_N(v_N,v_N)}{||v_N||_N^2} C_1^* N^{-d} \leq \lambda \leq \frac{a_N(v_N,v_N)}{||v_N||_N^2} C_2^* N^{-d} \ .$$

Moreover, we remind that (see the extension of (4.3.20) to the d-dimensional case)

$$(6.3.31) \qquad ||v_N||_{0,w}^2 \leq ||v_N||_N^2 \leq 2^d ||v_N||_{0,w}^2 \ .$$

Recalling that $a_N(\cdot,\cdot)$ is continuous and coercive with respect to the Chebyshev seminorm $|v|_{1,w} = ||\nabla v||_{0,w}$ (see Canuto, Hussaini, Quarteroni and Zang (1988), Theor. 11.2), using (6.3.27) and Poincaré inequality (4.3.53) (that also holds in the d-dimensional case: see, e.g., Canuto, Hussaini, Quarteroni and Zang (1988), Sect. 11.1.2) we have

$$\frac{\alpha}{C_\Omega} ||v_N||_{0,w}^2 \leq \alpha ||\nabla v_N||_{0,w}^2 \leq a_N(v_N,v_N) \leq \gamma ||\nabla v_N||_{0,w}^2 \leq \gamma C_3^* N^4 ||v_N||_{0,w}^2 \ .$$

Therefore from (6.3.30) and (6.3.31) it follows

$$(6.3.32) \qquad C_4^* N^{-d} \leq \lambda \leq C_5^* N^{4-d} \ ,$$

and, using (6.3.28), for each eigenvalue μ of A_{sp}^c we find

$$(6.3.33) \qquad K_1^* \leq \mu \leq K_2^* N^4 \ .$$

Hence (6.3.25) holds since

$$\chi_{sp}(A_{sp}^c) = \frac{\mu_{\max}(A_{sp}^c)}{\mu_{\min}(A_{sp}^c)} \ ,$$

and the same result is true for $\chi_{sp}(A_{sp})$.

We also point out that the eigenvalue problem: find $\mu \in \mathbb{R}$ and $\omega_N \in \mathbb{Q}_N^0$, $\omega_N \neq 0$, such that

$$(6.3.34) \qquad a_N(\omega_N,v_N) = \mu(\omega_N,v_N)_N \qquad \forall \, v_N \in \mathbb{Q}_N^0 \ ,$$

is equivalent to the one for A_{sp}^c. As a matter of fact, setting $\omega_N = \sum_j \xi_j \psi_j$ and taking $v_N = \psi_i$ (the i-th Lagrangian function), using discrete integration by parts as done in (6.2.23) we obtain from (6.3.34)

$$\mu\,\xi_i w_i = a_N(\omega_N, \psi_i) = -\Delta\omega_N(\mathbf{x}_i)w_i = -\sum_{j=1}^{N_h} \xi_j \Delta\psi_j(\mathbf{x}_i)w_i \ .$$

Thus, simplifying the coefficient w_i we obtain

$$A_{sp}^c \boldsymbol{\xi} = \mu\,\boldsymbol{\xi} \ .$$

Since this result is true for both the Chebyshev and the Legendre case, we notice that in the latter situation from (6.3.34) one finds at once

$$\frac{1}{3^d C_\Omega} \le \mu = \frac{a_N(\omega_N, \omega_N)}{\|\omega_N\|_N^2} = \frac{\|\nabla\omega_N\|_N^2}{\|\omega_N\|_N^2} \le 3^d C_3^* N^4 \ ,$$

having used (4.5.41), (6.3.27) and Poincaré inequality (6.1.18) for Chebyshev norms. This gives that $\chi(A_{sp}^c) = O(N^4)$ also when considering Legendre polynomials.

The spectral system (6.3.6) or (6.3.8) is generally faced by iterative methods, due to the topological structure of A_{sp} (or A_{sp}^c) that make unsuitable direct methods. (Indeed, a direct method that is based on successive diagonalizations of A_{sp}^c is also available in the case of a constant coefficient operator, see Haldenwang, Labrosse, Abboudi and Deville (1984).) The drawback is that the spectral condition number of both matrices grows like $O(N^4)$ (like $O(N^3)$ for A_{sp} in the Legendre case), and this makes the use of preconditioners compulsory.

Indeed, Orszag (1980) introduced as preconditioner the centered finite difference matrix, computed at the same Gauss-Lobatto grid-points. Later, the finite element matrix was proposed instead (Deville and Mund (1985), Canuto and Quarteroni (1985)), using bilinear elements on the grid induced by the Gauss-Lobatto nodes. Effective generalizations of the finite element preconditioner have been carried out later on by Canuto and Pietra (1991), Deville and Mund (1990) and Quarteroni and Zampieri (1992). For the Laplace equation, both these matrices lead to a preconditioned matrix whose condition number is independent of N. A theoretical analysis has been provided by Haldenwang, Labrosse, Abboudi and Deville (1984).

Spectral preconditioners for the case of the variable coefficient operator (6.1.1) that are based on a constant coefficient one are proposed and analyzed in Guillard and Desideri (1990).

In Table 6.3.2 (a) we report the spectral condition number of the Legendre preconditioned matrix $P^{-1}A_{sp}^c$ for several types of preconditioners P and values of the polynomial degree N. We still refer to the problem (6.1.28) in the domain $\Omega = (0,1)^2$. For the same problem we report in Table 6.3.2

(b) the condition number $\chi_1(P^{-1}A_{sp}^c) := \|P^{-1}A_{sp}^c\|_1\|(P^{-1}A_{sp}^c)^{-1}\|_1$ (see (2.1.22)). Table 6.3.3 reports the spectral condition number for the matrix $P^{-1}A_{sp}$. The preconditioners P for the non-symmetric matrix A_{sp}^c are: the diagonal one introduced in (2.5.10); the incomplete LU (ILU) decomposition (see Section 2.5); the finite difference matrix described above, together with its incomplete decompositions ILU and $RS5$ (see, e.g., Meijerink and van der Vorst (1981), and Canuto, Hussaini, Quarteroni and Zang (1988), Sect. 5.4.2); the matrix $A_{fe}^* := M^{-1}A_{fe}$, where A_{fe} is the finite element stiffness matrix and $M_{ij} := (\varphi_i, \varphi_j)$ is the finite element mass matrix (see Quarteroni and Zampieri (1992)). As for the symmetric matrix A_{sp}, preconditioners are still the diagonal matrix, the incomplete Cholesky decomposition (2.5.13) and the finite element stiffness matrix.

Table 6.3.2 (a). The spectral condition number $\chi_{sp}(P^{-1}A_{sp}^c)$ for the Poisson problem (6.1.28) for different types of preconditioners P

N				P			
	I	$\text{diag}(A_{sp}^c)$	$ILU(A_{sp}^c)$	$ILU(A_{fd})$	$RS5(A_{fd})$	A_{fd}	A_{fe}^*
4	6.909	5.898	1.357	1.690	1.517	1.554	1.465
8	59.501	27.208	2.851	3.836	2.511	1.945	1.352
12	261.451	64.399	5.245	7.129	3.297	2.101	1.355
16	778.872	117.557	8.691	11.705	3.979	2.185	1.359
20	1840.229	186.689	13.231	17.428	4.703	2.238	1.368

Table 6.3.2 (b). The condition number $\chi_1(P^{-1}A_{sp}^c)$ for the Poisson problem (6.1.28) for different types of preconditioners P

N				P		
	I	$\text{diag}(A_{sp}^c)$	$ILU(A_{sp}^c)$	$ILU(A_{fd})$	$RS5(A_{fd})$	A_{fd}
4	12.576	10.893	1.578	2.699	2.140	2.267
8	123.285	46.706	5.421	7.428	4.796	3.311
12	547.125	106.495	11.262	14.875	9.790	3.678
16	1629.664	190.290	19.087	25.143	22.359	3.912
20	3848.456	298.079	28.809	38.559	43.239	4.078

As an example of an operator which is not self-adjoint we consider the advection-diffusion problem

Table 6.3.3. The spectral condition number $\chi_{sp}(P^{-1}A_{sp})$ for the Poisson problem (6.1.28) for different types of preconditioners P

N	I	diag(A_{sp})	$IC(A_{sp})$	A_{fe}
			P	
4	6.019	6.068	1.357	3.517
8	28.954	28.217	2.851	5.473
12	71.333	66.097	5.245	6.115
16	143.664	119.922	8.691	6.440
20	261.369	189.751	13.231	6.635

(6.3.35)
$$\begin{cases} -\Delta u + \operatorname{div}(\mathbf{b}u) = f & \text{in } \Omega = (0,1)^2 \\ u = \varphi & \text{on } \partial\Omega \ , \end{cases}$$

which is approximated by a Legendre collocation method using $(N+1)^2$ grid-points. The associated matrix A_{sp}^c is non-symmetric. Table 6.3.4 refers to a test case in which $\mathbf{b}(x_1, x_2) = (1, x_1 + x_2)$. We report the condition number $\chi_1(P^{-1}A_{sp}^c)$ for several choices of the preconditioner P. For the same problem we display in Fig. 6.3.2 the spectrum distribution in the case $N = 16$ for several preconditioners

Table 6.3.4. The condition number $\chi_1(P^{-1}A_{sp}^c)$ for the advection-diffusion problem (6.3.35) for different types of preconditioners P

N	I	diag(A_{sp}^c)	$ILU(A_{sp}^c)$	$ILU(A_{fd})$	$RS5(A_{fd})$	A_{fd}
			P			
4	12.509	10.623	1.549	2.662	2.168	2.257
8	120.852	45.414	5.234	7.231	4.475	3.383
12	532.105	103.356	10.883	14.433	10.612	3.766
16	1580.752	184.506	18.401	24.302	24.813	3.978
20	3728.608	289.370	27.896	37.109	48.407	4.140

The iteration procedure for spectral collocation methods is typically the conjugate gradient method when A_{sp} is symmetric. Otherwise, for A_{sp}^c (or A_{sp} for operators which are not self-adjoint) the Richardson-Minimum Residual method, or the GMRES method, or else the Bi-CGSTAB method are effectively used.

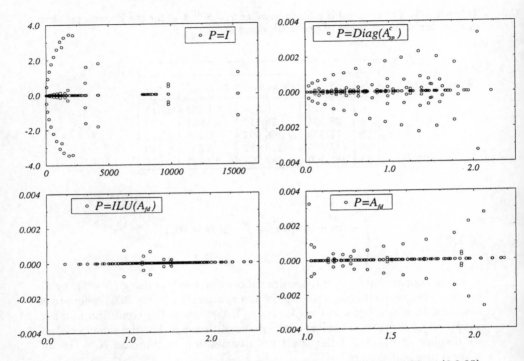

Fig. 6.3.2. The spectrum of $P^{-1}A_{sp}^c$ for the advection-diffusion problem (6.3.35)

For problem (6.1.28) we report in Fig. 6.3.3 the convergence history of the preconditioned iterations for the euclidean norm of the normalized residual. The Legendre spectral collocation method is applied, for both values $N = 12$ (Figs. 6.3.3 (a) and (c)) and $N = 24$ (Figs. 6.3.3 (b) and (d)). The iterative procedure is the Bi-CGSTAB method (see (2.6.42)-(2.6.53)) in the case of the non-symmetric matrix A_{sp}^c, and the CG method (see (2.4.36)-(2.4.41)) in the case of the matrix A_{sp}.

6.4 Domain Decomposition Methods

Domain decomposition methods allow for effective implementation of numerical techniques for partial differential equations on parallel architectures.

Any such method is based on the assumption that the given computational domain, say Ω, is partitioned into subdomains Ω_i, $i = 1, ..., M$, that may or may not overlap. Next, the original problem can be reformulated upon each subdomain Ω_i yielding a family of subproblems of reduced size which are coupled each others through the values of the unknown solution at subdomain interfaces.

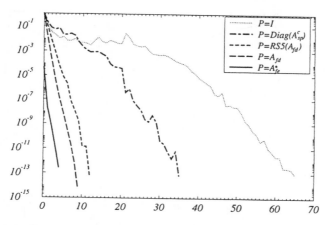

Fig. 6.3.3 (a). Convergence history of preconditioned Bi-CGSTAB iterations for the matrix A_{sp}^c for the Poisson problem (6.1.28) for $N = 12$

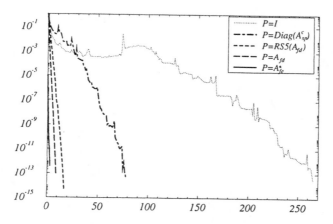

Fig. 6.3.3 (b). Convergence history of preconditioned Bi-CGSTAB iterations for the matrix A_{sp}^c for the Poisson problem (6.1.28) for $N = 24$

The interface coupling can be relaxed at the expense of introducing an iterative process among subdomains, yielding at each step independent subproblems upon subdomains. The latter can be solved simultaneously within a parallel computing environment.

We illustrate the basic aspects of domain decomposition methods in connection with the solution of the model Poisson problem (6.1.28). We aim at reviewing some well known methods (Sections 6.4.1 and 6.4.2), their principal

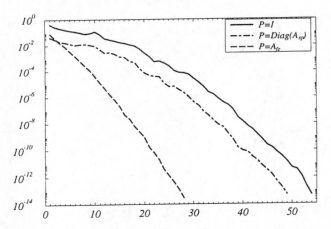

Fig. 6.3.3 (c). Convergence history of preconditioned CG iterations for the matrix A_{sp} for the Poisson problem (6.1.28) for $N = 12$

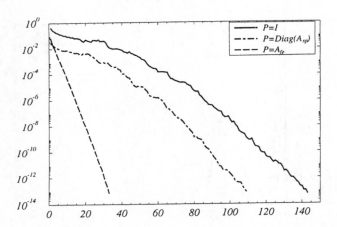

Fig. 6.3.3 (d). Convergence history of preconditioned CG iterations for the matrix A_{sp} for the Poisson problem (6.1.28) for $N = 24$

mathematical properties (Sections 6.4.3 and 6.4.4), and to provide a flavour of the capabilities of these approaches.

6.4.1 The Schwarz Method

The Schwarz alternating procedure (Schwarz (1869)) is the earliest example of domain decomposition approach in the context of partial differential equations. Let us decompose Ω in two overlapping subdomains Ω_1 and Ω_2

such that $\Omega = \Omega_1 \cup \Omega_2$, and denote by $\Gamma_1 := \partial\Omega_1 \cap \Omega_2$, $\Gamma_2 := \partial\Omega_2 \cap \Omega_1$, $\Omega_{1,2} := \Omega_1 \cap \Omega_2$ (see Fig. 6.4.1 for a pair of examples).

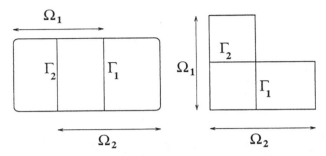

Fig. 6.4.1. Two examples of overlapping partitions of Ω

Let $\tilde{u}^0$ be an initialization function, that vanishes on $\partial\Omega$. The Schwarz method for the solution of (6.1.28) defines two sequences $\tilde{u}^{2k+1}$ and $\tilde{u}^{2k+2}$ ($k \geq 0$) by solving respectively:

(6.4.1)
$$\begin{cases} -\Delta\tilde{u}^{2k+1} = f & \text{in } \Omega_1 \\ \tilde{u}^{2k+1} = \tilde{u}^{2k} & \text{on } \Gamma_1 \\ \tilde{u}^{2k+1} = 0 & \text{on } \partial\Omega_1 \cap \partial\Omega \end{cases}$$

and

(6.4.2)
$$\begin{cases} -\Delta\tilde{u}^{2k+2} = f & \text{in } \Omega_2 \\ \tilde{u}^{2k+2} = \tilde{u}^{2k+1} & \text{on } \Gamma_2 \\ \tilde{u}^{2k+2} = 0 & \text{on } \partial\Omega_2 \cap \partial\Omega \ . \end{cases}$$

The above sequences converge to the solution u of (6.1.28). Precisely, there exist $C_1, C_2 \in (0,1)$ such that for all $k \geq 0$

(6.4.3) $\|u - \tilde{u}^{2k+1}\|_{L^\infty(\Omega_1)} \leq C_1^k C_2^k \|u - \tilde{u}^0\|_{L^\infty(\Gamma_1)}$,

(6.4.4) $\|u - \tilde{u}^{2k+2}\|_{L^\infty(\Omega_2)} \leq C_1^{k+1} C_2^k \|u - \tilde{u}^0\|_{L^\infty(\Gamma_2)}$,

The error reduction constants C_1 and C_2 can be rather close to one if the overlapping region $\Omega_{1,2}$ is thin. The proof of the above estimates can be obtained via maximum principle (see, e.g., Lions (1989)).

The Schwarz method (6.4.1), (6.4.2) can be given the following variational interpretation (Lions (1988)). Define at first for $k \geq 0$

$$u^{2k+1} := \begin{cases} \tilde{u}^{2k+1} & \text{in } \Omega_1 \\ \tilde{u}^{2k} & \text{in } \Omega \setminus \Omega_1 \end{cases} , \quad u^{2k+2} := \begin{cases} \tilde{u}^{2k+2} & \text{in } \Omega_2 \\ \tilde{u}^{2k+1} & \text{in } \Omega \setminus \Omega_2 \end{cases} .$$

These sequences satisfy

$$u^{2k+1} - u^{2k} = P_1(u - u^{2k}) \qquad \forall\, k \geq 0$$
$$u^{2k+2} - u^{2k+1} = P_2(u - u^{2k+1}) \qquad \forall\, k \geq 0 ,$$

or, equivalently,

$$u - u^{2k+1} = (I - P_1)(u - u^{2k}) \qquad \forall\, k \geq 0$$
$$u - u^{2k+2} = (I - P_2)(u - u^{2k+1}) \qquad \forall\, k \geq 0 .$$

Here I is the identity operator, P_i, $i = 1, 2$, is the orthogonal projection of $H_0^1(\Omega)$ onto V_i with respect to the bilinear form $(\nabla w, \nabla v)$ associated with the operator $-\Delta$, i.e., for any $w \in H_0^1(\Omega)$

$$(6.4.5) \qquad P_i w \in V_i \; : \; (\nabla(P_i w - w), \nabla v) = 0 \qquad \forall\, v \in V_i ,$$

and V_i is the closed subspace of $H_0^1(\Omega)$ obtained by extending by 0 each element of $H_0^1(\Omega_i)$ to Ω.

Setting $e^k := u - u^{2k}$ the above relations yield the error recursion formula:

$$(6.4.6) \qquad e^{k+1} = (I - P_2)(I - P_1)e^k = (I - Q)e^k \qquad \forall\, k \geq 0 ,$$

where

$$(6.4.7) \qquad\qquad\qquad Q := P_1 + P_2 - P_2 P_1 .$$

Based on (6.4.6), the Schwarz alternating method can therefore be regarded as an iterative procedure for the solution of the problem

$$(6.4.8) \qquad\qquad\qquad Qu = g$$

(for a suitable right hand side g).

The multiplicative term $P_2 P_1$ is the responsible for Q to be a polynomial operator of second degree, thus preventing method (6.4.1), (6.4.2) from being parallelizable. As a matter of fact, (6.4.1), (6.4.2) is indeed a sequential algorithm. For this reason, (6.4.1), (6.4.2) is nowadays recognized as the *multiplicative* form of the Schwarz alternating method.

The *additive* form (e.g., Matsokin and Nepomnyashchikh (1985), Dryja and Widlund (1987)), originates from replacing (6.4.8) by

$$(6.4.9) \qquad\qquad P^* u := (P_1 + P_2)u = g^* ,$$

where g^* is a new right hand side that is determined in order for (6.4.8) and (6.4.9) to have the same solution. It is easy to see that $g^* = g_1^* + g_2^*$, where $g_1^* \in V_1$, $g_2^* \in V_2$ and

$$(\nabla g_i^*, \nabla \varphi)_{\Omega_i} = (f, \varphi)_{\Omega_i} \qquad \forall\, \varphi \in H_0^1(\Omega_i) , \quad i = 1, 2 .$$

The finite dimensional realization of (6.4.9) is straightforward. Let V_h be a finite dimensional approximation of $V = H_0^1(\Omega)$; as before, we define $V_{i,h}$ to be the subspace of V_h consisting of functions which are 0 outside Ω_i, $i = 1, 2$. With this notation, instead of (6.4.9) we consider

$$(6.4.10) \qquad P_h^* u_h := (P_{1,h} + P_{2,h}) u_h = g_h^* \ ,$$

where $u_h \in V_h$, $g_h^* = g_{1,h}^* + g_{2,h}^*$ and $g_{i,h}^* \in V_{i,h}$ are the solution to the Dirichlet problem

$$(6.4.11) \qquad (\nabla g_{i,h}^*, \nabla \varphi_h)_{\Omega_i} = (f, \varphi_h)_{\Omega_i} \qquad \forall \, \varphi_h \in V_{i,h} \ , \quad i = 1, 2 \ .$$

The additive form of the Schwarz alternating method is then obtained by applying conjugate gradient iterations to the algebraic realization of (6.4.10). Since P_h^* is symmetric and positive definite with respect to the form $(\nabla w, \nabla v)$, the scalar product for the conjugate gradient method is the one induced from the stiffness matrix associated to $(\nabla w, \nabla v)$ (rather than the euclidean one).

For the analysis of these and other results about the Schwarz method we also refer, e.g., to Glowinski, Dinh and Périaux (1983), Bjørstad (1989), Dryja (1989), Dryja and Widlund (1990, 1992), Cai (1990), Kuznetsov (1991), Widlund (1992).

6.4.2 Iteration-by-Subdomain Methods Based on Transmission Conditions at the Interface

Assume now that Ω is divided into *non-overlapping* subdomains like in Fig. 6.4.2. We consider the simplest case in which we have only two subdomains Ω_1 and Ω_2, and denote by Γ their common boundary. We denote by $\mathbf{n}$ the normal direction on $\partial \Omega_1 \cap \Gamma$, oriented outward.

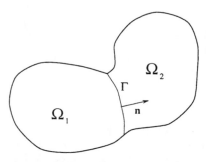

Fig. 6.4.2. Non-overlapping partition of Ω

Assuming f (and consequently u) smooth enough, and denoting by u_i the restriction of u to Ω_i, $i = 1, 2$, the Poisson problem (6.4.1) can be reformulated in the equivalent multidomain form:

$$(6.4.12) \qquad -\Delta u_1 = f \quad \text{in } \Omega_1$$

$$(6.4.13) \qquad u_1 = 0 \quad \text{on } \partial\Omega_1 \cap \partial\Omega$$

$$(6.4.14) \qquad u_1 = u_2 \quad \text{on } \Gamma$$

$$(6.4.15) \qquad \frac{\partial u_2}{\partial n} = \frac{\partial u_1}{\partial n} \quad \text{on } \Gamma$$

$$(6.4.16) \qquad u_2 = 0 \quad \text{on } \partial\Omega_2 \cap \partial\Omega$$

$$(6.4.17) \qquad -\Delta u_2 = f \quad \text{in } \Omega_2 \ .$$

The matching relationships (6.4.14) and (6.4.15) between u_1 and u_2 on Γ are the *transmission conditions*. Iteration-by-subdomain methods use such conditions iteratively to provide *boundary* conditions for u_1 and u_2, allowing therefore the solution of a problem in Ω_1 and another one in Ω_2.

This can be accomplished in several ways: some of them are presented below. The common framework is that two sequences of functions $\{u_1^k\}$, $\{u_2^k\}$ are generated starting from an initial guess u_1^0, u_2^0, and converge (hopefully) to u_1 and u_2, respectively.

(i) *A first approach: the Dirichlet/Neumann method*
Solve for each $k \geq 0$:

$$(6.4.18) \qquad \begin{cases} -\Delta u_1^{k+1} = f & \text{in } \Omega_1 \\[2mm] u_1^{k+1} = 0 & \text{on } \partial\Omega_1 \cap \partial\Omega \\[2mm] u_1^{k+1} = \lambda^k & \text{on } \Gamma \ , \end{cases}$$

then

$$(6.4.19) \qquad \begin{cases} -\Delta u_2^{k+1} = f & \text{in } \Omega_2 \\[2mm] u_2^{k+1} = 0 & \text{on } \partial\Omega_2 \cap \partial\Omega \\[2mm] \dfrac{\partial u_2^{k+1}}{\partial n} = \dfrac{\partial u_1^{k+1}}{\partial n} & \text{on } \Gamma \ , \end{cases}$$

with

$$(6.4.20) \qquad \lambda^k := \theta u_{2|\Gamma}^k + (1-\theta) u_{1|\Gamma}^k \ ,$$

θ being a positive acceleration parameter.

This method was considered, e.g., by Bjørstad and Widlund (1986) and Marini and Quarteroni (1989) in the framework of finite element approximations, and by Funaro, Quarteroni and Zanolli (1988) for spectral collocation approximations.

(ii) *A second approach: the Neumann/Neumann method*
In this case, being also given the initial values ψ_1^0 and ψ_2^0, for each $k \geq 0$ one has to solve

$$(6.4.21) \quad \begin{cases} -\Delta u_i^{k+1} = f & \text{in } \Omega_i \\[2mm] u_i^{k+1} = 0 & \text{on } \partial\Omega_i \cap \partial\Omega \\[2mm] u_i^{k+1} = \lambda^k & \text{on } \Gamma \end{cases}$$

and then

$$(6.4.22) \quad \begin{cases} -\Delta \psi_i^{k+1} = 0 & \text{in } \Omega_i \\[2mm] \psi_i^{k+1} = 0 & \text{on } \partial\Omega_i \cap \partial\Omega \\[2mm] \dfrac{\partial \psi_i^{k+1}}{\partial n} = \dfrac{\partial u_1^{k+1}}{\partial n} - \dfrac{\partial u_2^{k+1}}{\partial n} & \text{on } \Gamma \quad , \end{cases}$$

for $i = 1, 2$, with

$$(6.4.23) \qquad \lambda^k := u_{1|\Gamma}^k - \theta(\alpha_1 \psi_{1|\Gamma}^k - \alpha_2 \psi_{2|\Gamma}^k) \ .$$

Again, θ is a positive acceleration parameter, while α_1 and α_2 are two positive averaging coefficients.

This method was considered by Bourgat, Glowinski, Le Tallec and Vidrascu (1989); a former version was already investigated in Agoshkov and Lebedev (1985).

Remark 6.4.1 The iteration-by-subdomain methods that we have considered above can be extended to the case of a more general differential boundary value problem of the form

$$(6.4.24) \qquad \begin{cases} Lu = f & \text{in } \Omega \\[2mm] Bu = 0 & \text{on } \partial\Omega \quad , \end{cases}$$

where L is an elliptic partial differential operator, B is anyone of the boundary operator we have considered in Section 6.1.1, f is a given data, and u is the unknown solution. Both scalar and vector cases are admissible (in the latter case, (6.4.24) will represent a system of elliptic partial differential equations). The above methods can be formulated also in this general situation and convergence is assured under suitable assumptions (see, e.g., Bramble, Pasciak and Schatz (1986a, 1986b), Dryja and Widlund (1990), Glowinski and LeTallec (1990), Lions (1990), Quarteroni (1991), Quarteroni, Sacchi Landriani and Valli (1991)). $\qquad\qquad\qquad\qquad\qquad\qquad\qquad\qquad\qquad\qquad\square$

In view of a *parallel implementation*, a much better situation is the one in which Ω has been partitioned into many subdomains Ω_i, $i = 1, ..., M$ (see, e.g., Fig. 6.4.2, left). Each individual subproblem (6.4.18), (6.4.19)) (or else (6.4.21), (6.4.22)) can be solved by any kind of finite dimensional approximation method (such as finite elements, finite differences, spectral and collocation methods). The way of enforcing the differential equations depends upon which numerical method is being used. For instance, for finite element methods and for spectral Galerkin methods, equations as well as interface conditions are dealt with variationalwise, whereas they are enforced pointwise in the framework of finite differences and collocation methods. For these results we refer, e.g., to Gropp and Keyes (1988), Gropp (1992) and the references therein.

Remark 6.4.2 An approach different from those described above is the one based on the so-called *fictitious domain* method. The key idea is to imbed the given computational domain Ω into a larger one $\hat{\Omega}$ having a simple shape (e.g., a rectangle in two-dimensions), then reducing the original problem to a sequence of new problems set on $\hat{\Omega}$. This idea can be implemented in many different fashions. Pioneering works are those of the Russian school (see, e.g., Marchuk, Kuznetsov and Matsokin (1986) and references therein; see also Nepomnyaschikh (1992)). For more recent contributions see Glowinski, Pan and Périaux (1993, 1994), Ernst and Golub (1994).

The fictitious domain method has some strict connections with the *capacitance matrix* method, which has been proposed by Buzbee, Dorr, Georg and Golub (1971), Proskurowski and Widlund (1976) (see also O'Leary and Widlund (1979), Dryja (1982, 1983, 1984)).   □

6.4.3 The Steklov-Poincaré Operator

Domain decomposition methods can often be interpreted as iterative procedures for an *interface* equation which is associated with the given differential problem. This interface problem is handled by the Steklov-Poincaré operator that we are going to introduce (see also Agoshkov and Lebedev (1985), Agoshkov (1988), Quarteroni and Valli (1991a, 1991b)).

When properly devised, these iterative procedures intrinsically embody a preconditioner for the system induced on the interface unknowns. A distinguishing feature of a domain decomposition method is the property of optimality of such a preconditioner, i.e., its capability of generating a sequence that converges with a rate that doesn't depend on the size of the original system.

Let us bear in mind the model problem (6.1.28) together with its multidomain formulation (6.4.12)-(6.4.17) referred to the domain partition of Fig. 6.4.2. Let λ denote the unknown value of u on Γ. We consider the two Dirichlet problems:

$$(6.4.25) \qquad \begin{cases} -\Delta w_i = f & \text{in } \Omega_i \\[2mm] w_i = 0 & \text{on } \partial\Omega_i \cap \partial\Omega \\[2mm] w_i = \lambda & \text{on } \Gamma \quad, \end{cases}$$

for $i = 1, 2$. Using a linearity argument we can write

(6.4.26) $$w_i = H_i\lambda + T_i f \ ,$$

where

(6.4.27) $$\begin{cases} -\Delta H_i\lambda = 0 & \text{in } \Omega_i \\ H_i\lambda = 0 & \text{on } \partial\Omega_i \cap \partial\Omega \\ H_i\lambda = \lambda & \text{on } \Gamma \end{cases}$$

and

(6.4.28) $$\begin{cases} -\Delta T_i f = f & \text{in } \Omega_i \\ T_i f = 0 & \text{on } \partial\Omega_i \cap \partial\Omega \\ T_i f = 0 & \text{on } \Gamma \ . \end{cases}$$

For each $i = 1, 2$, $H_i\lambda$ is the *harmonic extension* of λ into Ω_i. Comparing (6.4.12)-(6.4.17) with (6.4.25) it follows that

(6.4.29) $$w_i = u_i \text{ for } i = 1, 2 \text{ if and only if } \frac{\partial w_1}{\partial n} = \frac{\partial w_2}{\partial n} \text{ on } \Gamma \ .$$

In turn this yields that the unknown value λ we are seeking on Γ has to satisfy

(6.4.30) $$S\lambda = \psi \quad \text{on } \Gamma \ ,$$

where

(6.4.31) $$\psi := \frac{\partial}{\partial n} T_2 f - \frac{\partial}{\partial n} T_1 f \ ,$$

and S is the Steklov-Poincaré operator, formally defined as

(6.4.32) $$S\mu := \frac{\partial}{\partial n} H_1\mu - \frac{\partial}{\partial n} H_2\mu \ .$$

The operator S acts between the space of trace functions $\Lambda := H_{00}^{1/2}(\Gamma) = \{v_{|\Gamma} \mid v \in H_0^1(\Omega)\}$ and its dual $\Lambda' = H_{00}^{-1/2}(\Gamma)$ (see Section 1.3 and Lions and Magenes (1968a) for the definition of these spaces). It can be noticed that starting from the weak formulation of (6.4.27) one is led to the following equivalent definition of the Steklov-Poincaré operator:

(6.4.33) $\langle S\gamma, \mu\rangle_{\Lambda',\Lambda} = (\nabla H_1\gamma, \nabla H_1\mu)_{\Omega_1} + (\nabla H_2\gamma, \nabla H_2\mu)_{\Omega_2} \quad \forall \ \gamma, \mu \in \Lambda$.

Therefore S is symmetric and positive definite, as

$$\langle S\mu, \mu\rangle_{\Lambda',\Lambda} = ||\nabla H_1\mu||_{0,\Omega_1}^2 + ||\nabla H_2\mu||_{0,\Omega_2}^2 > 0 \quad \forall \ \mu \in \Lambda \ .$$

The above procedure can be repeated at the finite dimensional level. In this case, as usual, let us denote by V_h a finite dimensional subspace of $V = H_0^1(\Omega)$, by $V_{i,h}$, $i = 1, 2$, and Λ_h its restrictions to Ω_i and Γ, respectively. Then for each $\mu_h \in \Lambda_h$ we denote by $H_{i,h}\mu_h \in V_{i,h}$ the approximate solutions of (6.1.25). The finite dimensional Steklov-Poincaré operator S_h reads:

$$
(6.4.34) \qquad
\begin{aligned}
\langle S_h \gamma_h, \mu_h \rangle_{\Lambda_h', \Lambda_h} &= (\nabla H_{1,h}\gamma_h, \nabla H_{1,h}\mu_h)_{\Omega_1} \\
&\quad + (\nabla H_{2,h}\gamma_h, \nabla H_{2,h}\mu_h)_{\Omega_2} \quad \forall \, \gamma_h, \mu_h \in \Lambda_h \ .
\end{aligned}
$$

Clearly $S_h = S_{1,h} + S_{2,h}$, with obvious use of notations.

The algebraic meaning of the finite dimensional operator S_h can be more easily caught by the help of a picture. We draw in Fig. 6.4.3 the finite element grid and distinguish among finite element nodes belonging to Γ ($\bullet$) and those of Ω_1 ($\square$) and Ω_2 ($\circ$). We denote the corresponding vector of the finite element unknowns by $\boldsymbol{\xi}_\Gamma$, $\boldsymbol{\xi}_1$ and $\boldsymbol{\xi}_2$, respectively, and their lengths by $N_{\Gamma,h}$, $N_{1,h}$ and $N_{2,h}$.

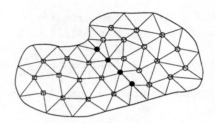

Fig. 6.4.3. Finite element triangulation and nodes

The finite element problem

$$\text{find } u_h \in V_h \ : \ (\nabla u_h, \nabla v_h) = (f, v_h) \quad \forall \, v_h \in V_h$$

can be written in the algebraic form

$$A\boldsymbol{\xi} = \mathbf{F}$$

(see Section 6.3 for notation). A is the $N_h \times N_h$ stiffness matrix, with $N_h = N_{\Gamma,h} + N_{1,h} + N_{2,h}$, and can be written in block form as

$$
\begin{pmatrix} A_{11} & 0 & A_{1\Gamma} \\ 0 & A_{22} & A_{2\Gamma} \\ A_{\Gamma 1} & A_{\Gamma 2} & A_{\Gamma\Gamma} \end{pmatrix}
\begin{pmatrix} \boldsymbol{\xi}_1 \\ \boldsymbol{\xi}_2 \\ \boldsymbol{\xi}_\Gamma \end{pmatrix}
=
\begin{pmatrix} \mathbf{F}_1 \\ \mathbf{F}_2 \\ \mathbf{F}_\Gamma \end{pmatrix} \ .
$$

Notice that $A_{\Gamma i} = A_{i\Gamma}^T$, $i = 1, 2$.

After eliminating $\boldsymbol{\xi}_1$ and $\boldsymbol{\xi}_2$ we obtain

$$(6.4.35) \qquad\qquad \Sigma\boldsymbol{\xi}_\Gamma = \mathbf{G}_\Gamma \ ,$$

with

(6.4.36)
$$\mathbf{G}_\Gamma := \mathbf{F}_\Gamma - A_{\Gamma 1} A_{11}^{-1} \mathbf{F}_1 - A_{\Gamma 2} A_{22}^{-1} \mathbf{F}_2$$

and

(6.4.37)
$$\Sigma := A_{\Gamma\Gamma} - A_{\Gamma 1} A_{11}^{-1} A_{1\Gamma} - A_{\Gamma 2} A_{22}^{-1} A_{2\Gamma} \ .$$

Σ is the *Schur complement* of the stiffness matrix A: it is precisely the algebraic counterpart of the discrete Steklov-Poincaré operator S_h. In turn, (6.4.35) is called *Schur complement system*. Σ is symmetric and positive definite, as it is the Schur complement of a symmetric and positive definite matrix. This is also easily seen from the relationship

(6.4.38)
$$\begin{aligned}(\Sigma\gamma_h, \mu_h) &= \langle S_h\gamma_h, \mu_h\rangle_{\Lambda'_h, \Lambda_h} \\ &= (\nabla H_{1,h}\gamma_h, \nabla H_{1,h}\mu_h)_{\Omega_1} + (\nabla H_{2,h}\gamma_h, \nabla H_{2,h}\mu_h)_{\Omega_2}\end{aligned}$$

for each $\gamma_h, \mu_h \in \Lambda_h$. Here $(\cdot, \cdot)$ is the euclidean scalar product of $\mathbb{R}^{N_{\Gamma,h}}$, and, given $\mu_h \in \Lambda_h$, $\boldsymbol{\mu}_h$ denotes the set of its values at the nodes on Γ.

As in the case of the operator S_h, we can clearly write $\Sigma = \Sigma_1 + \Sigma_2$. It turns out that either Σ_1 or Σ_2 can serve as *optimal preconditioner* of Σ. As a matter of fact, the spectral condition number of the preconditioned matrix $\Sigma_i^{-1}\Sigma$ satisfies

(6.4.39)
$$\chi_{sp}(\Sigma_i^{-1}\Sigma) \leq C \ ,$$

whereas $\chi_{sp}(\Sigma)$ is unbounded, as it grows like h^{-1}. Property (6.4.39) is a consequence of the so-called *extension theorem* which states that the norms of $S_{1,h}$ and $S_{2,h}$ are uniformly equivalent (for the proof see, e.g., Bramble, Pasciak and Schatz (1986a), Widlund (1987), Matsokin (1988) or Marini and Quarteroni (1989)).

6.4.4 The Connection Between Iteration-by-Subdomain Methods and the Schur Complement System

As we already noticed, we can reformulate the methods introduced in Section 6.4.2 within a finite dimensional framework. For either approach (i) or (ii), the iteration process can be regarded as operating on $\lambda_h := u_{h|\Gamma}$, the restriction of the finite dimensional solution to the interface Γ.

In this respect, the k-th step provides a new value λ_h^{k+1} that depends linearly on the previous one λ_h^k, and therefore can be regarded as an iterative procedure applied directly to the discrete Steklov-Poincaré equation. By analogy, identifying λ_h^k with the vector $\boldsymbol{\lambda}_h^k$ of its values at the nodes on Γ, we obtain a sequence that approximates the solution $\boldsymbol{\xi}_\Gamma$ of the Schur complement system (6.4.35).

analogy, identifying λ_h^k with the vector $\boldsymbol{\lambda}_h^k$ of its values at the nodes on Γ, we obtain a sequence that approximates the solution $\boldsymbol{\xi}_\Gamma$ of the Schur complement system (6.4.35).

More precisely, it can be shown that the first approach (Dirichlet/Neumann method) outlined in Section 6.4.2 yields the following iteration procedure for (6.4.35):

$$(6.4.40) \qquad P(\boldsymbol{\lambda}_h^{k+1} - \boldsymbol{\lambda}_h^k) = \theta(\mathbf{G}_\Gamma - \Sigma\boldsymbol{\lambda}_h^k) \ , \quad k \geq 0 \ ,$$

with $P = \Sigma_2$. Thus, the Dirichlet/Neumann method can be regarded as a mean to carry out Richardson iterations on (6.4.35), using the matrix Σ_2 as a *preconditioner* for Σ. At this stage, provided θ is properly chosen, we can infer that the error reduction factor at each iteration is $\rho = (\chi - 1)/(\chi + 1)$, with $\chi := \chi_{sp}(\Sigma_2^{-1}\Sigma)$. Thus, the convergence rate of $\boldsymbol{\lambda}_h^k$ to $\boldsymbol{\xi}_\Gamma$ is independent of the grid-size h.

A similar property can be established for the second approach considered in Section 6.4.2 (the Neumann/Neumann method); in that case the preconditioner is given by $P = (\alpha_1 \Sigma_1^{-1} + \alpha_2 \Sigma_2^{-1})^{-1}$ (see De Roeck and LeTallec (1990), LeTallec (1992), Roux (1992)).

Since P is symmetric and positive definite, conjugate gradient (rather than Richardson) iterations could also be used, based on the same kind of preconditioner. These approaches can still be interpreted as subdomain iteration methods which differ slightly from (6.4.18)-(6.4.19) and (6.4.21)-(6.4.22) (see Quarteroni and Sacchi Landriani (1989) and Bourgat, Glowinski, Le Tallec and Vidrascu (1989)).

Pioneering works on interface preconditioners were done by Dryja (1982, 1983, 1984), Golub and Mayers (1984), Bramble (1986), Chan (1987), Meurant (1988).

A very important issue is how to get optimality on preconditioners when partitioning the domain into many substructures. At this stage, a multilevel approach needs to be pursued, in order to ensure a fast propagation of information among subdomains, even in the case of local grid refinement. For some important contributions to this research field see, e.g., Bramble, Pasciak and Schatz (1986b, 1987, 1988, 1989), Tezduyar and Liou (1988), Bramble, Pasciak and Xu (1990), Smith (1990, 1991), Dryja and Widlund (1990), Chan and Mathew (1991), Tezduyar, Behr, Aliabadi, Mittal and Ray (1994).

The adaptation of domain decomposition techniques to initial-boundary value problems has also received great attention lately, as witnessed by Cai (1991), Dawson and Du (1991), Dryja (1991), Meurant (1991), Dawson and Dupont (1992), Kuznetsov (1994) and the references quoted therein.

7. Elliptic Problems: Approximation by Mixed and Hybrid Methods

In this Chapter, we consider some alternative formulations of second order linear elliptic equations, based on variational principles that differ from the classical one presented in the previous Chapter. We then introduce new numerical approaches that are tailored on the new formulations. More precisely, we first introduce the so-called equilibrium finite element methods through the minimization of the complementary energy. We then consider the Hellinger-Reissner principle, giving rise to both mixed and hybrid finite element methods. Finally, a more general analysis of saddle-point problems and their approximation via Lagrangian multipliers is provided.

7.1 Alternative Mathematical Formulations

Let us now consider the elliptic equation

$$(7.1.1) \qquad Lu := -\sum_{i,j=1}^{d} D_i(a_{ij}D_j u) = f \quad , \quad a_{ij} \in L^\infty(\Omega) \quad ,$$

supplemented with a homogeneous Dirichlet boundary condition. We have analyzed this problem in Section 6.1.1, introducing the variational formulation (6.1.8) which is equivalent to the minimization in $H_0^1(\Omega)$ of the energy functional

$$(7.1.2) \qquad J(v) := \frac{1}{2}\int_\Omega \sum_{i,j=1}^{d} a_{ij} D_i v D_j v - \int_\Omega fv$$

(under the restriction that coefficients a_{ij} are symmetric, i.e., $a_{ij} = a_{ji}$).

Let us now introduce the vector field **p** which has the following components

$$p_i = \sum_{j=1}^{d} a_{ij} D_j u \quad , \quad i = 1, ..., d \quad ,$$

where u is the solution to (7.1.1).

With respect to the new couple of unknowns $(u, \mathbf{p})$, the original problem (7.1.1) can be written as a first order system

$$(7.1.3) \quad \begin{cases} p_i = \sum_{j=1}^{d} a_{ij} D_j u \quad \text{in } \Omega \quad, \quad i = 1, ..., d \quad, \\ \operatorname{div} \mathbf{p} + f = 0 \qquad \text{in } \Omega \\ u_{|\partial\Omega} = 0 \quad . \end{cases}$$

The approximation methods we are going to present are based on this alternative formulation. The motivation is that in several physical situations one is more interested in finding an approximation of $\mathbf{p}$, which is at least as accurate as that of u.

In principle, once u_h has been calculated (e.g., either by a Galerkin or by a collocation method), one could construct $\mathbf{p}_h$ as

$$(\mathbf{p}_h)_i := \sum_{j=1}^{d} a_{ij} D_j u_h \quad,$$

the derivative of u_h being computed in some way consistent with the functional form of u_h. In this way, however, $\mathbf{p}_h$ does not satisfy the equilibrium relation $\operatorname{div} \mathbf{p} + f = 0$ exactly but only in a weak sense, and the two normal traces of $\mathbf{p}_h$ across an interelement boundary do not coincide. Furthermore, theoretical estimates, as well as numerical experience, outline that the error $\mathbf{p} - \mathbf{p}_h$ has a worse behavior than $u - u_h$. Thus it is better to construct a solution $(u_h, \mathbf{p}_h)$ by approximating directly (7.1.3).

7.1.1 The Minimum Complementary Energy Principle

To describe this method, let us assume for simplicity that $a_{ij} = a_{ji}$. First of all, due to the ellipticity assumption, for almost every $\mathbf{x} \in \Omega$ we can construct the inverse of the coefficient matrix, which will be denoted by $a^{ij}(\mathbf{x})$ and which is clearly symmetric and positive definite. Thus, we have

$$(7.1.4) \qquad \sum_{j=1}^{d} a_{ij}(\mathbf{x}) a^{js}(\mathbf{x}) = \delta_{is} \quad, \quad i, s = 1, ..., d \quad,$$

for almost every $\mathbf{x} \in \Omega$.

We are going to show that the solution $\mathbf{p}$ to (7.1.3) is given by the minimization of a suitable functional. In fact, set

$$(7.1.5) \qquad I(\mathbf{q}) := \frac{1}{2} \int_{\Omega} \sum_{i,j=1}^{d} a^{ij} q_i q_j \quad,$$

which is called the *complementary energy* functional, and define for each $f \in L^2(\Omega)$

(7.1.6) $$W^f := \{\mathbf{q} \in H(\text{div}; \Omega) \mid \text{div}\,\mathbf{q} + f = 0 \text{ in } \Omega\} \ .$$

The following theorem shows the equivalence of the minimization of the complementary energy $I(\mathbf{q})$ with the solution of its associated Euler equation. Moreover, it clarifies the relationship between the values $\mathbf{p}$ and u at which the minimum of $I(\mathbf{q})$ and of the energy functional $J(v)$, respectively, is attained.

Theorem 7.1.1 *Assume that $f \in L^2(\Omega)$ and $a_{ij} = a_{ji}$ for each $i, j = 1, ..., d$. Then, there exists a unique solution to the minimization problem*

(7.1.7) $$\text{find } \mathbf{p} \in W^f : I(\mathbf{p}) = \min_{\mathbf{q} \in W^f} I(\mathbf{q}) \ .$$

This problem is equivalent to:

(7.1.8) $$\text{find } \mathbf{p} \in W^f : \int_\Omega \sum_{i,j=1}^d a^{ij} p_j q_i = 0 \qquad \forall\, \mathbf{q} \in W^0 \ .$$

Moreover, the solution $\mathbf{p}$ is related to the solution u of (7.1.1) satisfying $u_{|\partial\Omega} = 0$ through

(7.1.9) $$p_i = \sum_{j=1}^d a_{ij} D_j u \ , \qquad i = 1, ..., d \ .$$

Proof. As $I(\mathbf{q})$ is a quadratic convex functional over the affine manifold W^f, the existence of a minimum follows from well-known results in convex analysis (see, e.g., Rockafellar (1970)).

To prove the equivalence of (7.1.7) and (7.1.8), let us first recall that the Gâteaux derivative of $I(\mathbf{q})$ at $\mathbf{p}$ is vanishing. Hence, for each $\mathbf{q} \in W^0$ we have

$$0 = \lim_{t \to 0} \frac{I(\mathbf{p} + t\mathbf{q}) - I(\mathbf{p})}{t} = \int_\Omega \sum_{i,j=1}^d a^{ij} p_j q_i \ .$$

Notice that $\mathbf{q} \in W^0$ implies $(\mathbf{p} + t\mathbf{q}) \in W^f$ for each $t \in \mathbb{R}$.

Conversely, let $\mathbf{p}$ be a solution to (7.1.8). Then, for each $\mathbf{q} \in W^0$

$$I(\mathbf{p} + \mathbf{q}) = \frac{1}{2} \int_\Omega \sum_{i,j=1}^d a^{ij}(p_i + q_i)(p_j + q_j)$$

$$= I(\mathbf{p}) + \int_\Omega \sum_{i,j=1}^d a^{ij} p_j q_i + \frac{1}{2} \int_\Omega \sum_{i,j=1}^d a^{ij} q_i q_j \geq I(\mathbf{p}) \ ,$$

where the latter inequality follows as the matrix $\{a^{ij}(\mathbf{x})\}$ is positive definite for almost every $\mathbf{x} \in \Omega$.

Moreover, the solution to (7.1.8) is unique. In fact, let $\mathbf{p}_1$ and $\mathbf{p}_2$ be two solutions, then $(\mathbf{p}_1 - \mathbf{p}_2) \in W^0$, and choosing $\mathbf{q} = \mathbf{p}_1 - \mathbf{p}_2$ in (7.1.8) yields $\mathbf{p}_1 = \mathbf{p}_2$.

Finally, let u be the solution of (7.1.1) with $u_{|\partial\Omega} = 0$. Defining $\mathbf{p}$ as in (7.1.9), we have

$$\operatorname{div} \mathbf{p} + f = 0 \quad \text{in } \mathcal{D}'(\Omega) \ ,$$

hence $\mathbf{p} \in W^f$. Furthermore, using (7.1.4) for each $\mathbf{q} \in W^0$ it follows

$$\int_\Omega \sum_{i,j=1}^d a^{ij} p_j q_i = \int_\Omega \sum_{i,j,s=1}^d a^{ij} a_{js} D_s u \, q_i = \int_\Omega \nabla u \cdot \mathbf{q}$$

$$= -\int_\Omega u \operatorname{div} \mathbf{q} + \langle \mathbf{q} \cdot \mathbf{n}, u \rangle_{\partial\Omega} = 0 \ ,$$

where, throughout this Chapter, $\langle \cdot, \cdot \rangle_\Sigma$ denotes the pairing between $H^{-1/2}(\Sigma)$ and $H^{1/2}(\Sigma)$ (see Section 1.3), Σ being a subset of the boundary $\partial\Omega$ which may vary from case to case.   □

Remark 7.1.1 If the coefficient matrix $\{a_{ij}\}$ is not symmetric, by proceeding as before we can prove that the solutions to the equations (7.1.1) and (7.1.8) are still related through (7.1.9). The only difference is that in this case (7.1.8) is no longer the Euler equation associated to the functional $I(\mathbf{q})$ defined in (7.1.5).   □

Let us now consider the case of different kind of boundary value problems. To start with, let us consider the *non-homogeneous Dirichlet* boundary value problem

$$\begin{cases} Lu = f & \text{in } \Omega \\ u_{|\partial\Omega} = \varphi & \text{on } \partial\Omega \ , \end{cases}$$

with $\varphi \in H^{1/2}(\partial\Omega)$. It is easily seen that the complementary energy functional becomes

$$I(\mathbf{q}) := \frac{1}{2} \int_\Omega \sum_{i,j=1}^d a^{ij} q_i q_j - \langle \mathbf{q} \cdot \mathbf{n}, \varphi \rangle_{\partial\Omega} \ .$$

The corresponding Euler equation is given by

$$\int_\Omega \sum_{i,j=1}^d a^{ij} p_j q_i - \langle \mathbf{q} \cdot \mathbf{n}, \varphi \rangle_{\partial\Omega} = 0 \quad \forall \, \mathbf{q} \in W^0 \ ,$$

or equivalently by

$$\int_\Omega \left(\sum_{i,j=1}^d a^{ij} p_j q_j - \mathbf{q} \cdot \nabla \tilde{\varphi} \right) = 0 \quad \forall \, \mathbf{q} \in W^0 \ ,$$

where $\tilde{\varphi} \in H^1(\Omega)$ is a suitable extension of φ satisfying $\tilde{\varphi}_{|\partial\Omega} = \varphi$.

If u is the solution to (7.1.1) with different boundary conditions, one has to formulate the minimization problem (7.1.7) in a different way. Here we confine ourselves to consider the *Neumann* boundary condition, i.e.,

$$\frac{\partial u}{\partial n_L} = g \quad \text{on} \quad \partial\Omega \ , \quad g \in L^2(\partial\Omega) \ .$$

This corresponds to $\mathbf{p} \cdot \mathbf{n} = g$ on $\partial\Omega$, hence we can write:

(7.1.10) $$\text{find } \mathbf{p} \in W^{f,g} : I(\mathbf{p}) = \min_{q \in W^{f,g}} I(\mathbf{q})$$

where

(7.1.11) $$W^{f,g} := \{\mathbf{q} \in H(\text{div}; \Omega) \mid \text{div}\,\mathbf{q} + f = 0 \text{ in } \Omega, \mathbf{q} \cdot \mathbf{n} = g \text{ on } \partial\Omega\} \ .$$

The right hand side f and the datum g must thus satisfy the compatibility condition

(7.1.12) $$\int_\Omega f + \int_{\partial\Omega} g = 0 \ .$$

Indeed, the coerciveness assumption (6.1.20) is not satisfied for the operator L introduced in (7.1.1). However, the existence and uniqueness up to a constant of a solution to the Neumann problem associated to (7.1.1) can be proven under the compatibility condition (7.1.12). (See, e.g., Grisvard (1985).)

Starting from the weak formulation (7.1.8), one can easily introduce suitable Galerkin approximations based on finite elements or other methods. For this, we refer the interested reader to, e.g., Fraeijs de Veubeke (1965), Thomas (1977), Raviart and Thomas (1979).

However, despite its apparent elementary formulation, problem (7.1.8) hides the trouble of constructing a solution satisfying the constraint $\text{div}\,\mathbf{p} + f = 0$, which may be quite complicated at a finite dimensional level. As an example, let us just recall that seeking a $\mathbf{p}$ such that, for each element K of a triangulation $\mathcal{T}_h$, $\mathbf{p}_{|K} \in H(\text{div}; K)$ doesn't guarantee that $\mathbf{p} \in H(\text{div}; \Omega)$; actually, one needs $(\mathbf{n} \cdot \mathbf{p}_{|K_1}) = (\mathbf{n} \cdot \mathbf{p}_{|K_2})$ over each interelement boundary $K_1 \cap K_2$ (see Proposition 3.2.2). Moreover, the constraint $\text{div}\,\mathbf{p} + f = 0$ must be satisfied in each K; in particular this requires that

$$\int_{\partial K} \mathbf{n} \cdot \mathbf{p}_{|K} + \int_K f = 0 \ .$$

This difficulty has fostered the introduction of other equivalent formulations, which are the object of the next Section.

7.1.2 Saddle-Point Formulations: Mixed and Hybrid Methods

The idea of *mixed methods* is based on the introduction of a Lagrangian multiplier to relax the constraint $\operatorname{div} \mathbf{p} + f = 0$. Still assuming that $a_{ij} = a_{ji}$, let us introduce the Lagrangian functional

$$(7.1.13) \qquad \mathcal{L}(\mathbf{q}, v) := I(\mathbf{q}) + \int_{\Omega} (\operatorname{div} \mathbf{q} + f) v \ ,$$

defined in $H(\operatorname{div}; \Omega) \times L^2(\Omega)$. We want to minimize over $\mathbf{q}$ and maximize over v, i.e., we look for $(\mathbf{p}, u) \in H(\operatorname{div}; \Omega) \times L^2(\Omega)$ which is the saddle-point of $\mathcal{L}(\cdot, \cdot)$:

$$(7.1.14) \qquad \mathcal{L}(\mathbf{p}, v) \leq \mathcal{L}(\mathbf{p}, u) \leq \mathcal{L}(\mathbf{q}, u) \quad \forall \ (\mathbf{q}, v) \in H(\operatorname{div}; \Omega) \times L^2(\Omega) \ .$$

If $(\mathbf{p}, u)$ is such a saddle-point, then, necessarily, the constraint $\operatorname{div} \mathbf{p} + f = 0$ is satisfied, otherwise taking $v = v_\tau = \tau(\operatorname{div} \mathbf{p} + f)$ would give

$$\lim_{\tau \to \infty} \mathcal{L}(\mathbf{p}, v_\tau) = \infty \ .$$

The characterization of the solution $(\mathbf{p}, u)$ to (7.1.8) and (7.1.1) as the saddle-point of the Lagrangian functional (7.1.13) takes the name of *Hellinger-Reissner principle*. Finite element methods based on this variational formulation will be considered in Section 7.2.

Remark 7.1.2 *(About terminology).* Any formulation stemming from the minimization of the energy $J(v)$ defined in (7.1.2) is called *primal*, whereas those arising from the minimization of the complementary energy $I(\mathbf{q})$ defined in (7.1.5) are said to be *dual*.

In any of the cases in which relaxation of constraints is accomplished throughout a saddle-point formulation, one generates the so-called *mixed* methods. Among the latter, we give the name *hybrid* to those methods which are based on partitioning the domain Ω into subsets, in which framework one of the unknowns is defined only at the interfaces.

According to this classification, the Hellinger-Reissner approach falls under the *dual mixed* case. $\square$

Let us now prove the existence of a unique saddle-point of the Lagrangian $\mathcal{L}(\mathbf{q}, v)$, and its characterization as the solution of a system of variational equations.

Theorem 7.1.2 *Assume that $f \in L^2(\Omega)$ and $a_{ij} = a_{ji}$ for each $i, j = 1, ..., d$. The saddle-point problem (7.1.14) has a unique solution $(\mathbf{p}, u) \in H(\operatorname{div}; \Omega) \times L^2(\Omega)$, which is also the solution to the equivalent problem*

(7.1.15) $$\int_\Omega \left(\sum_{i,j=1}^d a^{ij} p_j q_i + u \operatorname{div} \mathbf{q} \right) = 0 \quad \forall\, \mathbf{q} \in H(\operatorname{div}; \Omega)$$

(7.1.16) $$\int_\Omega (\operatorname{div} \mathbf{p} + f) v = 0 \quad \forall\, v \in L^2(\Omega) \ .$$

Moreover, $(\mathbf{p}, u)$ is the solution to (7.1.15), (7.1.16) if, and only if, u is the solution to (7.1.1) satisfying $u_{|\partial\Omega} = 0$ and $\mathbf{p}$ and u are related through (7.1.9).

Proof. The existence of a saddle-point solution follows by well-known results of convex minimization (see, e.g., Rockafellar (1970)). Since at $(\mathbf{p}, u)$ the Gateaux derivative of $\mathcal{L}(\cdot, \cdot)$ must vanish, one has

$$\lim_{t \to 0} \frac{\mathcal{L}(\mathbf{p} + t\mathbf{q}, u) - \mathcal{L}(\mathbf{p}, u)}{t} = 0 \quad \forall\, \mathbf{q} \in H(\operatorname{div}; \Omega)$$

$$\lim_{t \to 0} \frac{\mathcal{L}(\mathbf{p}, u + tv) - \mathcal{L}(\mathbf{p}, u)}{t} = 0 \quad \forall\, v \in L^2(\Omega) \ ,$$

and (7.1.15), (7.1.16) easily follow.

Conversely, let $(\mathbf{p}, u)$ be a solution to (7.1.15), (7.1.16). Then for each $\mathbf{q} \in H(\operatorname{div}; \Omega)$ we have

$$\mathcal{L}(\mathbf{p} + \mathbf{q}, u) = I(\mathbf{p}) + \int_\Omega \sum_{i,j=1}^d a^{ij} p_j q_i + \frac{1}{2} \int_\Omega \sum_{i,j=1}^d a^{ij} q_i q_j$$

$$+ \int_\Omega (\operatorname{div} \mathbf{p} + f) u + \int_\Omega u \operatorname{div} \mathbf{q} \geq \mathcal{L}(\mathbf{p}, u) \ ,$$

as $\{a^{ij}\}$ is a positive definite matrix. Moreover, for each $v \in L^2(\Omega)$ it holds

$$\mathcal{L}(\mathbf{p}, v) = I(\mathbf{p}) + \int_\Omega (\operatorname{div} \mathbf{p} + f) v = I(\mathbf{p}) \ .$$

Hence, $(\mathbf{p}, u)$ is a saddle-point of $\mathcal{L}(\cdot, \cdot)$.

Uniqueness of the solution to (7.1.15), (7.1.16) is proven as follows. Let $(\mathbf{p}_1, u_1)$ and $(\mathbf{p}_2, u_2)$ be two solutions; from (7.1.16) we have

$$\operatorname{div} \mathbf{p}_1 + f = 0 = \operatorname{div} \mathbf{p}_2 + f \ ,$$

hence $\operatorname{div}(\mathbf{p}_1 - \mathbf{p}_2) = 0$. Thus, choosing in (7.1.15) $\mathbf{q} = \mathbf{p}_1 - \mathbf{p}_2$ we find

$$0 = \int_\Omega \left[\sum_{i,j=1}^d a^{ij} (\mathbf{p}_1 - \mathbf{p}_2)_i (\mathbf{p}_1 - \mathbf{p}_2)_j + (u_1 - u_2) \operatorname{div}(\mathbf{p}_1 - \mathbf{p}_2) \right]$$

$$= \int_\Omega \sum_{i,j=1}^d a^{ij} (\mathbf{p}_1 - \mathbf{p}_2)_i (\mathbf{p}_1 - \mathbf{p}_2)_j \ ,$$

yielding $\mathbf{p}_1 = \mathbf{p}_2$. As a consequence (7.1.15) gives now

$$\int_\Omega (u_1 - u_2) \operatorname{div} \mathbf{q} = 0 \quad \forall\, \mathbf{q} \in H(\operatorname{div}; \Omega) \ .$$

Choosing $\mathbf{q} = \nabla\psi$, where ψ is the solution to

$$\begin{cases} \Delta\psi = u_1 - u_2 \\ \psi_{|\partial\Omega} = 0 \ , \end{cases}$$

we find at once $u_1 = u_2$.

Now let u be the solution to (7.1.1) with a homogeneous Dirichlet boundary condition, and define $\mathbf{p}$ through (7.1.9). For each $\mathbf{q} \in H(\operatorname{div}; \Omega)$ it holds

$$\int_\Omega \sum_{i,j=1}^d a^{ij} p_j q_i = \int_\Omega \nabla u \cdot \mathbf{q} = - \int_\Omega u \operatorname{div} \mathbf{q} \ ,$$

and moreover

$$\operatorname{div} \mathbf{p} + f = -Lu + f = 0 \ .$$

On the other side, if $(\mathbf{p}, u)$ is the solution to (7.1.15), (7.1.16), choosing $\mathbf{q} \in (\mathcal{D}(\Omega))^d$ we find

$$(7.1.17) \qquad D_i u = \sum_{j=1}^d a^{ij} p_j \in L^2(\Omega) \ , \quad i = 1, ..., d \ ,$$

in the distributional sense, hence $u \in H^1(\Omega)$. Moreover,

$$\sum_{s=1}^d a_{is} D_s u = \sum_{j,s=1}^d a_{is} a^{sj} p_j = p_i \ , \quad i = 1, ..., d \ ,$$

and consequently $-Lu = -\operatorname{div} \mathbf{p} = f$ in Ω. Finally, take now $\mathbf{q} \in H(\operatorname{div}; \Omega)$ in (7.1.15). If we integrate by parts, we find from (7.1.17)

$$0 = \int_\Omega \sum_{i=1}^d \left(\sum_{j=1}^d a^{ij} p_j - D_i u \right) q_i + \int_{\partial\Omega} u\,\mathbf{q} \cdot \mathbf{n} = \int_{\partial\Omega} u\,\mathbf{q} \cdot \mathbf{n} \ .$$

As $\mathbf{q}$ is arbitrary, it follows that $u_{|\partial\Omega} = 0$. $\qquad \square$

The second component u of the couple $(\mathbf{p}, u)$ is the *Lagrangian multiplier* associated to the constraint $\operatorname{div} \mathbf{p} + f = 0$. It is regarded as an $L^2(\Omega)$-function, thus we need low regularity for its approximation.

Remark 7.1.3 The symmetry assumption on the coefficients a_{ij} has not been used to obtain the equivalence between the homogeneous Dirichlet problem associated to (7.1.1) and problem (7.1.15), (7.1.16). $\qquad \square$

We now turn to the case of other boundary conditions. If we consider the *non-homogeneous Dirichlet* problem, similarly to what has been done in the previous Section when $u_{|\partial\Omega} = \varphi$, one has to modify (7.1.15) as follows

$$(7.1.18) \quad \int_\Omega \left(\sum_{i,j=1}^d a^{ij} p_j q_i + u \operatorname{div} \mathbf{q} \right) - \langle \mathbf{q} \cdot \mathbf{n}, \varphi \rangle_{\partial\Omega} = 0 \quad \forall\, \mathbf{q} \in H(\operatorname{div}; \Omega) \ ,$$

or

$$(7.1.19) \quad \int_\Omega \left[\sum_{i,j=1}^d a^{ij} p_j q_i + u \operatorname{div} \mathbf{q} - \operatorname{div}(\tilde\varphi \mathbf{q}) \right] = 0 \quad \forall\, \mathbf{q} \in H(\operatorname{div}; \Omega) \ ,$$

where $\tilde\varphi \in H^1(\Omega)$, $\tilde\varphi_{|\partial\Omega} = \varphi$.

If u is the solution to the *Neumann* problem

$$\begin{cases} Lu = f & \text{in } \Omega \\[2mm] \dfrac{\partial u}{\partial n_L} = g & \text{on } \partial\Omega \\[2mm] \int_\Omega u = 0 \ , \end{cases}$$

where f and g satisfy the compatibility condition (7.1.12), the Hellinger-Reissner principle reads: find $(\mathbf{p}^0, u) \in H_0(\operatorname{div}; \Omega) \times L_0^2(\Omega)$ such that

$$(7.1.20) \quad \int_\Omega \left(\sum_{i,j=1}^d a^{ij} p_j^0 q_i + u \operatorname{div} \mathbf{q} \right) = -\int_\Omega \sum_{i,j=1}^d a^{ij} p_j^g q_i \quad \forall\, \mathbf{q} \in H_0(\operatorname{div}; \Omega)$$

$$(7.1.21) \quad \int_\Omega (\operatorname{div} \mathbf{p}^0 + f) v = -\int_\Omega (\operatorname{div} \mathbf{p}^g) v \quad \forall\, v \in L_0^2(\Omega) \ .$$

Here we have set

$$H_0(\operatorname{div}; \Omega) := \left\{ \mathbf{q} \in H(\operatorname{div}; \Omega) \,|\, \mathbf{q} \cdot \mathbf{n} = 0 \text{ on } \partial\Omega \right\}$$

$$L_0^2(\Omega) := \left\{ u \in L^2(\Omega) \,|\, \int_\Omega u = 0 \right\}$$

and $\mathbf{p}^g$ is a suitable function in $H(\operatorname{div}; \Omega)$ such that $\mathbf{p}^g \cdot \mathbf{n} = g$ on $\partial\Omega$. In this case u and $\mathbf{p}^0$ are related through

$$p_i^0 = \sum_{j=1}^d a_{ij} D_j u - p_i^g \ , \quad i = 1, ..., d \ .$$

This relation takes the place of (7.1.9).

When considering the *mixed* boundary value problem

$$\begin{cases} Lu = f & \text{in } \Omega \\[2mm] u = 0 & \text{on } \Gamma_D \\[2mm] \dfrac{\partial u}{\partial n_L} = g & \text{on } \Gamma_N \quad, \end{cases}$$

the solution $(\mathbf{p}^0, u)$ and the test functions $(\mathbf{q}, v)$ must belong to the space $H_{0,\Gamma_N}(\text{div}; \Omega) \times L^2(\Omega)$, where

$$H_{0,\Gamma_N}(\text{div}; \Omega) := \big\{ \mathbf{q} \in H(\text{div}; \Omega) \,|\, \mathbf{q} \cdot \mathbf{n} = 0 \text{ on } \Gamma_N \big\} \quad.$$

In this case the function $\mathbf{p}^g$ appearing in (7.1.20), (7.1.21) belongs to $H(\text{div}; \Omega)$ and satisfies $\mathbf{p}^g \cdot \mathbf{n} = g$ on Γ_N, $\mathbf{p}^g \cdot \mathbf{n} = 0$ on Γ_D.

Finally, for the *Robin* boundary condition

$$\frac{\partial u}{\partial n_L} + \kappa u = g \text{ on } \partial\Omega \quad, \quad \kappa \in L^\infty(\partial\Omega) \quad,$$

one has to find $(\mathbf{p}, u) \in \mathcal{H}(\text{div}; \Omega) \times L^2(\Omega)$ such that

(7.1.22)
$$\int_\Omega \left(\sum_{i,j=1}^d a^{ij} p_j q_i + u \, \text{div}\, \mathbf{q} \right) + \int_{\partial\Omega} \kappa^{-1}(\mathbf{p} \cdot \mathbf{n})(\mathbf{q} \cdot \mathbf{n})$$
$$= \int_{\partial\Omega} \kappa^{-1} g \mathbf{q} \cdot \mathbf{n} \quad \forall\, \mathbf{q} \in \mathcal{H}(\text{div}; \Omega)$$

(7.1.23)
$$\int_\Omega (\text{div}\, \mathbf{p} + f) v = 0 \quad \forall\, v \in L^2(\Omega) \quad.$$

Here we have assumed $\kappa(x) \geq \kappa_* > 0$ almost everywhere on $\partial\Omega$, and we have set

(7.1.24)
$$\mathcal{H}(\text{div}; \Omega) := \big\{ \mathbf{q} \in H(\text{div}; \Omega) \,|\, \mathbf{q} \cdot \mathbf{n} \in L^2(\partial\Omega) \big\} \quad,$$

which is a Hilbert space with respect to the scalar product

(7.1.25)
$$(\mathbf{p}, \mathbf{q})_{\mathcal{H}(\text{div};\Omega)} := (\mathbf{p}, \mathbf{q})_{H(\text{div};\Omega)} + \int_{\partial\Omega} (\mathbf{p} \cdot \mathbf{n})(\mathbf{q} \cdot \mathbf{n}) \quad.$$

We come back now to the homogeneous Dirichlet problem (7.1.15), (7.1.16). As we remarked in Theorem 7.1.2, the component u of the solution to (7.1.15), (7.1.16) indeed belongs to $H^1(\Omega)$, being the solution to (7.1.1). Thus we can rewrite (7.1.15) and (7.1.16) as: find $(\mathbf{p}, u) \in (L^2(\Omega))^d \times H_0^1(\Omega)$ such that

$$(7.1.26) \qquad \int_\Omega \left(\sum_{i,j=1}^d a^{ij} p_j q_i - \nabla u \cdot \mathbf{q} \right) = 0 \quad \forall\, \mathbf{q} \in (L^2(\Omega))^d$$

$$(7.1.27) \qquad \int_\Omega \mathbf{p} \cdot \nabla v = \int_\Omega fv \quad \forall\, v \in H_0^1(\Omega) \ .$$

Eliminating p via (7.1.26) gives the usual homogeneous Dirichlet problem associated to (7.1.1). We will not dwell here on the formulation (7.1.26), (7.1.27), since it has been analyzed by Babuška, Oden and Lee (1977) (see also Ciarlet and Destuynder (1979)).

Remark 7.1.4 The non-homogeneous Dirichlet problem based on a primal formulation has been dealt with in Section 6.1.1 (see Remark 6.1.2). An alternative way, which is based on the relaxation of the boundary condition $u = \varphi$ on $\partial\Omega$, is provided by the following formulation: find $(u, \mu) \in H^1(\Omega) \times H^{-1/2}(\partial\Omega)$ such that

$$(7.1.28) \qquad \int_\Omega \sum_{i,j=1}^d a_{ij} D_j u D_i v - \langle \mu, v \rangle_{\partial\Omega} = \int_\Omega fv \quad \forall\, v \in H^1(\Omega)$$

$$(7.1.29) \qquad \langle \eta, u - \varphi \rangle_{\partial\Omega} = 0 \quad \forall\, \eta \in H^{-1/2}(\partial\Omega) \ .$$

These are the saddle-point equations for the Lagrangian functional $\mathcal{L}(v, \eta) = J(v) - \langle \eta, v - \varphi \rangle_{\partial\Omega}$. The Lagrangian multiplier μ has the meaning of $\frac{\partial u}{\partial n_L}$ on $\partial\Omega$. In this regard, (7.1.28), (7.1.29) is a *primal mixed* approach. This problem has been introduced and analyzed by Babuška (1973a). $\qquad \square$

Another alternative saddle-point formulation is the one which leads to the introduction of the so-called *hybrid methods*. To describe it, we need to introduce a partition $\mathcal{P}$ of $\overline{\Omega}$ into disjoint subsets K. For instance, we may think of a finite element triangulation $\mathcal{T}_h$ as described in Chapter 3. We then define the subspace Y of $L^2(\Omega)$ as follows

$$(7.1.30) \qquad Y := \{ v \in L^2(\Omega) \mid v_{|K} \in H^1(K) \quad \forall\, K \in \mathcal{P} \} \ .$$

It is easily noticed that the Sobolev space $H_0^1(\Omega)$ can be characterized as being the space of functions $v \in Y$ such that the following constraint holds:

$$(7.1.31) \qquad \sum_{K \in \mathcal{P}} \langle \mathbf{q} \cdot \mathbf{n}, v_{|K} \rangle_{\partial K} = 0 \quad \forall\, \mathbf{q} \in H(\mathrm{div}; \Omega) \ .$$

Thus, we are led into looking for a solution $(\mathbf{p}, u) \in H(\mathrm{div}; \Omega) \times Y$ to the following system:

$$(7.1.32) \qquad \sum_{K \in \mathcal{P}} \left(\int_K \sum_{i,j=1}^d a_{ij} D_j u D_i v - \langle \mathbf{p} \cdot \mathbf{n}, v_{|K} \rangle_{\partial K} \right) = \int_\Omega fv \quad \forall\, v \in Y$$

$$(7.1.33) \qquad \sum_{K \in \mathcal{P}} \langle \mathbf{q} \cdot \mathbf{n}, u_{|K} \rangle_{\partial K} = 0 \quad \forall\, \mathbf{q} \in H(\mathrm{div}; \Omega) \ .$$

Due to the characterization of $H_0^1(\Omega)$ given by (7.1.31), this problem is clearly equivalent to the homogeneous Dirichlet problem associated to (7.1.1). However, let us notice that $\mathbf{p}$ is not uniquely determined since only $\mathbf{p} \cdot \mathbf{n}$ appears in (7.1.32). On the other hand, this fact would allow one to replace $\mathbf{p} \cdot \mathbf{n}$ by λ and any $\mathbf{q} \cdot \mathbf{n}$ by μ, where λ and μ belong now to the space of functions that are traces on $\Gamma := \{\cup\, \partial K \mid K \in \mathcal{P}\}$ of the normal component of vector functions in $H(\mathrm{div};\Omega)$. This formulation is usually referred to as the *primal hybrid* formulation.

A last saddle-point formulation that we are interested in, is given by the *dual hybrid* formulation. In this case we introduce the affine subspace $\hat{W}^f$ of $(L^2(\Omega))^d$ defined as follows:

$$(7.1.34) \quad \hat{W}^f := \{\mathbf{q} \in (L^2(\Omega))^d \mid \mathrm{div}(\mathbf{q}_{|K}) + f = 0 \text{ in } L^2(K) \quad \forall\, K \in \mathcal{P}\}\ .$$

It is easily proven that the affine subspace W^f defined in (7.1.6) can be characterized as being the set of functions $\mathbf{q} \in \hat{W}^f$ such that

$$(7.1.35) \quad \sum_{K \in \mathcal{P}} \langle \mathbf{q}_{|K} \cdot \mathbf{n}, v\rangle_{\partial K} = 0 \quad \forall\, v \in H_0^1(\Omega)\ .$$

The presence of this constraint allows us to formulate the following problem: find $(\mathbf{p}, u) \in \hat{W}^f \times H_0^1(\Omega)$ such that

$$(7.1.36) \quad \int_\Omega \sum_{i,j=1}^d a^{ij} p_j q_i - \sum_{K \in \mathcal{P}} \langle \mathbf{q}_{|K} \cdot \mathbf{n}, u\rangle_{\partial K} = 0 \quad \forall\, \mathbf{q} \in \hat{W}^0$$

$$(7.1.37) \quad \sum_{K \in \mathcal{P}} \langle \mathbf{p}_{|K} \cdot \mathbf{n}, v\rangle_{\partial K} = 0 \quad \forall\, v \in H_0^1(\Omega)\ .$$

When $a_{ij} = a_{ji}$, $i, j = 1, ..., d$, this corresponds to the saddle-point of the Lagrangian functional

$$(7.1.38) \quad \mathcal{D}(\mathbf{q}, v) := I(\mathbf{q}) - \sum_{K \in \mathcal{P}} \langle \mathbf{q}_{|K} \cdot \mathbf{n}, v\rangle_{\partial K}\ .$$

The relationship (7.1.38) enlights the basic difference between the Hellinger-Reissner principle and the dual hybrid formulation. In the former, $\mathbf{q}$ is sought in the function space W^f (see (7.1.6)), which is an affine subspace of $H(\mathrm{div};\Omega)$. In the latter, the constraint $\mathbf{q} \in H(\mathrm{div};\Omega)$ has been relaxed as it is now enforced only locally on each $K \in \mathcal{P}$ (see (7.1.34)).

Clearly, u is not uniquely determined in $H_0^1(\Omega)$, as only its traces on ∂K have to satisfy (7.1.36). However, if $u \in H_0^1(\Omega)$ is the solution to (7.1.1) and $\mathbf{p}$ is given by (7.1.9), then $(\mathbf{p}, u)$ is a solution to (7.1.36), (7.1.37).

In Table 7.1.1 we summarize the various variational principles we have introduced so far in Chapters 6 and 7 for the homogeneous Dirichlet problem associated to (7.1.1). In the first column we enter the name of the method,

Table 7.1.1. Summary of the variational methods presented in Section 7.1

	Extremal problem	Functional
Displacement	$\min\limits_{v\in H_0^1(\Omega)} J_\Omega(v)$	$J_\Omega(v) := \frac{1}{2}\int_\Omega \sum_{i,j} a_{ij} D_i v D_j v$ $\quad - \int_\Omega f v$ (potential energy)
Equilibrium	$\min\limits_{\mathbf{q}\in W^f} I(\mathbf{q})$ $(W^f$ defined in (7.1.6))	$I(\mathbf{q}) := \frac{1}{2}\int_\Omega \sum_{i,j} a^{ij} q_i q_j$ (complementary energy)
Dual mixed	$\min\limits_{\mathbf{q}\in H(\mathrm{div};\Omega)} \max\limits_{v\in L^2(\Omega)} \mathcal{L}(\mathbf{q},v)$	$\mathcal{L}(\mathbf{q},v) := I(\mathbf{q})$ $\quad + \int_\Omega (\mathrm{div}\,\mathbf{q} + f) v$ (Hellinger-Reissner energy)
Primal mixed	$\min\limits_{v\in H^1(\Omega)} \max\limits_{\eta\in H^{-1/2}(\partial\Omega)} \mathcal{L}_*(v,\eta)$	$\mathcal{L}_*(v,\eta) := J_\Omega(v) - \langle\eta, v\rangle_{\partial\Omega}$
Dual hybrid	$\min\limits_{\mathbf{q}\in \hat{W}^f} \max\limits_{v\in H_0^1(\Omega)} \mathcal{D}(\mathbf{q},v)$ $(\hat{W}^f$ defined in (7.1.34))	$\mathcal{D}(\mathbf{q},v) := I(\mathbf{q})$ $\quad -\sum_{K\in\mathcal{P}}\langle\mathbf{q}_{|K}\cdot\mathbf{n}, v\rangle_{\partial K}$
Primal hybrid	$\min\limits_{v\in Y} \max\limits_{\mathbf{q}\in H(\mathrm{div};\Omega)} \mathcal{D}_*(\mathbf{q},v)$ $(Y$ defined in (7.1.30))	$\mathcal{D}_*(\mathbf{q},v) := \sum_{K\in\mathcal{P}} J_K(v)$ $\quad -\sum_{K\in\mathcal{P}}\langle\mathbf{q}\cdot\mathbf{n}, v_{|K}\rangle_{\partial K}$

in the second the minimization problem or the saddle-point problem under consideration, in the third one the functional employed.

In the following Sections we are going to introduce and analyze some approximation methods based on the variational formulations presented here.

We mainly focus on finite element methods for the dual mixed formulation (Hellinger-Reissner principle) (Section 7.2).

7.2 Approximation by Mixed Methods

Let us concentrate our attention on the dual mixed formulation (7.1.15), (7.1.16) associated with the homogeneous Dirichlet boundary value problem for the elliptic equation (7.1.1). A thorough analysis of other boundary value problems, of the primal mixed formulation as well as of primal and dual hybrid formulations can be found in Roberts and Thomas (1991) and Brezzi and Fortin (1991). A general theory of approximation for constrained problems (including mixed and hybrid methods) is also presented in Section 7.4 of this book.

7.2.1 Setting up and Analysis

Starting from (7.1.15), (7.1.16), we are led to consider the following discrete problem:

$$(7.2.1) \quad \begin{cases} \text{find } (\mathbf{p}_h, u_h) \in W_h \times Y_h : \\[2mm] \displaystyle\int_\Omega \left(\sum_{i,j=1}^d a^{ij} p_{h,j} q_{h,i} + u_h \operatorname{div} \mathbf{q}_h \right) = 0 \quad \forall\, \mathbf{q}_h \in W_h \\[4mm] \displaystyle\int_\Omega (\operatorname{div} \mathbf{p}_h + f) v_h = 0 \qquad\qquad\qquad \forall\, v_h \in Y_h \quad , \end{cases}$$

where $W_h \subset H(\operatorname{div}; \Omega)$ and $Y_h \subset L^2(\Omega)$ are suitable finite dimensional spaces.

Some remarks are now in order. First of all, $(7.2.1)_2$ does not necessarily imply that $\operatorname{div} \mathbf{p}_h + f = 0$ in Ω, hence $\mathbf{p}_h \notin W^f$ in general. A necessary condition for $\mathbf{p}_h$ to belong to W^f is clearly $f \in \operatorname{div} W_h$. This is also sufficient if $\operatorname{div} W_h \subset Y_h$, because in such a case we are allowed to choose $v_h = \operatorname{div} \mathbf{p}_h + f$ in $(7.2.1)_2$.

Secondly, the existence and uniqueness of a solution to (7.2.1) doesn't follow from the theory developed in the infinite dimensional case, where one needs barely to verify that the solution $u \in H_0^1(\Omega)$ to (7.1.1) and the corresponding $\mathbf{p}$ defined in (7.1.9) do satisfy (7.1.15), (7.1.16). In the current case, denoting by $V_h \subset H_0^1(\Omega)$ a suitable finite dimensional subspace, the solution to the Galerkin problem

$$(7.2.2) \qquad u_h \in V_h : \int_\Omega \sum_{i,j=1}^d a_{ij} D_j u_h D_i v_h = \int_\Omega f v_h \quad \forall\, v_h \in V_h \quad ,$$

together with the vector $\mathbf{p}_h$ defined by

$$(7.2.3) \qquad p_{h,i} := \sum_{j=1}^{d} a_{ij} D_j u_h \quad , \quad i = 1, ..., d \quad ,$$

do not satisfy (7.2.1), as $\mathbf{p}_h \notin H(\text{div}; \Omega)$ and $(7.2.1)_2$ has not even a meaning. Thus let us start by proving an existence theorem.

Theorem 7.2.1 *Assume that the spaces* W_h *and* Y_h *satisfy the following compatibility condition: there exists* $\beta_h > 0$ *such that*

$$(7.2.4) \qquad \begin{array}{c} \forall \, v_h \in Y_h \; \exists \; \mathbf{q}_h \in W_h \, , \mathbf{q}_h \neq \mathbf{0} : \\[6pt] \displaystyle\int_{\Omega} v_h \, \text{div} \, \mathbf{q}_h \geq \beta_h ||\mathbf{q}_h||_{H(\text{div};\Omega)} ||v_h||_0 \quad . \end{array}$$

Then (7.2.1) *has a unique solution.*

Proof. Since we are considering a finite dimensional problem, we have only to check uniqueness, namely that $f = 0$ implies $\mathbf{p}_h = \mathbf{0}$ and $u_h = 0$. To this aim, take $\mathbf{q}_h = \mathbf{p}_h$ in $(7.2.1)_1$: using $(7.2.1)_2$ (with $f = 0$) and the fact that $\{a^{ij}\}$ is positive definite we find at once $\mathbf{p}_h = \mathbf{0}$.

On the other hand, from $(7.2.1)_1$ we still obtain

$$\int_{\Omega} u_h \, \text{div} \, \mathbf{q}_h = 0 \quad \forall \, \mathbf{q}_h \in W_h$$

thus it follows that $u_h = 0$ owing to (7.2.4). $\qquad\square$

The compatibility condition (7.2.4) is also called *inf-sup* condition or *Ladyzhenskaya-Babuška-Brezzi* condition. In particular, notice that it implies that $\dim Y_h \leq \dim W_h$. Indeed, if this was not the case, there would exist $v_h \in Y_h$ orthogonal to the space $\{\text{div} \, \mathbf{q}_h \, | \, \mathbf{q}_h \in W_h\}$, contradicting (7.2.4).

Now let us come to the stability and convergence of the approximate solutions $(\mathbf{p}_h, u_h)$. We need to introduce the following finite dimensional subspace

$$(7.2.5) \qquad W_h^f := \left\{ \mathbf{q}_h \in W_h \, | \, \int_{\Omega} (\text{div} \, \mathbf{q}_h + f) v_h = 0 \quad \forall \, v_h \in Y_h \right\} \quad .$$

As we have already remarked, in general W_h^f is not contained in W^f. However, assuming that $W_h^0 \subset W^0$, we can easily prove that there exists a positive constant C_0 such that

$$(7.2.6) \qquad \int_{\Omega} \sum_{i,j=1}^{d} a^{ij} p_{h,j} p_{h,i} \geq C_0 ||\mathbf{p}_h||_{H(\text{div};\Omega)}^2 \quad \forall \, \mathbf{p}_h \in W_h^0 \quad ,$$

since the norms $|| \cdot ||_0$ and $|| \cdot ||_{H(\text{div};\Omega)}$ do coincide on W^0. We have already noticed that a sufficient condition for having $W_h^0 \subset W^0$ is that $\text{div} \, W_h \subset Y_h$.

We can state now the following results, which will be proven in a more general context in the following Section 7.4.1 (see Proposition 7.4.1).

Proposition 7.2.1 *Assume that W_h and Y_h satisfy the compatibility condition (7.2.4). Then for each $f \in L^2(\Omega)$ there exists a unique $\mathbf{q}_h^f \in (W_h^0)^\perp$ such that*

$$(7.2.7) \qquad \int_\Omega (\operatorname{div} \mathbf{q}_h^f + f) v_h = 0 \qquad \forall\, v_h \in Y_h$$

and

$$(7.2.8) \qquad ||\mathbf{q}_h^f||_{H(\operatorname{div};\Omega)} \leq \frac{1}{\beta_h} \sup_{\substack{v_h \in Y_h \\ v_h \neq 0}} \frac{\int_\Omega f v_h}{||v_h||_0} \ .$$

Furthermore, if $\mathbf{p}_h \in W_h$ satisfies

$$(7.2.9) \qquad \int_\Omega \sum_{i,j=1}^d a^{ij} p_{h,j} q_{h,i} = 0 \qquad \forall\, \mathbf{q}_h \in W_h^0 \ ,$$

then there exists a unique $u_h \in Y_h$ such that

$$(7.2.10) \qquad \int_\Omega u_h \operatorname{div} \mathbf{q}_h + \int_\Omega \sum_{i,j=1}^d a^{ij} p_{h,j} q_{h,i} = 0 \qquad \forall\, \mathbf{q}_h \in W_h$$

and

$$(7.2.11) \qquad ||u_h||_0 \leq \frac{1}{\beta_h} \sup_{\substack{\mathbf{q}_h \in W_h \\ \mathbf{q}_h \neq 0}} \frac{\int_\Omega \sum_{i,j=1}^d a^{ij} p_{h,j} q_{h,i}}{||\mathbf{q}_h||_{H(\operatorname{div};\Omega)}} \ .$$

Here we have denoted by $(W_h^0)^\perp$ the subspace of W_h orthogonal to W_h^0 with respect to the scalar product of $H(\operatorname{div};\Omega)$.

We are now in a position to state the main result of this Section.

Theorem 7.2.2 (Stability and convergence). *Assume that the spaces W_h and Y_h satisfy the compatibility condition (7.2.4) with a constant $\beta > 0$ independent of h, and moreover that $W_h^0 \subset W^0$. Then the solution $(\mathbf{p}_h, u_h)$ to (7.2.1) satisfies*

$$(7.2.12) \qquad ||\mathbf{p}_h||_{H(\operatorname{div};\Omega)} + ||u_h||_0 \leq C_1 ||f||_0 \ .$$

Moreover, the difference between $(\mathbf{p}_h, u_h)$ and the solution $(\mathbf{p}, u)$ to (7.1.15), (7.1.16) satisfies

$$(7.2.13) \qquad \begin{aligned} &||\mathbf{p} - \mathbf{p}_h||_{H(\operatorname{div};\Omega)} + ||u - u_h||_0 \\ &\qquad \leq C_2 \left\{ \inf_{\mathbf{q}_h \in W_h} ||\mathbf{p} - \mathbf{q}_h||_{H(\operatorname{div};\Omega)} + \inf_{v_h \in Y_h} ||u - v_h||_0 \right\} \ . \end{aligned}$$

The constants C_1 and C_2 depend increasingly on β^{-1}.

Proof. From (7.2.4) and Proposition 7.2.1, there exists a unique $\mathbf{q}_h^f \in (W_h^0)^\perp$ such that (7.2.7) and (7.2.8) hold. Hence, setting $\mathbf{p}_h = \mathbf{p}_h^0 + \mathbf{q}_h^f$ we can rewrite (7.2.1) as: find $(\mathbf{p}_h^0, u_h) \in W_h^0 \times Y_h$ such that

$$(7.2.14) \quad \int_\Omega \left(\sum_{i,j=1}^d a^{ij} p_{h,j}^0 q_{h,i} + u_h \operatorname{div} \mathbf{q}_h \right)$$

$$= -\int_\Omega \sum_{i,j=1}^d a^{ij} q_{h,j}^f q_{h,i} \quad \forall \, \mathbf{q}_h \in W_h ,$$

On the other hand, (7.2.14) is equivalent to: find $\mathbf{p}_h^0 \in W_h^0$ such that

$$(7.2.15) \qquad \int_\Omega \sum_{i,j=1}^d a^{ij} p_{h,j}^0 q_{h,i} = -\int_\Omega \sum_{i,j=1}^d a^{ij} q_{h,j}^f q_{h,i} \quad \forall \, \mathbf{q}_h \in W_h^0 .$$

In fact, (7.2.15) at once follows from (7.2.14), and the converse also holds due to Proposition 7.2.1.

Having assumed $W_h^0 \subset W^0$, (7.2.6) gives that the bilinear form

$$\int_\Omega \sum_{i,j=1}^d a^{ij} p_{h,j} q_{h,i}$$

is coercive in W_h^0. As a consequence of Lax-Milgram lemma (see Theorem 5.1.1), there exists a unique solution $\mathbf{p}_h^0 \in W_h^0$ to (7.2.15), and it satisfies

$$\|\mathbf{p}_h^0\|_{H(\operatorname{div};\Omega)} \le C \|\mathbf{q}_h^f\|_0 .$$

Thus, from (7.2.8)

$$(7.2.16) \qquad \|\mathbf{p}_h\|_{H(\operatorname{div};\Omega)} \le C \|\mathbf{q}_h^f\|_{H(\operatorname{div};\Omega)} \le \frac{C}{\beta} \|f\|_0 .$$

Moreover, from (7.2.11) it follows

$$\|u_h\|_0 \le \frac{C}{\beta} \|\mathbf{p}_h\|_0 ,$$

hence (7.2.12) holds.

To prove (7.2.13), let us start showing that

$$(7.2.17) \qquad \|\mathbf{p} - \mathbf{p}_h\|_{H(\operatorname{div};\Omega)} \le C \inf_{\mathbf{q}_h^* \in W_h^f} \|\mathbf{p} - \mathbf{q}_h^*\|_{H(\operatorname{div};\Omega)} .$$

By subtracting (7.1.15) and (7.2.1)$_1$, for each $\mathbf{q}_h \in W_h$, $\mathbf{q}_h^* \in W_h^f$, $v_h \in Y_h$ we have

$$\int_\Omega \left[\sum_{i,j=1}^d a^{ij}(p_{h,j} - q_{h,j}^*)q_{h,i} + (u_h - v_h)\operatorname{div}\mathbf{q}_h \right]$$

$$= \int_\Omega \left[\sum_{i,j=1}^d a^{ij}(p_j - q_{h,j}^*)q_{h,i} + (u - v_h)\operatorname{div}\mathbf{q}_h \right] .$$

Choosing $\mathbf{q}_h = (\mathbf{p}_h - \mathbf{q}_h^*) \in W_h^0$, from (7.2.6) and the fact that $W_h^0 \subset W^0$ it follows

$$||\mathbf{q}_h||_{H(\operatorname{div};\Omega)}^2 \leq C||\mathbf{q}_h||_0 ||\mathbf{p} - \mathbf{q}_h^*||_0 ,$$

i.e.,

$$||\mathbf{q}_h||_{H(\operatorname{div};\Omega)} \leq C||\mathbf{p} - \mathbf{q}_h^*||_0 .$$

Writing $\mathbf{p} - \mathbf{p}_h = \mathbf{p} - \mathbf{q}_h^* - \mathbf{q}_h$, (7.2.17) follows at once. Now let us show that

$$(7.2.18)\quad \inf_{\mathbf{q}_h^* \in W_h^f} ||\mathbf{p} - \mathbf{q}_h^*||_{H(\operatorname{div};\Omega)} \leq \left(1 + \frac{1}{\beta}\right) \inf_{\mathbf{q}_h \in W_h} ||\mathbf{p} - \mathbf{q}_h||_{H(\operatorname{div};\Omega)} .$$

From Proposition 7.2.1, for each $\mathbf{q}_h \in W_h$ there exists a unique $\mathbf{z}_h \in (W_h^0)^\perp$ such that

$$\int_\Omega v_h \operatorname{div}\mathbf{z}_h = \int_\Omega v_h \operatorname{div}(\mathbf{p} - \mathbf{q}_h) \quad \forall\, v_h \in Y_h ,$$

and

$$||\mathbf{z}_h||_{H(\operatorname{div};\Omega)} \leq \frac{1}{\beta}||\operatorname{div}(\mathbf{p} - \mathbf{q}_h)||_0 .$$

Setting $\mathbf{q}_h^* := \mathbf{z}_h + \mathbf{q}_h$, it follows that $\mathbf{q}_h^* \in W_h^f$, as $\operatorname{div}\mathbf{p} + f = 0$ in Ω. Furthermore,

$$||\mathbf{p} - \mathbf{q}_h^*||_{H(\operatorname{div};\Omega)} \leq ||\mathbf{p} - \mathbf{q}_h||_{H(\operatorname{div};\Omega)} + ||\mathbf{z}_h||_{H(\operatorname{div};\Omega)}$$

$$\leq \left(1 + \frac{1}{\beta}\right) ||\mathbf{p} - \mathbf{q}_h||_{H(\operatorname{div};\Omega)} ,$$

and (7.2.18) holds true.

Now let us estimate $||u - u_h||_0$. From the compatibility condition (7.2.4), for each $v_h \in Y_h$ we have

$$||u_h - v_h||_0 \leq \frac{1}{\beta} \sup_{\substack{\mathbf{q}_h \in W_h \\ \mathbf{q}_h \neq 0}} \frac{\int_\Omega (u_h - v_h)\operatorname{div}\mathbf{q}_h}{||\mathbf{q}_h||_{H(\operatorname{div};\Omega)}} .$$

On the other hand, by subtracting (7.1.15) and $(7.2.1)_1$ it follows

$$\int_\Omega (u_h - v_h)\operatorname{div}\mathbf{q}_h = \int_\Omega (u - v_h)\operatorname{div}\mathbf{q}_h + \int_\Omega \sum_{i,j=1}^d a^{ij}(\mathbf{p} - \mathbf{p}_h)_j q_{h,i} .$$

Thus,

$$||u_h - v_h||_0 \leq \frac{1}{\beta}(||u - v_h||_0 + C||\mathbf{p} - \mathbf{p}_h||_0) \ ,$$

and writing $u - u_h = u - v_h + v_h - u_h$, we conclude the proof of (7.2.13). $\quad\square$

Notice that the error estimate for $||\mathbf{p} - \mathbf{p}_h||_{H(\mathrm{div};\Omega)}$ depends on $1/\beta$, whereas the one for $||u - u_h||_0$ depends on $1/\beta^2$. This makes the approximation of u more sensitive to the smallness of the constant β in (7.2.4).

It is now necessary to construct a couple of subspaces W_h and Y_h such that the assumptions of Theorem 7.2.2 are satisfied. Moreover, we want to estimate the approximation errors appearing in (7.2.13). An example of such a choice is presented in the following Section 7.2.2.

7.2.2 An Example: the Raviart-Thomas Finite Elements

Let us assume that $\Omega \subset \mathbb{R}^d$, $d = 2, 3$, is a polygonal domain with Lipschitz boundary, and that $\mathcal{T}_h$ is a family of triangulations of $\overline{\Omega}$. For simplicity, we will restrict our attention to a reference polyhedron $\hat{K}$ which is the unit d-simplex (see (3.1.6)). However, analogous results also hold for $\hat{K} = [0,1]^d$ (see, e.g., Roberts and Thomas (1991)).

When considering the two-dimensional case, Raviart and Thomas (1977) proposed to choose as W_h and Y_h the following spaces ($k \geq 1$ is a given integer):

$$(7.2.19) \quad W_h = W_h^k := \{\mathbf{q}_h \in H(\mathrm{div};\Omega) \,|\, \mathbf{q}_{h|K} \in \mathbb{D}_k \ \ \forall \, K \in \mathcal{T}_h\}$$

$$(7.2.20) \quad Y_h = Y_h^{k-1} := \{v_h \in L^2(\Omega) \,|\, v_{h|K} \in \mathbb{P}_{k-1} \ \ \forall \, K \in \mathcal{T}_h\} \ ,$$

where

$$(7.2.21) \qquad\qquad \mathbb{D}_k := (\mathbb{P}_{k-1})^d \oplus \mathbf{x}\mathbb{P}_{k-1} \ .$$

(These spaces have already been introduced in (3.2.6), (3.5.7) and (3.2.2), respectively.) The degrees of freedom for the Raviart-Thomas finite elements are illustrated in Section 3.3.3. The analysis of the properties of these elements in the three-dimensional case has been carried out by Nédélec (1980).

We want to show that the assumptions in Theorem 7.2.2 are satisfied with the choice (7.2.19), (7.2.20). First of all, in Section 3.3.3 we have already noticed that $\mathrm{div}\, W_h^k \subset Y_h^{k-1}$, so that $W_h^0 \subset W^0$.

To establish that the compatibility condition (7.2.4) holds with a constant β independent of h, the following abstract lemma, due to Fortin (1977), will be useful.

Lemma 7.2.1 (Fortin). *Let us assume that there exists $\beta^* > 0$ such that the compatibility condition*

$$\forall\, v \in L^2(\Omega) \;\exists\; \mathbf{q} \in H(\operatorname{div}; \Omega), \mathbf{q} \neq \mathbf{0}:$$

(7.2.22)
$$\int_{\Omega} v \operatorname{div} \mathbf{q} \geq \beta^* ||\mathbf{q}||_{H(\operatorname{div};\Omega)} ||v||_0$$

is satisfied, and, moreover, that there exists an operator $\tau_h : H(\operatorname{div}; \Omega) \to W_h$ such that

(i) $\int_{\Omega} v_h \operatorname{div}(\mathbf{q} - \tau_h(\mathbf{q})) = 0 \quad \forall\, \mathbf{q} \in H(\operatorname{div}; \Omega),\; v_h \in Y_h$

(ii) $||\tau_h(\mathbf{q})||_{H(\operatorname{div};\Omega)} \leq C_*||\mathbf{q}||_{H(\operatorname{div};\Omega)} \quad \forall\, \mathbf{q} \in H(\operatorname{div}; \Omega)$,

where $C_* > 0$ doesn't depend on h. Then, the compatibility condition (7.2.4) is satisfied with $\beta = \beta^*/C_*$.

Proof. From (7.2.22), for any $v_h \in Y_h$ there exists $\mathbf{q}^* \in H(\operatorname{div}; \Omega), \mathbf{q}^* \neq \mathbf{0}$, such that

(7.2.23)
$$\int_{\Omega} v_h \operatorname{div} \mathbf{q}^* \geq \beta^* ||\mathbf{q}^*||_{H(\operatorname{div};\Omega)} ||v_h||_0 \; .$$

As we can assume $v_h \neq 0$, (7.2.23) gives in particular $\int_{\Omega} v_h \operatorname{div} \mathbf{q}^* \neq 0$, and consequently (i) yields $\tau_h(\mathbf{q}^*) \neq \mathbf{0}$.

On the other hand, from (i), (ii) and (7.2.23) we find

$$\int_{\Omega} v_h \operatorname{div}(\tau_h(\mathbf{q}^*)) = \int_{\Omega} v_h \operatorname{div} \mathbf{q}^* \geq \frac{\beta^*}{C_*} ||\tau_h(\mathbf{q}^*)||_{H(\operatorname{div};\Omega)} ||v_h||_0 \;,$$

and the thesis follows.   □

The compatibility condition (7.2.22) is the infinite dimensional counterpart of (7.2.4). Notice also that condition (i) in particular implies that there are no spurious zero-energy modes, i.e., elements $v_h^* \in Y_h$ such that $\int_{\Omega} v_h^* \operatorname{div} \mathbf{q}_h = 0$ for all $\mathbf{q}_h \in W_h$ but satisfying $\int_{\Omega} v_h^* \operatorname{div} \mathbf{q} \neq 0$ for some $\mathbf{q} \in W$.

Let us now verify that the assumptions of Lemma 7.2.1 hold for W_h^k and Y_h^{k-1} as in (7.2.19), (7.2.20). First of all, given $v \in L^2(\Omega),\, v \neq 0$ one can find $\mathbf{q} \in H(\operatorname{div}; \Omega),\, \mathbf{q} \neq \mathbf{0}$ such that $\operatorname{div} \mathbf{q} = v$. This can be done, for instance, by choosing $\mathbf{q} = \nabla \psi$, where ψ is the solution to

$$\begin{cases} \Delta \psi = v & \text{in } \Omega \\ \\ \psi_{|\partial\Omega} = 0 & \text{on } \partial\Omega \; . \end{cases}$$

With this choice, (7.2.22) is satisfied, as $||\mathbf{q}||_0 = ||\nabla\psi||_0 \leq C_\Omega^{1/2}||v||_0$ and $||\operatorname{div} \mathbf{q}||_0 = ||v||_0$, C_Ω being the constant appearing in the Poincaré inequality (1.3.2).

To construct the operator τ_h, let $\mathbf{q} \in H(\operatorname{div}; \Omega)$ and define

$$\phi := \begin{cases} \operatorname{div} \mathbf{q} & \text{in } \Omega \\ 0 & \text{in } B \setminus \Omega \end{cases},$$

where B is an open ball containing Ω. Clearly $\phi \in L^2(\Omega)$, therefore we can solve

$$\begin{cases} \Delta \psi_* = \phi & \text{in } B \\ \\ \psi_{*|\partial B} = 0 & \text{on } \partial B \end{cases}$$

and we obtain $\psi^* \in H^2(B)$, $\|\psi_*\|_{2,B} \leq C_B \|\phi\|_{0,B}$. Thus, defining $\mathbf{q}_* := (\nabla \psi_*)_{|\Omega}$, we have $\mathbf{q}_* \in (H^1(\Omega))^d$, $\operatorname{div} \mathbf{q}_* = \operatorname{div} \mathbf{q}$ in Ω and

$$\|\mathbf{q}_*\|_{1,\Omega} \leq C_B \|\operatorname{div} \mathbf{q}\|_{0,\Omega} .$$

We can now consider the interpolant of $\mathbf{q}_*$ in W_h, i.e., we define

$$(7.2.24) \qquad \tau_h(\mathbf{q}) := \omega_h^k(\mathbf{q}_*) ,$$

where ω_h^k is the interpolation operator introduced in (3.4.20). Notice that ω_h^k is defined in $(H^1(\Omega))^d$ and not in $H(\operatorname{div}; \Omega)$, and this is the reason for using the "regularizing" procedure above. The operator τ_h satisfies

$$\|\tau_h(\mathbf{q})\|_{H(\operatorname{div};\Omega)} = \|\omega_h^k(\mathbf{q}_*)\|_{H(\operatorname{div};\Omega)} \leq C \|\mathbf{q}_*\|_{1,\Omega} \leq C \|\operatorname{div} \mathbf{q}\|_{0,\Omega} ,$$

hence (ii) holds. Property (i) is a straightforward consequence of (3.4.23), i.e.,

$$(7.2.25) \qquad \operatorname{div}(\omega_h^k(\mathbf{q}^*)) = p_h^{k-1}(\operatorname{div} \mathbf{q}^*) = p_h^{k-1}(\operatorname{div} \mathbf{q}) ,$$

where p_h^{k-1} is the $L^2(\Omega)$-orthogonal projection onto Y_h^{k-1}.

We have thus proven that, when considering the Raviart-Thomas finite elements, the compatibility condition (7.2.4) is satisfied with a constant β independent of h. Hence, all assumptions of Theorem 7.2.2 hold true and the error estimate (7.2.13) is valid. On the other hand, we already know from Theorem 3.4.4 that, if $\mathcal{T}_h$ is a regular family of triangulations,

$$\inf_{\mathbf{q}_h \in W_h^k} \|\mathbf{p} - \mathbf{q}_h\|_{H(\operatorname{div};\Omega)} \leq C h^l (|\mathbf{p}|_{l,\Omega} + |\operatorname{div} \mathbf{p}|_{l,\Omega}) , \quad 1 \leq l \leq k$$

and that (see (3.5.24))

$$\inf_{v_h \in Y_h^{k-1}} \|u - v_h\|_0 \leq C h^l |u|_{l,\Omega} , \quad 1 \leq l \leq k .$$

The following error estimate therefore holds for $1 \leq l \leq k$:

$$(7.2.26) \quad \|\mathbf{p} - \mathbf{p}_h\|_{H(\operatorname{div};\Omega)} + \|u - u_h\|_0 \leq C h^l (|\mathbf{p}|_{l,\Omega} + |\operatorname{div} \mathbf{p}|_{l,\Omega} + |u|_{l,\Omega}) .$$

Remark 7.2.1 Following Falk and Osborn (1980) it can also be proven that

$$\|\mathbf{p} - \mathbf{p}_h\|_0 \leq C h^l \|\mathbf{p}\|_{l,\Omega} , \quad 1 \leq l \leq k .$$

Moreover,

$$||u - u_h||_0 \leq Ch^l ||u||_{l*,\Omega} \quad , \quad l^* = \max(2, l) \ , \ 1 \leq l \leq k \ ,$$

provided the regularity condition (6.2.12) holds.   □

Remark 7.2.2 Since $(C^\infty(\overline{\Omega}))^d$ is dense in $H(\text{div}; \Omega)$ (see, e.g., Girault and Raviart (1986), p. 27) and $C_0^\infty(\Omega)$ is dense in $L^2(\Omega)$, one can repeat the argument adopted in Theorem 6.2.1 to prove that $\mathbf{p}_h \to \mathbf{p}$ in $H(\text{div}; \Omega)$ and $u_h \to u$ in $L^2(\Omega)$ under the only assumption $\mathbf{p} \in H(\text{div}; \Omega)$, $u \in L^2(\Omega)$ or, equivalently, $u \in H^1(\Omega)$.   □

Remark 7.2.3 By proceeding in the same way as in the construction of the operator τ_h, it is easily seen that for each $v_h \in Y_h^{k-1}$ there exists $\mathbf{q}_h \in W_h^k$ such that $\text{div}\,\mathbf{q}_h = v_h$. (Just take v_h instead of $\text{div}\,\mathbf{q}$.) It is thus proven that $\text{div}\,W_h^k = Y_h^{k-1}$.   □

Remark 7.2.4 Let us emphasise that the commutativity property (7.2.25) is of great importance in the analysis of the Raviart-Thomas finite elements. In fact, it at once yields $\text{div}\,W_h^k \subset Y_h^{k-1}$, so that $W_h^k \subset W^0$; moreover, it is helpful in the construction of the operator τ_h in Lemma 3.1.

At the end of this Section we present another family of finite elements enjoying property (7.2.25).   □

Similar results hold for the other boundary value problems we have considered so far.

As we noticed in Section 7.1.2, the Neumann problem can be written in the form (7.1.20), (7.1.21), and the basic spaces are now $H_0(\text{div}; \Omega)$ and $L_0^2(\Omega)$. The compatibility condition (7.2.22) is satisfied, as for each $v \in L_0^2(\Omega)$ we can construct $\mathbf{q} \in H_0(\text{div}; \Omega)$ such that $\text{div}\,\mathbf{q} = v$ and $||\mathbf{q}||_{H(\text{div};\Omega)} \leq C||v||_0$ by solving the Neumann problem

$$\begin{cases} \Delta\psi = v & \text{in } \Omega \\ \dfrac{\partial\psi}{\partial n} = 0 & \text{on } \partial\Omega \ , \end{cases}$$

setting then $\mathbf{q} = \nabla\psi$.

To show that the discrete compatibility condition (7.2.4) holds uniformly with respect to h is a more delicate matter. However, it can be shown that the Raviart-Thomas spaces

$$W_{0,h}^k := \{\mathbf{q}_h \in H_0(\text{div}; \Omega) \,|\, \mathbf{q}_{h|K} \in \mathbb{D}_k \ \ \forall\, K \in \mathcal{T}_h\}$$
$$Y_{0,h}^k := \{v_h \in L_0^2(\Omega) \,|\, v_{h|K} \in \mathbb{P}_{k-1} \ \ \forall\, K \in \mathcal{T}_h\}$$

satisfy (7.2.4). A proof is reported in Roberts and Thomas (1991), and it is valid also for the problem with mixed boundary conditions.

When considering the Robin problem (7.1.22), (7.1.23), the only difference from the Dirichlet case stems from the use of the space $\mathcal{H}(\text{div}; \Omega)$ defined in (7.1.24) instead of $H(\text{div}; \Omega)$. Nonetheless, by proceeding as in the construction of the operator τ_h in (7.2.24), it is easily seen that for each $v \in L^2(\Omega)$ there exists $\mathbf{q} \in \mathcal{H}(\text{div}; \Omega)$ such that $\text{div}\,\mathbf{q} = v$ and $||\mathbf{q}||_{\mathcal{H}(\text{div};\Omega)} \leq C||v||_0$. The compatibility condition (7.2.22) is thus satisfied, and as before one can take as W_h and Y_h the Raviart-Thomas finite element spaces W_h^k and Y_h^{k-1}. (Just notice that W_h^k is indeed contained in $\mathcal{H}(\text{div}; \Omega)$.)

Now we are now going to introduce another family of finite element spaces which have been proposed for problem (7.2.1) by Brezzi, Douglas and Marini (1985) in dimension $d = 2$ and by Brezzi, Douglas, Duran and Fortin (1987) in dimension $d = 3$.

In this case the choice of W_h and Y_h is given by:

(7.2.27) $W_h = W_h^{*k} := \{\mathbf{q}_h \in H(\text{div}; \Omega) \,|\, \mathbf{q}_{h|K} \in (\mathbb{P}_{k-1})^d \ \forall K \in \mathcal{T}_h\}$

(7.2.28) $Y_h = Y_h^{k-2}$,

where $k \geq 2$. The degrees of freedom related to W_h^{*k} in the two-dimensional case are given on each triangle K by

$$\int_{F_K} \mathbf{n} \cdot \mathbf{q}\,\psi \quad , \quad \psi \in \mathbb{P}_{k-1}$$

$$\int_K \mathbf{q} \cdot \nabla\chi \quad , \quad \chi \in \mathbb{P}_{k-2} \ (k \geq 3)$$

$$\int_K \mathbf{q} \cdot \mathbf{w} \quad , \quad \mathbf{w} \in (\mathbb{P}_{k-1})^2 \ , \ \text{div}\,\mathbf{w} = 0 \text{ in } K \ , \ \mathbf{w} \cdot \mathbf{n} = 0 \text{ on } \partial K \ (k \geq 3) \ ,$$

where $\mathbf{q} \in (\mathbb{P}_{k-1})^2$ and F_K is any face of K. Hence, on each K there are $2k$ degrees of freedom less than for the Raviart-Thomas elements.

It has been proven that the commutativity property (7.2.25) still holds with Y_h^{k-1} replaced by Y_h^{k-2}, and the following estimates for the interpolation error are valid:

$$||\mathbf{q} - \boldsymbol{\omega}_h^{*k}(\mathbf{q})||_{0,\Omega} \leq Ch^l|\mathbf{q}|_{l,\Omega} \ , \ 1 \leq l \leq k \ ,$$

for each $\mathbf{q} \in H^l(\Omega)$, and

$$||\,\text{div}(\mathbf{q} - \boldsymbol{\omega}_h^{*k}(\mathbf{q}))||_{0,\Omega} \leq Ch^l|\,\text{div}\,\mathbf{q}|_{l,\Omega} \ , \ 1 \leq l \leq k - 1 \ ,$$

for each $\mathbf{q} \in H^1(\Omega)$ such that $\text{div}\,\mathbf{q} \in H^l(\Omega)$. Therefore, the following error estimate follows from Theorem 7.2.2 for $1 \leq l \leq k - 1$:

(7.2.29) $||\mathbf{p} - \mathbf{p}_h||_{H(\text{div};\Omega)} + ||u - u_h||_0 \leq Ch^l(|\mathbf{p}|_{l,\Omega} + |\,\text{div}\,\mathbf{p}|_{l,\Omega} + |u|_{l,\Omega})$.

Notice that the interpolation errors for the Raviart-Thomas and the Brezzi-Douglas-Marini elements are of the same order when measured in the

$L^2(\Omega)$-norm. If one is mainly interested in the approximation of the vector **p**, these latter elements are a viable choice.

With the aim of illustrating the accuracy of mixed methods, let us consider the Neumann problem

(7.2.30)
$$
\begin{cases}
-\operatorname{div}(\alpha\nabla u) = f & \text{in } \Omega = (0,1)^2 \\
\alpha\dfrac{\partial u}{\partial n} = 0 & \text{on } \partial\Omega \ ,
\end{cases}
$$

where the datum f satisfies the compatibility condition $\int_\Omega f = 0$. We triangulate Ω by a uniform grid using $2N^2$ triangles of equal size (the finite element grid spacing is $h = \sqrt{2}/N$). In Tables 7.2.1, 7.2.2 and 7.2.3 we report the relative error in the $L^2(\Omega)$-norm of the vector function $\mathbf{p} = \alpha\nabla u$.

Table 7.2.1 refers to the case in which $\alpha = 1$ and $f(x_1,x_2) = 2(1-x_1-x_2)$, and compares the errors obtained by using the mixed Raviart-Thomas finite elements based on the spaces (7.2.19), (7.2.20) with $k = 1$ (RT(1)) and the mixed Brezzi-Douglas-Marini finite elements based on the spaces (7.2.27), (7.2.28) with $k = 2$ (BDM(2)). We observe linear convergence for the Raviart-Thomas elements and quadratic convergence for the Brezzi-Douglas-Marini elements.

Table 7.2.1. Comparison between two different types of mixed finite element approximations for $\alpha = 1$

$N(=\sqrt{2}h^{-1})$	RT(1)	BDM(2)
10	0.93e-1	0.45e-2
20	0.47e-1	0.11e-2
40	0.23e-1	0.28e-3
60	0.16e-1	0.12e-3
80	0.12e-1	0.71e-4
100	0.93e-2	0.45e-4

Table 7.2.2 refers to a similar comparison between RT(1) and BDM(2), this time when the coefficient $\alpha = \alpha(\mathbf{x})$ is constant only with respect to x_2 and exhibits a variation with respect to x_1 (i.e., $\max_{x_1} \alpha(\mathbf{x})/\min_{x_1} \alpha(\mathbf{x})$) of the order of 10^4.

In Table 7.2.3 we report a calculation made using different values of N for a coefficient $\alpha(\mathbf{x})$ which behaves as before, with $\max_{x_1} \alpha(\mathbf{x})/\min_{x_1} \alpha(\mathbf{x})$ of the order of 10^i for $i = 0, 1, 2, 4, 8$.

At the end of the following Section we will compare the accuracy of these mixed finite elements versus the computational cost (for a suitable solution algorithm).

Table 7.2.2. Comparison between two different types of mixed finite element approximations when α exhibits a gradient of the order of 10^4

$N(=\sqrt{2}h^{-1})$	RT(1)	BDM(2)
10	0.21e-0	0.40e-1
20	0.93e-1	0.11e-1
40	0.45e-1	0.27e-2
60	0.30e-1	0.12e-2
80	0.22e-1	0.67e-3
100	0.18e-1	0.43e-3

Table 7.2.3. Comparison between two different types mixed finite element approximations when α exhibits a large gradient

i	RT(1) ($N=40$)	BDM(2) ($N=40$)	BDM(2) ($N=60$)	BDM(2) ($N=100$)
0	0.29e-1	0.11e-2	0.51e-3	0.20e-3
1	0.41e-1	0.23e-2	0.10e-2	0.37e-3
2	0.44e-1	0.26e-2	0.11e-2	0.41e-3
4	0.45e-1	0.27e-2	0.12e-2	0.43e-3
8	0.45e-1	0.28e-2	0.12e-2	0.43e-3

7.3 Some Remarks on the Algorithmic Aspects

In order to formulate the algebraic problem associated to (7.2.1), let us start by setting

$$(7.3.1) \qquad a(\mathbf{r}, \mathbf{q}) := \int_\Omega \sum_{s,t=1}^{d} a^{st} r_t q_s \ , \quad b(\mathbf{q}, v) := \int_\Omega v \operatorname{div} \mathbf{q} \ ,$$

and by choosing two bases $\{\varphi_j \,|\, j = 1, ..., N_h\}$ and $\{\psi_l \,|\, l = 1, ..., K_h\}$ of W_h and Y_h, respectively. Writing the solutions $\mathbf{p}_h$ and u_h as $\mathbf{p}_h(\mathbf{x}) = \sum_{j=1}^{N_h} p_j \varphi_j(\mathbf{x})$ and $u_h(\mathbf{x}) = \sum_{l=1}^{K_h} u_l \psi_l(\mathbf{x})$, (7.2.1) can be formulated algebraically as follows

$$(7.3.2) \qquad \begin{cases} A\mathbf{p} + B^T \mathbf{u} = 0 \\ \\ B\mathbf{p} \qquad\quad = -\mathbf{f} \ , \end{cases}$$

where

$$(7.3.3) \qquad A_{ij} := a(\boldsymbol{\varphi}_j, \boldsymbol{\varphi}_i) \ , \quad B_{li} := b(\boldsymbol{\varphi}_i, \psi_l) \ , \quad f_l := (f, \psi_l) \ .$$

The matrix

$$(7.3.4) \qquad S := \begin{pmatrix} A & B^T \\ B & 0 \end{pmatrix}$$

is non-singular if $\ker B^T = \mathbf{0}$, which is equivalent to require that the compatibility condition (7.2.4) holds. Since A is positive definite, under this assumption one could solve (7.3.2) by eliminating $\mathbf{p}$ from the first equation

$$(7.3.5) \qquad \mathbf{p} = -A^{-1}B^T\mathbf{u}$$

and then finding the solution $\mathbf{u}$ to

$$(7.3.6) \qquad BA^{-1}B^T\mathbf{u} = \mathbf{f} \ .$$

This procedure will be given the name of *mixed-Schur complement* algorithm. This procedure resembles the one used in the framework of the Stokes problem for eliminating the velocity field from the momentum equation (see Section 9.6.1).

For the sake of simplicity, let us assume from now on that the bilinear form $a(\cdot, \cdot)$ is symmetric. Then the matrix $R := BA^{-1}B^T$ is symmetric and positive definite but full (due to the presence of the inverse of A), hence (7.3.6) can be solved, e.g., by the conjugate gradient method described in Section 2.4.2. This procedure would require at each step the solution of the linear system associated to the matrix A, which is sparse but of large dimension N_h. Since the condition number of R can grow like h^{-2}, a preconditioning procedure is compulsory. In the numerical test that will be reported in Table 7.3.1, the matrix BB^T is used as a preconditioner of R. As such, it enjoys optimal spectral properties. However, it may be very much memory demanding when K_h is large.

A different solution algorithm is based on the observation that inverting the matrix A would be economical if the vector functions in W_h were not required to belong to $H(\text{div}; \Omega)$, allowing therefore their normal components to be discontinuous across the interelement boundaries. In such a case A would be a block-diagonal matrix, each block (corresponding to a triangle) being a small non-singular matrix.

The idea is therefore to resorting to a *mixed-Lagrangian* formulation, introducing suitable Lagrangian multipliers with the aim of relaxing the continuity assumption for the normal components at the interelement boundaries.

To give an example, let us introduce

$$(7.3.7) \qquad \tilde{W}_h := \{\mathbf{q}_h \in (L^2(\Omega))^2 \,|\, \mathbf{q}_{h|K} \in \mathbb{D}_1 \ \forall K \in \mathcal{T}_h\} \ ,$$

which is to the *discontinuous* first order Raviart-Thomas finite element space. Notice that $W_h \subset \tilde{W}_h$, and that an element $\mathbf{q}_h$ of $\tilde{W}_h$ belongs to W_h if and only if $\mathbf{q}_h \in H(\text{div}; \Omega)$.

Setting

$$(7.3.8) \qquad \Gamma_h := \bigcup_{K \in \mathcal{T}_h} \partial K \ , \quad \Gamma_h^i := \Gamma_h \cap \Omega \ ,$$

and denoting by $\mathcal{S}$ a side of the triangle K, the space of multipliers is defined as

$$(7.3.9) \qquad \Lambda_h := \{\mu_h \in L^2(\Gamma_h) \,|\, \mu_{h|\mathcal{S}} \in \mathbb{P}_0 \ \forall \, \mathcal{S} \in \Gamma_h^i,$$
$$\mu_{h|\mathcal{S}} = 0 \ \forall \, \mathcal{S} \in \Gamma_h \setminus \Gamma_h^i\} \ .$$

Furthermore, we introduce the bilinear form

$$(7.3.10) \qquad d(\mathbf{q}_h, \mu_h) := - \sum_{K \in \mathcal{T}_h} \int_{\partial K} \mu_h \, \mathbf{q}_{h|K} \cdot \mathbf{n}_K = \sum_{\mathcal{S} \in \Gamma_h^i} \int_{\mathcal{S}} \mu_h \, [\mathbf{q}_h \cdot \mathbf{n}]$$

for $\mathbf{q}_h \in \tilde{W}_h$, $\mu_h \in \Lambda_h$, where $\mathbf{n}_K$ is the unit outward normal vector on ∂K and $[\mathbf{q}_h \cdot \mathbf{n}]$ denotes the jump of the normal component of $\mathbf{q}_h$ across the side $\mathcal{S}$. Notice that an element $\mathbf{q}_h \in \tilde{W}_h$ satisfies $d(\mathbf{q}_h, \mu_h) = 0$ for each $\mu_h \in \Lambda_h$ if and only if $\mathbf{q}_h \in H(\text{div}; \Omega)$, i.e., $\mathbf{q}_h \in W_h$.

Since $b(\mathbf{q}_h, v_h)$ is not defined for $\mathbf{q}_h \in \tilde{W}_h$, we also set

$$(7.3.11) \qquad b_h(\mathbf{q}_h, v_h) := \sum_{K \in \mathcal{T}_h} \int_K v_h \, \text{div} \, \mathbf{q}_h$$

for each $\mathbf{q}_h \in \tilde{W}_h$, $\mu_h \in \Lambda_h$.

Instead of (7.3.2), we are led to consider the following problem: find $\tilde{\mathbf{p}}_h \in \tilde{W}_h$, $\tilde{u}_h \in Y_h$, $\lambda_h \in \Lambda_h$ such that

$$(7.3.12) \qquad \begin{cases} a(\tilde{\mathbf{p}}_h, \tilde{\mathbf{q}}_h) + b_h(\tilde{\mathbf{q}}_h, \tilde{u}_h) + d(\tilde{\mathbf{q}}_h, \lambda_h) = 0 & \forall \, \tilde{\mathbf{q}}_h \in \tilde{W}_h \\[2mm] b_h(\tilde{\mathbf{p}}_h, v_h) = -(f, v_h) & \forall \, v_h \in Y_h \\[2mm] d(\tilde{\mathbf{p}}_h, \mu_h) = 0 & \forall \, \mu_h \in \Lambda_h \ . \end{cases}$$

Clearly, a solution to (7.3.12) is also a solution to (7.2.1), i.e., $\tilde{\mathbf{p}}_h = \mathbf{p}_h$, $\tilde{u}_h = u_h$. In turn, the Lagrangian multiplier λ_h represents an approximation of the trace of the solution u on Γ_h. This value can be used together with $\tilde{u}_h$ for obtaining a better approximation of u (see, e.g., Arnold and Brezzi (1985)).

The algebraic restatement of (7.3.12) reads

$$(7.3.13) \qquad \begin{cases} \tilde{A}\mathbf{p} + B_h^T \mathbf{u} + D^T \boldsymbol{\lambda} = \mathbf{0} \\[2mm] B_h \mathbf{p} \qquad\qquad\quad = -\mathbf{f} \\[2mm] D\mathbf{p} \qquad\qquad\quad = \mathbf{0} \ , \end{cases}$$

where $\tilde{A}_{ij} := a(\tilde{\varphi}_j, \tilde{\varphi}_i)$, $(B_h)_{li} := b_h(\tilde{\varphi}_i, \psi_l)$, $D_{si} := d(\tilde{\varphi}_i, \chi_s)$ and $\{\tilde{\varphi}_j \mid j = 1, ..., \tilde{N}_h\}$ and $\{\chi_s \mid s = 1, ..., L_h\}$ are bases of $\tilde{W}_h$ and Λ_h, respectively.

The matrix

$$(7.3.14) \qquad\qquad \tilde{S} := \begin{pmatrix} \tilde{A} & B_h^T & D^T \\ B_h & 0 & 0 \\ D & 0 & 0 \end{pmatrix}$$

is symmetric, non-singular but fails to be positive definite. However, $\tilde{A}$ is a block-diagonal matrix, each block being a 3×3 matrix easy to invert. We can therefore eliminate $\mathbf{p}$ in (7.3.13) finding

$$(7.3.15) \qquad \begin{cases} B_h\tilde{A}^{-1}B_h^T\mathbf{u} + B_h\tilde{A}^{-1}D^T\boldsymbol{\lambda} = \mathbf{f} \\[2mm] D\tilde{A}^{-1}B_h^T\mathbf{u} + D\tilde{A}^{-1}D^T\boldsymbol{\lambda} = 0 \end{cases},$$

i.e., a linear system associated to a symmetric and positive definite matrix.

The unknown $\tilde{u}_h$ is not continuous, therefore another simple elimination procedure leads to a system where the only unknown is $\boldsymbol{\lambda}$. It reads

$$(7.3.16) \qquad\qquad\qquad \tilde{R}\boldsymbol{\lambda} = \mathbf{g} \ ,$$

where

$$(7.3.17) \qquad \begin{aligned} \tilde{R} &:= D\tilde{A}^{-1}D^T - D\tilde{A}^{-1}B_h^T(B_h\tilde{A}^{-1}B_h^T)^{-1}B_h\tilde{A}^{-1}D^T \\ \mathbf{g} &:= -D\tilde{A}^{-1}B_h^T(B_h\tilde{A}^{-1}B_h^T)^{-1}\mathbf{f} \ . \end{aligned}$$

The symmetric and positive definite matrix $\tilde{R}$ has dimension $L_h \times L_h$, where L_h is the number of internal sides of the triangulation, whereas the matrix $R = BA^{-1}B^T$ has dimension $K_h \times K_h$, K_h being the number of the triangles. Although in general L_h is larger than K_h, it is easier to apply a conjugate gradient procedure to the linear system associated to the matrix $\tilde{R}$ than to the one associated to R, as $\tilde{A}$ is a block-diagonal matrix with 3×3 blocks and $B_h\tilde{A}^{-1}B_h^T$ is diagonal.

Having determined the solution of system (7.3.16) for the Lagrange multiplier $\boldsymbol{\lambda}$, the mixed-Lagrangian algorithm furnishes $\mathbf{u}$ through (7.3.15)$_1$ and $\mathbf{p}$ through (7.3.13)$_1$.

Let us present now some numerical results. Table 7.3.1 refers to the solution of the boundary value problem (7.2.30) (with $\alpha = 1$ and $f(x_1, x_2) = 2(1 - x_1 - x_2)$), using the BDM(2) mixed finite elements. Here, the purpose is to compare the performance of the mixed-Schur complement (M-SC) algorithm with the one of the mixed-Lagrangian (M-L).

For several values of the grid-size we report both the number of CG-iterations (ITER) and the global CPU-time (in seconds) for solving system (7.3.6) (with preconditioner BB^T) in the M-SC approach, or system (7.3.16)

Table 7.3.1. Comparison between mixed-Lagrangian and mixed-Schur complement algorithms for the solution of problem (7.2.30) by the BDM(2) mixed finite elements

$N(=\sqrt{2}h^{-1})$	Mixed-Lagrangian		Mixed-Schur complement	
	ITER	CPU-time	ITER	CPU-time
10	30	0.08	7	0.23
20	38	0.42	7	0.84
40	69	3.03	8	3.59
60	85	8.32	7	7.58
80	111	16.18	7	12.96
100	139	37.36	7	22.35

(with preconditioner the incomplete Cholesky decomposition of $\tilde{R}$) in the M-L approach.

For the same problem we report in Fig. 7.3.1 the CPU-time (in seconds) versus the accuracy (still for the $L^2(\Omega)$-norm of the relative error $||\mathbf{p} - \mathbf{p}_h||_0/||\mathbf{p}||_0$) when using the RT(1) and the BDM(2) mixed finite elements. Computations for both elements have been performed by the M-SC algorithm.

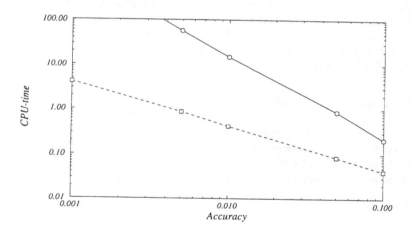

Fig. 7.3.1. CPU-time versus accuracy for both RT(1) (solid line) and BDM(2) (dashed line) mixed finite elements

7.4 The Approximation of More General Constrained Problems

In this Section we want to present an abstract theory, due to Babuška (1971) and Brezzi (1974), concerning existence and uniqueness of the solution to problems like (7.1.15), (7.1.16), as well as their internal approximation (7.2.1). As we already noticed, if the coefficient matrix $\{a^{ij}\}$ is symmetric, these problems are equivalent to the determination of the unique saddle-point of a Lagrangian functional related to a constrained problem. However, this symmetry assumption is not necessary for the analysis we are going to develop, and we are thus using the expressions "constrained problems" in a wider meaning, as the results presented here are also valid for more general problems.

In our presentation we will follow the theory developed by Brezzi (1974). This theory extends and embodies the one presented in Section 7.2 which was limited to the second order elliptic equation (7.1.1). Indeed, the analysis of the Galerkin approximation to the Stokes problem in the primitive variables $\mathbf{u}$ (the velocity) and p (the pressure) (see Chapter 9) will rely upon the abstract theory presented hereafter.

More generally, the numerical analysis of mixed approximations to several kind of boundary-value problems can be brought back to this abstract theory. An example is provided by boundary value problems for the biharmonic operator Δ^2 and its applications to plate bending problems, or to the Stokes and Navier-Stokes problems in the stream function formulation (see Section 10.4; see also Brezzi and Fortin (1991), Chap. IV). Another example arises from the mixed approximation of linear elasticity problems (see Brezzi and Fortin (1991), Chap. VII, and the references therein).

7.4.1 Abstract Formulation

Let us introduce the general framework for this analysis. Let X and M be two (real) Hilbert spaces, with norms $||\cdot||_X$ and $||\cdot||_M$, respectively. Let X' and M' be their dual spaces (the spaces of linear and continuous functional on X and M, respectively), and introduce two bilinear forms

$$(7.4.1) \qquad a(\cdot,\cdot): X \times X \to \mathbb{R} \quad , \quad b(\cdot,\cdot): X \times M \to \mathbb{R}$$

such that

$$(7.4.2) \qquad |a(w,v)| \le \gamma||w||_X||v||_X \quad , \quad |b(w,\mu)| \le \delta||w||_X||\mu||_M$$

for each $w, v \in X$ and $\mu \in M$.

We consider the following constrained problem:

$$
(7.4.3) \quad
\begin{cases}
\text{find } (u, \eta) \in X \times M : \\[2mm]
a(u, v) + b(v, \eta) = \langle l, v \rangle \quad \forall\, v \in X \\[2mm]
b(u, \mu) = \langle \sigma, \mu \rangle \qquad\quad \forall\, \mu \in M \ ,
\end{cases}
$$

where $l \in X'$ and $\sigma \in M'$, and $\langle \cdot, \cdot \rangle$ denotes the duality pairing between X' and X or M' and M.

It is useful to rewrite problem (7.4.3) in an operator form. We associate to $a(\cdot, \cdot)$ and $b(\cdot, \cdot)$ the operators $\mathsf{A} \in \mathcal{L}(X; X')$ and $\mathsf{B} \in \mathcal{L}(X; M')$ defined by

$$
(7.4.4) \qquad \langle \mathsf{A}w, v \rangle = a(w, v) \quad \forall\, w, v \in X
$$
$$
(7.4.5) \qquad \langle \mathsf{B}v, \mu \rangle = b(v, \mu) \quad \forall\, v \in X,\ \mu \in M \ .
$$

We denote by $\mathsf{B}^T \in \mathcal{L}(M; X')$ the adjoint operator of B, i.e., the operator defined by

$$
(7.4.6) \qquad \langle \mathsf{B}^T \mu, v \rangle = \langle \mathsf{B}v, \mu \rangle = b(v, \mu) \quad \forall\, v \in X,\ \mu \in M \ .
$$

Thus we can write (7.4.3) as: find $(u, \eta) \in X \times M$ such that

$$
(7.4.7) \qquad
\begin{cases}
\mathsf{A}u + \mathsf{B}^T \eta = l \quad \text{in } X' \\[2mm]
\mathsf{B}u = \sigma \qquad\quad \text{in } M' \ .
\end{cases}
$$

Now define the affine manifold

$$
X^\sigma := \{ v \in X \mid b(v, \mu) = \langle \sigma, \mu \rangle \ \forall\, \mu \in M \} \ .
$$

Clearly $X^0 = \ker \mathsf{B}$, and this is a closed subspace of X. We can now associate to problem (7.4.3) the following problem:

$$
(7.4.8) \qquad \text{find } u \in X^\sigma : a(u, v) = \langle l, v \rangle \quad \forall\, v \in X^0 \ .
$$

It is readily seen that if (u, η) is a solution to (7.4.3), then u is a solution to (7.4.8). We will introduce suitable conditions ensuring that the converse is also true, and that the solution to (7.4.8) does exist and is unique, thus constructing a solution to (7.4.3).

Before doing this, we need to prove an equivalence result concerning the bilinear form $b(\cdot, \cdot)$ and the operator B. Denote by $X_\sharp^0$ the polar set of X^0, i.e.,

$$
(7.4.9) \qquad X_\sharp^0 := \{ g \in X' \mid \langle g, v \rangle = 0 \ \forall\, v \in X^0 \} \ .
$$

Since $X^0 = \ker \mathsf{B}$, we can also write $X_\sharp^0 = (\ker \mathsf{B})_\sharp$. Let us decompose X as follows:

$$
X = X^0 \oplus (X^0)^\perp \ .
$$

Clearly, B is not an isomorphism from X onto M', as in general $\ker \mathsf{B} = X^0 \neq \{0\}$. We are going to introduce a condition which is equivalent to the fact that B is indeed an isomorphism from $(X^0)^\perp$ onto M' (and moreover B^T is an isomorphism from M onto $X_\sharp^0$).

Proposition 7.4.1 *The following statements are equivalent:*
a. *there exists a constant $\beta^* > 0$ such that the compatibility condition*

$$(7.4.10) \qquad \forall\, \mu \in M\ \exists\, v \in X\ ,\ v \neq 0 : b(v,\mu) \geq \beta^* ||v||_X ||\mu||_M$$

holds;
b. *the operator B^T is an isomorphism from M onto $X_\sharp^0$ and*

$$(7.4.11) \qquad ||B^T\mu||_{X'} := \sup_{\substack{v \in X \\ v \neq 0}} \frac{\langle B^T\mu, v\rangle}{||v||_X} \geq \beta^* ||\mu||_M \qquad \forall\, \mu \in M\ ;$$

c. *the operator B is an isomorphism from $(X^0)^\perp$ onto M' and*

$$(7.4.12) \qquad ||Bv||_{M'} := \sup_{\substack{\mu \in M \\ \mu \neq 0}} \frac{\langle Bv, \mu\rangle}{||\mu||_M} \geq \beta^* ||v||_X \qquad \forall\, v \in (X^0)^\perp\ .$$

Proof. First of all we show that *a.* and *b.* are equivalent. From the definition of B^T (see (7.4.6)) it is clear that (7.4.10) and (7.4.11) are equivalent. Thus, we have only to prove that B^T is an isomorphism from M onto $X_\sharp^0$. Clearly, (7.4.11) shows that B^T is an injective operator from M onto its range $\mathcal{R}(B^T)$, with a continuous inverse. Thus, $\mathcal{R}(B^T)$ is a closed subspace of X'. It remains to be proven that $\mathcal{R}(B^T) = X_\sharp^0$. The application of the Closed Range theorem (see, e.g., Yosida (1974), p. 205) gives at once

$$\mathcal{R}(B^T) = (\ker B)_\sharp = X_\sharp^0\ .$$

Let us now prove that *b.* and *c.* are equivalent. We first see that $X_\sharp^0$ can be identified with the dual of $(X^0)^\perp$. In fact, at each $g \in ((X^0)^\perp)'$ we associate $\hat{g} \in X'$ satisfying

$$\langle \hat{g}, v\rangle = \langle g, P^\perp v\rangle \qquad \forall\, v \in X\ ,$$

where $P^\perp$ is the orthogonal projection onto $(X^0)^\perp$. Clearly $\hat{g} \in X_\sharp^0$ and one verifies that $g \to \hat{g}$ is an isometric bijection from $((X^0)^\perp)'$ onto $X_\sharp^0$. As a consequence, B^T is an isomorphism from M onto $((X^0)^\perp)'$ satisfying

$$||(B^T)^{-1}||_{\mathcal{L}(X_\sharp^0; M)} \leq \frac{1}{\beta^*}$$

if, and only if, B is an isomorphism from $(X^0)^\perp$ onto M' satisfying

$$||B^{-1}||_{\mathcal{L}(M'; (X^0)^\perp)} \leq \frac{1}{\beta^*}\ .$$

The proof is now complete. $\qquad\qquad\qquad\qquad\qquad\qquad\qquad\qquad\qquad\quad \Box$

We are now in a position to prove the main theorem of this Section.

Theorem 7.4.1 *Assume that the bilinear form $a(\cdot,\cdot)$ satisfies (7.4.2) and is coercive on X^0, i.e., there exists a constant $\alpha > 0$ such that*

$$(7.4.13) \qquad a(v,v) \geq \alpha||v||_X^2 \qquad \forall\, v \in X^0 \ .$$

Assume, moreover, that the bilinear form $b(\cdot,\cdot)$ satisfies (7.4.2) and the compatibility condition (7.4.10) holds true. Then, for each $l \in X', \sigma \in M'$ there exists a unique solution u to (7.4.8), and a unique $\eta \in M$ such that (u,η) is the (unique) solution to (7.4.3). Furthermore, the map $(l,\sigma) \to (u,\eta)$ is an isomorphism from $X' \times M'$ onto $X \times M$, and

$$(7.4.14) \qquad ||u||_X \leq \frac{1}{\alpha}\left(||l||_{X'} + \frac{\alpha+\gamma}{\beta^*}||\sigma||_{M'}\right)$$

$$(7.4.15) \qquad ||\eta||_M \leq \frac{1}{\beta^*}\left[\left(1 + \frac{\gamma}{\alpha}\right)||l||_{X'} + \frac{\gamma(\alpha+\gamma)}{\alpha\beta^*}||\sigma||_{M'}\right] \ .$$

Proof. Uniqueness of the solution to problem (7.4.8) is a straightforward consequence of the coerciveness assumption (7.4.13).

Let us now prove the existence. As condition (7.4.10) is satisfied, from $c.$ in Proposition 7.4.1 there exists a unique $u^0 \in (X^0)^\perp$ such that $Bu^0 = \sigma$ and

$$(7.4.16) \qquad ||u^0||_X \leq \frac{1}{\beta^*}||\sigma||_{M'} \ .$$

Thus, we rewrite (7.4.8) as:

$$(7.4.17) \qquad \text{find } \tilde{u} \in X^0 : a(\tilde{u},v) = \langle l,v\rangle - a(u^0,v) \qquad \forall\, v \in X^0 \ ,$$

and define the solution u to (7.4.8) as $u = \tilde{u} + u^0$. The existence of a unique solution to (7.4.17) is assured by Lax-Milgram lemma (see Theorem 5.1.1), and, moreover,

$$(7.4.18) \qquad ||\tilde{u}||_X \leq \frac{1}{\alpha}(||l||_{X'} + \gamma||u^0||_X) \ .$$

Let us now consider problem (7.4.3). As (7.4.17) can be written in the form

$$\langle Au - l, v\rangle = 0 \qquad \forall\, v \in X^0 \ ,$$

we have $(Au - l) \in X_\sharp^0$. Moreover, from $b.$ in Proposition 7.4.1 we can find a unique $\eta \in M$ such that $Au - l = -B^T\eta$, i.e., (u,η) is a solution to (7.4.3), and satisfies

$$(7.4.19) \qquad ||\eta||_M \leq \frac{1}{\beta^*}||Au - l||_{X'} \ .$$

We have already noticed that each solution (u, η) to (7.4.3) gives a solution u to (7.4.8): also for problem (7.4.3) uniqueness thus holds.

Finally, summing up inequalities (7.4.16), (7.4.18) and (7.4.19) the thesis follows at once.   □

We are now in a position to consider the approximation of the abstract constrained problem (7.4.3). Let X_h and M_h be finite dimensional subspaces of X and M, respectively. We look for a solution to the following problem: being given $l \in X'$ and $\sigma \in M'$

(7.4.20)
$$\begin{cases} \text{find } (u_h, \eta_h) \in X_h \times M_h : \\[2mm] a(u_h, v_h) + b(v_h, \eta_h) = \langle l, v_h \rangle \quad \forall\, v_h \in X_h \\[2mm] b(u_h, \mu_h) = \langle \sigma, \mu_h \rangle \qquad\qquad \forall\, \mu_h \in M_h \quad . \end{cases}$$

As in the infinite dimensional case, we define the space

$$X_h^\sigma := \{ v_h \in X_h \,|\, b(v_h, \mu_h) = \langle \sigma, \mu_h \rangle \ \forall\, \mu_h \in M_h \} \quad .$$

Since M_h is, in general, a proper subspace of M, we notice that $X_h^\sigma \not\subset X^\sigma$.

The finite dimensional problem corresponding to (7.4.8) reads as follows:

(7.4.21) $\text{find } u_h \in X_h^\sigma : a(u_h, v_h) = \langle l, v_h \rangle \quad \forall\, v_h \in X_h^0 \ .$

Clearly, each solution (u_h, η_h) to (7.4.20) provides a solution u_h to (7.4.21). We are going to present suitable conditions to ensure that the converse is also true. Moreover, a sound stability and convergence analysis will follow.

7.4.2 Analysis of Stability and Convergence

We first state the finite dimensional counterpart of Theorem 7.4.1, which shows at the same time the stability of the approximate solutions (u_h, η_h).

Theorem 7.4.2 (Stability). *Assume that the bilinear form $a(\cdot, \cdot)$ satisfies (7.4.2) and is coercive on X_h^0, i.e., there exists a constant $\alpha_h > 0$ such that*

(7.4.22) $a(v_h, v_h) \geq \alpha_h ||v_h||_X^2 \quad \forall\, v_h \in X_h^0 \ .$

Assume, moreover, that the bilinear form $b(\cdot, \cdot)$ satisfies (7.4.2) and that the following compatibility condition holds: there exists $\beta_h > 0$ such that

(7.4.23) $\forall\, \mu_h \in M_h \ \exists\, v_h \in X_h, v_h \neq 0 : b(v_h, \mu_h) \geq \beta_h ||v_h||_X ||\mu_h||_M \ .$

Then for each $l \in X', \sigma \in M'$ there exists a unique solution (u_h, η_h) to (7.4.20), which satisfies (7.4.14) and (7.4.15) with α_h instead of α and β_h instead of β^*. In those cases in which both α_h and β_h in (7.4.22) and (7.4.23) are independent of h, this is a stability result for (u_h, η_h).

The proof of the preceding Theorem follows straightforwardly by taking X_h and M_h instead of X and M in Theorem 7.4.1, and noticing that

$$||l||_{X'_h} \leq ||l||_{X'} \quad , \quad ||\sigma||_{M'_h} \leq ||\sigma||_{M'} \ .$$

It must be observed, however, that (7.4.13) does not imply (7.4.22), as $X_h^0 \not\subset X^0$, and (7.4.10) does not imply (7.4.23), since, in general, X_h is a proper subspace of X.

As we have already noticed in Section 7.2.1, the compatibility condition (7.4.23) is also called *inf-sup* or *Ladyzhenskaya-Babuška-Brezzi* condition.

Let us now come to the convergence theorem, whose proof will be reported here even if it is very similar to that of Theorem 7.2.2.

Theorem 7.4.3 (Convergence). *Let the assumptions of Theorems 7.4.1 and 7.4.2 be satisfied. Then the solutions (u, η) and (u_h, η_h) to (7.4.3) and (7.4.20), respectively, satisfy the following error estimates*

$$(7.4.24) \quad ||u - u_h||_X \leq \left(1 + \frac{\gamma}{\alpha_h}\right) \inf_{v_h^* \in X_h^\sigma} ||u - v_h^*||_X + \frac{\delta}{\alpha_h} \inf_{\mu_h \in M_h} ||\eta - \mu_h||_M$$

$$
\begin{aligned}
(7.4.25) \quad ||\eta - \eta_h||_M \leq &\frac{\gamma}{\beta_h} \left(1 + \frac{\gamma}{\alpha_h}\right) \inf_{v_h^* \in X_h^\sigma} ||u - v_h^*||_X \\
&+ \left(1 + \frac{\delta}{\beta_h} + \frac{\gamma\delta}{\alpha_h\beta_h}\right) \inf_{\mu_h \in M_h} ||\eta - \mu_h||_M \ .
\end{aligned}
$$

Moreover, the following estimate holds

$$(7.4.26) \quad \inf_{v_h^* \in X_h^\sigma} ||u - v_h^*||_X \leq \left(1 + \frac{\delta}{\beta_h}\right) \inf_{v_h \in X_h} ||u - v_h||_X \ .$$

From (7.4.24)-(7.4.26) it follows that convergence is optimal, provided that (7.4.22) and (7.4.23) hold with constants independent of h.

Proof. Take $v_h \in X_h, v_h^* \in X_h^\sigma$ and $\mu_h \in M_h$. By subtracting (7.4.20) from (7.4.3)$_1$ it follows

$$a(u_h - v_h^*, v_h) + b(v_h, \eta_h - \mu_h) = a(u - v_h^*, v_h) + b(v_h, \eta - \mu_h) \ .$$

Choosing $v_h = (u_h - v_h^*) \in X_h^0$, from (7.4.22) and (7.4.2) we have

$$||u_h - v_h^*||_X \leq \frac{1}{\alpha_h}(\gamma ||u - v_h^*||_X + \delta ||\eta - \mu_h||_M) \ ,$$

and consequently (7.4.24) holds true.

To obtain (7.4.25), we start from the compatibility condition (7.4.23). For each $\mu_h \in M_h$ we find

$$||\eta_h - \mu_h||_M \le \frac{1}{\beta_h} \sup_{\substack{v_h \in X_h \\ v_h \ne 0}} \frac{b(v_h, \eta_h - \mu_h)}{||v_h||_X} .$$

On the other hand, by subtracting $(7.4.20)_2$ from $(7.4.3)_2$ it follows

$$b(v_h, \eta_h - \mu_h) = a(u - u_h, v_h) + b(v_h, \eta - \mu_h) .$$

From (7.4.2) we obtain

$$||\eta_h - \mu_h||_M \le \frac{1}{\beta_h} (\gamma ||u - u_h||_X + \delta ||\eta - \mu_h||_M) ,$$

which, together with (7.4.24), yields (7.4.25).

Finally, let us prove (7.4.26). For each $v_h \in X_h$, owing to (7.4.23) and Proposition 7.4.1 there exists a unique $z_h \in (X_h^0)^\perp$ such that

$$b(z_h, \mu_h) = b(u - v_h, \mu_h) \quad \forall \mu_h \in M_h$$

and

$$||z_h||_X \le \frac{\delta}{\beta_h} ||u - v_h||_X .$$

Setting $v_h^* := z_h + v_h$, we readily see that $v_h^* \in X_h^\sigma$, as $b(u, \mu_h) = \langle \sigma, \mu_h \rangle$ for each $\mu_h \in M_h$. Furthermore,

$$||u - v_h^*||_X \le ||u - v_h||_X + ||z_h||_X \le \left(1 + \frac{\delta}{\beta_h}\right) ||u - v_h||_X ,$$

and (7.4.26) follows. □

Remark 7.4.1 Looking at the proof of Theorem 7.2.2 (see in particular (7.2.17)), we notice that, under the condition $X_h^0 \subset X_h$, it is possible to estimate $||u - u_h||_X$ in terms of $\inf_{X_h} ||u - v_h||_X$ solely. Also, error estimate (7.4.24) holds even if the compatibility conditions (7.4.10) and (7.4.23) are not satisfied. □

Remark 7.4.2 *(Spurious modes)*. The compatibility condition (7.4.23) is necessary to achieve uniqueness of η_h. Actually, it can be written as:

if $\mu_h \in M_h$ and $b(v_h, \mu_h) = 0$ for each $v_h \in X_h$, then $\mu_h = 0$.

Thus, if (7.4.23) is not satisfied, there exists $\mu_h^* \in M_h$, $\mu_h^* \ne 0$, such that

$$b(v_h, \mu_h^*) = 0 \quad \forall v_h \in X_h .$$

As a consequence, if (u_h, η_h) solves (7.4.20), also $(u_h, \eta_h + \tau \mu_h^*), \tau \in \mathbb{R}$, is a solution to the same problem.

Any such element μ_h^* is called *spurious* (or *parasitic) mode*, as it cannot be detected by the numerical problem (7.4.20). It might thus generate instabilities for the method (see also Section 9.2.2). □

7.4.3 How to Verify the Uniform Compatibility Condition

In most applications, it is not difficult to verify that (7.4.22) holds uniformly with respect to h (see, e.g., Section 7.2.1 and Chapter 9). More difficult is to prove the compatibility condition (7.4.23) holds uniformly on h.

A first criterium in this direction has been proposed by Fortin (1977) (see Lemma 7.2.1). Here we report the general abstract formulation, whose proof follows the same guidelines of that presented before.

Lemma 7.4.1 (Fortin). *Assume that the compatibility condition (7.4.10) is satisfied, and, moreover, that there exists an operator $\tau_h : X \to X_h$ such that*
   (i) $b(v - \tau_h(v), \mu_h) = 0$ for each $v \in X$, $\mu_h \in M_h$
   (ii) $\|\tau_h(v)\|_X \leq C_*\|v\|_X$ for each $v \in X$,
where $C_* > 0$ doesn't depend on h. Then, the compatibility condition (7.4.23) is satisfied with $\beta = \beta^*/C_*$.

Another interesting case in related to the choice $X = (H_0^1(\Omega))^d, M = L_0^2(\Omega)$, $a(\mathbf{w}, \mathbf{v}) = \nu(\nabla\mathbf{w}, \nabla\mathbf{v})$, $b(\mathbf{v}, q) = -(q, \mathrm{div}\,\mathbf{v})$, $\nu > 0$, which are the functional spaces and the bilinear forms appearing in the variational formulation of the Stokes problem (see Chapter 9). In this case, Verfürth (1984b) has proven the following result:

Lemma 7.4.2 (Verfürth). *Let $\mathcal{T}_h$ be a quasi-uniform family of triangulations of $\overline{\Omega}$. Assume that $M_h \subset H^1(\Omega) \cap L_0^2(\Omega)$ and that there exists $\hat{\beta} > 0$ such that*

$$(7.4.27) \qquad \forall\, q_h \in M_h\ \exists\, \mathbf{v}_h \in X_h, \mathbf{v}_h \neq \mathbf{0} : b(\mathbf{v}_h, q_h) \geq \hat{\beta}\|\mathbf{v}_h\|_0\|q_h\|_1 \ .$$

Assume, moreover, that the there exists an operator $\mathbf{R}_h : X \to X_h$ and a constant $K > 0$ such that

$$(7.4.28) \qquad \|\mathbf{v} - \mathbf{R}_h(\mathbf{v})\|_0 + h\|\mathbf{v} - \mathbf{R}_h(\mathbf{v})\|_1 \leq Kh|\mathbf{v}|_1\ \forall\,\mathbf{v} \in X \ .$$

Then the compatibility condition (7.4.23) is satisfied.

Proof. First of all, recalling that from the inverse inequality (6.3.21) for each $\mathbf{v}_h \in X_h$ we have

$$\|\mathbf{v}_h\|_1 \leq Ch^{-1}\|\mathbf{v}_h\|_0 \ ,$$

condition (7.4.27) implies that

$$(7.4.29) \qquad \forall\, q_h \in M_h\ \exists\, \mathbf{v}_h \in X_h, \mathbf{v}_h \neq \mathbf{0} : b(\mathbf{v}_h, q_h) \geq K_1 h\|\mathbf{v}_h\|_1\|q_h\|_1 \ .$$

Let us now recall that for each $q_h \in M_h$ there exists $\mathbf{w} \in X$ such that $\mathrm{div}\,\mathbf{w} = -q_h$ and

$$(7.4.30) \qquad \|\mathbf{w}\|_1 \leq K_2\|q_h\|_0$$

(see, e.g., Girault and Raviart (1986), p. 24). Thus from (7.4.28) we find

(7.4.31)
$$\|\mathbf{R}_h(\mathbf{w})\|_1 \leq \|\mathbf{R}_h(\mathbf{w}) - \mathbf{w}\|_1 + \|\mathbf{w}\|_1$$
$$\leq (1 + K)\|\mathbf{w}\|_1 \leq K_2(1 + K)\|q_h\|_0 \ .$$

If q_h is such that

(7.4.32)
$$\|q_h\|_0 \leq KK_2 h\|q_h\|_1 \ ,$$

then (7.4.29) yields

(7.4.33)
$$b(\mathbf{v}_h, q_h) \geq \frac{K_1}{KK_2}\|\mathbf{v}_h\|_1\|q_h\|_0 \ .$$

On the contrary, if

(7.4.34)
$$\|q_h\|_0 > KK_2 h\|q_h\|_1 \ ,$$

from (7.4.28), (7.4.30) and (7.4.31) we have

(7.4.35)
$$b(\mathbf{R}_h(\mathbf{w}), q_h) = b(\mathbf{w}, q_h) + b(\mathbf{R}_h(\mathbf{w}) - \mathbf{w}, q_h)$$
$$\geq \|q_h\|_0^2 - \|\mathbf{R}_h(\mathbf{w}) - \mathbf{w}\|_0\|q_h\|_1$$
$$\geq \|q_h\|_0^2 - KK_2 h\|q_h\|_0\|q_h\|_1$$
$$\geq \frac{1}{K_2(1 + K)}(\|q_h\|_0 - KK_2 h\|q_h\|_1)\|\mathbf{R}_h(\mathbf{w})\|_1 \ ,$$

where the term $b(\mathbf{R}_h(\mathbf{w}) - \mathbf{w}, q_h)$ has been integrated by parts.

Putting together (7.4.29) and (7.4.35), it is readily proven that for each $q_h \in M_h$ satisfying (7.4.34) there exists $\mathbf{z}_h \in X_h$, $\mathbf{z}_h \neq \mathbf{0}$, such that

(7.4.36)
$$b(\mathbf{z}_h, q_h) \geq Q(h\|q_h\|_1)\|\mathbf{z}_h\|_1 \ ,$$

where

$$Q(\xi) := \max\left\{K_1\xi, \frac{1}{K_2(1 + K)}(\|q_h\|_0 - KK_2\xi)\right\} \quad , \quad \xi > 0 \ .$$

Finally, taking the minimum over $\{\xi > 0\}$ of $Q(\xi)$, from (7.4.33) and (7.4.36) it easily follows that (7.4.23) holds with $\beta = K_1 K_2^{-1}[K_1(1 + K) + K]^{-1}$. $\square$

Notice that (7.4.27) is different from (7.4.23) as the "wrong" norms occur at the right hand side. However, (7.4.27) has been proven to hold for some finite element spaces frequently used in the approximation of the Stokes problem (see for instance the Taylor-Hood or Bercovier-Pironneau elements in Section 9.3.2; the proof is reported, e.g., in Fortin (1993) and Bercovier and Pironneau (1979), respectively).

Furthermore, under suitable assumptions, the approximation property (7.4.28) is true for several choices of the space X_h (see for instance (3.5.22), or Clément (1975)).

Other viable criteria for checking the uniform compatibility condition (7.4.23) when considering the Stokes problem have been proposed by Boland and Nicolaides (1983) and Stenberg (1984) (see also Gunzburger (1989), pp. 16–21, and Girault and Raviart (1986), pp. 129–132). They essentially require to subdivide Ω into some macro-elements, and to verify the compatibility condition (7.4.23) only locally over each macro-element.

7.5 Complements

For the mixed method, error estimates of $u - u_h$ in the $L^\infty(\Omega)$-norm have been proven, e.g., by Scholz (1977), Douglas and Roberts (1985), Gastaldi and Nochetto (1987).

The extension of the results presented in this Chapter to the case of a general elliptic operator

$$Lw := -\sum_{i,j=1}^{d} D_i(a_{ij} D_j w) + \sum_{i=1}^{d} c_i D_i w + a_0 w$$

can give rise to some difficulties. The case of dual mixed formulation has been analyzed by Douglas and Roberts (1985).

A different generalization is concerned with problems that can be written in the form

$$\begin{cases} \text{find } (u, \eta) \in X \times M : \\[2mm] a(u, v) + b(v, \eta) = \langle l, v \rangle \quad \forall\, v \in X \\[2mm] b(u, \mu) - d(\eta, \mu) = \langle \sigma, \mu \rangle \quad \forall\, \mu \in M \quad, \end{cases}$$

where $d(\cdot, \cdot)$ is a continuous and non-negative bilinear form on $M \times M$. For the analysis of this problem we refer to Roberts and Thomas (1991) (see also Section 9.6.7).

Another extension of the theory of Brezzi is related to the problem

$$\begin{cases} \text{find } (u, \eta) \in X \times M : \\[2mm] a(u, v) + b_1(v, \eta) = \langle l, v \rangle \quad \forall\, v \in X \\[2mm] b_2(u, \mu) = \langle \sigma, \mu \rangle \quad \forall\, \mu \in M \quad, \end{cases}$$

where the bilinear form b_1 and b_2 are distinct (see Nicolaides (1982) and Bernardi, Canuto and Maday (1988), the latter being motivated by the Chebyshev approximation of the Stokes problem).

8. Steady Advection-Diffusion Problems

In this Chapter we consider second order linear elliptic equations that have an advective term which is much stronger than the diffusive one.

At first, the difficulties stemming from the use of the Galerkin method will be outlined on a simple one-dimensional finite difference scheme. The first remedy will be to resort to "upwind" schemes, which introduce a "numerical viscosity" that have the effect of stabilizing the computational algorithm. However, this procedure is only first order accurate.

More general and high order accurate methods of stabilization are described and analyzed in the two- and three-dimensional cases, for both the finite element and the spectral collocation approximation. Another Section is devoted to the presentation of some numerical results obtained by the use of the proposed methods.

Finally, an heterogeneous approach, based on a domain decomposition technique, is briefly illustrated.

The search of optimal stabilization methods for advection-dominated problems is still an active research area. The interested reader is advised to keep abreast of the latest literature.

8.1 Mathematical Formulation

Let us now consider the second order linear elliptic operator

$$(8.1.1) \qquad Lz := - \sum_{i,j=1}^{d} D_i(a_{ij} D_j z) + \sum_{i=1}^{d} D_i(b_i z) + a_0 z \ ,$$

and its associated bilinear form

$$(8.1.2) \qquad a(z,v) := \int_{\Omega} \left[\sum_{i,j=1}^{d} a_{ij} D_j z D_i v - \sum_{i=1}^{d} b_i z D_i v + a_0 z v \right] \ ,$$

that have been introduced in Section 6.1.

We assume that $a(\cdot,\cdot)$ is continuous and coercive in $V \times V$, V being a closed subspace of $H^1(\Omega)$, i.e., there exist constants $\gamma > 0$ and $\alpha > 0$ such that

$$(8.1.3) \qquad |a(z,v)| \leq \gamma ||z||_1 ||v||_1 \qquad \forall\, z, v \in V$$

$$(8.1.4) \qquad a(v,v) \geq \alpha ||v||_1^2 \qquad \forall\, v \in V \ .$$

As reported in Section 6.1.2, suitable assumptions on the coefficients a_{ij}, b_i and a_0 ensure that (8.1.3), (8.1.4) hold. For instance, if $V = H_0^1(\Omega)$, assuming that (6.1.19) holds with $\eta = 0$ we can choose α in the following way

$$(8.1.5) \qquad \alpha = \frac{\alpha_0}{1 + C_\Omega} \ ,$$

α_0 and C_Ω being the ellipticity and Poincaré constants, respectively (see Definition 6.1.1 and (1.3.2)). Moreover, γ can be expressed as

$$(8.1.6) \qquad \gamma = C_{d,\Omega} \left(\max_{1 \leq i,j \leq d} ||a_{ij}||_{L^\infty(\Omega)} + ||\mathbf{b}||_{L^\infty(\Omega)} + ||a_0||_{L^\infty(\Omega)} \right) \ ,$$

where $C_{d,\Omega}$ is a suitable constant depending upon the physical dimension d and the domain Ω.

We recall that under the above assumptions (8.1.3) and (8.1.4), the existence of a unique solution

$$(8.1.7) \qquad u \in V : a(u,v) = \mathcal{F}(v) \qquad \forall\, v \in V \ ,$$

$\mathcal{F}(\cdot)$ being a continuous linear functional on the Hilbert space V, is assured by Lax-Milgram lemma (see Theorem 5.1.1).

In Section 6.2.1 we have introduced the Galerkin method as a possible way to approximate the solution u to (8.1.7). However, in Section 6.1.4 we noticed that the Galerkin method can have a poor performance if the coerciveness constant α is small in comparison with the continuity constant γ. In particular, error estimate (6.2.8) can have, in front, a very large multiplicative constant C if the ellipticity constant α_0 is small with respect to $||\mathbf{b}||_{L^\infty(\Omega)}$ and/or to $||a_0||_{L^\infty(\Omega)}$.

The aim of the following Sections is to show that these kind of difficulties may really occur for the Galerkin method. Next, we describe some alternative ways for overcoming them.

In Section 8.2 we first start from a simple one-dimensional example which already shows where instabilities come from. Then, in Section 8.3 we provide a description of some of the most common remedies, especially in the framework of finite element and spectral approximation.

8.2 A One-Dimensional Example

Let us consider the one-dimensional advection-diffusion equation

$$\begin{cases} -\varepsilon D^2 u + b Du = 0 \ , \quad 0 < x < 1 \ , \\[2mm] u(0) = 0 \\[2mm] u(1) = 1 \ , \end{cases}$$

(8.2.1)

with $\varepsilon > 0$ and $b > 0$. Its exact solution is $u(x) = (e^{b/\varepsilon} - 1)^{-1}(e^{bx/\varepsilon} - 1)$, and it exhibits a boundary layer of width $O(\varepsilon/b)$ near to $x = 1$ if ε/b is small enough.

8.2.1 Galerkin Approximation and Centered Finite Differences

We start by approximating (8.2.1) by means of the Galerkin method with piecewise-linear finite elements over a uniform grid. Precisely, after choosing a positive integer M, we set $h := 1/M$, $x_j := jh$, $j = 0, ..., M$. We then approximate the non-homogeneous Dirichlet problem (8.2.1) as pointed out in the second half of Remark 6.2.2, taking $V_h = X_h^1 \cap H_0^1(0,1)$ and $V_h^* = \{v_h \in X_h^1 \, | \, v_h(0) = 0 \ , \ v_h(1) = 1\}$, with X_h^1 defined as in (3.2.4).

This leads to the following linear system

(8.2.2) $$A\boldsymbol{\xi} = \mathbf{F} \ ,$$

where

(8.2.3) $$A := \operatorname{tridiag}\left(-\frac{\varepsilon}{h} - \frac{b}{2} \, , \ \frac{2\varepsilon}{h} \, , \ -\frac{\varepsilon}{h} + \frac{b}{2}\right)$$

(8.2.4) $$\mathbf{F} := \left(0, ..., 0, \frac{\varepsilon}{h} - \frac{b}{2}\right)$$

(8.2.5) $$\boldsymbol{\xi} := \{u_h(x_j)\}_{j=1,...,M-1} \ .$$

Assuming that $2\varepsilon \neq bh$, it is easily seen that the solution to (8.2.2) is given by

(8.2.6) $$\xi_j = u_h(x_j) = \frac{\left(\dfrac{1 + \text{Pe}}{1 - \text{Pe}}\right)^j - 1}{\left(\dfrac{1 + \text{Pe}}{1 - \text{Pe}}\right)^M - 1} \ , \quad j = 1, ..., M - 1 \ ,$$

where

(8.2.7) $$\text{Pe} := \frac{bh}{2\varepsilon}$$

is the *Péclet number*. If $\text{Pe} > 1$ then $(1 + \text{Pe})/(1 - \text{Pe}) < 0$, and consequently the solution u_h exhibits an oscillatory behaviour. Clearly, being fixed ε and b, in principle it is always possible to choose the grid-size h small enough so that $\text{Pe} \leq 1$, thus avoiding oscillations (see Fig. 8.2.1). However, this is very often unpractical if ε is very small with respect to b, since one would obtain linear systems having too many unknowns (and actually it may be unfeasible in the higher dimensional case).

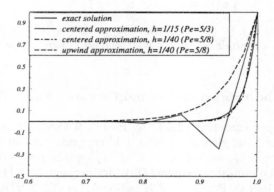

Fig. 8.2.1. Exact solution, centered and upwind approximation to (8.2.1), with $\varepsilon = 1/50$ and $b = 1$

Therefore, the first conclusion could be: if an advection-diffusion problem has a boundary layer and the mesh-size h is larger than ε/b, the Galerkin solution may be affected by oscillations. We will see in Section 8.3 how to overcome this problem.

The linear system (8.2.2) is equivalent to the one obtained by approximating (8.2.1) by *centered* finite differences. (This coincidence occurs as we are considering a uniform grid, and a constant advective velocity b.) In fact, the latter scheme reads:

$$(8.2.8) \quad \begin{cases} -\varepsilon \dfrac{u_{j+1} - 2u_j + u_{j-1}}{h^2} + b \dfrac{u_{j+1} - u_{j-1}}{2h} = 0 \ , \quad j = 1, ..., M-1 \\[2mm] u_0 = 0 \\[2mm] u_M = 1 \ , \end{cases}$$

where u_j is the approximation of $u(x_j)$. This is a system of linear difference equations whose solution is $u_j = \xi_j$, with ξ_j given by (8.2.6) (e.g., Isaacson and Keller (1966), Sect. 8.4). Therefore the Galerkin method with piecewise-linear polynomials is equivalent to the approximation by means of centered finite differences.

Remark 8.2.1 Instead of problem (8.2.1) let us consider, for a while, the reaction-diffusion problem

$$\begin{cases} -\varepsilon D^2 u + a_0 u = 0 \ , \quad 0 < x < 1 \\[2mm] u(0) = 0 \\[2mm] u(1) = 1 \ , \end{cases}$$

with $\varepsilon > 0$ and $a_0 > 0$. Its solution is $u(x) = \sinh(\sqrt{a_0}x/\sqrt{\varepsilon})/\sinh(\sqrt{a_0}/\sqrt{\varepsilon})$, and has a boundary layer of width $O(\sqrt{\varepsilon}/\sqrt{a_0})$ at $x = 1$ whenever ε/a_0 is small enough.

The centered finite difference scheme would approximate the exact solution $u(x_j)$ by

$$(8.2.9) \qquad \xi_j = (\tau_2^M - \tau_1^M)^{-1}(\tau_2^j - \tau_1^j) \ , \quad j = 1, ..., M - 1 \ ,$$

with $\tau_1 = 1 + \sigma - \sqrt{\sigma(2 + \sigma)}$ and $\tau_2 = 1 + \sigma + \sqrt{\sigma(2 + \sigma)}$, $\sigma := a_0 h^2 \varepsilon^{-1}/2$. Since $0 < \tau_1 < 1 < \tau_2$, for all value of the grid-size h, the sequence ξ_j is monotonically increasing, thus the finite difference scheme is oscillation-free for all values of h.

On the other hand, approximating (8.2.1) by the Galerkin method using piecewise-linear finite elements would provide the following set of equations

$$\varepsilon \sum_{i=1}^{M-1} \xi_i \int_0^1 D\varphi_i D\varphi_j + a_0 \sum_{i=1}^{M-1} \xi_i \int_0^1 \varphi_i \varphi_j = 0 \ , \quad j = 1, ..., M - 1 \ ,$$

where φ_i are the basis functions of V_h. After computing the integrals we obtain the following algebraic equations

$$\begin{cases} -\varepsilon \dfrac{\xi_{j+1} - 2\xi_j + \xi_{j-1}}{h} + a_0 h \dfrac{\xi_{j+1} + 4\xi_j + \xi_{j-1}}{6} = 0 \ , \quad j = 1, ..., M - 1 \\[2mm] \xi_0 = 0 \\[2mm] \xi_M = 1 \ . \end{cases}$$

Correspondingly, for $\sigma \neq 3$ we obtain a solution like (8.2.9), but with

$$\tau_1 = \frac{3 + 2\sigma - \sqrt{3\sigma(\sigma + 6)}}{3 - \sigma} \ , \quad \tau_2 = \frac{3 + 2\sigma + \sqrt{3\sigma(\sigma + 6)}}{3 - \sigma} \ .$$

These characteristic roots have the same sign of $3 - \sigma$; thus the finite solution is affected by oscillations, unless $\sigma < 3$, i.e., $h < (6\varepsilon/a_0)^{1/2}$.

This time the "responsible" for the numerical oscillation is the zero-order term $a_0 u$. As a matter of fact, if we replace the mass matrix $M_{ij} := \int_0^1 \varphi_i \varphi_j$ (i.e., the finite element representation of the identity operator) by its *lumped* form $\hat{M}_{ij} := (\sum_{k=1}^{M-1} M_{ik})\delta_{ij}$, we verify at once that $\hat{M} = I$ and therefore we reobtain exactly the finite difference scheme, thus an unconditionally stable solution. (For a discussion on the mass lumping procedure for time-dependent problems, see Section 11.4.) $\qquad\qquad \square$

8.2.2 Upwind Finite Differences and Numerical Diffusion

By noticing that in problem (8.2.1) transport occurs from left to right (as the advective coefficient b has a positive sign), one is led to consider backward (*upwind*) finite differences to approximate the term bDu.

The resulting scheme can be written as:

$$
(8.2.10) \quad
\begin{cases}
-\varepsilon \dfrac{\tilde{u}_{j+1} - 2\tilde{u}_j + \tilde{u}_{j-1}}{h^2} + b\dfrac{\tilde{u}_j - \tilde{u}_{j-1}}{h} = 0 \ , \quad j = 1, ..., M-1 \\[2ex]
\tilde{u}_0 = 0 \\[1.5ex]
\tilde{u}_M = 1 \ ,
\end{cases}
$$

and has the solution

$$
(8.2.11) \qquad \tilde{u}_j = \frac{(1 + 2\mathrm{Pe})^j - 1}{(1 + 2\mathrm{Pe})^M - 1} \ , \quad j = 1, ..., M-1 \ .
$$

This time $\tilde{u}_j$ doesn't oscillate any longer. Indeed, its value increases with j, as $1 + 2\mathrm{Pe} > 0$ for each value of ε, b and h (see Fig. 8.2.1).

It can be noticed that this upwind scheme introduces a local truncation error of order $O(h)$, whereas the centered difference is of order $O(h^2)$. (We warn the reader that the error arising from piecewise-linear finite elements decays like $O(h)$ in the H^1-norm (see (6.2.8)), but like $O(h^2)$ in the maximum norm (see Ciarlet (1968)). This is consistent with the fact that piecewise-linear finite elements coincide with centered finite differences on a uniform grid in one dimension, provided the boundary value problem at hand has constant coefficients.)

Despite this lack of accuracy the upwind scheme is indeed better suited for the approximation of the solution to (8.2.1) when the Péclet number Pe is larger than 1.

This fact can also be explained in a different way. The upwind approximation of Du can be written as

$$
\frac{\tilde{u}_j - \tilde{u}_{j-1}}{h} = \frac{\tilde{u}_{j+1} - \tilde{u}_{j-1}}{2h} - \frac{h}{2}\frac{\tilde{u}_{j+1} - 2\tilde{u}_j + \tilde{u}_{j-1}}{h^2} \ ,
$$

therefore it corresponds to a centered difference approximation of the regularized operator $D - \frac{h}{2}D^2$. In other words, it introduces a numerical dissipation that can be regarded as a direct discretization of the artificial viscous term $-\frac{h}{2}D^2u$. This is frequently called *numerical diffusion* (or *numerical viscosity*). This also explains from a different point of view why the upwind difference scheme is only first order accurate.

It is possible to interpret the upwind difference scheme as a Petrov-Galerkin approximation of (8.2.1). This method is based on the fact that the spaces of trial and test functions are different, and usually denoted by W_h and V_h, respectively. In the approximation of (8.2.1), the space of trial functions can be chosen as $W_h = X_h^1 \cap H_0^1(0,1)$; the corresponding affine subspace $W_h^* = \{v_h \in X_h^1 \,|\, v_h(0) = 0\,, v_h(1) = 1\}$ accounts for the non-homogeneous Dirichlet condition. To define the space of test function V_h, consider the piecewise-quadratic function

(8.2.12)
$$\sigma_2(x) := \begin{cases} -3x(1+x) & ,\ -1 \le x \le 0 \\ -3x(1-x) & ,\ 0 \le x \le 1 \\ 0 & ,\ |x| \ge 1 \end{cases}.$$

We can then introduce the finite dimensional space V_h as

(8.2.13)
$$V_h = \operatorname{span}\{\psi_1, ..., \psi_{M-1}\}\ ,$$

where

(8.2.14)
$$\psi_j(x) := \varphi_j(x) + \sigma_2(h^{-1}x - j)\ ,\quad j = 1, ..., M-1\ ,$$

φ_j being the basis function of W_h corresponding to the node x_j.

It is easily verified (see, e.g., Mitchell and Griffiths (1980), pp. 221–223) that the linear system associated to (8.2.10) is equivalent to the one associated to the Petrov-Galerkin problem

$$\text{find } u_h \in W_h^* : \int_0^1 (\varepsilon Du_h Dv_h + bDu_h v_h) = 0\ \forall\, v_h \in V_h\ .$$

8.2.3 Spectral Approximation

In this Section we consider a Chebyshev collocation approximation of the advection-diffusion problem (8.2.1). We will see that, although oscillations are indeed present if the advective term is dominant, a sort of maximum principle holds for the spectral solution. These results are due to Canuto (1988).

For the sake of simplicity, let us deal with the same Dirichlet problem as before in the interval (-1,1), i.e.,

(8.2.15)
$$\begin{cases} -\varepsilon D^2 u + bDu = 0\ ,\ -1 < x < 1 \\ u(-1) = 0 \\ u(1) = 1\ . \end{cases}$$

If we define for $N \ge 2$

(8.2.16)
$$V_N^* := \{v_N \in \mathbb{P}_N \,|\, v_N(-1) = 0\,,\ v_N(1) = 1\}\ ,$$

and

(8.2.17)
$$V_N = \mathbb{P}_N^0 := \{v_N \in \mathbb{P}_N \,|\, v_N(-1) = v_N(1) = 0\}\ ,$$

the spectral Chebyshev problem reads:

$$
\text{(8.2.18)} \qquad \text{find } u_N \in V_N^* : \varepsilon \left(D u_N, \frac{D(v_N w)}{w} \right)_N
$$
$$
+ b(D u_N, v_N)_N = 0 \quad \forall\, v_N \in V_N \;,
$$

where $(\cdot, \cdot)_N$ is the discrete scalar product introduced in (4.2.4) for the Chebyshev nodes and weights defined in (4.3.12) and (4.3.13), respectively.

We notice that (8.2.18) embodies both Galerkin and collocation cases. As a matter of fact, since $D(v_N w)w^{-1} \in \mathbb{P}_{N-1}$ (as $v_N = (1 - x^2)p_{N-2}$, with $p_{N-2} \in \mathbb{P}_{N-2}$), we have from (4.2.6) that

$$
\left(D u_N, \frac{D(v_N w)}{w} \right)_N = \left(D u_N, \frac{D(v_N w)}{w} \right)_w
$$

and

$$
(D u_N, v_N)_N = (D u_N, v_N)_w \;.
$$

Hence (8.2.18) can be regarded as a Galerkin method for (8.2.15) with respect to the scalar product $(\cdot, \cdot)_w$. Moreover, since

$$
\left(D u_N, \frac{D(v_N w)}{w} \right)_w = -(D^2 u_N, v_N)_w = -(D^2 u_N, v_N)_N
$$

(still using (4.2.6)), we deduce from (8.2.18) that u_N satisfies

$$
-\varepsilon D^2 u_N + b u_N = 0 \quad \text{at} \quad x_j = \cos \frac{\pi j}{N} \;, \quad j = 1, ..., N-1 \;,
$$

i.e., a collocation problem.

The exact solution to (8.2.15)

$$
u(x) = (e^{2b/\varepsilon} - 1)^{-1}[e^{b(x+1)/\varepsilon} - 1]
$$

is an increasing function satisfying the maximum principle $0 \le u(x) \le 1$, while this is not the case for the approximate solution $u_N(x)$, as a Gibbs phenomenon occurs at $x = 1$ if ε/b is very small compared to $1/N$.

In other words, the maximum principle in the *physical space* does not hold for spectral collocation (or Galerkin) approximation. However, a sort of maximum principle may hold in the *frequency space*. In fact, it can be proven that the Chebyshev coefficients of the spectral solution

$$
u_N(x) = \sum_{m=0}^{N} \hat{u}_m T_m(x)
$$

are such that $\hat{u}_m > 0$ for $1 \le m \le N$ ($T_m(x)$ being the Chebyshev polynomials introduced in (4.3.2)).

A remarkable consequence of this fact is that

$$u_N(x) = \sum_{m=0}^{N} \hat{u}_m T_m(x) \leq \hat{u}_0 + \sum_{m=1}^{N} \hat{u}_m |T_m(x)|$$

(8.2.19)

$$\leq \sum_{m=0}^{N} \hat{u}_m = \sum_{m=0}^{N} \hat{u}_m T_m(1) = u_N(1) = 1$$

for any $x \in [-1, 1]$, thus $u_N(x)$ is uniformly bounded from above by the boundary data.

A more precise result can be derived if ε/b is small enough compared to $1/N^2$, i.e., if

(8.2.20)
$$\frac{\varepsilon}{b} \leq C_0 N^{-2}$$

holds for a small constant $C_0 > 0$. Notice that under this condition a Gibbs phenomenon indeed occurs at $x = 1$. In this case, if N is odd, one obtains $\hat{u}_m > 0$ for $0 \leq m \leq N$, and consequently, by proceeding as in the proof of (8.2.19), it follows that $|u_N(x)| \leq 1$ for $-1 \leq x \leq 1$. On the other hand, if N is even, one finds $\hat{u}_0 < 0$ and $\hat{u}_m > 0$ for $1 \leq m \leq N$, thus $-1 - 2|\hat{u}_0| \leq u_N(x) \leq 1$ for $-1 \leq x \leq 1$. Note, however, that in this case $|\hat{u}_0|$ depends on ε, b and N (precisely, it behaves like $N^{-2}b/\varepsilon$), hence $u_N(x)$ is not uniformly bounded from below. In particular, when ε/b is very small compared to $1/N$, this result suggests that spectral approximation of advection-dominated problems may exhibit a strong sensitivity to the parity of N.

8.3 Stabilization Methods

We now return to the general d-dimensional problem

(8.3.1)
$$\begin{cases} Lu = f & \text{in } \Omega \\ u = 0 & \text{on } \partial\Omega \end{cases},$$

where $d = 2, 3$, L is defined in (8.1.1) and, for the sake of simplicity, Ω is a polygonal domain.

The Galerkin approximation of (8.3.1) (stated in the form (8.1.7)) would read:

(8.3.2)   find $u_h \in V_h$: $a(u_h, v_h) = (f, v_h)$ $\forall v_h \in V_h$,

where V_h is a suitable finite dimensional subspace of $V = H_0^1(\Omega)$. As seen in the preceding Section, this method can fail to furnish a satisfactory approximate solution if the ellipticity constant α_0 is small with respect to $||\mathbf{b}||_{L^\infty(\Omega)}$ and the mesh size h.

An instance is provided in Fig. 8.3.1, where we report the Galerkin approximation based on piecewise-linear finite elements for the advection-diffusion equation

$$(8.3.3) \qquad -\varepsilon \Delta u + \operatorname{div}(\mathbf{b}u) = 0 \quad \text{in} \quad \Omega = (0,1)^2 \ ,$$

where $\varepsilon = 10^{-3}$, $\mathbf{b} = (0,1)$ and the triangulation is made by 30×30 uniformly distributed nodes. It is clearly seen that severe oscillations arise across the layer.

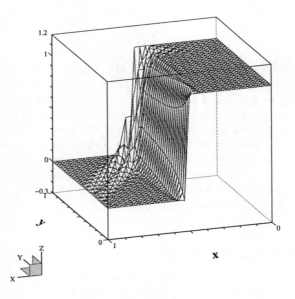

Fig. 8.3.1. Numerical results obtained by the Galerkin method

Notice that when considering problem (8.3.3) the advective term does not play any role in the stability estimate. Therefore nothing can be controlled a-priori independently of the parameter ε.

We have shown before that a possible cure is to resort to a Petrov-Galerkin approximation, which, in the one-dimensional case, turns out to be equivalent to using an upwind finite difference scheme. Unfortunately, the accuracy of this method is affected by the introduction of a numerical viscosity of order $O(h)$.

In the following Sections, we review some of the methods that have been proposed for approximating the d-dimensional problem (8.3.1): at first, we present the artificial diffusion method, which is first order accurate; then we focus on some stabilization methods which are strongly consistent (see (8.3.6)) and high order accurate.

8.3.1 The Artificial Diffusion Method

The classical *artificial diffusion* method is a d-dimensional generalization of the upwind finite difference scheme, and introduces a numerical diffusion of order $O(h)$ acting in all directions.

The idea behind this method is easily explained: as the oscillations in the Galerkin method appear when the Péclet number Pe is large, i.e., the ellipticity constant α_0 is small compared with $||\mathbf{b}||_{L^\infty(\Omega)}$ and h, to avoid these difficulties one is led to consider a modified problem with a larger ellipticity constant. For instance, adding to L the operator $\mathcal{L}$ such that

$$\mathcal{L}z := -hQ\Delta z \ , \quad Q := ||\mathbf{b}||_{L^\infty(\Omega)} \ ,$$

the ellipticity constant becomes $\alpha_0^* := \alpha_0 + hQ$, and the Galerkin method applied to this new operator leads to an approximate solution which is not affected by oscillations any longer.

We notice that the new solution u_h^* satisfies:

$$(8.3.4) \qquad u_h^* \in V_h \ : \ a_h(u_h^*, v_h) = (f, v_h) \quad \forall \, v_h \in V_h \ ,$$

where

$$a_h(z, v) := a(z, v) + hQ(\nabla z, \nabla v) \quad \forall \, z, v \in V \ ,$$

i.e., a generalized Galerkin problem (see Section 5.5).

This procedure has a clear drawback. Since the additional term $-hQ\Delta z$ is $O(h)$, we obtain in particular that

$$(8.3.5) \qquad \sup_{\substack{v_h \in V_h \\ v_h \neq 0}} \frac{|a_h(u, v_h) - (f, v_h)|}{||v_h||_1} = O(h) \ ,$$

u being the solution to (8.3.1). This means that the scheme is *consistent*, but with first order accuracy only. As a consequence, the artificial diffusion method cannot be better than first order accurate, no matter how large the polynomial degree k of the finite element space is.

In the lowest degree case of piecewise-linear finite elements ($k = 1$), the above stabilization would not prevent to keep first order accuracy for the H^1-norm of the error. However, it damps the maximum norm of the error from $O(h^2|\log h|)$ (see Remark 6.2.3) to $O(h)$.

The stabilizing term is therefore non-compatible with the optimal order guaranteed by the polynomial approximation. This loss of accuracy is even more dramatic when spectral methods are used.

The error estimate for the artificial diffusion method can be obtained by applying Proposition 5.5.1. Precisely, we have

$$||u - u_h^*||_1 \le C \left(\inf_{v_h \in V_h} ||u - v_h||_1 + h||u||_1 \right) \ ,$$

where $C = C(\alpha, \gamma, Q)$ (α and γ being respectively the coerciveness and continuity constants of the bilinear form $a(\cdot, \cdot)$).

Likewise the one-dimensional upwind finite difference case, stability is thus achieved at the expense of accuracy. Typical results exhibit excessive diffusive behaviour, as already shown in Fig. 8.2.1 for the one-dimensional case. In particular, in the two-dimensional case one can notice a diffusion not only along the streamlines (i.e., the lines parallel to the direction $\mathbf{b}(\mathbf{x})$), but also in the direction perpendicular to $\mathbf{b}(\mathbf{x})$. This undesirable behavior is usually called *crosswind diffusion*, and it is witnessed in Fig. 8.3.2.

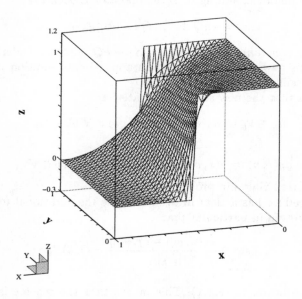

Fig. 8.3.2. Numerical results obtained by the artificial diffusion method

One could avoid it by adding to L the operator $\hat{\mathcal{L}}$ such that

$$\hat{\mathcal{L}}z := -\frac{h}{Q} \operatorname{div}[(\mathbf{b} \cdot \nabla z)\mathbf{b}] \ ,$$

which is not coercive in $H_0^1(\Omega)$ but satisfies

$$\int_\Omega (\hat{\mathcal{L}}v)v = \frac{h}{Q} \int_\Omega (\mathbf{b} \cdot \nabla v)^2 \geq 0 \qquad \forall \, v \in H_0^1(\Omega) \ .$$

This procedure, which was proposed by Raithby (1976) for finite difference schemes and by Hughes and Brooks (1979) in a finite element context, is called *streamline upwind* method and generates an artificial diffusion which pollutes the approximate solution only in the direction of the streamlines.

However, it still introduces an additional term of first order, hence it is only $O(h)$-accurate.

Remark 8.3.1 Several other upwind-type finite element methods have been proposed in the literature. In the following Section we are going to present a family of stabilization methods recently proposed by different authors. For a theoretical presentation and an analysis of the computational efficiency of earlier schemes, we refer to Ikeda (1983), Tabata (1986) and the references therein.   □

8.3.2 Strongly Consistent Stabilization Methods for Finite Elements

In this Section we describe some stabilization methods which don't deteriorate the inherent accuracy of the polynomial subspace. All these methods fit under the generalized Galerkin form (5.5.1), and share the desirable property that

$$(8.3.6) \qquad \mathcal{A}_h(u, v_h) - \mathcal{F}_h(v_h) = 0 \quad \forall \, v_h \in V_h \ ,$$

provided u is the exact solution of problem (8.3.1).

This property, which strengthens the consistency property (8.3.5), will be referred to as *strong consistency*, and allows the stabilized method to mantain the optimal accuracy which is inherent to the choice of the finite dimensional subspace.

We will see that these methods provide non-oscillatory solutions even when the Péclet number is large; some numerical diffusion is still present, but it mainly affects the solution in the streamline direction, sensibly reducing the crosswind diffusion.

Moreover, a suitable version of these methods can also be applied to different problems in order to cure different kinds of pathologies. For instance, when considering the approximation of the Stokes problem, the onset of instability is not due to large advection but rather to the presence of spurious pressure modes. The latter may arise when a suitable compatibility condition between the approximating subspaces for velocity and pressure is not fulfilled (see Sections 9.2 and 9.4).

Let us consider the operator L introduced in (8.1.1) and split it into its symmetric and skew-symmetric parts (with respect to the scalar product in $L^2(\Omega)$), i.e.,

$$(8.3.7) \qquad L = L_S + L_{SS} \ ,$$

where

$$(8.3.8) \qquad L_S z := - \sum_{i,j=1}^{d} D_i(a_{ij}^S D_j z) + \left(\frac{1}{2} \operatorname{div} \mathbf{b} + a_0 \right) z$$

$$(8.3.9) \qquad L_{SS} z := - \sum_{i,j=1}^{d} D_i(a_{ij}^{SS} D_j z) + \frac{1}{2} \operatorname{div}(\mathbf{b}z) + \frac{1}{2} \mathbf{b} \cdot \nabla z \ ,$$

and

$$(8.3.10) \qquad a_{ij}^S := \frac{a_{ij} + a_{ji}}{2} \ , \quad a_{ij}^{SS} := \frac{a_{ij} - a_{ji}}{2} \ , \quad i,j = 1, ..., d \ .$$

Then we introduce a regular family of triangulations $\mathcal{T}_h$ of $\overline{\Omega}$ (see Definition 3.4.1) and we consider a finite dimensional subspace $V_h \subset V = H_0^1(\Omega)$ which is made by piecewise-smooth functions. As an example, one can choose $V_h = X_h^k \cap H_0^1(\Omega)$, $k \geq 1$, where X_h^k is the finite element space defined in (3.2.4) or (3.2.5).

In this Chapter we deal with the *advection-dominated case*, i.e., we assume that for each $K \in \mathcal{T}_h$ the local Péclet function $\mathrm{Pe}_K(\mathbf{x})$ is larger than 1, i.e.,

$$(8.3.11) \qquad \mathrm{Pe}_K(\mathbf{x}) := \frac{|\mathbf{b}(\mathbf{x})| h_K}{2\alpha_0(\mathbf{x})} > 1 \quad \forall \, \mathbf{x} \in K \ ,$$

where

$$(8.3.12) \qquad \alpha_0(\mathbf{x}) := \min_{\boldsymbol{\xi} \in \mathbb{R}^d \setminus 0} \frac{\sum_{i,j=1}^{d} a_{ij}(\mathbf{x}) \xi_i \xi_j}{|\boldsymbol{\xi}|^2} \ , \quad \mathbf{x} \in \overline{\Omega} \ .$$

Here, and in the sequel, the coefficients $a_{ij}(\mathbf{x})$ and $\mathbf{b}(\mathbf{x})$ are supposed to be continuous in $\overline{\Omega}$. Notice that from the ellipticity assumption (see Definition 6.1.1) it follows $\alpha_0(\mathbf{x}) \geq \alpha_0 > 0$ for every $\mathbf{x} \in \overline{\Omega}$. Moreover, (8.3.11) implies, in particular, that $\inf_\Omega |\mathbf{b}(\mathbf{x})| > 0$.

We are thus in a position to introduce the *stabilized* approximation problem to (8.3.1). Define

$$\mathcal{L}_h^{(\rho)}(z_h, v_h) := \sum_{K \in \mathcal{T}_h} \delta \left(Lz_h, \frac{h_K}{|\mathbf{b}|} (L_{SS} + \rho L_S) v_h \right)_K$$

$$\varphi_h^{(\rho)}(v_h) := \sum_{K \in \mathcal{T}_h} \delta \left(f, \frac{h_K}{|\mathbf{b}|} (L_{SS} + \rho L_S) v_h \right)_K \ ,$$

where $z_h, v_h \in V_h$, $(\cdot, \cdot)_K$ denotes the $L^2(K)$-scalar product and the parameters $\delta > 0$ and $\rho \in \mathbb{R}$ have to be chosen. Then we consider the problem:

$$(8.3.13) \qquad \begin{aligned} &\text{find } u_h \in V_h : a(u_h, v_h) + \mathcal{L}_h^{(\rho)}(u_h, v_h) \\ &\qquad\qquad = (f, v_h) + \varphi_h^{(\rho)}(v_h) \quad \forall \, v_h \in V_h \ . \end{aligned}$$

We further denote by $a_h^{(\rho)}(\cdot, \cdot)$ the stabilized bilinear form above, i.e.,

(8.3.14) $a_h^{(\rho)}(z_h, v_h) := a(z_h, v_h) + \mathcal{L}_h^{(\rho)}(z_h, v_h) \quad \forall\, z_h, v_h \in V_h$,

and by $\mathcal{F}_h^{(\rho)}(\cdot)$ the perturbed continuous linear functional

$$\mathcal{F}_h^{(\rho)}(v_h) := (f, v_h) + \varphi_h^{(\rho)}(v_h) \quad \forall\, v_h \in V_h\ .$$

A first important remark is in order. Since the exact solution u satisfies $Lu = f \in L^2(\Omega)$, $a_h^{(\rho)}(u, v_h)$ is defined for each $v_h \in V_h$. Moreover, *strong consistency* (in the sense of (8.3.6)) holds for problem (8.3.13) for any choice of δ and ρ. The problem now is to find suitable values of these parameters in such a way that also stability and convergence are achieved, with respect to a suitable norm.

In Section 8.4 we will analyze the stabilized problem (8.3.13) for three different choices of the parameter ρ. These methods have been proposed by different researchers. However, following Baiocchi and Brezzi (1992), we have preferred to present them in a unified way. We consider $\rho = 0$, which corresponds to the *SUPG (Streamline Upwind/Petrov-Galerkin) method* proposed in Hughes and Brooks (1982) and Brooks and Hughes (1982) (this method has been also called the *Streamline Diffusion method*, see, e.g., Johnson (1987)); $\rho = 1$, which gives the *GALS (Galerkin/Least-Squares) method* introduced by Hughes, Franca and Hulbert (1989); $\rho = -1$, which leads to the method proposed by Franca, Frey and Hughes (1992), extending to advection-diffusion equations an idea used by Douglas and Wang (1989) in the approximation of the Stokes problem. We call this last choice *DWG (Douglas-Wang/Galerkin) method*. Let us notice however that the three cases above do coincide when $\frac{1}{2} \operatorname{div} \mathbf{b} + a_0 = 0$ and piecewise-linear polynomials are considered as test functions.

Remark 8.3.2 As we said above, the acronym SUPG stands for Streamline Upwind/Petrov-Galerkin. However, as pointed out by Hughes (1987), this name doesn't seem to be really appropriate. Indeed, SUPG is a special case of (8.3.13), which is a generalized Galerkin method.   □

Remark 8.3.3 The *diffusion-dominated case*, i.e., $\operatorname{Pe}_K(\mathbf{x}) \leq 1$ for each $\mathbf{x} \in K$ and for each $K \in \mathcal{T}_h$, can be satisfactory treated by the standard Galerkin method. Nonetheless, we can also use the stabilization approach defined in (8.3.13), provided that the stabilization parameter δ is chosen as a function of Pe_K, namely $\delta = O(\operatorname{Pe}_K)$ (see also Remark 8.4.2).   □

We conclude this Section showing some numerical results. At first we consider an approximation based on the stabilized GALS method using piecewise-quadratic finite elements for the advection-diffusion equation (8.3.3) (with $\varepsilon = 10^{-5}$). The triangulation is made by 30×30 nodes uniformly distributed. Different choices of the advective field $\mathbf{b}$ are made, as well as different boundary values for u are considered. From case to case, the values of

b and of the boundary data can be directly desumed from the illustrations we are going to show. Fig. 8.3.3 (a) and (b) show that numerical dissipation can only be added in the direction of the streamlines: there is no dissipation at all in (a), while the case (b) is diffusive since streamlines cross the discontinuity line. We also notice the presence of some under- and over-shooting in the latter case, a situation that often arises for these stabilized methods (see also Fig. 8.3.3 (c)). The test problem referred to in Fig. 8.3.3 (c) and (d) is virtually the same. However, (d) shows that there might be a severe grid-orientation effect that induces numerical diffusion even when **b** is parallel to the discontinuity line.

The non-monotonicity of the scheme, which is enlightened by the under- and over-shooting phenomena near the layer, can be cured at the expense of introducing a further stabilizing term in (8.3.13). This latter term is a shock-capturing nonlinear viscosity term (see Hughes, Mallet and Mizukami (1986), Galeão and Dutra do Carmo (1988), Shakib (1988)).

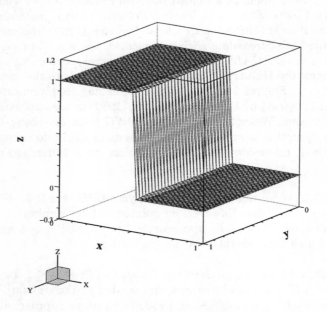

Fig. 8.3.3 (a). Numerical results obtained by the GALS method

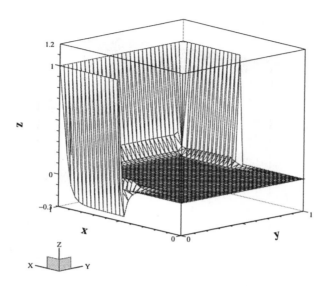

Fig. 8.3.3 (b). Numerical results obtained by the GALS method

8.3.3 Stabilization by Bubble Functions

Before analyzing stability and convergence of the methods presented in the previous Section, we want to propose a different interpretation of them, due originally to Brezzi, Bristeau, Franca, Mallet and Rogé (1992). Further remarks and improvements can be found in Baiocchi, Brezzi and Franca (1993). For a similar approach in a different context, also see Remark 9.4.2.

Let us restrict, for simplicity, to the two-dimensional case, i.e., $d = 2$. We start by introducing the finite dimensional space

$$(8.3.15) \qquad V_h^b := [X_h^k \cap H_0^1(\Omega)] \oplus B \ ,$$

where X_h^k is defined in (3.2.4), and B is a (finite dimensional) space of *bubble functions* having exactly one degree of freedom in each element $K \in \mathcal{T}_h$. For instance, we can choose the cubic bubble functions, namely,

$$(8.3.16) \quad B = B_3 := \{v_B \in H_0^1(\Omega) \,|\, v_{B|K} = c\,b_K \ , \ c \in \mathbb{R} \ , \ b_K = \lambda_1\lambda_2\lambda_3\} \ ,$$

where λ_i, $i = 1, 2, 3$, are linear polynomials, each one vanishing on one side of K and taking value 1 at the opposite vertex. The function b_K, which is an example of bubble function, is thus positive in the internal part of K and vanishes on ∂K.

Now consider the Galerkin approximation of (8.3.1) in V_h^b:

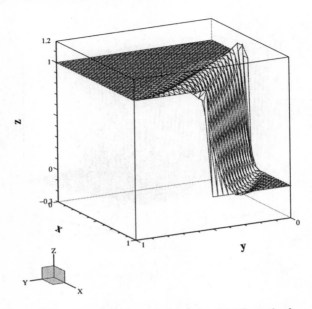

Fig. 8.3.3 (c). Numerical results obtained by the GALS method

$$
(8.3.17) \quad
\begin{aligned}
& \text{find } (u_h + u_B) \in V_h^b : a(u_h + u_B, v_h + v_B) \\
& \qquad\qquad = (f, v_h + v_B) \quad \forall\, (v_h + v_B) \in V_h^b \;.
\end{aligned}
$$

We want to reframe (8.3.17) as a stabilized Galerkin method in $X_h^k \cap H_0^1(\Omega)$, solving with respect to u_B. First of all, we can write u_B on each K as

$$
(8.3.18) \qquad u_{B|K} = c_{b,K} b_K \;.
$$

Choosing as test function in (8.3.17) $v_h = 0$ and $v_B \in B$ such that

$$
v_B = \begin{cases} b_K & \text{in } K \\ 0 & \text{on } \Omega \setminus K \end{cases},
$$

we find

$$
a(u_h + u_B, v_B) = a_K(u_h + c_{b,K} b_K, b_K) \;,
$$

having defined by $a_K(\cdot, \cdot)$ the restriction of $a(\cdot, \cdot)$ to the triangle K. Thus (8.3.17) becomes

$$
a_K(u_h, b_K) + c_{b,K} a_K(b_K, b_K) = (f, b_K)_K \;,
$$

and, since $u_{h|K} \in X_h^k$ and b_K vanishes on ∂K, we can integrate by parts in the first term, obtaining

$$
(8.3.19) \qquad c_{b,K} = \frac{(f - L u_h, b_K)_K}{a_K(b_K, b_K)} \;.
$$

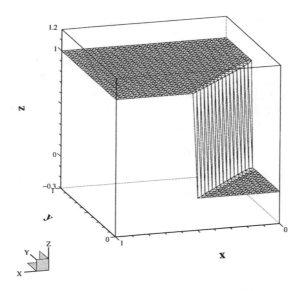

Fig. 8.3.3 (d). Numerical results obtained by the GALS method

We now choose as test function in (8.3.17) any $v_h \in X_h^k \cap H_0^1(\Omega)$ and $v_B = 0$, finding

$$a(u_h, v_h) + \sum_{K \in \mathcal{T}_h} c_{b,K} a_K(b_K, v_h) = (f, v_h) \ .$$

After we integrate by parts, the term $a_K(b_K, v_h)$ produces

$$a_K(b_K, v_h) = (b_K, L_S v_h - L_{SS} v_h)_K$$

(see (8.3.8) and (8.3.9) for notations), and similarly

$$a_K(b_K, b_K) = (L_S b_K, b_K)_K \ .$$

Using (8.3.19) we can finally write (8.3.17) as

(8.3.20) $a(u_h, v_h) + a_B(f; u_h, v_h) = (f, v_h) \quad \forall\, v_h \in X_h^k \cap H_0^1(\Omega) \ ,$

where

(8.3.21) $a_B(f; u_h, v_h) := \sum_{K \in \mathcal{T}_h} \dfrac{(Lu_h - f, b_K)_K (L_{SS} v_h - L_S v_h, b_K)_K}{(L_S b_K, b_K)_K} \ .$

This finite dimensional approximation to (8.3.1) is strongly consistent, and differs from the usual Galerkin one due to the presence of the additional term $a_B(f; u_h, v_h)$. Moreover, it resembles the DWG method introduced in (8.3.13) for $\rho = -1$. In fact, there the additional term takes the form

$$(8.3.22) \qquad \sum_{K \in \mathcal{T}_h} \delta \left(Lu_h - f, \frac{h_K}{|\mathbf{b}|}(L_{SS}v_h - L_S v_h) \right)_K .$$

We now want to show that, at least in a simplified case, (8.3.21) can indeed be presented in the form (8.3.22). Consider piecewise-linear finite elements (i.e., $k = 1$ in (3.2.4)) and suppose that $\mathbf{b}(\mathbf{x})$ is a constant vector, $f(\mathbf{x})$ is constant, $a_0(\mathbf{x}) = 0$ and $a_{ij}(\mathbf{x}) = \varepsilon \delta_{ij}$, where $\varepsilon > 0$ is a constant and δ_{ij} is the Kronecker symbol. In such a case we have

$$L_S z = -\varepsilon \Delta z \ , \quad L_{SS} z = \mathbf{b} \cdot \nabla z \ ,$$

and SUPG, GALS and DWG methods are all coincident. Choose, moreover, the space B of bubble functions as in (8.3.16). Note that the bubble function b_K satisfies

$$(8.3.23) \qquad \int_K b_K = C_1 \operatorname{meas}(K) = C_{1,K} h_K^2 \ , \quad \int_K |\nabla b_K|^2 = C_{2,K} \ ,$$

where constants $C_{1,K}$ and $C_{2,K}$ are independent of h_K.

Begin by computing $(Lu_h - f, b_K)_K$. Recalling that $u_{h|K} \in \mathbb{P}_1$, from (8.3.23) we have

$$(Lu_h - f, b_K)_K = (\mathbf{b} \cdot \nabla u_h - f, b_K)_K = (\mathbf{b} \cdot \nabla u_{h|K} - f) \int_K b_K$$
$$= (\mathbf{b} \cdot \nabla u_{h|K} - f) C_{1,K} h_K^2 \ .$$

Analogously,

$$(L_{SS}v_h - L_S v_h, b_K)_K = (\mathbf{b} \cdot \nabla v_h, b_K)_K = \mathbf{b} \cdot \nabla v_{h|K} C_1 \operatorname{meas}(K) \ .$$

Finally, from (8.3.23) it holds

$$(L_S b_K, b_K)_K = -\varepsilon (\Delta b_K, b_K)_K = \varepsilon C_{2,K} \ .$$

Therefore, we can write

$$a_B(f; u_h, v_h) = \sum_{K \in \mathcal{T}_h} \frac{(\mathbf{b} \cdot \nabla u_{h|K} - f)\mathbf{b} \cdot \nabla v_{h|K} C_1 C_{1,K} h_K^2 \operatorname{meas}(K)}{\varepsilon C_{2,K}}$$
$$= \sum_{K \in \mathcal{T}_h} \frac{C_1 C_{1,K} h_K^2}{\varepsilon C_{2,K}} (L_{SS} u_h - f, L_{SS} v_h)_K \ .$$

From definition (8.3.11) we can finally obtain

$$a_B(f; u_h, v_h)$$
$$(8.3.24) \qquad = \sum_{K \in \mathcal{T}_h} \frac{2 C_1 C_{1,K} \operatorname{Pe}_K}{C_{2,K}} \left(Lu_h - f, \frac{h_K}{|\mathbf{b}|}(L_{SS}v_h - L_S v_h) \right)_K \ .$$

Comparing this with (8.3.13) we find out that the additional term stemming from the presence of the bubble function is nothing but the stabilizing term of the DWG method, for a suitable choice of the constant $\delta = \delta_K$ on each element $K \in \mathcal{T}_h$. (Recall again that, under our assumptions, we have $L_S u_h = L_S v_h = 0$; hence SUPG, GALS and DWG methods are coincident.)

Indeed, (8.3.24) is not yet satisfactory for the approximation of an advection-dominated problem, as the constant $\delta_K := 2C_{2,K}^{-1} C_1 C_{1,K} \mathrm{Pe}_K$ depends on the local Péclet number. However, by proceeding in a slightly different way, one could choose other types of bubble functions in order to recover (8.3.13) with a parameter δ_K which is independent of Pe_K (see Brezzi, Bristeau, Franca, Mallet and Rogé (1992)).

8.3.4 Stabilization Methods for Spectral Approximation

In this Section we consider two examples of stabilization methods for spectral collocation approximation. The former shows that a standard spectral collocation scheme can be stabilized by adding extra trial/test functions with local support (typically, bubble functions). The latter modifies the usual Galerkin approach by introducing an additional term like in SUPG, GALS or DWG methods, by adapting the procedures presented in Section 8.3.2 to the spectral collocation case.

Let us focus, for simplicity, on the two-dimensional homogeneous Dirichlet problem

$$(8.3.25) \qquad \begin{cases} Lu := -\varepsilon \Delta u + \mathrm{div}(\mathbf{b}u) + a_0 u = f & \text{in } \Omega \\ \\ u = 0 & \text{on } \partial \Omega \end{cases},$$

where $f \in L^2(\Omega)$ and $\Omega = (-1,1)^2$. The bilinear form $a(\cdot,\cdot)$ associated to (8.3.25) is defined in (8.1.2) (with $a_{ij} = \varepsilon \delta_{ij}$), and is assumed to be continuous and coercive, i.e., to satisfy (8.1.3) and (8.1.4). The associated variational problem reads:

$$(8.3.26) \qquad \text{find } u \in H_0^1(\Omega) \ : \ a(u,v) = (f,v) \quad \forall \, v \in H_0^1(\Omega) \ .$$

For each integer $N \geq 2$, the Legendre Gauss-Lobatto nodes $\mathbf{x}_{ij} = (x_i, x_j)$, $i, j = 0, ..., N$, induce a triangulation $\mathcal{T}_h$ on $\overline{\Omega}$, given by the N^2 rectangular cells $K_{ij} := I_i \times I_j$ whose vertices are four neighbouring nodes of the Gauss-Lobatto grid. Here, we have set $I_i := [x_{i+1}, x_i]$; recall that the Legendre nodes are usually ordered from right toward the left, see Section 4.4.2. We have moreover denoted by w_{ij} the weights of the Legendre Gauss-Lobatto quadrature formula (see (4.5.38)).

Let $V_N := \mathbb{Q}_N^0$ be the space of polynomials of degree less than or equal to N in each variable x and y, vanishing on $\partial \Omega$. Similarly, let V_h be the finite element space $X_h^1 \cap H_0^1(\Omega)$, where X_h^1 is defined in (3.2.5). The space of bubble functions is denoted by

(8.3.27) $B := \{v_B \in H_0^1(\Omega) \,|\, v_{B|K_{ij}} = c\,b_{ij} \, , \; c \in \mathbb{R} \, , \; i, j = 0, ..., N-1\}$,

where, for each i, j, b_{ij} is a function, uniquely determined, vanishing at the boundary of K_{ij}. The easiest example is provided by $b_{ij} \in \mathbb{Q}_2$, $b_{ij}(x, y) = p_i(x)p_j(y)$, p_i and p_j being two parabolas vanishing at the endpoints of the intervals I_i and I_j, respectively, and taking value 1 at the midpoints. A more useful choice is the one outlined in (8.3.33) below.

Following Canuto (1994) we define

(8.3.28) $V_N^b := V_N \oplus B$,

and we introduce the Galerkin approximation to (8.3.26) in V_N^b:

(8.3.29)   find $u_N^b \in V_N^b \; : \; a(u_N^b, v_N^b) = (f, v_N^b) \quad \forall \, v_N^b \in V_N^b$.

This problem clearly has a unique solution u_N^b, which satisfies

(8.3.30) $\|u_N^b\|_1 \leq \dfrac{1}{\alpha}\|f\|_0 \quad \forall \, N \geq 1$

(8.3.31) $\|u - u_N^b\|_1 \leq \dfrac{\gamma}{\alpha} \inf_{v_N^b \in V_N^b} \|u - v_N^b\|_1 \leq \dfrac{\gamma}{\alpha} \inf_{v_N \in V_N} \|u - v_N\|_1$,

as $V_N \subset V_N^b$. This shows that the accuracy of this Galerkin approximation is at least as good as the one of the standard Galerkin method in V_N.

However, we have already noticed that the spectral Galerkin method is difficult to implement (see Section 6.2.1). Therefore, we are more interested in introducing a spectral collocation approximation to (8.3.26). For this purpose, let us assume that $f \in C^0(\overline{\Omega})$, and for simplicity let $\mathbf{b}$ be a constant vector and $a_0 = a_0(\mathbf{x}) \geq 0$.

The following spectral collocation method with bubble correction is proposed in Canuto (1994): find $u_N^b = u_N + u_b \in V_N^b$ such that

(8.3.32) $\begin{cases} (Lu_N, v_N)_N + a(u_b, \pi_h^1(v_N)) = (f, v_N)_N \quad \forall \, v_N \in V_N \\[2mm] (\pi_h^1(Lu_N), v_b) + a(u_b, v_b) = (\pi_h^1(f), v_b) \quad \forall \, v_b \in B \end{cases}$.

Here $(\cdot, \cdot)_N$ is the discrete scalar product introduced in (4.5.39), and $\pi_h^1 : C^0(\overline{\Omega}) \to V_h$ is the finite element interpolation operator defined in (3.4.1).

The derivation of (8.3.32) can be motivated as follows: split (8.3.29) into two sets of equations (one for u_N with the test function v_N, the other for u_b with the test function v_b), according to the decomposition (8.3.28). Then, replace the bilinear form $a(u_N, v_N)$ and the term (f, v_N) by their discrete counterparts (involving $(\cdot, \cdot)_N$ instead of $(\cdot, \cdot)$). As for the remaining terms, the algebraic polynomials, as well as f, are approximated by their finite element interpolants.

The introduction of these interpolation corrections aims at obtaining a more sparse global matrix. In fact, the block $a(u_b, \pi_h^1(v_N))$ is banded; the other block $(\pi_h^1(Lu_N), v_b)$ is still full, since Lu_N depends on all the grid-values of u_N. However, it depends on the grid-values of Lu_N in a banded way, since π_h^1 is a local operator. It follows that the bubble correction u_b can be expressed element-by-element in terms of the neighbouring grid-values of Lu_N, using (8.3.32)$_2$, and thus eliminated from (8.3.32)$_1$.

It remains to indicate how to construct the space of bubble functions B. Let $b^*(x)$ be a one-dimensional bubble function on the interval $(0, 1)$. By this we mean any non-negative function of $H_0^1(0, 1)$ (not identically vanishing). Then, on each rectangular cell $K_{ij} = (x_{i+1}, x_i) \times (x_{j+1}, x_j)$ let us set

$$(8.3.33) \qquad b_{ij}(x, y) := b^*\left(\frac{x - x_{i+1}}{x_i - x_{i+1}}\right) b^*\left(\frac{y - x_{j+1}}{x_j - x_{j+1}}\right) .$$

The space B is now defined according to (8.3.27).

Setting

$$c^* := \frac{\|b^*\|_0}{\|Db^*\|_0} ,$$

problem (8.3.32) has a unique solution provided c^* is small enough (independently of N). In that case, one also has

$$\|u_N^b\|_1 \le C\|f\|_{C^0(\overline{\Omega})} ,$$

and

$$\|u - u_N^b\|_1 \le C(N^{1-s}\|u\|_s + N^{-r}\|f\|_r) ,$$

assuming that the exact solution u belongs to $H^s(\Omega)$ and f belongs to $H^r(\Omega)$ ($s \ge 1, r \ge 2$).

A different approach, more germane to those introduced in Section 8.3.2 for finite elements, has been proposed by Pasquarelli and Quarteroni (1994). The approximate problem now reads: find $u_N \in \mathbb{Q}_N^0$ such that

$$(8.3.34) \qquad \begin{aligned} (L_N u_N, v_N + \delta(L_{N,SS} + \rho L_{N,S})v_N)_N \\ = (f, v_N + \delta(L_{N,SS} + \rho L_{N,S})v_N)_N \quad \forall\, v_N \in \mathbb{Q}_N^0 , \end{aligned}$$

where $\delta > 0$ is a parameter, and ρ may take the value $\rho = 0$ (SUPG method), $\rho = 1$ (GALS method) or $\rho = -1$ (DWG method). Here, L_N denotes the following pseudo-spectral approximation of the operator L:

$$(8.3.35) \qquad L_N z := -\varepsilon\Delta z + \frac{1}{2}[\operatorname{div} I_N(\mathbf{b}z) + \mathbf{b} \cdot \nabla z] + \left(\frac{1}{2}\operatorname{div}\mathbf{b} + a_0\right) z .$$

(Clearly, $L_N z = Lz$ if $\mathbf{b}$ is constant and $z \in \mathbb{Q}_N$.) Moreover, $L_{N,S}$ and $L_{N,SS}$ are its symmetric and skew-symmetric part with respect to the scalar product $(\cdot, \cdot)_N$, respectively, i.e.,

$$(8.3.36) \qquad L_{N,S}z := -\varepsilon \Delta z + \left(\frac{1}{2}\operatorname{div}\mathbf{b} + a_0\right)z$$

$$(8.3.37) \qquad L_{N,SS}z := \frac{1}{2}[\operatorname{div} I_N(\mathbf{b}z) + \mathbf{b}\cdot\nabla z] \ .$$

As before, let us denote by $\mathbf{x}_{ij} = (x_i, x_j)$ and $w_{ij} = w_i w_j$, $i,j = 0, ..., N$, the Legendre Gauss-Lobatto nodes and weights, respectively. The collocation interpretation of (8.3.34) reads:

$$(8.3.38) \qquad \begin{cases} L_N u_N(\mathbf{x}_{ij}) \\ \quad + \delta \sum_{l,k=0}^{N} L_N u_N(\mathbf{x}_{lk})\,(L_{N,SS} + \rho L_{N,S})\psi_{ij}(\mathbf{x}_{lk})\,\dfrac{w_{lk}}{w_{ij}} \\ \quad = \delta \sum_{l,k=0}^{N} f(\mathbf{x}_{lk})\,(L_{N,SS} + \rho L_{N,S})\psi_{ij}(\mathbf{x}_{lk})\,\dfrac{w_{lk}}{w_{ij}} \\ \qquad + f(\mathbf{x}_{ij}) \qquad\qquad \forall\, i,j \text{ such that } \mathbf{x}_{ij} \in \Omega \\ u_N(\mathbf{x}_{ij}) = 0 \qquad\qquad \forall\, i,j \text{ such that } \mathbf{x}_{ij} \in \partial\Omega \ , \end{cases}$$

where $\psi_{ij} \in \mathbb{Q}_N^0$ is the Lagrangian basis function associated to the internal node $\mathbf{x}_{ij}$.

For all i,j, the function $(L_{N,SS} + \rho L_{N,S})\psi_{ij}$ is a polynomial that is different from 0 at all collocation points. This entails that all equations (except those at the boundary) depend on the values of the polynomial $L_N u_N$ at all collocation nodes. In particular, one obtains a full matrix instead of the usual block matrix (see Sections 6.3.1 and 6.3.3).

With regard to the parameter δ, it can either vary at each collocation node, or be taken constant. In the latter case, a suitable scaling is

$$\delta = \frac{\hat{c}}{\varepsilon N^4} \ ,$$

where $\hat{c} > 0$ is a constant to be fixed independently of ε and N.

8.4 Analysis of Strongly Consistent Stabilization Methods

In this Section we analyze the stabilization methods defined in (8.3.13). Although they can be regarded as generalized Galerkin methods, they do not fall into the abstract framework of Proposition 5.5.1, as the associated bilinear form $a_h^{(\rho)}(\cdot,\cdot)$ does not satisfy the continuity requirement (5.5.5). On the other hand, the results proven in Theorem 5.5.1 do not imply convergence in the present situation.

For the sake of simplicity, in the sequel we assume that there exist constants μ_0 and μ_1 such that:

(8.4.1) $$0 < \mu_0 \leq \mu(\mathbf{x}) := \frac{1}{2} \operatorname{div} \mathbf{b}(\mathbf{x}) + a_0(\mathbf{x}) \leq \mu_1 \ , \ \mathbf{x} \in \Omega \ .$$

Let us notice however that this condition can be relaxed if the flow field $\mathbf{b}(\mathbf{x})$ has no closed streamlines. In that case, a weaker condition suffices, and it disappears in the limit $a_{ij} \to 0$ (see, e.g., Johnson, Nävert and Pitkäranta (1984)).

Assume moreover that the coefficients a_{ij} take the simple form

$$a_{ij}(\mathbf{x}) = \varepsilon \delta_{ij} \ , \quad i, j = 1, ..., d \ , \ \mathbf{x} \in \Omega \ ,$$

$\varepsilon > 0$ being a constant. Consequently

$$L_S z = -\varepsilon \Delta z + \mu z$$
$$L_{SS} z = \frac{1}{2} \operatorname{div}(\mathbf{b} z) + \frac{1}{2} \mathbf{b} \cdot \nabla z \ ,$$

and we find at once that

(8.4.2) $$a(v_h, v_h) = \varepsilon ||\nabla v_h||^2_{0,\Omega} + ||\mu^{1/2} v_h||^2_{0,\Omega} \quad \forall \, v_h \in V_h \ .$$

Now, let us start with the stability analysis for the SUPG method. First of all, we want an estimate of $a_h^{(0)}(v_h, v_h)$ from below.

Proposition 8.4.1 *Assume that we are dealing with the advection-dominated case, i.e.,*

(8.4.3) $$\frac{|\mathbf{b}(\mathbf{x})| h_K}{2\varepsilon} > 1 \quad \forall \, \mathbf{x} \in K \ .$$

If the parameter δ and the mesh-size h_K satisfy

(8.4.4) $$0 < \delta \leq C_0^{-1} \ , \quad \delta \, h_K \leq \frac{|\mathbf{b}(\mathbf{x})|}{2\mu(\mathbf{x})} \quad \forall \, \mathbf{x} \in K \ ,$$

where C_0 is the constant appearing in the inverse inequality (8.4.7) below, the bilinear form $a_h^{(0)}(\cdot, \cdot)$ associated to the SUPG method satisfies the coerciveness inequality

(8.4.5)
$$a_h^{(0)}(v_h, v_h) \geq \frac{\varepsilon}{2} ||\nabla v_h||^2_{0,\Omega} + \frac{1}{2} ||\mu^{1/2} v_h||^2_{0,\Omega}$$
$$+ \frac{1}{2} \sum_{K \in \mathcal{T}_h} \delta \left(\frac{h_K}{|\mathbf{b}|} L_{SS} v_h, L_{SS} v_h \right)_K \quad \forall \, v_h \in V_h \ .$$

Proof. Being satisfied (8.4.2), we concentrate our attention on the stabilizing term, which can be rewritten as

$$(8.4.6) \quad \sum_{K \in \mathcal{T}_h} \delta \left(-\varepsilon \Delta v_h, \frac{h_K}{|\mathbf{b}|} L_{SS} v_h \right)_K +$$

$$+ \sum_{K \in \mathcal{T}_h} \delta \left(\mu v_h, \frac{h_K}{|\mathbf{b}|} L_{SS} v_h \right)_K + \sum_{K \in \mathcal{T}_h} \delta \left(L_{SS} v_h, \frac{h_K}{|\mathbf{b}|} L_{SS} v_h \right)_K .$$

Consider the first term. The following inverse inequality

$$(8.4.7) \quad \sum_{K \in \mathcal{T}_h} h_K^2 \int_K |\Delta v_h|^2 \leq C_0 \|\nabla v_h\|_{0,\Omega}^2 \quad \forall \, v_h \in V_h$$

holds (it can be proven by proceeding as in Proposition 6.3.2). Furthermore, from (8.4.3) it follows that

$$(8.4.8) \quad \frac{\varepsilon^2}{|\mathbf{b}(\mathbf{x})| h_K} \leq \frac{\varepsilon}{2} \quad \forall \, \mathbf{x} \in \overline{\Omega} .$$

Therefore,

$$(8.4.9) \quad \left| \sum_{K \in \mathcal{T}_h} \delta \left(-\varepsilon \Delta v_h , \frac{h_K}{|\mathbf{b}|} L_{SS} v_h \right)_K \right|$$

$$\leq \frac{1}{4} \sum_{K \in \mathcal{T}_h} \delta \left(\frac{h_K}{|\mathbf{b}|} L_{SS} v_h, L_{SS} v_h \right)_K$$

$$+ \sum_{K \in \mathcal{T}_h} \delta \varepsilon^2 \left(\frac{h_K}{|\mathbf{b}|} \Delta v_h, \Delta v_h \right)_K$$

$$\leq \frac{1}{4} \sum_{K \in \mathcal{T}_h} \delta \left(\frac{h_K}{|\mathbf{b}|} L_{SS} v_h, L_{SS} v_h \right)_K + \frac{C_0 \delta \varepsilon}{2} \|\nabla v_h\|_{0,\Omega}^2 .$$

As for the second term in (8.4.6), using (8.4.1) it can be estimated as follows:

$$(8.4.10) \quad \left| \sum_{K \in \mathcal{T}_h} \delta \left(\mu v_h, \frac{h_K}{|\mathbf{b}|} L_{SS} v_h \right)_K \right|$$

$$\leq \frac{1}{4} \sum_{K \in \mathcal{T}_h} \delta \left(\frac{h_K}{|\mathbf{b}|} L_{SS} v_h, L_{SS} v_h \right)_K$$

$$+ \sum_{K \in \mathcal{T}_h} \delta \left(\frac{h_K}{|\mathbf{b}|} \mu^2 v_h, v_h \right)_K .$$

From (8.4.2), (8.4.6), (8.4.9) and (8.4.10) we can conclude that

$$a_h^{(0)}(v_h, v_h) \geq \varepsilon \left(1 - \frac{C_0 \delta}{2}\right) ||\nabla v_h||_0^2$$

(8.4.11)
$$+ \frac{1}{2} \sum_{K \in \mathcal{T}_h} \delta \left(\frac{h_K}{|\mathbf{b}|} L_{SS} v_h, L_{SS} v_h\right)_K$$

$$+ \sum_{K \in \mathcal{T}_h} \left(\mu \left[1 - \delta \frac{h_K}{|\mathbf{b}|} \mu\right] v_h, v_h\right)_K .$$

If we now choose δ and h_K as indicated in (8.4.4), the stability inequality (8.4.5) follows. $\qquad\square$

Notice that condition (8.4.4) on h_K is actually not restrictive, as we are considering advection-dominated problems for which $|\mathbf{b}(\mathbf{x})|$ is large enough.

From (8.4.5) we can deduce the stability result for the solution to problem (8.3.13). In fact, let us define

(8.4.12)
$$||v_h||_{SUPG}^2 := \varepsilon ||\nabla v_h||_{0,\Omega}^2 + ||\mu^{1/2} v_h||_{0,\Omega}^2$$
$$+ \sum_{K \in \mathcal{T}_h} \delta \left(\frac{h_K}{|\mathbf{b}|} L_{SS} v_h, L_{SS} v_h\right)_K .$$

If u_h is a solution to (8.3.13) with $\rho = 0$ we find

$$\frac{1}{2} ||u_h||_{SUPG}^2 \leq a_h^{(0)}(u_h, u_h) = (f, u_h) + \sum_{K \in \mathcal{T}_h} \delta \left(f, \frac{h_K}{|\mathbf{b}|} L_{SS} u_h\right)_K$$

$$\leq ||\mu^{-1/2} f||_{0,\Omega}^2 + \sum_{K \in \mathcal{T}_h} \delta \left(\frac{h_K}{|\mathbf{b}|} f, f\right)_K + \frac{1}{4} ||u_h||_{SUPG}^2 .$$

We conclude that there exists C, independent of h, such that

(8.4.13)
$$||u_h||_{SUPG} \leq C ||f||_{0,\Omega} .$$

Let us further stress that this stability estimate improves the one that can be obtained for the Galerkin method, owing to the presence at the left hand side of the additional term

$$\sum_{K \in \mathcal{T}_h} \delta \left(\frac{h_K}{|\mathbf{b}|} L_{SS} u_h, L_{SS} u_h\right)_K .$$

In particular, this provides a control for the derivative of u_h in the streamline direction $\mathbf{b}(\mathbf{x})$.

We now consider the GALS method. From (8.4.2) we easily obtain:

Proposition 8.4.2 For any $\delta > 0$ the bilinear form $a_h^{(1)}(\cdot, \cdot)$ associated to the GALS method satisfies

$$a_h^{(1)}(v_h, v_h)$$

(8.4.14)
$$= \varepsilon||\nabla v_h||_{0,\Omega}^2 + ||\mu^{1/2}v_h||_{0,\Omega}^2 + \sum_{K \in \mathcal{T}_h} \delta \left(\frac{h_K}{|\mathbf{b}|}Lv_h, Lv_h\right)_K$$

$$=: ||v_h||_{GALS}^2 \quad \forall \, v_h \in V_h \ .$$

This property of coerciveness leads to a stability estimate. In fact, if u_h is a solution to (8.3.13) with $\rho = 1$ we have

$$||u_h||_{GALS}^2 = a_h^{(1)}(u_h, u_h) = (f, u_h) + \sum_{K \in \mathcal{T}_h} \delta \left(f, \frac{h_K}{|\mathbf{b}|}Lu_h\right)_K$$

$$\leq ||\mu^{-1/2}f||_{0,\Omega}^2 + \sum_{K \in \mathcal{T}_h} \delta \left(\frac{h_K}{|\mathbf{b}|}f, f\right)_K + \frac{1}{4}||u_h||_{GALS}^2 \ ,$$

hence

(8.4.15)
$$||u_h||_{GALS} \leq C||f||_{0,\Omega} \ .$$

Notice that (8.4.15) holds for any choice of the parameter $\delta > 0$.

For the sake of completeness, we also present the proof of the stability for the DWG method.

Proposition 8.4.3 *Assume that* (8.4.3) *holds. If the parameter δ and the mesh-size h_K satisfy*

$$0 < \delta \leq (2C_0)^{-1} \ , \quad \delta \, h_K \leq \frac{|\mathbf{b}(\mathbf{x})|}{4\mu(\mathbf{x})} \quad \forall \, \mathbf{x} \in K \ ,$$

where C_0 is the constant appearing in the inverse inequality (8.4.7), *then the bilinear form $a_h^{(-1)}(\cdot, \cdot)$ associated to the DWG method satisfies the coerciveness inequality*

(8.4.16)
$$a_h^{(-1)}(v_h, v_h) \geq \frac{\varepsilon}{2}||\nabla v_h||_{0,\Omega}^2 + \frac{1}{2}||\mu^{1/2}v_h||_{0,\Omega}^2$$
$$+ \sum_{K \in \mathcal{T}_h} \delta \left(\frac{h_K}{|\mathbf{b}|}L_{SS}v_h, L_{SS}v_h\right)_K \quad \forall \, v_h \in V_h \ .$$

Proof. The stabilizing term can be rewritten as

$$\sum_{K \in \mathcal{T}_h} \delta \left(L_{SS}v_h + L_S v_h, \frac{h_K}{|\mathbf{b}|}(L_{SS}v_h - L_S v_h)\right)_K$$

$$= \sum_{K \in \mathcal{T}_h} \delta \left(\frac{h_K}{|\mathbf{b}|}L_{SS}v_h, L_{SS}v_h\right)_K - \sum_{K \in \mathcal{T}_h} \delta \left(\frac{h_K}{|\mathbf{b}|}L_S v_h, L_S v_h\right)_K \ .$$

Owing to (8.4.7), the second term can be estimated in the following way:

$$\sum_{K \in \mathcal{T}_h} \delta \left(\frac{h_K}{|\mathbf{b}|} L_S v_h, L_S v_h \right)_K$$

$$= \sum_{K \in \mathcal{T}_h} \delta \varepsilon^2 \left(\frac{h_K}{|\mathbf{b}|} \Delta v_h, \Delta v_h \right)_K + 2 \sum_{K \in \mathcal{T}_h} \delta \left(\frac{h_K}{|\mathbf{b}|} \varepsilon \Delta v_h, \mu v_h \right)_K$$

$$+ \sum_{K \in \mathcal{T}_h} \delta \left(\frac{h_K}{|\mathbf{b}|} \mu^2 v_h, v_h \right)_K$$

$$\leq 2 C_0 \varepsilon^2 \sum_{K \in \mathcal{T}_h} \delta \left(\frac{1}{|\mathbf{b}| h_K} \nabla v_h, \nabla v_h \right)_K + 2 \sum_{K \in \mathcal{T}_h} \delta \left(\frac{h_K}{|\mathbf{b}|} \mu^2 v_h, v_h \right)_K.$$

By proceeding in the same way as for the SUPG case, and taking into account (8.4.11), we therefore find

$$(8.4.17) \qquad \begin{aligned} a_h^{(-1)}(v_h, v_h) &\geq \varepsilon(1 - C_0 \delta) \|\nabla v_h\|_{0,\Omega}^2 \\ &+ \sum_{K \in \mathcal{T}_h} \delta \left(\frac{h_K}{|\mathbf{b}|} L_{SS} v_h, L_{SS} v_h \right)_K \\ &+ \sum_{K \in \mathcal{T}_h} \left(\mu \left[1 - 2\delta \frac{h_K}{|\mathbf{b}|} \mu \right] v_h, v_h \right)_K. \end{aligned}$$

The thesis follows from the assumptions on δ and h_K. $\square$

Clearly, this result at once entails the stability estimate (8.4.13) for the solution u_h to (8.3.13) with $\rho = -1$.

We are now going to provide an error estimate for the solution u_h to (8.3.13). For the sake of simplicity, we only consider the GALS case, corresponding to the choice $\rho = 1$ in (8.3.13). The proof of the convergence of the SUPG and DWG methods can be performed in a similar way. The interested reader can refer to Johnson, Nävert and Pitkäranta (1984) and Franca, Frey and Hughes (1992), respectively.

Proposition 8.4.4 *Assume that the space V_h satisfies the following approximation property: for each $v \in V \cap H^{k+1}(\Omega)$ there exists $\hat{v}_h \in V_h$ such that*

$$(8.4.18) \qquad \begin{aligned} \|v - \hat{v}_h\|_{0,K} + h_K \|\nabla(v - \hat{v}_h)\|_{0,K} + h_K^2 \|D^2(v - \hat{v}_h)\|_{0,K} \\ \leq C h_K^{k+1} |v|_{k+1,K} \end{aligned}$$

for each $K \in \mathcal{T}_h$. Assume moreover (8.4.3) is satisfied and that $0 < \delta \leq 2C_0^{-1}$, where C_0 is the constant appearing in the inverse inequality (8.4.7). Then the difference between the solution u of (8.1.7) and u_h of (8.3.13) (for $\rho = 1$) satisfies the following estimate

$$(8.4.19) \qquad ||u_h - u||_{GALS} \le Ch^{k+1/2}|u|_{k+1,\Omega} \ ,$$

provided that $u \in H^{k+1}(\Omega)$.

Proof. Let us write the error as $u_h - u = \sigma_h - \eta$, where

$$(8.4.20) \qquad \sigma_h := u_h - \hat{u}_h \ , \quad \eta := u - \hat{u}_h \ ,$$

and $\hat{u}_h \in V_h$ is the function related to u satisfying (8.4.18). If, for instance, $V_h = X_h^k \cap H_0^1(\Omega)$, then we can choose $\hat{u}_h = \pi_h^k(u)$, π_h^k being the finite element interpolation operator defined in (3.4.1) (see (3.4.14)).

We start by evaluating $||\sigma_h||_{GALS}$. From the strong consistency (8.3.6) we have

$$a_h^{(1)}(\sigma_h, \sigma_h) = a_h^{(1)}(u_h - u + \eta, \sigma_h) = a_h^{(1)}(\eta, \sigma_h) \ .$$

Therefore, using (8.4.14), we find

$$(8.4.21) \qquad ||\sigma_h||_{GALS}^2 = a_h^{(1)}(\eta, \sigma_h) \quad \forall \ \hat{u}_h \in V_h \ .$$

The term at the right hand side can be written as follows:

$$(8.4.22) \qquad \begin{aligned} &a_h^{(1)}(\eta, \sigma_h) \\ &= \varepsilon \int_\Omega \nabla\eta \cdot \nabla\sigma_h - \int_\Omega \eta \mathbf{b} \cdot \nabla\sigma_h + \int_\Omega a_0 \eta \sigma_h \\ &\quad + \sum_{K \in \mathcal{T}_h} \delta \left(L\eta, \frac{h_K}{|\mathbf{b}|} L\sigma_h \right)_K \\ &= \varepsilon(\nabla\eta, \nabla\sigma_h) - \sum_{K \in \mathcal{T}_h} (\eta, L\sigma_h)_K + 2(a_0\eta, \sigma_h) \\ &\quad + \sum_{K \in \mathcal{T}_h} (\eta, -\varepsilon\Delta\sigma_h)_K + \sum_{K \in \mathcal{T}_h} \delta \left(L\eta, \frac{h_K}{|\mathbf{b}|} L\sigma_h \right)_K \ . \end{aligned}$$

Let us estimate each term in (8.4.22). We first obtain:

$$(8.4.23) \qquad \varepsilon(\nabla\eta, \nabla\sigma_h) \le \frac{\varepsilon}{4}||\nabla\sigma_h||_{0,\Omega}^2 + \varepsilon||\nabla\eta||_{0,\Omega}^2 \ .$$

Next, we find

$$(8.4.24) \qquad \begin{aligned} -\sum_{K \in \mathcal{T}_h} (\eta, L\sigma_h)_K \le &\frac{1}{4} \sum_{K \in \mathcal{T}_h} \delta \left(\frac{h_K}{|\mathbf{b}|} L\sigma_h, L\sigma_h \right)_K \\ &+ \sum_{K \in \mathcal{T}_h} \left(\frac{|\mathbf{b}|}{\delta h_K} \eta, \eta \right)_K \ . \end{aligned}$$

For the third term we have

$$(8.4.25) \qquad 2(a_0\eta, \sigma_h) \le \frac{1}{2}||\mu^{1/2}\sigma_h||_{0,\Omega}^2 + 2||\mu^{-1/2}a_0\eta||_{0,\Omega}^2 \ .$$

Due to (8.4.8), the fourth term can be estimated in the following way:

$$\sum_{K \in \mathcal{T}_h} (\eta, -\varepsilon \Delta \sigma_h)_K \le \frac{1}{4} \sum_{K \in \mathcal{T}_h} \delta \varepsilon^2 \left(\frac{h_K}{|\mathbf{b}|} \Delta \sigma_h, \Delta \sigma_h \right)_K$$

(8.4.26)
$$+ \sum_{K \in \mathcal{T}_h} \left(\frac{|\mathbf{b}|}{\delta h_K} \eta, \eta \right)_K$$

$$\le \frac{\delta C_0 \varepsilon}{8} \|\nabla \sigma_h\|_{0,\Omega}^2 + \sum_{K \in \mathcal{T}_h} \left(\frac{|\mathbf{b}|}{\delta h_K} \eta, \eta \right)_K .$$

Finally, for the last term we can provide the following bound

$$\sum_{K \in \mathcal{T}_h} \delta \left(L\eta, \frac{h_K}{|\mathbf{b}|} L\sigma_h \right)_K \le \frac{1}{4} \sum_{K \in \mathcal{T}_h} \delta \left(\frac{h_K}{|\mathbf{b}|} L\sigma_h, L\sigma_h \right)_K$$

(8.4.27)
$$+ \sum_{K \in \mathcal{T}_h} \delta \left(\frac{h_K}{|\mathbf{b}|} L\eta, L\eta \right)_K .$$

In conclusion, as $0 < \delta \le 2C_0^{-1}$ from (8.4.21), (8.4.23)-(8.4.27) we obtain

$$\frac{1}{2} \|\sigma_h\|_{GALS}^2 \le \varepsilon \|\nabla \eta\|_{0,\Omega}^2 + 2 \sum_{K \in \mathcal{T}_h} \left(\frac{|\mathbf{b}|}{\delta h_K} \eta, \eta \right)_K$$

(8.4.28)
$$+ 2\|\mu^{-1/2} a_0 \eta\|_{0,\Omega}^2 + \sum_{K \in \mathcal{T}_h} \delta \left(\frac{h_K}{|\mathbf{b}|} L\eta, L\eta \right)_K .$$

Applying now (8.4.18) to the right hand side of (8.4.28) and using (8.4.8), we finally find

(8.4.29) $$\|\sigma_h\|_{GALS} \le C h^{k+1/2} |u|_{k+1,\Omega} , \quad u \in H^{k+1}(\Omega) ,$$

where C does not depend either on h or on Pe_K (C may however depend on several other quantities, such as μ_0 and μ_1 in (8.4.1), the constant C_0 in (8.4.7), $\|\mathbf{b}\|_{L^\infty(\Omega)}$ and $\|a_0\|_{L^\infty(\Omega)}$).

An analogous estimate clearly holds for $\|\eta\|_{GALS}$, too. Thus, by triangular inequality, we obtain the error estimate (8.4.19).   □

In particular, looking at the definition of the norm $\| \cdot \|_{GALS}$ in (8.4.14), this result shows that the estimate for the streamline derivative $|\mathbf{b}|^{-1}\mathbf{b} \cdot \nabla(u_h - u)$ with respect to the norm of $L^2(\Omega)$ is optimal, namely, it is of order k uniformly with respect to Pe_K. On the other hand, the norm of $u_h - u$ in $L^2(\Omega)$ is $O(h^{k+1/2})$, hence suboptimal by a power $1/2$. Finally, the $L^2(\Omega)$-norm of $\nabla(u_h - u)$ is not better than $O(h^k)$, as, in view of (8.4.3), ε is smaller than Ch.

Remark 8.4.1 Johnson, Nävert and Pitkäranta (1984) have derived localization results for non-smooth solutions having sharp internal and boundary layers. These results show that an error estimate like (8.4.19) holds outside a small neighbourhood of the layers. Notice that the standard Galerkin method for (8.3.1) does not allow similar local estimates, as the perturbing effect due to sharp layers may propagate in any direction.   □

Remark 8.4.2 For the sake of simplicity, in this Section we have dealt with the globally advection-dominated case (8.3.11). However, a method like (8.3.13) can be used also in the diffusion-dominated case or in a mixed case, substituting the function $\delta \operatorname{Pe}_K(\mathbf{x})$ to $\delta h_K |\mathbf{b}(\mathbf{x})|^{-1}$ in the stabilizing term when the local Péclet function satisfies $\operatorname{Pe}_K(\mathbf{x}) \leq 1$ (see, e.g., Hughes, Franca and Hulbert (1989), Franca, Frey and Hughes (1992)). In that case the method turns out to be $O(h^k)$ accurate in the GALS norm.   □

8.5 Some Numerical Results

The stiffness matrix A associated to the bilinear form $a_h^{(\rho)}(\cdot, \cdot)$ is positive definite (provided (8.4.1) holds) but not symmetric. A viable solution algorithm can therefore be based on iterative procedures like those described in Section 2.6. As the condition number of A scales dramatically with the grid-size h, using a preconditioner becomes unavoidable, at least for small h.

We present a few pictures related to the advection-diffusion problem (8.3.3) with $\mathbf{b} = (1, 1)$ and $\varepsilon \ll 1$ to be specified in the sequel. The boundary data for the solution u is equal to 1 on the upper side $\{x_2 = 1\}$, to 0 on $\{x_1 = 1\}$ and $\{x_2 = 0\}$ and finally on the side $\{x_1 = 0\}$ it takes the value 0 for $0 \leq x_2 \leq 1/3$, 1/2 for $1/3 < x_2 < 2/3$ and 1 for $2/3 \leq x_2 \leq 1$. The solution exhibits a double internal layer across the characteristic lines issuing from the discontinuity points on the side $\{x_1 = 0\}$. The domain is triangulated into $2n^2$ triangles of equal measure. Piecewise-quadratic finite elements have been used and the Bi-CGSTAB iterative procedure (see (2.6.42)-(2.6.53)) is applied. All pictures report the euclidean norm of the normalized residual versus the number of iterations.

Figs. 8.5.1 and 8.5.2 refer to the calculation corresponding to the value $\varepsilon = 10^{-2}$ and different values of n. The latter is obtained by the GALS method, while the former by the Galerkin one. In both cases the ILU preconditioner (see Section 2.5) has been used.

Figs. 8.5.3 and 8.5.4 report the convergence history when $n = 40, \varepsilon = 10^{-3}$ and different kinds of preconditioners are used: the diagonal preconditioner, the S.O.R. preconditioner or the ILU preconditioner. Fig. 8.5.3 refers to a Galerkin approximation while Fig. 8.5.4 to a GALS approximation.

Finally, Fig. 8.5.5 shows the convergence history for the GALS approximation, still using Bi-CGSTAB iterations with ILU preconditioner, for $n = 40$ and three different values of ε. The interesting conclusion is that the method converges very quickly and its rate is practically independent of ε.

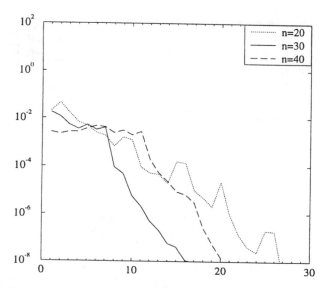

Fig. 8.5.1. Convergence history of the Galerkin method when $\varepsilon = 10^{-2}$ and different values of the grid-size are used

8.6 The Heterogeneous Method

Here, we consider another approach which can be utilized for the approximation of advection-diffusion problems like (8.3.1) where advection phenomena are dominant.

To be precise, let us suppose that the operator L associated with (8.3.1) is the one given in (8.1.1), that we rewrite in the following way

$$Lz = L_1 z + L_2 z \quad ,$$

where we have set

(8.6.1)
$$L_2 z := -\sum_{i,j=1}^{d} D_i(a_{ij} D_j z)$$

(8.6.2)
$$L_1 z := \sum_{i=1}^{d} D_i(b_i z) + a_0 z \quad .$$

We assume that the function $\alpha_0(\mathbf{x})$ defined in (8.3.12) satisfies

(8.6.3)
$$0 < \alpha_0 \leq \alpha_0(\mathbf{x}) \leq \alpha^* \quad \text{a.e. in } \Omega$$

and that (6.1.20) holds, so that the bilinear form $a(\cdot, \cdot)$ associated to L is coercive in $H_0^1(\Omega)$. We assume, moreover, that the constant α^* is very small compared to the size of $\mathbf{b}$, say

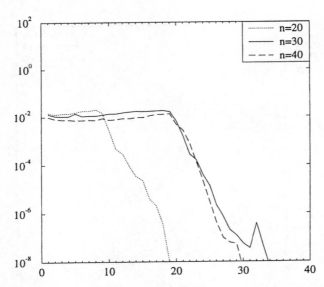

Fig. 8.5.2. Convergence history of the GALS method when $\varepsilon = 10^{-2}$ and different values of the grid-size are used

$$\hat{\alpha} := \frac{\alpha^*}{||\mathbf{b}||_{L^\infty(\Omega)}} \ll 1 \ .$$

As a consequence, we can expect boundary or internal layers of width proportional to $\hat{\alpha}$ or $\sqrt{\hat{\alpha}}$, respectively.

For the sake of definiteness, let us concentrate on the case of boundary layers. If their location is approximatively known in advance (which is very often the case), one may consider the complete operator L only in a neighbourhood of the layer (i.e., where the effects of diffusive terms depending on $\alpha_0(\mathbf{x})$ are really relevant). In the remaining region, the *reduced* operator L_1 can be used instead of L, getting rid of the contribution due to viscosity.

The operator L_1 is a first-order hyperbolic one, which falls under the family of operators considered in Chapter 14. Let us just notice that, in several situations, the determination of a solution to the hyperbolic problem $L_1 u = f$ requires less computational effort than the one needed to solve the complete problem $Lu = f$. Moreover, far from the boundary layer it is often possible to employ a coarse mesh for the hyperbolic problem without affecting accuracy, thus limiting the use of a fine one near the sharp layer.

Clearly, this approach requires that the computational domain Ω is split into two disjoint regions Ω_1 and Ω_2 with $\Omega_1 \cap \Omega_2 = \emptyset$ and $\overline{\Omega_1} \cup \overline{\Omega_2} = \overline{\Omega}$. Thus, suitable matching conditions between the solution of the complete problem in Ω_2 and the one of the reduced problem in Ω_1 need to be prescribed on the interface $\Gamma := \partial\Omega_1 \cap \partial\Omega_2$.

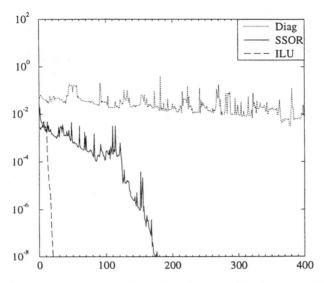

Fig. 8.5.3. Convergence history of the Galerkin method when $\varepsilon = 10^{-3}$, $n = 40$ and different types of preconditioners are used

For the sake of simplicity, let us restrict to the homogeneous Dirichlet problem (8.3.1). In Gastaldi, Quarteroni and Sacchi Landriani (1990) the following *coupled problem* has been proposed:

$$(8.6.4) \quad \begin{cases} Lu_2 = f & \text{in } \Omega_2 \\[2mm] L_1 u_1 = f & \text{in } \Omega_1 \\[2mm] u_2 = u_1 & \text{on } \Gamma^{in} \\[2mm] \dfrac{\partial u_2}{\partial n_L} = -\mathbf{b} \cdot \mathbf{n}\, u_1 & \text{on } \Gamma \\[2mm] u_2 = 0 & \text{on } \partial\Omega_2 \cap \partial\Omega \\[2mm] u_1 = 0 & \text{on } \partial\Omega_1^{in} \cap \partial\Omega \ , \end{cases}$$

where $\mathbf{n} = \mathbf{n}(\mathbf{x})$ is the unit outward normal vector on $\partial\Omega_1$, and $\frac{\partial u_2}{\partial n_L}$ denotes the conormal derivative of u_2 with respect to $\mathbf{n}$, i.e.,

$$\frac{\partial u_2}{\partial n_L} = \sum_{i,j=1}^{d} a_{ij}\, D_j u_2\, n_i - \mathbf{b} \cdot \mathbf{n}\, u_2 \ .$$

Finally,

$$\partial\Omega_1^{in} := \{\mathbf{x} \in \partial\Omega_1 \,|\, \mathbf{b}(\mathbf{x}) \cdot \mathbf{n}(\mathbf{x}) < 0\}$$

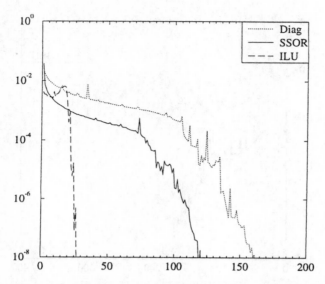

Fig. 8.5.4. Convergence history of the GALS method when $\varepsilon = 10^{-3}$, $n = 40$ and different types of preconditioners are used

is the inflow boundary of the domain Ω_1, and $\Gamma^{in} := \partial \Omega_1^{in} \cap \Gamma$ (see Fig. 8.6.1).

The matching condition $(8.6.4)_3$ states that continuity only occurs on Γ^{in}, while $(8.6.4)_4$ enforces the flux balance on the whole interface Γ.

The coupled problem (8.6.4) can be seen as a domain decomposition formulation for the solution of the complete equation (8.3.1) (see Remark 6.4.1). Due to the use of different operators, these approaches are referred to as *heterogeneous* domain decomposition methods (see, e.g., Quarteroni, Pasquarelli and Valli (1992)).

Notice that the interface conditions in (8.6.4) are the limit of those pertaining the fully elliptic problem

$$(8.6.5) \qquad\qquad L_\varepsilon u_\varepsilon = f \quad \text{in} \ \Omega \ ,$$

where $\varepsilon > 0$

$$(8.6.6) \qquad L_\varepsilon z := - \sum_{i,j=1}^{d} D_i(a_{ij}^\varepsilon D_j z) + \sum_{i=1}^{d} D_i(b_i z) + a_0 z \ ,$$

and

$$(8.6.7) \qquad\qquad a_{ij}^\varepsilon(\mathbf{x}) := \begin{cases} a_{ij}(\mathbf{x}) & , \ \mathbf{x} \in \Omega_2 \\ \varepsilon & , \ \mathbf{x} \in \Omega_1 \end{cases} .$$

In fact, in this case the solution u_ε to $L_\varepsilon u_\varepsilon = f$ in Ω satisfies on Γ

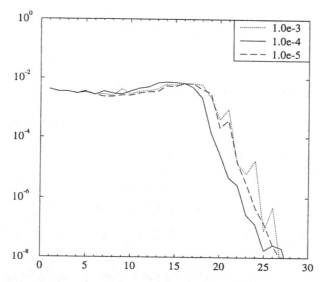

Fig. 8.5.5. Convergence history of the GALS method when $n = 40$, for different values of ε

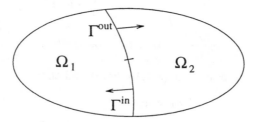

Fig. 8.6.1. Partition of the domain Ω

$$(8.6.8) \qquad \begin{cases} u_{\varepsilon|\Omega_2} = u_{\varepsilon|\Omega_1} \\[2mm] \dfrac{\partial u_{\varepsilon|\Omega_2}}{\partial n_L} = \varepsilon \dfrac{\partial u_{\varepsilon|\Omega_1}}{\partial n} - \mathbf{b} \cdot \mathbf{n}\, u_{\varepsilon|\Omega_1} \ . \end{cases}$$

In the limiting process, the continuity of the solution is mantained only on Γ^{in}. However, it can be proven that the size of the jump on $\Gamma \setminus \Gamma^{in}$ is as small as the value of $\max_\Gamma \alpha_0(\mathbf{x})$.

Now, let us come to the finite dimensional approximation of (8.6.4). We first of all introduce the bilinear forms

$$(8.6.9) \qquad a_2(z, v) := \int_{\Omega_2} \left[\sum_{i,j=1}^d a_{ij} D_j z D_i v - \sum_{i=1}^d b_i z D_i v + a_0 z v \right]$$

$$(8.6.10) \quad a_1(z,v) := \int_{\Omega_1} \left[-\sum_{i=1}^{d} b_i z D_i v + a_0 z v \right] .$$

We then rewrite (8.6.4) as: find $u_2 \in H^1_{\partial\Omega_2 \setminus \Gamma}(\Omega_2)$, $u_1 \in L^2(\Omega_1)$ such that

$$(8.6.11) \quad \begin{cases} a_2(u_2, v_2) - (\mathbf{b} \cdot \mathbf{n}\, u_2, v_2)_{\Gamma^{in}} \\ \qquad = (f, v_2)_{\Omega_2} + (\mathbf{b} \cdot \mathbf{n}\, u_1, v_2)_{\Gamma^{out}} \quad \forall\, v_2 \in H^1_{\partial\Omega_2 \setminus \Gamma}(\Omega_2) \\ a_1(u_1, v_1) + (\mathbf{b} \cdot \mathbf{n}\, u_1, v_1)_{\partial\Omega_1^{out}} \\ \qquad = (f, v_1)_{\Omega_1} - (\mathbf{b} \cdot \mathbf{n}\, u_2, v_1)_{\Gamma^{in}} \quad \forall\, v_1 \in H^1(\Omega_1) \quad, \end{cases}$$

where $\Gamma^{out} := \Gamma \setminus \Gamma^{in}$, $\partial\Omega_1^{out} := \partial\Omega_1 \setminus \partial\Omega_1^{in}$ and

$$H^1_{\partial\Omega_2 \setminus \Gamma}(\Omega_2) := \{ v \in H^1(\Omega_2) \,|\, v = 0 \text{ on } \partial\Omega_2 \setminus \Gamma \}$$

(see Section 1.3).

Notice that through $(8.6.11)_1$ we are imposing that $(8.6.4)_1$ is satisfied and moreover that

$$(8.6.12) \qquad \sum_{i,j=1}^{d} a_{ij}(D_j u_2) n_i = 0 \quad \text{on } \Gamma^{in}$$

$$(8.6.13) \qquad \frac{\partial u_2}{\partial n_L} = -\mathbf{b} \cdot \mathbf{n}\, u_1 \quad \text{on } \Gamma^{out} ,$$

whereas $(8.6.11)_2$ implies that $(8.6.4)_2$ holds and that

$$(8.6.14) \qquad u_1 = u_2 \quad \text{on } \Gamma^{in} .$$

The interface conditions (8.6.12)-(8.6.14) are clearly equivalent to $(8.6.4)_2$ and $(8.6.4)_3$.

Notice, moreover, that from $(8.6.11)_2$ u_1 indeed satisfies $\text{div}(\mathbf{b} u_1) \in L^2(\Omega_1)$, thus the trace $\mathbf{b} \cdot \mathbf{n}\, u_1$ on $\partial\Omega_1$ is well defined (see Theorem 1.3.2).

The solution of (8.6.11) can be effectively obtained by an iteration-by-subdomain procedure, which alternates the solution of the hyperbolic equation and of the elliptic one within the respective regions. This strategy is useful in view of the numerical computations, since it permits to use different solvers (and even different spatial discretizations) in the two regions.

Let us present a finite dimensional approximation of (8.6.11) based on the Galerkin method. If $V_{2,h}$ and $V_{1,h}$ are suitable finite dimensional subspaces of $H^1_{\partial\Omega_2 \setminus \Gamma}(\Omega_2)$ and $H^1(\Omega_1)$, we can write the discretized problem as: find $u_{2,h} \in V_{2,h}$ and $u_{1,h} \in V_{1,h}$ such that

$$(8.6.15) \quad \begin{cases} a_2(u_{2,h}, v_{2,h}) - (\mathbf{b} \cdot \mathbf{n}\, u_{2,h}, v_{2,h})_{\Gamma^{in}} \\ \qquad = (f, v_{2,h})_{\Omega_2} + (\mathbf{b} \cdot \mathbf{n}\, u_{1,h}, v_{2,h})_{\Gamma^{out}} \quad \forall\, v_{2,h} \in V_{2,h} \\ a_1(u_{1,h}, v_{1,h}) + (\mathbf{b} \cdot \mathbf{n}\, u_{1,h}, v_{1,h})_{\partial\Omega_1^{out}} \\ \qquad = (f, v_{1,h})_{\Omega_1} - (\mathbf{b} \cdot \mathbf{n}\, u_{2,h}, v_{1,h})_{\Gamma^{in}} \quad \forall\, v_{1,h} \in V_{1,h} \quad . \end{cases}$$

An effective iterative procedure for the solution of this problem reads as follows: being given any initial guess $u_{2,h}^0 \in V_{2,h}$, for $k \geq 1$ first solve

$$(8.6.16) \quad \begin{aligned} a_1(u_{1,h}^k, v_{1,h}) &+ (\mathbf{b} \cdot \mathbf{n}\, u_{1,h}^k, v_{1,h})_{\partial\Omega_1^{out}} \\ &= (f, v_{1,h})_{\Omega_1} - (\mathbf{b} \cdot \mathbf{n}\, u_{2,h}^{k-1}, v_{1,h})_{\Gamma^{in}} \quad \forall\, v_{1,h} \in V_{1,h} \quad , \end{aligned}$$

then solve

$$(8.6.17) \quad \begin{aligned} a_2(u_{2,h}^k, v_{2,h}) &- (\mathbf{b} \cdot \mathbf{n}\, u_{2,h}^k, v_{2,h})_{\Gamma^{in}} \\ &= (f, v_{2,h})_{\Omega_2} + (\mathbf{b} \cdot \mathbf{n}\, u_{1,h}^k, v_{2,h})_{\Gamma^{out}} \quad \forall\, v_{2,h} \in V_{2,h} \quad . \end{aligned}$$

Both these problems are solvable, as the associated matrices are positive definite.

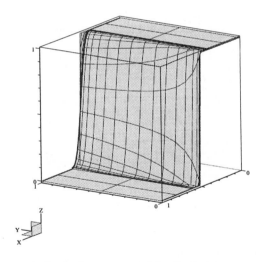

Fig. 8.6.2. The numerical solution computed by the heterogeneous method for the first test problem with an internal layer

The convergence of this iterative scheme has been proven in the two-dimensional case by Gastaldi, Quarteroni and Sacchi Landriani (1990) for the spectral collocation method, and the rate of convergence turns out to be independent of the discretization parameter h. An acceleration parameter

$\theta_k > 0$ can be also utilized, substituting $u_{2,h}^{k-1}$ at the right hand side of (8.6.16) by $\lambda_h^{k-1} := \theta_k u_{2,h}^{k-1} + (1 - \theta_k)u_{1,h}^{k-1}$ (having chosen the additional initial value $u_{1,h}^0 \in V_{1,h}$, too).

A thorough presentation of the practical performance of this method can be found in Quarteroni, Pasquarelli and Valli (1992) and Frati, Pasquarelli and Quarteroni (1993).

A different approach to the elliptic-hyperbolic coupling, which is called the χ-*formulation*, has been proposed by Brezzi, Canuto and Russo (1989) (see also Canuto and Russo (1993) and the references therein).

A couple of numerical results obtained by the heterogeneous method (subdividing $\Omega \subset \mathbb{R}^2$ in more than two domains) are reported below. Fig. 8.6.2 is the counterpart of Fig. 8.3.3 (a). The domain Ω has been subdivided into nine subdomains. The complete elliptic problem (with $\varepsilon = 10^{-3}$) is solved only in the three domains which contain the internal layer. In the remaining subdomains ε is set to 0. The numerical solution is achieved by the Legendre spectral collocation method, using 874 collocation points altogether.

Fig. 8.6.3 refers to the case in which the advective field $\mathbf{b}$ is given by (x_1, x_2), $\varepsilon = 10^{-4}$ and Ω is subdivided into nine subdomains: one large internal "hyperbolic" subdomain, and eight thin domains next to the boundary, in which the whole elliptic problem is solved. The collocation points we have used are 9 in the "hyperbolic" domain, 49 in each corner domain and 21 in anyone of the remaining four "boundary" domains.

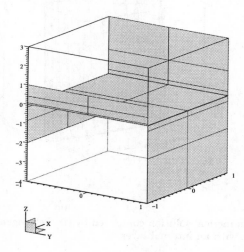

Fig. 8.6.3. The numerical solution computed by the heterogeneous method for the second test problem having four boundary layers

9. The Stokes Problem

This Chapter is devoted to the Stokes system of differential equations, that can be written as

(9.1)
$$\begin{cases} a_0\mathbf{u} - \nu\Delta\mathbf{u} + \nabla p = \mathbf{f} & \text{in } \Omega \subset \mathbb{R}^d \ , \ d = 2,3 \\[2mm] \text{div } \mathbf{u} = 0 & \text{in } \Omega \\[2mm] \mathbf{u} = 0 & \text{on } \partial\Omega \ , \end{cases}$$

where $a_0 \geq 0$ and $\nu > 0$ are given constants, $\mathbf{f} : \Omega \to \mathbb{R}^d$ is a given function, while $\mathbf{u} : \Omega \to \mathbb{R}^d$ and $p : \Omega \to \mathbb{R}$ are the problem unknowns.

Equations (9.1) can be obtained as part of the solution process for the Navier-Stokes equations of fluid mechanics, for a flow confined in Ω (for instance, adopting semi-implicit schemes or else splitting methods for time-advancing). In such cases, almost invariably a_0 plays the role of the inverse of the time-step. A complete derivation of the Navier-Stokes equations is given in Section 13.1. Another instance is the case of slow motion of fluids with very high viscosity (low Reynolds number), in which case $a_0 = 0$, and $\mathbf{f}$ represents the external forces (per unit mass). In both cases, p represents the ratio between the flow pressure and density, and $\mathbf{u}$ the velocity.

In (9.1) we are considering the classical homogeneous Dirichlet boundary condition. Other conditions involving the normal derivative of $\mathbf{u}$ or a linear combination between the latter and p on $\partial\Omega$ are also admissible, as we will point out in Section 10.1.1.

9.1 Mathematical Formulation and Analysis

We assume that $\mathbf{f} \in (L^2(\Omega))^d$, and set $\mathcal{V} := \{\mathbf{v} \in \mathcal{D}(\Omega) | \text{div } \mathbf{v} = 0\}$ and $V := (H_0^1(\Omega))^d$, which is a Hilbert space. From the first equation of (9.1) and Green formulae (5.1.2) and (5.1.5) we obtain

(9.1.1) $\qquad a_0(\mathbf{u}, \mathbf{v}) + \nu(\nabla\mathbf{u}, \nabla\mathbf{v}) - (p, \text{div } \mathbf{v}) = (\mathbf{f}, \mathbf{v}) \quad \forall \, \mathbf{v} \in (\mathcal{D}(\Omega))^d$

and therefore

(9.1.2) $\qquad a_0(\mathbf{u}, \mathbf{v}) + \nu(\nabla \mathbf{u}, \nabla \mathbf{v}) = (\mathbf{f}, \mathbf{v}) \qquad \forall\, \mathbf{v} \in \mathcal{V}$.

We define the space

(9.1.3) $\qquad\qquad\qquad V_{\mathrm{div}} := \{\mathbf{v} \in V \,|\, \mathrm{div}\,\mathbf{v} = 0\}$,

which is a closed subspace of V, and, by virtue of Poincaré inequality (1.3.2), it is a Hilbert space for the norm

(9.1.4) $\qquad\qquad\qquad\qquad ||\mathbf{v}|| := ||\nabla \mathbf{v}||_0$.

It follows that the bilinear form

(9.1.5) $\qquad a(\mathbf{w}, \mathbf{v}) := a_0(\mathbf{w}, \mathbf{v}) + \nu(\nabla \mathbf{w}, \nabla \mathbf{v})$, $\mathbf{w}, \mathbf{v} \in V$,

is coercive over $V_{\mathrm{div}} \times V_{\mathrm{div}}$. Since, in addition, the application $\mathbf{v} \to (\mathbf{f}, \mathbf{v})$ is linear and continuous over V_{div}, we conclude by virtue of the Lax-Milgram lemma that the problem

(9.1.6) $\qquad$ find $\mathbf{u} \in V_{\mathrm{div}} : a(\mathbf{u}, \mathbf{v}) = (\mathbf{f}, \mathbf{v}) \qquad \forall\, \mathbf{v} \in V_{\mathrm{div}}$

admits a unique solution.

We will make use of the following result.

Lemma 9.1.1 *Let Ω be a bounded domain in $\mathbb{R}^d$ with a Lipschitz continuous boundary, and let L be an element of V' (i.e., a linear continuous functional on V). Then L vanishes identically on V_{div} if and only if there exists a function $p \in L^2(\Omega)$ such that*

(9.1.7) $\qquad\qquad L(\mathbf{v}) = (p, \mathrm{div}\,\mathbf{v}) \qquad \forall\, \mathbf{v} \in V$.

Further, (9.1.7) defines a unique function p up to an additive constant.

The proof of this lemma, that we omit for the sake of brevity, is reported, e.g., in Girault and Raviart (1986), p. 22.

We are now able to prove

Theorem 9.1.1 *Let Ω be a bounded domain in $\mathbb{R}^d$ with a Lipschitz continuous boundary, and for each $\mathbf{f} \in (L^2(\Omega))^d$ let $\mathbf{u}$ be the solution to (9.1.6). Then there exists a function $p \in L^2(\Omega)$, which is unique up to an additive constant, such that*

(9.1.8) $\qquad a(\mathbf{u}, \mathbf{v}) - (p, \mathrm{div}\,\mathbf{v}) = (\mathbf{f}, \mathbf{v}) \qquad \forall\, \mathbf{v} \in V$.

Proof. Let $\mathbf{u}$ be the solution to (9.1.6), and define

$$L(\mathbf{v}) := a(\mathbf{u}, \mathbf{v}) - (\mathbf{f}, \mathbf{v}) \qquad \forall\, \mathbf{v} \in V$$.

By (9.1.6) L satisfies the assumptions of Lemma 9.1.1; thus there exists $p \in L^2(\Omega)$ such that the couple $(\mathbf{u}, p)$ is a solution to (9.1.8).   □

Let us present now a different approach. If we define the Hilbert space $Q = L_0^2(\Omega)$, with

$$(9.1.9) \qquad L_0^2(\Omega) := \left\{ q \in L^2(\Omega) \mid \int_\Omega q = 0 \right\}$$

and the bilinear form

$$(9.1.10) \qquad b(\mathbf{v}, q) := -(q, \operatorname{div} \mathbf{v}) \ , \quad \mathbf{v} \in V \ , \ q \in Q \ ,$$

from Theorem 9.1.1 we deduce that the problem

$$(9.1.11) \qquad \begin{cases} \text{find } \mathbf{u} \in V \ , \ p \in Q : \\[2mm] a(\mathbf{u}, \mathbf{v}) + b(\mathbf{v}, p) = (\mathbf{f}, \mathbf{v}) \quad \forall \, \mathbf{v} \in V \\[2mm] b(\mathbf{u}, q) = 0 \qquad\qquad\quad \forall \, q \in Q \end{cases}$$

has a unique solution (the pressure is still defined up to an additive constant).

It remains to show that the solution to (9.1.11) is also a solution to (9.1). This can be done using a classical density argument. Since $(\mathcal{D}(\Omega))^d$ is dense in V (i.e., V is the closure of $(\mathcal{D}(\Omega))^d$ with respect to the $(H^1(\Omega))^d$-norm), we can take any $\mathbf{v} \in (\mathcal{D}(\Omega))^d$ in (9.1.11) and find, integrating by parts, the first equation of (9.1) satisfied in the sense of distributions. In fact, such an equation holds almost everywhere as $\mathbf{f} \in (L^2(\Omega))^d$. Finally, the last equation of (9.1.11) ensures that $\operatorname{div} \mathbf{u} = 0$ almost everywhere, as $\operatorname{div} \mathbf{u} \in Q$ and it is orthogonal to all functions of Q.

From now on, we will refer to (9.1.11) as the *weak formulation* of the Stokes equations.

Remark 9.1.1 *(An alternative choice of the pressure space).* From (9.1) and the preceding analysis, it is clear that the pressure p is a function of $L^2(\Omega)$ which is defined up to an additive constant only. A way to remove this uncertainty is to look for a pressure with a vanishing average in Ω, i.e., belonging to the space $L_0^2(\Omega)$ defined in (9.1.9). This choice is suitable, especially in view of approximating (9.1.11) by spectral methods. Indeed, to find a discrete pressure with zero average one simply needs to disregard its lowest Fourier coefficient, i.e., the one associated with the constant polynomial (see Chapter 4).

However, for finite dimensional approximations of *local* type, such as those based upon finite elements or else finite differences, the global condition on the average may be unsuitable to deal with. In such cases, the pressure space Q can be replaced by $L^2(\Omega) \setminus \mathbb{R}$, so that, at the discrete level, one can, for instance, look for a pressure whose value at a given point of the domain Ω is a priori prescribed. $\qquad\square$

Remark 9.1.2 *(Saddle-point formulation).* Consider the Lagrangian functional

$$(9.1.12) \qquad \mathcal{L}(\mathbf{v}, q) := \frac{1}{2} a(\mathbf{v}, \mathbf{v}) + b(\mathbf{v}, q) - (\mathbf{f}, \mathbf{v}) \ , \quad \mathbf{v} \in V \ , \ q \in Q \ .$$

The solution $(\mathbf{u}, p)$ of the Stokes problem (9.1.11) is a saddle-point of the above functional, i.e.,

$$(9.1.13) \qquad\qquad \mathcal{L}(\mathbf{u}, p) = \min_{\mathbf{v} \in V} \max_{q \in Q} \mathcal{L}(\mathbf{v}, q)$$

and conversely. For the proof see, e.g., Temam (1984), Glowinski and Le Tallec (1989), and also Section 7.1.2. In this regard, p is the Lagrangian multiplier associated with the divergence-free constraint. $\qquad\square$

9.2 Galerkin Approximation

The analysis carried out in the previous Section introduces a couple of different weak formulations for the Stokes problem (9.1). The first one is given in (9.1.6): solving that problem provides the velocity field $\mathbf{u}$. Next, the pressure p can be recovered from (9.1.8) or any other method that allows for the determination of the pressure once the velocity is available. The other formulation is that given in (9.1.11), which in turns is equivalent to the saddle-point problem (9.1.13).

The two formulations suggest two different paths to solve numerically the Stokes problem. Let us start from (9.1.6) which is simpler to analyze, although most part of our discussion will be devoted to the numerical approximation to (9.1.11). Following the guidelines of Section 5.2, the Galerkin approximation of (9.1.6) reads

$$(9.2.1) \qquad \text{find } \mathbf{u}_h \in V_{\text{div},h} : a(\mathbf{u}_h, \mathbf{v}_h) = (\mathbf{f}, \mathbf{v}_h) \qquad \forall \ \mathbf{v}_h \in V_{\text{div},h} \ ,$$

where $\{V_{\text{div},h}\}$ is a family of finite dimensional subspaces of V_{div} satisfying the consistency assumption

$$(9.2.2) \qquad \forall \ \mathbf{v} \in V_{\text{div}} \quad \inf_{\mathbf{v}_h \in V_{\text{div},h}} \|\mathbf{v} - \mathbf{v}_h\| \to 0 \ \text{ as } \ h \to 0 \ .$$

A straightforward application of Theorem 5.2.1 yields existence and uniqueness of the solution to (9.2.1); such a solution is stable and converges to the solution $\mathbf{u}$ of (9.1.6). To be precise, there exists a constant C independent of h such that

(9.2.3) $$||\mathbf{u} - \mathbf{u}_h|| \le C \inf_{\mathbf{v}_h \in V_{\mathrm{div},h}} ||\mathbf{u} - \mathbf{v}_h|| \; .$$

Despite the mathematical elegance of this result, problem (9.2.1) is scarcely used in the practice to approximate the velocity field. The reason is twofold. Firstly, it is quite difficult to find internal approximations $V_{\mathrm{div},h}$ to the space (9.1.3) (i.e., a finite dimensional space of divergence-free vector valued functions), which are "rich enough" to find from (9.2.3) a good convergence behaviour with respect to h. Secondly, even when the former task has been successfully accomplished, the construction of a basis for $V_{\mathrm{div},h}$ may be very difficult in practice, although some successful examples have been provided in the literature, especially for spectral approximations (see, e.g., Moser, Moin and Leonard (1983)). In finite elements, it can be useful to modify the finite dimensional space in (9.2.1), as we show in Remark 9.2.1.

We now turn to (9.1.11). This problem has a familiar appearance, as it was encountered in Chapter 7 already. Following the guidelines of Section 7.4, we introduce two families of finite dimensional subspaces $V_h \subset V$ and $Q_h \subset Q$ depending on h. Then, we approximate (9.1.11) with the discrete problem

(9.2.4)
$$\begin{cases} \text{find } \mathbf{u}_h \in V_h \;, \; p_h \in Q_h : \\[2mm] a(\mathbf{u}_h, \mathbf{v}_h) + b(\mathbf{v}_h, p_h) = (\mathbf{f}, \mathbf{v}_h) \quad \forall \, \mathbf{v}_h \in V_h \\[2mm] b(\mathbf{u}_h, q_h) = 0 \qquad\qquad\qquad \forall \, q_h \in Q_h \;\; . \end{cases}$$

For the analysis of this problem, let us set

(9.2.5) $$Z_h := \{\mathbf{v}_h \in V_h |\; (q_h, \mathrm{div}\,\mathbf{v}_h) = 0 \text{ for each } q_h \in Q_h\} \; .$$

This is the space of discretely divergence-free functions associated with the finite dimensional spaces. The bilinear form $a(\cdot, \cdot)$ turns out to be coercive in Z_h as it is coercive in V and Z_h is a subspace of V, i.e., there exists $\alpha > 0$ such that

(9.2.6) $$a_0\,||\mathbf{v}_h||_0^2 + \nu\,||\nabla \mathbf{v}_h||_0^2 \ge \alpha\,||\mathbf{v}_h||^2 \; \forall \, \mathbf{v}_h \in Z_h \; .$$

Moreover, still using Poincaré inequality (1.3.2), there exist $\gamma > 0$ and $\delta > 0$ such that

(9.2.7) $$|a_0(\mathbf{w}, \mathbf{v}) + \nu(\nabla \mathbf{w}, \nabla \mathbf{v})| \le \gamma\,||\mathbf{w}||\,||\mathbf{v}|| \quad \forall \, \mathbf{w}, \mathbf{v} \in V \; ,$$

i.e., the bilinear form $a(\cdot, \cdot)$ is continuous over $V \times V$, and

(9.2.8) $$|(q, \mathrm{div}\,\mathbf{v})| \le \delta\,||\mathbf{v}||\,||q||_0 \quad \forall \, \mathbf{v} \in V \;, \; q \in Q \; ,$$

which in turn expresses the continuity of the bilinear form $b(\cdot, \cdot)$ over $V \times Q$.

Moreover, let us assume that the spaces V_h and Q_h enjoy the following property: there exists $\beta > 0$ such that

(9.2.9) $\forall\, q_h \in Q_h\ \exists\, \mathbf{v}_h \in V_h\ ,\ \mathbf{v}_h \neq \mathbf{0} : (q_h, \operatorname{div} \mathbf{v}_h) \geq \beta\,||\mathbf{v}_h||\,||q_h||_0$.

This is called *compatibility* (or *inf-sup* or *Ladyzhenskaya-Babuška-Brezzi) condition.*

Under these assumptions, Theorem 7.4.2 yields existence and uniqueness for the solution to (9.2.4). Further, this solution satisfies

(9.2.10) $||\mathbf{u}_h|| \leq \dfrac{1}{\alpha}\,||\mathbf{f}||_{V'}$, $||p_h||_0 \leq \dfrac{1}{\beta}\left(1 + \dfrac{\gamma}{\alpha}\right)||\mathbf{f}||_{V'}$.

These estimates state that the solution is stable if β is independent of h. When the latter condition is not true, or, even worse, (9.2.9) doesn't hold, the approximation (9.2.4) is said to be unstable.

The abstract convergence result given by Theorem 7.4.3 yields in the present case

(9.2.11) $||\mathbf{u} - \mathbf{u}_h|| \leq \left(1 + \dfrac{\gamma}{\alpha}\right) \inf_{\mathbf{v}_h \in Z_h} ||\mathbf{u} - \mathbf{v}_h|| + \dfrac{\delta}{\alpha} \inf_{q_h \in Q_h} ||p - q_h||_0$

and

(9.2.12)
$$||p - p_h||_0 \leq \dfrac{\gamma}{\beta}\left(1 + \dfrac{\gamma}{\alpha}\right) \inf_{\mathbf{v}_h \in Z_h} ||\mathbf{u} - \mathbf{v}_h||$$
$$+ \left(1 + \dfrac{\delta}{\beta} + \dfrac{\gamma\delta}{\alpha\beta}\right) \inf_{q_h \in Q_h} ||p - q_h||_0 \ .$$

Because
$$\inf_{\mathbf{v}_h \in Z_h} ||\mathbf{u} - \mathbf{v}_h|| \leq \left(1 + \dfrac{\delta}{\beta}\right) \inf_{\mathbf{v}_h \in V_h} ||\mathbf{u} - \mathbf{v}_h|| \ ,$$

due to (7.4.26), the convergence estimate is optimal provided β is independent of h.

Remark 9.2.1 *(Non-conforming approximation).* With the aim of removing the difficulty of construction of subspaces to $V_{\operatorname{div},h}$, we could consider, instead of (9.2.1), the problem

(9.2.13)   find $\mathbf{u}_h \in Z_h : a(\mathbf{u}_h, \mathbf{v}_h) = (\mathbf{f}, \mathbf{v}_h)$ $\forall\, \mathbf{v}_h \in Z_h$

(see (9.2.5)). Since, generally, Z_h is not a subspace of V_{div} (as a matter of fact, the elements of Z_h are not necessarily divergence-free), (9.2.1) would provide a *non-conforming* approximation to (9.1.6).

The main interest of using a non-conforming finite dimensional space Z_h is that it might be simpler to construct a set of basis functions for it. Some examples are provided in Hecht (1981). A proof of convergence can still be provided (as indicated in Remark 5.5.1), if we assume that the consistency property (9.2.2) does hold even with Z_h instead of $V_{\operatorname{div},h}$.   □

9.2.1 Algebraic Form of the Stokes Problem

Let us begin with problem (9.2.1). For any fixed h, let N_h^0 denote the dimension of the subspace $V_{\mathrm{div},h}$ and let $\{\boldsymbol{\varphi}_j^0 \,|\, j = 1, ..., N_h^0\}$ be a basis for $V_{\mathrm{div},h}$. Setting

$$\mathbf{u}_h(\mathbf{x}) = \sum_{j=1}^{N_h^0} \xi_j \boldsymbol{\varphi}_j^0(\mathbf{x}) \ ,$$

the linear system associated with (9.2.1) becomes

$$A^0 \boldsymbol{\xi} = \mathbf{f} \quad \text{with} \quad f_i := (\mathbf{f}, \boldsymbol{\varphi}_i^0) \ , \ A_{ij}^0 := a(\boldsymbol{\varphi}_j^0, \boldsymbol{\varphi}_i^0) \ .$$

The matrix A^0 is symmetric and positive definite.

A less simple situation occurs for problem (9.2.4). Now we denote by N_h and K_h the dimensions of V_h and Q_h, respectively, and by $\{\boldsymbol{\varphi}_j \,|\, j = 1, ..., N_h\}$ and $\{\psi_l \,|\, l = 1, ..., K_h\}$ the bases for V_h and Q_h, respectively. If we set

$$\mathbf{u}_h(\mathbf{x}) = \sum_{j=1}^{N_h} u_j \boldsymbol{\varphi}_j(\mathbf{x}) \ , \ p_h(\mathbf{x}) = \sum_{l=1}^{K_h} p_l \psi_l(\mathbf{x})$$

the linear system associated with (9.2.4) takes the following form

$$(9.2.14) \qquad \begin{pmatrix} A & B^T \\ B & 0 \end{pmatrix} \begin{pmatrix} \mathbf{u} \\ \mathbf{p} \end{pmatrix} = \begin{pmatrix} \mathbf{f} \\ 0 \end{pmatrix} \ ,$$

where

$$(9.2.15) \qquad A_{ij} := a(\boldsymbol{\varphi}_j, \boldsymbol{\varphi}_i) \ , \ B_{li} := b(\boldsymbol{\varphi}_i, \psi_l) \ , \ f_i := (\mathbf{f}, \boldsymbol{\varphi}_i) \ .$$

A is an $N_h \times N_h$ symmetric and positive definite matrix, while B is a rectangular $K_h \times N_h$ matrix. Using an algebraic argument it can easily be seen that the compatibility condition (9.2.9) holds if, and only if, $\ker B^T = \mathbf{0}$. In this case, the global matrix

$$(9.2.16) \qquad S := \begin{pmatrix} A & B^T \\ B & 0 \end{pmatrix} \ ,$$

whose dimension is $N_h + K_h$, is non-singular. As a matter of fact, in this case from (9.2.14) we obtain

$$(9.2.17) \qquad \mathbf{u} = A^{-1}(\mathbf{f} - B^T \mathbf{p}) \ ,$$
$$(9.2.18) \qquad B A^{-1} B^T \mathbf{p} = B A^{-1} \mathbf{f} \ .$$

Since the $K_h \times K_h$ symmetric *pressure-matrix*

$$(9.2.19) \qquad R := B A^{-1} B^T$$

is positive definite if $\ker B^T = \mathbf{0}$, it follows that $\mathbf{p}$ and then $\mathbf{u}$ are uniquely defined by (9.2.18) and (9.2.17). This in turn ensures that S is non-singular.

We also notice that S is symmetric but indefinite (thus eigenvalues of S are real but with variable sign).

We would also like to notice that system (9.2.14) could be replaced by an equivalent one obtained by changing the sign to the lowest K_h rows, i.e., replacing S with

$$(9.2.20) \qquad S' := \begin{pmatrix} A & B^T \\ -B & 0 \end{pmatrix}$$

This matrix is no longer symmetric, but it is positive definite, thus all its eigenvalues have positive real part.

The constants δ and β of (9.2.8) and (9.2.9) are related to the generalized eigenvalue problem

$$BA^{-1}B^T \omega^k = \lambda_k J \omega^k \ ,$$

where $J_{lm} := (\psi_l, \psi_m)$ is the pressure mass matrix. Precisely, one has (see, e.g., Fortin (1993))

$$\beta = \sqrt{\lambda_{\min}} \ , \quad \delta \geq \sqrt{\lambda_{\max}} \ .$$

9.2.2 Compatibility Condition and Spurious Pressure Modes

As seen in the previous Section, the compatibility condition (9.2.9) ensures that $\ker B^T = 0$, yielding the non-singularity of the global matrix S. Condition (9.2.9) means that the space V_h of discrete velocities is sufficiently rich compared with the one Q_h of discrete pressures.

Another remarkable implication of (9.2.9) is that no spurious pressure mode is allowed. By definition, a *spurious mode* is any function p_h^* such that

$$(9.2.21) \qquad p_h^* \in Q_h \ , \quad (p_h^*, \operatorname{div} v_h) = 0 \quad \text{for all } v_h \in V_h \ .$$

Any such function would give $b(v_h, p_h + c\, p_h^*) = b(v_h, p_h)$ for all $v_h \in V_h$ and all $c \in \mathbb{R}$. Therefore, (9.2.4) would define a discrete pressure only up to those functions which are linear combinations of spurious modes. In particular, this can easily yield the instability of the pressure calculation.

Now it is clear that if the compatibility condition (9.2.9) is satisfied, the only possible spurious mode is the trivial one $p_h^* = 0$. Indeed, if p_h^* is any function of Q_h different than zero, (9.2.9) ensures the existence of a function $v_h \in V_h$ such that $(p_h^*, \operatorname{div} v_h) > 0$.

A critical question can be raised now: what happens to the numerical scheme if the finite dimensional spaces don't satisfy the compatibility condition? We deduce from the previous analysis that the velocity field can be obtained in a stable and convergent way, as both (9.2.11) and the first inequality in (9.2.10) do not depend on the constant β. As a matter of fact, only the pressure field can be contaminated by the spurious modes. When a good discrete pressure p_h is also desired, the obvious remedy is to operate a choice of Q_h and V_h that satisfies (9.2.9). If this is not the case, however, several kind of remedies can be introduced.

A canonical approach is to operate a *stabilization* on the finite dimensional system (9.2.4) which amounts to relax the incomprimibility constraint on $\mathbf{u}_h$. This can be accomplished in various ways that can be rigorously motivated from a thorough mathematical analysis. This issue is addressed in Section 9.4.

Another approach consists of *filtering* the spurious modes out of the computed pressure. This is typically accomplished within an iterative procedure after each step, and requires a-priori knowledge of the structure of the spurious mode vector space. We refer for that to Section 9.5.

Finally, other approaches are also workable within specific frameworks of approximation. We mention, for instance, the use of *macro-elements* in the framework of finite element approximations. Such a technique is illustrated in Section 9.3.

9.2.3 Divergence-Free Property and Locking Phenomena

In general, the discrete continuity condition

$$(9.2.22) \qquad b(\mathbf{u}_h, q_h) = 0 \qquad \forall \, q_h \in Q_h$$

doesn't necessarily imply that

$$(9.2.23) \qquad \operatorname{div} \mathbf{u}_h = 0 \ .$$

As a matter of fact, (9.2.22) simply entails that $\mathbf{u}_h \in Z_h$ (see (9.2.5)), but this is not sufficient to conclude with (9.2.23), as we have pointed out in Remark 9.2.1. However, (9.2.23) holds if

$$(9.2.24) \qquad \operatorname{div} V_h \subset Q_h \ .$$

In fact, (9.2.23) follows at once taking $q_h = \operatorname{div} \mathbf{u}_h$ in (9.2.22). In the finite element approximations of the Stokes problem the property (9.2.24) is almost never verified.

The richer Q_h is, the more likely (9.2.24) will hold true. However, going along this direction may conflict with the fulfillment of the compatibility condition (9.2.9). In this respect, an extreme situation may occur when Q_h is exceedingly large compared with V_h so that the condition (9.2.22) overconstrains $\mathbf{u}_h$ at such a level that not only (9.2.23) holds, but even

$$\mathbf{u}_h = \mathbf{0} \ .$$

This characteristic behaviour is known as the *locking* phenomenon for the velocity field. We will encounter an example in Section 9.3.1.

9.3 Finite Element Approximation

Here , we adopt the notation of Chapter 3. In particular, we restrict ourselves to the case of a polygonal domain Ω, and $\mathcal{T}_h$ denotes a triangulation of $\overline{\Omega}$ into polyhedra K, that for brevity are called elements, whose diameter is at most h.

Let us define

$$(9.3.1) \qquad Y_h^k := \{\varphi_h \in L^2(\Omega)| \ \varphi_{h|K} \in \mathbb{P}_k \quad \forall \, K \in \mathcal{T}_h\} \ , \quad k \geq 0 \ ,$$

when each K is either a triangle or a tetrahedron, or

$$(9.3.2) \qquad Y_h^k := \{\varphi_h \in L^2(\Omega)| \ \varphi_{h|K} \circ T_K \in \mathbb{Q}_k \quad \forall \, K \in \mathcal{T}_h\} \ , \quad k \geq 0 \ ,$$

if each K is a parallelogram or a parallelepiped. (One could also consider general quadrilaterals or hexahedrons; however, similarly to what we have done in Chapter 3, we limit ourselves to the simplest case.) For both cases we set (accordingly with (3.2.4), (3.2.5))

$$(9.3.3) \qquad X_h^k = Y_h^k \cap C^0(\overline{\Omega}) \ , \quad k \geq 1 \ .$$

Clearly, a space like $Y_h^m \cap L_0^2(\Omega)$ for some $m \geq 0$, or $X_h^m \cap L_0^2(\Omega)$ for $m \geq 1$, is a good candidate for the pressure space Q_h; on the other hand, $(X_h^k \cap H_0^1(\Omega))^d$ can be used for V_h, for some $k \geq 1$.

Here, we present some possible combinations of these spaces, and for each one we will indicate the degrees of freedom for each velocity component and the pressure on the master element.

We point out that for the velocity spaces, the degrees of freedom need to be frozen on the boundary in order to match the homogeneous Dirichlet condition on $\mathbf{u}_h$. Concerning the pressure, the fact that we are dealing with functions whose average is 0, or that vanish at a given point $\mathbf{x}^*$ of $\overline{\Omega}$, should result in a reduction by one of the global number of degrees of freedom. When the latter option is adopted, choosing the discontinuous space Y_h^m has the nice feature that on each element K (indeed, excepting the one K^* containing $\mathbf{x}^*$) the divergence of $\mathbf{u}_h$ has vanishing average, i.e., the mass is conserved there. This property follows immediately from $(9.2.4)_2$ as in this case Q_h contains all functions that are piecewise constant in $\overline{\Omega} \setminus K^*$ and vanish in K^*.

9.3.1 Discontinuous Pressure Finite Elements

In this Section, we confine ourselves to the two-dimensional case. We begin considering

$$(9.3.4) \quad Q_h = Y_h^m \cap L_0^2(\Omega) \ , \quad m \geq 0 \quad , \quad V_h = (X_h^k \cap H_0^1(\Omega))^2 \ , \quad k \geq 1 \ .$$

The simplest situation occurs when $m = 0$ and $k = 1$ (piecewise-constant pressures and linear, or bilinear, velocities). The degrees of freedom are those indicated in Fig. 9.3.1. The symbol • denotes the values of the velocity components, whilst □ will refer to the value of the pressure.

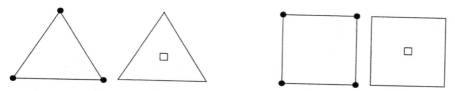

Fig. 9.3.1. Piecewise-constant pressure and piecewise-linear (or bilinear) velocities

These kind of elements don't satisfy the compatibility condition (9.2.9); further, a locking phenomenon may occur. This is, for instance, the case when the triangulation is made by N triangles having an edge on $\partial\Omega$ and a vertex that is common to each of them (see the Fig. 9.3.2). Since $\mathbf{u}_h$ has only two degrees of freedom (the two components at the only internal vertex), whereas p_h has $N - 1$ degrees of freedom (one less than the number of elements since p_h has vanishing mean value), the condition (9.2.22) yields $\mathbf{u}_h \equiv \mathbf{0}$.

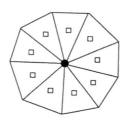

Fig. 9.3.2. Triangulation yielding the locking phenomenon for $\mathbf{u}_h$

A partial remedy to this situation is to adopt the so-called cross-grid triangulation. This is obtained by subdividing each triangle K (a *macroelement*) into three subtriangles having the center of gravity as common vertex, then using a constant pressure over K and piecewise-linear velocities upon each subtriangle that are continuous over $\overline{\Omega}$ (see Fig. 9.3.3). Such a choice eliminates the locking phenomenon, although it doesn't fulfill the compatibility condition (9.2.9) (see, e.g., Gunzburger (1989), p. 28).

Also, the piecewise $\mathbb{Q}_1$-$\mathbb{Q}_0$ elements displayed at the right hand of Fig. 9.3.1 don't pass the compatibility test. If, for instance, Ω is a rectangle, with sides parallel to the cartesian frame and subdivided into rectangles, then it can be shown that the checkerboard mode $p_h^* \in Q_h$, satisfying $p_h^* = 1$ or

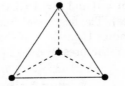 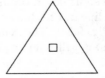

Fig. 9.3.3. (Cross-grid $\mathbb{P}_1$)-$\mathbb{P}_0$ elements

$p_h^* = -1$ alternatively on adjacent elements, is in fact a spurious mode for the pressure (see Girault and Raviart (1986), p. 160).

This mode can be filtered out resorting to the cross-grid triangulation illustrated in the following Fig. 9.3.4, left. From any parallelogram K of T_h we obtain four triangles by means of the two diagonals of K (the macroelement). The associated spaces are stable. Moreover, the following error estimate holds (see Brezzi and Fortin (1991), Sect. VI.5.4)

$$(9.3.5) \qquad ||\mathbf{u} - \mathbf{u}_h|| + ||p - p_h||_0 \leq Ch(||\mathbf{u}||_2 + ||p||_1) \ ,$$

where $|| \cdot ||_s$ is the Sobolev norm of order s (see Section 1.2) and h is the maximum diameter of the macroelements. The error estimate (9.3.5) follows from (9.2.11), (9.2.12), Theorem 3.4.2 and estimate (3.5.24). The implicit assumption is that both norms at the right hand side make sense.

Fig. 9.3.4. (Cross-grid $\mathbb{P}_1$)-$\mathbb{Q}_0$ elements (left) and (cross-grid $\mathbb{Q}_1$)-$\mathbb{Q}_0$ elements (right)

Another stabilization can be carried out using the macroelements of Fig. 9.3.4, right. Each parallelogram is subdivided into four parallelograms sharing the center of gravity as common vertex, then $\mathbb{Q}_1$ polynomials are used over each subelement. This approximation is stable and linearly convergent (see Gunzburger (1989), p. 30).

Another remedy to instability is the use of a richer space for the velocity approximation. In this respect, let us notice that the elements $\mathbb{P}_2$-$\mathbb{P}_0$ and $\mathbb{Q}_2$-$\mathbb{Q}_0$ are stable and both converge linearly (Fortin (1972)).

To obtain quadratic accuracy, one has to approximate the pressure with piecewise-linear polynomials. However, the $\mathbb{P}_2$-$\mathbb{P}_1$ element (which corresponds to take $k = 2$ and $m = 1$ in (9.3.4)) is still unstable (see, e.g., Brezzi and Fortin (1991), p. 231). It can be stabilized by resorting to the elements introduced by Crouzeix and Raviart (1973), based on discontinuous piecewise $\mathbb{P}_1$

pressures and piecewise $\mathbb{P}_2^b(K)$ velocities, where

$$(9.3.6) \qquad \mathbb{P}_k^b(K) := \{v = p_k + b \,|\, p_k \in \mathbb{P}_k \, ,$$
$$b \text{ is a cubic bubble function on } K\} \, ,$$

$k \geq 1$. A cubic *bubble function* b on a triangle K is a cubic polynomial that vanishes on the edges of K. Any such function can therefore be represented as a multiple of

$$(9.3.7) \qquad b_K := \lambda_1 \lambda_2 \lambda_3$$

where $\{\lambda_i \,|\, i = 1, 2, 3\}$ are linear polynomials, each one vanishing on one side of K and taking value 1 at the opposite vertex, and which is determined by the value that it attains at the center of gravity of K. The Crouzeix-Raviart elements are stable and converge quadratically. We report in Fig. 9.3.5 (left) their degrees of freedom.

A higher order generalization of the Crouzeix-Raviart elements is based upon discontinuous piecewise $\mathbb{P}_{k-1}$ pressures, and piecewise $\mathbb{P}_k \oplus B_{k+1}(K)$ velocities, where on each triangle K

$$(9.3.8) \qquad B_{k+1}(K) := \{v \in \mathbb{P}_{k+1} \,|\, v = p_{k-2}\, b_K \, , \ p_{k-2} \in \mathbb{P}_{k-2}\} \, , \ k \geq 2 \, ,$$

and b_K is defined in (9.3.7). B_{k+1} is therefore the space of bubble functions of degree less than or equal to $k + 1$ on K. These elements satisfy the compatibility condition and converge as $O(h^k)$. (For the proofs, see, e.g., Girault and Raviart (1986), p. 139.)

Fig. 9.3.5. Crouzeix-Raviart elements (left) and Boland-Nicolaides elements (right, $k = 2$)

Another good example is that based on parallelograms, with discontinuous piecewise $\mathbb{Q}_{k-1}$ pressures ($k \geq 1$) and piecewise $\mathbb{P}_k$ velocities on both triangles which are obtained dividing any parallelogram through one of its diagonals. See Fig. 9.3.5 (right) for an example with $k = 2$. These elements, introduced by Boland and Nicolaides (1983), satisfy the compatibility condition, and converge as $O(h^k)$ (see also Gunzburger (1989), p. 36).

Among other elements that are stable, we mention the $\mathbb{Q}_2$-$\mathbb{P}_1$ on parallelepipedal decompositions (pressures are therefore discontinuous). See Girault and Raviart (1986), p. 156.

9.3.2 Continuous Pressure Finite Elements

For simplicity, also in this Section we consider the two-dimensional case. When the finite element pressure is asked to be a continuous function, it is natural to consider the spaces

$$(9.3.9) \quad Q_h = X_h^m \cap L_0^2(\Omega) \, , \, m \geq 1 \, , \, V_h = (X_h^k \cap H_0^1(\Omega))^2 \, , \quad k \geq 1 \, .$$

To start with, let us mention that the so-called *equal interpolation* choice, the one based on the same polynomial degree for both velocity and pressure, i.e., $m = k$, yields an unstable approximation (see Sani, Gresho, Lee and Griffiths (1981), Sani, Gresho, Lee, Griffiths and Engelman (1981) and Brezzi and Fortin (1991), p. 210).

On the contrary, the elements proposed by Taylor and Hood (1973), corresponding to the choice $m = 1$, $k = 2$ in (9.3.9), are stable, and converge with the optimal rate, i.e.,

$$(9.3.10) \quad ||\mathbf{u} - \mathbf{u}_h|| + ||p - p_h||_0 \leq Ch^s(||\mathbf{u}||_{s+1} + ||p||_s) \, , \quad s = 1, 2 \, ,$$

provided the solution is regular enough so that the norms at the right hand side make sense (see, e.g., Girault and Raviart (1986), p. 176). This holds for both triangular and parallelepipedal elements. Higher-order Taylor-Hood elements (corresponding to the choice $m = k-1$, $k \geq 3$, in (9.3.9)) have been analyzed by Brezzi and Falk (1991). We show in Fig. 9.3.6 the associated degrees of freedom for the case $m = 1$, $k = 2$.

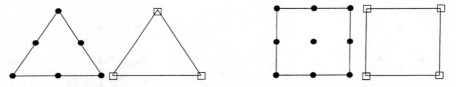

Fig. 9.3.6. Low-order Taylor-Hood triangles (left) and parallelograms (right)

Another possibility is to start with a triangulation $\mathcal{T}_h$ made by triangles K, take piecewise-linear pressures on each K, and piecewise-linear velocities over each of the three subtriangles of K sharing the center of gravity of K as common vertex (see Fig. 9.3.7 left). This choice, that will be indicated for the sake of brevity (cross-grid $\mathbb{P}_1$)-$\mathbb{P}_1$, satisfies the compatibility condition, and the corresponding finite element solution converges linearly with respect to h (see Pironneau (1988), p. 107).

A similar idea forms the basis of the so-called $(\mathbb{P}_1$ iso $\mathbb{P}_2)$-$\mathbb{P}_1$ elements (due to Bercovier-Pironneau (1979); see Fig. 9.3.7, right). The pressure space is still the same, while velocities are linear over each of the four subtriangles of K, obtained by joining the midpoints of the edges of K. This discretization is stable and linearly convergent. The reason of its name relies on the

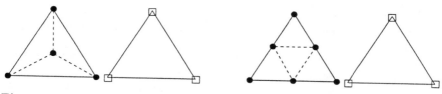

Fig. 9.3.7. (Cross-grid $\mathbb{P}_1$)-$\mathbb{P}_1$ (left) and ($\mathbb{P}_1$ iso $\mathbb{P}_2$)-$\mathbb{P}_1$ (right) elements

observation that the degrees of freedom are the same as those of the $\mathbb{P}_2$-$\mathbb{P}_1$ Taylor-Hood element (see Fig. 9.3.6, left). In a similar manner, one can consider the ($\mathbb{Q}_1$ iso $\mathbb{Q}_2$)-$\mathbb{Q}_1$ elements when T_h is made of parallelograms.

We conclude with the so-called *mini* element (introduced by Arnold, Brezzi and Fortin (1984)): pressures are still piecewise $\mathbb{P}_1$, while velocities are piecewise $\mathbb{P}_1^b$ (see (9.3.6)). This choice is stable, and leads to a linear convergence (see Fig. 9.3.8). The mini element is very much similar to (cross-grid $\mathbb{P}_1$)-$\mathbb{P}_1$ element of Fig. 9.3.7, left. Indeed, in the latter element the bubble function b on K, which is a cubic polynomial vanishing over ∂K, is replaced by a function $\psi \in C^0(K)$, still vanishing on ∂K and which is linear upon each one of the three subtriangles merging at the center of gravity of K. The two elements enjoy the same stability and convergence properties.

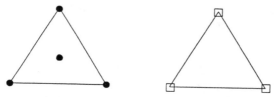

Fig. 9.3.8. The mini element

Many other finite element spaces have been used in the applications, such as, e.g., elements with non-conforming velocities and those based on reduced integration. We refer the interested reader to Girault and Raviart (1986) and Brezzi and Fortin (1991).

9.4 Stabilization Procedures

When the approximation (9.2.4) of the saddle-point problem (9.1.11) is based on a couple of spaces V_h, Q_h that don't satisfy the compatibility condition (9.2.9), a procedure aiming at stabilizing the discrete solution (which otherwise may be affected by spurious pressure modes) can be accomplished according to several criteria.

Throughout Section 9.3 we have already seen that the use of cross-grid triangulations can sometimes stabilize finite elements that are a priori unstable on general triangulation.

Another way of stabilizing without necessarily resorting to any special triangulation consists of slightly modifying (9.2.4), and consequently the associated algebraic system (9.2.14). Often, these methods aim at relaxing the incompressibility constraint, thus modifying the second set of equations of (9.2.4). In abstract form, the stabilized problem reads:

(9.4.1)
$$\begin{cases} \text{find } \mathbf{u}_h \in V_h \,, \ p_h \in Q_h : \\[2mm] a(\mathbf{u}_h, \mathbf{v}_h) + b(\mathbf{v}_h, p_h) = (\mathbf{f}, \mathbf{v}_h) \quad \forall \, \mathbf{v}_h \in V_h \\[2mm] b(\mathbf{u}_h, q_h) = \Phi_h(q_h) \qquad\qquad \forall \, q_h \in Q_h \,, \end{cases}$$

where the stabilization term Φ_h depends linearly on the test function q_h, but it may also depend on $\mathbf{u}_h$, p_h, $\mathbf{f}$ and h. Below, for the sake of simplicity, we confine ourselves to the case of a two-dimensional problem.

A first example was provided from Brezzi and Pitkäranta (1984) taking

(9.4.2)
$$\Phi_h(q_h) := \sum_{K \in \mathcal{T}_h} h_K^2 \int_K \nabla p_h \cdot \nabla q_h \,,$$

where, as usual, $\mathcal{T}_h$ is the finite element triangulation made by triangles K whose diameter is denoted by h_K. We have seen in Section 9.3.2 that the choice (9.3.9) with $m = k = 1$ (i.e., based on continuous piecewise-linear polynomials for both pressure and velocity) is unstable if $\Phi_h \equiv 0$. On the contrary, (9.4.1) with (9.4.2) yields a stable and linearly convergent approximation. Heuristically, owing to the presence of the term $\Phi_h(q_h)$, (9.4.1) can be regarded, at least in the case where one substitutes h_K with h in (9.4.2), as an approximation to the modified continuity equation

(9.4.3)
$$\operatorname{div} \mathbf{u} = h^2 \Delta p$$

with a homogeneous Neumann boundary condition on the pressure p. For the analysis of this approach, see also Brezzi and Douglas (1988).

Later on, Hughes, Franca and Balestra (1986) proposed to use

(9.4.4)
$$\Phi_h(q_h) := \delta \sum_K h_K^2 \int_K (a_0 \mathbf{u}_h - \nu \Delta \mathbf{u}_h + \nabla p_h - \mathbf{f}) \cdot \nabla q_h$$

for continuous pressure spaces, and they proved that the method is stable when $0 < \delta \leq C_I$ (C_I being a suitable constant). This approach is similar to the *Streamline Upwind/Petrov-Galerkin* (SUPG) method that has been considered in Section 8.3.2 for the steady advection-diffusion equation. If the spaces are chosen as in (9.3.9) with $m = k$, then both velocity and pressure converge like $O(h^k)$. In this case, at the differential level the continuity equation (when $h_K = h$ for all $K \in \mathcal{T}_h$) is modified as follows

$$\operatorname{div} \mathbf{u} = \delta h^2 \operatorname{div}(a_0 \mathbf{u} - \nu \varDelta \mathbf{u} + \nabla p - \mathbf{f}) \ .$$

However, the right hand side is identically zero from (9.1). In practice, the parameter δ cannot be too small otherwise the parasitic modes will not be eliminated. On the other hand, a large δ yields a poor approximation for the pressure field near to the boundary $\partial \Omega$. We also note that (9.4.4) yields a non-symmetric perturbation to (9.2.4).

Another possibility is to use

(9.4.5)
$$\Phi_h(q_h) := \delta \sum_K h_K^2 \left\{ \int_K (a_0 \mathbf{u}_h + \nabla p_h - \mathbf{f}) \cdot \nabla q_h \right.$$
$$\left. - \int_{\partial K \cap \partial \Omega} \nu \varDelta \mathbf{u}_h \cdot \mathbf{n} q_h \right\}$$

instead of (9.4.4) (see Brezzi and Douglas (1988)). The transformation is based on the remark that, for a divergence free vector $\mathbf{u}$, (9.4.4) and (9.4.5) do coincide (provided $\varDelta \mathbf{u}$ is continuous). This follows using Green formula (5.1.5). The interest of (9.4.5) versus (9.4.4) stems from the property that with (9.4.5) the perturbation on the matrix (9.2.16) concerns the degrees of freedom of $\mathbf{u}_h$ associated with the boundary elements solely.

A more general approach entails a perturbation of both continuity and momentum equations. The stabilized problem reads

(9.4.6)
$$\begin{cases} \text{find } \mathbf{u}_h \in V_h \ , \ p_h \in Q_h : \\[2mm] a(\mathbf{u}_h, \mathbf{v}_h) + b(\mathbf{v}_h, p_h) = (\mathbf{f}, \mathbf{v}_h) - \Psi_h^{(\rho)}(\mathbf{v_h}) \quad \forall \ \mathbf{v}_h \in V_h \\[2mm] b(\mathbf{u}_h, q_h) = \Phi_h(q_h) \hspace{3.2cm} \forall \ q_h \in Q_h \ , \end{cases}$$

where $\Phi_h(q_h)$ is as in (9.4.4),

$$\Psi_h^{(\rho)}(\mathbf{v}_h) := \delta \sum_K h_K^2 \int_K (a_0 \mathbf{u}_h - \nu \varDelta \mathbf{u}_h + \nabla p_h - \mathbf{f}) \cdot (\rho a_0 \mathbf{v}_h - \rho \nu \varDelta \mathbf{v}_h)$$

ρ is a parameter (typically, ρ takes the values -1, 0, 1) and $\delta > 0$.

For $\rho = 0$, this method coincides with the SUPG method (9.4.4). When $\rho = -1$ this method has been given the name of *Galerkin/Least-Squares* (GALS) (Hughes and Franca (1987)), and it leads to a symmetric system. The case $\rho = 1$ has been proposed by Douglas and Wang (1989): it yields a system which is no longer symmetric, however it is non-negative definite and stable under more general conditions. As a matter of fact, the following result holds (Franca and Stenberg (1991)):

Theorem 9.4.1 *Assume that $V_h = (X_h^k \cap H_0^1(\Omega))^2$ and $Q_h = Y_h^m \cap L_0^2(\Omega)$, for some $k \geq 2$ and $m \geq 0$. For the SUPG and GALS methods (i.e., $\rho = 0$ and $\rho = -1$, respectively) assume further that*

(9.4.7)
$$0 < \delta < C_I \ ,$$

where C_I is a suitable constant. Then (9.4.6) provides a stable solution, and moreover the following optimal convergence result holds

(9.4.8)
$$||\mathbf{u} - \mathbf{u}_h|| + ||p - p_h||_0 \leq C(h^k ||\mathbf{u}||_{k+1} + h^{m+1} ||p||_{m+1}) \ .$$

If $k = 1$, then the same result holds again provided $Q_h = X_h^m \cap L_0^2(\Omega)$ for some $m \geq 1$, so that $Q_h \subset C^0(\overline{\Omega})$.

The constant C_I in (9.4.7) can be very small. Indeed, C_I is precisely the constant showing up in the inverse inequality (6.3.21).

It is worthwhile to notice that some finite elements that are unstable for problem (9.2.4) become stable for the perturbed problem (9.4.6). Among these, let us mention the case of $\mathbb{P}_1$-$\mathbb{P}_1$ and $\mathbb{Q}_1$-$\mathbb{Q}_1$ elements (see Fig. 9.4.1)

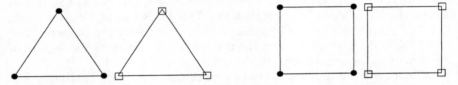

Fig. 9.4.1. Examples of finite elements fulfilling the assumptions of Theorem 9.4.1

Remark 9.4.1 For piecewise-linear (or -bilinear) velocities we have that $\Delta \mathbf{v}_{h|K} = 0$ for all $K \in \mathcal{T}_h$. Thus in this case, under the further assumption that $a_0 = 0$, all methods considered in this Section (except (9.4.2)) coincide with one another. □

The previous theorem doesn't apply to the case of the lowest order elements

(9.4.9)
$$V_h = (X_h^1 \cap H_0^1(\Omega))^2 \quad Q_h = Y_h^0 \cap L_0^2(\Omega)$$

(see Fig. 9.3.1) which, on the other hand, are not free of spurious modes on general triangulations (see Section 9.3.1). In this situation, a stabilized method reads:

(9.4.10)
$$\begin{cases} \text{find } \mathbf{u}_h \in V_h \ , \ p_h \in Q_h : \\[2mm] a(\mathbf{u}_h, \mathbf{v}_h) + b(\mathbf{v}_h, p_h) = (\mathbf{f}, \mathbf{v}_h) \qquad \forall \ \mathbf{v}_h \in V_h \\[2mm] b(\mathbf{u}_h, q_h) = \delta \sum_{\sigma \in \Gamma_h} h_\sigma \int_\sigma [p_h]_\sigma [q_h]_\sigma \quad \forall \ q_h \in Q_h \ , \end{cases}$$

where $\delta > 0$ is a suitable constant, Γ_h is the set of all edges σ of the triangulation except for those belonging to the boundary $\partial \Omega$, and h_σ is the length of σ. Finally, for any $q_h \in Q_h$, $[q_h]_\sigma$ denotes its jump across σ. This method yields a stable solution for all $\delta > 0$, which converges linearly, i.e.,

$$\|\mathbf{u} - \mathbf{u}_h\| + \|p - p_h\|_0 \leq Ch(\|\mathbf{u}\|_2 + \|p\|_1)$$

(Kechkar and Silvester (1992)).

The algebraic restatement of (9.4.10) yields the symmetric non-singular system

(9.4.11)
$$\begin{pmatrix} A & B^T \\ B & -\delta D \end{pmatrix} \begin{pmatrix} \mathbf{u} \\ \mathbf{p} \end{pmatrix} = \begin{pmatrix} \mathbf{f} \\ \mathbf{0} \end{pmatrix}$$

where D is a symmetric and non-negative definite matrix whose entries are

(9.4.12)
$$D_{ij} := \sum_{\sigma \in \Gamma_h} h_\sigma \int_\sigma [\psi_i]_\sigma [\psi_j]_\sigma .$$

Algebraically, the matrix in (9.4.11), say S_δ, can be decomposed as

$$S_\delta = \begin{pmatrix} A & 0 \\ B & I \end{pmatrix} \begin{pmatrix} A^{-1} & 0 \\ 0 & -BA^{-1}B^T - \delta D \end{pmatrix} \begin{pmatrix} A & B^T \\ 0 & I \end{pmatrix} .$$

Therefore, by the Sylvester law of inertia the number of positive and negative eigenvalues of S_δ coincide with the one of its second factor: the first N_h are positive and the latter K_h negative. Let us remember that if $\ker B^T \neq \mathbf{0}$ the matrix S in (9.2.16) has N_h positive eigenvalues while the latter K_h can be either negative or null.

A simplification in (9.4.10) occurs if we consider a triangulation made by *macroelements*, each of them being the union of four triangles or quadrilaterals (see Fig. 9.4.2). Then the sum over $\sigma \in \Gamma_h$ in (9.4.10) can be restricted to those edges that are internal to each macroelement. The reduced scheme enjoys the same stability and convergence properties as (9.4.10). The advantage is that now D has a narrower band.

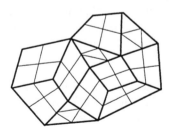

Fig. 9.4.2. A macroelement partition of Ω

Remark 9.4.2 *(Stabilization and bubble functions).* It has been pointed out in Baiocchi, Brezzi and Franca (1993) that the enrichment of the finite element space by summation of bubble functions results in a stabilization approach that is formally similar to those we have investigated above (see also

Pierre (1988, 1989), Bank and Welfert (1990)). A similar remark also applies to the stabilization methods introduced in Section 8.3.2 for advection-diffusion problems (see Section 8.3.3).

In an abstract fashion, we can formulate a finite dimensional approach that makes use of a space of bubble functions as follows. Let us set

$$\mathcal{V}_h^b := \mathcal{U}_h \oplus \mathcal{B}$$

where $\mathcal{U}_h$ is a standard finite element space, while $\mathcal{B}$ is a space of bubble functions. For instance, for the mini element of Fig. 9.3.8 we have

$$(9.4.13) \quad \mathcal{U}_h = (X_h^1 \cap H_0^1(\Omega))^2 \times (X_h^1 \cap L_0^2(\Omega)), \quad \mathcal{B}_{|K} = (B_3(K))^2 \times \{0\}$$

(see (9.3.8)). Then, let us denote by $\mathcal{A}(\cdot, \cdot)$ the linear form associated with the boundary value problem. The finite dimensional problem can be formulated as follows

$$(9.4.14) \quad \begin{aligned} \text{find } \mathbf{u}_h + \mathbf{u}_B \in \mathcal{V}_h^b : \mathcal{A}(\mathbf{u}_h + \mathbf{u}_B, \mathbf{v}_h + \mathbf{v}_B) \\ = \mathcal{F}(\mathbf{v}_h + \mathbf{v}_B) \ \forall \ \mathbf{v}_h \in \mathcal{U}_h, \mathbf{v}_B \in \mathcal{B} \ . \end{aligned}$$

In particular, taking $\mathbf{v}_h = \mathbf{0}$ yields

$$\mathcal{A}(\mathbf{u}_h + \mathbf{u}_B, \mathbf{v}_B) = \mathcal{F}(\mathbf{v}_B) \ \forall \ \mathbf{v}_B \in \mathcal{B} \ ,$$

and by computing $\mathbf{u}_B$ (element by element) we obtain

$$(9.4.15) \quad \mathbf{u}_B = \boldsymbol{\Phi}(\mathbf{f}, \mathbf{u}_h)$$

for a suitable functional $\boldsymbol{\Phi}$. Now, taking $\mathbf{v}_B = \mathbf{0}$ in (9.4.14) it follows

$$(9.4.16) \quad \mathcal{A}(\mathbf{u}_h, \mathbf{v}_h) = \mathcal{F}(\mathbf{v}_h) - \mathcal{A}(\boldsymbol{\Phi}(\mathbf{f}, \mathbf{u}_h), \mathbf{v}_h) \qquad \forall \ \mathbf{v}_h \in \mathcal{U}_h \ .$$

This problem is nothing but a Galerkin approximation formulated over the original space $\mathcal{U}_h$. The second term at the right hand side can be regarded as a perturbation that aims at stabilizing the solution $\mathbf{u}_h$. As a particular case, the effect produced by the use of bubble functions for the mini element can be regarded as a GALS approximation that makes use of the space $\mathcal{U}_h$. $\quad\square$

Remark 9.4.3 The stabilized problem (9.4.6) can be cast in a form similar to the one introduced in Section 8.3.2 for advection-diffusion equations. In fact, if we write the Stokes operator and the right hand side as

$$(9.4.17) \qquad L(\mathbf{w}, r) := \begin{pmatrix} a_0 \mathbf{w} - \nu \Delta \mathbf{w} + \nabla r \\ \text{div } \mathbf{w} \end{pmatrix} \ , \quad \mathbf{f}^* := \begin{pmatrix} \mathbf{f} \\ 0 \end{pmatrix} \ ,$$

the symmetric and skew-symmetric part of L with respect to the scalar product in $L^2(\Omega)$ are given by

$$(9.4.18) \qquad L_S(\mathbf{w}, r) := \begin{pmatrix} a_0\mathbf{w} - \nu\Delta\mathbf{w} \\ 0 \end{pmatrix} \;, \quad L_{SS}(\mathbf{w}, r) := \begin{pmatrix} \nabla r \\ \mathrm{div}\,\mathbf{w} \end{pmatrix} \;.$$

Then we can introduce the following stabilized problem: find $(\mathbf{u}_h, p_h) \in V_h \times Q_h$ such that

$$(9.4.19) \qquad \begin{aligned} a(\mathbf{u}_h, \mathbf{v}_h) &+ b(\mathbf{v}_h, p_h) - b(\mathbf{u}_h, q_h) \\ &+ \delta \sum_{K \in \mathcal{T}_h} \left(L(\mathbf{u}_h, p_h) - \mathbf{f}^*, h_K^2 (L_{SS} + \rho L_S)(\mathbf{v}_h, q_h) \right)_K \\ &= (\mathbf{f}, \mathbf{v}_h) \end{aligned}$$

for each $(\mathbf{v}_h, q_h) \in V_h \times Q_h$. This essentially corresponds to (9.4.6) (indeed, only an additional stabilizing term

$$\delta \sum_{K \in \mathcal{T}_h} h_K^2 (\mathrm{div}\,\mathbf{u}_h, \mathrm{div}\,\mathbf{v}_h)_K$$

has been introduced at the left hand side of (9.4.19)).

Notice that for advection-diffusion equations the GALS method is the one stemming from the choice $\rho = 1$, and not from $\rho = -1$. The reason of this name "switching" when passing to the Stokes problem is that GALS method was originally introduced starting from the Stokes operator written in the symmetric form

$$L^*(\mathbf{w}, r) := \begin{pmatrix} a_0\mathbf{w} - \nu\Delta\mathbf{w} + \nabla r \\ -\,\mathrm{div}\,\mathbf{w} \end{pmatrix}$$

(see Hughes and Franca (1987)), then subtracting at the left hand side

$$\delta \sum_{K \in \mathcal{T}_h} h_K^2 (a_0\mathbf{u}_h - \nu\Delta\mathbf{u}_h + \nabla p_h - \mathbf{f}, a_0\mathbf{v}_h - \nu\Delta\mathbf{v}_h + \nabla q_h)_K \;,$$

the "least-squares" term. This corresponds to the choice $\rho = -1$ in (9.4.19). Indeed, for the Stokes problem it would be more appropriate to give the name of "least-squares method" to the Douglas-Wang method (9.4.6) with $\rho = 1$.

$\square$

9.5 Approximation by Spectral Methods

These approximations are characterized by the fact that both test and trial functions are invariably algebraic polynomials of high degree. For the sake of simplicity, we assume that Ω is the domain $(-1, 1)^d$, $d = 2, 3$; the case of a more general domain demands for a domain decomposition approach (see, e.g., Section 6.4).

Let us begin with the approximation of problems (9.1.6), taking

$$(9.5.1) \qquad V_N := (\mathbb{Q}_N^0)^d \;, \quad V_{\mathrm{div},N} := \{\mathbf{v}_N \in V_N | \, \mathrm{div}\,\mathbf{v}_N = 0\} \;,$$

where, as usual, $\mathbb{Q}_N^0$ denotes the space of algebraic polynomials of degree less than or equal to N in each variable x_i, $i = 1, ..., d$, vanishing over $\partial\Omega$. In the current case, the Galerkin approximation (9.2.1) reads

$$(9.5.2) \qquad \text{find } \mathbf{u}_N \in V_{\text{div},N} : \quad a(\mathbf{u}_N, \mathbf{v}_N) = (\mathbf{f}, \mathbf{v}_N) \qquad \forall\, \mathbf{v}_N \in V_{\text{div},N} \ .$$

This problem has a unique solution, which is stable, and satisfies the convergence result (9.2.3) (with the obvious change of notation). In order to deduce a rate of convergence with respect to N^{-1}, we need to estimate the infimum at the right hand side of (9.2.3). For this, we have the following result that was proven by Sacchi Landriani and Vandeven (1989).

Theorem 9.5.1 *For any $s \geq 1$ there exists a constant C independent of N such that for all $\mathbf{v} \in (H^s(\Omega))^d \cap V_{\text{div}}$*

$$(9.5.3) \qquad \inf_{\mathbf{v}_N \in V_{\text{div},N}} ||\mathbf{v} - \mathbf{v}_N|| \leq C N^{1-s} ||\mathbf{v}||_s \ .$$

Proof. Let us confine ourselves to the two-dimensional case ($d = 2$). First of all, we notice that, for any $\mathbf{v} \in V_{\text{div}}$, there exists a *stream function* $\psi \in H^2(\Omega)$ such that

$$(9.5.4) \qquad \mathbf{v} = \text{rot}\,\psi := (\frac{\partial\psi}{\partial x_2}, -\frac{\partial\psi}{\partial x_1}) \quad \text{in } \Omega \ .$$

Further, if $\mathbf{v} \in (H^s(\Omega))^2$ for some $s \geq 1$, then $\psi \in H^{s+1}(\Omega)$ and

$$(9.5.5) \qquad ||\psi||_{s+1} \leq C||\mathbf{v}||_s \ .$$

Since $\mathbf{v} = \mathbf{0}$ on $\partial\Omega$, $\nabla\psi = 0$ on $\partial\Omega$; moreover, since ψ is determined up to an additive constant, we can enforce that it vanishes at one boundary point, so that we can infer that it vanishes identically on $\partial\Omega$. Thus $\psi \in H_0^2(\Omega)$, where

$$(9.5.6) \qquad H_0^2(\Omega) = \left\{ \varphi \in H^2(\Omega) \,|\, \varphi = \frac{\partial\varphi}{\partial n} = 0 \text{ on } \partial\Omega \right\} \ .$$

On the latter space it is possible to define an operator $P_{2,N}^0$, by associating to any function of $H_0^2(\Omega)$ its orthogonal projection, with respect to the scalar product of (9.5.6), upon the polynomials of degree less than or equal to N that vanish over $\partial\Omega$ together with their normal derivatives. By an argument similar to that used to derive error estimates for $P_{1,N}^0$ (see Section 4.5.2), it can be proven that

$$(9.5.7) \qquad ||\psi - P_{2,N}^0\psi||_2 \leq C N^{1-s} ||\psi||_{s+1} \leq C N^{1-s} ||\mathbf{v}||_s \ , \quad s \geq 1$$

(the latter inequality follows from (9.5.5)). Taking $\mathbf{v}_N = \text{rot}(P_{2,N}^0\psi)$, and noticing that $\mathbf{v}_N \in V_{\text{div},N}$, we conclude with (9.5.3). $\qquad\square$

It now follows from (9.2.3) and (9.5.3) that

$$(9.5.8) \qquad ||\mathbf{u} - \mathbf{u}_N|| \leq CN^{1-s}||\mathbf{u}||_s \ ,$$

provided $\mathbf{u} \in (H^s(\Omega))^d \cap V_{\mathrm{div}}$ for some $s \geq 1$. Once $\mathbf{u}_N$ is available, the problem is how to recover a pressure field p_N with the same optimal accuracy. This is not an easy task, that can be, however, accomplished in some cases. We refer, in particular, to the method of Moser, Moin and Leonard (1983) and its extensions (see Pasquarelli, Quarteroni and Sacchi Landriani (1987) and Pasquarelli (1991)). We should bear in mind, however, that this approach entails the use of a basis for $V_{\mathrm{div},N}$ made by linear combinations of Chebyshev polynomials (see Section 4.5.1). As a consequence, in (9.5.2) every integral is replaced by a weighted one using the Chebyshev weight function (see Section 4.5.4).

In general, the difficulty of generating a basis for $V_{\mathrm{div},N}$ and the one of recovering the pressure field suggests to approximate the saddle-point problem (9.1.11) directly. In turn, the latter can be solved by a spectral Galerkin method, or else by a generalized Galerkin method, which is often equivalent to a spectral collocation method.

9.5.1 Spectral Galerkin Approximation

Let Q_N and V_N be two polynomial subspaces of $L_0^2(\Omega)$ and $(H_0^1(\Omega))^d$, respectively, and then consider the following spectral approximation to (9.1.11):

$$(9.5.9) \qquad \begin{cases} \text{find } \mathbf{u}_N \in V_N \ , \ p_N \in Q_N : \\[2mm] a(\mathbf{u}_N, \mathbf{v}_N) + b(\mathbf{v}_N, p_N) = (\mathbf{f}, \mathbf{v}_N) \quad \forall \, \mathbf{v}_N \in V_N \\[2mm] b(\mathbf{u}_N, q_N) = 0 \qquad\qquad\qquad \forall \, q_N \in Q_N \ . \end{cases}$$

Typically, the space V_N is the one defined in (9.5.1); this guarantees an error estimate like (9.5.8) for the velocity field. Correspondingly, we would like to take $Q_N = M_N$, with $M_N := \mathbb{Q}_N \cap L_0^2(\Omega)$. Unfortunately, this choice is unsuitable because the couple V_N, M_N would not satisfy the compatibility condition (9.2.9). For this reason, we need to take a proper subspace of M_N as Q_N.

To begin with, let us define

$$(9.5.10) \qquad S_N := \{p_N^* \in \mathbb{Q}_N \mid b(\mathbf{v}_N, p_N^*) = 0 \ \ \forall \, \mathbf{v}_N \in V_N \} \ ,$$

i.e., the subspace of $\mathbb{Q}_N$ made by all spurious pressure modes (see (9.2.21)). Also, let us define

$$(9.5.11) \qquad D_N := \{\varphi_N \in \mathbb{Q}_N \mid \exists \, \mathbf{v}_N \in V_N \ , \ \mathrm{div}\,\mathbf{v}_N = \varphi_N \} \ ,$$

which is the image of $\mathbb{Q}_N$ through the divergence operator. Clearly, D_N is the orthogonal subspace of S_N with respect to the $L^2(\Omega)$ scalar product. The space S_N can be fully characterized. For the time being, notice that we can take, as pressure space Q_N, any *supplementary space* to S_N in $\mathbb{Q}_N$, i.e., any $Q_N \subset \mathbb{Q}_N$ such that

$$(9.5.12) \qquad \dim S_N + \dim Q_N = \dim \mathbb{Q}_N \ , \quad S_N \cap Q_N = \{0\} \ .$$

The following result (proven, e.g., in Bernardi and Maday (1992), pp. 126–127) characterizes the space of spurious modes for two-dimensional approximations.

Theorem 9.5.2 *Assume that $\Omega = (-1,1)^2$. The space S_N has dimension 8, and a basis for it is given by the polynomials*

$$(9.5.13) \qquad \begin{array}{l} 1 \ , \ L_N(x_1) \ , \ L_N(x_2) \ , \ L_N(x_1)L_N(x_2) \ , \ L_N'(x_1)L_N'(x_2) \ , \\ L_N'(x_1)L_{N+1}'(x_2) \ , \ L_{N+1}'(x_1)L_N'(x_2) \ , \ L_{N+1}'(x_1)L_{N+1}'(x_2) \ , \end{array}$$

where $\{L_k \mid k = 0, 1, ...\}$ are the Legendre polynomials introduced in Section 4.4.1.

Proof. It is easy to show that any function in (9.5.13) belongs to S_N. Conversely, let us prove that any function $p_N^* \in S_N$ can be obtained by a linear combination of the above polynomials. We notice that p_N^* must satisfy

$$(9.5.14) \qquad \int_\Omega p_N^* \left(\frac{\partial v_{N,1}}{\partial x_1} + \frac{\partial v_{N,2}}{\partial x_2} \right) d\mathbf{x} = 0 \qquad \forall \, \mathbf{v}_N = (v_{N,1}, v_{N,2}) \in V_N \ .$$

First taking $\mathbf{v}_N = (\varphi_{mn}, 0)$ and then $\mathbf{v}_N = (0, \varphi_{mn})$ with

$$\varphi_{mn}(x_1, x_2) = (1 - x_1^2)(1 - x_2^2)L_m'(x_1)L_n'(x_2) \ , \quad 1 \le m, n \le N-1 \ ,$$

we deduce from (9.5.14)

$$(9.5.15) \qquad \begin{aligned} & \int_\Omega p_N^*(x)L_m(x_1)(1 - x_2^2)L_n'(x_2) \, d\mathbf{x} \\ & = \int_\Omega p_N^*(x)(1 - x_1^2)L_m'(x_1)L_n(x_2) \, d\mathbf{x} = 0 \ . \end{aligned}$$

We have used the Sturm-Liouville formula

$$(9.5.16) \qquad ((1 - \xi^2)L_k'(\xi))' = -k(k+1)L_k(\xi), \ -1 < \xi < 1 \ ,$$

for Legendre polynomials (see (4.4.2)). Recalling that (see (6.2.20))

$$(9.5.17) \qquad \varphi_k(\xi) := (1 - \xi^2)L_k'(\xi) \ , \quad 1 \le k \le N-1 \ ,$$

is a basis for the space $\mathbb{P}_N^0$ of one dimensional polynomials of degree less than or equal to N vanishing at $\xi = \pm 1$. From (9.5.16) and (4.4.6) it follows that the orthogonal subspace to $\mathbb{P}_N^0$ in $\mathbb{P}_N$ is generated by $L_N'(\xi)$ and $L_{N+1}'(\xi)$. Thus, from (9.5.15) we deduce that p_N^* can be expressed either as

$$p_N^*(x_1, x_2) = a_0(x_2) + a_N(x_2)L_N(x_1)$$
$$+ b_N(x_1)L_N'(x_2) + b_{N+1}(x_1)L_{N+1}'(x_2)$$

or

$$p_N^*(x_1, x_2) = c_0(x_1) + c_N(x_1)L_N(x_2)$$
$$+ d_N(x_2)L_N'(x_1) + d_{N+1}(x_2)L_{N+1}'(x_1) \ ,$$

where a_0, a_N,..., d_N, d_{N+1} are algebraic polynomials of degree less than or equal to N. Therefore, p_N^* can be expressed as a linear combination of the polynomials (9.5.13).

It remains to show that the eight polynomials (9.5.13) are linearly independent. For that, it is sufficient to verify the linear independency of the polynomials $\{1, L_N, L_N', L_{N+1}'\}$. To this aim, let us assume that

$$\alpha_0 + \alpha_N L_N + \beta_N L_N' + \beta_{N+1} L_{N+1}' = 0$$

Owing to the integral equation

$$\int_{-1}^{\xi} L_k(\eta) \, d\eta = \frac{1}{2k+1}(L_{k+1}(\xi) - L_{k-1}(\xi)) \ , \quad -1 \leq \xi \leq 1 \ , \quad k \geq 1 \ ,$$

we can replace L_{N+1}' with $(2N+1)L_N + L_{N-1}'$. Since L_N and L_N' have different parity, we conclude easily that $\alpha_0 = \alpha_N = \beta_N = \beta_{N+1} = 0$. $\quad\square$

Equivalently, the above theorem states that the space D_N has codimension 8 in $\mathbb{Q}_N$.

The existence of spurious modes that are different from the constant functions reveals that the compatibility condition can't be satisfied taking $Q_N = \mathbb{Q}_N$ as pressure space. Of course, this condition holds taking $Q_N = D_N$; however, the constant β in (9.2.9) is not uniformly bounded from below with respect to N. A crucial step to prove this property relies on the following result (for the proof see Bernardi and Maday (1992), p. 130).

Lemma 9.5.1 *For any $q_N \in D_N$ there exists one polynomial $\mathbf{v}_N \in V_N$ such that*

(9.5.18) $$\operatorname{div} \mathbf{v}_N = q_N$$

and

(9.5.19) $$||\mathbf{v}_N|| \leq CN||q_N||_0 \ .$$

Owing to this lemma, for all $q_N \in D_N$ it is possible to find $\mathbf{v}_N \in V_N$ such that

$$|b(\mathbf{v}_N, q_N)| = \left| \int_\Omega q_N \operatorname{div} \mathbf{v}_N \, d\mathbf{x} \right| = \int_\Omega q_N^2 \, d\mathbf{x} \geq C \, N^{-1}||\mathbf{v}_N|| \, ||q_N||_0 \ .$$

Thus, (V_N, D_N) satisfy the compatibility condition with a constant β larger than $C N^{-1}$. This result cannot be improved, in the sense that it is not possible to find a constant β satisfying condition (9.2.9) that goes to zero slower than N^{-1} as N goes to infinity. Thus, we conclude that we have

$$(9.5.20) \qquad\qquad \beta = CN^{-1}$$

for the spectral Galerkin approximation to the Stokes problem when

$$V_N = (\mathbb{Q}_N^0)^2 \quad \text{and} \quad Q_N = D_N \ .$$

We need, however, to notice that choosing $Q_N = D_N$ as pressure space is not convenient from the point of view of consistency. Indeed, the fact that D_N is orthogonal, for instance, to the polynomial $L_N'(x_1)L_N'(x_2)$ implies that some low degree polynomials cannot belong to D_N. As a consequence, D_N is unsuitable to approximate all functions of the physical pressure space $L_0^2(\Omega)$.

In this respect, a more convenient choice, indicated in Bernardi, Maday and Metivet (1987), is to take Q_N as the space of all polynomials of $\mathbb{Q}_N$ which are orthogonal to the following polynomials

$$1, \ L_N(x_1), \ L_N(x_2), \ L_N(x_1)L_N(x_2), \ A_{N-1}(x_1)A_{N-1}(x_2),$$
$$A_{N-1}(x_1)A_N(x_2), \ A_N(x_1)A_{N-1}(x_2), \ A_N(x_1)A_N(x_2) \ ,$$

where

$$A_{N-1} := L_N' - P_{[\lambda N]}L_N' \ , \quad A_N := P_{N-1}L_{N+1}' - P_{[\lambda N]}L_{N+1}' \ .$$

For any $0 < \lambda < 1$, $[\lambda N]$ denotes the integral part of λN, and $P_{[\lambda N]}$ is the orthogonal projection operator upon the space of one-dimensional polynomials of degree less than or equal to $[\lambda N]$ with respect to the scalar product defined in (4.4.5).

This choice looks rather complicated; however, it is suitable from the theoretical point of view, as the compatibility condition is satisfied with a constant β that behaves the same as in (9.5.20). Further, the space Q_N chosen with the above criterion is rich enough to include all polynomials of degree less than or equal to $[\lambda N]$ with zero average. Thus Q_N enjoys an optimal convergence property. Indeed, using (9.2.12) and keeping in mind (9.5.20), we deduce the following error estimate for the pressure

$$(9.5.21) \qquad \|p - p_N\|_0 \le CN^{2-s}(\|\mathbf{u}\|_s + \|p\|_{s-1}) \ , \quad s \ge 2 \ ,$$

provided the exact solution $(\mathbf{u}, p)$ has the smoothness required from the right hand side. Let us notice that (9.5.21) is optimal up to a factor N which is precisely the inverse of the constant β.

9.5.2 Spectral Collocation Approximation

Most often, collocation methods are preferred to the Galerkin method for the spectral approximation to the Stokes equations. Besides other reasons, let us mention the difficulty of calculating the right hand side of (9.5.9), which is overcome with the collocation approach as all integrals are replaced by Gaussian numerical integration.

Set $\Omega = (-1,1)^2$, let N be a positive integer, and denote by Ω_N the $(N+1)^2$ Legendre Gauss-Lobatto nodes (see (4.5.38)), by Ξ_N the subset of Ω_N of the $(N-1)^2$ nodes belonging to the interior of Ω, and by Σ_N the subset of the $4N$ nodes lying on the boundary $\partial\Omega$. The collocation problem that approximates (9.1) when $\Omega = (-1,1)^2$ reads

$$(9.5.22) \quad \begin{cases} a_0\mathbf{u}_N - \nu\Delta\mathbf{u}_N + \nabla p_N = \mathbf{f} & \text{at } \mathbf{x} \in \Xi_N \\[2mm] \text{div } \mathbf{u}_N = 0 & \text{at } \mathbf{x} \in \Omega_N \setminus R_N \\[2mm] \mathbf{u}_N = 0 & \text{at } \mathbf{x} \in \Sigma_N \quad, \end{cases}$$

where R_N is a set of eight nodes. Four of them are the corners of $\overline{\Omega}$; the others, say $\{\mathbf{x}^1, \mathbf{x}^2, \mathbf{x}^3, \mathbf{x}^4\}$, can be taken in any way so that the Gram matrix $\phi_i(\mathbf{x}^j)$, $i,j = 1,...,4$, is non-singular, where $\phi_1(\mathbf{x}) = 1$, $\phi_2(\mathbf{x}) = L_N(x_1)$, $\phi_3(\mathbf{x}) = L_N(x_2)$, $\phi_4(\mathbf{x}) = L_N(x_1)L_N(x_2)$.

Using the discrete scalar product (4.5.39) and property (4.5.40), the collocation problem (9.5.22) can be equivalently restated as

$$(9.5.23) \quad \begin{cases} \text{find } \mathbf{u}_N \in V_N, \ p_N \in Q_N : \\[2mm] a_N(\mathbf{u}_N, \mathbf{v}_N) + b_N(\mathbf{v}_N, p_N) = (\mathbf{f}, \mathbf{v}_N)_N & \forall\, \mathbf{v}_N \in V_N \\[2mm] b_N(\mathbf{u}_N, q_N) = 0 & \forall\, q_N \in Q_N \quad. \end{cases}$$

Symbol explanation is as follows. V_N is the space defined in (9.5.1), while Q_N is a space of dimension $(N+1)^2 - 8$, which is supplementary to the space S_N^c spanned by the functions

$$(9.5.24) \quad \begin{array}{l} 1, \ L_N(x_1), \ L_N(x_2), \ L_N(x_1)L_N(x_2), \ L_N'(x_1)L_N'(x_2), \\[2mm] L_N'(x_1)x_2 L_N'(x_2), \ x_1 L_N'(x_1)L_N'(x_2), \ x_1 L_N'(x_1)x_2 L_N'(x_2) \quad. \end{array}$$

Moreover,

$$(9.5.25) \quad \begin{array}{l} a_N(\mathbf{w}_N, \mathbf{v}_N) := a_0(\mathbf{w}_N, \mathbf{v}_N)_N + \nu(\nabla\mathbf{w}_N, \nabla\mathbf{v}_N)_N \\[2mm] b_N(\mathbf{v}_N, q_N) := -(q_N, \text{div }\mathbf{v}_N)_N \quad, \end{array}$$

where $(\cdot, \cdot)_N$ is the discrete scalar product defined in (4.4.14) and $\mathbf{w}_N, \mathbf{v}_N \in V_N$, $q_N \in Q_N$.

Problem (9.5.23) therefore appears as a generalized Galerkin approximation to (9.1.11). As in the Galerkin case, Q_N contains $\mathbb{Q}_{[\lambda N]}$ for $0 < \lambda < 1$.

For such an approach, a theorem of stability and convergence that generalizes Theorems 7.4.2 and 7.4.3 can be proven in a straightforward way. In particular, in the current case, the compatibility condition reads

$$(9.5.26) \quad \forall \, q_N \in Q_N \;\; \exists \, \mathbf{v}_N \in V_N \;:\; (q_N, \operatorname{div} \mathbf{v}_N)_N \geq \beta \, \|\mathbf{v}_N\| \, \|q_N\|_0 \;.$$

With the above choice of finite dimensional subspaces, (9.5.23) is free of spurious pressure modes, i.e., of functions p_N^* such that

$$(9.5.27) \quad\quad p_N^* \in Q_N \;,\; (p_N^*, \operatorname{div} \mathbf{v}_N)_N = 0 \quad \forall \, \mathbf{v}_N \in V_N \;.$$

By analogy with the Galerkin case, it can be shown that (9.5.26) is satisfied with a constant β that behaves as in (9.5.20). Thus, an error estimate similar to (9.5.8) and (9.5.21) can also be obtained in this case. The difference comes from the presence of an extra term involving the right hand side $\mathbf{f}$. Precisely, the new estimates read

$$(9.5.28) \quad\quad \|\mathbf{u} - \mathbf{u}_N\| \leq C N^{1-s} \|\mathbf{u}\|_s + N^{-r} \|\mathbf{f}\|_r$$

and

$$(9.5.29) \quad\quad \|p - p_N\|_0 \leq C N^{2-s} (\|\mathbf{u}\|_s + \|p\|_{s-1}) + N^{1-r} \|\mathbf{f}\|_r \;,$$

for some $s \geq 2$ and $r \geq 2$ for which the norms at the right hand side make sense. Also this approximation can be shown to enjoy the property that $\operatorname{div} \mathbf{u}_N = 0$ identically.

The algorithmic aspects of this method are discussed in Bernardi and Maday (1992), pp. 153–155.

9.5.3 Spectral Generalized Galerkin Approximation

This method, which is due to Maday, Meiron, Patera and Rønquist (1993), is no longer equivalent to a collocation method. As usual, we take $V_N = (\mathbb{Q}_N^0)^2$, while this time $Q_N = \mathbb{Q}_{N-2} \cap L_0^2(\Omega)$. The spectral problem reads

$$(9.5.30) \quad \begin{cases} \text{find } \mathbf{u}_N \in V_N, \; p_N \in Q_N : \\[2mm] a_N(\mathbf{u}_N, \mathbf{v}_N) + b(\mathbf{v}_N, p_N) = (\mathbf{f}, \mathbf{v}_N)_N \quad \forall \, \mathbf{v}_N \in V_N \\[2mm] b(\mathbf{u}_N, q_N) = 0 \hspace{3.2cm} \forall \, q_N \in Q_N \;, \end{cases}$$

where $a_N(\cdot, \cdot)$ is defined in (9.5.25) and $b(\cdot, \cdot)$ in (9.1.10). Owing to the exactness of the Legendre Gauss-Lobatto integration formula, and the fact that the degree of the polynomials of Q_N is now less than or equal to $N - 2$, we deduce

$$b(\mathbf{v}_N, q_N) = -(\operatorname{div} \mathbf{v}_N, q_N) = b_N(\mathbf{v}_N, q_N) \quad \forall \, \mathbf{v}_N \in V_N \;,\; q_N \in Q_N \;.$$

Owing to this property we obtain from (9.5.30) that:

$$a_0 \mathbf{u}_N - \nu \Delta \mathbf{u}_N + \nabla p_N = \mathbf{f} \quad \text{at} \quad \mathbf{x} \in \Xi_N$$
$$\mathbf{u}_N = 0 \qquad\qquad\qquad \text{at} \quad \mathbf{x} \in \Sigma_N \ .$$

Moreover, for any $\mathbf{x} = (\xi_i, \xi_j) \in \Xi_N$ we obtain that $\operatorname{div} \mathbf{u}_N(\mathbf{x})$ turns out to be a suitable linear combination of the values of $\operatorname{div} \mathbf{u}_N$ at the four points $(-1, \xi_j)$, $(1, \xi_j)$, $(\xi_i, -1)$, $(\xi_i, 1)$. This explains why (9.5.30) fails to be a complete collocation method.

The interest of this method is that the pressure space Q_N is now easy to characterize, and, most of all, it is free of spurious modes. Furthermore it can be proven that

$$(9.5.31) \quad \forall \, q_N \in Q_N \ \exists \, \mathbf{v}_N \in V_N \ : \quad b(\mathbf{v}_N, q_N) \geq C N^{-1/2} \|\mathbf{v}_N\| \, \|q_N\|_0 \ ,$$

i.e., the compatibility condition holds with a constant β that behaves like $N^{-1/2}$. Due to this fact, under the usual smoothness assumptions the error estimate takes the following form

$$(9.5.32) \quad \begin{aligned} \|\mathbf{u} - \mathbf{u}_N\| &+ N^{-1/2} \|p - p_N\|_0 \\ &\leq C \big[N^{1-s} (\|\mathbf{u}\|_s + \|p\|_{s-1}) + N^{-r} \|\mathbf{f}\|_r \big] \ , \end{aligned}$$

which is preferable to (9.5.29).

9.6 Solving the Stokes System

Here, we deal with the solution to the Stokes system (9.2.14). Some of the methods we are going to consider are actually based on a restatement of the differential system. The latter is then accordingly approximated, therefore yielding a numerical solution that doesn't necessarily coincide with that of (9.2.14).

We recall that the matrix S is symmetric but indefinite; it is non-singular if $\ker B^T = \mathbf{0}$, which is the case if the compatibility condition (9.2.9) holds. We report, hereafter, those solution techniques that have deserved broad attention in the literature. For a more complete survey see, e.g., Pironneau (1988), Gunzburger (1989), Quartapelle (1993) and Fortin (1993) for finite element approximations and Canuto, Hussaini, Quarteroni and Zang (1988), Boyd (1989) and Rønquist (1988) for spectral approximations.

We warn the reader that in this Section we are using the same notation for the vector function $\mathbf{u} = \mathbf{u}(\mathbf{x})$, the solution to the infinite dimensional problem (9.1), and the finite dimensional vector $\mathbf{u} \in \mathbb{R}^{N_h}$, the solution to the linear system (9.2.14).

9.6.1 The Pressure-Matrix Method

This method is based on the elimination procedure (9.2.17)-(9.2.18), and consists of generating an independent linear system for the pressure after elimination of the velocity vector. The resulting problem reads

$$(9.6.1) \qquad\qquad\qquad R\mathbf{p} = \mathbf{g}$$

where R is the $K_h \times K_h$ pressure-matrix given in (9.2.19) and $\mathbf{g} = BA^{-1}\mathbf{f}$. We have already noticed that R is symmetric and, when $\ker B^T = \mathbf{0}$, positive definite.

If $a_0 = 0$ in (9.1), then R is well-conditioned. In fact, νR is an approximation of the operator $\operatorname{div} \varDelta^{-1}\nabla$, which behaves like the identity operator. Thus νR is close to the variational equivalent of the identity, i.e., the pressure mass matrix $J := (\psi_l, \psi_m)_{lm}$ (see (9.2.15)). More precisely, denoting with $\chi_{sp}(R)$ the spectral condition number of R (see (2.4.8)), it can be proven that there exist two constants C_0 and C_1 such that

$$(9.6.2) \qquad\qquad\qquad C_0\beta^{-1} \le \chi_{sp}(R) \le C_1\beta^{-1}$$

where β is the constant of the compatibility condition (9.2.9) (see Fortin and Pierre (1992)). If β is independent of h, then $\chi_{sp}(R)$ is independent of h too.

In any such case, the conjugate gradient method can be efficiently used for the solution of (9.6.1), as its convergence rate is independent of the dimension of R (see Section 2.4.2). For each conjugate gradient iteration, the key step is the residual evaluation

$$\mathbf{r}^k = \mathbf{g} - R\mathbf{p}^k \ ,$$

which can be accomplished as follows:

(i)   compute first $\mathbf{q}^k = B^T\mathbf{p}^k$;
(ii)   then solve $A\mathbf{y}^k = \mathbf{q}^k$;
(iii) finally compute $\mathbf{r}^k = \mathbf{g} - B\mathbf{y}^k$.

Points (i) and (iii) are trivial as they simply require matrix-vector multiplications. Instead, point (ii) entails the solution of two linear systems with the same matrix, each of them being a Poisson problem for one velocity component. Since A is symmetric and positive definite, step (ii) can be faced by a direct method (through a Cholesky decomposition of A) or again by conjugate gradient inner iterations (see Chapter 2). In the latter case, a preconditioner for A oughts to be used, as the condition number of A grows with the dimension N_h of A (see (6.3.17) and (6.3.25)). Further, let us also notice that this second way to solve step (ii) can be fairly expensive, since at each step of the outer iteration on R it would be necessary to iterate until convergence for computing the vector $\mathbf{y}^k$. We will see in Section 9.6.7 a different approach which doesn't present this drawback.

For approximations based on spectral methods, the constant β of the compatibility condition may in some cases depend on N, the degree of the polynomial solution (see (9.5.20) and (9.5.31)). However, even in these cases the system (9.6.1) is faced by conjugate gradient outer iterations without preconditioning (or with the matrix J/ν as a preconditioner).

When a_0 is different from 0 in (9.1) (let us recall that this case occurs, e.g., for time-dependent problems) the spectral condition number of R is no longer independent of the matrix dimension. Indeed, in this case $\chi_{sp}(R)$ is proportional to the spectral condition number of A, which grows like N_h in the finite element case and like N_h^2 in the case of spectral approximations. Using a preconditioner P for system (9.6.1) is therefore in order. Here, let us review some examples that have been suggested in this respect.

To start with, let us notice that from (9.1) it follows (formally!) that the pressure p satisfies

$$\begin{cases} -\Delta p = -\operatorname{div} \mathbf{f} & \text{in } \Omega \\[2mm] \dfrac{\partial p}{\partial n} = \mathbf{f} \cdot \mathbf{n} + \nu \Delta \mathbf{u} \cdot \mathbf{n} - a_0 \mathbf{u} \cdot \mathbf{n} & \text{on } \partial\Omega \end{cases} .$$

Since $\mathbf{u} = \mathbf{0}$ on $\partial\Omega$, by neglecting $\nu \Delta \mathbf{u} \cdot \mathbf{n}$ when ν is very small we see that p satisfies a Poisson problem with Neumann boundary conditions on $\partial\Omega$. Thus it looks reasonable to define P as the $K_h \times K_h$ matrix associated with the Laplace operator with Neumann condition.

This choice, that was proposed by Benqué, Ibler, Keramsi and Labadie (1980), has been generalized later on by Cahouet and Chabard (1988) who suggested to take P as a suitable approximation of the operator $(\nu I^{-1} - a_0 \Delta^{-1})^{-1}$, still with Neumann boundary condition. At the discrete level, this corresponds to take as a preconditioner the matrix

$$P = (\nu J^{-1} + a_0 H^{-1})^{-1} ,$$

where $H := BM^{-1}B^T$ and $M_{ij} := (\varphi_i, \varphi_j)$ is the velocity mass matrix (see (9.2.15)). With this choice, $P^{-1}R$ turns out to be an approximation to the operator

$$-(\nu I^{-1} - a_0 \Delta^{-1}) \operatorname{div}(a_0 I - \nu\Delta)^{-1}\nabla$$

which behaves like the identity when $\nu = 0$ or $a_0 = 0$. Experimental results show that the spectral condition number of $P^{-1}R$ is independent of the finite element mesh size, at least for those finite element approximations that satisfy the compatibility condition (9.2.9) (see Cahouet and Chabard (1988)).

A slightly more general approach is to take an approximation to $(\mu_1 I^{-1} - \mu_2 \Delta^{-1})^{-1}$, where μ_1, μ_2 are chosen according to the kind of boundary conditions that are prescribed for the Stokes problem.

9.6.2 The Uzawa Method

This method is inspired by the following iteration procedure on the original problem (9.1). Let p^0 be given; for any $k \geq 0$ solve

(9.6.3)
$$a_0 \mathbf{u}^{k+1} - \nu \Delta \mathbf{u}^{k+1} = \mathbf{f} - \nabla p^k$$

with $\mathbf{u}^{k+1} = \mathbf{0}$ on $\partial \Omega$, and then

(9.6.4)
$$p^{k+1} - p^k = -\rho \operatorname{div} \mathbf{u}^{k+1} \quad .$$

The positive real number ρ is an acceleration parameter. If ρ is chosen suitably, this procedure is nothing but the gradient method applied to the dual of saddle-point functional (9.1.12). Also, notice that (9.6.4) can be regarded as a backward Euler approximation of the pseudo-evolutionary continuity equation

$$\frac{\partial p}{\partial t} + \operatorname{div} \mathbf{u} = 0 \quad ,$$

with ρ playing the role of a time-step Δt.

At the discrete level, the Uzawa scheme is indeed the preconditioned Richardson method (see Section 2.4.1) applied to equation (9.6.1), with the pressure mass matrix J as a preconditioner.

The convergence of (9.6.3)-(9.6.4) is achieved for $0 < \rho < 2\nu$ (for a proof, see, e.g., Temam (1984)). However, it is generally slow, and can be accelerated using in (9.6.4) a preconditioner P for the pressure-matrix R of the same type as those considered throughout the previous section. The algebraic restatement of the preconditioned procedure (9.6.3) reads

(9.6.5)
$$\begin{cases} A\mathbf{u}^{k+1} = \mathbf{f} - B^T \mathbf{p}^k \\[2mm] P(\mathbf{p}^{k+1} - \mathbf{p}^k) = \rho B \mathbf{u}^{k+1} \quad . \end{cases}$$

The convergence rate of this iterative method is usually independent of the mesh size h.

9.6.3 The Arrow-Hurwicz Method

A more complicated scheme (but requiring the same computational effort of the Uzawa method for each iteration) is defined as follows. Let p_0 and $\mathbf{u}_0$ be given, for $k \geq 0$ solve

(9.6.6)
$$\beta_0 \mathbf{u}^{k+1} - \Delta \mathbf{u}^{k+1} = \beta_0 \mathbf{u}^k - \Delta \mathbf{u}^k + \rho(\mathbf{f} - \nabla p^k - a_0 \mathbf{u}^k + \nu \Delta \mathbf{u}^k) \quad ,$$

with $\mathbf{u}^{k+1} = \mathbf{0}$ on $\partial \Omega$, and then

(9.6.7)
$$p^{k+1} - p^k = -\frac{\rho}{\sigma} \operatorname{div} \mathbf{u}^{k+1} \quad .$$

The parameters $\beta_0 \geq 0$, $\rho > 0$ and $\sigma > 0$ are chosen to ensure convergence. A simple choice is often given by $\beta_0 = a_0/\nu$. A proof of convergence in the simplified case $\beta_0 = a_0 = 0$ can be found, e.g., in Temam (1984). The proof in the general case follows a similar procedure and convergence is achieved for $0 < \rho < \min(2\beta_0/a_0, 2\sigma\nu/(1 + \sigma\nu^2))$.

Let us also notice that other types of symmetric and coercive preconditioning operators could be used instead of $\beta_0 I - \Delta$ (typically, at the discrete level, suitable symmetric and positive definite preconditioners P for the matrix A/ν).

9.6.4 Penalty Methods

A different approach is based on the so-called penalty method, where the functional to be minimized is modified in the following way

$$(9.6.8) \qquad \mathcal{I}_\varepsilon(\mathbf{v}) := \frac{a_0}{2} \int_\Omega |\mathbf{v}|^2 + \frac{\nu}{2} \int_\Omega |\nabla \mathbf{v}|^2 + \frac{1}{2\varepsilon} \int_\Omega |\operatorname{div} \mathbf{v}|^2 - (\mathbf{f}, \mathbf{v}) \ ,$$

$\varepsilon > 0$, and the minimum is now taken over $(H_0^1(\Omega))^d$.

This corresponds to solve

$$(9.6.9) \qquad \begin{cases} a_0 \mathbf{u}_\varepsilon - \nu \Delta \mathbf{u}_\varepsilon - \dfrac{1}{\varepsilon} \nabla \operatorname{div} \mathbf{u}_\varepsilon = \mathbf{f} & \text{in } \Omega \\[2mm] \mathbf{u}_\varepsilon = 0 & \text{on } \partial\Omega \ , \end{cases}$$

or, equivalently

$$(9.6.10) \qquad \begin{cases} a_0 \mathbf{u}_\varepsilon - \nu \Delta \mathbf{u}_\varepsilon + \nabla p_\varepsilon = \mathbf{f} & \text{in } \Omega \\[2mm] \mathbf{u}_\varepsilon = 0 & \text{on } \partial\Omega \\[2mm] \varepsilon p_\varepsilon + \operatorname{div} \mathbf{u}_\varepsilon = 0 & \text{in } \Omega \ . \end{cases}$$

The constraint $\operatorname{div} \mathbf{u} = 0$ is no more satisfied, but it can be proven that u_ε and p_ε converge as $\varepsilon \to 0$ to the solution of (9.1), with first order accuracy (see, e.g., Bercovier (1978), Temam (1984)). It must be noticed, however, that the spectral condition number of the matrix corresponding to (9.6.9) behaves like $1/\varepsilon$, and that recovering the pressure p_ε requires the additional solution of a linear system associated to the pressure mass matrix J.

A different penalty method reads as follows: for each $\varepsilon > 0$ find the solution $(\mathbf{u}_\varepsilon^*, p_\varepsilon^*)$ to

$$(9.6.11) \qquad \begin{cases} a_0 \mathbf{u}_\varepsilon^* - \nu \Delta \mathbf{u}_\varepsilon^* + \nabla p_\varepsilon^* = \mathbf{f} & \text{in } \Omega \\[2mm] \mathbf{u}_\varepsilon^* = 0 & \text{on } \partial\Omega \\[2mm] -\varepsilon \Delta p_\varepsilon^* + \operatorname{div} \mathbf{u}_\varepsilon^* = 0 & \text{in } \Omega \\[2mm] \dfrac{\partial p_\varepsilon^*}{\partial n} = 0 & \text{on } \partial\Omega \ . \end{cases}$$

A Galerkin finite element approximation of (9.6.11) would lead to the problem (9.4.1), (9.4.2) with h_K^2 replaced by ε in (9.4.2). Its analysis can be found in Brezzi and Pitkäranta (1984), and yields the error estimate

$$\|\mathbf{u} - \mathbf{u}_h\|_1 + \|p - p_h\|_0 = O(h) \quad,$$

having chosen V_h and Q_h as in (9.3.9) with $k = m = 1$.

We underline again that penalty methods can be seen indeed as suitable stabilization procedures for the Stokes system (see Section 9.4). However, these schemes are not strongly consistent (in the sense made precise in (8.3.6)), since the exact solution satisfies div $\mathbf{u} = 0$.

9.6.5 The Augmented-Lagrangian Method

This is a combination of the Uzawa method and the penalty method (9.6.10). Still in differential form, the augmented-Lagrangian iterations read: being given p^0, for any $k \geq 0$ solve the system:

$$(9.6.12) \qquad a_0 \mathbf{u}^{k+1} - \nu \Delta \mathbf{u}^{k+1} + \nabla p^{k+1} = \mathbf{f}$$

$$(9.6.13) \qquad p^{k+1} - p^k = -\rho \, \text{div} \, \mathbf{u}^{k+1}$$

with boundary conditions on $\mathbf{u}^{k+1}$; ρ is still an acceleration parameter.

The algebraic form of (9.6.13) reads

$$(9.6.14) \qquad J(\mathbf{p}^{k+1} - \mathbf{p}^k) = \rho B \mathbf{u}^{k+1} \quad,$$

where J is the pressure mass matrix, i.e., $J_{lm} = (\psi_l, \psi_m)$ (see (9.2.15)) for finite element or spectral Galerkin approximations, while J is the identity matrix for the spectral collocation method. Since J is in any case non-singular, $\mathbf{p}^{k+1}$ can formally be eliminated and replaced into (9.6.12) to get the system

$$(9.6.15) \qquad \left(A + \rho B^T J^{-1} B\right) \mathbf{u}^{k+1} = \mathbf{f} - B^T \mathbf{p}^k \quad.$$

This is a symmetric and positive definite system. Once it has been solved (e.g., by a conjugate gradient method with preconditioning) the pressure $\mathbf{p}^{k+1}$ can be recovered from (9.6.14).

The convergence of (9.6.12), (9.6.13) is achieved for any $\rho > 0$, and is very fast if ρ is very large (see Fortin and Pierre (1992)). However, it must be noticed that the spectral condition number of the matrix $(A + \rho B^T J^{-1} B)$ in (9.6.15) increases as far as ρ gets large (see Fortin and Glowinski (1983), p. 15). (Here it is assumed that ker $B \neq 0$, which is a necessary condition for finding non-vanishing velocity approximation.) Sometimes, the augmented-Lagrangian is used as preconditioner for the conjugate gradient method applied to problem (9.2.14) (see Fortin (1989)).

For a thorough presentation of the argumented-Lagrangian method we refer the reader to Glowinski and Le Tallec (1989).

9.6.6 Methods Based on Pressure Solvers

If the divergence operator is applied to the momentum equation in (9.1) we obtain

$$(9.6.16) \qquad \Delta p = \operatorname{div} \mathbf{f} \quad \text{in } \Omega$$

owing to the fact the $\mathbf{u}$ is divergence free. The Poisson problem (9.6.16) demands for a boundary condition on p or on its normal derivative. In a formal way, taking the normal component of the momentum equation on $\partial \Omega$ yields

$$(9.6.17) \qquad \frac{\partial p}{\partial n} = \mathbf{f} \cdot \mathbf{n} + \nu \Delta \mathbf{u} \cdot \mathbf{n} \quad \text{on } \partial \Omega$$

provided these terms make sense on $\partial \Omega$.

The above relation furnishes a Neumann condition for problem (9.6.16), which unfortunately relates $\mathbf{u}$ and p to one another. A possible way for decoupling $\mathbf{u}$ and p on $\partial \Omega$ is to resort to the following iterative procedure: p^0 is given, and for each $k \geq 0$ solve

$$(9.6.18) \qquad \begin{cases} a_0 \mathbf{u}^{k+1} - \nu \Delta \mathbf{u}^{k+1} = \mathbf{f} - \nabla p^k & \text{in } \Omega \\ \mathbf{u}^{k+1} = 0 & \text{on } \partial \Omega \quad, \end{cases}$$

then

$$(9.6.19) \qquad \begin{cases} -\Delta p^{k+1} = -\operatorname{div} \mathbf{f} \\ \dfrac{\partial p^{k+1}}{\partial n} = \mathbf{f} \cdot \mathbf{n} + \nu \Delta \mathbf{u}^{k+1} \cdot \mathbf{n} & \text{on } \partial \Omega \quad. \end{cases}$$

Another possibility is to endow (9.6.16) with a Dirichlet condition on the pressure. First of all, let us remark that if $(\mathbf{u}, p)$ is a classical solution to (9.1), then both conditions (9.6.16) and

$$(9.6.20) \qquad \operatorname{div} \mathbf{u} = 0 \quad \text{on } \partial \Omega$$

are satisfied. Conversely, if $(\mathbf{u}, p)$ is a classical solution of the momentum equation $(9.1)_1$ and of $(9.1)_3$, (9.6.16) and (9.6.20), then it is at once verified that

$$\begin{cases} a_0 \operatorname{div} \mathbf{u} - \nu \Delta \operatorname{div} \mathbf{u} = 0 & \text{in } \Omega \\ \operatorname{div} \mathbf{u} = 0 & \text{on } \partial \Omega \quad, \end{cases}$$

and therefore necessarily $\operatorname{div} \mathbf{u} = 0$ in Ω. Thus $(\mathbf{u}, p)$ would be a classical solution to the Stokes problem (9.1). Unfortunately, (9.6.20) is not an admissible boundary condition for the Poisson problem (9.6.16). One may overcome this difficulty by a different approach that we are going to describe.

The idea is to look for an unknown value λ on $\partial\Omega$ such that, if $p = p(\lambda)$ is the solution to the problem (9.6.16) with boundary condition

$$(9.6.21) \qquad\qquad p = \lambda \quad \text{on } \partial\Omega \ ,$$

then the solution $\mathbf{u} = \mathbf{u}(\lambda)$ to the Dirichlet problem

$$(9.6.22) \qquad \begin{cases} a_0\mathbf{u}(\lambda) - \nu\Delta\mathbf{u}(\lambda) = \mathbf{f} - \nabla p(\lambda) & \text{in } \Omega \\ \mathbf{u}(\lambda) = 0 & \text{on } \partial\Omega \end{cases}$$

satisfies (9.6.20). If this happens, in view of the previous derivation we would conclude that $(\mathbf{u}, p)$ is actually a solution to (9.1). This procedure can be implemented in several ways. We present two of them. The *influence matrix* method and the *integral condition* method.

Let us begin from the former. Let $H(\mu)$ be the harmonic extension of the boundary value μ, i.e., the solution to

$$(9.6.23) \qquad \begin{cases} \Delta H(\mu) = 0 & \text{in } \Omega \\ H(\mu) = \mu & \text{on } \partial\Omega \ . \end{cases}$$

The solution $p(\lambda)$ to (9.6.16), (9.6.21) can be written as $p(\lambda) = H(\lambda) + p_0$, where p_0 solves

$$(9.6.24) \qquad \begin{cases} \Delta p_0 = \operatorname{div}\mathbf{f} & \text{in } \Omega \\ p_0 = 0 & \text{on } \partial\Omega \ . \end{cases}$$

Similarly, the solution $\mathbf{u}(\lambda)$ to (9.6.22) can be split as $\mathbf{u}(\lambda) = \hat{\mathbf{u}}(\lambda) + \mathbf{u}_0$, where

$$(9.6.25) \qquad \begin{cases} a_0\hat{\mathbf{u}}(\lambda) - \nu\Delta\hat{\mathbf{u}}(\lambda) = -\nabla H(\lambda) & \text{in } \Omega \\ \hat{\mathbf{u}}(\lambda) = 0 & \text{on } \partial\Omega \end{cases}$$

and

$$(9.6.26) \qquad \begin{cases} a_0\mathbf{u}_0 - \nu\Delta\mathbf{u}_0 = \mathbf{f} - \nabla p_0 & \text{in } \Omega \\ \mathbf{u}_0 = 0 & \text{on } \partial\Omega \ . \end{cases}$$

Let us further introduce the function $\psi(\lambda)$ solution to

$$(9.6.27) \qquad \begin{cases} \Delta\psi(\lambda) = \operatorname{div}\mathbf{u}(\lambda) & \text{in } \Omega \\ \psi(\lambda) = 0 & \text{on } \partial\Omega \ . \end{cases}$$

Applying the divergence operator to $(9.6.25)_1$, from (9.6.16) and (9.6.27) it easily follows that

$$a_0 \Delta\psi(\lambda) - \nu\Delta^2\psi(\lambda) = 0 \quad \text{in } \Omega \ .$$

As a consequence, the condition $\frac{\partial\psi(\lambda)}{\partial n} = 0$ on $\partial\Omega$, if satisfied, would imply $\psi(\lambda) = \operatorname{div}\mathbf{u}(\lambda) = 0$ in Ω.

By proceeding formally, using the Green formula (1.3.3) and (9.6.23) and (9.6.27), for each boundary function μ in a suitable class Λ we have

$$\int_{\partial\Omega} \frac{\partial\psi(\lambda)}{\partial n}\, \mu = \int_\Omega \operatorname{div}[\nabla\psi(\lambda)\, H(\mu)]$$

$$= \int_\Omega \Delta\psi(\lambda)\, H(\mu) + \int_\Omega \nabla\psi(\lambda)\, \nabla H(\mu)$$

$$= \int_\Omega \operatorname{div}\mathbf{u}(\lambda)\, H(\mu) - \int_\Omega \psi(\lambda)\, \Delta H(\mu) = \int_\Omega \operatorname{div}\mathbf{u}(\lambda)\, H(\mu) \ .$$

We are then led to looking for that function $\lambda \in \Lambda$ such that

(9.6.28)
$$\int_\Omega \operatorname{div}\mathbf{u}(\lambda)\, H(\mu) = 0 \quad \forall\, \mu \in \Lambda \ .$$

Using again the Green formula (1.3.3), equation (9.6.28) can be equivalently reformulated as

(9.6.29)
$$\text{find } \lambda \in \Lambda \ : \ \mathcal{A}(\lambda, \mu) = \mathcal{F}(\mu) \quad \forall\, \mu \in \Lambda \ ,$$

where

$$\mathcal{A}(\eta, \mu) := a_0(\hat{\mathbf{u}}(\eta), \hat{\mathbf{u}}(\mu)) + \nu(\nabla\hat{\mathbf{u}}(\eta), \nabla\hat{\mathbf{u}}(\mu)) \ ,$$
$$\mathcal{F}(\mu) := -a_0(\mathbf{u}_0, \hat{\mathbf{u}}(\mu)) - \nu(\nabla\mathbf{u}_0, \nabla\hat{\mathbf{u}}(\mu)) \ .$$

The form $\mathcal{A}(\cdot, \cdot)$ is symmetric. Moreover, if Λ is suitably chosen, $\mathcal{A}(\cdot, \cdot)$ can be proven to be continuous and coercive on $\Lambda \times \Lambda$ (see Glowinski and Pironneau (1979), or Girault and Raviart (1986), p. 183). Hence Lax-Milgram lemma ensures the existence of a unique solution λ of (9.6.29).

From the numerical point of view, the size of problem (9.6.29) is the number of grid-points lying on $\partial\Omega$. This problem can be solved, for instance, by the conjugate gradient method. Once λ is available on $\partial\Omega$ the Poisson problem (9.6.16) can be approximated, and finally $\mathbf{u}$ can be recovered solving the Poisson problem (9.6.22), which, from the algebraic point of view, corresponds to two linear systems with the same matrix, one for each velocity component.

A detailed analysis and an efficient implementation of this method can be found in Glowinski and Pironneau (1979) (see also Chinosi and Comodi (1991)) for finite element approximations, in Kleiser and Schumann (1980), Canuto and Sacchi Landriani (1986) and Sacchi Landriani (1987) in the framework of spectral approximations.

The name influence matrix method is borrowed from the name of the matrix transforming the vector of λ at the grid-points on $\partial\Omega$ into the one of $\operatorname{div}\mathbf{u}(\lambda)$ at the same points.

The integral condition method has been proposed by Quartapelle and Napolitano (1986). The name originates from the fact that in this formulation the pressure is required to satisfy a condition of integral character, which, at some extent, plays the role of a pressure boundary condition. Its derivation is as follows (for definiteness, let us consider the three-dimensional case). Introducing the operator

$$(9.6.30) \qquad \mathbf{Curl}\,\mathbf{v} := (D_2 v_3 - D_3 v_2, D_3 v_1 - D_1 v_3, D_1 v_2 - D_2 v_1) \ ,$$

one easily verifies that

$$-\Delta \mathbf{v} = \mathbf{Curl}\,\mathbf{Curl}\,\mathbf{v} - \nabla \operatorname{div} \mathbf{v} \ .$$

Therefore the following Green formula holds:

$$(9.6.31) \qquad
\begin{aligned}
-\int_\Omega \Delta \mathbf{w} \cdot \mathbf{v} &= \int_\Omega (\operatorname{div} \mathbf{w} \operatorname{div} \mathbf{v} + \mathbf{Curl}\,\mathbf{w} \cdot \mathbf{Curl}\,\mathbf{v}) \\
&\quad - \int_{\partial\Omega} [(\operatorname{div} \mathbf{w})\,\mathbf{v} \cdot \mathbf{n} + (\mathbf{Curl}\,\mathbf{w}) \cdot \mathbf{n} \times \mathbf{v}] \ .
\end{aligned}$$

Let $\mathbf{K}(\mu)$ be the solution to

$$(9.6.32) \qquad
\begin{cases}
a_0 \mathbf{K}(\mu) - \nu \Delta \mathbf{K}(\mu) = \mathbf{0} & \text{in } \Omega \\[2mm]
\mathbf{K}(\mu) \cdot \mathbf{n} = \mu & \text{on } \partial\Omega \\[2mm]
\mathbf{K}(\mu) \times \mathbf{n} = \mathbf{0} & \text{on } \partial\Omega \ .
\end{cases}$$

Using (9.6.31) and (9.6.32) the solution $(\mathbf{u}, p)$ to (9.1) satisfies

$$(9.6.33) \quad \int_\Omega \nabla p \cdot \mathbf{K}(\mu) = \int_\Omega (\mathbf{f} - a_0 \mathbf{u} + \nu \Delta \mathbf{u}) \cdot \mathbf{K}(\mu) = \int_\Omega \mathbf{f} \cdot \mathbf{K}(\mu) \quad \forall \mu \in \Lambda \ .$$

As this integral relation has to be satisfied for each boundary function μ, it can be viewed as a boundary condition for the pressure. Starting from (9.6.33) we can formulate a variational problem whose solution provides the correct boundary value for the pressure. Having split $p(\lambda)$ as before, we seek $\lambda \in \Lambda$ such that

$$(9.6.34) \qquad \int_\Omega \nabla H(\lambda) \cdot \mathbf{K}(\mu) = \int_\Omega (\mathbf{f} - \nabla p_0) \cdot \mathbf{K}(\mu) \quad \forall \mu \in \Lambda \ .$$

This problem replaces (9.6.29); notice that the bilinear form at the left hand side is not symmetric.

Using (9.6.25) and (9.6.26) for rewriting $\nabla H(\lambda)$ and ∇p_0 and integrating by parts the terms containing $\hat{\mathbf{u}}(\lambda)$ and $\mathbf{u}_0$, from (9.6.31) it follows that the vector field $\mathbf{u}(\lambda)$ corresponding to the solution λ to (9.6.34) satisfies

$$(9.6.35) \qquad \nu \int_{\partial\Omega} \operatorname{div} \mathbf{u}(\lambda)\,\mu = 0 \quad \forall \mu \in \Lambda \ ,$$

i.e., div $\mathbf{u}(\lambda) = 0$ on $\partial\Omega$.

An extensive analysis of the integral condition method, of its effective implementation, as well as a comparison with the influence matrix method can be found in Quartapelle (1993), Sects. 5.3-5.5.

9.6.7 A Global Preconditioning Technique

A drawback of conjugate gradient iterations applied to the pressure-matrix R (or else of the Uzawa method, which corresponds to preconditioned Richardson iterations for R) is that at each step of the iterative procedure one needs to compute the action of R, thus, in particular, the action of A^{-1}.

If a double iteration is used, i.e., an outer conjugate gradient procedure is applied to (9.6.1) and an inner iteration is employed to solve the system related to A, a preconditioner for A (say, P_0) oughts to be used, as A is ill-conditioned. Following this procedure, it happens that the overall solution process can be somewhat costly, as at each step of the outer iteration it is necessary to iterate until convergence the inner one.

It is possible, however, to transform problem (9.2.14) into an equivalent one, still based on the unknowns $(\mathbf{u}, \mathbf{p})$, which turns out to be symmetric and positive definite with respect to a suitable scalar product, and which can be thus faced by the conjugate gradient method. The advantage now is that we only have to consider a single-level iteration procedure, and moreover it can be shown that it just requires at each step the action of the preconditioner P_0^{-1}. In many applications, the amount of work to solve (9.2.14) with this single-level iteration is comparable to that required for one evaluation of A^{-1} in the double iteration procedure.

The use of a preconditioner P_0 for the matrix A is also a characteristic feature of a Arrow-Hurwicz like method (see Section 9.6.3). However, this would require the selection of suitable iteration parameters ρ and σ, and it is not clear how to choose them in an optimal way.

In contrast, the application of the conjugate gradient method to the reformulated problem (9.6.38) (that we are going to introduce here below) does not require to find criteria for the determination of any parameter and furnishes in a natural way an optimally converging scheme. Here, we report this technique, that has been proposed by Bramble and Pasciak (1988), and which also applies to the system

(9.6.36)
$$\begin{cases} A\mathbf{u} + B^T\mathbf{p} = \mathbf{f} \\ B\mathbf{u} - D\mathbf{p} = \mathbf{g} \ , \end{cases}$$

under the assumptions that A is symmetric and positive definite, ker $B^T = 0$ and D is symmetric and non-negative definite. System (9.6.36) is slightly more general than (9.2.14), and it includes the case of stabilization methods that we have discussed in Section 9.4.

Let P_0 be a convenient symmetric and positive definite preconditioner of A (e.g., its incomplete Cholesky decomposition, or a finite element preconditioner if A is obtained by a spectral approximation), satisfying moreover

$$(9.6.37) \qquad K_1(P_0 \mathbf{v}, \mathbf{v}) \le (A\mathbf{v}, \mathbf{v}) \le K_2(P_0 \mathbf{v}, \mathbf{v}) \quad \forall \, \mathbf{v} \in \mathbb{R}^{N_h}$$

with $K_1 > 1$. As a consequence, $((A - P_0)\mathbf{v}, \mathbf{v}) \ge (K_1 - 1)|\mathbf{v}|^2$ for each $\mathbf{v} \in \mathbb{R}^{N_h}$. From (9.6.36) we deduce

$$P_0^{-1} A\mathbf{u} + P_0^{-1} B^T \mathbf{p} = P_0^{-1} \mathbf{f}$$

and also

$$\begin{aligned} 0 &= B P_0^{-1}(A\mathbf{u} + B^T \mathbf{p} - \mathbf{f}) \\ &= B P_0^{-1}(A - P_0)\mathbf{u} + B\mathbf{u} + B P_0^{-1} B^T \mathbf{p} - B P_0^{-1} \mathbf{f} \ . \end{aligned}$$

Replacing $B\mathbf{u}$ by $D\mathbf{p} + \mathbf{g}$ we find

$$(9.6.38) \qquad S^* \begin{pmatrix} \mathbf{u} \\ \mathbf{p} \end{pmatrix} = \begin{pmatrix} P_0^{-1} \mathbf{f} \\ B P_0^{-1} \mathbf{f} - \mathbf{g} \end{pmatrix} \ ,$$

where

$$S^* := \begin{pmatrix} P_0^{-1} A & P_0^{-1} B^T \\ B P_0^{-1}(A - P_0) & D + B P_0^{-1} B^T \end{pmatrix} \ .$$

Introducing the scalar product

$$\left[\begin{pmatrix} \mathbf{w} \\ \mathbf{r} \end{pmatrix}, \begin{pmatrix} \mathbf{v} \\ \mathbf{q} \end{pmatrix} \right] := ((A - P_0)\mathbf{w}, \mathbf{v}) + (\mathbf{r}, \mathbf{q}) \ ,$$

it can be seen that S^* is symmetric and positive definite with respect to $[\cdot, \cdot]$; more precisely, one finds

$$\left[S^* \begin{pmatrix} \mathbf{v} \\ \mathbf{q} \end{pmatrix}, \begin{pmatrix} \mathbf{v} \\ \mathbf{q} \end{pmatrix} \right] \ge c_0 \left[\begin{pmatrix} \mathbf{v} \\ (D + B A^{-1} B^T)\mathbf{q} \end{pmatrix}, \begin{pmatrix} \mathbf{v} \\ \mathbf{q} \end{pmatrix} \right]$$

for a suitable constant $c_0 > 0$. Moreover, the spectral condition number $\chi_{sp}(S^*)$ satisfies

$$K_3^{-1} \chi_{sp}(\hat{S}) \le \chi_{sp}(S^*) \le K_3 \chi_{sp}(\hat{S}) \ ,$$

where $K_3 > 0$ is related to the constants K_1 and K_2 in (9.6.37), and

$$\hat{S} := \begin{pmatrix} I & 0 \\ 0 & D + B A^{-1} B^T \end{pmatrix} \ .$$

Thus the conjugate gradient method can be successfully applied to (9.6.38), yielding at each step the solution of linear systems associated to the matrix P_0, which can be carried out by direct methods.

Notice finally that, in applications, the selected preconditioner P_0 may not satisfy condition (9.6.37) with $K_1 > 1$. A scaling procedure is thus in order;

the optimal scaling factor η^* can be determined, for instance, by estimating the lowest eigenvalue η_1 of $P_0^{-1}A$ (recall that, by Theorem 2.5.1, the largest possible constant K_1 in (9.6.37) is given indeed by η_1), choosing then η^* smaller than but close enough to η_1.

9.7 Complements

In the last years, the stabilization methods described in Section 9.4 have been intensively used for numerical computations. An account is given in Tezduyar (1992).

In the spectral collocation case, another approach can be based on the use of a *staggered grid*: the usual Gauss-Lobatto nodes are employed to represent velocities while Gauss nodes (internal to Ω) are associated with the pressure field (see, e.g., Bernardi and Maday (1992)).

When the Chebyshev (rather than the Legendre) collocation points are used, in the matrix S associated to the Stokes operator the non-diagonal blocks are no longer transposed to one another. This is due to the fact that the ∇ operator is not the adjoint of $(-\operatorname{div})$ when taken with respect to the Chebyshev discrete scalar product. Therefore, in this case the analysis of the collocation approximation of the Stokes problem has to be performed using the theory developed by Bernardi, Canuto and Maday (1988) (see also Section 7.5).

A global preconditioning technique (different from the one presented in Section 9.6.7) consists of using the following block-preconditioner for system (9.6.36)

$$P = \begin{pmatrix} P_0 & 0 \\ 0 & Q_0 \end{pmatrix} \; ,$$

where P_0 is still a preconditioner of A, while Q_0 is a preconditioner of the pressure mass matrix J. If P_0 is an optimal preconditioner of A and Q_0 of J, then P turns out to be an optimal preconditioner for (9.6.36), provided that there exists $\kappa > 0$, independent of h, such that

$$\kappa(J\mathbf{q}, \mathbf{q}) \leq ((D + BA^{-1}B^T)\mathbf{q}, \mathbf{q}) \quad \forall\, \mathbf{q} \in \mathbb{R}^{K_h}$$

(see Wathen and Silvester (1993), Silvester and Wathen (1994)).

We conclude mentioning the possibility of facing the Stokes system (9.2.14) directly by a multi-grid method. The interested reader is referred to Verfürth (1984a) and Wittum (1987, 1988) for finite element approximations and to Heinrichs (1992, 1993) for discretizations based on spectral methods.

10. The Steady Navier-Stokes Problem

This Chapter addresses the steady Navier-Stokes equations, which describe the motion, independent of time, of a homogeneous incompressible fluid. A complete derivation of these equations will be provided in Section 13.1.

After introducing the weak form of the problem, we focus on the approximation of branches of non-singular solutions by finite elements and spectral methods. Next, we analyze several iterative procedures to solve the system of nonlinear equations. We finish by considering the formulation of the Navier-Stokes equations in term of the stream function and vorticity variables.

10.1 Mathematical Formulation

The Navier-Stokes equations provide a model of the flow motion of an homogeneous incompressible Newtonian fluid. In the steady case they read

(10.1.1) $$-\nu\Delta\mathbf{u} + (\mathbf{u}\cdot\nabla)\mathbf{u} + \nabla p = \mathbf{f} \quad \text{in } \Omega$$

(10.1.2) $$\text{div}\,\mathbf{u} = 0 \quad \text{in } \Omega \ ,$$

where we have set

$$(\mathbf{u}\cdot\nabla)\mathbf{u} := \sum_{i=1}^{d} u_i D_i \mathbf{u} \ .$$

Moreover, Ω is a bounded domain of $\mathbb{R}^d$, $d = 2, 3$, with a Lipschitz continuous boundary $\partial\Omega$, and we retain the notations of Chapter 9. Therefore, $\mathbf{u}$ denotes the velocity of the fluid, p the ratio between its pressure and density, $\mathbf{f}$ is the external force field per unit mass and $\nu > 0$ is the constant kinematic viscosity. The derivation of equations (10.1.1), (10.1.2) is presented in Section 13.1.

The above equations need to be supplemented by some boundary conditions. For the sake of simplicity, we consider the following homogeneous Dirichlet condition:

(10.1.3) $$\mathbf{u} = 0 \quad \text{on } \partial\Omega \ ,$$

which describes a fluid confined into a domain Ω whose boundary is fixed. Other boundary conditions are admissible as well, and account for different kind of physical situations. A short description is given in Section 10.1.1 below.

The Navier-Stokes system differs from the Stokes one due to the presence of the nonlinear convective term $(\mathbf{u} \cdot \nabla)\mathbf{u}$. If we consider the two Hilbert spaces $V = (H_0^1(\Omega))^d$ and $Q = L_0^2(\Omega)$ as in Section 9, the weak formulation of (10.1.1)-(10.1.3) can be stated as follows: given $\mathbf{f} \in (L^2(\Omega))^d$,

(10.1.4)
$$\begin{cases} \text{find } \mathbf{u} \in V , \ p \in Q : \\[2mm] a(\mathbf{u}, \mathbf{v}) + c(\mathbf{u}; \mathbf{u}, \mathbf{v}) + b(\mathbf{v}, p) = (\mathbf{f}, \mathbf{v}) \quad \forall \, \mathbf{v} \in V \\[2mm] b(\mathbf{u}, q) = 0 \hspace{3.5cm} \forall \, q \in Q \ . \end{cases}$$

We recall that
$$a : V \times V \to \mathbb{R} , \ b : V \times Q \to \mathbb{R}$$

are the bilinear forms $a(\mathbf{w}, \mathbf{v}) := \nu(\nabla \mathbf{w}, \nabla \mathbf{v})$ and $b(\mathbf{v}, q) := -(q, \text{div } \mathbf{v})$, where $(\cdot, \cdot)$ denotes the scalar product in $L^2(\Omega)$ or $(L^2(\Omega))^d$. Besides, $c : V \times V \times V \to \mathbb{R}$ is defined by

(10.1.5) $$c(\mathbf{w}; \mathbf{z}, \mathbf{v}) := \int_\Omega [(\mathbf{w} \cdot \nabla)\mathbf{z}] \cdot \mathbf{v} = \sum_{i,j=1}^d \left(w_j \frac{\partial z_i}{\partial x_j}, v_i \right) \ ,$$

and is the trilinear form associated with the nonlinear convective term.

At first, we notice that the form $c(\cdot; \cdot, \cdot)$ is continuous on $(H^1(\Omega))^d$. Indeed, by using Hölder inequality it is easy to see that

$$\left| \int_\Omega w_j \frac{\partial z_i}{\partial x_j} v_i dx \right| \leq ||w_j||_{L^4(\Omega)} \left\| \frac{\partial z_i}{\partial x_j} \right\|_0 ||v_i||_{L^4(\Omega)} \ .$$

Since $H^1(\Omega) \subset L^4(\Omega)$ for $d = 2, 3$ (see Theorem 1.3.4), owing to the Poincaré inequality (1.3.2) we conclude that there exists a constant $\hat{C} > 0$ such that

(10.1.6) $$|c(\mathbf{w}; \mathbf{z}, \mathbf{v})| \leq \hat{C} |\mathbf{w}|_1 |\mathbf{z}|_1 |\mathbf{v}|_1 \quad \forall \, \mathbf{w}, \mathbf{z}, \mathbf{v} \in (H_0^1(\Omega))^d \ .$$

(See Section 1.2 for the definition of norms and seminorms in Sobolev spaces.) In particular, for any fixed $\mathbf{w} \in V$, the map $\mathbf{v} \to c(\mathbf{w}; \mathbf{w}, \mathbf{v})$ is linear and continuous on V, i.e., it is an element of V'.

Alternatively to (10.1.4) we can formulate the Navier-Stokes problem as

(10.1.7) $$\text{find } \mathbf{u} \in V_{\text{div}} : \ a(\mathbf{u}, \mathbf{v}) + c(\mathbf{u}; \mathbf{u}, \mathbf{v}) = (\mathbf{f}, \mathbf{v}) \quad \forall \, \mathbf{v} \in V_{\text{div}} \ ,$$

where V_{div} is the subspace of V of divergence-free functions introduced in (9.1.3). Clearly, problems (10.1.4) and (10.1.7) provide the nonlinear generalizations of problems (9.1.11) and (9.1.6), respectively.

If $(\mathbf{u}, p)$ is a solution to (10.1.4), then $\mathbf{u}$ is a solution to (10.1.7). The converse is also true in the sense stated by the following result.

Lemma 10.1.1 *Let* $\mathbf{u}$ *be a solution to problem* (10.1.7). *Then there exists a unique* $p \in Q$ *such that* $(\mathbf{u}, p)$ *is a solution of problem* (10.1.4).

Proof. The map $\mathbf{v} \to a(\mathbf{u}, \mathbf{v}) + c(\mathbf{u}; \mathbf{u}, \mathbf{v}) - (\mathbf{f}, \mathbf{v})$ belongs to V', and vanishes on V_{div}. Therefore, the thesis follows from Lemma 9.1.1. $\qquad\square$

Let us come now to the proof of the existence and uniqueness of a solution to (10.1.7). At first define the space

$$(10.1.8) \qquad H_{\mathrm{div}} := \left\{ \mathbf{v} \in (L^2(\Omega))^d \,|\, \mathrm{div}\,\mathbf{v} = 0 \text{ in } \Omega , \ \mathbf{v} \cdot \mathbf{n} = 0 \text{ on } \partial\Omega \right\} ,$$

where $\mathbf{n}$ is the unit outward normal vector on $\partial\Omega$. Further, as norm in V_{div} choose the $H^1(\Omega)$-seminorm $|\cdot|_1$.

Theorem 10.1.1 *Let* $\mathbf{f} \in H_{\mathrm{div}}$ *with*

$$(10.1.9) \qquad \frac{\|\mathbf{f}\|_0}{\nu^2} < \frac{1}{\hat{C}C_\Omega^{1/2}} ,$$

where $\hat{C} > 0$ *is the constant appearing in* (10.1.6) *and* $C_\Omega > 0$ *is the constant of the Poincaré inequality* (1.3.2). *Then there exists a unique solution* $\mathbf{u} \in V_{\mathrm{div}}$ *to problem* (10.1.7).

Proof. For each $\mathbf{w} \in V_{\mathrm{div}}$ let us define the bilinear form

$$(10.1.10) \qquad \mathcal{A}_{\mathbf{w}}(\mathbf{z}, \mathbf{v}) := a(\mathbf{z}, \mathbf{v}) + c(\mathbf{w}; \mathbf{z}, \mathbf{v}) ,$$

which is clearly continuous in $V_{\mathrm{div}} \times V_{\mathrm{div}}$. Besides, it is also coercive, as integrating by parts one obtains for each $\mathbf{v} \in V_{\mathrm{div}}$

$$
\begin{aligned}
(10.1.11) \qquad \mathcal{A}_{\mathbf{w}}(\mathbf{v}, \mathbf{v}) &= \nu|\mathbf{v}|_1^2 + \int_\Omega \sum_{i,j=1}^d w_i v_j D_i v_j \\
&= \nu|\mathbf{v}|_1^2 + \frac{1}{2} \int_\Omega \sum_{i=1}^d w_i D_i(|\mathbf{v}|^2) \\
&= \nu|\mathbf{v}|_1^2 - \frac{1}{2} \int_\Omega \mathrm{div}\,\mathbf{w}|\mathbf{v}|^2 + \frac{1}{2} \int_{\partial\Omega} \mathbf{w} \cdot \mathbf{n}|\mathbf{v}|^2 = \nu|\mathbf{v}|_1^2 .
\end{aligned}
$$

Applying Lax-Milgram lemma (see Theorem 5.1.1) for each $\mathbf{w} \in V_{\mathrm{div}}$ there exists a unique solution $\mathbf{z} \in V_{\mathrm{div}}$ to

$$(10.1.12) \qquad \mathcal{A}_{\mathbf{w}}(\mathbf{z}, \mathbf{v}) = (\mathbf{f}, \mathbf{v}) \qquad \forall\, \mathbf{v} \in V_{\mathrm{div}} .$$

A fixed point of the nonlinear map $\Phi : \mathbf{w} \to \mathbf{z}$ is clearly a solution (10.1.7). To prove the existence of such a fixed point, let us start by showing that any solution to (10.1.12) (or else any solution to (10.1.7)) is contained in a fixed ball of V_{div}. Taking $\mathbf{v} = \mathbf{z}$ in (10.1.12), from (10.1.11) it follows

$$|\mathbf{z}|_1^2 \leq \frac{1}{\nu}||\mathbf{f}||_0\,||\mathbf{z}||_0 \leq \frac{C_\Omega^{1/2}}{\nu}||\mathbf{f}||_0\,|\mathbf{z}|_1 \ ,$$

where C_Ω is the constant in the Poincaré inequality (1.3.2). We have therefore obtained the a-priori estimate

$$(10.1.13) \qquad\qquad |\mathbf{z}|_1 \leq \frac{C_\Omega^{1/2}}{\nu}||\mathbf{f}||_0 \ .$$

Define the closed subset $\mathcal{K} \subset V_{\mathrm{div}}$ as

$$(10.1.14) \qquad \mathcal{K} := \left\{ \mathbf{v} \in V_{\mathrm{div}} \mid |\mathbf{v}|_1 \leq \frac{C_\Omega^{1/2}}{\nu}||\mathbf{f}||_0 \right\} \ .$$

Owing to (10.1.13) we derive that $\Phi(V_{\mathrm{div}}) \subset \mathcal{K}$.

We claim that the map $\Phi : \mathcal{K} \to \mathcal{K}$ is a contraction, i.e., for each $\mathbf{w}, \overline{\mathbf{w}} \in \mathcal{K}$, it satisfies

$$(10.1.15) \qquad |\Phi(\mathbf{w}) - \Phi(\overline{\mathbf{w}})|_1 = |\mathbf{z} - \overline{\mathbf{z}}|_1 \leq \kappa|\mathbf{w} - \overline{\mathbf{w}}|_1$$

for a suitable constant κ satisfying $0 < \kappa < 1$. As a matter of fact, taking the difference between the equations satisfied by $\mathbf{z}$ and $\overline{\mathbf{z}}$ one finds

$$\nu(\nabla(\mathbf{z} - \overline{\mathbf{z}}), \nabla\mathbf{v}) + c(\mathbf{w}; \mathbf{z}, \mathbf{v}) - c(\overline{\mathbf{w}}; \overline{\mathbf{z}}, \mathbf{v}) = 0 \qquad \forall\,\mathbf{v} \in V_{\mathrm{div}} \ .$$

Choosing $\mathbf{v} = \mathbf{z} - \overline{\mathbf{z}}$, this last equation can be rewritten as

$$\nu|\mathbf{z} - \overline{\mathbf{z}}|_1^2 + c(\mathbf{w}; \mathbf{z} - \overline{\mathbf{z}}, \mathbf{z} - \overline{\mathbf{z}}) + c(\mathbf{w} - \overline{\mathbf{w}}; \overline{\mathbf{z}}, \mathbf{z} - \overline{\mathbf{z}}) = 0 \ .$$

Now notice that by integration by parts

$$c(\mathbf{w}; \mathbf{z} - \overline{\mathbf{z}}, \mathbf{z} - \overline{\mathbf{z}}) = \frac{1}{2} \int_\Omega \sum_{i=1}^{d} w_i D_i(|\mathbf{z} - \overline{\mathbf{z}}|^2) = 0 \ ,$$

as $\mathbf{w} \in V_{\mathrm{div}}$. Therefore, using (10.1.6) we have

$$\nu|\mathbf{z} - \overline{\mathbf{z}}|_1^2 = -c(\mathbf{w} - \overline{\mathbf{w}}; \overline{\mathbf{z}}, \mathbf{z} - \overline{\mathbf{z}}) \leq \hat{C}|\mathbf{w} - \overline{\mathbf{w}}|_1\,|\overline{\mathbf{z}}|_1\,|\mathbf{z} - \overline{\mathbf{z}}|_1 \ ,$$

hence

$$\nu|\mathbf{z} - \overline{\mathbf{z}}|_1 \leq \frac{\hat{C}C_\Omega^{1/2}}{\nu}||\mathbf{f}||_0\,|\mathbf{w} - \overline{\mathbf{w}}|_1 \ ,$$

since $\overline{\mathbf{z}} \in \mathcal{K}$. Assuming (10.1.9), inequality (10.1.15) follows choosing the constant $\kappa = \hat{C}C_\Omega^{1/2}||\mathbf{f}||_0/\nu^2$.

The existence of a unique fixed point $\mathbf{u} = \Phi(\mathbf{u}) \in \mathcal{K}$ (hence a solution to (10.1.7)) is now a consequence of the Banach contraction theorem. Since any other solution to (10.1.7) belongs to $\mathcal{K}$ and is a fixed point of Φ, uniqueness is also proven. $\qquad\square$

Notice that it is not restrictive to assume $\mathbf{f} \in H_{\mathrm{div}}$. As a matter of fact, the so-called Helmholtz decomposition principle states that the space $(L^2(\Omega))^d$ can be decomposed into the direct sum of H_{div} and $\mathcal{G}$, where

(10.1.16)
$$\mathcal{G} := \{\mathbf{v} \in (L^2(\Omega))^d \,|\, \mathbf{v} = \nabla q \,,\, q \in H^1(\Omega)\}$$

(see, e.g., Temam (1984), p. 15). Further, by integration by parts it is easily verified that

(10.1.17)
$$(\nabla q, \mathbf{v}) = 0 \quad \forall \, q \in H^1(\Omega) \,,\, \mathbf{v} \in H_{\mathrm{div}} \,,$$

hence the gradient component of the external force field $\mathbf{f}$ does not play any role in (10.1.7).

We remark that for each $\mathbf{f} \in H_{\mathrm{div}}$ a solution (not necessarily unique) to (10.1.7) does exist (see, e.g., Temam (1984), p. 164). The smallness condition (10.1.9) is necessary for proving uniqueness. As a matter of fact, it is known (see, e.g., Temam (1984), Chap. II, Sect. 4) that the solution may not be unique when ν is small with respect to $\mathbf{f}$ for problems similar to (10.1.7) (e.g., the Taylor problem for flows between rotating infinite cylinders, the Benard problem for thermo-conductive fluids).

10.1.1 Other Kind of Boundary Conditions

The no-slip condition (10.1.3) is the correct boundary condition for a viscous fluid contained in a rigid vessel, as the viscous effects force the particles of the fluid to be adherent at the boundary. In different physical situations other boundary conditions have to be imposed. Let us briefly present some of them.

A *non-friction* condition is given by

(10.1.18)
$$\sum_{j=1}^{d}[-p\delta_{ij} + \nu(D_i u_j + D_j u_i)]n_j = g_i \quad \text{on } \partial\Omega \,,\quad i = 1,...,d \,,$$

where δ_{ij} is the Kronecker tensor (i.e., $\delta_{ij} = 0$ for $i \neq j$ and $\delta_{ii} = 1$), $\mathbf{n}$ denotes the unit outward normal vector on $\partial\Omega$ and $\mathbf{g}$ is a given vector. Typically, one has $\mathbf{g} = -p_e\mathbf{n}$, where p_e denotes the external pressure (for instance, $p_e = 0$). This condition can be seen as an outflow condition on a fictitious boundary (say, the exit of a tube). Mathematically, it corresponds to a Neumann condition for the operator

(10.1.19)
$$\sum_{j=1}^{d} D_j[p\delta_{ij} - \nu(D_i u_j + D_j u_i)] = D_i p - \nu\Delta u_i$$

(recall that $\mathbf{u}$ satisfies the incompressibility constraint div $\mathbf{u} = 0$). The tensor $T_{ij} := \rho^*[-p\delta_{ij} + \nu(D_i u_j + D_j u_i)]$, where $\rho^* > 0$ is the (constant) density, is the stress tensor associated to the motion of homogeneous incompressible flows (see Section 13.1). The Stokes problem subjected to this boundary condion has been studied by Kreĭn and Laptev (1968).

A *slip* boundary condition can be expressed as follows:

$$(10.1.20) \quad \mathbf{u} \cdot \mathbf{n} = 0 \quad \text{on } \partial\Omega$$

$$(10.1.21) \quad \sum_{i,j=1}^{d} \nu(D_i u_j + D_j u_i) n_j \tau_i^k = g^k \quad \text{on } \partial\Omega \ , \quad k = 1, ..., d-1 \ ,$$

where $\tau^k = \tau^k(\mathbf{x})$ are linearly independent tangent vectors at $\mathbf{x} \in \partial\Omega$ and g^k are given data, $k = 1, ..., d - 1$. In this case the fluid cannot leave the domain Ω, but the particles can slip on the boundary. This boundary value problem can be seen as a simplified model for the motion of a fluid limited by a free surface. The Stokes problem with this slip boundary condition has been considered by Solonnikov and Ščadilov (1973), and the approximation of the solution of both the Stokes and the nonlinear Navier-Stokes systems has been studied by Conca (1984) and Verfürth (1987).

The *free boundary* problem, i.e., the motion of a fluid which is not confined in a rigid vessel but can move freely, can be formulated as follows. Firstly, the non-friction condition (10.1.18) has to be satisfied on $\partial\Omega$, which is an unknown surface to be determined. To close the system, another boundary condition is thus required. It can be expressed by imposing that each particle of the fluid lying on the boundary remains on $\partial\Omega$ during the motion. In the stationary case this corresponds to impose the slip condition (10.1.20). If surface tension is taken into consideration, (10.1.18) has to be substituted by

$$(10.1.22) \quad \sum_{j=1}^{d} [-p\delta_{ij} + \nu(D_i u_j + D_j u_i)] n_j$$

$$= g_i + 2\sigma H n_i \quad \text{on } \partial\Omega \ , \quad i = 1, ..., d \ ,$$

where $\sigma > 0$ is the surface tension coefficients and H is the mean curvature of $\partial\Omega$. An existence theorem for the free-boundary problem for the Navier-Stokes equations has been proven by Bemelmans (1981).

Other boundary conditions have been considered by Pironneau (1986) and Bègue, Conca, Murat and Pironneau (1988). For simplicity, let us limit our presentation to the three-dimensional case $d = 3$. They proved existence and uniqueness theorem for the Navier-Stokes equations subjected to both these two sets of conditions:

$$(10.1.23) \quad \mathbf{u} \times \mathbf{n} = \mathbf{g}_1 \ , \quad p + \frac{1}{2}|\mathbf{u}|^2 = g_2 \quad \text{on } \partial\Omega$$

or else

$$(10.1.24) \quad \mathbf{u} \cdot \mathbf{n} = g_1 \ , \quad \mathbf{Curl}\,\mathbf{u} \times \mathbf{n} = \mathbf{g}_2 \quad \text{on } \partial\Omega$$

(the operator **Curl** has been introduced in (9.6.30)). These boundary conditions can be used for modeling the flow in a tube (using (10.1.23) at the inflow and the outflow) or the external flow around an obstacle (using again (10.1.23) on the artificial computational boundary far from the obstacle). The finite element approximation of these problems has been considered by Girault (1988, 1990), who in the second paper also analyzed the following boundary condition:

$$(10.1.25) \quad \mathbf{u} \cdot \mathbf{n} = 0 \;, \quad \mathbf{Curl}\,\mathbf{u} \cdot \mathbf{n} = 0 \;, \quad \mathbf{Curl}\,\mathbf{Curl}\,\mathbf{u} \cdot \mathbf{n} = 0 \quad \text{on } \partial\Omega \;.$$

We point out that all the above sets of boundary conditions are suitable for the Stokes problem (9.1) as well. The only exception is represented by (10.1.23), which should become

$$(10.1.26) \qquad \mathbf{u} \times \mathbf{n} = \mathbf{g}_1 \;, \quad p = g_2 \quad \text{on } \partial\Omega \;.$$

For additional remarks on admissible boundary conditions we refer to Pironneau (1988), pp. 123–127.

10.1.2 An Abstract Formulation

Both problems (10.1.4) and (10.1.7) can be regarded as particular cases of the following class of problems:

$$(10.1.27) \qquad \text{find } (\lambda, w) \in \Lambda \times W \; : \; F(\lambda, w) := w + TG(\lambda, w) = 0 \;,$$

where $T \in \mathcal{L}(Y; W)$, G is a C^2-mapping form $\Lambda \times W$ into Y, W and Y are two Banach spaces and Λ is a connected subset of $\mathbb{R}^+ := \{\mu \in \mathbb{R} \,|\, \mu > 0\}$. Here we make use of the notation $\mathcal{L}(A; B)$ to indicate the space of continuous linear mappings between the Banach spaces A and B (see Section 1.2 (ii)).

We begin with problem (10.1.4), and set:

$$(10.1.28) \qquad W = V \times Q \;, \; Y = V' \;, \; \Lambda = \mathbb{R}^+ \;.$$

Next we define the linear operator T as follows: given $\mathbf{f}^* \in V'$ we denote by

$$(10.1.29) \qquad T\mathbf{f}^* := (\mathbf{u}^*, p^*) \in V \times Q$$

the solution of Stokes problem with homogeneous Dirichlet condition

$$(10.1.30) \qquad \begin{cases} (\nabla \mathbf{u}^*, \nabla \mathbf{v}) + b(\mathbf{v}, p^*) = \langle \mathbf{f}^*, \mathbf{v} \rangle & \forall\, \mathbf{v} \in V \\[2mm] b(\mathbf{u}^*, q) = 0 & \forall\, q \in Q \;. \end{cases}$$

A C^∞-mapping from $\Lambda \times W$ into Y defined by

$$(10.1.31) \qquad G : (\mu, z) \to G(\mu, z) = \mu \left(\sum_{j=1}^{d} v_j \frac{\partial \mathbf{v}}{\partial x_j} - \mathbf{f} \right) \;,$$

is associated to the datum $\mathbf{f} \in (L^2(\Omega))^d$. Here $z = (\mathbf{v}, q) \in W$ (the map G is indeed independent of q). The following result can now be stated easily.

Lemma 10.1.2 *The pair* $(\mathbf{u}, p) \in (H_0^1(\Omega))^d \times L_0^2(\Omega)$ *is a solution of problem* (10.1.4) *if, and only if,* $\lambda = 1/\nu$, $w = (\mathbf{u}, p/\nu)$ *is a solution of* (10.1.27), *where* W *and* Y *are defined in* (10.1.28), T *is defined by* (10.1.29), (10.1.30), *and* G *by* (10.1.31).

In a similar manner, we can formulate (10.1.7) as (10.1.27). In such a case

$$(10.1.32) \qquad W = V_{\mathrm{div}} \ , \ Y = (V_{\mathrm{div}})' \ , \ \Lambda = \mathbb{R}^+ \ ,$$

while the operator T is defined as follows. For any $\mathbf{f}^* \in (V_{\mathrm{div}})'$, denote by $T\mathbf{f}^* := \mathbf{u}^* \in V_{\mathrm{div}}$ the solution of

$$(10.1.33) \qquad P\Delta\mathbf{u}^* = \mathbf{f}^* \quad \text{in } \Omega \ ,$$

i.e.,

$$(10.1.34) \qquad (\nabla\mathbf{u}^*, \nabla\mathbf{v}) = \langle \mathbf{f}^*, \mathbf{v} \rangle \quad \forall \, \mathbf{v} \in V_{\mathrm{div}} \ .$$

Finally, for any $\mathbf{f} \in (L^2(\Omega))^d$ the operator $G : \Lambda \times W \to Y$ is defined by

$$(10.1.35) \qquad G(\mu, \mathbf{v}) := \mu P \left(\sum_{j=1}^{d} v_j \frac{\partial \mathbf{v}}{\partial x_j} - \mathbf{f} \right) \ .$$

10.2 Finite Dimensional Approximation

In this Section we deal with the case of problem (10.1.4) that in Section 10.1.2 has been reformulated as: given $\lambda \in \Lambda$,

$$(10.2.1) \qquad \text{find } w(\lambda) \in W \ : \ F(\lambda, w(\lambda)) := w(\lambda) + TG(\lambda, w(\lambda)) = 0 \ ,$$

where T is a linear continuous map from Y and W, G is a C^2-mapping form $\Lambda \times W$ into Y, W and Y are two Banach spaces and Λ is a connected subset of $\mathbb{R}^+ := \{\mu \in \mathbb{R} \,|\, \mu > 0\}$.

We first propose a general approximation method for solving (10.2.1), and illustrate its properties in an abstract fashion. This is done in Section 10.2.1. Next, we consider two different approximation methods that are specifically devised for Navier-Stokes equations, and show how they fit under the general framework described in Section 10.2.1. In particular, in Section 10.2.2 we investigate mixed approximation methods which consist in discretizing directly both momentum and continuity equations. We also make a few comments about approximations based on projection upon divergence-free subspaces. In Section 10.2.3 we address approximations based on spectral collocation methods.

We say that $\{(\lambda, w(\lambda)) \, | \, \lambda \in \Lambda\}$ is a *branch of solutions* of (10.2.1) if $\lambda \to w(\lambda)$ is a continuous function from Λ into W and $F(\lambda, w(\lambda)) = 0$. Our analysis will be confined to the approximation of a *branch of non-singular solutions* of (10.2.1). By that we mean that on the "curve" $\{(\lambda, w(\lambda)) \, | \, \lambda \in \Lambda\}$ the Fréchet derivative $D_w F$ of the map F with respect to w is an isomorphism of W. This happens, for instance, if there exists a constant $\alpha > 0$ such that

$$\|D_w F(\lambda, w(\lambda))v\|_W = \|v + T D_w G(\lambda, w(\lambda))v\|_W$$
$$\geq \alpha \|v\|_W \quad \text{for all } v \in W, \ \lambda \in \Lambda .$$

The symbol $D_w F(\lambda_0, w_0)$ denotes the Fréchet derivative of F with respect to the variable w, computed at the point (λ_0, w_0).

In particular, we notice that if the Reynolds number is small enough (namely, the viscosity ν satisfies (10.1.9)) and $(\mathbf{u}, p)$ is the unique solution of problem (10.1.4), then $\lambda = 1/\nu$, $w = (\mathbf{u}, p/\nu)$ is a non-singular solution of (10.1.27) (for a proof, see, e.g., Girault and Raviart (1986), p. 300).

The Reynolds number Re is a non-dimensional number and is defined as follows (see Landau and Lifshitz (1959), Sect. 19)

$$(10.2.2) \qquad\qquad \mathrm{Re} := \frac{l \, |\mathbf{v}^*|}{\nu} ,$$

where l is a characteristic length of the domain Ω and $\mathbf{v}^*$ a typical velocity of the flow. Other definitions are also possible, involving different measures of the data. Looking back at Theorem 10.1.1, we could for instance define the Reynolds number as

$$\mathrm{Re} := \frac{l^{3/2 - d/4} \, \|\mathbf{f}\|_0^{1/2}}{\nu} .$$

The Reynolds number in the form (10.2.2) expresses the ratio of convection to diffusion. It has no absolute meaning: it enables problems with the same geometry and similar data (boundary and source terms) to be compared. In practice, the larger is Re the more difficult is the problem to handle.

10.2.1 An Abstract Approximate Problem

Let $\{W_h \, | \, h > 0\}$ be a family of finite dimensional subspaces of W. The approximation method for problem (10.2.1) that we are going to consider has the following abstract form: given $\lambda \in \Lambda$

$$(10.2.3) \quad \text{find } w_h(\lambda) \in W_h \ : \ F_h(\lambda, w_h(\lambda)) := w_h(\lambda) + T_h G(\lambda, w_h(\lambda)) = 0 .$$

The discrete linear operator $T_h : Y \to W_h$ is an approximation of the linear operator T. We will make a set of assumptions on T_h that guarantee the existence of a branch of solutions of (10.2.3) and its convergence as $h \to 0$ to a branch of non-singular solutions of (10.2.1).

To begin with, we make some additional assumptions on G and T. We will denote by $||| \cdot |||$ the norm of a bilinear functional $\mathcal{F} : X_1 \times X_2 \to X_3$, X_1, X_2 and X_3 Banach spaces, i.e.,

$$|||\mathcal{F}||| := \sup_{\substack{v_1 \in X_1 \\ v_1 \neq 0}} \sup_{\substack{v_2 \in X_2 \\ v_2 \neq 0}} \frac{||\mathcal{F}(v_1, v_2)||_{X_3}}{||v_1||_{X_1} ||v_2||_{X_2}} \ .$$

We suppose that

(10.2.4)   there exists a Banach space $H \subset Y$, with continuous imbedding, such that G is a C^2-map from $\Lambda \times W$ into H .

Moreover, we require that $D^2 G$ is bounded from every bounded subset of $\Lambda \times W$ into H. This means that there exists a locally bounded function $\Phi : \mathbb{R}^+ \to \mathbb{R}^+$ such that

(10.2.5) $|||D^2 G(\mu, v)||| \leq \Phi(\mu + ||v||_W)$ for all $\mu \in \Lambda$, $v \in W$.

Finally, we assume that

(10.2.6) T is a compact operator from H into W

(for the definition of compact operator between Banach spaces we refer to Section 1.3). Since we have supposed that $T \in \mathcal{L}(Y; W)$, (10.2.6) is satisfied when the inclusion $H \subset Y$ is compact.

The following result states the conditions under which problem (10.2.3) has solutions that are stable and converge to those of problem (10.2.1).

Theorem 10.2.1 *Suppose that* (10.2.4)-(10.2.6) *are satisfied. Assume moreover that:*

a.   the operators $T_h \in \mathcal{L}(Y; W)$ satisfy

(10.2.7)
$$\lim_{h \to 0} ||(T - T_h)\psi||_W = 0 \quad \forall \, \psi \in Y$$
$$\lim_{h \to 0} ||T - T_h||_{\mathcal{L}(H;W)} = 0 \ ;$$

b.   there exists a linear operator $\Pi_h : W \to W_h$ satisfying

(10.2.8) $$\lim_{h \to 0} ||v - \Pi_h v||_W = 0 \quad \forall \, v \in W \ .$$

Then there exists a neighbourhood Θ of the origin in W and, for h small enough, a unique C^2-mapping $\lambda \in \Lambda \to w_h(\lambda) \in W_h$ such that, for all $\lambda \in \Lambda$,

(10.2.9) $$F_h(\lambda, w_h(\lambda)) = 0 \quad and \quad w_h(\lambda) - w(\lambda) \in \Theta \ .$$

Further, the following convergence inequality holds

(10.2.10)
$$||w(\lambda) - w_h(\lambda)||_W \leq C\{||w(\lambda) - \Pi_h w(\lambda)||_W$$
$$+ ||(T - T_h)G(\lambda, w(\lambda))||_W\}$$

where $C > 0$ is independent of both h and λ.

Notice that $(10.2.7)_2$ follows from $(10.2.7)_1$ provided that the inclusion $H \subset Y$ is compact.

A proof of Theorem 10.2.1 is given in Girault and Raviart (1986), pp. 307–308. A slightly different version was formerly proven in Brezzi, Rappaz and Raviart (1980). A more general result, in which also the nonlinear map G is approximated by a suitable G_h, has been obtained by Maday and Quarteroni (1982b).

The convergence estimate (10.2.10) involves all the ingredients of the finite dimensional approximation. As a matter of fact, the first term of the right hand side measures the distance of the subspace W_h from the solution $w(\lambda)$ and the second term depends on how well the linear operator T is approximated by T_h.

10.2.2 Approximation by Mixed Finite Element Methods

An approximation to the Navier-Stokes equations (10.1.1)-(10.1.3) can be devised by applying a Galerkin method to (10.1.4). With this aim, let $\{V_h \,|\, h > 0\}$ be a family of subspaces of V, and $\{Q_h \,|\, h > 0\}$ a family of subspaces of Q. For the sake of exposition, we may think of Q_h and V_h as finite element spaces, although what we are developing can be easily adapted to the case of spectral Galerkin approximations as well. We therefore assume that Q_h and V_h are defined as in Section 9.3 and that they are *compatible spaces*, i.e., satisfy the compatibility condition (9.2.9). Besides, the following approximation properties are supposed to hold: there exist two operators $\mathbf{r}_h : (H^1(\Omega))^d \to V_h$ and $s_h : L^2(\Omega) \to Q_h$ such that

$$(10.2.11) \qquad \begin{aligned} \|\mathbf{v} - \mathbf{r}_h(\mathbf{v})\|_1 &\leq Ch^{l_1}\|\mathbf{v}\|_{l_1+1} \quad \forall\, \mathbf{v} \in (H^{l_1+1}(\Omega))^d \\ \|q - s_h(q)\|_0 &\leq Ch^{l_2+1}\|q\|_{l_2+1} \quad \forall\, q \in H^{l_2+1}(\Omega)) \ , \end{aligned}$$

for certain integers $l_1 \geq 1$, $l_2 \geq 0$. For instance, if the finite dimensional subspaces are given by $V_h = (X_h^k \cap H_0^1(\Omega))^d$ and $Q_h = (Y_h^m \cap L_0^2(\Omega)$ as in (9.3.4), we can choose $\mathbf{r}_h = \pi_h^k$, the interpolation operator over V_h, and $s_h = p_h^m$, the L^2-orthogonal projection over Q_h, respectively. With this choice, (10.2.11) holds with $l_1 = k$ and $l_2 = m$.

Then for each $h > 0$ we consider the problem:

$$(10.2.12) \qquad \begin{cases} \text{find } \mathbf{u}_h \in V_h \,, \ p_h \in Q_h \,: \\[4pt] a(\mathbf{u}_h, \mathbf{v}_h) + c(\mathbf{u}_h; \mathbf{u}_h, \mathbf{v}_h) + b(\mathbf{v}_h, p_h) = (\mathbf{f}, \mathbf{v}_h) \quad \forall\, \mathbf{v}_h \in V_h \\[4pt] b(\mathbf{u}_h, q_h) = 0 \hspace{4.2cm} \forall\, q_h \in Q_h \,, \end{cases}$$

which is nothing but the nonlinear generalization of the finite dimensional Stokes problem (9.2.4).

Problem (10.2.12) can be represented in the form (10.2.3). Indeed, define $W = V \times Q$, $Y = V' = (H^{-1}(\Omega))^d$ and $\Lambda = \mathbb{R}^+$ as in (10.1.28), T as in (10.1.29), (10.1.30) and G as in (10.1.31). Furthermore, let us set $W_h = V_h \times Q_h$ and define $T_h : (H^{-1}(\Omega))^d \to W_h$ as follows: for any $\mathbf{f}^* \in (H^{-1}(\Omega))^d$, $T_h \mathbf{f}^* := (\mathbf{u}_h^*, p_h^*) \in V_h \times Q_h$ is such that

(10.2.13)
$$\begin{cases} (\nabla \mathbf{u}_h^*, \nabla \mathbf{v}_h) + b(\mathbf{v}_h, p_h^*) = \langle \mathbf{f}^*, \mathbf{v}_h \rangle & \forall \ \mathbf{v}_h \in V_h \\ b(\mathbf{u}_h^*, q_h) = 0 & \forall \ q_h \in Q_h \ . \end{cases}$$

In other words, (10.2.13) is the finite dimensional approximation to the Stokes problem (10.1.30). If we set $w_h := (\mathbf{u}_h, p_h/\nu)$ we deduce from (10.2.12) that

$$w_h = -T_h G(1/\nu, w_h) \ ,$$

or, equivalently,

$$F_h(\lambda, w_h) := w_h + T_h G(\lambda, w_h) = 0$$

with $\lambda = 1/\nu$.

Hereafter, we briefly outline how this approximation matches the requirements of Theorem 10.2.1. A complete proof can be found, e.g., in Girault and Raviart (1986), Chap. IV, Sect. 4.1 and 4.2.

First of all, by applying the convergence theory presented in Section 9.2 and proceeding as in Proposition 6.2.1, it can be concluded that

$$\lim_{h \to 0} \{ \|\mathbf{u}^* - \mathbf{u}_h^*\|_1 + \|p^* - p_h^*\|_0 \} = 0$$

i.e., for all $\mathbf{f}^* \in (H^{-1}(\Omega))^d$ we have

(10.2.14)
$$\lim_{h \to 0} \|(T - T_h) \mathbf{f}^*\|_W = 0 \ .$$

Therefore, $(10.2.7)_1$ is satisfied.

Now, we wish to find a space H verifying (10.2.4), (10.2.5) and, moreover, compactly imbedded in Y (so that also (10.2.6) and $(10.2.7)_2$ would be satisfied). We claim that a good candidate is, for instance, $H = (L^{3/2}(\Omega))^d$. As a matter of fact, from Theorem 1.3.5 we know that the inclusion $L^{3/2}(\Omega) \subset H^{-1}(\Omega)$ is compact, hence H is compactly imbedded into Y. Moreover, we notice that, when $d \leq 3$, owing to Theorem 1.3.4 the space $H_0^1(\Omega)$ is continuously imbedded into $L^6(\Omega)$. Therefore, using the Hölder inequality (1.2.8), for each $\mathbf{v}, \mathbf{w} \in V$ we have

$$\sum_{j=1}^d \left(v_j \frac{\partial \mathbf{w}}{\partial x_j} + w_j \frac{\partial \mathbf{v}}{\partial x_j} \right) \in (L^{3/2}(\Omega))^d$$

and (10.2.4), (10.2.5) hold, provided we assume $\mathbf{f} \in (L^{3/2}(\Omega))^d$.

Finally, let us define for each $v = (\mathbf{v}, q) \in V \times Q$

$$\Pi_h v := (\mathbf{P}_{V_h}(\mathbf{v}), P_{Q_h}(q)) \ ,$$

where $\mathbf{P}_{V_h}$ and P_{Q_h} are the orthogonal projections over V_h and Q_h with respect to the scalar product of $(H^1(\Omega))^d$ and $L^2(\Omega)$, respectively. By proceeding as in Proposition 6.2.1, using (10.2.11) we can conclude that property (10.2.8) is satisfied.

In view of Theorem 10.2.1 we conclude that, for h small enough, there exists a unique branch of non-singular solutions of (10.2.12). Moreover, for what concerns the convergence estimate (10.2.10), in the present situation we obtain

$$(10.2.15) \quad \begin{aligned} &||\mathbf{u}(\lambda) - \mathbf{u}_h(\lambda)||_1 + ||p(\lambda) - p_h(\lambda)||_0 \\ &\qquad \leq C(h^{l_1}||\mathbf{u}(\lambda)||_{l_1+1} + h^{l_2+1}||p(\lambda)||_{l_2+1}) \ . \end{aligned}$$

As a matter of fact,

$$\begin{aligned} ||w(\lambda) - \Pi_h w(\lambda)||_W^2 &= ||\mathbf{u}(\lambda) - \mathbf{P}_{V_h}(\mathbf{u}(\lambda))||_1^2 + ||p(\lambda) - P_{Q_h}(p(\lambda))||_0^2 \\ &\leq ||\mathbf{u}(\lambda) - \mathbf{r}_h(\mathbf{u}(\lambda))||_1^2 + ||p(\lambda) - s_h(p(\lambda))||_0^2 \end{aligned}$$

and the right hand side can be bounded making use of (10.2.11). Moreover $||(T-T_h)G(\lambda, w(\lambda))||_W$ is nothing but the error arising from the finite element approximation to a Stokes problem whose right hand side is $G(\lambda, w(\lambda))$. It can therefore be bounded by the same right hand side of (10.2.15), owing to the results of Section 9.3 on the Stokes problem.

In the case in which the finite dimensional subspace is $V_{\mathrm{div},h} \subset V_{\mathrm{div}}$, by generalizing what has been done in Section 9.2 for the Stokes problem we can find the following Galerkin approximation to problem (10.1.7):

$$\text{find } \mathbf{u}_h \in V_{\mathrm{div},h} \ : \ a(\mathbf{u}_h, \mathbf{v}_h) + c(\mathbf{u}_h; \mathbf{u}_h, \mathbf{v}_h) = (\mathbf{f}, \mathbf{v}_h) \quad \forall \ v_h \in V_{\mathrm{div},h} \ .$$

It is easy to see that this problem can be formulated as (10.2.3), and (10.1.7) as (10.1.27). The analysis can therefore be carried out using Theorem 10.2.1. The interested reader can refer to Girault and Raviart (1986), Chap. IV, Sect. 4.3.

10.2.3 Approximation by Spectral Collocation Methods

When the Navier-Stokes equations are approximated by the spectral collocation method, or else by finite elements with numerical integrations, the result is a system like (10.2.12) in which all integrals are a-priori replaced by convenient quadrature formulae.

As an example, we suppose that $\Omega = (-1, 1)^2$ and adopt the $\mathbb{Q}_N$-$\mathbb{Q}_{N-2}$ representation considered in Section 9.5.3, therefore $V_N := (\mathbb{Q}_N^0)^2$ and $Q_N := \mathbb{Q}_{N-2} \cap L_0^2(\Omega)$. If the momentum equations are collocated at the $(N-1)^2$ internal nodes of the Legendre Gauss-Lobatto formula, then we are left with the problem:

$$(10.2.16) \quad \begin{cases} \text{find } \mathbf{u}_N \in V_N \,, \ p_N \in Q_N \ : \\[1mm] a_N(\mathbf{u}_N, \mathbf{v}_N) + c_N(\mathbf{u}_N; \mathbf{u}_N, \mathbf{v}_N) + b(\mathbf{v}_N, p_N) \\[1mm] \hspace{2.5cm} = (\mathbf{f}, \mathbf{v}_N)_N \qquad \forall \ \mathbf{v}_N \in V_N \\[2mm] b(\mathbf{u}_N, q_N) = 0 \hspace{2cm} \forall \ q_N \in Q_N \ , \end{cases}$$

which generalizes (9.5.30) to the nonlinear Navier-Stokes equations. We have set:

$$a_N(\mathbf{u}, \mathbf{v}) := \nu(\nabla \mathbf{u}, \nabla \mathbf{v})_N$$
$$c_N(\mathbf{u}; \mathbf{u}, \mathbf{v}) := ((\mathbf{u} \cdot \nabla)\mathbf{u}, \mathbf{v})_N \ ,$$

where, as usual, $(\varphi, \psi)_N := \sum_{i,j=0}^{N} \varphi(\mathbf{x}_{ij})\psi(\mathbf{x}_{ij})w_{ij}$ is the Legendre Gauss-Lobatto quadrature formula that approximates the integral $(\varphi, \psi) = \int_\Omega \varphi \psi$ (see (4.5.39)). We notice that

$$b(\mathbf{v}_N, q_N) = -(q_N, \operatorname{div} \mathbf{v}_N)_N$$

for all $\mathbf{v}_N \in V_N$ and $q_N \in Q_N$, due to the exactness of the Gauss-Lobatto integration formula (see (4.5.40)).

Defining $W_N := V_N \times Q_N$, problem (10.2.16) can be easily reformulated as:

$$\text{find } w_N \in W_N \ : \ F_N(\lambda, w_N) := w_N + T_N G_N(\lambda, w_N) = 0 \ ,$$

provided we set $\lambda = 1/\nu$, $w_N = (\mathbf{u}_N, p_N/\nu)$ and define T_N and G_N as follows.

For any continuous vector function $\mathbf{f}^*$, $T_N \mathbf{f}^* := (\mathbf{u}_N^*, p_N^*) \in V_N \times Q_N$ is such that

$$(10.2.17) \quad \begin{cases} (\nabla \mathbf{u}_N^*, \nabla \mathbf{v}_N)_N + b(\mathbf{v}_N, p_N^*) = (\mathbf{f}^*, \mathbf{v}_N)_N \quad \forall \ \mathbf{v}_N \in V_N \\[2mm] b(\mathbf{u}_N^*, q_N) = 0 \hspace{2.5cm} \forall \ q_N \in Q_N \ . \end{cases}$$

In other words, $T_N \mathbf{f}^*$ provides the spectral collocation approximation to $T\mathbf{f}^* := (\mathbf{u}^*, p^*)$ which is a solution to the following Stokes problem:

$$\begin{cases} -\Delta \mathbf{u}^* + \nabla p^* = \mathbf{f}^* & \text{in } \Omega \\[2mm] \operatorname{div} \mathbf{u}^* = 0 & \text{in } \Omega \\[2mm] \mathbf{u}^* = 0 & \text{on } \partial\Omega \ . \end{cases}$$

We know from (9.5.32) that T_N converges to T (the exact Stokes resolvent) with spectral accuracy provided $\mathbf{f}^*$ and $T\mathbf{f}^*$ are smooth.

Concerning G_N, for each $\mathbf{v}_N, \mathbf{w}_N \in V_N$, $q_N \in Q_N$ and $\mathbf{f} \in C^0([-1,1]^2)$ we define

$$(10.2.18) \quad \langle G_N(\mu, z_N), \mathbf{w}_N \rangle := \mu[c_N(\mathbf{v}_N; \mathbf{v}_N, \mathbf{w}_N) - (\mathbf{f}, \mathbf{w}_N)_N] \ ,$$

where $z_N := (\mathbf{v}_N, q_N)$ and $\langle \cdot, \cdot \rangle$ denotes the duality pairing between V and V'. Since $(\varphi, \psi)_N$ converges to (φ, ψ) with spectral accuracy (see (4.4.24)), it follows that G_N is clearly a good approximation to the operator G defined in (10.1.31), as the latter acts as follows

$$(10.2.19) \qquad \langle G(\mu, z), \mathbf{w} \rangle = \mu[c(\mathbf{v}; \mathbf{v}, \mathbf{w}) - (\mathbf{f}, \mathbf{w})] \quad \forall \, \mathbf{w} \in V \ ,$$

where we have set $z := (\mathbf{v}, q)$.

As G_N is different from G, the convergence proof should be carried out by applying the generalization of Theorem 10.2.1 due to Maday and Quarteroni (1982b). This kind of analysis, although in slightly different contexts, can be found in Maday and Quarteroni (1982a, 1982b).

10.3 Numerical Algorithms

We briefly address the main algorithms that can be used to solve the non-linear problem (10.2.3). It is worthwhile mentioning, however, that often the solution to the stationary problem (10.1.1)-(10.1.3) is regarded as the steady-state solution to a time-dependent Navier-Stokes problem. In such a case, any good approximation method considered in Chapter 13 is suitable to lead to an approximate solution for (10.1.1)-(10.1.3).

In this Section we attack problem (10.2.3) directly. For simplicity of notation, we drop the subscript h everywhere, hence we refer to problem (10.1.27).

10.3.1 Newton Methods and the Continuation Method

If we have to solve the abstract problem (10.1.27), where

$$(10.3.1) \qquad F : \Lambda \times W \to W$$

is a differentiable mapping and, fixed $\lambda \in \Lambda$, we look for an isolated non-singular solution $w = w(\lambda)$, an efficient way is based on the following Newton algorithm.

We start from an initial guess w^0 and construct the sequence $\{w^n\}$ in W by

$$(10.3.2) \qquad D_w F(\lambda, w^n)(w^{n+1} - w^n) = -F(\lambda, w^n) \ .$$

At each step one should therefore solve a linear differential problem associated with the linear operator $D_w F(\lambda, w^n)$.

Since the latter varies with n, the above process can be too costly. A cheaper alternative is to replace (10.3.2) by

$$(10.3.3) \qquad D_w F(\lambda, w^0)(w^{n+1} - w^n) = -F(\lambda, w^n) \ .$$

Both methods converge, provided $D_w F(\lambda, v)$ is Lipschitz continuous with respect to v in a suitable ball $S(w; \delta)$. More precisely, assume that there exists $K > 0$ such that

$$
(10.3.4) \quad
\begin{aligned}
&\|D_w F(\lambda, v) - D_w F(\lambda, v^*)\|_{\mathcal{L}(W;W)} \\
&\qquad \leq K\|v - v^*\|_W \quad \forall\, v, v^* \in S(w; \delta) \ .
\end{aligned}
$$

Then there exists a δ' with $0 < \delta' \leq \delta$ such that, for each initial value w^0 in $S(w; \delta')$, Newton algorithm (10.3.2) determines a unique sequence $\{w^n\} \in S(w; \delta')$ that converges quadratically to w, i.e.,

$$
(10.3.5) \qquad \exists\, C_1 > 0 \ : \ \|w^{n+1} - w\|_W \leq C_1 \|w^n - w\|_W^2 \ .
$$

Similarly, there exists a δ'' with $0 < \delta'' \leq \delta$ such that for each initial guess w^0 in $S(w; \delta'')$ algorithm (10.3.3) determines a unique sequence $\{w^n\} \in S(u; \delta'')$ that converges linearly to w, i.e.,

$$
(10.3.6) \qquad \exists\, 0 < C_2 < 1 \ : \ \|w^{n+1} - w\|_W \leq C_2 \|w^n - w\|_W \ .
$$

We sketch the proof of these results, referring to Girault and Raviart (1986), pp. 362–367, for the details. We first define the constant $\gamma = \gamma(\lambda)$ as

$$
\gamma(\lambda) := \|[D_w F(\lambda, w(\lambda))]^{-1}\|_{\mathcal{L}(W;W)} \ , \quad \lambda \in \Lambda \ .
$$

Such a constant exists since $w = w(\lambda)$ is a non-singular solution. Let $v \in S(w; \delta')$, with $\delta' \leq \delta$ to be chosen in a suitable way. We set

$$
\begin{aligned}
\mathcal{B} :&= [D_w F(\lambda, w(\lambda))]^{-1} [D_w F(\lambda, w(\lambda)) - D_w F(\lambda, v)] \\
&= I - [D_w F(\lambda, w(\lambda))]^{-1} D_w F(\lambda, v) \ .
\end{aligned}
$$

From (10.3.4) it follows

$$
\|\mathcal{B}\|_{\mathcal{L}(W;W)} \leq \gamma K \delta' \ ,
$$

therefore if δ' satisfies $\gamma K \delta' < 1$ the operator $I - \mathcal{B}$ is invertible, and $\|I - \mathcal{B}\|_{\mathcal{L}(W;W)} \leq 1/(1 - \gamma K \delta')$. We can thus conclude that $[D_w F(\lambda, v)]^{-1} = (I - \mathcal{B})^{-1}[D_w F(\lambda, w(\lambda))]^{-1}$, and

$$
(10.3.7) \qquad \|[D_w F(\lambda, v)]^{-1}\|_{\mathcal{L}(W;W)} \leq \frac{\gamma}{1 - \gamma K \delta'} \quad \forall\, v \in S(w; \delta') \ .
$$

Now, we show that when $w^0 \in S(w; \delta')$ (with the δ' chosen above), then (10.3.2) defines a sequence $w^n \in S(w; \delta')$ that converges to w. This can be done proceeding by induction. As a matter of fact, suppose that w^n belongs to $S(w; \delta')$; then $[D_w F(\lambda, w^n)]^{-1}$ exists and

$$
\begin{aligned}
w^{n+1} - w &= w^n - w + [D_w F(\lambda, w^n)]^{-1}(F(\lambda, w) - F(\lambda, w^n)) \\
&= [D_w F(\lambda, w^n)]^{-1} \\
&\quad \times [F(\lambda, w) - F(\lambda, w^n) - D_w F(\lambda, w^n)(w - w^n)] \\
&= [D_w F(\lambda, w^n)]^{-1} \\
&\quad \times \int_0^1 [D_w F(\lambda, w^n + t(w - w^n)) - D_w F(\lambda, w^n)](w - w^n)\,dt \ .
\end{aligned}
$$

From (10.3.4) and (10.3.7) we deduce that

$$||w^{n+1} - w||_W \le \frac{\gamma K}{2(1 - \gamma K \delta')}||w^n - w||_W^2 \ .$$

Since $||w^n - w||_W \le \delta'$, we also obtain

$$||w^{n+1} - w||_W \le \frac{\gamma K \delta'}{2(1 - \gamma K \delta')}||w^n - w||_W \ .$$

Choosing δ' such that $\gamma K \delta' < 2/3$ it follows $\gamma K \delta' / 2(1 - \gamma K \delta') < 1$, therefore w^{n+1} belongs to $S(w; \delta')$ and the last two inequalities guarantee that the sequence converges quadratically.

The convergence proof for the algorithm (10.3.3) is similar. Proceeding as before we obtain in the current case

$$w^{n+1} - w = [D_w F(\lambda, w^0)]^{-1}$$

$$\times \int_0^1 [D_w F(\lambda, w^n + t(w - w^n)) - D_w F(\lambda, w^0)](w - w^n)dt \ .$$

Assuming that w^n belongs to $S(w; \delta'')$ we deduce

$$||w^{n+1} - w||_W \le 2\gamma K(||w^n - w^0||_W + ||w^n - w||_W)||w^n - w||_W$$
$$\le 6\gamma K \delta''||w^n - w||_W \ .$$

Taking $\delta'' < 1/(6\gamma K)$ we find that w^{n+1} belongs to $S(w; \delta'')$ and further that the sequence $\{w^n\}$ generated by (10.3.3) converges linearly to w.

When applied to the Navier-Stokes problem (10.1.4), the Newton method (10.3.2) reads (remember that $\lambda = 1/\nu$ in this case): find $(\mathbf{u}^{n+1}, p^{n+1}) \in V \times Q$ such that:

$$(10.3.8) \quad \begin{cases} a(\mathbf{u}^{n+1}, \mathbf{v}) + c(\mathbf{u}^{n+1}; \mathbf{u}^n, \mathbf{v}) + c(\mathbf{u}^n; \mathbf{u}^{n+1}, \mathbf{v}) + b(\mathbf{v}, p^{n+1}) \\ \qquad = c(\mathbf{u}^n; \mathbf{u}^n, \mathbf{v}) + (\mathbf{f}, \mathbf{v}) \qquad\qquad\qquad \forall \, \mathbf{v} \in V \\ \\ b(\mathbf{u}^{n+1}, q) = 0 \qquad\qquad\qquad\qquad\qquad\qquad \forall \, q \in Q \ . \end{cases}$$

Likewise, the variant (10.3.3) reads: find $(\mathbf{u}^{n+1}, p^{n+1}) \in V \times Q$ such that:

$$(10.3.9) \quad \begin{cases} a(\mathbf{u}^{n+1}, \mathbf{v}) + c(\mathbf{u}^{n+1}; \mathbf{u}^0, \mathbf{v}) + c(\mathbf{u}^0; \mathbf{u}^{n+1}, \mathbf{v}) + b(\mathbf{v}, p^{n+1}) \\ \qquad = c(\mathbf{u}^0 - \mathbf{u}^n; \mathbf{u}^n, \mathbf{v}) + c(\mathbf{u}^n; \mathbf{u}^0, \mathbf{v}) + (\mathbf{f}, \mathbf{v}) \qquad \forall \, \mathbf{v} \in V \\ \\ b(\mathbf{u}^{n+1}, q) = 0 \qquad\qquad\qquad\qquad\qquad\qquad\qquad \forall \, q \in Q \ . \end{cases}$$

A few comments are in order. First of all, the new iterate $(\mathbf{u}^{n+1}, p^{n+1})$ is independent of p^n. Moreover, since $D^2 G(\lambda, w)$ is constant, $D_w F(\lambda, w)$ is obviously Lipschitz continuous. Therefore, if $(\mathbf{u}, p)$ is a non-singular solution of (10.1.4) (equivalently, of (10.1.1)-(10.1.3)), for any $\mathbf{u}^0$ sufficiently close to

u, the scheme (10.3.8) (respectively, (10.3.9)) determines a unique sequence that converges quadratically (respectively, linearly) to $(\mathbf{u}, p)$.

A major drawback of both algorithms (10.3.2) and (10.3.3) is that the initial guess needs to be near the exact solution. When this solution is a point of a branch of non-singular solutions, i.e., $w = w(\lambda)$ for a fixed λ, and a neighbouring solution is known, say $w(\lambda - \Delta\lambda)$ for some small increment $\Delta\lambda$, then the latter value can be used to provide the initial guess for the iterative procedure. The resulting procedure is known as the *continuation method*. We describe it hereby. By formally differentiating (10.1.27) we obtain

$$D_w F(\lambda, w(\lambda)) \frac{dw(\lambda)}{d\lambda} + D_\lambda F(\lambda, w(\lambda)) = 0 \quad \forall \lambda \in \Lambda$$

and therefore

(10.3.10)
$$\frac{dw(\lambda)}{d\lambda} = -\psi(\lambda) \ ,$$

where

$$\psi(\lambda) = [D_w F(\lambda, w(\lambda))]^{-1} D_\lambda F(\lambda, w(\lambda)) \ .$$

The ordinary differential equation (10.3.10) can be solved by any single-step method (such as, e.g., Euler, Runge-Kutta) or by an explicit multi-step method. If, for the sake of simplicity, the forward Euler method is used with a step-length $\Delta\lambda$, assuming that $w(\lambda - \Delta\lambda)$ is known we obtain:

(10.3.11)
$$\overline{w}(\lambda) = w(\lambda - \Delta\lambda) - \psi(\lambda - \Delta\lambda)\Delta\lambda \ .$$

This means that $\overline{w}(\lambda)$ is defined by

$$D_w F(\lambda - \Delta\lambda, w(\lambda - \Delta\lambda))(\overline{w}(\lambda) - w(\lambda - \Delta\lambda))$$
$$= -D_\lambda F(\lambda - \Delta\lambda, w(\lambda - \Delta\lambda))\Delta\lambda \ .$$

Since the exact value is

$$w(\lambda) = w(\lambda - \Delta\lambda) - \int_{\lambda-\Delta\lambda}^{\lambda} \psi(\mu) d\mu \ ,$$

subtracting (10.3.11) from the previous equation gives

$$w(\lambda) - \overline{w}(\lambda) = -\int_{\lambda-\Delta\lambda}^{\lambda} [\psi(\mu) - \psi(\lambda - \Delta\lambda)] d\mu$$
$$= -\int_{\lambda-\Delta\lambda}^{\lambda} \psi'(\xi_\mu)(\mu - \lambda + \Delta\lambda) d\mu \ .$$

Therefore,

$$||w(\lambda) - \overline{w}(\lambda)||_W \le \frac{(\Delta\lambda)^2}{2} \max_{\lambda-\Delta\lambda \le \eta \le \lambda} ||\psi'(\eta)||_W \ .$$

If $\Delta\lambda$ is small enough, $\overline{w}(\lambda)$ belongs to $S(w(\lambda); \delta')$ and can therefore serve as initial guess w^0 for the Newton iterates.

If we apply the above continuation method to the Navier-Stokes problem, and set for ease of notation:

$$\lambda := 1/\nu \ , \ \lambda^* := \lambda - \Delta\lambda \ ,$$

$$\mathbf{u} := \mathbf{u}(\lambda) \ , \ p := p(\lambda) \ , \ \mathbf{u}^* := \mathbf{u}(\lambda - \Delta\lambda) \ , \ p^* := p(\lambda - \Delta\lambda)$$

$$\delta\mathbf{u} := \overline{\mathbf{u}}(\lambda) - \mathbf{u}^* \ , \ \delta p = \overline{p}(\lambda) - p^* \ ,$$

we obtain the following problem: find $\delta\mathbf{u} \in V$, $\delta p \in Q$ such that:

(10.3.12)

$$\begin{cases} \dfrac{1}{\lambda^*}(\nabla\delta\mathbf{u}, \nabla\mathbf{v}) + c(\mathbf{u}^*; \delta\mathbf{u}, \mathbf{v}) + c(\delta\mathbf{u}; \mathbf{u}^*, \mathbf{v}) \\ \qquad\qquad + \dfrac{1}{\lambda^*}b(\mathbf{v}, \delta p) \\ \qquad = \dfrac{\Delta\lambda}{\lambda^*}[(\mathbf{f}, \mathbf{v}) - c(\mathbf{u}^*; \mathbf{u}^*, \mathbf{v})] \qquad \forall \, \mathbf{v} \in V \\ b(\delta\mathbf{u}, q) = 0 \qquad\qquad\qquad\qquad \forall \, q \in Q \ . \end{cases}$$

This problem has the same complexity as one iteration of Newton algorithm.

Remark 10.3.1 For solving (10.1.27), instead of (10.3.2) or (10.3.3) one could use a more general *quasi-Newton* method. This would read:

$$H(\lambda, w^n)(w^{n+1} - w^n) = -F(\lambda, w^n) \ ,$$

where H is a convenient approximation of the Jacobian $D_w F$. In this respect, the algorithm (10.3.3) falls under this cathegory.

Another way is to start from the Newton method but at each step n solve the linear problem (10.3.2) by GMRES iterations (see Section 2.6.2)). More precisely, (10.3.2) first needs to be reformulated as (10.3.8), then approximated in space for each n to yield a linear system of the form

(10.3.13)

$$J_n \begin{pmatrix} \mathbf{u}^{n+1} \\ \mathbf{p}^{n+1} \end{pmatrix} = \begin{pmatrix} \mathbf{g}^n \\ 0 \end{pmatrix} \ ,$$

where J_n is the matrix associated to the Jacobian $D_w F(\lambda, w^n)$ and $\mathbf{g}^n$ is the vector associated to the right hand side of the discretized momentum equation. With obvious notation, system (10.3.13) can be given the abstract form $A\mathbf{x} = \mathbf{b}$, then the GMRES iterates (2.6.20) can be started on, taking $\mathbf{x}^0 = \begin{pmatrix} \mathbf{u}^n \\ \mathbf{p}^n \end{pmatrix}$. At convergence they provide the solution of (10.3.13). For a detailed description see Brown and Saad (1990). $\qquad\qquad\qquad\square$

10.3.2 An Operator-Splitting Algorithm

Another iterative approach consists in seeking the solution to (10.1.1), (10.1.2) as the limit of a pseudo-time advancing scheme. For instance, at each step the equations can be split into a linear Stokes problem and a nonlinear convection-diffusion equation for the velocity field, following a procedure similar to the one used for the Peaceman-Rachford scheme (5.7.7). In differential form, starting from $(\mathbf{u}^0, p^0)$ we construct a sequence $(\mathbf{u}^n, p^n)$ by

$$
(10.3.14) \quad
\begin{cases}
-\eta\nu\Delta\mathbf{u}^{n+1/2} + \alpha_n\mathbf{u}^{n+1/2} + \nabla p^{n+1/2} \\
\qquad = \mathbf{f} - (\mathbf{u}^n\cdot\nabla)\mathbf{u}^n + (1-\eta)\nu\Delta\mathbf{u}^n + \alpha_n\mathbf{u}^n \quad \text{in } \Omega \\[2mm]
\operatorname{div}\mathbf{u}^{n+1/2} = 0 \qquad\qquad\qquad\qquad\qquad\qquad\quad \text{in } \Omega \\[2mm]
\mathbf{u}^{n+1/2} = 0 \qquad\qquad\qquad\qquad\qquad\qquad\quad\; \text{on } \partial\Omega
\end{cases}
$$

$$
(10.3.15) \quad
\begin{cases}
-(1-\eta)\nu\Delta\mathbf{u}^{n+1} + (\mathbf{u}^{n+1}\cdot\nabla)\mathbf{u}^{n+1} + \alpha_n\mathbf{u}^{n+1} \\
\qquad = \mathbf{f} - \nabla p^{n+1/2} + \eta\nu\Delta\mathbf{u}^{n+1/2} + \alpha_n\mathbf{u}^{n+1/2} \quad \text{in } \Omega \\[2mm]
\mathbf{u}^{n+1} = 0 \qquad\qquad\qquad\qquad\qquad\qquad\qquad\;\; \text{on } \partial\Omega \quad,
\end{cases}
$$

where $0 < \eta < 1$. The parameters α_n need to be chosen conveniently in order to speed up the convergence.

In general, the proof of the convergence of such a scheme is not an easy matter. In this respect, knowing that the exact stationary solution to be approximated is stable can provide a justification of the method (see, e.g., Heywood and Rannacher (1986a)).

The approach (10.3.14, (10.3.15) was introduced by Glowinski (1984), p. 255. It has then been extended to a three-step algorithm of Strang type (see (5.7.11)), in which the first two steps reproduce (10.3.14) and (10.3.15), while the third one is again based on a Stokes problem like (10.3.14) (see Bristeau, Glowinski and Périaux (1987)).

The Stokes problem (10.3.14) can be faced as discussed in Chapter 9. On the other hand, (10.3.15) is a nonlinear convection-diffusion system of the form

$$
(10.3.16) \quad
\begin{cases}
-\nu\Delta\mathbf{v} + (\mathbf{v}\cdot\nabla)\mathbf{v} + \alpha^*\mathbf{v} = \mathbf{g} \quad \text{in } \Omega \\[2mm]
\mathbf{v} = 0 \qquad\qquad\qquad\qquad\quad \text{on } \partial\Omega \quad.
\end{cases}
$$

It is easy to see that this problem can also be given the form $F(\nu, \mathbf{v}) = 0$, and analyzed as done in the abstract framework of Sections 10.1 and 10.2. Its solution can be accomplished in several ways, e.g., by a gradient algorithm (see Girault and Raviart (1986), p. 361) or a conjugate gradient algorithm (see Glowinski, Mantel, Périaux, Perrier and Pironneau (1982)). In both cases, each iteration would require solving four Dirichlet problems for the vector operator $-\nu\Delta\mathbf{v} + \alpha^*\mathbf{v}$.

10.4 Stream Function-Vorticity Formulation of the Navier-Stokes Equations

The velocity $\mathbf{u}$ and the pressure p are usually called the primitive variables for the Navier-Stokes equations. It is possible to rewrite the problem introducing different unknowns (non-primitive variables). A typical example is the following one.

If $\Omega \subset \mathbb{R}^2$ is a simply-connected domain, a well-known result states that a vector function $\mathbf{u} \in (L^2(\Omega))^2$ satisfies $\mathrm{div}\,\mathbf{u} = 0$ if, and only if, there exists a function ψ in $H^1(\Omega)$, called *stream function*, such that

$$(10.4.1) \qquad \mathbf{u} = \mathbf{curl}\,\psi := \left(\frac{\partial \psi}{\partial x_2}, -\frac{\partial \psi}{\partial x_1} \right) \ .$$

Clearly, ψ is defined only up to an additive constant. For the proof see, e.g., Girault and Raviart (1986), p. 37.

Meanwhile, we define the *vorticity* associated with $\mathbf{u} \in (H^1(\Omega))^2$ to be the scalar field $\omega \in L^2(\Omega)$ such that

$$(10.4.2) \qquad \omega := \mathrm{curl}\,\mathbf{u} = \frac{\partial u_2}{\partial x_1} - \frac{\partial u_1}{\partial x_2} \ .$$

The two operators $\mathbf{curl}$ and curl are formally adjoint operators in the sense that

$$(10.4.3) \qquad \begin{aligned} \int_\Omega \mathbf{v} \cdot \mathbf{curl}\,\varphi \, d\mathbf{x} - \int_\Omega \varphi \, \mathrm{curl}\,\mathbf{v}\, d\mathbf{x} &= \int_{\partial\Omega} \varphi \,(v_1 n_2 - v_2 n_1)\, d\gamma \\ &= \int_{\partial\Omega} \varphi \,(\mathbf{v} \times \mathbf{n})_3 \, d\gamma \ , \end{aligned}$$

provided all operations are justified (this is the case if, e.g., $\varphi \in H^1(\Omega)$ and $\mathbf{v} \in (H^1(\Omega))^2$). Here, $\mathbf{n}$ is the unit outward normal vector on $\partial\Omega$.

Notice that the operators $\mathbf{curl}$ and curl can be viewed in a unified way if we look at a three-dimensional problem, and we identify any scalar function φ with the vector function $(0, 0, \varphi)$. With this understanding, one easily verifies that

$$((\mathbf{curl}\,\varphi)_1, (\mathbf{curl}\,\varphi)_2, 0) = \mathbf{Curl}\,(0, 0, \varphi) \ , \quad (0, 0, \mathrm{curl}\,\mathbf{v}) = \mathbf{Curl}\,(v_1, v_2, 0) \ ,$$

where, as in (9.6.30), for each vector function $\mathbf{v} = (v_1, v_2, v_3)$ we have defined

$$\mathbf{Curl}\,\mathbf{v} := (D_2 v_3 - D_3 v_2, D_3 v_1 - D_1 v_3, D_1 v_2 - D_2 v_1) \ .$$

Finally, it is verified at once that for any sufficiently smooth scalar function φ and vector function $\mathbf{v}$ we have

$$(10.4.4) \qquad \begin{cases} \mathrm{div}\,\mathbf{curl}\,\varphi = 0 \ , & \mathrm{div}\,\mathbf{Curl}\,\mathbf{v} = 0 \\ \mathrm{curl}\,\nabla\varphi = 0 \ , & \mathbf{Curl}\,\nabla\varphi = \mathbf{0} \\ \mathrm{curl}\,\mathbf{curl}\,\varphi = -\Delta\varphi \ , & \mathbf{Curl}\,\mathbf{Curl}\,\mathbf{v} = -\Delta\mathbf{v} + \nabla\,\mathrm{div}\,\mathbf{v} \ . \end{cases}$$

Now let **u** denote the solution to the Navier-Stokes equations (10.1.1)-(10.1.3), and let ψ and ω be the associated stream function and vorticity, respectively. From (10.4.1) and (10.4.2) it follows easily that

$$(10.4.5) \qquad\qquad -\Delta\psi = \omega \ .$$

Furthermore, applying the curl operator to the momentum equation (10.1.1) and using the continuity equation (10.1.2) as well as (10.4.4), we obtain

$$(10.4.6) \qquad\qquad -\nu\Delta\omega + \mathbf{curl}\,\psi \cdot \nabla\omega = \mathrm{curl}\,\mathbf{f} \ .$$

Note that the pressure is not needed, and that the incompressibility constraint is avoided.

The boundary conditions that accompany (10.4.5) and (10.4.6) are

$$(10.4.7) \qquad\qquad \psi = 0 \ , \ \frac{\partial\psi}{\partial n} = 0 \ \ \text{on } \partial\Omega \ ,$$

which are equivalent to $\mathbf{u} = \mathbf{0}$ on $\partial\Omega$. The subtlety of the stream function-vorticity formulation is that there are no physical boundary conditions for the vorticity.

It is also possible to reduce the problem to only one unknown, namely the stream function. In fact, the elimination of the vorticity in (10.4.6) by means of (10.4.5) gives the fourth-order equation

$$(10.4.8) \qquad\qquad \nu\Delta^2\psi - \mathbf{curl}\,\psi \cdot \nabla\Delta\psi = \mathrm{curl}\,\mathbf{f} \ .$$

This elliptic equation, supplemented by the boundary conditions (10.4.7), can be reformulated in a variational way as follows:

$$(10.4.9) \qquad \begin{aligned} &\text{find } \psi \in H_0^2(\Omega) \ : \ \nu(\Delta\psi, \Delta\varphi) + (\Delta\psi\,\mathbf{curl}\,\psi, \nabla\varphi) \\ &\qquad\qquad\qquad = (\mathbf{f}, \mathrm{curl}\,\varphi) \quad \forall\,\varphi \in H_0^2(\Omega) \ . \end{aligned}$$

Once a solution ψ is available, the vorticity ω can be explicitly recovered from (10.4.5).

The associated bilinear form $\nu(\Delta\psi, \Delta\varphi)$ is continuous and coercive in $H_0^2(\Omega)$ and the trilinear one $(\Delta\psi\,\mathbf{curl}\,\psi, \nabla\varphi)$ is a compact perturbation. Problem (10.4.9) can therefore be refrased in the abstract form (10.1.27). For its approximation the theory based on the methods of Section 10.2 can be applied.

An approximation of (10.4.9) based on standard conforming finite elements is not so easy to implement, due to the fact that piecewise-polynomial functions must have the first derivatives continuous across interelement boundaries in order to belong to $H_0^2(\Omega)$. An approach based on mixed finite elements for the stream function-vorticity formulation (10.4.5), (10.4.6) is more suitable (see, e.g., Brezzi and Fortin (1991), Chap. VI; see also Gunzburger and Peterson (1988) and Gunzburger (1989)). A spectral approximation would provide an optimal estimate of the error $||\psi - \psi_N||_{H^2(\Omega)}$ using the convergence analysis for the biharmonic problem (for this, see Bernardi and Maday (1992), Chap. V).

The extension of the stream function-vorticity approach to three-dimensional problems requires a more careful analysis. In that case, assuming that the domain Ω is simply-connected and that $\partial\Omega$ is connected, the condition $\operatorname{div}\mathbf{u} = 0$ is equivalent to the existence of a vector stream function $\boldsymbol\psi$ such that $\mathbf{Curl}\,\boldsymbol\psi = \mathbf{u}$ and $\operatorname{div}\boldsymbol\psi = 0$. Proceeding as in the two-dimensional case, it is easily seen that the stream function formulation (10.4.9) has to be replaced by

(10.4.10)
$$\text{find }\boldsymbol\psi \in \boldsymbol\Psi \;:\; \nu(\Delta\boldsymbol\psi, \Delta\boldsymbol\varphi) - (\Delta\boldsymbol\psi \times \mathbf{Curl}\,\boldsymbol\psi, \mathbf{Curl}\,\boldsymbol\varphi)$$
$$= (\mathbf{f}, \mathbf{Curl}\,\boldsymbol\varphi) \quad \forall\,\boldsymbol\varphi \in \boldsymbol\Psi \;,$$

where

$$\boldsymbol\Psi := \left\{\boldsymbol\varphi \in (L^2(\Omega))^3 \mid \operatorname{div}\boldsymbol\varphi \in H^1(\Omega)\,,\ \mathbf{Curl}\,\boldsymbol\varphi \in (H_0^1(\Omega))^3\,,\ (\boldsymbol\varphi\times\mathbf{n})_{|\partial\Omega} = 0\right\}\,.$$

This formulation is not equivalent to (10.1.7), as it also is necessary to prove that the solution $\boldsymbol\psi$ to (10.4.10) satisfies $\operatorname{div}\boldsymbol\psi = 0$, so that, setting $\mathbf{u} = \mathbf{Curl}\,\boldsymbol\psi$ and defining the vector vorticity as $\boldsymbol\omega := \mathbf{Curl}\,\mathbf{u}$, it follows

$$\boldsymbol\omega = \mathbf{Curl}\,\mathbf{u} = \mathbf{Curl}\,\mathbf{Curl}\,\boldsymbol\psi = -\Delta\boldsymbol\psi$$

(see (10.4.4)). We do not dwell on this formulation here, referring the interested reader to Girault and Raviart (1986), pp. 294–297. We also refer to Quartapelle (1993), for a comprehensive presentation of Navier-Stokes equations in non-primitive variables.

10.5 Complements

A review paper on theoretical and computational issues which are of interest in the approximation of Navier-Stokes equations is the one by Gresho (1991).

All along this Chapter, we have been dealing with regular branches of solutions to the Navier-Stokes equations. For the approximation of singular points (bifurcation and return points) we refer to the abstract theory developed in Brezzi, Rappaz and Raviart (1981a, 1981b) and also to the monograph of Crouzeix and Rappaz (1990).

The continuation method that has been mentioned in Section 10.3.1 is investigated in Reinhart (1980), in connection with the solution of nonlinear partial differential equations. For a more general presentation, see, e.g., Allgower and Georg (1990, 1993).

The application of stabilization methods as those reported in Sections 8.3 and 9.4 has been proposed and analyzed for the Navier-Stokes equations by Brooks and Hughes (1982), Behr, Franca and Tezduyar (1993), Zhou and Feng (1993).

The stream function-vorticity formulation of the Navier-Stokes equations has often been used in flow computations, also combined with stabilization techniques (see, e.g., Tezduyar, Glowinski and Liou (1988), Tezduyar, Liou and Ganjoo (1990) and the references therein).

11. Parabolic Problems

This Chapter deals with linear second-order parabolic equations. The typical example is given by the heat equation, which is the non-stationary counterpart of the Laplace equation.

First we present a short review of the theory concerning existence and uniqueness of the solution to several initial-boundary value problems. A semi-discrete approximation will be then introduced by discretizing with respect to the space variable, using either finite element or spectral methods. Finally, a total discretization procedure, based on finite difference schemes for the time derivative, will also be presented. A sound stability and convergence analysis will be provided for all these schemes.

The algorithmic aspects of these methods will be shortly analyzed in the last Section of this Chapter.

11.1 Initial-Boundary Value Problems and Weak Formulation

Let us assume that Ω is a bounded domain in $\mathbb{R}^d$, $d = 2, 3$, with Lipschitz boundary. Consider a second order differential operator L given by

$$(11.1.1) \qquad Lw := - \sum_{i,j=1}^{d} D_i(a_{ij} D_j w) + \sum_{i=1}^{d} [D_i(b_i w) + c_i D_i w] + a_0 w \ .$$

Let us recall that L is *elliptic* if there exists a constant $\alpha_0 > 0$ such that

$$(11.1.2) \qquad \sum_{i,j=1}^{d} a_{ij}(\mathbf{x}) \xi_i \xi_j \geq \alpha_0 |\boldsymbol{\xi}|^2 \ ,$$

for each $\boldsymbol{\xi} \in \mathbb{R}^d$ and for almost every $\mathbf{x} \in \Omega$.

Definition 11.1.1 The operator

$$(11.1.3) \qquad \qquad \frac{\partial}{\partial t} + L$$

is said to be parabolic if L is elliptic.

We warn the reader that we are only considering the case of a time-independent operator L (namely, a_{ij}, b_i, c_i and a_0 don't depend on t). A classical example is provided by the heat equation $\frac{\partial u}{\partial t} - \Delta u = f$, in which case $L = -\Delta$. Most of the results here presented can be extended to the case of a time-dependent operator (see, e.g., Thomée (1984), Fujita and Suzuki (1991)).

We are interested in the same boundary value problems considered in Chapter 6 (Dirichlet, Neumann, mixed, Robin) for the operator (11.1.1). Let us indicate by $Bu = g$ any of these boundary conditions.

Due to the presence of the time derivative operator, an initial condition has finally to be imposed to determine the solution u.

We are therefore dealing with the following initial-boundary value problem:

$$(11.1.4) \quad \begin{cases} \dfrac{\partial u}{\partial t} + Lu = f & \text{in } Q_T := (0,T) \times \Omega \\[2mm] Bu = g & \text{on } \Sigma_T := (0,T) \times \partial\Omega \\[2mm] u_{|t=0} = u_0 & \text{on } \Omega \ , \end{cases}$$

where $f = f(t,\mathbf{x})$, $g = g(t,\mathbf{x})$ and $u_0 = u_0(\mathbf{x})$ are given data.

As in the elliptic case considered before, a weak formulation can be provided. Let V be a closed subspace of $H^1(\Omega)$ such that

$$H_0^1(\Omega) \subset V \subset H^1(\Omega) \ .$$

We have already pointed out that each one of the boundary conditions we are considering requires a specific choice of V (see Section 6.1.1). Moreover, let us recall that we denote by $L^2(0,T;V)$ the space

$$L^2(0,T;V) := \left\{ u : (0,T) \to V \mid u \text{ is measurable and } \int_0^T \|u(t)\|_1^2 \, dt < \infty \right\} \ ,$$

and similarly we define $C^0([0,T]; L^2(\Omega))$ and other functional spaces for space-time functions (see also Section 1.2).

The weak formulation of (11.1.4) for the homogeneous Dirichlet boundary condition reads as follows: given $f \in L^2(Q_T)$ and $u_0 \in L^2(\Omega)$, find $u \in L^2(0,T;V) \cap C^0([0,T]; L^2(\Omega))$ such that

$$(11.1.5) \quad \begin{cases} \dfrac{d}{dt}(u(t),v) + a(u(t),v) = (f(t),v) & \forall\, v \in V \\[2mm] u(0) = u_0 \ , \end{cases}$$

where $V = H_0^1(\Omega)$, $(\cdot, \cdot)$ denotes the scalar product in $L^2(\Omega)$, the bilinear form $a(\cdot, \cdot)$ is defined in (6.1.2), and the above equation has to be intended in the sense of distributions in $(0, T)$.

For the other boundary conditions, one has to modify the choice of V and both the left and right hand sides of (11.1.5) according to (6.1.11), (6.1.14), (6.1.16).

11.1.1 Mathematical Analysis of Initial-Boundary Value Problems

Several methods can be employed to prove the existence and uniqueness of a solution to (11.1.5). We are going to present one of them, based on the Faedo-Galerkin method and suitable energy estimates.

Let us first assume that the bilinear form $a(\cdot, \cdot)$ is continuous and, moreover, *weakly coercive* in V, i.e., there exist two constants $\alpha > 0$ and $\lambda \geq 0$ such that

$$(11.1.6) \qquad a(v, v) + \lambda ||v||_0^2 \geq \alpha ||v||_1^2 \qquad \forall \, v \in V \ .$$

This is also called Gårding inequality. Very often, it is satisfied taking $\lambda = 0$, namely the bilinear form $a(\cdot, \cdot)$ is coercive. This is, e.g., the case of the heat equation $\frac{\partial u}{\partial t} - \Delta u = f$ with homogeneous Dirichlet boundary condition.

In its general form, the inequality (11.1.6) is satisfied for all the boundary value problems we are dealing with, provided that, for each $i, j = 1, ..., d$, the coefficients a_{ij}, b_i, c_i and a_0 of the operator L belong to $L^\infty(\Omega)$. In fact, using (11.1.2) one finds

$$a(v, v) = \int_\Omega \left[\sum_{i,j} a_{ij} D_i v D_j v - \sum_i (b_i - c_i) v D_i v + a_0 v^2 \right]$$

$$\geq \alpha_0 ||Dv||_0^2 - ||\mathbf{b} - \mathbf{c}||_{L^\infty(\Omega)} ||Dv||_0 ||v||_0 - ||a_0||_{L^\infty(\Omega)} ||v||_0^2 \ .$$

Recalling that for each $\varepsilon > 0$

$$||Dv||_0 ||v||_0 \leq \varepsilon ||Dv||_0^2 + \frac{1}{4\varepsilon} ||v||_0^2 \ ,$$

it follows that (11.1.6) holds for $a(w, v)$, choosing for instance

$$\lambda > C \left(\frac{1}{\alpha_0} ||\mathbf{b} - \mathbf{c}||_{L^\infty(\Omega)}^2 + ||a_0||_{L^\infty(\Omega)} \right) \ ,$$

where $C = C(d, \Omega)$ is a suitable constant.

When considering the Robin problem (6.1.15), the bilinear form is given by $a(w, v) + (\kappa w, v)_{\partial\Omega}$ (see (6.1.16)). Assumption (11.1.6) is still valid for this bilinear form. As a matter of fact, by the trace theorem and an interpolation result (see Theorems 1.3.1 and 1.3.7) for each $\varepsilon > 0$ one has

$$||v||_{0,\partial\Omega}^2 \leq \varepsilon ||Dv||_0^2 + C_\varepsilon ||v||_0^2 \ ,$$

thus

$$(\kappa v, v)_{\partial\Omega} \geq -\frac{\alpha_0}{2}||Dv||_0^2 - \frac{C}{\alpha_0}||\kappa||_{L^\infty(\partial\Omega)}^2||v||_0^2 \ .$$

In this case it is sufficient to choose

$$\lambda > C\left(\frac{1}{\alpha_0}||\kappa||_{L^\infty(\partial\Omega)}^2 + \frac{1}{\alpha_0}||\mathbf{b} - \mathbf{c}||_{L^\infty(\Omega)}^2 + ||a_0||_{L^\infty(\Omega)}\right) \ .$$

Notice, moreover, that also the continuity of the bilinear form $a(\cdot, \cdot)$ is always easily verified under the above assumptions on the coefficients.

Before coming to the existence and uniqueness theorem, let us notice that, if we introduce the change of variable $u_\lambda(t, \mathbf{x}) := e^{-\lambda t}u(t, \mathbf{x})$, where u is the solution of (11.1.4), the new unknown u_λ satisfies

$$\frac{\partial u_\lambda}{\partial t} + Lu_\lambda + \lambda u_\lambda = e^{-\lambda t}f \quad \text{in } Q_T \ .$$

If (11.1.6) holds for $a(w, v)$, the bilinear form $a_\lambda(w, v) := a(w, v) + \lambda(w, v)$ associated to this last problem is coercive, i.e., it satisfies (11.1.6) with $\lambda = 0$. Therefore, if we replace f with $e^{-\lambda t}f$ and L with $L + \lambda I$, I being the identity operator, without loosing generality we can assume that the bilinear form associated to the initial-boundary value problem (11.1.4) satisfies (11.1.6) with $\lambda = 0$.

This will be always assumed in the sequel of this Chapter. However, it is worthy to notice that the estimates we will prove are valid for the auxiliary unknown $u_\lambda(t, \mathbf{x})$ (or its approximations), and that the corresponding estimates for the solution $u(t, \mathbf{x})$ show an extra multiplicative factor $e^{\lambda t}$.

Let us now prove the existence theorem. We notice that hereafter all norms refer to the space variables, i.e., $||\cdot||_k$ is the norm in the Sobolev space $H^k(\Omega)$ for $k \geq 0$.

Theorem 11.1.1 *Assume that the bilinear form $a(\cdot, \cdot)$ is continuous in $V \times V$ and that (11.1.6) is satisfied with $\lambda = 0$. Then, given $f \in L^2(Q_T)$ and $u_0 \in L^2(\Omega)$, there exists a unique solution $u \in L^2(0, T; V) \cap C^0([0, T]; L^2(\Omega))$ to (11.1.5). Moreover, $\frac{\partial u}{\partial t} \in L^2(0, T; V')$ and the energy estimate*

$$(11.1.7) \qquad ||u(t)||_0^2 + \alpha \int_0^t ||u(\tau)||_1^2 \leq ||u_0||_0^2 + \frac{1}{\alpha}\int_0^t ||f(\tau)||_0^2$$

holds for each $t \in [0, T]$.

Proof. We employ the so-called Faedo-Galerkin method, and construct an approximate sequence solving suitable finite dimensional problems.

Since V is a closed subspace of $H^1(\Omega)$, it is a separable Hilbert space. Let $\{\phi_j\}_{j \geq 1}$ be a complete orthonormal basis in V and define $V^N :=$ span $\{\phi_1, ..., \phi_N\}$. Consider the approximate problem: for each $t \in [0, T]$ find $u^N(t) \in V^N$ such that:

$$(11.1.8) \quad \begin{cases} \dfrac{d}{dt}(u^N(t), \phi_j) + a(u^N(t), \phi_j) \\ \qquad = (f(t), \phi_j) \,, \ \forall \, j = 1, ..., N \,, \quad t \in (0, T) \\ u^N(0) = u_0^N := P_N(u_0) = \displaystyle\sum_{s=1}^{N} \rho_s \phi_s \,, \end{cases}$$

where P_N is the orthogonal projection in $L^2(\Omega)$ on V^N, hence the vector ρ is the solution of the linear system $(M\rho)_j = (u_0, \phi_j)$, M being the mass matrix $M_{js} := (\phi_j, \phi_s)$. Since $\{\phi_j\}$, $j = 1, ..., N$, is a basis for V^N, the equation in (11.1.8) is indeed satisfied for each $v^N \in V^N$.

Writing

$$u^N(t) = \sum_{s=1}^{N} c_s^N(t) \phi_s \,,$$

(11.1.8) is equivalent to solving

$$(11.1.9) \quad \begin{cases} M \dfrac{d}{dt} \mathbf{c}^N(t) + A \mathbf{c}^N(t) = \mathbf{F}(t) \\ M \mathbf{c}^N(0) = \mathbf{c}_0 \,, \end{cases}$$

where for $i, j = 1, ..., N$

$$A_{ij} := a(\phi_j, \phi_i) \,, \quad F_i(t) := (f(t), \phi_i) \,, \quad c_{0,i} := (u_0, \phi_i) \,.$$

Since M is positive definite, we find a unique solution $\mathbf{c}^N$ to (11.1.9). As $\mathbf{F} \in L^2(0, T)$, it follows $\mathbf{c}^N \in H^1(0, T)$, i.e., $u^N \in H^1(0, T; V)$. Choosing $u^N(t)$ in (11.1.8) as a test function, we have

$$\left(\frac{d}{dt} u^N(t), u^N(t) \right) + a(u^N(t), u^N(t)) = (f(t), u^N(t)) \,,$$

and therefore, owing to (11.1.6),

$$\frac{1}{2} \frac{d}{dt} \|u^N(t)\|_0^2 + \alpha \|u^N(t)\|_1^2 \leq \|f(t)\|_0 \|u^N(t)\|_0 \,.$$

Integrating over $(0, \tau)$, $\tau \in (0, T]$, we obtain

$$(11.1.10) \qquad \|u^N(\tau)\|_0^2 + \alpha \int_0^\tau \|u^N(t)\|_1^2 \leq \|u_0\|_0^2 + \frac{1}{\alpha} \int_0^\tau \|f(t)\|_0^2 \,.$$

The sequence u^N is thus bounded in $L^\infty(0, T; L^2(\Omega)) \cap L^2(0, T; V)$. Hence, we can select a subsequence (still denoted by u^N) which converges in the weak* topology of $L^\infty(0, T; L^2(\Omega))$ and weakly in $L^2(0, T; V)$ (see, e.g., Yosida (1974), pp. 137 and 126). This means that there exists $u \in L^\infty(0, T; L^2(\Omega)) \cap L^2(0, T; V)$ such that

$$\int_0^T (u^N(t), \varphi(t)) \to \int_0^T (u(t), \varphi(t)) \quad \text{as } N \to \infty$$

for each $\varphi \in L^1(0, T; L^2(\Omega))$ and

$$\int_0^T (\nabla u^N(t), \Phi(t)) \to \int_0^T (\nabla u(t), \Phi(t)) \quad \text{as } N \to \infty$$

for each $\Phi \in L^2(0, T; L^2(\Omega))$.

In order to pass to the limit in (11.1.8), take $\Psi \in C^1([0, T])$ with $\Psi(T) = 0$. By multiplying (11.1.8) by Ψ and integrating by parts over $(0, T)$ the first term at the left hand side, we find

$$\int_0^T \left(\frac{du^N}{dt}(t), \phi_j \right) \Psi(t) = -\int_0^T (u^N(t), \phi_j) \frac{d\Psi}{dt}(t) - (u_0^N, \phi_j)\Psi(0) \ .$$

Since u_0^N converges in $L^2(\Omega)$ to u_0, by choosing arbitrarily N_0 and passing to the limit in (11.1.8) we finally obtain

(11.1.11)
$$-\int_0^T (u(t), \phi_j) \frac{d\Psi}{dt}(t) - (u_0, \phi_j)\Psi(0) + \int_0^T a(u(t), \phi_j)\Psi(t)$$
$$= \int_0^T (f(t), \phi_j)\Psi(t) \ , \quad \forall j = 1, ..., N_0 \ .$$

Since the linear combinations of ϕ_j are dense in V, we can also write (11.1.11) for each $v \in V$. Moreover, taking $\Psi \in \mathcal{D}(0, T)$, (11.1.11) is nothing else than (11.1.5). We have thus constructed a solution $u \in L^\infty(0, T; L^2(\Omega)) \cap L^2(0, T; V)$ of problem (11.1.5).

It remains only to show that $u(0) = u_0$ and $\frac{\partial u}{\partial t} \in L^2(0, T; V')$. Let us recall, in fact, that from the latter result it follows that $u \in L^2(0, T; V) \cap H^1(0, T; V')$, therefore $u \in C^0([0, T]; L^2(\Omega))$ (see Lions and Magenes (1968a), p. 23). Let us begin with proving that $u(0) = u_0$. First of all, from (11.1.5) we have that $(u(t), v) \in H^1(0, T)$, hence it is a continuous function on $[0, T]$. Now multiply (11.1.5) by $\Psi \in C^1([0, T])$ with $\Psi(T) = 0$. After integrating by parts one has

$$-\int_0^T (u(t), v) \frac{d\Psi}{dt}(t) - (u(0), v)\Psi(0) + \int_0^T a(u(t), v)\Psi(t)$$
$$= \int_0^T (f(t), v)\Psi(t) \quad \forall v \in V \ ;$$

thus, taking $\Psi(0) = 1$

$$(u(0) - u_0, v) = 0 \quad \forall v \in V \ .$$

This implies that $u(0) = u_0$.

Finally, from (11.1.5) it also follows that

$$(11.1.12) \qquad\qquad \frac{\partial u}{\partial t} + Lu = f$$

in the sense of distributions on Q_T. Since $Lu \in L^2(0, T; V')$, we find $\frac{\partial u}{\partial t} \in L^2(0, T; V')$.

The uniqueness of the solution is a consequence of (11.1.7), which is indeed an *a-priori* estimate for any solution $u \in L^2(0, T; V) \cap H^1(0, T; V')$ to (11.1.5). Rewrite (11.1.5) as

$$\left\langle \frac{\partial u}{\partial t}(t), v \right\rangle + a(u(t), v) = (f(t), v) \qquad \forall\, v \in V \ ,$$

where $\langle \cdot, \cdot \rangle$ denotes the duality pairing between V' and V (see Section 1.2 (ii)). Taking $v = u(t)$, it follows (see, for instance, Lions and Magenes (1968a); Temam (1984), p. 260)

$$\frac{1}{2} \frac{d}{dt} \|u(t)\|_0^2 + a(u(t), u(t)) = (f(t), u(t))$$

and thus (11.1.7) is obtained by proceeding as in the proof of (11.1.10). □

Remark 11.1.1 The same result is obtained if the assumption on f is weakened to $f \in L^2(0, T; V')$, as one can easily verify by following the proof presented above. □

Remark 11.1.2 An alternative approach to prove the existence of a solution is based on the semigroup theory, in which case the solution u to (11.1.4) is formally given by

$$(11.1.13) \qquad\qquad u(t) = e^{-tA} u_0 + \int_0^t e^{-(t-s)A} f(s)\, ds \ .$$

Here, A represents a suitable "realization" of L with the associated boundary condition $Bu = 0$ on $\partial\Omega$. We notice that, in order to give a meaning to (11.1.3), the assumptions on u_0 and f must be somehow strenghtened. We are not going to present further details concerning this approach, referring the interested reader to Pazy (1983). □

Remark 11.1.3 *(The symmetric case).* If the bilinear form $a(w, v)$ is coercive and symmetric, i.e., it satisfies (11.1.6) with $\lambda = 0$ and $a(w, v) = a(v, w)$ for each $w, v \in V$, it is possible to costruct the solution to (11.1.5) as a series

$$(11.1.14) \qquad u(t) = \sum_{i=1}^{\infty} \left\{ (u_0, \omega_i) e^{-\lambda_i t} + \int_0^t (f(s), \omega_i) e^{-\lambda_i(t-s)}\, ds \right\} \ .$$

Here, $\lambda_i \in \mathbb{R}$ and $\omega_i \in V$, $\omega_i \neq 0$, are the eigenvalues and eigenvectors of the bilinear form $a(\cdot, \cdot)$, i.e., λ_i and ω_i satisfy for each i

$$a(\omega_i, v) = \lambda_i(\omega_i, v) \quad \forall\, v \in V .$$

Since the inclusion $V \subset L^2(\Omega)$ is compact (see Theorem 1.3.5), the eigenvalues $\{\lambda_i\}$ form a non-decreasing sequence bounded from below by the constant α and unbounded from above. Moreover, $\{\omega_i\}$ provides a complete orthonormal basis in $L^2(\Omega)$, and $\{\lambda_i^{-1/2}\omega_i\}$ is a complete orthonormal basis in V with respect to the scalar product induced by $a(\cdot, \cdot)$ (see, e.g., Brezis (1983), pp. 97–98, Raviart and Thomas (1983), pp. 137–138). $\qquad\square$

Having shown the existence of a unique solution, we will give some additional a-priori estimates which will be useful in the sequel.

Proposition 11.1.1 *Assume* (11.1.6) *is satisfied with* $\lambda = 0$, *and that* $f \in L^2(Q_T)$, $u_0 \in V$, $a_{ij}, b_i \in C^1(\overline{\Omega})$ *and* $c_i, a_0 \in L^\infty(\Omega)$. *Then the solution* u *to* (11.1.5) *belongs to* $L^\infty(0,T;V) \cap H^1(0,T;L^2(\Omega))$ *and satisfies the energy estimate*

$$(11.1.15) \quad \sup_{t\in(0,T)} \|u(t)\|_1^2 + \int_0^T \left\|\frac{\partial u}{\partial t}(t)\right\|_0^2 \le C_\alpha \left(\|u_0\|_1^2 + \int_0^T \|f(t)\|_0^2 \right) ,$$

where $C_\alpha > 0$ *is a constant independent of* T.

Proof. We will proceed in a formal way, because a rigorous proof would require considering the approximate problem (11.1.8), finding the estimate (11.1.15) for the sequence u^N, and passing to the limit. The limit function u obtained is obviously the solution to (11.1.5), and satisfies (11.1.15) due to the properties of the weak convergence.

Notice moreover that, when $a_{ij} \in C^1(\overline{\Omega})$, it is not restrictive assuming $a_{ij} = a_{ji}$ for each $i,j = 1, ..., d$. In fact, one can write

$$\sum_{i,j=1}^d D_i(a_{ij}D_ju) = \sum_{i,j=1}^d D_i \left[\left(\frac{a_{ij} + a_{ji}}{2}\right) D_ju \right]$$

$$+ \sum_{i,j=1}^d \left[D_i \left(\frac{a_{ij} - a_{ji}}{2}\right) \right] D_ju .$$

Clearly, the coefficients $a_{ij}^S := (a_{ij} + a_{ji})/2$ still satisfy the ellipticity assumption (11.1.2), with the same constant $\alpha_0 > 0$. Further, the vector $\mathbf{c}^S$ defined by

$$c_j^S := \sum_{i=1}^d \left[D_i \left(\frac{a_{ij} - a_{ji}}{2}\right) \right] , \quad j = 1, ..., d ,$$

belongs to $C^0(\overline{\Omega})$ (and satisfies div $\mathbf{c}^S = 0$). We can therefore assume in the sequel that $a_{ij} = a_{ji}$.

Multiply now $(11.1.4)_1$ by $\frac{\partial u}{\partial t}$ and integrate on Ω. Since $\frac{\partial u}{\partial t} = 0$ on Σ_T, for almost each $t \in [0, T]$ we have

$$\int_\Omega \left[\left(\frac{\partial u}{\partial t} \right)^2 + Lu \frac{\partial u}{\partial t} \right]$$

(11.1.16)
$$= \int_\Omega \left[\left(\frac{\partial u}{\partial t} \right)^2 + \sum_{i,j} a_{ij} D_j u D_i \frac{\partial u}{\partial t} \right.$$

$$\left. - \sum_i \left(b_i u D_i \frac{\partial u}{\partial t} - c_i D_i u \frac{\partial u}{\partial t} \right) + a_0 u \frac{\partial u}{\partial t} \right]$$

$$= \int_\Omega f \frac{\partial u}{\partial t} \ .$$

Integrating by parts one finds

$$\int_\Omega \sum_{i,j} a_{ij} D_j u D_i \frac{\partial u}{\partial t} = \frac{1}{2} \frac{d}{dt} \int_\Omega \sum_{i,j} a_{ij} D_i u D_j u$$

and

$$\int_\Omega \sum_i b_i u D_i \frac{\partial u}{\partial t} = - \int_\Omega \operatorname{div}(\mathbf{b} u) \frac{\partial u}{\partial t} \ .$$

Integrating in time (11.1.16) on $(0, t)$, from the ellipticity assumption we have

$$\|u(t)\|_1^2 + \int_0^t \left\| \frac{\partial u}{\partial t}(\tau) \right\|_0^2$$

$$\leq C \left[\|u_0\|_1^2 + \int_0^t (\|u(\tau)\|_1 + \|f(\tau)\|_0) \left\| \frac{\partial u}{\partial t}(\tau) \right\|_0 \right] \ .$$

This yields

$$(11.1.17) \quad \|u(t)\|_1^2 + \int_0^t \left\| \frac{\partial u}{\partial t}(\tau) \right\|_0^2 \leq C \left[\|u_0\|_1^2 + \int_0^t (\|u(\tau)\|_1^2 + \|f(\tau)\|_0^2) \right] \ ,$$

hence, by employing (11.1.7), the energy estimate (11.1.15) follows at once.

$\square$

Remark 11.1.4. Let us show how to modify Proposition 11.1.1 when the boundary condition is the (homogeneous) Neumann one. In such a case we have $V = H^1(\Omega)$. As before, let us assume that $a_{ij} = a_{ji}$ for each $i, j = 1, ..., d$. Estimate (11.1.17) is obtained in the same way, only noting that

$$- \int_\Omega \sum_i b_i u D_i \frac{\partial u}{\partial t} = \int_\Omega \operatorname{div}(\mathbf{b} u) \frac{\partial u}{\partial t} - \int_{\partial \Omega} \mathbf{b} \cdot \mathbf{n} u \frac{\partial u}{\partial t}$$

$$= \int_\Omega \operatorname{div}(\mathbf{b} u) \frac{\partial u}{\partial t} - \frac{1}{2} \frac{d}{dt} \int_{\partial \Omega} \mathbf{b} \cdot \mathbf{n} u^2 \ .$$

By integrating on $(0, t)$, the trace theorem and an interpolation result (see Theorems 1.3.1 and 1.3.7) yield

$$\left| \int_{\partial \Omega} \mathbf{b} \cdot \mathbf{n} u^2(t) \right| \leq \varepsilon ||u(t)||_1^2 + C_\varepsilon ||u(t)||_0^2 \ ,$$

thus (11.1.17) holds with the additional term $C||u(t)||_0^2$ at the right hand side. This last term can be estimated by means of (11.1.7), so that (11.1.15) is obtained.

A similar procedure can be employed when considering the (homogeneous) Robin case. $\qquad \square$

Remark 11.1.5 Proposition 11.1.1 still holds (with suitable modifications) when considering the non-homogeneous case for the Neumann and Robin problems. In fact, assuming $\partial \Omega \in C^2$ and $g \in L^2(0,T; H^{1/2}(\partial \Omega)) \cap H^{1/4}(0,T; L^2(\partial \Omega))$ (and moreover $\kappa \in C^1(\partial \Omega)$ for the Robin problem), a proof of this result can be found, e.g., in Lions and Magenes (1968b), p. 34. Clearly, one has also to add the norm of the boundary datum g to the norms of u_0 and f at the right hand side of (11.1.15). $\qquad \square$

A simple consequence of Proposition 11.1.1 is given by

Corollary 11.1.1 *Assume that the solution u obtained in Proposition 11.1.1 satisfies*

$$(11.1.18) \qquad ||u(t)||_2^2 \leq C(||Lu(t)||_0^2 + ||u(t)||_1^2) \quad a.e. \ in \ [0,T] \ .$$

Then $u \in L^2(0,T; H^2(\Omega)) \cap H^1(0,T; L^2(\Omega)) \cap C^0([0,T]; V)$ and satisfies the estimate

$$(11.1.19) \qquad \begin{aligned} \max_{t \in [0,T]} ||u(t)||_1^2 + \int_0^T & \left(\left|\left| \frac{\partial u}{\partial t}(t) \right|\right|_0^2 + ||u(t)||_2^2 \right) \\ & \leq C_\alpha \left(||u_0||_1^2 + \int_0^T ||f(t)||_0^2 \right) \ . \end{aligned}$$

Proof. Estimate (11.1.19) follows from (11.1.18), (11.1.15) and (11.1.7), since $Lu = f - \frac{\partial u}{\partial t}$. Moreover, by interpolation one has that

$$L^2(0,T; H^2(\Omega)) \cap H^1(0,T; L^2(\Omega)) \subset C^0([0,T]; H^1(\Omega))$$

(see Theorem 1.3.8 or Lions and Magenes (1968a), p. 23). $\qquad \square$

Remark 11.1.6 Assumption (11.1.18) is satisfied for the homogeneous Dirichlet problem if $\partial \Omega \in C^2$ (see, e.g., Lions and Magenes (1968a), p. 166) or if Ω is a plane convex polygonal domain (see Grisvard (1976)). Indeed, under these assumptions, (11.1.18) is also satisfied for the (homogeneous) Neumann and Robin problem (assuming moreover, in this last case $\kappa \in C^1(\partial\Omega)$). On the contrary, (11.1.18) in general does not hold for the mixed problem. $\square$

11.2 Semi-Discrete Approximation

A first step towards the approximation of the solution to (11.1.5) entails the discretization of the space variable only. This leads to a system of ordinary differential equations, whose solution $u_h(t)$ is an approximation of the exact solution for each $t \in [0, T]$.

All methods described in Chapters 5 and 6 can be employed to fulfill this aim. Here, as in the sequel of this Chapter, we focus on the finite element method (Section 11.2.1) and the spectral collocation method (Section 11.2.2).

In Section 11.2.1 we will prove an error estimate with respect to the norms of $C^0([0, T]; L^2(\Omega))$ and $L^2(0, T; H^1(\Omega))$ for piecewise-linear polynomials (see Proposition 11.2.1). Moreover, assuming that the solution u is smooth enough, an optimal estimate for higher order elements will be provided (see Proposition 11.2.2), and in Remark 11.2.3 the error estimate for a "rough" initial datum $u_0 \in L^2(\Omega)$ will be presented. All these results are obtained by means of the energy method.

Similar results are also presented in Section 11.2.2 for the spectral collocation case.

11.2.1 The Finite Element Case

The variational formulation (11.1.5) naturally leads to a semi-discrete problem by approximating the space V by a finite dimensional space V_h.

The proof of Theorem 11.1.1 provides an instance of such a procedure: the so-called Faedo-Galerkin method, consisting in choosing $V_h = V^N :=$ span $\{\phi_1, ..., \phi_N\}$, $\{\phi_j\}$ being a complete orthonormal basis of V.

Now we want to present other choices of V_h, based on the finite element method. The semi-discrete approximate problem reads as follows: given $f \in L^2(Q_T)$ and $u_{0,h} \in V_h$, a suitable approximation of the initial datum $u_0 \in L^2(\Omega)$, for each $t \in [0, T]$ find $u_h(t) \in V_h$ such that

$$(11.2.1) \quad \begin{cases} \dfrac{d}{dt}(u_h(t), v_h) + a(u_h(t), v_h) \\ \qquad = (f(t), v_h) \quad \forall\, v_h \in V_h \ , \quad t \in (0, T) \\ u_h(0) = u_{0,h} \ . \end{cases}$$

Here we have assumed that Ω is a polygonal domain with Lipschitz boundary, and we are considering, for simplicity, the homogeneous Dirichlet boundary condition, i.e., V_h is chosen as in (6.2.4). When dealing with other boundary conditions, the space V_h must be chosen as in (6.2.5)-(6.2.7); moreover, the bilinear form and the linear functional at the right hand side must be modified as indicated in (6.1.10), (6.1.13) and (6.1.16).

Problem (11.2.1) is a system of ordinary differential equations. Writing $u_h(t) = \sum_j \xi_j(t)\varphi_j$, where $\{\varphi_j\}$, $j = 1, ..., N_h$, is a basis of V_h, and $u_{0,h} = \sum_j \xi_{0,j}\varphi_j$, problem (11.2.1) can be written as

$$(11.2.2) \qquad \begin{cases} M\dfrac{d}{dt}\boldsymbol{\xi}(t) + A\boldsymbol{\xi}(t) = \mathbf{F}(t) \\[2mm] \boldsymbol{\xi}(0) = \boldsymbol{\xi}_0 \ , \end{cases}$$

where

$$M_{ij} = (M_{fe})_{ij} := (\varphi_i, \varphi_j) \ , \ A_{ij} = (A_{fe})_{ij} := a(\varphi_j, \varphi_i) \ ,$$
$$F_i(t) = (F_{fe})_i(t) := (f(t), \varphi_i) \ , \qquad i, j = 1, ..., N_h \ .$$

Since M_{fe} is positive definite, there exists a unique solution $\boldsymbol{\xi}(t)$ of (11.2.2).

Repeating the proof of Theorem 11.1.1, we see that the solution u_h satisfies an energy estimate like (11.1.7), provided $u_{0,h}$ converge to u_0 in $L^2(\Omega)$. This proves the stability of the method.

We are now going to prove the convergence of u_h to u in a suitable topology, and give an estimate of the order of convergence.

Proposition 11.2.1 *Let $\mathcal{T}_h$ be a regular family of triangulations and assume that piecewise-linear or -bilinear finite elements are used. Assume moreover that (11.1.6) (with $\lambda = 0$) and (11.1.18) are satisfied and that $f \in L^2(Q_T)$, $u_0 \in V$, $a_{ij}, b_i \in C^1(\overline{\Omega})$, $c_i, a_0 \in L^\infty(\Omega)$. Then the solutions u and u_h to (11.1.5) and (11.2.1), respectively, satisfy*

$$(11.2.3) \quad \begin{aligned} \|u(t) - u_h(t)\|_0^2 &+ \alpha \int_0^t \|u(\tau) - u_h(\tau)\|_1^2 \\ &\leq \|u_0 - u_{0,h}\|_0^2 + C_{\alpha,\gamma}h^2 \left(\|u_{0,h}\|_1^2 + \|u_0\|_1^2 + \int_0^t \|f(\tau)\|_0^2\right) \end{aligned}$$

for each $t \in [0, T]$. Here α is the coerciveness constant in (11.1.6), γ is the continuity constant of the bilinear form $a(\cdot, \cdot)$, and $C_{\alpha,\gamma} > 0$ is a suitable constant independent of h.

Proof. For each $t \in [0, T]$ define $e_h(t) := u(t) - u_h(t)$; from (11.1.5) and (11.2.1) one finds

$$(11.2.4) \quad \begin{aligned} \frac{d}{dt}(e_h(t), v_h) &+ a(e_h(t), v_h) \\ &= \left(\frac{\partial e_h}{\partial t}(t), v_h\right) + a(e_h(t), v_h) = 0 \qquad \forall\, v_h \in V_h \ . \end{aligned}$$

For almost any fixed t, choose $v_h = u_h(t) - w_h$, $w_h \in V_h$ in (11.2.4). For each $\varepsilon > 0$ and for almost any $t \in [0, T]$ we find

(11.2.5)
$$\frac{1}{2} \frac{d}{dt}(e_h(t), e_h(t)) + a(e_h(t), e_h(t))$$
$$= \left(\frac{\partial e_h}{\partial t}(t), u(t) - w_h \right) + a(e_h(t), u(t) - w_h)$$
$$\leq \left\| \frac{\partial e_h}{\partial t}(t) \right\|_0 \|u(t) - w_h\|_0 + \gamma \|e_h(t)\|_1 \|u(t) - w_h\|_1$$
$$\leq \left\| \frac{\partial e_h}{\partial t}(t) \right\|_0 \|u(t) - w_h\|_0 + \frac{\gamma^2}{4\varepsilon} \|u(t) - w_h\|_1^2 + \varepsilon \|e_h(t)\|_1^2 .$$

From Corollary 11.1.1, it follows that $u \in L^2(0, T; H^2(\Omega))$. Hence, choosing for almost any $t \in [0, T]$ $w_h = \pi_h^k(u(t))$ (see (3.4.1)), we find

$$\|u(t) - \pi_h^k(u(t))\|_0^2 + h^2 \|u(t) - \pi_h^k(u(t))\|_1^2 \leq C h^4 \|u(t)\|_2^2 .$$

Integrating (11.2.5) in $(0, t)$ and choosing $\varepsilon = \alpha/2$ yields

$$\|e_h(t)\|_0^2 + \alpha \int_0^t \|e_h(\tau)\|_1^2 \leq \|u_0 - u_{0,h}\|_0^2$$
$$+ C_{\alpha,\gamma} h^2 \int_0^t \left(\left\| \frac{\partial u}{\partial t}(\tau) \right\|_0^2 + \left\| \frac{\partial u_h}{\partial t}(\tau) \right\|_0^2 + \|u(\tau)\|_2^2 \right) .$$

As in (11.1.15) we have

$$\int_0^t \left\| \frac{\partial u_h}{\partial t}(\tau) \right\|_0^2 \leq C_\alpha \left(\|u_{0,h}\|_1^2 + \int_0^t \|f(\tau)\|_0^2 \right) ,$$

therefore the thesis follows from (11.1.19).   □

Let us notice that the procedure employed in the proof above leads to an optimal error estimate with respect to the norm of $L^2(0, T; H^1(\Omega))$ and for piecewise-linear polynomials only. A different method is presented in the following Proposition 11.2.2, yielding an optimal error estimate in the norm of $C^0([0, T]; L^2(\Omega))$ for any degree of the approximating piecewise-polynomials, provided that the solution u is smooth enough.

Remark 11.2.1 If $u_{0,h} \in V_h$ is chosen in such a way that

(11.2.6) $\|u_0 - u_{0,h}\|_0 + h\|u_0 - u_{0,h}\|_1 \leq Ch\|u_0\|_1$,

then the error $u - u_h$ is $O(h)$ in the space $C^0([0, T]; L^2(\Omega)) \cap L^2(0, T; H^1(\Omega))$. For instance, we can take $u_{0,h} = P_{1,h}^k(u_0)$, where $P_{1,h}^k$ is the projection on V_h with respect to the scalar product of $H^1(\Omega)$. Assuming that T_h is quasi-uniform, another possible choice is given by $u_{0,h} = P_h^k(u_0)$, where P_h^k is the

$L^2(\Omega)$-projection onto V_h (see (3.5.3)). With both these choices (11.2.6) is satisfied, provided that Ω is a plane convex polygonal domain (see Section 3.5 and in particular (3.5.22)). However, in practice these choices are not the easiest to be implemented. If we know that $u_0 \in H^2(\Omega)$, it is better to take $u_{0,h} = \pi_h^k(u_0)$, where π_h^k is the finite element interpolation operator. $\square$

Assuming more regularity on the data and, furthermore, that suitable compatibility conditions between the initial datum and the boundary data are satisfied at $t = 0$ on $\partial\Omega$, the solution u to (11.1.5) is indeed more regular. This in principle implies that the convergence of u_h to u is of higher order. As an example, let us just prove the following error estimate, first obtained by Wheeler (1973):

Proposition 11.2.2 *Let $\mathcal{T}_h$ be a regular family of triangulations. Assume that (11.1.6) is satisfied with $\lambda = 0$ and that the solution $\varphi(r)$ of the adjoint problem*

$$\varphi(r) \in V : a(v, \varphi(r)) = (r, v) \qquad \forall \, v \in V$$

satisfies $\varphi(r) \in H^2(\Omega)$ when $r \in L^2(\Omega)$. Assume moreover that $u_0 \in H^{k+1}(\Omega)$, $k \geq 1$, and that the solution u to (11.1.5) is such that $\frac{\partial u}{\partial t} \in L^1(0, T; H^{k+1}(\Omega))$. Then, using piecewise-polynomials of degree less than or equal to k in the definition of the finite element space V_h in (6.2.4), for each $t \in [0, T]$ the solution u_h to (11.2.1) satisfies

$$\|u(t) - u_h(t)\|_0 \leq \|u_0 - u_{0,h}\|_0$$

(11.2.7)
$$+ Ch^{k+1} \left(\|u_0\|_{k+1} + \int_0^t \left\| \frac{\partial u}{\partial t}(\tau) \right\|_{k+1} \right) ,$$

where $C > 0$ is a suitable constant independent of h.

Proof. We make use of the elliptic "projection" operator $\Pi_{1,h}^k$, which is defined as follows for each $v \in V$:

(11.2.8) $\Pi_{1,h}^k(v) \in V_h : a(\Pi_{1,h}^k(v), v_h) = a(v, v_h) \qquad \forall \, v_h \in V_h$.

As (11.1.6) is satisfied with $\lambda = 0$, the existence of such a solution is ensured by the Lax-Milgram lemma. If $a(\cdot, \cdot)$ is symmetric, then $\Pi_{1,h}^k$ is really an orthogonal projection operator onto V_h, with respect to the scalar product $a(\cdot, \cdot)$. If $\mathcal{T}_h$ is a regular family of triangulations and for each $r \in L^2(\Omega)$ the solution $\varphi(r)$ of the adjoint problem belongs to $H^2(\Omega)$, by proceeding as in Section 3.5 one has

(11.2.9) $\|v - \Pi_{1,h}^k(v)\|_0 + h\|v - \Pi_{1,h}^k(v)\|_1 \leq Ch^{k+1}|v|_{k+1} \quad \forall \, v \in H^{k+1}(\Omega)$.

For any fixed $t \in [0, T]$ let us write $u_h(t) - u(t) = w_1(t) + w_2(t)$, where

(11.2.10) $w_1(t) := u_h(t) - \Pi_{1,h}^k(u(t))$, $w_2(t) := \Pi_{1,h}^k(u(t)) - u(t)$.

From (11.2.9), w_2 is easily estimated in the following way:

$$||w_2(t)||_0 \leq Ch^{k+1}||u(t)||_{k+1}$$

(11.2.11)

$$\leq Ch^{k+1}\left(||u_0||_{k+1} + \int_0^t \left|\left|\frac{\partial u}{\partial t}(\tau)\right|\right|_{k+1}\right) \ .$$

On the other hand, from (11.1.5), (11.2.1) and (11.2.8) for each $v_h \in V_h$ we have

$$\left(\frac{\partial w_1}{\partial t}(t), v_h\right) + a(w_1(t), v_h)$$

$$= \left(\frac{\partial u_h}{\partial t}(t), v_h\right) + a(u_h(t), v_h)$$

$$- \left(\frac{\partial}{\partial t}[\Pi_{1,h}^k(u(t))], v_h\right) - a(\Pi_{1,h}^k(u(t)), v_h)$$

$$= \left(\frac{\partial u}{\partial t}(t), v_h\right) - \left(\frac{\partial}{\partial t}[\Pi_{1,h}^k(u(t))], v_h\right) = \left(-\frac{\partial w_2}{\partial t}(t), v_h\right) \ .$$

Choosing $v_h = w_1(t)$ it follows

$$\frac{1}{2}\frac{d}{dt}||w_1(t)||_0^2 + a(w_1(t), w_1(t)) = -\left(\frac{\partial w_2}{\partial t}(t), w_1(t)\right) \ .$$

Using again (11.1.6), we have

(11.2.12)
$$\frac{1}{2}\frac{d}{dt}||w_1(t)||_0^2 + \alpha||w_1(t)||_1^2 \leq -\left(\frac{\partial w_2}{\partial t}(t), w_1(t)\right) \ .$$

Thus, integrating over $(0, t)$

$$||w_1(t)||_0 \leq ||w_1(0)||_0 + \int_0^t \left|\left|\frac{\partial w_2}{\partial t}(\tau)\right|\right|_0 \ .$$

Since $\frac{\partial}{\partial t}$ commutes with $\Pi_{1,h}^k$, the thesis follows from (11.2.9) and (11.2.11).

$\square$

In a similar way one can obtain an error estimate with respect to the norm $L^2(0, T; H^1(\Omega))$. Precisely, assuming that $u_0 \in H^k(\Omega)$, $u \in L^2(0, T; H^{k+1}(\Omega))$ and $\frac{\partial u}{\partial t} \in L^2(0, T; H^k(\Omega))$, it is easily shown that

$$\left(\alpha \int_0^T ||u(\tau) - u_h(\tau)||_1^2\right)^{1/2} = O(h^k) \ .$$

Remark 11.2.2 If we had taken advantage of the presence of the term $\alpha||w_1(t)||_1^2$ at the left hand side of (11.2.12) we would have obtained

$$\frac{d}{dt}||w_1(t)||_0 + \alpha||w_1(t)||_0 \le \left|\left|\frac{\partial w_2}{\partial t}(t)\right|\right|_0$$

(notice that we have bounded from below $||w_1(t)||_1$ with $||w_1(t)||_0$), i.e.,

$$\frac{d}{dt}[\exp(\alpha t)||w_1(t)||_0] \le \exp(\alpha t)\left|\left|\frac{\partial w_2}{\partial t}(t)\right|\right|_0 .$$

Integrating over $(0,t)$ we find

$$||w_1(t)||_0 \le ||w_1(0)||_0 \exp(-\alpha t)$$

(11.2.13)
$$+ \int_0^t \exp[-\alpha(t-\tau)]\left|\left|\frac{\partial w_2}{\partial t}(\tau)\right|\right|_0 ,$$

thus for each $t \in [0,T]$

$$||u(t) - u_h(t)||_0 \le ||u_0 - u_{0,h}||_0 \exp(-\alpha t)$$

$$+ Ch^{k+1}\left\{||u_0||_{k+1} \exp(-\alpha t) + ||u(t)||_{k+1}\right.$$

$$\left. + \int_0^t \exp[-\alpha(t-\tau)]\left|\left|\frac{\partial u}{\partial t}(\tau)\right|\right|_{k+1}\right\} .$$

This shows, in particular, that the effect of the initial error $u_0 - u_{0,h}$ tends exponentially to 0 as t goes to ∞.

For further results of this type, the interested reader can refer to Thomée (1984) and Fujita and Suzuki (1991). □

Remark 11.2.3 If the initial datum is not regular, it is possible to prove optimal error estimates depending on a negative power of t. Assuming, for simplicity, that $f = 0, \lambda = 0$ in (11.1.6) and that a_{ij}, b_i, c_i, a_0 in (11.1.1) are smooth enough, $i, j = 1, ..., d$, a typical result reads as follows

$$(11.2.14) \qquad ||u(t) - u_h(t)||_0 \le Ch^{k+1}t^{-(k+1)/2}||u_0||_0 , \quad t > 0 .$$

As usual, $k \ge 1$ refers to the degree of the polynomials $\mathbb{P}_k$ used in defining V_h, and $u_{0,h} = P_h^k(u_0)$ is the $L^2(\Omega)$-orthogonal projection of u_0 onto V_h. For a proof of (11.2.14) and of other sharp error estimates for a non-smooth initial datum see, e.g., Bramble, Schatz, Thomée and Wahlbin (1977), Huang and Thomée (1981), Sammon (1982), Luskin and Rannacher (1982a) and Lasiecka (1984). See also Thomée (1984), Fujita and Suzuki (1991).

In particular, estimate (11.2.14) shows that, being $f = 0$, the error $u(t) - u_h(t)$ decays to zero as t goes to infinity even if u_0 is not smooth. An exponential decay can be recovered by proceeding as in Remark 11.2.2. □

Remark 11.2.4 (L^∞-error estimate). It is possible to obtain a result similar to (11.2.7) also when considering the error estimate in the $L^\infty(\Omega)$-norm. In fact, assume that the bilinear form $a(\cdot, \cdot)$ is symmetric and coercive (i.e., (11.1.6) holds with $\lambda = 0$). In the case of a two-dimensional domain and piecewise-linear finite elements, setting $u_{0,h} = \Pi_{1,h}^1(u_0)$, in Schatz, Thomée and Wahlbin (1980) the following error estimate is proven for each $t \in [0, T]$:

$$\|u(t) - u_h(t)\|_{L^\infty(\Omega)} \leq Ch^2 |\log h|^2 \left[\|u_0\|_{W^{2,\infty}(\Omega)} + \int_0^t \left\| \frac{\partial u}{\partial t}(\tau) \right\|_{W^{2,\infty}(\Omega)} \right].$$

Let us notice that a logarithmic factor also appears in the analogous estimate for elliptic equations (see Remark 6.2.3). □

11.2.2 The Case of Spectral Methods

Now we consider the semi-discrete approximation of problem (11.1.5), based upon algebraic polynomial approximations in space.

As done in the finite element context, we confine to the case in which $u(t) = 0$ on $\partial\Omega$ for all $t \geq 0$ (homogeneous Dirichlet boundary condition); moreover, for simplicity we only consider the two-dimensional case. Therefore we set $\Omega = (-1, 1)^2$ and denote by V_N the space of algebraic polynomials of degree less than or equal to N with respect to each coordinate x_i, $i = 1, 2$, that vanish on $\partial\Omega$. This space was already introduced in (6.2.14).

Instead of (11.2.1) we consider the following problem. Let $u_{0,N} \in V_N$ be a convenient approximation of u_0. For each $t \in [0, T]$, we look for a function $u_N(t) \in V_N$ that satisfies

$$(11.2.15) \quad \begin{cases} \dfrac{d}{dt}(u_N(t), v_N) + a(u_N(t), v_N) \\ \qquad = (f(t), v_N) \qquad \forall\, v_N \in V_N \ , \quad t \in (0, T) \\ u_N(0) = u_{0,N} \ . \end{cases}$$

This is the spectral Galerkin semi-discrete approximation to (11.1.5).

A more effective (and flexible) approach is the one based on the spectral collocation approximation. Let us consider, for simplicity, the Legendre case, referring to Section 6.2.2 for the modifications necessary to deal with the Chebyshev one. Formally speaking, this can be easily formulated by replacing (11.2.15) with

$$(11.2.16) \quad \begin{cases} \dfrac{d}{dt}(u_N(t), v_N)_N + a_N(u_N(t), v_N) \\ \qquad = (f(t), v_N)_N \qquad \forall\, v_N \in V_N \ , \quad t \in (0, T) \\ u_N(0) = u_{0,N} \ . \end{cases}$$

As explained in detail in Section 6.2.2, this method is based on the idea of splitting the operator L into its symmetric and skew-symmetric part, replacing any exact scalar product $(\varphi, \psi) = \int_\Omega \varphi\psi$ with numerical integration based upon Legendre Gauss-Lobatto formulae, i.e., (following (4.5.38), (4.5.39), but using a single sub-index for both the nodes and the weights $\mathbf{x}_i$ and w_i) with

$$(\varphi, \psi)_N = \sum_{i=1}^{(N+1)^2} \varphi(\mathbf{x}_i)\, \psi(\mathbf{x}_i)\, w_i \ .$$

Therefore the approximate bilinear form a_N reads

$$a_N(z, v) := \sum_{i,j=1}^{2} (a_{ij} D_j z, D_i v)_N - \frac{1}{2} \sum_{i=1}^{2} [(b_i z, D_i v)_N - (b_i D_i z, v)_N]$$
$$+ \left(\left(\frac{1}{2} \operatorname{div} \mathbf{b} + a_0 \right) z, v \right)_N \ .$$

In order to simplify our presentation, we have assumed $c_i = 0$, $i = 1, 2$, and in the sequel we suppose that $u_{0,N}$ as well as $f(t)$ (for each t) are continuous.

With this understanding, (11.2.16) can be restated collocationwise as follows:

(11.2.17)
$$\begin{cases} \dfrac{d}{dt} u_N(t) + L_N u_N(t) = f(t) & \text{at } \mathbf{x}_i \in \Omega \\[2mm] u_N(t) = 0 & \text{at } \mathbf{x}_i \in \partial\Omega \\[2mm] u_N(0) = u_{0,N} & \text{at } \mathbf{x}_i \in \overline{\Omega} \end{cases}$$

for each $t \in (0, T)$, where L_N is the pseudo-spectral approximation of the operator L defined in (6.2.40).

Showing the equivalence between (11.2.17) and (11.2.16) is an easy matter: since it involves space-like arguments solely, it can be carried out as done in Section 6.2.2.

The algebraic reformulation of the above problems still takes the form (11.2.2), having denoted by φ_j a basis for V_N, $j = 1, ..., (N-1)^2 =: N_h$. For the spectral Galerkin method such a basis can be chosen, e.g., as indicated in (6.2.20).

In the spectral collocation case, a more suitable choice is to take $\varphi_j = \psi_j$, the Lagrangian function associated to the collocation node $\mathbf{x}_j$, i.e., $\psi_j \in V_N$ and satisfies $\psi_j(\mathbf{x}_i) = \delta_{ij}$ for $i, j = 1, ..., N_h$ (see Section 6.3.1). Then, following the notation of Section 6.3.1, we have

(11.2.18)
$$M_{ij} = (M_{sp})_{ij} := (\psi_i, \psi_j)_N \ , \quad A_{ij} = (A_{sp})_{ij} := a_N(\psi_j, \psi_i) \ ,$$
$$F_i(t) = (F_{sp})_i(t) := (f(t), \psi_i)_N \ , \quad i, j = 1, ..., N_h \ .$$

We notice that M_{sp} is diagonal (which is a nice feature) as

$$M_{sp} = \operatorname{diag}(w_1, ..., w_{N_h}) =: D \ ,$$

where w_i denotes the Gaussian coefficient associated with $\mathbf{x}_i$, while $(F_{sp})_i(t) = f(t, \mathbf{x}_i)w_i$. Furthermore, generalizing (6.3.11) to the case of the operator L, if we define the matrix A_{sp}^c associated to the strong form of the collocation problem as

$$(A_{sp}^c)_{ij} := (L_N \psi_j)(\mathbf{x}_i) \quad , \quad i, j = 1, ..., N_h \quad ,$$

it can be shown that $A_{sp} = D A_{sp}^c$ (as in (6.3.12)). Moreover, in the strong collocation case the mass matrix, say M_{sp}^c, is the identity matrix.

Since in both the Galerkin and the collocation case M is a positive definite matrix, the existence of a unique solution can be inferred straigthforwardly.

The stability analysis of the spectral approximations doesn't differ substantially from the one carried out in Section 11.2.1 for the finite element case, whereas the convergence analysis does, as it exploits a different kind of approximation theory.

In the spectral Galerkin case (11.2.15), taking $v_N = u_N(t)$ at each t and assuming that the form $a(\cdot, \cdot)$ satisfies (11.1.6) with $\lambda = 0$ yields

$$(11.2.19) \qquad \|u_N(t)\|_0^2 + \alpha \int_0^t \|u_N(\tau)\|_1^2 \le \|u_{0,N}\|_0^2 + \frac{1}{\alpha} \int_0^t \|f(\tau)\|_0^2$$

for each $t \in [0, T]$. If, for instance, $u_{0,N}$ is the L^2-projection of u_0 upon V_N, then $\|u_{0,N}\|_0 \le \|u_0\|_0$, thus (11.2.19) provides stability.

In the spectral collocation case, the procedure is about the same, however it yields:

$$\frac{1}{2} \frac{d}{dt} \|u_N(t)\|_N^2 + a_N(u_N(t), u_N(t)) = (f(t), u_N(t))_N \quad .$$

Let us consider, for simplicity, the Legendre collocation case. If we assume that there exists a constant $\alpha_N > 0$ such that

$$(11.2.20) \qquad a_N(v_N, v_N) \ge \alpha_N \|v_N\|_1^2 \quad \forall \, v_N \in V_N \quad ,$$

then, from the equality above and (4.5.41), we obtain

$$(11.2.21) \qquad \begin{aligned} \|u_N(t)\|_N^2 &+ \alpha_N \int_0^t \|u_N(\tau)\|_1^2 \\ &\le \|u_{0,N}\|_N^2 + \frac{9}{\alpha_N} \int_0^t \|f(\tau)\|_N^2 \end{aligned}$$

for each $t \in [0, T]$, where $\|v\|_N = \sqrt{(v, v)_N}$ is the discrete norm. Now, we remember that (see (6.2.27))

$$\|v\|_N \le 2\|v\|_{C^0(\overline{\Omega})} \quad \forall \, v \in C^0(\overline{\Omega}) \quad .$$

Then, if we assume $u_0 \in C^0(\overline{\Omega})$ and we take $u_{0,N} = I_N u_0$ (the interpolant of u_0 at the collocation nodes $\{\mathbf{x}_i\}$), we deduce from (11.2.21) that

$$\|u_N(t)\|_N^2 + \alpha_N \int_0^t \|u_N(\tau)\|_1^2$$

(11.2.22)

$$\leq 4 \left(\|u_0\|_{C^0(\overline{\Omega})}^2 + \frac{9}{\alpha_N} \int_0^t \|f(\tau)\|_{C^0(\overline{\Omega})}^2 \right) \quad .$$

If it happens that α_N is bounded from below uniformly with respect to N, i.e.,

(11.2.23) $\exists \, \overline{\alpha} > 0$ such that $\alpha_N \geq \overline{\alpha} \quad \forall \, N$,

then from (11.2.22) it follows that

$$\|u_N(t)\|_N^2 + \overline{\alpha} \int_0^t \|u_N(\tau)\|_1^2$$

(11.2.24)

$$\leq 4 \left(\|u_0\|_{C^0(\overline{\Omega})}^2 + \frac{9}{\overline{\alpha}} \int_0^t \|f(\tau)\|_{C^0(\overline{\Omega})}^2 \right) \quad .$$

Since $\|u_N(t)\|_N \geq \|u_N(t)\|_0$ (see (4.5.41)), the above inequality implies uniform stability for the spectral collocation solution.

Before concluding, let us notice that assumptions (11.2.20), (11.2.23) are satisfied for the bilinear form associated to the operator $L + \lambda I$, λ a suitable positive constant, if the coefficients a_{ij}, b_i and a_0 and moreover div $\mathbf{b}$ are continuous functions in $\overline{\Omega}$. (For simplicity, we have assumed here that $c_i = 0$ for $i = 1, 2$.) As a matter of fact, in such a case

$$a_N(v_N, v_N) \geq \alpha_0 \|\nabla v_N\|_N^2 - \left(\frac{1}{2} \| \text{div } \mathbf{b}\|_{C^0(\overline{\Omega})} + \|a_0\|_{C^0(\overline{\Omega})} \right) \|v_N\|_N^2 \ ,$$

α_0 being the ellipticity constant of L (see (11.1.2)). Therefore, setting $\lambda := \frac{1}{2}\| \text{div } \mathbf{b}\|_{C^0(\overline{\Omega})} + \|a_0\|_{C^0(\overline{\Omega})}$, the modified bilinear form $a_N(z, v) + \lambda(z, v)_N$ (which is associated to the operator $L + \lambda I$) satisfies (11.2.20), (11.2.23) with $\overline{\alpha}$ which depends solely on the Poincaré constant C_Ω (see (1.3.2)).

About the proof of convergence, we restrict ourselves to the spectral Galerkin case (11.2.15). In order to exploit all the potential accuracy of polynomial approximation, we make the assumption that the data and the solution to problem (11.1.5) are smooth as much as needed. In particular, we suppose that

(11.2.25) $u_0 \in H^s(\Omega)$ and $\dfrac{\partial u}{\partial t} \in L^1(0, T; H^s(\Omega))$ for some $s \geq 1$.

We then follow the guidelines of the proof of Proposition 11.2.2, adapting the notation (and concepts) to the spectral Galerkin framework. First, for any $v \in V = H_0^1(\Omega)$, let us denote by $\Pi_{1,N} v \in V_N$ its elliptic "projection" which is defined as follows:

(11.2.26) $a(\Pi_{1,N} v, v_N) = a(v, v_N) \quad \forall \, v_N \in V_N$.

If $a(\cdot,\cdot)$ is symmetric, $\Pi_{1,N}$ is indeed an orthogonal projection operator with respect to the scalar product $a(\cdot,\cdot)$. In any case, the existence and uniqueness of such a $\Pi_{1,N}v$ is ensured from the Lax-Milgram lemma as $a(\cdot,\cdot)$ is coercive and continuous on V. Let us observe that if $a(z,v) = (\nabla z, \nabla v)$ (i.e., we are dealing with the heat equation, where $Lz = -\Delta z$), then $\Pi_{1,N}$ coincides with the orthogonal projection operator $P_N^{1,0}$ that has been introduced in (4.5.33). In any case, a proof similar to the one leading to (4.5.37) yields the estimate

$$(11.2.27) \quad N||u(t) - \Pi_{1,N}u(t)||_0 + ||u(t) - \Pi_{1,N}u(t)||_1$$
$$\leq CN^{1-s}||u(t)||_s \quad \forall\, t \in [0,T] \;.$$

We now set $u_N(t) - u(t) = w_1(t) + w_2(t)$, where $w_1(t) := u_N(t) - \Pi_{1,N}u(t)$, and $w_2(t) := \Pi_{1,N}u(t) - u(t)$. Owing to (11.2.27) we have for each $t \in [0,T]$:

$$(11.2.28) \quad ||w_2(t)||_0 \leq CN^{-s}||u(t)||_s \leq CN^{-s}\left(||u_0||_s + \int_0^t \left\| \frac{\partial u}{\partial t}(\tau) \right\|_s \right) \;.$$

On the other hand, also in the spectral case w_1 satisfies an inequality like (11.2.13). From (11.2.28) and (11.2.13), and proceeding as done in Remark 11.2.2 we obtain for any $t \in [0,T]$:

$$||u(t) - u_N(t)||_0 \leq ||u_0 - u_{0,N}||_0 \exp(-\alpha t)$$

$$(11.2.29) \qquad\qquad + CN^{-s}\Big\{ ||u_0||_s \exp(-\alpha t) + ||u(t)||_s$$

$$+ \int_0^t \exp[-\alpha(t-\tau)] \left\| \frac{\partial u}{\partial t}(\tau) \right\|_s \Big\} \;.$$

As in the finite element case, the influence of the initial error (that here behaves as $CN^{-s}||u_0||_s$) decays exponentially to 0 as t increases.

A convergence in the norm of $L^2(0,T;H^1(\Omega))$ for the error $u - u_N$ can also be proven. Indeed, taking advantage of the presence of the term $\alpha||w_1(t)||_1^2$ in (11.2.12), one can deduce that

$$\left(\alpha \int_0^T ||u(\tau) - u_N(\tau)||_1^2 \right)^{1/2} = O(N^{1-s}) \;,$$

under the assumptions $u_0 \in H^{s-1}(\Omega)$, $u \in L^2(0,T;H^s(\Omega))$ and $\frac{\partial u}{\partial t} \in L^2(0,T;H^{s-1}(\Omega))$.

11.3 Time-Advancing by Finite Differences

In order to obtain a full discretization of (11.1.5), we consider a uniform mesh for the time variable t and define

$$(11.3.1) \qquad t_n := n\Delta t \quad , \quad n = 0, 1, ..., \mathcal{N} \quad ,$$

$\Delta t > 0$ being the time-step, and $\mathcal{N} := [T/\Delta t]$, the integral part of $T/\Delta t$. Next, we replace the time derivative by means of suitable difference quotients, thus constructing a sequence $u_h^n(\mathbf{x})$ that approximates the exact solution $u(t_n, \mathbf{x})$.

Let us describe this procedure on a general system of ordinary differential equations

$$(11.3.2) \qquad \begin{cases} \dfrac{d\mathbf{y}}{dt}(t) = \boldsymbol{\Psi}(t, \mathbf{y}(t)) \quad , \quad t \in (0, T) \\[2mm] \mathbf{y}(0) = \mathbf{y}_0 \quad . \end{cases}$$

Several approaches can be used, such as multi-step methods, Runge-Kutta methods or methods based on rational approximations of the exponential (see, e.g., Gear (1971), Crouzeix and Mignot (1984), Thomée (1984), Butcher (1987), Lambert (1991)). For simplicity we confine ourselves to the so-called θ-method (see Section 5.6.2), which consists in replacing (11.3.2) by the following scheme: find $\mathbf{y}^n$ such that

$$(11.3.3) \qquad \begin{cases} \dfrac{1}{\Delta t}(\mathbf{y}^{n+1} - \mathbf{y}^n) \\[2mm] \quad = \theta\boldsymbol{\Psi}(t_{n+1}, \mathbf{y}^{n+1}) + (1-\theta)\boldsymbol{\Psi}(t_n, \mathbf{y}^n) \ , \ n = 0, 1, ..., \mathcal{N} - 1 \ , \\[2mm] \mathbf{y}^0 = \mathbf{y}_0 \quad . \end{cases}$$

Here, $\theta \in [0, 1]$ is a parameter. When $\theta = 0$ or $\theta = 1$, this scheme is called *forward Euler* method or *backward Euler* method, respectively; for $\theta = 1/2$ it is called *Crank-Nicolson* method.

For $\theta \geq 1/2$ the scheme is A-stable (see, e.g., Lambert (1991), p. 244); for $\theta = 1/2$, however, it may suffer from unexpected instabilities especially when rough perturbations are introduced in the data.

In the following Sections we analyze the θ-scheme for both the finite element and the spectral case. For finite elements, a stability and convergence result is first presented in the general situation (see Theorems 11.3.1 and 11.3.2), and then an alternative proof is provided in the case of a symmetric problem (see Theorems 11.3.3 and 11.3.4). Analogous results for the spectral method will be stated in Section 11.3.2.

11.3.1 The Finite Element Case

Applying the θ-scheme to the semi-discrete approximation (11.2.1), one is left with the following problem: find $u_h^n \in V_h$ such that

$$(11.3.4) \quad \begin{cases} \dfrac{1}{\Delta t}(u_h^{n+1} - u_h^n, v_h) + a(\theta u_h^{n+1} + (1-\theta)u_h^n, v_h) \\ \qquad = (\theta f(t_{n+1}) + (1-\theta)f(t_n), v_h) \qquad \forall\, v_h \in V_h \\[2mm] u_h^0 = u_{0,h} \end{cases}$$

for each $n = 0, 1, ..., \mathcal{N} - 1$, having assumed that f is everywhere defined on $[0, T]$.

At each time step, one has to solve the linear system

$$(11.3.5) \qquad (M + \theta \Delta t A)\xi^{n+1} = \eta^n \ ,$$

where η^n is known from the previous steps, the matrices M and A are defined in (11.2.2) and

$$u_h^{n+1} = \sum_{j=1}^{N_h} \xi_j^{n+1}\varphi_j \ ,$$

φ_j being the basis functions of V_h.

Assuming that (11.1.6) is satisfied with $\lambda = 0$, the matrix $(M + \theta \Delta t A)$ is positive definite. In such a case (11.3.5) has a unique solution. Moreover, $(M + \theta \Delta t A)$ is symmetric if the bilinear form is symmetric, i.e., $a(z, v) = a(v, z)$ for each $z, v \in V$. Let us also notice that in general (11.3.5) is a genuine linear system, namely, the matrix $(M + \theta \Delta t A)$ is not diagonal. We will see in Section 11.4 how to resort to a diagonal system when considering the forward Euler method (i.e., $\theta = 0$).

Now let us turn to the proof of stability of u_h^n. It is useful to introduce the following notation: for any function $\phi \in L^2(\Omega)$, define

$$(11.3.6) \qquad ||\phi||_{-1,h} := \sup_{\substack{v_h \in V_h \\ v_h \neq 0}} \frac{(\phi, v_h)}{||v_h||_1} \ ,$$

which is a norm on V_h (clearly, $||\phi||_{-1,h} \leq ||\phi||_0$ for each $\phi \in L^2(\Omega)$).

We are going to prove that the θ-scheme (11.3.4) is unconditionally stable with respect to the $L^2(\Omega)$-norm provided that $1/2 \leq \theta \leq 1$. On the contrary, in the case $0 \leq \theta < 1/2$ we have to assume that a stability condition is satisfied (see (11.3.7) here below).

Theorem 11.3.1 (Stability). *Assume that (11.1.6) is satisfied with $\lambda = 0$ and that the map $t \to ||f(t)||_0$ is bounded in $[0, T]$. When $0 \leq \theta < 1/2$, assume, moreover, that $\mathcal{T}_h$ is a quasi-uniform family of triangulations and that the following restriction on the time-step*

$$(11.3.7) \qquad \Delta t \left(1 + C_3 h^{-2}\right) < \frac{2\alpha}{(1 - 2\theta)\gamma^2}$$

is satisfied. Here, C_3 is the constant appearing in the inverse inequality (6.3.21), while α and γ are the coerciveness and continuity constants, respectively. Then u_h^n defined in (11.3.4) satisfies

$$(11.3.8) \qquad ||u_h^n||_0 \leq C_\theta \left(||u_{0,h}||_0 + \sup_{t \in [0,T]} ||f(t)||_0 \right), \quad n = 0, 1, ..., \mathcal{N},$$

where the constant $C_\theta > 0$ is a non-decreasing function of α^{-1}, γ and T, and is independent of $\mathcal{N}$, Δt and h.

Proof. Take $v_h = \theta u_h^{n+1} + (1 - \theta)u_h^n$ in (11.3.4). It is easily verified that

$$\frac{1}{2}||u_h^{n+1}||_0^2 - \frac{1}{2}||u_h^n||_0^2 + \left(\theta - \frac{1}{2}\right)||u_h^{n+1} - u_h^n||_0^2$$
$$+ \Delta t \, a(\theta u_h^{n+1} + (1 - \theta)u_h^n, \theta u_h^{n+1} + (1 - \theta)u_h^n)$$
$$= \Delta t(\theta f(t_{n+1}) + (1 - \theta)f(t_n), \theta u_h^{n+1} + (1 - \theta)u_h^n).$$

By the coerciveness assumption (11.1.6), for each $0 < \epsilon \leq 1$ we find

$$(11.3.9) \qquad \begin{aligned} ||u_h^{n+1}||_0^2 - ||u_h^n||_0^2 &+ (2\theta - 1)||u_h^{n+1} - u_h^n||_0^2 \\ &+ 2(1 - \epsilon)\alpha \Delta t ||\theta u_h^{n+1} + (1 - \theta)u_h^n||_1^2 \\ &\leq \frac{\Delta t}{2\epsilon\alpha}||\theta f(t_{n+1}) + (1 - \theta)f(t_n)||_{-1,h}^2. \end{aligned}$$

When $1/2 \leq \theta \leq 1$, the left hand side is larger than $||u_h^{n+1}||_0^2 - ||u_h^n||_0^2$, and in particular we can set $\epsilon = 1$. On the contrary, when $0 \leq \theta < 1/2$ we proceed as follows: choosing $v_h = u_h^{n+1} - u_h^n$ in (11.3.4) we find

$$(11.3.10) \qquad \begin{aligned} ||u_h^{n+1} &- u_h^n||_0^2 \\ &= -\Delta t \, a(\theta u_h^{n+1} + (1 - \theta)u_h^n, u_h^{n+1} - u_h^n) \\ &\quad + \Delta t(\theta f(t_{n+1}) + (1 - \theta)f(t_n), u_h^{n+1} - u_h^n) \\ &\leq \gamma \Delta t ||\theta u_h^{n+1} + (1 - \theta)u_h^n||_1 \, ||u_h^{n+1} - u_h^n||_1 \\ &\quad + \Delta t ||\theta f(t_{n+1}) + (1 - \theta)f(t_n)||_{-1,h} \, ||u_h^{n+1} - u_h^n||_1. \end{aligned}$$

By means of the inverse inequality (6.3.21), we finally have

$$(11.3.11) \qquad \begin{aligned} ||u_h^{n+1} - u_h^n||_0 &\leq \Delta t(1 + C_3 h^{-2})^{1/2}\big[\gamma ||\theta u_h^{n+1} + (1 - \theta)u_h^n||_1 \\ &\quad + ||\theta f(t_{n+1}) + (1 - \theta)f(t_n)||_{-1,h}\big]. \end{aligned}$$

Setting for each $\eta > 0$

$$\kappa_\eta := \left[2(1 - \epsilon)\alpha - (1 - 2\theta)\gamma(\gamma + \eta)\Delta t(1 + C_3 h^{-2})\right],$$

it follows

$$\|u_h^{n+1}\|_0^2 - \|u_h^n\|_0^2 + \Delta t\, \kappa_\eta \,\|\theta u_h^{n+1} + (1-\theta)u_h^n\|_1^2$$
$$\leq C_{\epsilon,\eta}\, \Delta t(1 + \Delta t\, h^{-2})\|\theta f(t_{n+1}) + (1-\theta)f(t_n)\|_{-1,h}^2\ .$$

Choosing ϵ and η small enough, due to (11.3.7) we have $\kappa_\eta > 0$ and moreover $1 + \Delta t\, h^{-2} \leq C_*$ for a suitable $C_* > 0$, therefore

$$(11.3.12) \qquad \|u_h^{n+1}\|_0^2 - \|u_h^n\|_0^2 \leq C_{\epsilon,\eta}\, \Delta t\, \|\theta f(t_{n+1}) + (1-\theta)f(t_n)\|_{-1,h}^2\ .$$

Now let m be a fixed index, $1 \leq m \leq \mathcal{N}$. In both cases we have considered above ($0 \leq \theta < 1/2$ or $1/2 \leq \theta \leq 1$), summing up from $n = 0$ to $n = m - 1$ we find

$$\|u_h^m\|_0^2 \leq \|u_{0,h}\|_0^2 + C\Delta t \sum_{n=0}^{m-1} \|\theta f(t_{n+1}) + (1-\theta)f(t_n)\|_{-1,h}^2$$

and the thesis follows. $\qquad\qquad\qquad\qquad\qquad\qquad\qquad\qquad\qquad\qquad \square$

A slightly different stability result is proven in Section 12.2.

Now we turn to the convergence analysis. We start proving an error estimate between the semi-discrete solution $u_h(t_n)$ and the fully-discrete one u_h^n for any fixed h.

Theorem 11.3.2 (Convergence). *Assume that* (11.1.6) *is satisfied with* $\lambda = 0$ *and that* $\frac{\partial u_h}{\partial t}(0) \in L^2(\Omega)$, $f \in L^2(Q_T)$ *with* $\frac{\partial f}{\partial t} \in L^2(Q_T)$. *When* $0 \leq \theta < 1/2$, *assume, moreover, that* $\mathcal{T}_h$ *is a quasi-uniform family of triangulations and that the time-step restriction* (11.3.7) *is satisfied. Then the functions* u_h^n *and* $u_h(t)$ *defined in* (11.3.4) *and in* (11.2.1), *respectively, satisfy*

$$(11.3.13) \qquad \|u_h^n - u_h(t_n)\|_0 \leq C_\theta \Delta t \left(\left\| \frac{\partial u_h}{\partial t}(0) \right\|_0^2 + \int_0^T \left\| \frac{\partial f}{\partial t}(s) \right\|_0^2 \right)^{1/2}$$

for each $n = 0, 1, ..., \mathcal{N}$.

When $\theta = 1/2$, *under the additional assumptions* $\frac{\partial^2 f}{\partial t^2} \in L^2(Q_T)$ *and* $\frac{\partial^2 u_h}{\partial t^2}(0) \in L^2(\Omega)$, *the following estimate also holds:*

$$\|u_h^n - u_h(t_n)\|_0$$
$$(11.3.14)$$
$$\leq C(\Delta t)^2 \left(\left\| \frac{\partial^2 u_h}{\partial t^2}(0) \right\|_0^2 + \int_0^T \left\| \frac{\partial^2 f}{\partial t^2}(s) \right\|_0^2 \right)^{1/2}$$

for each $n = 0, 1, ..., \mathcal{N}$. *The constant* $C_\theta > 0$ *and* $C > 0$ *are non-decreasing functions of* α^{-1}, γ *and* T, *and are independent of* $\mathcal{N}$, Δt *and* h.

Proof. The semi-discrete solution $u_h(t)$ satisfies (11.2.1), which can be rewritten as

$$\frac{1}{\Delta t}(u_h(t_{n+1}) - u_h(t_n), v_h)$$

$$+ a(\theta u_h(t_{n+1}) + (1-\theta)u_h(t_n), v_h)$$

(11.3.15)

$$= \left(\frac{u_h(t_{n+1}) - u_h(t_n)}{\Delta t} - \theta \frac{\partial u_h}{\partial t}(t_{n+1}) - (1-\theta)\frac{\partial u_h}{\partial t}(t_n), v_h\right)$$

$$+ (\theta f(t_{n+1}) + (1-\theta)f(t_n), v_h) \ .$$

The first term at the right hand side of (11.3.15) is equal to

$$-\frac{1}{\Delta t}\left(\int_{t_n}^{t_{n+1}} [s - (1-\theta)t_{n+1} - \theta t_n]\frac{\partial^2 u_h}{\partial t^2}(s)ds, v_h\right) \ .$$

Hence, defining $e_h^n := u_h^n - u_h(t_n)$, from (11.3.4) and (11.3.15) it follows

$$\frac{1}{\Delta t}(e_h^{n+1} - e_h^n, v_h) + a(\theta e_h^{n+1} + (1-\theta)e_h^n, v_h)$$

(11.3.16)

$$= \frac{1}{\Delta t}\left(\int_{t_n}^{t_{n+1}} [s - (1-\theta)t_{n+1} - \theta t_n]\frac{\partial^2 u_h}{\partial t^2}(s)ds, v_h\right)$$

for each $v_h \in V_h$. Now taking $v_h = \theta e_h^{n+1} + (1-\theta)e_h^n$ in (11.3.16), by proceeding as in the proof of Theorem 11.3.1, for each $0 < \epsilon \le 1$ we find

$$||e_h^{n+1}||_0^2 - ||e_h^n||_0^2 + (2\theta - 1)||e_h^{n+1} - e_h^n||_0^2$$

$$+ 2(1-\epsilon)\alpha\Delta t||\theta e_h^{n+1} + (1-\theta)e_h^n||_1^2$$

$$\le \frac{(\Delta t)^2}{2\epsilon\alpha}\int_{t_n}^{t_{n+1}} \left|\left|\frac{\partial^2 u_h}{\partial t^2}(s)\right|\right|_{-1,h}^2 \ .$$

It is now necessary to distinguish between the case $0 \le \theta < 1/2$ and $1/2 \le \theta \le 1$. In the latter, the left hand side is larger than $||e_h^{n+1}||_0^2 - ||e_h^n||_0^2$, while in the former one has to proceed as in the proof of Theorem 11.3.1 for evaluating $||e_h^{n+1} - e_h^n||_0^2$. Taking into account (11.3.7), one finally obtains

$$||e_h^{n+1}||_0^2 - ||e_h^n||_0^2 \le C(\Delta t)^2 \int_{t_n}^{t_{n+1}} \left|\left|\frac{\partial^2 u_h}{\partial t^2}(s)\right|\right|_{-1,h}^2 \ .$$

Therefore, by fixing an index m, $1 \le m \le \mathcal{N}$, and summing up from $n = 0$ to $n = m - 1$, as $e_h^0 = 0$ we finally obtain

$$||e_h^m||_0^2 \le C(\Delta t)^2 \int_0^{t_m} \left|\left|\frac{\partial^2 u_h}{\partial t^2}(s)\right|\right|_{-1,h}^2 \ .$$

Differentiating (11.2.1) with respect to t, we have

(11.3.17) $$\left(\frac{\partial^2 u_h}{\partial t^2}(t), v_h\right) + a\left(\frac{\partial u_h}{\partial t}(t), v_h\right) = \left(\frac{\partial f}{\partial t}(t), v_h\right) \quad \forall \ v_h \in V_h \ ,$$

hence

$$\left\|\frac{\partial^2 u_h}{\partial t^2}(t)\right\|_{-1,h} \leq C\left(\left\|\frac{\partial u_h}{\partial t}(t)\right\|_1 + \left\|\frac{\partial f}{\partial t}(t)\right\|_{-1,h}\right) \ .$$

Moreover, choosing for each fixed $t \in [0, T]$ $v_h = \frac{\partial u_h}{\partial t}(t)$ in (11.3.17) and proceeding as in the proof of Theorem 11.1.1, it follows

$$\int_0^{t_m} \left\|\frac{\partial u_h}{\partial t}(s)\right\|_1^2 \leq C\left(\left\|\frac{\partial u_h}{\partial t}(0)\right\|_0^2 + \int_0^{t_m} \left\|\frac{\partial f}{\partial t}(s)\right\|_{-1,h}^2\right) \ ,$$

hence (11.3.13) holds.

Now take $\theta = 1/2$. One easily verifies that the first term at the right hand side of (11.3.15) is equal to

$$\frac{1}{2\Delta t}\left(\int_{t_n}^{t_{n+1}} (t_{n+1} - s)(t_n - s)\frac{\partial^3 u_h}{\partial t^3}(s)ds, v_h\right) \ .$$

By proceeding as before, one finds

$$\|e_h^m\|_0^2 \leq C(\Delta t)^4 \int_0^{t_m} \left\|\frac{\partial^3 u_h}{\partial t^3}(s)\right\|_{-1,h}^2 \ .$$

Differentiating (11.3.17) with respect to t and applying the previous argument to the equation thus obtained, estimate (11.3.14) easily follows.   □

Remark 11.3.1 The term $\|\frac{\partial u_h}{\partial t}(0)\|_0^2$ appearing in (11.3.13) can be estimated with respect to the data of the problem. For instance, take $u_{0,h} = \Pi_{1,h}^k(u_0)$ and assume that $u_0 \in H^2(\Omega) \cap V$ (the elliptic "projection" $\Pi_{1,h}^k$ is defined in (11.2.8)). Then, choosing $v_h = \frac{\partial u_h}{\partial t}(0)$ in (11.2.1) we find

$$\left\|\frac{\partial u_h}{\partial t}(0)\right\|_0^2 = -a\left(u_{0,h}, \frac{\partial u_h}{\partial t}(0)\right) + \left(f(0), \frac{\partial u_h}{\partial t}(0)\right)$$

$$= -a\left(u_0, \frac{\partial u_h}{\partial t}(0)\right) + \left(f(0), \frac{\partial u_h}{\partial t}(0)\right)$$

$$= \left(-Lu_0 + f(0), \frac{\partial u_h}{\partial t}(0)\right) \ .$$

Hence,

$$\left\|\frac{\partial u_h}{\partial t}(0)\right\|_0 \leq C(\|u_0\|_2 + \|f(0)\|_0) \ ,$$

as $\|\Pi_{1,h}^k(u_0)\|_1 \leq C\|u_0\|_1$ by (11.2.8).

If T_h is a quasi-uniform family of triangulations, for any initial datum $u_{0,h} \in V_h$ using the inverse inequality (6.3.21) yields

$$\left\|\frac{\partial u_h}{\partial t}(0)\right\|_0 \leq C(h^{-1}\|u_{0,h} - u_0\|_1 + \|u_0\|_2 + \|f(0)\|_0) \ .$$

Taking for instance $u_{0,h} = \pi_h^k(u_0)$, the finite element interpolation of u_0, again produces an estimate uniform with respect to h.   □

A proof of convergence that doesn't make use of the error estimate for the semi-discrete approximation can be performed as follows (notice that a similar procedure is also employed in the proof of Theorem 12.2.2). Setting $\eta_h^n := u_h^n - \Pi_{1,h}^k(u(t_n))$, one verifies that it satisfies the scheme (11.3.32) below. Taking there the test function $v_h = \theta\eta_h^{n+1} + (1-\theta)\eta_h^n$ and proceeding as in the proof of Theorem 11.3.1, one easily finds

$$||u_h^n - \Pi_{1,h}^k(u(t_n))||_0^2 \leq ||u_{0,h} - \Pi_{1,h}^k(u_0)||_0^2$$

$$+ C_\theta \int_0^{t_n} \left|\left|(I - \Pi_{1,h}^k)\frac{\partial u}{\partial t}(s)\right|\right|_{-1,h}^2$$

$$+ C_\theta(\Delta t)^2 \int_0^{t_n} \left|\left|\frac{\partial^2 u}{\partial t^2}(s)\right|\right|_{-1,h}^2 ,$$

or, when $\theta = 1/2$,

$$||u_h^n - \Pi_{1,h}^k(u(t_n))||_0^2 \leq ||u_{0,h} - \Pi_{1,h}^k(u_0)||_0^2$$

$$+ C \int_0^{t_n} \left|\left|(I - \Pi_{1,h}^k)\frac{\partial u}{\partial t}(s)\right|\right|_{-1,h}^2$$

$$+ C(\Delta t)^4 \int_0^{t_n} \left|\left|\frac{\partial^3 u}{\partial t^3}(s)\right|\right|_{-1,h}^2 .$$

From these results an optimal error estimate follows, under suitable assumptions, by proceeding as in the following Corollary 11.3.1.

Now we consider the *symmetric* case, i.e., $a(z,v) = a(v,z)$ for each $z, v \in V$. (Notice that this is never the case if either $b_i \neq 0$, or $c_i \neq 0$, or $a_{ij} \neq a_{ji}$ for some index i and j.) Under this assumption, it is possible to analyze stability and convergence taking advantage of the structure of the eigenvalues of the bilinear form $a(\cdot, \cdot)$. We still assume that (11.1.6) holds with $\lambda = 0$, and we recall that V is compactly embedded in $L^2(\Omega)$ since Ω is bounded (see Theorem 1.3.5). Thus, there exists a non-decreasing sequence of eigenvalues $\alpha \leq \mu_1 \leq \mu_2 \leq ...$ for the bilinear form $a(\cdot, \cdot)$, i.e.,

$$(11.3.18) \qquad \omega_j \in V , \ \omega_j \neq 0 \ : \ a(\omega_j, v) = \mu_j(\omega_j, v) \qquad \forall \ v \in V .$$

The corresponding eigenfunctions $\{\omega_i\}$ form a complete orthonormal basis in $L^2(\Omega)$ (see Remark 11.1.3).

In an analogous way, when considering the finite dimensional problem in V_h, we find a sequence of eigenvalues $\alpha \leq \mu_{1,h} \leq \mu_{2,h} \leq ... \leq \mu_{N_h,h}$ and a $L^2(\Omega)$-orthonormal basis of eigenvectors $\omega_{i,h} \in V_h$, $i = 1, ..., N_h$. Any function v_h in V_h can thus be expanded with respect to the system $\omega_{i,h}$ as

(11.3.19)
$$v_h = \sum_{i=1}^{N_h} (v_h, \omega_{i,h}) \omega_{i,h} \quad,$$

and

$$\|v_h\|_0^2 = \sum_{i=1}^{N_h} (v_h, \omega_{i,h})^2 \quad.$$

In particular, we have

(11.3.20)
$$u_h^n = \sum_{i=1}^{N_h} u_i^n \omega_{i,h} \quad, \quad u_i^n := (u_h^n, \omega_{i,h}) \quad.$$

Moreover, let f_h^n be the $L^2(\Omega)$-orthogonal projection of $\theta f(t_{n+1}) + (1-\theta) f(t_n)$ onto V_h, i.e., $f_h^n \in V_h$ and

(11.3.21)
$$(f_h^n, v_h) = (\theta f(t_{n+1}) + (1-\theta) f(t_n), v_h) \quad \forall v_h \in V_h \quad,$$

and set

(11.3.22)
$$f_h^n = \sum_{i=1}^{N_h} f_i^n \omega_{i,h} \quad, \quad f_i^n := (f_h^n, \omega_{i,h}) \quad.$$

We are now in a position to prove the stability result.

Theorem 11.3.3 (Stability). *Let the bilinear form $a(\cdot, \cdot)$ be symmetric and coercive, i.e., (11.1.6) is satisfied with $\lambda = 0$. Assume that the map $t \to \|f(t)\|_0$ is bounded in $[0, T]$. Moreover, when $0 \le \theta < 1/2$ let the time-step Δt satisfy*

(11.3.23)
$$\Delta t \, \mu_{N_h, h} \le \frac{2}{1 - 2\theta} \quad.$$

Then, the solution u_h^n of the fully-discrete problem (11.3.4) satisfies

(11.3.24)
$$\|u_h^n\|_0 \le \|u_{0,h}\|_0 + T \sup_{t \in [0,T]} \|f(t)\|_0 \quad, \quad n = 0, 1, ..., \mathcal{N} \quad.$$

Proof. Equation (11.3.4) is equivalent to

(11.3.25)
$$\frac{1}{\Delta t} (u_i^{n+1} - u_i^n) + \mu_{i,h} [\theta u_i^{n+1} + (1-\theta) u_i^n] = f_i^n \quad, \quad i = 1, ..., N_h \quad,$$

for each $n = 0, 1, ..., \mathcal{N} - 1$. We can rewrite (11.3.25) as

(11.3.26)
$$u_i^{n+1} = \frac{1 - (1-\theta) \Delta t \mu_{i,h}}{1 + \theta \Delta t \mu_{i,h}} u_i^n + \frac{\Delta t}{1 + \theta \Delta t \mu_{i,h}} f_i^n \quad.$$

The stability condition (11.3.23) yields

(11.3.27)
$$\left| \frac{1 - (1 - \theta)\Delta t \mu_{i,h}}{1 + \theta \Delta t \mu_{i,h}} \right| \leq 1 \; .$$

Notice that this condition is always satisfied if $1/2 \leq \theta \leq 1$. Hence, taking the absolute value of (11.3.26) we have

(11.3.28)
$$|u_i^m| \leq |u_i^0| + \frac{\Delta t}{1 + \theta \Delta t \mu_{i,h}} \sum_{n=0}^{m-1} |f_i^n| \; , \quad m \geq 1 \; .$$

Recalling that $\|u_h^n\|_0^2 = \sum_i |u_i^n|^2$, the thesis thus follows from (11.3.28) and the Minkowski inequality. $\square$

The stability condition (11.3.23) entails a relation between the mesh size h and the time step Δt. In fact, when $\mathcal{T}_h$ is a quasi-uniform family of triangulations, the greatest eigenvalue $\mu_{N_h,h}$ of the bilinear form $a(\cdot,\cdot)$ is $O(h^{-2})$ (see (6.3.24)), hence (11.3.23) states

(11.3.29)
$$\Delta t \, h^{-2} \leq \frac{C}{1 - 2\theta} \; ,$$

where C is a constant independent of h and Δt.

Now let us turn to the convergence analysis. We will slightly modify the procedure followed in Theorem 11.3.2 to obtain an explicit estimate of the difference $u_h^n - u(t_n)$. First of all, we will give an estimate of $\eta_h := u_h^n - \Pi_{1,h}^k(u(t_n))$, where as usual $\Pi_{1,h}^k(v)$ is the elliptic "projection" defined in (11.2.8). From this result the error estimate will follow at once, without making use of the results obtained in Proposition 11.2.2 for the semi-discrete error $\|u(t_n) - u_h(t_n)\|_0$.

Theorem 11.3.4 (Convergence). *Let the bilinear form $a(\cdot,\cdot)$ be symmetric and coercive. Assume that $u_0 \in V$ and that the solution u to (11.1.5) satisfies $\frac{\partial u}{\partial t} \in L^1(0,T;V)$ and $\frac{\partial^2 u}{\partial t^2} \in L^1(0,T;L^2(\Omega))$. Moreover, when $0 \leq \theta < 1/2$ let the time-step restriction (11.3.23) also be satisfied. Then u_h^n defined in (11.3.4) satisfies*

(11.3.30)
$$\|u_h^n - u(t_n)\|_0 \leq \|(I - \Pi_{1,h}^k)u(t_n)\|_0 + \|u_{0,h} - \Pi_{1,h}^k(u_0)\|_0$$
$$+ \int_0^{t_n} \left\| (I - \Pi_{1,h}^k)\frac{\partial u}{\partial t}(s) \right\|_0 + C_\theta^* \Delta t \int_0^{t_n} \left\| \frac{\partial^2 u}{\partial t^2}(s) \right\|_0$$

for each $n = 0, 1, ..., \mathcal{N}$, where $C_\theta^* := \max(\theta, 1 - \theta)$.
When $\theta = 1/2$, under the additional assumption $\frac{\partial^3 u}{\partial t^3} \in L^1(0,T;L^2(\Omega))$, the following estimate also holds:

(11.3.31)
$$\|u_h^n - u(t_n)\|_0 \leq \|(I - \Pi_{1,h}^k)u(t_n)\|_0 + \|u_{0,h} - \Pi_{1,h}^k(u_0)\|_0$$
$$+ \int_0^{t_n} \left\| (I - \Pi_{1,h}^k)\frac{\partial u}{\partial t}(s) \right\|_0 + \frac{(\Delta t)^2}{8} \int_0^{t_n} \left\| \frac{\partial^3 u}{\partial t^3}(s) \right\|_0$$

for each $n = 0, 1, ..., \mathcal{N}$.

Proof. Since $u \in C^0([0, T]; V)$, we can consider $\Pi_{1,h}^k(u(t_n))$. It is easily seen that the difference $\eta_h^n := u_h^n - \Pi_{1,h}^k(u(t_n))$ satisfies

$$(11.3.32) \quad \frac{1}{\Delta t}(\eta_h^{n+1} - \eta_h^n, v_h) + a(\theta \eta_h^{n+1} + (1-\theta)\eta_h^n, v_h) = (\varepsilon_h^n, v_h) \quad \forall v_h \in V_h \ ,$$

where $\varepsilon_h^n \in V_h$ is defined by the relation

$$
\begin{aligned}
(11.3.33) \quad (\varepsilon_h^n, v_h) &= (\theta f(t_{n+1}) + (1-\theta)f(t_n), v_h) \\
&\quad - \frac{1}{\Delta t}(\Pi_{1,h}^k(u(t_{n+1}) - u(t_n)), v_h) \\
&\quad - a(\theta u(t_{n+1}) + (1-\theta)u(t_n), v_h) \quad \forall v_h \in V_h \ .
\end{aligned}
$$

The discrete function η_h^n thus satisfies a scheme like (11.3.4). By proceeding as in Theorem 11.3.3, we find

$$(11.3.34) \quad ||\eta_h^n||_0 \leq ||\eta_h^0||_0 + \Delta t \sum_{n=0}^{m-1} ||\varepsilon_h^n||_0 \ .$$

Let us now estimate $||\varepsilon_h^n||_0$. From (11.1.5) we can write

$$(f(t), v_h) - a(u(t), v_h) = \left(\frac{\partial u}{\partial t}(t), v_h\right) \quad \forall v_h \in V_h \ ,$$

and consequently, as $\frac{\partial}{\partial t}$ commutes with $\Pi_{1,h}^k$:

$$
\begin{aligned}
(11.3.35) \quad (\varepsilon_h^n, v_h) &= \left(\theta \frac{\partial u}{\partial t}(t_{n+1}) + (1-\theta)\frac{\partial u}{\partial t}(t_n) - \frac{u(t_{n+1}) - u(t_n)}{\Delta t}, v_h\right) \\
&\quad + \frac{1}{\Delta t}\left(\int_{t_n}^{t_{n+1}} (I - \Pi_{1,h}^k)\frac{\partial u}{\partial t}(s)ds, v_h\right) \ .
\end{aligned}
$$

As in the proof of Theorem 11.3.2, the first term at the right hand side can be written in the following way:

$$\frac{1}{\Delta t}\left(\int_{t_n}^{t_{n+1}} [s - (1-\theta)t_{n+1} - \theta t_n]\frac{\partial^2 u}{\partial t^2}(s)ds, v_h\right) \ .$$

Hence,

$$||\varepsilon_h^n||_0 \leq C_\theta^* \int_{t_n}^{t_{n+1}} \left|\left|\frac{\partial^2 u}{\partial t^2}(s)\right|\right|_0 + \frac{1}{\Delta t}\int_{t_n}^{t_{n+1}} \left|\left|(I - \Pi_{1,h}^k)\frac{\partial u}{\partial t}(s)\right|\right|_0 \ ,$$

and (11.3.30) easily follows.

To prove (11.3.31), one has only to remark that the first term at the right hand side of (11.3.35) can be written as

$$-\frac{1}{2\Delta t}\left(\int_{t_n}^{t_{n+1}}(t_{n+1}-s)(t_n-s)\frac{\partial^3 u}{\partial t^3}(s)ds,v_h\right)\ .$$

The result thus follows as in the preceding case. $\square$

A simple consequence of this result, based on the estimate (11.2.9), is stated in the following

Corollary 11.3.1 *Let the assumptions of Theorem 11.3.4 be satisfied. Assume moreover that the approximation property (11.2.9) holds and that $u_0 \in H^{k+1}(\Omega)$ and $\frac{\partial u}{\partial t} \in L^1(0,T;H^{k+1}(\Omega))$. Then, for each $n = 0,1,...,\mathcal{N}$*

$$\|u_h^n - u(t_n)\|_0$$

(11.3.36)
$$\leq \|u_{0,h} - u_0\|_0 + Ch^{k+1}\left(|u_0|_{k+1} + \int_0^{t_n}\left|\frac{\partial u}{\partial t}(s)\right|_{k+1}\right)$$
$$+ C_\theta^* \Delta t \int_0^{t_n}\left\|\frac{\partial^2 u}{\partial t^2}(s)\right\|_0\ .$$

In the particular case $\theta = 1/2$, if we make the further assumption that $\frac{\partial^3 u}{\partial t^3} \in L^1(0,T;L^2(\Omega))$, then for each $n = 0,1,...,\mathcal{N}$

$$\|u_h^n - u(t_n)\|_0$$

(11.3.37)
$$\leq \|u_{0,h} - u_0\|_0 + Ch^{k+1}\left(|u_0|_{k+1} + \int_0^{t_n}\left|\frac{\partial u}{\partial t}(s)\right|_{k+1}\right)$$
$$+ \frac{(\Delta t)^2}{8}\int_0^{t_n}\left\|\frac{\partial^3 u}{\partial t^3}(s)\right\|_0\ .$$

The constant $C > 0$ is independent of $\mathcal{N}$, Δt and h.

We also notice that, when $1/2 < \theta \leq 1$, one can obtain an exponential decay of $\|u_h^n - \Pi_{1,h}^k(u(t_n))\|_0$ (see, e.g., Raviart and Thomas (1983), p. 181).

Remark 11.3.2. As in Remark 11.2.3, it is possible to prove optimal error estimates even when the initial datum is not regular, i.e., $u_0 \in L^2(\Omega)$.

For instance, let us consider, for simplicity, the case $f = 0$ and assume that the coefficients of the operator L are smooth enough. Choosing $u_{0,h} = P_h^k(u_0)$, P_h^k being the $L^2(\Omega)$-orthogonal projection onto V_h, we obtain the estimate

(11.3.38) $$\|u_h^n - u(t_n)\|_0 \leq C\left(\frac{h^{k+1}}{t_n^{(k+1)/2}} + \frac{\Delta t}{t_n}\right)\|u_0\|_0\ ,\quad n \geq 1\ ,$$

for both the backward Euler ($\theta = 1$) and the forward Euler ($\theta = 0$) schemes (see Luskin and Rannacher (1982a), Huang and Thomée (1982), Thomée (1984)).

When considering the Crank-Nicolson case ($\theta = 1/2$), one has to add some backward Euler steps to the scheme to obtain

$$(11.3.39) \qquad ||u_h^n - u(t_n)||_0 \leq C \left(\frac{h^{k+1}}{t_n^{(k+1)/2}} + \frac{(\Delta t)^2}{t_n^2} \right) ||u_0||_0 \ , \quad n \geq 1$$

(see Luskin and Rannacher (1982b)).   □

An alternative approach to the full discretizazion of (11.1.5) by a finite difference scheme in the time variable is based on a Galerkin method also in time. This approach is called *discontinuous Galerkin* method, and was proposed for parabolic problems by Jamet (1978).

Let us introduce the space

$$(11.3.40) \qquad \begin{aligned} W_h^{\Delta t} := \{ v : [0,T] &\to V_h \ | \\ & v_{|(t_n, t_{n+1})} \in \mathbb{P}_q^n(V_h) \ , \ n = 0, 1, ..., \mathcal{N} - 1 \} \ , \end{aligned}$$

where $q \geq 0$, $t_n := n\Delta t$, $\Delta t := T/\mathcal{N}$ and

$$(11.3.41) \quad \mathbb{P}_q^n(V_h) := \left\{ v : (t_n, t_{n+1}) \to V_h \ | \ v(\tau) = \sum_{s=0}^q v_s \tau^s \ , \ v_s \in V_h \right\} \ .$$

Let us notice that an element of $W_h^{\Delta t}$ is not required to be continuous at the points t_n, $n = 1, ..., \mathcal{N} - 1$. Thus we introduce the notation

$$(11.3.42) \qquad v_+^n := \lim_{\tau \to 0^+} v(t_n + \tau) \ , \quad v_-^n := \lim_{\tau \to 0^-} v(t_n + \tau)$$

and

$$(11.3.43) \qquad [v^n] := v_+^n - v_-^n \ .$$

The discontinuous Galerkin method for (11.1.5) reads as follows:

$$(11.3.44) \qquad \text{find } u \in W_h^{\Delta t} : B(u,v) = G(v) \qquad \forall \ v \in W_h^{\Delta t} \ ,$$

where

$$(11.3.45) \qquad \begin{aligned} B(z,v) := & \sum_{n=0}^{\mathcal{N}-1} \int_{t_n}^{t_{n+1}} \left[\left(\frac{\partial z}{\partial t}(t), v(t) \right) + a(z(t), v(t)) \right] \\ & + \sum_{n=1}^{\mathcal{N}-1} ([z^n], v_+^n) + (z_+^0, v_+^0) \end{aligned}$$

and

$$(11.3.46) \qquad G(v) := \int_0^T (f(t), v(t)) + (u_0, v_+^0) \ .$$

Since the test functions can be chosen independently on each time-interval $[t_n, t_{n+1}]$, (11.3.44) may also be formulated as follows: given $U_{-1}(0) := u_0$, for each $n = 0, 1, ..., \mathcal{N} - 1$ find $U_n : [t_n, t_{n+1}] \to V_h$ such that $U_n \in \mathbb{P}_q^n(V_h)$ and

$$
\begin{aligned}
(11.3.47) \quad & \int_{t_n}^{t_{n+1}} \left[\left(\frac{\partial U_n}{\partial t}(t), v(t) \right) + a(U_n(t), v(t)) \right] + (U_n(t_n), v(t_n)) \\
& = \int_{t_n}^{t_{n+1}} (f(t), v(t)) + (U_{n-1}(t_n), v(t_n)) \quad \forall \, v \in \mathbb{P}_q^n(V_h) \ .
\end{aligned}
$$

The continuity of u at each point t_n, $n = 1, ..., \mathcal{N} - 1$, is therefore imposed only in a weak sense.

Assuming $\lambda = 0$ in (11.1.6), the existence of a unique solution to (11.3.47) is a straightforward consequence of the energy estimate

$$
||U_n(t_{n+1})||_0^2 + \alpha \int_{t_n}^{t_{n+1}} ||U_n(t)||_1^2 \le ||U_{n-1}(t_n)||_0^2 + \frac{1}{\alpha} \int_{t_n}^{t_{n+1}} ||f(t)||_0^2 \ ,
$$

which is obtained choosing $v = U_n$ in (11.3.47). Taking $v = u$ in (11.3.44) one can also obtain the global stability estimate

$$
||u_-^{n+1}||_0^2 + \sum_{m=0}^{n} ||[u^m]||_0^2 + \alpha \int_0^{t_{n+1}} ||u(t)||_1^2 \le ||u_0||_0^2 + \frac{1}{\alpha} \int_0^{t_{n+1}} ||f(t)||_0^2
$$

for each $n = 0, 1, ..., \mathcal{N} - 1$ (here we have set $u_-^0 := u_0$).

When $q = 0$, U_n is constant on $[t_n, t_{n+1}]$ and it can be noticed that (11.3.47) reduces to

$$
(U_n - U_{n-1}, v) + \Delta t \, a(U_n, v) = \left(\int_{t_n}^{t_{n+1}} f(t), v \right) \quad \forall \, v \in V_h \ ,
$$

which is the backward Euler discretization of (11.2.1), with a different right hand side.

A thorough analysis of the stability and convergence properties of the discontinuous Galerkin method can be found in Thomée (1984), Eriksson, Johnson and Thomée (1985) (see also Dupont (1982)). Let us just recall that the method is formally of order $2q + 1$ at the nodal points t_n, and of order $q + 1$ in the global interval $[0, T]$.

11.3.2 The Case of Spectral Methods

The time-advancing θ-scheme applied to the spectral Galerkin problem (11.2.15) reads: find $u_N^n \in V_N$ such that

$$(11.3.48) \quad \begin{cases} \dfrac{1}{\Delta t}(u_N^{n+1} - u_N^n, v_N) + a(\theta u_N^{n+1} + (1-\theta)u_N^n, v_N) \\ \qquad = (\theta f(t_{n+1}) + (1-\theta)f(t_n), v_N) \qquad \forall\, v_N \in V_N \\[2mm] u_N^0 = u_{0,N} \end{cases}$$

for each $n = 0, 1, ..., \mathcal{N} - 1$.

The same kind of formulation holds for the spectral collocation method (11.2.16) also, provided that we operate the change of notation: $(\cdot, \cdot) \to (\cdot, \cdot)_N$ and $a(\cdot, \cdot) \to a_N(\cdot, \cdot)$ (we refer to Section 11.2.2 for explanation of notation in the Legendre case). Applying the θ-scheme directly on the pointwise version (11.2.17) yields the problem: for $n = 0, 1, ..., \mathcal{N} - 1$ find $u_N^n \in \mathbb{Q}_N$ such that

$$(11.3.49) \quad \begin{cases} \dfrac{1}{\Delta t}(u_N^{n+1} - u_N^n) + \theta L_N u_N^{n+1} + (1-\theta)L_N u_N^n \\ \qquad = \theta f(t_{n+1}) + (1-\theta)f(t_n) \qquad \text{at } \mathbf{x}_{jk} \in \Omega \\[2mm] u_N^{n+1} = 0 \qquad\qquad\qquad\qquad\qquad \text{at } \mathbf{x}_{jk} \in \partial\Omega \\[2mm] u_N^0 = u_{0,N} \qquad\qquad\qquad\qquad\quad\; \text{at } \mathbf{x}_{jk} \in \overline{\Omega} \;\; . \end{cases}$$

In both cases (11.3.48) and (11.3.49), at each time level t_{n+1} one is left with a problem like (11.3.5). In the collocation case (11.3.49), however, M is the identity matrix I, making the forward Euler scheme ($\theta = 0$) a fully explicit one.

The stability analysis for problem (11.3.48) can be carried out by reproducing the proof of Theorem 11.3.1. Under the same assumptions on θ, Δt and f, and supposing that the stability condition

$$(11.3.50) \qquad \Delta t\, (1 + C_3^* N^4) < \frac{2\alpha}{(1-2\theta)\gamma^2}$$

holds for $0 \le \theta < 1/2$ (here C_3^* is the constant appearing in the inverse inequality (6.3.27)), the result reads

$$(11.3.51) \quad \|u_N^n\|_0 \le C_\theta \left(\|u_{0,N}\|_0 + \sup_{t\in[0,T]} \|f(t)\|_0 \right), \quad n = 0, 1, ..., \mathcal{N} \;,$$

where $C_\theta > 0$ is a non-decreasing function of α^{-1}, γ and T, which is however independent of $\mathcal{N}$, Δt and N.

A similar result can be easily proven for the Legendre spectral collocation problem also. The only difference arises from the presence of the discrete norm $\| \cdot \|_N$ that replaces the L^2-norm $\| \cdot \|_0$ (see also Canuto, Hussaini, Quarteroni and Zang (1988), Sect. 12.2.2). Assuming that $a_N(\cdot, \cdot)$ satisfies (11.2.20) and (11.2.23), and moreover that

$$a_N(z_N, v_N) \le \overline{\gamma}\,\|z_N\|_1\,\|v_N\|_1 \quad \forall\, z_N, v_N \in V_N \;,$$

the stability condition for $0 \le \theta < 1/2$ now reads

$$\Delta t \left(1 + C_3^* N^4\right) < \frac{2\overline{\alpha}}{(1 - 2\theta)\overline{\gamma}^2} \ .$$

This condition (as well as (11.3.50)) is more severe than the one needed for the finite element case (see (11.3.7)). We will comment on this issue at the end of the Section.

Also the convergence analysis doesn't present any extra-difficulty compared with the one developed for finite elements. It can therefore be proven that, under the assumptions of Theorem 11.3.2 on θ, Δt, f and u_N (and requiring that the stability condition (11.3.50) holds when $0 \le \theta < 1/2$), the difference between the fully-discrete and semi-discrete spectral Galerkin solutions to problems (11.3.48) and (11.2.15), respectively, satisfies for each $n = 0, 1, ..., \mathcal{N}$:

$$
\begin{aligned}
&||u_N^n - u_N(t_n)||_0 \\
(11.3.52) \quad &\le C_\theta(\Delta t)^p \left(\left\| \frac{\partial^p u_N}{\partial t^p}(0) \right\|_0^2 + \int_0^T \left\| \frac{\partial^p f}{\partial t^p}(t) \right\|_0^2 \right)^{1/2} ,
\end{aligned}
$$

where $p = 2$ if $\theta = 1/2$, and $p = 1$ otherwise. Clearly, we need to assume that the norms at the right hand side are finite. The constant $C_\theta > 0$ has the same parameter dependence as the one that appears in (11.3.51).

As pointed out in Remark 11.3.1, in order for (11.3.52) to be effective, the norm of the time derivative of u_N at $t = 0$ needs to be bounded uniformly with respect to N. Indeed, this is easily proven if $u_{0,N}$ is set equal to $\Pi_{1,N} u_0$, the elliptic projection upon V_N of the exact initial value u_0 (see (11.2.26)). In such a case, taking $v_N = \frac{\partial u_N}{\partial t}(0)$ in (11.2.15), and repeating the arguments used in Remark 11.3.1, we obtain

$$\left\| \frac{\partial u_N}{\partial t}(0) \right\|_0 \le C(||u_0||_2 + ||f(0)||_0) \ ,$$

provided the data at the initial time u_0 and $f(0)$ have the required smoothness.

The usual modifications occur for the analysis of the spectral collocation problem. A direct estimate of $||u_N^n - \Pi_{1,N} u(t_n)||_0$ can also be found in Canuto, Hussaini, Quarteroni and Zang (1988), Sect. 12.2.2.

As done for finite elements in Theorems 11.3.3 and 11.3.4, a different approach can also be followed when the bilinear form $a(\cdot, \cdot)$ is coercive and symmetric. Denote now by $\mu_{j,N}^* \in \mathbb{R}$ (respectively, $\omega_{j,N}^* \in V_N$, $\omega_{j,N}^* \ne 0$) the eigenvalues (respectively, eigenvectors) of the finite dimensional eigenvalue problem

$$(11.3.53) \qquad a(\omega_{j,N}^*, v_N) = \mu_{j,N}^*(\omega_{j,N}^*, v_N) \quad \forall\, v_N \in V_N \ ,$$

$j = 1, ..., N_h := (N-1)^2$. Clearly, $\alpha \leq \mu_{1,N}^* \leq \cdots \leq \mu_{N_h,N}^*$, and $(\omega_{j,N}^*, \omega_{k,N}^*) = \delta_{kj}$ for $j, k = 1, ..., N_h$.

In a similar manner, in the Legendre collocation case we can define $\mu_{j,N} \in \mathbb{R}$ and $\omega_{j,N} \in V_N$, $\omega_{j,N} \neq 0$, by

$$(11.3.54) \qquad a_N(\omega_{j,N}, v_N) = \mu_{j,N}(\omega_{j,N}, v_N)_N \quad \forall \, v_N \in V_N \ .$$

At this regard, we observe that in the Legendre case $a_N(\cdot, \cdot)$ is symmetric if $a(\cdot, \cdot)$ is symmetric. Actually, $a_N(\cdot, \cdot)$ is obtained from $a(\cdot, \cdot)$ barely by replacing any scalar product $(\cdot, \cdot)$ with $(\cdot, \cdot)_N$, and this operation is symmetry-preserving. Assuming that (11.2.20) and (11.2.23) hold, from (4.4.16) it clearly follows that $3^{-d} \overline{\alpha} \leq \mu_{1,N} \leq \cdots \leq \mu_{N_h,N}$; further, orthogonality now holds for the discrete scalar product, namely, $(\omega_{j,N}, \omega_{k,N})_N = \delta_{kj}$ for $j, k = 1, ..., N_h$.

The spectral analysis carried out in the proof of Theorem 11.3.3 can be repeated (with straightforward modifications) for both the spectral Galerkin and the spectral collocation problems introduced in this Section. The stability result reads

$$(11.3.55) \qquad |||u_N^n||| \leq |||u_{0,N}||| + T \sup_{t \in [0,T]} |||f(t)||| \ , \quad n = 0, 1, ..., \mathcal{N} \ ,$$

provided

$$(11.3.56) \qquad \Delta t \, \hat{\mu}_N \leq \frac{2}{1 - 2\theta}$$

when $0 \leq \theta < 1/2$. Here, for the sake of brevity, we have denoted by $\hat{\mu}_N$ the eigenvalue $\mu_{N_h,N}^*$ in the Galerkin case, and $\mu_{N_h,N}$ in the collocation case. Similarly, $||| \cdot ||| = || \cdot ||_0$ for the Galerkin problem, whereas $||| \cdot ||| = || \cdot ||_N$ for the collocation problem.

Both cases share the (unpleasant) property that $\hat{\mu}_N = O(N^4)$. (This is shown in Section 6.3.3 for the collocation case and can be proven by proceeding as in Section 6.3.2 for the Galerkin case.) Therefore inequality (11.3.56) takes the form

$$(11.3.57) \qquad \Delta t \, N^4 \leq \frac{2\hat{C}}{1 - 2\theta} \ ,$$

where $\hat{C}$ is a suitable constant (precisely, $\hat{C} := \sup\{N^4/\hat{\mu}_N \mid N \geq 1\}$).

This stability limit for the time-step Δt is much more severe than its finite-element counterpart. This makes explicit time-advancing schemes (such as the forward Euler one, corresponding to $\theta = 0$) often unsuitable for the time discretization of parabolic equations when using spectral methods for space variables. As a matter of fact, for advection-diffusion equations a fairly common approach is to adopt a semi-implicit scheme, which consists of evaluating implicitly the second order term (diffusion) and explicitly the lower order ones (advection and reaction). This procedure weakens the stability limit (11.3.57) at the espense of solving a linear diffusion system at each step. An account is given in Canuto, Hussaini, Quarteroni and Zang (1988); see also the following Section 12.2.

The convergence analysis for spectral approximations in the symmetric case quite closely resembles the one developed in the finite element case, and yields the same kind of results. For brevity, we don't report it here.

Remark 11.3.3 A fully spectral discretization of initial-boundary value problems can be accomplished if a spectral (rather than a finite difference) technique is applied for the discretization of time derivative in (11.2.15) or (11.2.16). This amounts to looking for a function u_N which is not only an algebraic polynomial with respect to the space variables, but for any fixed $\mathbf{x} \in \Omega$ is an algebraic polynomial also with respect to the time variable t.

This approach has the advantage of yielding spectral accuracy both in space and time. The disadvantage is that the discrete problem is now fully coupled, hence advancing stepwise along the temporal direction as in the finite difference approach is no longer possible in this context. Among references addressing the issue of spectral approximations in time we mention Morchoisne (1979), Tal-Ezer (1986, 1989) and Ierley, Spencer and Worthing (1992).   □

Remark 11.3.4 For the spatial approximation, Chebyshev (rather than Legendre) expansions are frequently used, as pointed out, e.g., in Chapter 6. In the spectral Galerkin case, this amounts to replacing the scalar product $(\cdot, \cdot)$ with the weighted one $(\cdot, \cdot)_w$. In the collocation case, the difference with the Legendre case is that now the Chebyshev Gauss-Lobatto points are used as collocation nodes.

Obviously, the stability and convergence analysis is harder in the Chebyshev case, due to the presence of the Chebyshev weight function $w(\mathbf{x}) = [(1 - x_1)(1 - x_2)]^{-1/2}$. In particular, the latter entails the lack of symmetry of the bilinear form even in the case of a self-adjoint spatial operator L. An in-depth analysis can be found in Gottlieb and Orszag (1977), Canuto, Hussaini, Quarteroni and Zang (1988), Boyd (1989) and Funaro (1992).   □

Remark 11.3.5 A sharp temporal stability analysis for spectral approximations based on the concept of pseudo-eigenvalues is provided in Reddy and Trefethen (1990).   □

11.4 Some Remarks on the Algorithmic Aspects

Whenever an implicit time-advancing method is adopted, at each time-level a linear system of the form (11.3.5) needs to be solved. This occurs for both finite element and spectral collocation approximation.

When A is symmetric and positive definite, so is the matrix $E := M + \theta \Delta t A$. However, the spectral condition number of E is generally much smaller than that of A, especially when the time-step Δt is small.

In Table 11.4.1 we report the spectral condition number of $E_{fe} := M_{fe} + \theta \Delta t A_{fe}$ for several values of h (the grid-size) and $\theta \Delta t$. (The finite element stiffness matrix A_{fe} and mass matrix M_{fe} are defined in (11.2.2).) Results refer to the discretization of the heat equation

$$(11.4.1) \quad \begin{cases} \dfrac{\partial u}{\partial t} - \Delta u = 0 & \text{in } Q_T := (0,T) \times \Omega \\[2mm] u = 0 & \text{on } \Sigma_T := (0,T) \times \partial\Omega \\[2mm] u_{|t=0} = u_0 & \text{on } \Omega \ , \end{cases}$$

with $\Omega = (0,1)^2$, and space discretization carried out with piecewise-bilinear finite elements on a uniform mesh.

Table 11.4.1. The spectral condition number of the finite element matrix E_{fe}

N	\multicolumn{6}{c}{$\theta\Delta t$}
	.5	.1	.05	.01	.005	.001
4	2.92	2.28	1.84	1.50	2.13	3.57
8	11.69	8.67	6.59	2.44	1.55	3.17
12	26.40	19.42	14.63	5.10	2.94	1.96
16	47.00	34.48	25.90	8.83	4.97	1.68
20	73.50	53.84	40.39	13.64	7.59	1.90
24	105.88	77.51	58.11	19.52	10.79	2.59

We recall that in the spectral collocation case M is diagonal. As a matter of fact, following the discussion carried out throughout Section 6.3.1, we can implement the collocation case in two fashions. In the former we have (see (11.2.18))

$$(11.4.2) \quad M_{ij} = (M_{sp})_{ij} := (\psi_i, \psi_j)_N \ , \quad A_{ij} = (A_{sp})_{ij} := a_N(\psi_j, \psi_i) \ ,$$

thus $M = D := \text{diag}(w_i)$. In the latter case we have

$$(11.4.3) \quad M = M_{sp}^c = I \ , \quad A = A_{sp}^c = D^{-1} A_{sp} \ .$$

This is obtained by multiplying to the left M and A in (11.4.2) by M_{sp}^{-1}, resorting therefore to classical (differential) form of the collocation method. We recall, moreover, that A_{sp}^c is not symmetric (see (6.3.12)), but has positive and real eigenvalues.

We report in Table 11.4.2 the condition number $\chi_1(E_{sp}^c)$ (see (2.1.25)) of the Legendre spectral evolution matrix $E_{sp}^c = I + \theta \Delta t A_{sp}^c$ for several values of N and $\theta \Delta t$.

Table 11.4.2. The condition number $\chi_1(E_{sp}^c)$ of the spectral collocation matrix E_{sp}^c

N	$\theta \Delta t$					
	.5	.1	.05	.01	.005	.001
4	11.44	8.46	3.95	2.64	1.85	1.17
8	110.54	77.20	28.38	15.23	8.24	2.58
12	489.49	340.38	120.47	63.60	31.95	7.79
16	1458.93	1012.49	361.17	187.08	92.675	20.55
20	3444.97	2389.97	851.16	440.09	217.25	45.34
24	6993.40	4851.04	1726.54	892.05	439.33	89.46

When $\theta = 0$, the solution $\boldsymbol{\xi}^{n+1}$ is explicitly given in terms of $\boldsymbol{\xi}^n$ throughout (11.3.5). This is a highly desirable feature of the spectral collocation method, which unfortunately is not shared by the finite element method, since $M_{ij} = (M_{fe})_{ij} = (\varphi_i, \varphi_j)$, $\{\varphi_j\}$ being the finite element basis such that $\varphi_j(\mathbf{a}_i) = \delta_{ij}$ for any grid-point $\mathbf{a}_i$. For this reason, in common practice it is usual to approximate M_{fe} by a diagonal matrix. This is the so-called *lumping* process. We illustrate this method when considering the finite dimensional subspace $V_h = X_h^1 \cap H_0^1(\Omega)$, X_h^1 defined in (3.2.4).

A first way to introduce the lumping process is to substitute the mass matrix M by the matrix $\hat{M}$ given by

$$(11.4.4) \qquad \hat{M}_{ij} := \left(\sum_{k=1}^{N_h} M_{ik} \right) \delta_{ij} = \left[\sum_{k=1}^{N_h} (\varphi_i, \varphi_k) \right] \delta_{ij} \ ,$$

i.e., $\hat{M}$ is the diagonal matrix having the element $\hat{M}_{ii}$ equal to the sum of the elements of M on the i-th row.

In order to verify that $\hat{M}$ is indeed non-singular and to establish suitable error estimates it is useful to interpret this procedure from another point of view. Let us introduce the quadrature formula (trapezoidal rule)

$$(11.4.5) \qquad \int_K \psi \simeq \frac{1}{3} \operatorname{meas}(K) \sum_{j=1}^{3} \psi(\mathbf{b}_{j,K}) \ ,$$

where K is any element of the triangulation T_h and $\mathbf{b}_{j,K}$, $j = 1, 2, 3$, are the vertices of K. We can define the approximate scalar product $(\cdot, \cdot)_h$ as

$$(11.4.6) \qquad (\varphi, \psi)_h := \frac{1}{3} \sum_{K \in T_h} \left[\text{meas}(K) \sum_{j=1}^{3} \varphi(\mathbf{b}_{j,K}) \psi(\mathbf{b}_{j,K}) \right] .$$

Now we are in a position to formulate the problem: find $\hat{u}_h(t) \in V_h$ such that

$$(11.4.7) \qquad \begin{cases} \dfrac{d}{dt}(\hat{u}_h(t), v_h)_h + a(\hat{u}_h(t), v_h) \\ \qquad = (f(t), v_h) \quad \forall\, v_h \in V_h \ , \ t \in [0, T] \\ \hat{u}_h(0) = u_{0,h} \ , \end{cases}$$

where $V_h = X_h^1 \cap H_0^1(\Omega)$ is the space of piecewise-linear functions vanishing on $\partial \Omega$. If $\hat{u}_h$ is written as

$$\hat{u}_h(t) = \sum_{j=1}^{N_h} \hat{\xi}_j(t) \varphi_j \ ,$$

where φ_j is the Lagrangian basis function associated to the node $\mathbf{a}_j$, system (11.4.7) reads as follows

$$\sum_{j=1}^{N_h} (\varphi_i, \varphi_j)_h \frac{d\hat{\xi}_j}{dt}(t) + \sum_{j=1}^{N_h} a(\varphi_j, \varphi_i) \hat{\xi}_j(t) = (f(t), \varphi_i) \quad \forall\, i = 1, ..., N_h \ .$$

The corresponding mass matrix is given by $(\varphi_i, \varphi_j)_h$, whereas the stiffness matrix is unchanged, i.e., $A_{ij} = a(\varphi_j, \varphi_i)$.

We claim that

$$(11.4.8) \qquad \hat{M}_{ij} = (\varphi_i, \varphi_j)_h$$

(thus, in particular, $\hat{M}$ is non-singular). To prove this fact, notice at first that trivially $(\varphi_i, \varphi_j)_h = 0$ if $i \neq j$, as in that case $\varphi_i \varphi_j$ vanishes at each node of T_h. Moreover, one easily verifies that (φ_i, φ_k) is non-zero only if the nodes $\mathbf{a}_i$ and $\mathbf{a}_k$ belong to the same triangle K. In this case, a simple calculation shows that

$$\int_K \varphi_i \varphi_k = \begin{cases} \dfrac{1}{12} \text{meas}(K) \ , & i \neq k \ , \\ \dfrac{1}{6} \text{meas}(K) \ , & i = k \ . \end{cases}$$

For each pair $\mathbf{a}_i$, $\mathbf{a}_k$, there are exactly two triangles K containing $\mathbf{a}_i$ and $\mathbf{a}_k$; thus, denoting by D_i the union of the triangles having $\mathbf{a}_i$ as a vertex, it follows

$$\sum_{k \neq i} (\varphi_i, \varphi_k) = \sum_{k \neq i} \sum_{K \in \mathcal{T}_h} \int_K \varphi_i \varphi_k = \frac{1}{6} \operatorname{meas}(D_i)$$

$$\|\varphi_i\|_0^2 = \sum_{K \in \mathcal{T}_h} \int_K \varphi_i^2 = \frac{1}{6} \operatorname{meas}(D_i) \ .$$

Hence

$$\hat{M}_{ii} := \sum_{k=1}^{N_h} (\varphi_i, \varphi_k) = \frac{1}{3} \operatorname{meas}(D_i) \ ,$$

and trivially we also have

$$(\varphi_i, \varphi_i)_h = \frac{1}{3} \sum_{K \in \mathcal{T}_h} \left[\operatorname{meas}(K) \sum_{j=1}^{3} \varphi_i^2(\mathbf{b}_{j,K}) \right] = \frac{1}{3} \operatorname{meas}(D_i) \ ,$$

therefore the proof of (11.4.8) is complete.

It is worthwhile to notice that optimal error estimates like those proven, for instance, in (11.2.7) and in Corollary 11.3.1 also hold for the "lumped" problem (11.4.7) (see Raviart (1973), and also Thomée (1984), pp. 170–179).

For the solution of the system (11.3.5) we refer to the algorithms described throughout Chapter 2. It is useful to point out that a Cholesky or *LU* decomposition (when used) can be carried out only once, as E doesn't change along the time. Also, let us stress the fact that iterative procedures are fast to converge as E is nicely conditioned, and, at each time-level, an accurate initial guess is provided by the numerical solution available from the previous step.

11.5 Complements

Besides the paper by Wheeler (1973), other early results on the Galerkin finite element approximation of parabolic equations are the ones by Douglas and Dupont (1970), Bramble and Thomée (1974) and Zlámal (1974). Another approach, based on high order accurate rational approximations of the exponential function, is the one by Baker, Bramble and Thomée (1977).

Optimal error estimates in non-isotropic Sobolev spaces of fractional order for the Galerkin approximation of parabolic problems have been obtained by Hackbusch (1981b), Baiocchi and Brezzi (1983) and Tomarelli (1984).

The error estimate (11.3.36) is not well suited for being used in choosing the time-step adaptively. For a detailed analysis of this important issue, we refer to Johnson (1987), Sect. 8.4, Eriksson, Estep, Hansbo and Johnson (1994), and the references therein.

The approximation of parabolic equations based on the mixed finite element approch has been proposed and analyzed by Johnson and Thomée (1981).

12. Unsteady Advection-Diffusion Problems

In this Chapter we consider linear second-order parabolic equations with advective terms dominating over the diffusive ones. The corresponding steady case was considered in Chapter 8.

At first we recall the mathematical formulation of parabolic problems, and underline the difficulties stemming from strong advection (and/or small diffusion). Next we introduce a family of time-advancing finite difference schemes (explicit, implicit or semi-implicit), and analyze their stability properties, improving some results obtained in Chapter 11. Then we consider the so-called discontinuous Galerkin method, which accounts for the stabilization procedures for space discretization that have been introduced in Section 8.3.2. We continue with operator-splitting methods, which yield a decoupling between advection and diffusion. We finish with the so-called characteristic Galerkin method, which is based on the replacement of the advective part of the equation by total differentiation along characteristics.

12.1 Mathematical Formulation

We assume that Ω is a bounded domain in $\mathbb{R}^d$, $d = 2, 3$, with Lipschitz boundary, and consider the parabolic initial-boundary value problem: for each $t \in [0, T]$ find $u(t)$ such that

(12.1.1)
$$\begin{cases} \dfrac{\partial u}{\partial t} + Lu = f & \text{in } Q_T := (0, T) \times \Omega \\[2mm] Bu = 0 & \text{on } \Sigma_T := (0, T) \times \partial\Omega \\[2mm] u = u_0 & \text{on } \Omega, \text{ for } t = 0 \quad, \end{cases}$$

where L is the second-order elliptic operator

(12.1.2)
$$Lw := -\varepsilon \Delta w + \sum_{i=1}^{d} D_i(b_i w) + a_0 w$$

and the boundary operator B denotes anyone of the boundary conditions considered in Chapter 6 (Dirichlet, Neumann, mixed, Robin).

For the purpose of our analysis, we are considering here the case in which $\varepsilon \ll ||\mathbf{b}||_{L^\infty(\Omega)}$. In practice this situation is more difficult to treat, giving rise to instability phenomena if the spatial grid-size h is not small enough. Similar problems also appear in the steady case, and in Chapter 8 we have considered some methods to overcome these difficulties.

The "simplified" form considered in (12.1.2) is a perfect alias of the most general situation in which L is given by (8.1.1), whenever the diffusion coefficients a_{ij} are significantly smaller then the advective ones b_i, $i,j = 1, ..., d$. With no loss of generality, we further suppose that $\mathbf{b}$ is normalized to $||\mathbf{b}||_{L^\infty(\Omega)} = 1$.

We assume here that there exist two positive constants μ_0 and μ_1 such that

$$(12.1.3) \qquad 0 < \mu_0 \le \mu(\mathbf{x}) := \frac{1}{2} \operatorname{div} \mathbf{b}(\mathbf{x}) + a_0(\mathbf{x}) \le \mu_1$$

for almost every $\mathbf{x} \in \Omega$. With this condition the bilinear form

$$(12.1.4) \qquad a(w,v) := \int_\Omega \left[\varepsilon \nabla w \cdot \nabla v - \sum_{i=1}^{d} b_i w D_i v + a_0 wv \right] \ ,$$

is coercive in the Hilbert space $H_0^1(\Omega)$, i.e., it satisfies (5.1.14) with a constant $\alpha = \min\{\varepsilon, (\varepsilon + C_\Omega \mu_0)/(1 + C_\Omega)\}$, where C_Ω is the constant in Poincaré inequality (1.3.2).

Condition (12.1.3) is not restrictive, as, by means of the change of variable $u_\lambda(t, \mathbf{x}) := e^{-\lambda t} u(t, \mathbf{x})$, we can deal with the auxiliary bilinear form $a_\lambda(w,v) := a(w,v) + \lambda(w,v)$, where $(\cdot, \cdot)$ denotes the scalar product in $L^2(\Omega)$. Its coefficients always satisfy (12.1.3) provided that both a_0 and div $\mathbf{b}$ belong to $L^\infty(\Omega)$ and λ is large enough.

In the sequel we will consider, for simplicity, the homogeneous Dirichlet boundary condition. The parabolic advection-diffusion problem (12.1.1) can be reformulated in a weak fashion as follows: given $f \in L^2(Q_T)$ and $u_0 \in L^2(\Omega)$, find $u \in L^2(0, T; V) \cap C^0([0, T]; L^2(\Omega))$ such that

$$(12.1.5) \qquad \begin{cases} \dfrac{d}{dt}(u(t), v) + a(u(t), v) = (f(t), v) & \forall \, v \in V \\[2mm] u(0) = u_0 \ , \end{cases}$$

where $V = H_0^1(\Omega)$ and the equation has to be intended in the sense of distributions in $(0, T)$. Next, its semi-discrete (continuous in time) approximation reads as (11.2.1).

The stability and convergence analysis carried out on problem (11.2.1), however, reveals that for case (12.1.2) with $\varepsilon \ll 1$ some difficulties could arise. Indeed, looking for instance at the proof of Proposition 11.2.1 it is readily seen that the constant $C_{\alpha,\gamma}$ appearing there behaves like γ^2/α, where γ is the continuity constant of the form $a(\cdot, \cdot)$. After rewriting the bilinear form as

$$(12.1.6) \quad a(w,v) = \int_{\Omega} \left[\varepsilon \nabla w \cdot \nabla v - \frac{1}{2} \sum_{i=1}^{d} (b_i w D_i v - b_i v D_i w) + \mu w v \right] \ ,$$

it can be easily seen that $a(\cdot, \cdot)$ satisfies (5.1.13) and the continuity constant γ can be estimated by

$$\gamma \leq \varepsilon + 1 + \mu_1 \ .$$

We conclude that the error estimate (11.2.3) (or else (11.2.7)) for the semi-discrete approximation deteriorates if ε is small. A similar situation was already observed in the steady case (see Section 8.1).

Clearly, this difficulties also arise for the fully-discrete approximation given by the θ-scheme (11.3.4). In fact, it is easily seen that also in this case the multiplicative constants appearing in the error estimates obtained in Theorem 11.3.2 and Corollary 11.3.1 depend increasingly on ε^{-1}.

A final remark concerns the stability condition (11.3.7) obtained for the θ-scheme in Theorem 11.3.1. It reads as follows:

$$\Delta t \leq \frac{Ch^2 \alpha}{(1-2\theta)\gamma^2} \ .$$

For a small ε this becomes

$$(12.1.7) \qquad\qquad \Delta t \leq \frac{Ch^2 \varepsilon}{(1-2\theta)(\varepsilon^2 + 1 + \mu_1^2)} \ .$$

We notice that this restriction, obtained by an energy analysis, is not fully satisfactory. As a matter of fact, from a von Neumann stability analysis one would expect a condition on Δt which, for fixed h, becomes more favourable as far as ε decreases.

The aim of this Chapter is two-fold. First, improve some results obtained in Chapter 11, enlightening the dependence of stability and convergence estimates on the small diffusion coefficient ε. Indeed, we will begin the following Section by showing via a sharper analysis that stability holds even if the time-step Δt satisfies a condition weaker than (12.1.7). We also introduce a semi-implicit scheme which is not subjected to time-step restrictions. Secondly, we introduce other time-advancing methods that were not considered in Chapter 11 and that are better suited for the approximation of parabolic problems where advection is dominant.

12.2 Time-Advancing by Finite Differences

As done in (11.2.1), we consider the following semi-discrete (continuous in time) approximation of the advection-diffusion initial-boundary value problem (12.1.5):

$$(12.2.1) \quad \begin{cases} \dfrac{d}{dt}(u_h(t), v_h) + a(u_h(t), v_h) = (f(t), v_h) \; \forall v_h \in V_h \;, \quad t \in (0, T) \\[3mm] u_h(0) = u_{0,h} \;. \end{cases}$$

Here $V_h \subset H_0^1(\Omega)$ is a suitable finite dimensional space and $u_{0,h} \in V_h$ an approximation of the initial datum u_0.

In this respect, we have noticed in Chapter 8 that the Galerkin approximation for the steady problem using the classical finite element space (6.2.4) is not satisfactory when advection is much stronger than diffusion, as the approximate solution is highly oscillatory unless the mesh-size h is comparable with ε. In order to avoid these difficulties, in the rest of this Section we will focus on the case in which diffusion is dominant. More appropriate approaches for the advection-dominated case will be presented and discussed in Sections 12.3, 12.4 and 12.5.

12.2.1 A Sharp Stability Result for the θ-scheme

In Section 11.3 we have analyzed a family of time-advancing methods based on the so-called θ-scheme:

$$(12.2.2) \quad \begin{cases} \dfrac{1}{\Delta t}(u_h^{n+1} - u_h^n, v_h) + a(\theta u_h^{n+1} + (1 - \theta)u_h^n, v_h) \\[3mm] \qquad = (\theta f(t_{n+1}) + (1 - \theta)f(t_n), v_h) \; \forall v_h \in V_h \\[3mm] u_h^0 = u_{0,h} \end{cases}$$

for each $n = 0, 1, ..., \mathcal{N} - 1$, where $0 \le \theta \le 1$, $\Delta t := T/\mathcal{N}$ is the time step, $\mathcal{N}$ is a positive integer, and $u_{0,h} \in V_h$ is a suitable approximation of the initial datum u_0. We remind that this includes the schemes: forward Euler ($\theta = 0$), backward Euler ($\theta = 1$) and Crank-Nicolson ($\theta = 1/2$).

At each time step one has to solve a linear system associated to the matrix $M + \theta \Delta t A$, where the stiffness matrix A and the mass matrix M are defined as in (11.2.2), i.e.,

$$(12.2.3) \qquad A_{ij} := a(\varphi_j, \varphi_i) \;, \quad M_{ij} := (\varphi_i, \varphi_j) \;,$$

$\{\varphi_j \,|\, j = 1, ..., N_h\}$ being a basis of V_h. Notice that the forward Euler scheme is not explicit, unless the mass matrix M is lumped to a diagonal one (see Section 11.4).

Now we prove a new stability result for these schemes, that improves the one of Theorem 11.3.1.

Proposition 12.2.1 *Assume that (12.1.3) holds and that the map $t \to \|f(t)\|_0$ is bounded in $[0, T]$. When $0 \le \theta < 1/2$, assume, moreover, that T_h is a quasi-uniform family of triangulations and that the time-step restriction*

$$(12.2.4) \qquad \Delta t \le \frac{h^2}{(1 - 2\theta)C^*} \min\left\{\frac{1}{\varepsilon}, \frac{2\mu_0}{1 + h^2\mu_1^2}\right\}$$

holds, C^* being the constant appearing in the inequality (12.2.10) here below. Then, the solution u_h^n of (12.2.2) satisfies

$$(12.2.5) \qquad \|u_h^n\|_0 \le \|u_{0,h}\|_0 + C^{**}\sqrt{\frac{T}{\varepsilon}} \max_{t \in [0,T]} \|f(t)\|_0 \ ,$$

where the positive constant C^{**} depends only on Ω.

Proof. First of all notice that, owing to (12.1.3), we can assume that the constant λ appearing in (11.1.6) is 0. Focussing on the case $0 \le \theta < 1/2$, from Theorem 11.3.1 we deduce

$$(12.2.6) \qquad \begin{aligned} &\frac{1}{2}\|u_h^{n+1}\|_0^2 - \frac{1}{2}\|u_h^n\|_0^2 + \Delta t\, a(u_{h,\theta}^{n+1}, u_{h,\theta}^{n+1}) \\ &= \Delta t\,(f_\theta^{n+1}, u_{h,\theta}^{n+1}) + \left(\frac{1}{2} - \theta\right)\|u_h^{n+1} - u_h^n\|_0^2 \ , \end{aligned}$$

where we have used the simplified notation

$$u_{h,\theta}^{n+1} := \theta u_h^{n+1} + (1 - \theta)u_h^n \ , \quad f_\theta^{n+1} := \theta f(t_{n+1}) + (1 - \theta)f(t_n) \ .$$

Owing to (12.1.3) we obtain

$$(12.2.7) \qquad a(u_{h,\theta}^{n+1}, u_{h,\theta}^{n+1}) \ge \varepsilon \|\nabla u_{h,\theta}^{n+1}\|_0^2 + \mu_0 \|u_{h,\theta}^{n+1}\|_0^2 \ .$$

Moreover there exists a constant $\hat{C}_1 > 0$, only dependent on Ω, such that

$$(12.2.8) \qquad \Delta t\,(f_\theta^{n+1}, u_{h,\theta}^{n+1}) \le \hat{C}_1 \frac{\Delta t}{\varepsilon}\|f_\theta^{n+1}\|_{-1,h}^2 + \Delta t \frac{\varepsilon}{2}\|\nabla u_{h,\theta}^{n+1}\|_0^2$$

(the norm $\|\cdot\|_{-1,h}$ has been defined in (11.3.6)). Finally, we know that

$$(12.2.9) \qquad \begin{aligned} &\|u_h^{n+1} - u_h^n\|_0^2 \\ &= -\Delta t\, a(u_{h,\theta}^{n+1}, u_h^{n+1} - u_h^n) + \Delta t\,(f_\theta^{n+1}, u_h^{n+1} - u_h^n) \\ &=: \hat{a} + \hat{f} \ . \end{aligned}$$

The first term at the right hand side can be estimated using (12.1.6) and providing a strict bound for each one of the terms. This yields

$$\hat{a} \leq \Delta t\,\varepsilon \|\nabla u_{h,\theta}^{n+1}\|_0 \|\nabla u_h^{n+1} - \nabla u_h^n\|_0 + \frac{\Delta t}{2}\|u_{h,\theta}^{n+1}\|_0\|\nabla u_h^{n+1} - \nabla u_h^n\|_0$$

$$+ \Delta t\left(\frac{1}{2}\|\nabla u_{h,\theta}^{n+1}\|_0 + \mu_1\|u_{h,\theta}^{n+1}\|_0\right)\|u_h^{n+1} - u_h^n\|_0 \ .$$

By means of the inverse inequality (6.3.21) we finally have

$$\hat{a} \leq \Delta t\left[C_3^{1/2}h^{-1}\varepsilon\|\nabla u_{h,\theta}^{n+1}\|_0 + (C_3^{1/2}h^{-1} + \mu_1)\|u_{h,\theta}^{n+1}\|_0\right]\|u_h^{n+1} - u_h^n\|_0 \ .$$

Using again (6.3.21), the second term at the right hand side in (12.2.9) satisfies

$$\hat{f} \leq \hat{C}_2\Delta t\,h^{-1}\|f_\theta^{n+1}\|_{-1,h}\|u_h^{n+1} - u_h^n\|_0 \ .$$

In conclusion, from (12.2.9) we deduce

(12.2.10)
$$\|u_h^{n+1} - u_h^n\|_0^2 \leq C^*(\Delta t)^2 \left[h^{-2}\varepsilon^2\|\nabla u_{h,\theta}^{n+1}\|_0^2 \right.$$
$$+ h^{-2}(1 + h^2\mu_1^2)\|u_{h,\theta}^{n+1}\|_0^2$$
$$\left. + h^{-2}\|f_\theta^{n+1}\|_{-1,h}^2\right] \ ,$$

where $C^* > 0$ only depends on Ω. If the time-step Δt satisfies (12.2.4) we can conclude that

$$\|u_h^{n+1}\|_0^2 - \|u_h^n\|_0^2 \leq \hat{C}_3\frac{\Delta t}{\varepsilon}\|f_\theta^{n+1}\|_{-1,h}^2 \ ,$$

where $\hat{C}_3 > 0$ only depends on Ω. Now the stability estimate (12.2.5) follows by proceeding as in Theorem 11.3.1.   □

Stability condition (12.2.4) is less restrictive when ε becomes small, and it is mainly dictated by the advection coefficient $\mathbf{b}$ and by a_0 through the constants μ_0 and μ_1 in (12.1.3). Despite this fact, it might not be the best possible one yet, at least in some particular cases. As a matter of fact, a von Neumann stability analysis for the forward-in-time ($\theta = 0$) backward-in-space finite difference scheme applied to the one-dimensional advection-diffusion operator $Lu := -\varepsilon u'' + bu' + a_0u$, ε, b and a_0 positive constants, on the whole real line $\mathbb{R}$ yields the stability condition

$$\Delta t \leq \frac{h^2}{2\varepsilon + bh}$$

(see, e.g., Strikwerda (1989), pp. 129–132). For small ε, the time-step Δt behaves like $O(h)$, whereas in (12.2.4) it is $O(h^2)$ for each $\varepsilon > 0$.

12.2.2 A Semi-Implicit Scheme

Aiming at avoiding stability restrictions, without resorting to a fully implicit approximation, one is led to consider the so-called *semi-implicit* time-discretization approach. This consists in evaluating the principal part of the operator L (i.e., its second order term $-\varepsilon\Delta$) at the time level t_{n+1}, whereas the remaining parts are considered at the time t_n. More precisely, this scheme reads

(12.2.11)
$$\begin{cases} \dfrac{1}{\Delta t}(u_h^{n+1} - u_h^n, v_h) + \varepsilon(\nabla u_h^{n+1}, \nabla v_h) - \sum_{i=1}^{d}(b_i u_h^n, D_i v_h) \\ \qquad\qquad + (a_0 u_h^n, v_h) = (f(t_{n+1}), v_h) \qquad \forall\, v_h \in V_h \\ u_h^0 = u_{0,h} \end{cases}$$

for each $n = 0, 1, ..., \mathcal{N} - 1$. This scheme is expected to enjoy better stability properties than the one previously analyzed. In fact, we can prove:

Theorem 12.2.1 *Assume that the map $t \to ||f(t)||_0$ is bounded in $[0, T]$. Then the semi-implicit approximation scheme (12.2.11) is unconditionally stable on any finite time interval $[0, T]$ and the solution u_h^n satisfies*

(12.2.12)
$$||u_h^n||_0 \le \left(||u_{0,h}||_0 + C\sqrt{\dfrac{T}{\varepsilon}} \max_{t \in [0,T]} ||f(t)||_0 \right)$$
$$\times \exp\left[\dfrac{CT}{\varepsilon}(1 + ||a_0||^2_{L^\infty(\Omega)}) \right]$$

for each $n = 0, 1, ..., \mathcal{N}$, where the constant $C > 0$ only depends on Ω.

Proof. By proceeding essentially as in the proof of Theorem 11.3.1, the choice $v_h = u_h^{n+1}$ in (12.2.11) produces

$$\frac{1}{2}||u_h^{n+1}||_0^2 - \frac{1}{2}||u_h^n||_0^2 + \frac{1}{2}||u_h^{n+1} - u_h^n||_0^2 + \varepsilon\Delta t||\nabla u_h^{n+1}||_0^2$$
$$\le C\Delta t(1 + ||a_0||_{L^\infty(\Omega)})\,||u_h^n||_0\,||\nabla u_h^{n+1}||_0$$
$$+ C\Delta t||f(t_{n+1})||_{-1,h}\,||\nabla u_h^{n+1}||_0 \ .$$

It follows easily that

$$||u_h^{n+1}||_0^2 - ||u_h^n||_0^2 \le \frac{C}{\varepsilon}\Delta t(1 + ||a_0||^2_{L^\infty(\Omega)})\,||u_h^n||_0^2$$
$$+ \frac{C}{\varepsilon}\Delta t||f(t_{n+1})||^2_{-1,h} \ .$$

Let now m be a fixed index, $1 \le m \le \mathcal{N}$. Summing over n from 0 to $m - 1$ we find

$$||u_h^m||_0^2 \leq ||u_{0,h}||_0^2 + \frac{C}{\varepsilon}\Delta t \left(1 + ||a_0||_{L^\infty(\Omega)}^2\right) \sum_{n=0}^{m-1} ||u_h^n||_0^2$$

$$+ \frac{C}{\varepsilon}\Delta t \sum_{n=0}^{m-1} ||f(t_{n+1})||_{-1,h}^2 \ .$$

Using the discrete Gronwall inequality (see Lemma 1.4.1) and recalling that $||\phi||_{-1,h} \leq ||\phi||_0$ for each $\phi \in L^2(\Omega)$ gives (12.2.12). $\qquad\square$

The stability estimate (12.2.12) deteriorates exponentially with respect to ε^{-1}. However, numerical experiences show that, at least in some particular cases, the bound given in (12.2.12) is too pessimistic, and that the dependence on ε^{-1} is linear rather than exponential (see Bressan and Quarteroni (1986)).

Now we turn to the convergence analysis for the semi-implicit problem (12.2.11). We have the following

Theorem 12.2.2 *Assume that $u_0 \in H_0^1(\Omega)$ and that the solution u to (11.1.5) is such that $\frac{\partial u}{\partial t} \in L^2(0, T; H_0^1(\Omega))$ and $\frac{\partial^2 u}{\partial t^2} \in L^2(0, T; L^2(\Omega))$. Then u_h^n defined in (12.2.11) satisfies for each $n = 0, 1, ..., \mathcal{N}$*

$$||u_h^n - u(t_n)||_0 \leq ||(I - \Pi_{1,h}^k)u(t_n)||_0 + \exp(C_\varepsilon t_n)$$

(12.2.13)
$$\times \left\{ ||u_{0,h} - \Pi_{1,h}^k(u_0)||_0^2 + C_\varepsilon \int_0^{t_n} \left\|(I - \Pi_{1,h}^k)\frac{\partial u}{\partial t}(s)\right\|_0^2 \right.$$

$$\left. + C_\varepsilon(\Delta t)^2 \int_0^{t_n} \left(\left\|\frac{\partial u}{\partial t}(s)\right\|_1^2 + \left\|\frac{\partial^2 u}{\partial t^2}(s)\right\|_0^2 \right) \right\}^{1/2},$$

where $\Pi_{1,h}^k$ is the elliptic "projection" operator defined in (11.2.8) and $C_\varepsilon > 0$ is a non-decreasing function of ε^{-1}.

Proof. From our assumptions it follows that $u \in C^0([0, T]; H_0^1(\Omega))$, hence the operator $\Pi_{1,h}^k(u(t_n))$ is defined for each $n = 0, 1, ..., \mathcal{N}$. Set $\eta_h^n := u_h^n - \Pi_{1,h}^k(u(t_n))$ and observe that for each $v_h \in V_h$ and $n = 0, 1, ..., \mathcal{N} - 1$

$$\frac{1}{\Delta t}(\eta_h^{n+1} - \eta_h^n, v_h) + \varepsilon(\nabla \eta_h^{n+1}, \nabla v_h)$$

$$- \sum_{i=1}^d (b_i \eta_h^n, D_i v_h) + (a_0 \eta_h^n, v_h)$$

$$= (f(t_{n+1}), v_h) - \frac{1}{\Delta t}(\Pi_{1,h}^k(u(t_{n+1}) - u(t_n)), v_h)$$

(12.2.14)
$$- \varepsilon(\nabla u(t_{n+1}), \nabla v_h) + \sum_{i=1}^d (b_i u(t_{n+1}), D_i v_h)$$

$$- (a_0 u(t_{n+1}), v_h) - \varepsilon(\nabla u(t_n), \nabla v_h)$$

$$+ \sum_{i=1}^d (b_i u(t_n), D_i v_h) - (a_0 u(t_n), v_h)$$

$$+ \varepsilon(\nabla \Pi_{1,h}^k(u(t_n)), \nabla v_h) - \sum_{i=1}^d (b_i \Pi_{1,h}^k(u(t_{n+1})), D_i v_h)$$

$$+ (a_0 \Pi_{1,h}^k(u(t_{n+1})), v_h) \ .$$

On the other hand, the solution u satisfies

$$\left(\frac{\partial u}{\partial t}(t_{n+1}), v_h\right) = (f(t_{n+1}), v_h) - \varepsilon(\nabla u(t_{n+1}), \nabla v_h)$$

$$+ \sum_{i=1}^d (b_i u(t_{n+1}), D_i v_h) - (a_0 u(t_{n+1}), v_h) \ .$$

Let us further recall that the definition of $\Pi_{1,h}^k(\cdot)$ reads

$$-\varepsilon(\nabla u(t_n), \nabla v_h) + \sum_{i=1}^d (b_i u(t_n), D_i v_h) - (a_0 u(t_n), v_h)$$

$$= -\varepsilon(\nabla \Pi_{1,h}^k(u(t_n)), \nabla v_h) + \sum_{i=1}^d (b_i \Pi_{1,h}^k(u(t_n)), D_i v_h)$$

$$- (a_0 \Pi_{1,h}^k(u(t_n)), v_h) \ .$$

Since $\frac{\partial}{\partial t}$ commutes with $\Pi_{1,h}^k$, (12.2.14) can be finally rewritten as

$$\frac{1}{\Delta t}(\eta_h^{n+1} - \eta_h^n, v_h) + \varepsilon(\nabla \eta_h^{n+1}, \nabla v_h)$$

$$- \sum_{i=1}^d (b_i \eta_h^n, D_i v_h) + (a_0 \eta_h^n, v_h) = (\varepsilon_h^n, v_h) \ ,$$

where $\varepsilon_h^n \in V_h$ is defined by the relation

$$
\begin{aligned}
(\varepsilon_h^n, v_h) = & \left(\frac{\partial u}{\partial t}(t_{n+1}) - \frac{u(t_{n+1}) - u(t_n)}{\Delta t}, v_h \right) \\
& + \frac{1}{\Delta t} \left(\int_{t_n}^{t_{n+1}} (I - \Pi_{1,h}^k) \frac{\partial u}{\partial t}(s) ds, v_h \right) \\
& + \sum_{i=1}^{d} (b_i \Pi_{1,h}^k (u(t_n) - u(t_{n+1})), D_i v_h) \\
& - (a_0 \Pi_{1,h}^k (u(t_n) - u(t_{n+1})), v_h) \ .
\end{aligned}
$$

(12.2.15)

Therefore η_h^n satisfies a scheme like (12.2.11), and consequently

(12.2.16)
$$
\begin{aligned}
\|\eta_h^n\|_0^2 \leq & \left(\|u_{0,h} - \Pi_{1,h}^k(u_0)\|_0^2 + \frac{C}{\varepsilon} \Delta t \sum_{n=0}^{m-1} \|\varepsilon_h^n\|_{-1,h}^2 \right) \\
& \times \exp \left[\frac{C t_n}{\varepsilon} (1 + \|a_0\|_{L^\infty(\Omega)}^2) \right] \ .
\end{aligned}
$$

A suitable bound for $\|\varepsilon_h^n\|_{-1,h}$ is easily found. In fact, the first two terms at the right hand side of (12.2.15) can be estimated by proceeding as in the proof of Theorem 11.3.2. Moreover, we have

$$
\sum_{i=1}^{d} (b_i \Pi_{1,h}^k (u(t_n) - u(t_{n+1})), D_i v_h) \leq \|\Pi_{1,h}^k (u(t_n) - u(t_{n+1}))\|_0 \|v_h\|_1 \ .
$$

Therefore we conclude

(12.2.17)
$$
\begin{aligned}
\|\varepsilon_h^n\|_{-1,h}^2 \leq C \Big[& \Delta t \int_{t_n}^{t_{n+1}} \left\| \frac{\partial^2 u}{\partial t^2}(s) \right\|_{-1,h}^2 \\
& + \frac{1}{\Delta t} \int_{t_n}^{t_{n+1}} \left\| (I - \Pi_{1,h}^k) \frac{\partial u}{\partial t}(s) \right\|_{-1,h}^2 \\
& + (1 + \|a_0\|_{L^\infty(\Omega)}^2) \|\Pi_{1,h}^k (u(t_{n+1}) - u(t_n))\|_0^2 \Big] \ .
\end{aligned}
$$

Finally, since the operator $\Pi_{1,h}^k$ is uniformly bounded in $H_0^1(\Omega)$, we obtain

(12.2.18)
$$
\begin{aligned}
\|\Pi_{1,h}^k (u(t_{n+1}) - u(t_n))\|_0^2 & \leq \Delta t \int_{t_n}^{t_{n+1}} \left\| \Pi_{1,h}^k \frac{\partial u}{\partial t}(s) \right\|_0^2 \\
& \leq \Delta t \frac{\gamma^2}{\alpha^2} \int_{t_n}^{t_{n+1}} \left\| \frac{\partial u}{\partial t}(s) \right\|_1^2 ,
\end{aligned}
$$

α and γ being the coerciveness and continuity constants of the form $a(\cdot, \cdot)$, respectively. The thesis follows now from (12.2.16)-(12.2.18). $\square$

From (12.2.13) we can easily obtain an error estimate with respect to h and Δt by proceeding as in Corollary 11.3.1. Under the additional assumptions $u_0 \in H^{k+1}(\Omega)$ and $\frac{\partial u}{\partial t} \in L^2(0, T; H^{k+1}(\Omega))$, the final result reads

$$||u_h^n - u(t_n)||_0 = O(h^{k+1} + \Delta t) \ .$$

The main motivation for dealing with the diffusive part implicitly and the advective one explicitly can be ascribed to the following facts. As proven in Theorem 12.2.1, for finite time integration the scheme is inconditionally stable. Besides, at each step one needs to solve a linear system with a *symmetric* matrix, although the original problem was not self-adjoint. Further, when applied to problems with nonlinear advection, the semi-implicit scheme provides an effective linearization procedure, as it still yields a system with a symmetric matrix, which moreover is the same at each time-level.

12.3 The Discontinuous Galerkin Method for Stabilized Problems

We have already shown in Chapter 8 that a pure Galerkin approach for space discretization can perform poorly when dealing with advection-dominated problems. In that case a stabilization procedure is therefore in order, and, if we choose to operate on the space variables solely, it can be devised on the ground of what has been done in Chapter 8 for steady advection-diffusion problems. In particular, one can resort to using in (12.2.1) the space V_h introduced in (8.3.15), which is obtained from $X_h^k \cap H_0^1(\Omega)$ by adding a bubble function in each element $K \in \mathcal{T}_h$. In other words,

$$V_h = V_h^b := [X_h^k \cap H_0^1(\Omega)] \oplus B \ ,$$

B being a finite dimensional space of bubble functions (see, for instance, (8.3.16)). Unfortunately, the resulting scheme tends to be over-diffusive across sharp layers.

Then we could decide to follow a different approach, resorting to stabilization techniques like those presented in Section 8.3.2, thus choosing $V_h = X_h^k \cap H_0^1(\Omega)$, $k \geq 1$, where X_h^k is the finite element space defined in (3.2.4) or (3.2.5), but replacing in (12.2.1) the bilinear form $a(\cdot, \cdot)$ and the right hand side by more involved ones. For instance, we could consider the following "stabilized" semi-discrete problem:

$$(12.3.1) \quad \begin{cases} \dfrac{d}{dt}(u_h(t), v_h) + a(u_h(t), v_h) \\ \qquad + \displaystyle\sum_{K \in \mathcal{T}_h} \delta \left(\dfrac{\partial u_h}{\partial t}(t) + L u_h(t), \dfrac{h_K}{|\mathbf{b}|}(L_{SS} v + \rho L_S v) \right)_K \\ = (f(t), v_h) \\ \qquad + \displaystyle\sum_{K \in \mathcal{T}_h} \delta \left(f(t), \dfrac{h_K}{|\mathbf{b}|}(L_{SS} v + \rho L_S v) \right)_K \quad \forall\, v_h \in V_h \\ u_h(0) = u_{0,h} \ , \end{cases}$$

for each $t \in (0, T)$. In practical computations, however, this last scheme is in general not satisfactory, especially for solutions having large derivatives in time. As a matter of fact, the fully-discrete backward Euler scheme applied to (12.3.1) turns out to be unstable for $\Delta t/h$ too close to 0, which inhibits the possibility of achieving optimal accuracy in time.

A more efficient approach, though its computational cost is in general fairly high, is based on the discontinuous Galerkin method, which makes use of a finite element formulation for both space and time variables. Its distinguishing feature is that trial functions are continuous in space but not necessarily in time (see Section 11.3.1 for a presentation of the method and notations).

Before introducing the numerical approximation of (12.1.5), as done in Section 8.3.2 we denote by L_S and L_{SS} the symmetric and skew-symmetric part of the operator L, respectively, i.e.,

$$(12.3.2) \qquad L_S z := -\varepsilon \Delta z + \left(\frac{1}{2} \operatorname{div} \mathbf{b} + a_0 \right) z$$

$$(12.3.3) \qquad L_{SS} z := \frac{1}{2} \operatorname{div}(\mathbf{b}z) + \frac{1}{2}\mathbf{b} \cdot \nabla z \ .$$

Introduce the notation

$$\mathcal{M}_h^{(\rho)}(z, v)(t)$$
$$:= \sum_{K \in \mathcal{T}_h} \delta \left(\frac{\partial z}{\partial t}(t) + L z(t), \frac{h_K}{|\mathbf{b}|} \left(\frac{\partial v}{\partial t}(t) + L_{SS} v(t) + \rho L_S v(t) \right) \right)_K$$

$$\psi_h^{(\rho)}(v)(t) := \sum_{K \in \mathcal{T}_h} \delta \left(f(t), \frac{h_K}{|\mathbf{b}|} \left(\frac{\partial v}{\partial t}(t) + L_{SS} v(t) + \rho L_S v(t) \right) \right)_K$$

for each $z, v \in W_h^{\Delta t}$. The discontinuous Galerkin method reads: given $U_{-1}(0) := u_0$, for each $n = 0, 1, ..., \mathcal{N} - 1$ find $U_n : [t_n, t_{n+1}] \to V_h$ such that $U_n \in \mathbb{P}_q^n(V_h)$ and satisfies

$$\int_{t_n}^{t_{n+1}} \left[\left(\frac{\partial U_n}{\partial t}(t), v(t) \right) + a(U_n(t), v(t)) \right] + (U_n(t_n), v(t_n))$$

$$+ \int_{t_n}^{t_{n+1}} \mathcal{M}_h^{(\rho)}(U_n, v)(t)$$

$$(12.3.4)$$

$$= \int_{t_n}^{t_{n+1}} (f(t), v(t)) + (U_{n-1}(t_n), v(t_n))$$

$$+ \int_{t_n}^{t_{n+1}} \psi_h^{(\rho)}(v)(t) \qquad \forall\, v \in \mathbb{P}_q^n(V_h) \ .$$

As usual, $\delta > 0$ and $\rho \in \mathbb{R}$ are parameters, and for simplicity we have assumed $\inf_\Omega |\mathbf{b}(\mathbf{x})| > 0$, which is justified in the globally advection-dominated case (see (8.3.11)). The choices $\rho = 1$, $\rho = 0$ and $\rho = -1$ correspond, respectively, to the GALS, SUPG and DWG methods introduced in Section 8.3.2.

For each $n = 0, 1, ..., \mathcal{N}-1$ problem (12.3.4) is equivalent to a linear system of algebraic equations, which has a unique solution provided the parameter δ is suitably chosen. In fact, taking $v = U_n$ and proceeding as in Propositions 8.4.1, 8.4.2 and 8.4.3, one obtains an energy estimate for U_n. For instance, in the GALS case we have for each stabilization parameter $\delta > 0$

$$
\begin{aligned}
(12.3.5) \quad & \|U_n(t_{n+1})\|_0^2 + 2\varepsilon \int_{t_n}^{t_{n+1}} \|\nabla U_n(t)\|_0^2 + \mu_0 \int_{t_n}^{t_{n+1}} \|U_n(t)\|_0^2 \\
& + \int_{t_n}^{t_{n+1}} \sum_{K \in \mathcal{T}_h} \delta \left\| \frac{h_K^{1/2}}{|\mathbf{b}|^{1/2}} \left(\frac{\partial U_n}{\partial t}(t) + L U_n(t) \right) \right\|_{0,K}^2 \\
& \leq \|U_{n-1}(t_n)\|_0^2 + \frac{1}{\mu_0} \int_{t_n}^{t_{n+1}} \|f(t)\|_0^2 \\
& + \int_{t_n}^{t_{n+1}} \sum_{K \in \mathcal{T}_h} \delta \left\| \frac{h_K^{1/2}}{|\mathbf{b}|^{1/2}} f(t) \right\|_{0,K}^2 ,
\end{aligned}
$$

where $\mu_0 > 0$ is defined in (12.1.3). As usual, $\| \cdot \|_0$ denotes the norm of $L^2(\Omega)$ and $\| \cdot \|_{0,K}$ the one of $L^2(K)$.

A global estimate on the time interval $[0, T]$ can also be obtained starting from a formulation like the one given in (11.3.44). For each $n = 0, 1, ..., \mathcal{N}-1$ (having set $u_-^0 := u_0$) this result reads

$$
\begin{aligned}
(12.3.6) \quad & \|u_-^{n+1}\|_0^2 + \sum_{m=0}^{n} \| [u^m] \|_0^2 \\
& + 2\varepsilon \int_0^{t_{n+1}} \|\nabla u(t)\|_0^2 + \mu_0 \int_0^{t_{n+1}} \|u(t)\|_0^2 \\
& + \int_0^{t_{n+1}} \sum_{K \in \mathcal{T}_h} \delta \left\| \frac{h_K^{1/2}}{|\mathbf{b}|^{1/2}} \left(\frac{\partial u}{\partial t}(t) + L u(t) \right) \right\|_{0,K}^2 \\
& \leq \|u_0\|_0^2 + \frac{1}{\mu_0} \int_0^{t_{n+1}} \|f(t)\|_0^2 \\
& + \int_0^{t_{n+1}} \sum_{K \in \mathcal{T}_h} \delta \left\| \frac{h_K^{1/2}}{|\mathbf{b}|^{1/2}} f(t) \right\|_{0,K}^2 .
\end{aligned}
$$

Here, $u_{|(t_m, t_{m+1})} = U_m$, thus $[u^m] = U_m(t_m) - U_{m-1}(t_m)$ for each $m = 0, 1, ..., \mathcal{N} - 1$. The extra stability term appearing in (12.3.6) for $\delta > 0$ witnesses the better properties of this method when compared with the pure Galerkin one (i.e., when $\delta = 0$).

The convergence analysis is presented in Hughes, Franca and Mallet (1987) for $\rho = 0$ and Hughes, Franca and Hulbert (1989) for $\rho = 1$. The method is $O(h^{k+1/2})$-accurate with respect to the norm defined at the left hand side of (12.3.6) (however, when $\rho = 0$ the operator L has to be replaced by its skew-symmetric part L_{SS}).

Notice that one could also choose a different finite dimensional space V_h and/or a different polynomial degree q on each interval $[t_n, t_{n+1}]$.

An even more general approach proposed by Johnson, Nävert and Pitkä-ranta (1984) is based on a space-time triangulation of any strip $S_n := [t_n, t_{n+1}] \times \Omega$. In this case for each n one chooses a finite element subspace $V_h^{(n)}$ of

$$H_*^1(S_n) := \{v \in H^1(S_n) \,|\, v = 0 \text{ on } [t_n, t_{n+1}] \times \partial\Omega\}$$

based on space-time elements of size less than or equal to h. The triangulations relative to adjoining strips do not need to match on the common time level. The problem still reads as (12.3.4), but now both the solution U_n and the test function v belong to $V_h^{(n)}$, and $\mathcal{T}_h$ denotes a family of triangulations of S_n.

12.4 Operator-Splitting Methods

Another effective approach for finding the approximate solution of a non-stationary advection-diffusion equation is based on an operator-splitting strategy. Let us still consider problem (12.1.1). The operator L can be written as

$$(12.4.1) \qquad\qquad L = L_1 + L_2 \ ,$$

where L_i, $i = 1, 2$, are suitable operators. For instance, a typical splitting for the advection-diffusion case is given by

$$(12.4.2) \quad L_1 u = -\varepsilon \Delta u + \eta\mu u \ , \quad L_2 u = \frac{1}{2}\operatorname{div}(\mathbf{b}u) + \frac{1}{2}\mathbf{b} \cdot \nabla u + (1-\eta)\mu u \ ,$$

where μ is defined in (12.1.3) and $0 \leq \eta \leq 1$. With this choice, we have separate diffusion from advection. In particular, if $\eta = 1$ we have $L_1 = L_S$ and $L_2 = L_{SS}$, the symmetric and skew-symmetric part of the operator L with respect to the scalar product in $L^2(\Omega)$ (see (12.3.2) and (12.3.3)).

The basic idea of an operator-splitting procedure is as follows: starting from $u^n(\mathbf{x})$, an approximation of $u(t_n, \mathbf{x})$, construct $u^{n+1}(\mathbf{x})$ through two or more intermediate values, each one obtained by solving a boundary value problem related to only one of the operators L_i (see Section 5.7 for a general presentation of these methods).

We start focussing on the Peaceman-Rachford scheme: given $u^0 := u_0$, for each $n = 0, 1, ..., \mathcal{N} - 1$ solve at first

$$(12.4.3) \qquad \frac{u^{n+1/2} - u^n}{\tau} + L_1 u^{n+1/2} + L_2 u^n = f(t_{n+1/2}) \quad \text{in } \Omega \ ,$$

then

(12.4.4) $$\frac{u^{n+1} - u^{n+1/2}}{\tau} + L_1 u^{n+1/2} + L_2 u^{n+1} = f(t_{n+1/2}) \quad \text{in } \Omega \ ,$$

where $\tau = \Delta t/2$ and $t_{n+1/2} = t_n + \Delta t/2$.

Clearly, suitable boundary conditions have to be added to both (12.4.3) and (12.4.4). In general, the choice of which conditions impose heavily depends on the properties of the operators L_1 and L_2, as well as on the prescribed boundary operator B. Considering, for instance, the homogeneous Dirichlet case (i.e., $Bu = u$ in (12.1.1)), and assuming that the operators L_1 and L_2 are those introduced in (12.4.2), the boundary condition to be prescribed for (12.4.3) is clearly $u = 0$ on $\partial\Omega$. On the other hand, (12.4.4) is a steady advection problem (see Section 5.1; see also Chapter 14). Therefore, the homogeneous Dirichlet condition has to be imposed only on the inflow boundary, i.e., on $\partial\Omega^{in} := \{ \mathbf{x} \in \partial\Omega \,|\, \mathbf{b}(\mathbf{x}) \cdot \mathbf{n}(\mathbf{x}) < 0 \}$, where $\mathbf{n}$ is the unit outward normal vector on $\partial\Omega$. With this choice problem (12.4.4) has a unique solution.

In this way, still assuming that (12.1.3) is satisfied, we have

(12.4.5) $$(L_1 v, v) = \int_{\Omega} (\varepsilon|\nabla v|^2 + \eta \mu v^2) \geq \alpha ||v||_1^2 \quad \forall \, v \in H_0^1(\Omega)$$

with $\alpha := \min\{\varepsilon, \eta\mu_0\}$, and

$$(L_2 v, v) = \int_{\Omega} \left\{ \frac{1}{2}[\mathrm{div}(\mathbf{b}v)v + (\mathbf{b} \cdot \nabla v)v] + (1-\eta)\mu v^2 \right\}$$

(12.4.6) $$= \frac{1}{2} \int_{\partial\Omega \setminus \partial\Omega^{in}} \mathbf{b} \cdot \mathbf{n} \, v^2 + (1-\eta) \int_{\Omega} \mu v^2$$

$$\geq (1-\eta)\mu_0 ||v||_0^2 \quad \forall \, v \in H_{\partial\Omega^{in}}^1(\Omega) \ ,$$

where, for a subset $\Sigma \subset \partial\Omega$, we have defined $H_{\Sigma}^1(\Omega) := \{v \in H^1(\Omega) \,|\, v = 0 \text{ on } \Sigma\}$ (see Section 1.3). For splittings based on non-negative operators, stability and convergence results as $\Delta t \to 0$ are proven, e.g., in Yanenko (1971) and Marchuk (1990). For nonlinear operators, we refer to Lions and Mercier (1979).

Notice that any other fractional-step scheme introduced in Section 5.7 can be used instead of the Peaceman-Rachford one.

In view of the effective resolution of the problem, for each n one is led to approximate both (12.4.3) and (12.4.4) by a finite dimensional approximation. The convergence of the discrete fractional-step scheme is ensured provided the approximate operators, say $L_{1,h}$ and $L_{2,h}$, satisfy properties (12.4.5) and (12.4.6) on all functions v belonging to the appropriate finite dimensional spaces.

The splitting procedure described above can also be used at an algebraic level, i.e., when applied to a semi-discrete approximation like (12.2.1). In this case, the problem reads

$$(12.4.7) \qquad \begin{cases} M \dfrac{d\boldsymbol{\xi}}{dt}(t) + A\boldsymbol{\xi}(t) = \mathbf{F}(t) \ , \quad t \in (0,T) \\[2mm] \boldsymbol{\xi}(0) = \boldsymbol{\xi}_0 \ , \end{cases}$$

where the matrix A and M are defined in (12.2.3), $F_i(t) := (f(t), \varphi_i)$, and we have set

$$u_h(t, \mathbf{x}) = \sum_{j=1}^{N_h} \xi_j(t) \varphi_j(\mathbf{x}) \ , \quad u_{0,h}(\mathbf{x}) = \sum_{j=1}^{N_h} \xi_{0,j} \varphi_j(\mathbf{x}) \ ,$$

$\{\varphi_j \,|\, j = 1, ..., N_h\}$ being a basis of V_h. Let us recall that the matrix M is symmetric and positive definite.

The operator-splitting (12.4.1), (12.4.2) induces the matrix-splitting $A = A_1 + A_2$, where

$$(12.4.8) \qquad (A_1)_{ij} := a_1(\varphi_j, \varphi_i) \ , \quad (A_2)_{ij} := a_2(\varphi_j, \varphi_i)$$

for $i, j = 1, ..., N_h$, and

$$(12.4.9) \quad a_1(w, v) := \varepsilon \int_\Omega \nabla w \cdot \nabla v + \int_\Omega \eta \mu w v$$

$$(12.4.10) \quad a_2(w, v) := -\frac{1}{2} \int_\Omega \sum_{k=1}^{d} b_k(w D_k v - v D_k w) + \int_\Omega (1 - \eta) \mu w v \ .$$

We have to notice that the boundary treatment is straightforward in this framework, as the original matrix A already accounts for the boundary conditions.

This approach can be called *algebraic-splitting*, and provides the ground for implementing fractional-step schemes. For instance, the Peaceman-Rachford method in this case reads: take $\boldsymbol{\xi}^0 := \boldsymbol{\xi}_0$, then solve for $n = 0, 1, ..., \mathcal{N} - 1$

$$(12.4.11) \qquad \begin{cases} M \dfrac{\boldsymbol{\xi}^{n+1/2} - \boldsymbol{\xi}^n}{\tau} + A_1 \boldsymbol{\xi}^{n+1/2} + A_2 \boldsymbol{\xi}^n = \mathbf{F}(t_{n+1/2}) \\[2mm] M \dfrac{\boldsymbol{\xi}^{n+1} - \boldsymbol{\xi}^{n+1/2}}{\tau} + A_1 \boldsymbol{\xi}^{n+1/2} + A_2 \boldsymbol{\xi}^{n+1} = \mathbf{F}(t_{n+1/2}) \ , \end{cases}$$

where $\tau = \Delta t / 2$. (We recall that $u_h^n(\mathbf{x}) = \sum_{j=1}^{N_h} \xi_j^n \varphi_j(\mathbf{x})$ is the function approximating $u(t_n, \mathbf{x})$.)

It is easily seen that A_1 is positive definite for each $0 \le \eta \le 1$, while A_2 is positive definite for $0 \le \eta < 1$ and non-negative definite for $\eta = 1$. Hence both $\boldsymbol{\xi}^{n+1/2}$ and $\boldsymbol{\xi}^{n+1}$ can be uniquely determined from (12.4.11).

Now we can prove a stability result for the Peaceman-Rachford scheme.

Proposition 12.4.1 *Let M be the symmetric and positive definite matrix in (12.4.11) and introduce the scalar product and the associated norm*

(12.4.12) $$(\boldsymbol{\eta}, \boldsymbol{\xi})_M := (M\boldsymbol{\eta}, \boldsymbol{\xi}) \quad , \quad |\boldsymbol{\eta}|_M^2 = (M\boldsymbol{\eta}, \boldsymbol{\eta})$$

for $\boldsymbol{\eta}, \boldsymbol{\xi} \in \mathbb{R}^{N_h}$. Assume that $M^{-1}A_1$ and $M^{-1}A_2$ commute, i.e., that

(12.4.13) $$M^{-1}A_1 M^{-1}A_2 = M^{-1}A_2 M^{-1}A_1$$

and set $u_h^n(\mathbf{x}) = \sum_{j=1}^{N_h} \xi_j^n \varphi_j(\mathbf{x})$, where $\boldsymbol{\xi}^n$ is the Peaceman-Rachford sequence constructed in (12.4.11). Then

(12.4.14) $$||u_h^n||_0 \leq ||u_{0,h}||_0 + t_n \max_{t \in [0,T]} ||f(t)||_0 \quad .$$

Proof. Let us start from the recurrence relation

(12.4.15) $$\boldsymbol{\xi}^{n+1} = \mathcal{B}\boldsymbol{\xi}^n + \Delta t\, \mathcal{G} M^{-1}\mathbf{F}(t_{n+1/2}) \quad ,$$

where

(12.4.16) $$\mathcal{B} := (I + \tau M^{-1}A_2)^{-1}(I - \tau M^{-1}A_1)(I + \tau M^{-1}A_1)^{-1}(I - \tau M^{-1}A_2)$$

and

(12.4.17) $$\mathcal{G} := (I + \tau M^{-1}A_2)^{-1}(I + \tau M^{-1}A_1)^{-1} \quad .$$

From (12.4.15) we derive

(12.4.18) $$\boldsymbol{\xi}^n = \mathcal{B}^n \boldsymbol{\xi}_0 + \Delta t \sum_{k=1}^{n} \mathcal{B}^{n-k} \mathcal{G} M^{-1}\mathbf{F}(t_{k-1/2}) \quad ,$$

therefore

(12.4.19) $$|\boldsymbol{\xi}^n|_M \leq |\mathcal{B}^n \boldsymbol{\xi}_0|_M + \Delta t \sum_{k=1}^{n} |\mathcal{B}^{n-k} \mathcal{G} M^{-1}\mathbf{F}(t_{k-1/2})|_M \quad .$$

From (12.4.13) we can write

$$\mathcal{B} = \hat{\mathcal{B}} := (I - \tau M^{-1}A_1)(I + \tau M^{-1}A_1)^{-1}(I - \tau M^{-1}A_2)(I + \tau M^{-1}A_2)^{-1} \quad .$$

We notice now that for each $0 \leq \eta \leq 1$ the matrix $M^{-1}A_1$ is positive definite with respect to the scalar product $(\cdot, \cdot)_M$, whereas $M^{-1}A_2$ is positive definite for $0 \leq \eta < 1$ and non-negative definite for $\eta = 1$, still with respect to $(\cdot, \cdot)_M$. Therefore, by proceeding as in Proposition 2.3.1 we obtain

$$||\hat{\mathcal{B}}||_M := \max_{\substack{\boldsymbol{\eta} \in \mathbb{R}^{N_h} \\ \boldsymbol{\eta} \neq 0}} \frac{|\hat{\mathcal{B}}\boldsymbol{\eta}|_M}{|\boldsymbol{\eta}|_M} < 1 \quad .$$

In a similar way we prove that $||\mathcal{G}||_M < 1$, and we conclude that

(12.4.20) $$|\boldsymbol{\xi}^n|_M \leq |\boldsymbol{\xi}_0|_M + t_n \max_{t \in [0,T]} |M^{-1}\mathbf{F}(t)|_M \quad , \quad n = 0, 1, ..., \mathcal{N} \quad .$$

Finally, it is easily seen that

$$|\boldsymbol{\xi}^n|_M = ||u_h^n||_0 \quad , \quad |M^{-1}\mathbf{F}(t)|_M = ||P_h^k(f(t))||_0 \leq ||f(t)||_0 \quad ,$$

where P_h^k is the orthogonal projection operator on V_h with respect to the $L^2(\Omega)$-scalar product. $\square$

The commutativity relation (12.4.13) is rather restrictive, and in general is not satisfied when A_1 and A_2 are the matrices given in (12.4.8)-(12.4.10). However, if we introduce the auxiliary unknown

(12.4.21) $\boldsymbol{\zeta}^n := (I + \tau M^{-1}A_2)\boldsymbol{\xi}^n \quad ,$

relation (12.4.15) becomes

(12.4.22) $\boldsymbol{\zeta}^{n+1} = \hat{B}\boldsymbol{\zeta}^n + \Delta t(I + \tau M^{-1}A_1)^{-1}M^{-1}\mathbf{F}(t_{n+1/2}) \quad .$

We can thus repeat the procedure above, obtaining (12.4.20) for $\boldsymbol{\zeta}^n$ instead of $\boldsymbol{\xi}^n$. This corresponds to a stability estimate for $\boldsymbol{\xi}^n$ with respect to the vector norm associated to the symmetric and positive definite matrix

$$\hat{M} := M + \tau(A_2 + A_2^T) + \tau^2 A_2^T M^{-1} A_2 \quad .$$

For the convergence analysis of the Peaceman-Rachford method we refer to Yanenko (1971) and Marchuk (1990).

Notice that the matrix A_1 defined in (12.4.8), (12.4.9) is the one corresponding to the discretization of the homogeneous Dirichlet problem for the operator L_1 defined in (12.4.2). On the contrary, the matrix A_2 defined in (12.4.8), (12.4.10), which has dimension N_h, is not the discrete counterpart of the operator L_2 with Dirichlet condition on $\partial\Omega^{in}$. In fact, in the latter case the number of degrees of freedom is equal to the number of the internal nodes N_h plus the nodes on $\partial\Omega \setminus \partial\Omega^{in}$, as we are allowed to assign the boundary condition on $\partial\Omega^{in}$ solely.

On the base of this remark, we can conclude that if we apply first the fractional-step procedure (12.4.3), (12.4.4), and then approximate the two boundary value problems thus obtained, the resulting scheme doesn't necessarily coincide with the one that we would obtain taking first the semidiscretization (12.4.7) of the original problem and then advancing in time by the fractional-step (12.4.11).

Furthermore, we have to warn the reader that the treatment of more general boundary conditions in the framework of operator-splitting procedures needs to be done carefully, in order to ensure consistency with the original boundary value problem. In some cases also the boundary conditions have to be split. For instance, when considering the Neumann problem

$$\begin{cases} Lu = f & \text{in } \Omega \\[2mm] \varepsilon\dfrac{\partial u}{\partial n} - \mathbf{b}\cdot\mathbf{n}\,u = g & \text{on } \partial\Omega \quad , \end{cases}$$

one could choose $a_1(\cdot,\cdot)$ as in (12.4.9) and

$$a_2(w,v) := -\frac{1}{2}\int_\Omega \sum_{k=1}^d b_k(wD_kv - vD_kw) + \int_\Omega (1-\eta)\mu\, w\, v - \frac{1}{2}\int_{\partial\Omega} \mathbf{b}\cdot\mathbf{n}\, w\, v \ .$$

The matrix A_1 now corresponds to the discretization of the differential operator L_1 associated to the boundary operator $\varepsilon\frac{\partial}{\partial n}$ on $\partial\Omega$. On the other hand, A_2 corresponds to the discretization of L_2 with Dirichlet boundary condition on $\partial\Omega^{in}$ (imposed in a *weak* sense).

12.5 A Characteristic Galerkin Method

This method stems from considering the non-stationary advection-diffusion equation from a Lagrangian (instead of Eulerian) point of view, and can be traced back to Pironneau (1982), Douglas and Russell (1982), Ewing, Russell and Wheeler (1984).

At first we define the characteristic lines associated to a vector field $\mathbf{b} = \mathbf{b}(t,\mathbf{x})$. Being given $\mathbf{x} \in \overline{\Omega}$ and $s \in [0,T]$, they are the vector functions $\mathbf{X} = \mathbf{X}(t; s, x)$ such that

(12.5.1)
$$\begin{cases} \dfrac{d\mathbf{X}}{dt}(t; s, \mathbf{x}) = \mathbf{b}(t, \mathbf{X}(t; s, \mathbf{x})) \ , \quad t \in (0, T) \\[2mm] \mathbf{X}(s; s, \mathbf{x}) = \mathbf{x} \ . \end{cases}$$

The existence and uniqueness of the characteristic lines for each choice of s and $\mathbf{x}$ holds under suitable assumptions on $\mathbf{b}$, for instance $\mathbf{b}$ continuos in $[0,T] \times \overline{\Omega}$ and Lipschitz continuous in $\overline{\Omega}$, uniformly with respect to $t \in [0,T]$ (see, e.g., Hartman (1973)).

From a geometric point of view, $\mathbf{X}(t; s, \mathbf{x})$ provides the position at time t of a particle which has been driven by the field $\mathbf{b}$ and that occupied the position $\mathbf{x}$ at the time s. The uniqueness result gives in particular that

(12.5.2)
$$\mathbf{X}(t; s, \mathbf{X}(s; \tau, \mathbf{x})) = \mathbf{X}(t; \tau, \mathbf{x})$$

for each $t, s, \tau \in [0,T]$ and $\mathbf{x} \in \overline{\Omega}$. Hence $\mathbf{X}(t; s, \mathbf{X}(s; t, \mathbf{x})) = \mathbf{X}(t; t, \mathbf{x}) = \mathbf{x}$, i.e., for fixed t and s, the inverse function of $\mathbf{x} \rightarrow \mathbf{X}(s; t, \mathbf{x})$ is given by $\mathbf{y} \rightarrow \mathbf{X}(t; s, \mathbf{y})$.

Therefore, defining

(12.5.3)
$$\overline{u}(t, \mathbf{y}) := u(t, \mathbf{X}(t; 0, \mathbf{y}))$$

(or, equivalently, $u(t, \mathbf{x}) = \overline{u}(t, \mathbf{X}(0, t, \mathbf{x}))$ from (12.5.1) it follows that

$$\frac{\partial \overline{u}}{\partial t}(t, \mathbf{y}) = \frac{\partial u}{\partial t}(t, \mathbf{X}(t; 0, \mathbf{y})) + \sum_{i=1}^{d} D_i u(t, \mathbf{X}(t; 0, \mathbf{y}))\frac{dX_i}{dt}(t; 0, \mathbf{y})$$

(12.5.4)

$$= \left(\frac{\partial u}{\partial t} + \mathbf{b} \cdot \nabla u\right)(t, \mathbf{X}(t; 0, \mathbf{y})) \ .$$

Following the same notation introduced in (12.5.3), we can rewrite the non-stationary advection-diffusion equation as

(12.5.5) $$\frac{\partial \overline{u}}{\partial t} - \varepsilon \overline{\Delta u} + (\overline{\operatorname{div} \mathbf{b}} + \overline{a_0})\overline{u} = \overline{f} \quad \text{in} \quad Q_T \ .$$

We are thus led to discretize (12.5.5). The time derivative is approximated by the backward Euler scheme, i.e.,

(12.5.6) $$\frac{\partial \overline{u}}{\partial t}(t_{n+1}, \mathbf{y}) \cong \frac{\overline{u}(t_{n+1}, \mathbf{y}) - \overline{u}(t_n, \mathbf{y})}{\Delta t} \ .$$

If we set $\mathbf{y} = \mathbf{X}(0; t_{n+1}, \mathbf{x})$, from (12.5.3) we obtain

$$\frac{\partial \overline{u}}{\partial t}(t_{n+1}, \mathbf{X}(0; t_{n+1}, \mathbf{x})) \cong \frac{u(t_{n+1}, \mathbf{x}) - u(t_n, \mathbf{X}(t_n; t_{n+1}, \mathbf{x}))}{\Delta t} \ .$$

Denoting by $\mathbf{X}^n(\mathbf{x})$ a suitable approximation of $\mathbf{X}(t_n; t_{n+1}, \mathbf{x})$, $n = 0, 1, ...,$ $\mathcal{N} - 1$, we can finally write the following implicit discretization scheme for problem (12.5.5): set $u^0 := u_0$, then for $n = 0, 1, ..., \mathcal{N} - 1$ solve

(12.5.7) $$\frac{u^{n+1} - u^n \circ \mathbf{X}^n}{\Delta t} - \varepsilon \Delta u^{n+1} + [\operatorname{div} \mathbf{b}(t_{n+1}) + a_0]u^{n+1}$$
$$= f(t_{n+1}) \quad \text{in} \quad \Omega \ .$$

Clearly, a boundary condition has to be imposed on $\partial \Omega$: for simplicity, let us consider the homogeneous Dirichlet condition $u^{n+1}_{|\partial\Omega} = 0$.

One can typically choose a backward Euler scheme also for discretizing

(12.5.8) $$\frac{d\mathbf{X}}{dt}(t; t_{n+1}, \mathbf{x}) = \mathbf{b}(t, \mathbf{X}(t; t_{n+1}, \mathbf{x})) \ .$$

This produces the following approximation of $\mathbf{X}(t_n; t_{n+1}, \mathbf{x})$:

(12.5.9) $$\mathbf{X}^n_{(1)}(\mathbf{x}) := \mathbf{x} - \mathbf{b}(t_{n+1}, \mathbf{x})\Delta t \ .$$

Notice that $\mathbf{X}^n_{(1)}$ is a second order approximation of $\mathbf{X}(t_n; t_{n+1}, \mathbf{x})$ since we are integrating (12.5.8) on the time interval (t_n, t_{n+1}), which has length Δt.

A more accurate scheme is provided by the second order Runge-Kutta scheme

(12.5.10) $$\mathbf{X}^n_{(2)}(\mathbf{x}) := \mathbf{x} - \mathbf{b}\left(t_{n+1/2}, \mathbf{x} - \mathbf{b}(t_{n+1}, \mathbf{x})\frac{\Delta t}{2}\right)\Delta t \ ,$$

which gives a third order approximation of $\mathbf{X}(t_n; t_{n+1}, \mathbf{x})$.

It is necessary to verify that $\mathbf{X}^n_{(i)}(\mathbf{x}) \in \Omega$ for each $\mathbf{x} \in \overline{\Omega}$, $i = 1, 2$, so that we can compute $\mathbf{u}^n \circ \mathbf{X}^n_{(i)}$. For simplicity, let us assume that $\mathbf{b}(t, \mathbf{x}) = \mathbf{0}$ for each $t \in [0, T]$ and $\mathbf{x} \in \partial\Omega$. As a consequence, $\mathbf{X}^n_{(i)}(\mathbf{x}) = \mathbf{x}$ for $\mathbf{x} \in \partial\Omega$, $i = 1, 2$. If we denote by $\mathbf{x}^* \in \partial\Omega$ the point having minimal distance from $\mathbf{x} \in \Omega$, we have

$$|\mathbf{X}^n_{(1)}(\mathbf{x}) - \mathbf{x}| = |\mathbf{b}(t_{n+1}, \mathbf{x})| \, \Delta t = |\mathbf{b}(t_{n+1}, \mathbf{x}) - \mathbf{b}(t_{n+1}, \mathbf{x}^*)| \, \Delta t$$
$$\leq |\mathbf{b}(t_{n+1})|_{Lip(\overline{\Omega})} \, |\mathbf{x} - \mathbf{x}^*| \, \Delta t \ ,$$

where

$$|\mathbf{g}|_{Lip(\overline{\Omega})} := \sup_{\substack{\mathbf{x}_1, \mathbf{x}_2 \in \overline{\Omega} \\ \mathbf{x}_1 \neq \mathbf{x}_2}} \frac{|\mathbf{g}(\mathbf{x}_1) - \mathbf{g}(\mathbf{x}_2)|}{|\mathbf{x}_1 - \mathbf{x}_2|} \ .$$

Assuming that

$$(12.5.11) \qquad \max_{t \in [0,T]} |\mathbf{b}(t)|_{Lip(\overline{\Omega})} \, \Delta t < 1 \ ,$$

it follows at once that $\mathbf{X}^n_{(1)}(\mathbf{x}) \in \Omega$ for each $\mathbf{x} \in \Omega$ and each $n = 0, 1, ..., \mathcal{N}-1$. A similar results holds for $\mathbf{X}^n_{(2)}(\mathbf{x})$.

From now on we will consider the second order approximation (12.5.9), referring to Pironneau (1988), pp. 86–90, for higher order schemes based on (12.5.10). If we suppose that

$$(12.5.12) \qquad \operatorname{div} \mathbf{b}(t, \mathbf{x}) + a_0(\mathbf{x}) \geq 0$$

for each $t \in [0, T]$ and almost every $\mathbf{x} \in \Omega$, stability is easily proven. In fact, multiplying (12.5.7) by u^{n+1} and integrating over Ω one obtains

$$(12.5.13) \qquad \begin{aligned} ||u^{n+1}||_0^2 &+ \varepsilon \Delta t ||\nabla u^{n+1}||_0^2 \\ &\leq (||u^n \circ \mathbf{X}^n_{(1)}||_0 + \Delta t ||f(t_{n+1})||_0) ||u^{n+1}||_0 \ . \end{aligned}$$

From (12.5.11) it also follows that the map $\mathbf{X}^n_{(1)}$ is injective. Therefore, we can introduce the change of variable $\mathbf{y} = \mathbf{X}^n_{(1)}(\mathbf{x})$, and setting $\mathbf{Y}^n_{(1)}(\mathbf{y}) := (\mathbf{X}^n_{(1)})^{-1}(\mathbf{y})$ we have

$$(12.5.14) \quad ||u^n \circ \mathbf{X}^n_{(1)}||_0^2 = \int_{\mathbf{X}^n_{(1)}(\Omega)} u^n(\mathbf{y})^2 |\det(\operatorname{Jac} \mathbf{X}^n_{(1)}) \circ \mathbf{Y}^n_{(1)}(\mathbf{y})|^{-1} \, d\mathbf{y} \ .$$

On the other hand,

$$|\det(\operatorname{Jac} \mathbf{X}^n_{(1)})(\mathbf{x})| \geq 1 - \Delta t \, C_1 \, ||\operatorname{Jac} \mathbf{b}(t_{n+1})||_{L^\infty(\Omega)} > 0$$

for almost every $\mathbf{x} \in \Omega$, provided that

$$(12.5.15) \qquad \mu_1^* \, \Delta t \leq C_2 \ ,$$

where

$$\mu_1^* := \max_{t\in[0,T]} ||\text{Jac } \mathbf{b}(t)||_{L^\infty(\Omega)}$$

and $C_1 > 0$, $0 < C_2 < C_1^{-1}$ are suitable constants. Therefore, possibly choosing a smaller constant C_2 in (12.5.15), from (12.5.14) one has

(12.5.16) $$||u^n \circ \mathbf{X}_{(1)}^n||_0^2 \le (1 + \Delta t\, C_3\, \mu_1^*)\, ||u^n||_0^2\ .$$

Notice that condition (12.5.15) implies (12.5.11) (if C_2 is small enough). From (12.5.13) we finally obtain for each $n = 0, 1, ..., \mathcal{N} - 1$

(12.5.17)
$$\begin{aligned}
(||u^{n+1}||_0^2 &+ \varepsilon\Delta t||\nabla u^{n+1}||_0^2)^{1/2} \\
&\le (1 + C_3\mu_1^*\Delta t)^{1/2}\, ||u^n||_0 + \Delta t||f(t_{n+1})||_0 \\
&\le (1 + C_3\mu_1^*\Delta t)^{\frac{n+1}{2}}\, ||u_0||_0 \\
&\qquad + \Delta t \sum_{k=1}^{n+1}(1 + C_3\mu_1^*\Delta t)^{\frac{n+1-k}{2}}\, ||f(t_k)||_0 \\
&\le \left(||u_0||_0 + t_{n+1} \max_{t\in[0,T]} ||f(t)||_0\right) \exp\left(\frac{C_3}{2}\mu_1^* t_{n+1}\right)\ .
\end{aligned}$$

Notice that the $L^2(\Omega)$-stability holds independently of ε.

The convergence of u^n to $u(t_n)$ is proven in a similar way. Defining the error function $\epsilon^n := u(t_n) - u^n$, from (12.5.7) we obtain for $n = 0, 1, ..., \mathcal{N} - 1$

(12.5.18)
$$\begin{aligned}
\frac{\epsilon^{n+1} - \epsilon^n \circ \mathbf{X}_{(1)}^n}{\Delta t} &- \varepsilon\Delta\epsilon^{n+1} + [\text{div } \mathbf{b}(t_{n+1}) + a_0]\epsilon^{n+1} \\
&= \frac{u(t_{n+1}) - u(t_n) \circ \mathbf{X}_{(1)}^n}{\Delta t} \\
&\qquad - \frac{\partial u}{\partial t}(t_{n+1}) - \mathbf{b}(t_{n+1}) \cdot \nabla u(t_{n+1}) \quad \text{in } \Omega\ .
\end{aligned}$$

Since $\mathbf{X}_{(1)}^n(\mathbf{x}) - \mathbf{X}(t_n; t_{n+1}, \mathbf{x}) = O((\Delta t)^2)$, it is easily shown that the right hand side in (12.5.18) is $O(\Delta t)$. Therefore convergence follows from (12.5.17) applied to ϵ^n, recalling that $\epsilon^0 = 0$.

The method we have described above also applies to the fully-discretized problem obtained from (12.5.7) by using, for example, the finite element method. The resulting scheme would read: given $u_h^0 := u_{0,h} \in V_h$, for each $n = 0, 1, ..., \mathcal{N} - 1$ find $u_h^{n+1} \in V_h$ such that

(12.5.19)
$$\begin{aligned}
\frac{1}{\Delta t}(u_h^{n+1} &- u_h^n \circ \mathbf{X}^n, v_h) \\
&+ \varepsilon(\nabla u_h^{n+1}, \nabla v_h) + ([\text{div } \mathbf{b}(t_{n+1}) + a_0]u_h^{n+1}, v_h) \\
&= (f(t_{n+1}), v_h) \quad \forall v_h \in V_h\ ,
\end{aligned}$$

where $(\cdot, \cdot)$ denotes the $L^2(\Omega)$ scalar product and V_h is a suitable finite element subspace of $V = H_0^1(\Omega)$ (the usual changes occur for boundary conditions different from the homogeneous Dirichlet one). Assuming that (12.5.12) and (12.5.15) hold and that $\mathbf{b}(t)_{|\partial\Omega} = \mathbf{0}$ for each $t \in [0,T]$, stability can be proven exactly as before. Further, if V_h is given by $X_h^1 \cap H_0^1(\Omega)$, the space of continuous and piecewise-linear polynomials vanishing on $\partial\Omega$, it can be shown that $||u(t_n) - u_h^n||_0 = 0(h + \Delta t + \frac{h^2}{\Delta t})$ (see Pironneau (1988), pp. 86–89).

We conclude this Section by noting that some additional problems have to be faced while implementing this method. In fact, one has to compute the integrals $(u_h^n \circ \mathbf{X}^n, v_h)$, and this is usually accomplished by means of a quadrature formula. It turns out that the effects of this quadrature procedure can give rise to instability phenomena (see Morton, Priestley and Süli (1988)). Moreover, numerical integration requires the knowledge of the value of $u_h^n \circ \mathbf{X}^n$ at some nodal points. This means that, for any fixed node $\mathbf{x}_k$, it is necessary to know which triangle $K \in T_h$ contains the point $\mathbf{X}^n(\mathbf{x}_k)$. Several remarks on the implementation of the method can be found in Pironneau (1988), pp. 91–93 (see also Priestley (1993)).

The implementation of this idea in the framework of spectral methods is described, e.g., in Ho, Maday, Patera and Rønquist (1990) and Süli and Ware (1992) for the incompressible Navier-Stokes equations and in Süli and Ware (1991) for advection-dominated elliptic problems and hyperbolic problems.

13. The Unsteady Navier-Stokes Problem

In this Chapter we turn our attention towards unsteady viscous flows, especially in the incompressible case. We therefore consider the time-dependent counterpart of the Navier-Stokes problem (10.1.1)-(10.1.3), that reads

$$(13.1) \quad \begin{cases} \dfrac{\partial \mathbf{u}}{\partial t} - \nu \Delta \mathbf{u} + (\mathbf{u} \cdot \nabla)\mathbf{u} + \nabla p = \mathbf{f} & \text{in } Q_T := (0, T) \times \Omega \\[2mm] \operatorname{div} \mathbf{u} = 0 & \text{in } Q_T \\[2mm] \mathbf{u} = 0 & \text{on } \Sigma_T := (0, T) \times \partial\Omega \\[2mm] \mathbf{u}_{|t=0} = \mathbf{u}_0 & \text{on } \Omega \ , \end{cases}$$

where $\mathbf{f} = \mathbf{f}(t, \mathbf{x})$ and $\mathbf{u}_0 = \mathbf{u}_0(\mathbf{x})$ are given data, Ω is an open bounded domain of $\mathbb{R}^d$, with $d = 2, 3$, and $\partial\Omega$ is its boundary. One remarkable feature of (13.1) is the absence of an equation containing $\frac{\partial p}{\partial t}$. Indeed, in (13.1) the pressure p appears as a Lagrange multiplier associated to the divergence-free constraint $\operatorname{div} \mathbf{u} = 0$.

To begin with, we derive in Section 13.1 the equations describing the motion of a general fluid, either incompressible or compressible. In Section 13.2 we introduce the weak formulation of problem (13.1), and comment on the existence, uniqueness and behaviour of solutions. Next we turn to the numerical approximation. Most effective numerical methods can be defined through a separation between temporal and spatial discretization. For this reason, most arguments we have developed in Chapter 10 for the stationary problem are relevant to the time-dependent case. Concerning the temporal discretization, we review some of the most popular finite differencing schemes and operator-splitting algorithms. We restrict our attention to the case of laminar flow problems.

For specific monographs on the subject we refer to Ladyzenskaya (1969), Lions (1969), Temam (1983, 1984) and Kreiss and Lorenz (1989) for the mathematical analysis of Navier-Stokes equations, to Temam (1984), Thomasset (1981), Peyret and Taylor (1983), Fletcher (1988a, 1988b), Canuto, Hussaini, Quarteroni and Zang (1988), Gunzburger (1989), Pironneau (1988) and Quartapelle (1993) for the analysis of numerical approximation methods.

13.1 The Navier-Stokes Equations for Compressible and Incompressible Flows

The equations describing the motion of a general fluid are derived from three conservation principles: momentum, mass and energy. Denoting by $\mathbf{u}$ the velocity of the fluid, $\rho > 0$ its density and e its internal energy per unit mass, these equations read (see, e.g., Landau and Lifshitz (1959), Sects. 15, 1, 49):

$$(13.1.1) \qquad \frac{\partial(\rho \mathbf{u})}{\partial t} + \mathbf{div}\,(\rho\,\mathbf{u} \otimes \mathbf{u}) = \mathbf{div}\,\mathbf{T} + \rho \mathbf{f}$$

$$(13.1.2) \qquad \frac{\partial \rho}{\partial t} + \mathrm{div}(\rho \mathbf{u}) = 0$$

$$(13.1.3) \qquad \frac{\partial\left(\rho e + \frac{1}{2}\rho|\mathbf{u}|^2\right)}{\partial t} + \mathrm{div}\left[\left(\left(\rho e + \frac{1}{2}\rho|\mathbf{u}|^2\right)\mathbf{u}\right)\right]$$
$$= \mathrm{div}(\mathbf{T}\cdot\mathbf{u}) - \mathrm{div}\,\mathbf{q} + \rho\mathbf{f}\cdot\mathbf{u} + \rho r \ .$$

Here we have denoted by $\mathbf{u} \otimes \mathbf{u}$ the tensor $\{u_i u_j\}$, $i, j = 1, ..., d$, $\mathbf{T} = \mathbf{T}(\mathbf{u}, \hat{p})$ the stress tensor, $\hat{p}$ the pressure, $\mathbf{q} = \mathbf{q}(\vartheta)$ the heat flux, $\vartheta > 0$ the (absolute) temperature, and finally by $\mathbf{f}$ and r the external force field per unit mass and the heat supply per unit mass per unit time, respectively. Moreover, we have used the notation

$$(\mathbf{div}\,\mathbf{T})_i := \sum_{j=1}^{d} D_j T_{ij} \quad , \quad (\mathbf{T}\cdot\mathbf{u})_i := \sum_{j=1}^{d} T_{ij} u_j \quad , \quad i = 1, ..., d \quad .$$

By using (13.1.2) we can rewrite (13.1.1) and (13.1.3) in the non-conservative form:

$$(13.1.4) \qquad \rho\left[\frac{\partial \mathbf{u}}{\partial t} + (\mathbf{u}\cdot\nabla)\mathbf{u}\right] = \mathbf{div}\,\mathbf{T} + \rho\mathbf{f}$$

$$(13.1.5) \qquad \rho\left[\frac{\partial e}{\partial t} + \mathbf{u}\cdot\nabla e\right] = \mathbf{T} : \mathbf{D} - \mathrm{div}\,\mathbf{q} + \rho r \quad ,$$

where $\mathbf{D} = \mathbf{D}(\mathbf{u})$ is the deformation tensor

$$D_{ij} := \frac{1}{2}(D_j u_i + D_i u_j)$$

and we have used the notation

$$\mathbf{T} : \mathbf{D} := \sum_{i,j=1}^{d} T_{ij} D_{ij} \ .$$

Having written these conservation laws, we are now in a position to introduce the constitutive equations which characterize the motion of any particular fluid. We will limit our presentation to those physical situations in which the following constitutive equations hold:

(13.1.6) $$T_{ij} = \left[-\hat{p} + \left(\zeta - \frac{2\mu}{d}\right) \text{div } \mathbf{u}\right]\delta_{ij} + 2\mu D_{ij}$$

(13.1.7) $$\mathbf{q} = -\chi\nabla\vartheta \ ,$$

μ and ζ being the shear and bulk viscosity coefficients, respectively, and χ the heat conductivity coefficient. The symbol δ_{ij} denotes the Kronecker tensor, i.e., $\delta_{ii} = 1$ and $\delta_{ij} = 0$ for $i \neq j$, $i,j = 1,...,d$.

To close the system some other relations have to be imposed, furnished by thermodynamic principles. At this stage we have to distinguish between compressible and incompressible flows, as the thermodynamic assumptions characterizing these types of flows are different.

13.1.1 Compressible Flows

Choosing as thermodynamic unknowns the density ρ and the temperature ϑ, in order to close the system we add the following state equations

(13.1.8) $$\hat{p} = P(\rho, \vartheta)$$

(13.1.9) $$e = E(\rho, \vartheta)$$

(13.1.10) $$\mu = \mu^*(\rho, \vartheta) \ , \quad \zeta = \zeta^*(\rho, \vartheta) \ , \quad \chi = \chi^*(\rho, \vartheta) \ ,$$

where P, E, μ^*, ζ^* and χ^* are known functions subjected to the thermodynamic restrictions (Clausius-Duhem inequalities)

(13.1.11) $$\mu^* \geq 0 \ , \quad \zeta^* \geq 0 \ , \quad \chi^* \geq 0 \ .$$

For physical reasons, it holds $E_\vartheta > 0$ (this latter condition is a general attribute of all real materials), and $P_\rho > 0$ (this condition is generally true for one-phase fluids). Moreover, from the well-known relation (see, e.g., Landau and Lifshitz (1959), Sect. 6)

(13.1.12) $$dE = \vartheta dS - Pd(\rho^{-1}) \ , \quad S \text{ specific entropy} \ ,$$

E and P must satisfy the compatibility condition

(13.1.13) $$E_\rho = \rho^{-2}(P - \vartheta P_\vartheta) \ .$$

We can finally rewrite equations (13.1.4) and (13.1.5) as:

(13.1.14)
$$\rho\left[\frac{\partial \mathbf{u}}{\partial t} + (\mathbf{u} \cdot \nabla)\mathbf{u}\right] = -\nabla P + \sum_{j=1}^{d} D_j(\mu^* D_j\mathbf{u} + \mu^*\nabla u_j)$$
$$+ \nabla\left[\left(\zeta^* - \frac{2\mu^*}{d}\right)\text{div } \mathbf{u}\right] + \rho\mathbf{f}$$

$$\rho E_\vartheta \left(\frac{\partial \vartheta}{\partial t} + \mathbf{u} \cdot \nabla \vartheta \right) = -\vartheta P_\vartheta \operatorname{div} \mathbf{u} + \frac{\mu^*}{2} \sum_{i,j=1}^{d} (D_i u_j + D_j u_i)^2$$

(13.1.15)

$$+ \left(\zeta^* - \frac{2\mu^*}{d} \right) (\operatorname{div} \mathbf{u})^2 + \operatorname{div}(\chi^* \nabla \vartheta) + \rho r \quad.$$

When the viscosity μ^* is greater than 0, then equations (13.1.14), (13.1.2) and (13.1.15) are called *Navier-Stokes* equations for compressible flows. The case $\mu^* = 0$, $\zeta^* = 0$ gives the *Euler* equations for compressible flows.

13.1.2 Incompressible Flows

In this case the pressure $\hat{p}$ is no longer related to thermodynamic unknowns, and must be determined through the momentum equation. The state equation (13.1.8) is replaced by the requirement that any given amount of fluid does not change its volume along the motion. This physical assumption reads

(13.1.16) $\operatorname{div} \mathbf{u} = 0 \quad.$

Moreover, now the thermodynamic relations (13.1.9), (13.1.10) reduce to

(13.1.17) $e = E(\vartheta) \quad, \quad \mu = \mu^*(\vartheta) \quad, \quad \chi = \chi^*(\vartheta) \quad.$

Notice that, owing to (13.1.6), the bulk viscosity ζ doesn't appear in the stress tensor any longer.

Finally, the specific entropy S is related to E and ϑ through

(13.1.18) $dE = \vartheta dS \,,$

which substitutes (13.1.12).

Therefore we rewrite equations (13.1.4), (13.1.3) and (13.1.5) as:

(13.1.19) $\rho \left[\frac{\partial \mathbf{u}}{\partial t} + (\mathbf{u} \cdot \nabla) \mathbf{u} \right] = -\nabla \hat{p} + \sum_{j=1}^{d} D_j(\mu^* D_j \mathbf{u} + \mu^* \nabla u_j) + \rho \mathbf{f}$

(13.1.20) $\frac{\partial \rho}{\partial t} + \mathbf{u} \cdot \nabla \rho = 0$

$$\rho E_\vartheta \left(\frac{\partial \vartheta}{\partial t} + \mathbf{u} \cdot \nabla \vartheta \right) = \frac{\mu^*}{2} \sum_{i,j=1}^{d} (D_i u_j + D_j u_i)^2$$

(13.1.21)

$$+ \operatorname{div}(\chi^* \nabla \vartheta) + \rho r \quad.$$

The *Navier-Stokes* system for incompressible flows is given by equations (13.1.19)-(13.1.21), (13.1.16) when $\mu^* > 0$, whereas the *Euler* system for incompressible flows is obtained as a particular case setting $\mu^* = 0$. The latter case will not be addressed in this book. The interested reader can refer to Marchioro and Pulvirenti (1993).

A flow for which $\rho = \rho(t)$ (i.e., the density doesn't depend on the space variable $\mathbf{x}$) is called *homogeneous*. It is easily verified that a flow is incompressible and homogeneous if and only if its density ρ is constant. Further, if an incompressible flow is homogeneous at the initial time, i.e., the initial density ρ_0 is constant, say $\rho_0(\mathbf{x}) = \rho^* > 0$ for each $\mathbf{x} \in \overline{\Omega}$, owing to (13.1.20) it follows that $\rho(t, \mathbf{x}) = \rho^*$ for each $(t, \mathbf{x}) \in \overline{Q_T}$. Therefore equation (13.1.20) can be eliminated from the system governing the motion of a homogeneous incompressible flow. Moreover, if μ^* is constant the velocity and the pressure are independent of ϑ, and equation (13.1.21) can be solved separately after (13.1.19), (13.1.16). Under both these assumptions the Navier-Stokes system takes the well-known form (13.1), with $\nu := \mu^*/\rho^*$ and $p := \hat{p}/\rho^*$.

13.2 Mathematical Formulation and Behaviour of Solutions

As in Chapter 10 we introduce the Hilbert spaces: H_{div}, V_{div}, $V = (H_0^1(\Omega))^d$ and $Q = L_0^2(\Omega)$ (see (10.1.8), (9.1.3) and (9.1.9)). By generalizing (10.1.7) we easily see that the weak formulation of (13.1) reads: given $\mathbf{f} \in L^2(0, T; H_{\mathrm{div}})$ and $\mathbf{u}_0 \in H_{\mathrm{div}}$, find $\mathbf{u} \in L^2(0, T; V_{\mathrm{div}}) \cap L^\infty(0, T; H_{\mathrm{div}})$ such that

(13.2.1)
$$\begin{cases} \dfrac{d}{dt}(\mathbf{u}(t), \mathbf{v}) + a(\mathbf{u}(t), \mathbf{v}) + c(\mathbf{u}(t); \mathbf{u}(t), \mathbf{v}) \\ \qquad\qquad\qquad = (\mathbf{f}(t), \mathbf{v}) \qquad \forall\, \mathbf{v} \in V_{\mathrm{div}} \\ \mathbf{u}(0) = \mathbf{u}_0 \ . \end{cases}$$

The above equation has to be intended in the sense of distributions in $(0, T)$. The symbol $(\cdot, \cdot)$ denotes the scalar product in $(L^2(\Omega))^d$; for the definition of the bilinear forms $a(\cdot, \cdot)$ and $b(\cdot, \cdot)$ and the trilinear form $c(\cdot; \cdot, \cdot)$ we refer to (9.1.5), (9.1.10) and (10.1.5), respectively.

The existence of a solution to (13.2.1) has been proven by Leray (1934) and Hopf (1951). Uniqueness is still an open problem in the three-dimensional case, whereas for $d = 2$ the solution $\mathbf{u}$ has been shown to belong to $C^0([0, T]; H_{\mathrm{div}})$ and to be unique (Ladyzhenskaya (1958, 1959), Lions and Prodi (1959)).

Any solution to (13.2.1) also satisfies the following energy estimate:

(13.2.2)
$$\sup_{t \in (0,T)} ||u(t)||_0^2 + \nu \int_0^T ||\nabla u(t)||_0^2 \leq ||u_0||_0^2 + \frac{C_\Omega}{\nu} \int_0^T ||f(t)||_0^2 \ ,$$

where C_Ω is the constant of the Poincaré inequality (1.3.2).

Under additional assumptions it is also possible to prove the existence of more regular solutions. Precisely, in the two-dimensional case, assuming that the boundary $\partial\Omega$ is a regular manifold and $\mathbf{u}_0 \in V_{\text{div}}$, the solution $\mathbf{u}$ to (13.2.1) belongs to $L^2(0,T;(H^2(\Omega))^2) \cap C^0([0,T];V_{\text{div}})$. In the three-dimensional case an analogous result holds, although only locally in time. More precisely, there exists $T^* \in (0,T]$ (but not necessarily $T^* = T$) such that the solution $\mathbf{u}$ to (13.2.1) belongs to $L^2(0,T^*;(H^2(\Omega))^3) \cap C^0([0,T^*];V_{\text{div}})$, and is unique in that class (see Prodi (1962)). It must be noticed, however, that the existence of a global-in-time unique solution is assured also in the three-dimensional case, provided the data $\mathbf{f}$ and $\mathbf{u}_0$ are small compared to the viscosity ν.

The proofs of these results can be found, e.g., in Temam (1984), Chap. III. Other results concerning the existence of more regular solutions in $\overline{Q_T}$ and the necessity of non-local compatibility conditions for the data on $\partial\Omega$ at $t = 0$ can be found in Heywood (1980) and Temam (1982).

An alternative weak formulation of (13.1), which is related to (10.1.4) instead of (10.1.7), is as follows: find $\mathbf{u}(t) \in V$ and $p(t) \in Q$ such that for almost every $t \in (0,T)$

(13.2.3)
$$\begin{cases} \dfrac{d}{dt}(\mathbf{u}(t),\mathbf{v}) + a(\mathbf{u}(t),\mathbf{v}) + c(\mathbf{u}(t);\mathbf{u}(t),\mathbf{v}) \\ \qquad\qquad\qquad + b(\mathbf{v},p(t)) = (\mathbf{f}(t),\mathbf{v}) \quad \forall\, \mathbf{v} \in V \\[2mm] b(\mathbf{u}(t),q) = 0 \qquad\qquad\qquad\qquad\qquad\quad \forall\, q \in Q \\[2mm] \mathbf{u}(0) = \mathbf{u}_0 \ . \end{cases}$$

Clearly, any solution of this problem is also a solution to (13.2.1). The converse is true, provided the solution to (13.2.1) is regular enough to apply Lemma 9.1.1. For instance, when $\mathbf{u} \in L^2(0,T;(H^2(\Omega))^2) \cap C^0([0,T];V_{\text{div}})$ it is possible to determine a pressure p which belongs to $L^2(0,T;H^1(\Omega))$.

13.3 Semi-Discrete Approximation

The space discretization of (13.2.1) can be accomplished exactly as described in Chapter 10 for the stationary case. At first, we choose a finite dimensional subspace of the divergence free subspace V_{div} (see (9.1.3)). Denoting it by $V_{\text{div},h}$, we have the reduced, one-field problem: for each $t \in [0,T]$ seek $\mathbf{u}_h(t,\cdot) \in V_{\text{div},h}$ such that

(13.3.1)
$$\begin{cases} \dfrac{d}{dt}(\mathbf{u}_h(t),\mathbf{v}_h) + a(\mathbf{u}_h(t),\mathbf{v}_h) + c(\mathbf{u}_h(t);\mathbf{u}_h(t),\mathbf{v}_h) \\ \qquad\qquad = (\mathbf{f}(t),\mathbf{v}_h) \quad \forall\, \mathbf{v}_h \in V_{\text{div},h} \ , \quad t \in (0,T) \\[2mm] \mathbf{u}_h(0) = \mathbf{u}_{0,h} \ , \end{cases}$$

where $\mathbf{u}_{0,h} \in V_{\mathrm{div},h}$ is an approximation to the initial data $\mathbf{u}_0$. This is therefore a Galerkin approximation to (13.2.1).

In turn, when approximating problem (13.2.3) we consider two subspaces $V_h \subset V$ and $Q_h \subset Q$ and for each $t \in [0,T]$ we seek $\mathbf{u}_h(t,\cdot) \in V_h$ and $p_h(t,\cdot) \in Q_h$ such that

$$(13.3.2) \quad \begin{cases} \dfrac{d}{dt}(\mathbf{u}_h(t),\mathbf{v}_h) + a(\mathbf{u}_h(t),\mathbf{v}_h) + c(\mathbf{u}_h(t);\mathbf{u}_h(t),\mathbf{v}_h) \\ \qquad\qquad + b(\mathbf{v}_h,p_h(t)) = (\mathbf{f}(t),\mathbf{v}_h) \qquad \forall\, \mathbf{v}_h \in V_h\ ,\ t \in (0,T) \\[2mm] b(\mathbf{u}_h(t),q_h) = 0 \qquad\qquad\qquad\quad \forall\, q_h \in Q_h\ ,\ t \in (0,T) \\[2mm] \mathbf{u}_h(0) = \mathbf{u}_{0,h}\ , \end{cases}$$

with $\mathbf{u}_{0,h} \in V_h$.

The analysis of the semi-discrete scheme (13.3.2) has been provided in a series of papers by Heywood and Rannacher (1982, 1986a, 1988, 1990) (see also Okamoto (1982) and Bernardi and Raugel (1985)). In these papers, the scheme (13.3.2) has been modified into another one in which the trilinear form $c(\mathbf{w};\mathbf{z},\mathbf{v})$ is replaced by

$$\tilde{c}(\mathbf{w};\mathbf{z},\mathbf{v}) := \frac{1}{2}[c(\mathbf{w};\mathbf{z},\mathbf{v}) - c(\mathbf{w};\mathbf{v},\mathbf{z})]\ .$$

This is due to stability purposes. As a matter of fact, one sees at once that the new trilinear form $\tilde{c}(\cdot;\cdot,\cdot)$ is skew-symmetric, i.e., $\tilde{c}(\mathbf{w};\mathbf{v},\mathbf{v}) = 0$ for each $\mathbf{v} \in V$. This enables us to use the energy method: for each fixed t, taking in (13.3.3) $\mathbf{v}_h = \mathbf{u}_h(t)$ provides the stability estimate

$$(13.3.3) \qquad ||\mathbf{u}_h(t)||_0^2 + \nu \int_0^t ||\nabla \mathbf{u}_h(\tau)||_0^2 \le ||\mathbf{u}_{0,h}||_0^2 + \frac{C_\Omega}{\nu} \int_0^t ||\mathbf{f}(\tau)||_0^2\ ,$$

which is the discrete counterpart of (13.2.2). At the continuous level, the introduction of the new trilinear form $\tilde{c}(\cdot;\cdot,\cdot)$ corresponds to the addition of the term $\frac{1}{2}(\mathrm{div}\,\mathbf{u})\mathbf{u}$ at the left hand side of $(13.1)_1$. However, this perturbation is consistent with the Navier-Stokes equations, as the exact velocity field is divergence free. The price we pay with the modified scheme derives from the necessity of approximating the extra convective term, too.

Let us come now to the convergence analysis for (13.3.2). Assume that V_h and Q_h are a couple of finite element spaces that satisfy the compatibility condition (9.2.9). Further, we make the assumptions that for all $\mathbf{v} \in V$ and $q \in Q$

$$\inf_{\mathbf{v}_h \in V_h} ||\mathbf{v} - \mathbf{v}_h||_1 + \inf_{q_h \in Q_h} ||q - q_h||_0 = O(h)\ .$$

Let us notice that all the choices of spaces V_h and Q_h described in Section 9.3 satisfy these assumptions.

In this context, the following error estimate can be proven:

(13.3.4) $||\mathbf{u}(t) - \mathbf{u}_h(t)||_0 \leq C_1(t)h^2$, $||p(t) - p_h(t)||_0 \leq C_2(t)h$.

If the data satisfy appropriate regularity assumptions, then over any time interval $(0, T)$ in which the Dirichlet integral $||\nabla \mathbf{u}||_0$ remains uniformly bounded one has

(13.3.5) $C_1(t) \leq K e^{Kt}$, $C_2(t) \leq K\tau(t)^{-1/2} e^{Kt}$

with $\tau(t) := \min(t, 1)$.

Due to the presence of the exponential, these estimates are virtually meaningless for large values of the time variable. However, if the data of the problem are small enough it can be proven that C_1 and C_2 are indeed uniformly bounded as $t \to \infty$. Finally, in the particular case $\mathbf{u}_0 = \mathbf{f}(0, \cdot) = \mathbf{0}$ the function C_2 remains bounded as $t \to 0$. On the other hand, notice that, if the exact solution is not supposed to be stable, the exponential growth of C_1 and C_2 has to be expected, since there can be exponential growth in the difference between initially neighboring exact solutions.

The error estimate (13.3.4) is obtained without assuming any special regularity of the solution $(\mathbf{u}, p)$ as $t \to 0$, except that $\mathbf{u}_0 \in V_{\text{div}}$. This is an important point since the regularity of the solution of an initial-boundary value problem up to $t = 0$ holds only if suitable compatibility conditions for the data on $\partial\Omega$ at $t = 0$ are satisfied. While for parabolic equations these compatibility conditions are of local type, in the case of Navier-Stokes problem they read as follows. If the $H^3(\Omega)$-norm of $\mathbf{u}(t)$ (or equivalently the $H^1(\Omega)$-norm of $\frac{\partial \mathbf{u}}{\partial t}(t)$) remains bounded up to $t = 0$, the external force field $\mathbf{f}$ at $t = 0$ and the initial velocity $\mathbf{u}_0$ must be such that

(13.3.6) $\nabla p_0 = \mathbf{f}(0, \cdot) + \nu \Delta \mathbf{u}_0$ on $\partial\Omega$,

where p_0 is the solution (defined up to an additive constant) to the Neumann problem

$$\begin{cases} \Delta p_0 = -\operatorname{div}[(\mathbf{u}_0 \cdot \nabla)\mathbf{u}_0] & \text{in } \Omega \\[2mm] \dfrac{\partial p_0}{\partial n} = \nu \Delta \mathbf{u}_0 \cdot \mathbf{n} & \text{on } \partial\Omega \end{cases}$$

(remind that we have assumed that $\mathbf{f}(0, \cdot) \in H_{\text{div}}$). Because of its non-local nature, condition (13.3.6) is virtually uncheckable for general data (excepting when $\mathbf{u}_0 = \mathbf{f}(0, \cdot) = \mathbf{0}$).

Still requiring the compatibility condition $\mathbf{u}_0 \in V_{\text{div}}$ (and nothing more), the error estimate (13.3.4) can be improved to

(13.3.7) $||\mathbf{u}(t) - \mathbf{u}_h(t)||_0 \leq C_1(t)h^k$, $||p(t) - p_h(t)||_0 \leq C_2(t)h^{k-1}$

for $k = 2, ..., 5$, provided the finite dimensional spaces V_h and Q_h have been chosen with the corresponding approximability property. Here the functions C_1 and C_2 behave like

(13.3.8) $C_1(t) \leq K\tau(t)^{-(k-2)/2} e^{Kt}$, $C_2(t) \leq K\tau(t)^{-(k-1)/2} e^{Kt}$.

A further improvement is obtained assuming that the exact solution is stable in a suitable sense. In this case, it is reasonable to expect a uniform bound for $C_1(t)$ and $C_2(t)$ as $t \to \infty$. In fact, this result can be proven under the assumption that the solution $\mathbf{u}$ is *exponentially* stable. A precise definition of this concept is provided in Heywood and Rannacher (1986a, 1986b). Roughly, it can be described by saying that there is a fixed length of time during which any sufficiently small perturbation, starting at any time $t_0 \geq 0$, will decay to half of its original size. It must be noticed that exponential stability is a weaker requirement than "universal energy" stability, i.e., that all perturbations decay monotonically and exponentially in the $L^2(\Omega)$-norm, regardless of their initial size. In particular, the Reynolds number of the flow (see (10.2.2)) is not required to be so small to guarantee this latter property. In the papers quoted above, more general definitions of stability are also considered, and they appear sufficiently general to apply to such phenomena as Taylor cells and von Kármán vortex shedding. In correspondence, several error estimates are provided.

Results of this type are also valid for the fully-discrete approximation of (13.2.3) obtained by time-advancing schemes based on a finite difference approximation of the time-derivative. The analysis of the forward Euler, backward Euler and Crank-Nicolson schemes is reported in Rannacher (1982), Heywood and Rannacher (1986a) and Heywood and Rannacher (1990), respectively. See also Section 13.4 below.

An approach slightly more general than (13.3.2) can be formulated as follows: for each $t \in [0, T]$ find $\mathbf{u}_h(t, \cdot) \in V_h$ and $p_h(t, \cdot) \in Q_h$ such that

$$(13.3.9) \quad \begin{cases} \dfrac{d}{dt}(\mathbf{u}_h(t), \mathbf{v}_h)_h + a_h(\mathbf{u}_h(t), \mathbf{v}_h) + c_h(\mathbf{u}_h(t); \mathbf{u}_h(t), \mathbf{v}_h) \\ \qquad + b_{1,h}(\mathbf{v}_h, p_h(t)) = (\mathbf{f}(t), \mathbf{v}_h)_h \quad \forall\, \mathbf{v}_h \in V_h \ , \quad t \in (0, T) \\[2mm] b_{2,h}(\mathbf{u}_h(t), q_h) = 0 \qquad\qquad\qquad \forall\, q_h \in Q_h \ , \quad t \in (0, T) \\[2mm] \mathbf{u}_h(0) = \mathbf{u}_{0,h} \ , \end{cases}$$

where $(\cdot, \cdot)_h$ is an approximation of $(\cdot, \cdot)$, $a_h(\cdot, \cdot)$ of $a(\cdot, \cdot)$, $c_h(\cdot; \cdot, \cdot)$ of $c(\cdot; \cdot, \cdot)$, and both $b_{1,h}(\cdot, \cdot)$ and $b_{2,h}(\cdot, \cdot)$ are approximations of $b(\cdot, \cdot)$. The *generalized Galerkin method* (13.3.9) is encountered, e.g., whenever numerical integration is used in the framework of finite elements or spectral methods. In the latter case, a special mention deserves the spectral collocation method that has been discussed in Sections 9.5.2 and 10.2.3 in the framework of the steady problem.

Remark 13.3.1 The theory previously developed is suitable for moderately low Reynolds numbers (see (10.2.2)). For high Reynolds numbers the convective term might induce numerical oscillations if not properly treated (see Sections 8.2 and 8.3). It should therefore be approximated by an upwind procedure (see, e.g., Girault and Raviart (1986), pp. 336–352) or by resorting to stabilization methods like those of Section 8.3 (see, e.g., Brooks and Hughes (1982), Johnson and Saranen (1986), Hansbo and Szepessy (1990), Franca and Frey (1992)).   □

The algebraic form of problem (13.3.2) is easily derived, proceeding as done in Section 9.2.1. Denoting by $\{\varphi_j \,|\, j = 1, ..., N_h\}$ and $\{\psi_l \,|\, l = 1, ..., K_h\}$ the bases of V_h and Q_h, respectively, and considering the expansions

$$\mathbf{u}_h(t, \mathbf{x}) = \sum_{j=1}^{N_h} u_j(t)\varphi_j(\mathbf{x}) \quad , \quad p_h(t, \mathbf{x}) = \sum_{l=1}^{K_h} p_l(t)\psi_l(\mathbf{x}) \quad ,$$

we obtain the following system of nonlinear equations:

$$(13.3.10) \quad \begin{cases} M\dfrac{d\mathbf{u}}{dt}(t) + A\mathbf{u}(t) + C(\mathbf{u}(t))\mathbf{u}(t) + B^T\mathbf{p}(t) = \mathbf{f}(t) \quad , \ t \in (0, T) \\[2mm] B\mathbf{u}(t) = 0 \qquad\qquad\qquad\qquad\qquad\qquad\qquad , \ t \in (0, T) \\[2mm] \mathbf{u}(0) = \mathbf{u}_0 \ . \end{cases}$$

Here we have set:

$$\begin{aligned} & M_{ij} := (\varphi_i, \varphi_j) \ , \ A_{ij} := a(\varphi_j, \varphi_i) \\[2mm] (13.3.11) \qquad & (C(\mathbf{w}))_{ij} := \sum_{m=1}^{N_h} w_m c(\varphi_m; \varphi_j, \varphi_i) \\[2mm] & B_{li} := b(\varphi_i, \psi_l) \ , \ f_i(t) := (\mathbf{f}(t), \varphi_i) \ . \end{aligned}$$

The above is a system of differential algebraic equations (see Brenan, Campbell and Petzold (1989)). We recall that the mass matrix M is in diagonal form, e.g., in the case of the spectral collocation method, or when the finite element method is used with a lumping procedure (see Section 11.4).

13.4 Time-Advancing by Finite Differences

Problem (13.3.10) can be advanced in time by suitable finite difference schemes. For instance, if we use the single-step θ-*scheme* (introduced in Section 5.6.2) we obtain at each time level $t_{n+1} = (n+1)\Delta t$, $n = 0, 1, ..., \mathcal{N} - 1$, the system:

$$(13.4.1) \quad \begin{cases} M\dfrac{\mathbf{u}^{n+1} - \mathbf{u}^n}{\Delta t} + A\mathbf{u}_\theta^{n+1} + C(\mathbf{u}_\theta^{n+1})\mathbf{u}_\theta^{n+1} + B^T\mathbf{p}_\theta^{n+1} = \hat{\mathbf{f}}_\theta^{n+1} \\[2mm] B\mathbf{u}^{n+1} = 0 \ , \end{cases}$$

with $\mathbf{u}_\theta^{n+1} := \theta\mathbf{u}^{n+1} + (1 - \theta)\mathbf{u}^n$, $\mathbf{p}_\theta^{n+1} := \theta\mathbf{p}^{n+1} + (1 - \theta)\mathbf{p}^n$, and $\hat{\mathbf{f}}_\theta^{n+1} :=$ $\mathbf{f}(\theta t_{n+1} + (1 - \theta)t_n)$ for $0 \leq \theta \leq 1$, and M, A, $C(\mathbf{w})$ and B are defined in

(13.3.11). This scheme is second order accurate with respect to Δt if $\theta = 1/2$ while it is only first order accurate for all other values of θ. (Of course, accuracy is measured in H^1-norm for each velocity component and L^2-norm for the pressure.)

A convenient way of solving the problem is to resort to the θ-indexed variables. To this aim, we rewrite for $\theta \neq 0$ the continuity equation as

$$B\mathbf{u}_\theta^{n+1} = \begin{cases} (1-\theta)B\mathbf{u}^0 & \text{if } n = 0 \\ 0 & \text{if } n \geq 1 \end{cases} ,$$

and, similarly, we set

$$\frac{1}{\Delta t}M(\mathbf{u}^{n+1} - \mathbf{u}^n) = \frac{1}{\theta\Delta t}M(\mathbf{u}_\theta^{n+1} - \mathbf{u}^n) .$$

Then directly solve $\mathbf{u}_\theta^{n+1}$ and $\mathbf{p}_\theta^{n+1}$, and set

$$\mathbf{u}^{n+1} = \frac{1}{\theta}(\mathbf{u}_\theta^{n+1} - (1-\theta)\mathbf{u}^n) , \quad \mathbf{p}^{n+1} = \frac{1}{\theta}(\mathbf{p}_\theta^{n+1} - (1-\theta)\mathbf{p}^n) .$$

We have to notice that whenever $\theta \neq 1$ this scheme requires a value $\mathbf{p}^0$ for the pressure at the initial time $t_0 = 0$ which is not prescribed for the problem (13.1).

When $\theta = 1$ (*backward Euler* case) at each time-level we obtain a problem that reads

$$(13.4.2) \qquad \begin{cases} \left(A + \dfrac{M}{\Delta t}\right)\mathbf{u} + C(\mathbf{u})\mathbf{u} + B^T\mathbf{p} = \mathbf{G} \\ B\mathbf{u} = 0 , \end{cases}$$

where $\mathbf{G}$ is known. This is like the problem encountered in Chapter 10 for the discretization of the steady Navier-Stokes problem, and therefore can be faced by a Newton method or any one of the methods illustrated in Section 10.4. If Δt is small, a good starting guess is the solution provided from the previous time step.

A second order *backward differentiation* scheme is obtained by setting $\theta = 1$ in (13.4.1) but replacing the first-order backward difference $(\mathbf{u}^{n+1} - \mathbf{u}^n)/\Delta t$ by the two-step, second order one $(3\mathbf{u}^{n+1} - 4\mathbf{u}^n + \mathbf{u}^{n-1})/2\Delta t$ (see (5.6.17)). This scheme, however, requires larger storage compared with the single-step, second order Crank-Nicolson method. Moreover, it needs an additional second order initialization $\mathbf{u}^1$.

The fully implicit scheme requires the solution of a nonlinear system at every time step. One way to avoid it is to resort to *semi-implicit* methods. For example, linearizing (13.4.1) by the Newton method and performing only one iteration at each time step provides a semi-implicit method. A similar method is given by:

$$(13.4.3) \quad \begin{cases} \dfrac{1}{\theta \Delta t} M \mathbf{u}_\theta^{n+1} + A \mathbf{u}_\theta^{n+1} + C(\mathbf{u}^n) \mathbf{u}_\theta^{n+1} + B^T \mathbf{p}_\theta^{n+1} \\ \qquad\qquad\qquad\qquad\qquad = \hat{\mathbf{f}}_\theta^{n+1} + \dfrac{1}{\theta \Delta t} M \mathbf{u}^n \\ B \mathbf{u}_\theta^{n+1} = \begin{cases} (1 - \theta) B \mathbf{u}^0 & \text{if } n = 0 \\ 0 & \text{if } n \geq 1 \end{cases} \end{cases} .$$

Other semi-implicit approaches entail a fully explicit treatment of the nonlinear term. An instance is provided by the following second order scheme which approximates the linear terms by the Crank-Nicolson method and the nonlinear (convective) one by the explicit Adams-Bashforth method. It reads (see (5.6.16)):

$$(13.4.4) \quad \begin{cases} \dfrac{2}{\Delta t} M \mathbf{u}^{n+1} + A \mathbf{u}^{n+1} + B^T \mathbf{p}^{n+1} = \dfrac{2}{\Delta t} M \mathbf{u}^n + \mathbf{f}^{n+1} + \mathbf{f}^n \\ \qquad\qquad - A \mathbf{u}^n - B^T \mathbf{p}^n - 3 C(\mathbf{u}^n) \mathbf{u}^n + C(\mathbf{u}^{n-1}) \mathbf{u}^{n-1} \\ B \mathbf{u}^{n+1} = 0 \ , \end{cases}$$

for $n = 1, 2, ..., \mathcal{N} - 1$, having chosen a suitable second order initialization $\mathbf{u}^1$. This discretization is second order accurate with respect to Δt, provided that the data and the solution are smooth enough with respect to the time-variable.

The substantial difference between (13.4.3) and (13.4.4) is that in the latter case we obtain a linear system whose associated matrix doesn't change with n. Indeed we have

$$(13.4.5) \quad \begin{cases} \left(A + \dfrac{2}{\Delta t} M \right) \mathbf{u} + B^T \mathbf{p} = \mathbf{H} \\ B \mathbf{u} = 0 \ , \end{cases}$$

for $\mathbf{u} = \mathbf{u}^{n+1}$, $\mathbf{p} = \mathbf{p}^{n+1}$ and a suitable right hand side $\mathbf{H}$ that depends on known values at time t_n and t_{n-1}. This system is like the one (9.2.14) associated with the Stokes problem (9.1), and can therefore be solved by the algorithms illustrated in Section 9.6.

The semi-implicit scheme (13.4.4) is very often used when the spatial approximation is based on the spectral collocation method, as the latter can take advantage of explicit evaluation of the nonlinear terms. In this context, the method is stable under the condition $\Delta t = O(N^{-2})$, N being the polynomial degree of the velocity field.

Another semi-implicit approach that enjoies the same kind of property (a linear system that doesn't change at each step) can be formulated as follows (see Gunzburger (1989), p. 128)

$$(13.4.6) \quad \begin{cases} \dfrac{3}{2\Delta t} M\mathbf{u}^{n+1} + A\mathbf{u}^{n+1} + B^T\mathbf{p}^{n+1} \\ \qquad = \mathbf{f}^{n+1} - C(\mathbf{u}^*)\mathbf{u}^* + \dfrac{2}{\Delta t}M\mathbf{u}^n - \dfrac{1}{2\Delta t}M\mathbf{u}^{n-1} \\ B\mathbf{u}^{n+1} = 0 \ , \end{cases}$$

where $\mathbf{u}^* := 2\mathbf{u}^n - \mathbf{u}^{n-1}$ and $\mathbf{u}^{-1} := \mathbf{u}^0$. This scheme is based on a backward difference formula which is second order accurate in time. However, it is only conditionally stable; for instance, in the finite element case it has been proven in Baker, Dougalis and Karakashian (1982) that it must be submitted to the restriction $\Delta t = O(h^{4/5})$.

13.5 Operator-Splitting Methods

A very classical fractional-step method for solving problem (13.2.1) was introduced by Chorin (1967, 1968) and Temam (1969). We confine to the two-dimensional case $(d = 2)$ and for $\mathbf{w}, \mathbf{z}, \mathbf{v} \in V = (H_0^1(\Omega))^2$ we define

$$a_j(\mathbf{w}, \mathbf{v}) := \nu \int_\Omega D_j\mathbf{w} D_j\mathbf{v}$$

$$\tilde{c}_j(\mathbf{w}; \mathbf{z}, \mathbf{v}) := \frac{1}{2}\sum_{i=1}^2 \int_\Omega w_j[v_i D_j z_i - z_i D_j v_i] \ ,$$

where $j = 1, 2$. Moreover, we split the external force field as $\mathbf{f} = \sum_{j=1}^2 \mathbf{f}_j$.

The fractional-step method reads: for each $n = 0, 1, ..., \mathcal{N}-1$ find $\mathbf{u}_h^{n+1/3} \in V_h$ such that

$$(13.5.1) \quad \begin{aligned} \frac{1}{\Delta t}(\mathbf{u}_h^{n+1/3} - \mathbf{u}_h^n, \mathbf{v}_h) + a_1(\mathbf{u}_h^{n+1/3}, \mathbf{v}_h) \\ + \tilde{c}_1(\mathbf{u}_h^n; \mathbf{u}_h^{n+1/3}, \mathbf{v}_h) = (\mathbf{f}_{1,*}^{n+1/2}, \mathbf{v}_h) \quad \forall \mathbf{v}_h \in V_h \ , \end{aligned}$$

where $(\cdot, \cdot)$ is the scalar product in $(L^2(\Omega))^2$ and for each vector function $\mathbf{g} \in (L^2(Q_T))^2$ we have set

$$\mathbf{g}_*^{n+1/2} := \frac{1}{\Delta t}\int_{t_n}^{t_{n+1}} \mathbf{g}(t)dt \ .$$

Then find $\mathbf{u}_h^{n+2/3} \in V_h$ such that

$$(13.5.2) \quad \begin{aligned} \frac{1}{\Delta t}(\mathbf{u}_h^{n+2/3} - \mathbf{u}_h^{n+1/3}, \mathbf{v}_h) + a_2(\mathbf{u}_h^{n+2/3}, \mathbf{v}_h) \\ + \tilde{c}_2(\mathbf{u}_h^{n+1/3}; \mathbf{u}_h^{n+2/3}, \mathbf{v}_h) = (\mathbf{f}_{2,*}^{n+1/2}, \mathbf{v}_h) \quad \forall \mathbf{v}_h \in V_h \ . \end{aligned}$$

Finally, $\mathbf{u}_h^{n+1} \in W_h$ is the unique solution to the problem

(13.5.3) $(\mathbf{u}_h^{n+1}, \mathbf{w}_h) = (\mathbf{u}_h^{n+2/3}, \mathbf{w}_h)$ $\forall \, \mathbf{w}_h \in W_h$,

where W_h is a subspace of V_h which constitutes a suitable approximation of V_{div} (see (9.1.3)). Notice that in general W_h is not required to be a subspace of V_{div}.

Existence and uniqueness for both (13.5.1) and (13.5.2) follow from positiveness, as $\tilde{c}_j(\mathbf{z}; \mathbf{v}, \mathbf{v}) = 0$ for each $\mathbf{z}, \mathbf{v} \in V$, $j = 1, 2$. On the other hand, (13.5.3) states that $\mathbf{u}_h^{n+1}$ is the L^2-orthogonal projection of $\mathbf{u}_h^{n+2/3}$ onto W_h. In other words, (13.5.3) means that

$$\mathbf{u}_h^{n+1} = P_h \mathbf{u}_h^{n+2/3}$$

with $P_h : V_h \rightarrow W_h$ is the orthogonal projection operator with respect to the scalar product of $(L^2(\Omega))^2$. For such a reason the fractional-step scheme (13.5.1)-(13.5.3) is known as the *projection method*.

This scheme is unconditionally stable in the norm of $(L^2(\Omega))^2$. Moreover, if the time step-restriction $\Delta t = O(h^2)$ is assumed to hold, one has

$$\Delta t \sum_{n=0}^{\mathcal{N}-1} ||\mathbf{u}_h^{n+1}||_1^2 \leq C \ .$$

Under the latter restriction the scheme is also convergent (see Temam (1969, 1984)).

A slightly different approach, that makes use of two (rather than three) steps, consists in solving: find $\mathbf{u}_h^{n+1/2} \in V_h$ solution of the nonlinear elliptic convection-diffusion problem

(13.5.4)
$$\frac{1}{\Delta t}(\mathbf{u}_h^{n+1/2} - \mathbf{u}_h^n, \mathbf{v}_h) + a(\mathbf{u}_h^{n+1/2}, \mathbf{v}_h)$$
$$+ \tilde{c}(\mathbf{u}_h^{n+1/2}; \mathbf{u}_h^{n+1/2}, \mathbf{v}_h) = (\mathbf{f}_*^{n+1/2}, \mathbf{v}_h) \quad \forall \, \mathbf{v}_h \in V_h \ ,$$

where $\tilde{c} := \tilde{c}_1 + \tilde{c}_2$, and then find $\mathbf{u}_h^{n+1} \in W_h$ such that

(13.5.5) $(\mathbf{u}_h^{n+1}, \mathbf{w}_h) = (\mathbf{u}_h^{n+1/2}, \mathbf{w}_h)$ $\forall \, \mathbf{w}_h \in W_h$,

i.e., $\mathbf{u}_h^{n+1} = P_h \mathbf{u}_h^{n+1/2}$. The existence of a unique solution to (13.5.4) can be proven as done in Sections 10.1 and 10.2. (For a solution of this problem based on a least squares approach see Glowinski (1984), Chap. VII.)

In its continuous version this method amounts to defining two sequences of vector functions $\mathbf{u}^{n+1/2}$, $\mathbf{u}^{n+1}$ and a sequence of scalar functions q^{n+1} recursively given as: $\mathbf{u}^0 := \mathbf{u}_0$, and for $n = 0, 1, ..., \mathcal{N} - 1$

(13.5.6)
$$\begin{cases} \dfrac{1}{\Delta t}(\mathbf{u}^{n+1/2} - \mathbf{u}^n) - \nu \Delta \mathbf{u}^{n+1/2} + (\mathbf{u}^{n+1/2} \cdot \nabla)\mathbf{u}^{n+1/2} \\[2mm] \qquad\qquad + \dfrac{1}{2}(\mathrm{div}\, \mathbf{u}^{n+1/2})\mathbf{u}^{n+1/2} = \mathbf{f}_*^{n+1/2} \qquad \text{in } \Omega \\[4mm] \mathbf{u}^{n+1/2} = 0 \qquad\qquad\qquad\qquad\qquad\qquad\qquad \text{on } \partial\Omega \end{cases}$$

$$\begin{cases} \dfrac{1}{\Delta t}(\mathbf{u}^{n+1} - \mathbf{u}^{n+1/2}) + \nabla q^{n+1} = 0 & \text{in } \Omega \\[2mm] \operatorname{div} \mathbf{u}^{n+1} = 0 & \text{in } \Omega \\[2mm] \mathbf{u}^{n+1} \cdot \mathbf{n} = 0 & \text{on } \partial\Omega \end{cases}$$

(13.5.7)

where $\mathbf{n}$ is the unit outward normal vector on $\partial\Omega$. As we have already pointed out in Section 13.3, the presence of the term $\frac{1}{2}(\operatorname{div} \mathbf{u}^{n+1/2})\mathbf{u}^{n+1/2}$ in (13.5.6) is due to stability purposes.

Equations (13.5.7) stem from the fact that $\mathbf{u}^{n+1}$ is the L^2-orthogonal projection of $\mathbf{u}^{n+1/2}$ upon the space H_{div} (defined in (10.1.8)). The last two equations follow straightforwardly, while the existence of a scalar q^{n+1} is a consequence of the so-called Helmholtz decomposition principle. This states that any function $\mathbf{v} \in (L^2(\Omega))^2$ can be uniquely represented as $\mathbf{v} = \mathbf{w} + \nabla q$, where $\mathbf{w} \in H_{\operatorname{div}}$ and $q \in H^1(\Omega)$. Clearly, $\int_\Omega \mathbf{z} \cdot \nabla q = 0$ for each $\mathbf{z} \in H_{\operatorname{div}}$, and therefore $\mathbf{w} = P_{\operatorname{div}}\mathbf{v}$, where P_{div} is the orthogonal projection operator from $(L^2(\Omega))^2$ onto H_{div}. In turn, since $\mathbf{w} \in H_{\operatorname{div}}$, q turns out to be the solution to the Neumann problem:

$$\begin{cases} \Delta q = \operatorname{div} \mathbf{u} & \text{in } \Omega \\[2mm] \dfrac{\partial q}{\partial n} = \mathbf{u} \cdot \mathbf{n} & \text{on } \partial\Omega \end{cases}$$

which defines q up to an additive constant.

A question that naturally arises is: how to recover the pressure function if one is interested in approximating the pressure too? In Temam (1969) it is proven that the scalar function $q^{n+1}(\mathbf{x})$ does approximate the pressure $p(t_{n+1}, \mathbf{x})$, although in a very weak sense. As a matter of fact, from (13.5.6) and (13.5.7) one can easily infer that

(13.5.8) $$\frac{\partial q^{n+1}}{\partial n} = \nabla q^{n+1} \cdot \mathbf{n} = 0 \quad \text{on } \partial\Omega ,$$

since both $\mathbf{u}^{n+1} \cdot \mathbf{n}$ and $\mathbf{u}^{n+1/2} \cdot \mathbf{n}$ vanish on $\partial\Omega$. Besides, still from (13.5.7) we deduce

(13.5.9)
$$\begin{aligned} \Delta q^{n+1} &= \frac{1}{\Delta t} \operatorname{div} \mathbf{u}^{n+1/2} \\ &= \operatorname{div}(\mathbf{f}_*^{n+1/2} + \nu\Delta\mathbf{u}^{n+1/2}) - \sum_{i,j} D_j u_i^{n+1/2} D_i u_j^{n+1/2} \\ &\quad - \frac{1}{2}(\operatorname{div}\mathbf{u}^{n+1/2})^2 - \frac{3}{2}\mathbf{u}^{n+1/2} \cdot \nabla \operatorname{div} \mathbf{u}^{n+1/2} \quad \text{in } \Omega , \end{aligned}$$

the latter equality being obtained from (13.5.6). Therefore, the function q^{n+1} satisfies a Poisson problem with a *homogeneous* Neumann condition. On the other hand, it is known that the exact pressure p satisfies the *non-homogeneous* Neumann boundary value problem

$$(13.5.10) \qquad \begin{cases} \Delta p = \operatorname{div} \mathbf{f} - \sum_{i,j} D_j u_i \, D_i u_j & \text{in } \Omega \\[2mm] \dfrac{\partial p}{\partial n} = (\mathbf{f} + \nu \Delta \mathbf{u}) \cdot \mathbf{n} & \text{on } \partial\Omega \quad . \end{cases}$$

The discrepancy between the right-hand sides of equations (13.5.8)-(13.5.9) and (13.5.10) is responsible for the poor convergence of $q^{n+1}(\mathbf{x})$ to $p(t_{n+1}, \mathbf{x})$ which is experimented in practical computations (for additional comments on this topic, see Temam (1991a)). Nevertheless, the method represents correctly the velocity field in many flow problems of physical interest (see, e.g., Gresho and Chan (1990)). If one wishes to approximate both the pressure and the velocity satisfactorily in the whole domain Ω, the projection method needs to be modified. Examples are provided, e.g., in Kim and Moin (1985), Orszag, Israeli and Deville (1986), Gresho (1990), Gresho and Chan (1990). A different interpretation, which is based on performing a block-LU decomposition of the algebraic problem resulting from a fully-discrete approximation of (13.1), has been proposed by Perot (1993).

A thorough analysis of the projection method and of its relation with stabilization methods like (9.6.11) (with $\varepsilon = \Delta t$) is given in Rannacher (1992). For $n = 0, 1, ..., \mathcal{N} - 1$ the scheme is rewritten in the equivalent form

$$(13.5.11) \qquad \begin{cases} \dfrac{1}{\Delta t}(\mathbf{u}^{n+1/2} - \mathbf{u}^{n-1/2}) - \nu \Delta \mathbf{u}^{n+1/2} + (\mathbf{u}^{n+1/2} \cdot \nabla)\mathbf{u}^{n+1/2} \\[2mm] \qquad\qquad + \dfrac{1}{2}(\operatorname{div} \mathbf{u}^{n+1/2})\mathbf{u}^{n+1/2} + \nabla q^n = \mathbf{f}_*^{n+1/2} & \text{in } \Omega \\[2mm] \mathbf{u}^{n+1/2} = 0 & \text{on } \partial\Omega \end{cases}$$

$$(13.5.12) \qquad \begin{cases} \Delta q^{n+1} = \dfrac{1}{\Delta t}\operatorname{div} \mathbf{u}^{n+1/2} & \text{in } \Omega \\[2mm] \dfrac{\partial q^{n+1}}{\partial n} = 0 & \text{on } \partial\Omega \end{cases}$$

(here, the initializations $\mathbf{u}^{-1/2} := \mathbf{u}_0$ and $q^0 = 0$ are also needed). The velocity and the pressure are proven to be convergent at the first order with respect to Δt in the norms of $(L^2(\Omega))^2$ and of the dual space of $H^1(\Omega) \cap L_0^2(\Omega)$, respectively. Further, with respect to the norms of $(H^1(\Omega))^2$ and $L^2(\Omega)$ the convergence is at the order of $(\Delta t)^{1/2}$. It is worthy to notice that the same convergence results are also obtained when the nonlinear convection-diffusion equation $(13.5.11)_1$ is linearized, substituting the term $(\mathbf{u}^{n+1/2} \cdot \nabla)\mathbf{u}^{n+1/2}$ with $(\mathbf{u}^n \cdot \nabla)\mathbf{u}^{n+1/2}$, where $\mathbf{u}^n := P_{\mathrm{div}}\mathbf{u}^{n-1/2}$. Moreover, the stabilizing term $\frac{1}{2}(\operatorname{div} \mathbf{u}^{n+1/2})\mathbf{u}^{n+1/2}$ can be omitted without affecting the validity of the result.

The analysis of Rannacher also indicates that in the interior of the domain Ω the pressure q^n is indeed a reasonable approximation of the exact pressure $p(t_n, \cdot)$, as the effects of the non-physical Neumann boundary condition decay exponentially with respect to $\operatorname{dist}(\mathbf{x}, \partial\Omega)\sqrt{\nu\Delta t}$.

Convergence results similar to those described above have been obtained by Shen (1992) for another modification of the projection method.

A different operator-splitting method is the one presented in Temam (1983), p. 94. As usual, we denote by V_h a finite dimensional subspace of V and by W_h a subspace of V_h which constitutes a suitable approximation of V_{div}. We advance from t_n to t_{n+1} by first finding $\mathbf{u}_h^{n+1/2} \in W_h$ such that

(13.5.13)
$$\frac{1}{\Delta t}(\mathbf{u}_h^{n+1/2} - \mathbf{u}_h^n, \mathbf{w}_h) + \frac{1}{2}a(\mathbf{u}_h^{n+1/2}, \mathbf{w}_h)$$
$$= (\mathbf{f}_*^{n+1/2}, \mathbf{w}_h) \quad \forall\, \mathbf{w}_h \in W_h \ ,$$

and then seeking $\mathbf{u}_h^{n+1} \in V_h$ that satisfies

(13.5.14)
$$\frac{1}{\Delta t}(\mathbf{u}_h^{n+1} - \mathbf{u}_h^{n+1/2}, \mathbf{v}_h) + \frac{1}{2}a(\mathbf{u}_h^{n+1}, \mathbf{v}_h)$$
$$+ \tilde{c}(\mathbf{u}_h^{n+1}; \mathbf{u}_h^{n+1}, \mathbf{v}_h) = 0 \quad \forall\, \mathbf{v}_h \in V_h \ .$$

For each n problem (13.5.13) can be regarded as a non-conforming approximation of the linear problem (9.1.6). The existence and uniqueness of a solution $\mathbf{u}_h^{n+1/2}$ follow from the Lax-Milgram lemma. On the other hand, (13.5.14) is a nonlinear elliptic convection-diffusion problem, whose analysis can be carried out as done in Sections 10.1 and 10.2. This operator-splitting method allows the separate treatment of the difficulties related to nonlinearity and to the incompressibility constraint. Unconditional stability and convergence are proven in Temam (1983), Sect. 13. This approach, however, doesn't furnish an approximation of the pressure.

Another approach is based on the following Peaceman-Rachford type formulae (see (5.7.7); see also Glowinski (1984), Chap. VII) applied to (13.2.3). We first solve the Stokes problem: find $(\mathbf{u}_h^{n+1/2}, p_h^{n+1/2}) \in V_h \times Q_h$ such that

(13.5.15)
$$\begin{cases} \dfrac{2}{\Delta t}(\mathbf{u}_h^{n+1/2} - \mathbf{u}_h^n, \mathbf{v}_h) + \eta\, a(\mathbf{u}_h^{n+1/2}, \mathbf{v}_h) \\ \qquad\qquad + b(\mathbf{v}_h, p_h^{n+1/2}) \\ \qquad = (\mathbf{f}^{n+1/2}, \mathbf{v}_h) - (1 - \eta)\, a(\mathbf{u}_h^n, \mathbf{v}_h) \\ \qquad\quad - c(\mathbf{u}_h^n; \mathbf{u}_h^n, \mathbf{v}_h) \qquad\qquad \forall\, \mathbf{v}_h \in V_h \\ b(\mathbf{u}_h^{n+1/2}, q_h) = 0 \qquad\qquad\qquad \forall\, q_h \in Q_h \ , \end{cases}$$

where $0 < \eta < 1$, $\mathbf{u}_h^0 := \mathbf{u}_{0,h} \in V_h$ and $\mathbf{f}^{n+\theta} := \mathbf{f}((n + \theta)\Delta t)$, $0 \le \theta \le 1$. Then we consider the nonlinear elliptic convection-diffusion problem: find $\mathbf{u}_h^{n+1} \in V_h$ such that

(13.5.16)
$$\frac{2}{\Delta t}(\mathbf{u}_h^{n+1} - \mathbf{u}_h^{n+1/2}, \mathbf{v}_h) + (1 - \eta)\, a(\mathbf{u}_h^{n+1}, \mathbf{v}_h)$$
$$+ c(\mathbf{u}_h^{n+1}; \mathbf{u}_h^{n+1}, \mathbf{v}_h)$$
$$= (\mathbf{f}^{n+1/2}, \mathbf{v}_h) - \eta\, a(\mathbf{u}_h^{n+1/2}, \mathbf{v}_h)$$
$$- b(\mathbf{v}_h, p_h^{n+1/2}) \quad \forall\, \mathbf{v}_h \in V_h \ .$$

A three-step variant of (13.5.15), (13.5.16), which is based on a Strang type splitting (see (5.7.11)), is defined as follows. Set $\mathbf{u}_h^0 := \mathbf{u}_{0,h} \in V_h$ and take $0 < \vartheta < 1/2$; we obtain $(\mathbf{u}_h^{n+\vartheta}, p_h^{n+\vartheta}) \in V_h \times Q_h$ by solving the Stokes problem:

$$(13.5.17) \quad \begin{cases} \dfrac{1}{\vartheta \Delta t}(\mathbf{u}_h^{n+\vartheta} - \mathbf{u}_h^n, \mathbf{v}_h) + \eta\, a(\mathbf{u}_h^{n+\vartheta}, \mathbf{v}_h) \\ \qquad + b(\mathbf{v}_h, p_h^{n+\vartheta}) \\ \qquad = (\mathbf{f}^n, \mathbf{v}_h) - (1 - \eta)\, a(\mathbf{u}_h^n, \mathbf{v}_h) \\ \qquad \quad - c(\mathbf{u}_h^n; \mathbf{u}_h^n, \mathbf{v}_h) \qquad\qquad \forall\, \mathbf{v}_h \in V_h \\[4pt] b(\mathbf{u}_h^{n+\vartheta}, q_h) = 0 \qquad\qquad\qquad\qquad \forall\, q_h \in Q_h \quad . \end{cases}$$

Then, we look for $\mathbf{u}_h^{n+1-\vartheta} \in V_h$ solution of the nonlinear elliptic convection-diffusion problem:

$$(13.5.18) \quad \begin{aligned} \frac{1}{(1 - 2\vartheta)\Delta t}(\mathbf{u}_h^{n+1-\vartheta} &- \mathbf{u}_h^{n+\vartheta}, \mathbf{v}_h) + (1 - \eta)\, a(\mathbf{u}_h^{n+1-\vartheta}, \mathbf{v}_h) \\ &+ c(\mathbf{u}_h^{n+1-\vartheta}; \mathbf{u}_h^{n+1-\vartheta}, \mathbf{v}_h) \\ &= (\mathbf{f}^{n+\vartheta}, \mathbf{v}_h) - \eta\, a(\mathbf{u}_h^{n+\vartheta}, \mathbf{v}_h) \\ &\quad - b(\mathbf{v}_h, p_h^{n+\vartheta}) \quad \forall\, \mathbf{v}_h \in V_h \quad . \end{aligned}$$

Finally, we find $(u_h^{n+1}, p_h^{n+1}) \in V_h \times Q_h$ by solving another Stokes problem:

$$(13.5.19) \quad \begin{cases} \dfrac{1}{\vartheta \Delta t}(\mathbf{u}_h^{n+1} - \mathbf{u}_h^{n+1-\vartheta}, \mathbf{v}_h) + \eta\, a(\mathbf{u}_h^{n+1}, \mathbf{v}_h) \\ \qquad + b(\mathbf{v}_h, p_h^{n+1}) \\ \qquad = (\mathbf{f}^{n+1-\vartheta}, \mathbf{v}_h) - (1 - \eta)\, a(\mathbf{u}_h^{n+1-\vartheta}, \mathbf{v}_h) \\ \qquad \quad - c(\mathbf{u}_h^{n+1-\vartheta}; \mathbf{u}_h^{n+1-\vartheta}, \mathbf{v}_h) \qquad \forall\, \mathbf{v}_h \in V_h \\[4pt] b(\mathbf{u}_h^{n+1}, q_h) = 0 \qquad\qquad\qquad\qquad\quad \forall\, q_h \in Q_h \quad . \end{cases}$$

This scheme has been analyzed by Klouček and Rys (1994); under suitable assumptions on the exact solution, the time-step Δt and the mesh-size h, it is stable and convergent.

Other splitting methods are presented in Yanenko (1971) and Marchuk (1990). Fractional-step methods for spectral approximations can be found in Canuto, Hussaini, Quarteroni and Zang (1988).

13.6 Other Approaches

The characteristic Galerkin method extends to the Navier-Stokes equations (13.1) the idea that was implemented for advection-diffusion equations in

Section 12.5. Being given $\mathbf{x} \in \overline{\Omega}$ and $s \in [0, T]$, we define the characteristic flow $\mathbf{X} = \mathbf{X}(t; s, \mathbf{x})$ through

(13.6.1)
$$\begin{cases} \dfrac{d\mathbf{X}}{dt}(t; s, \mathbf{x}) = \mathbf{u}(t, \mathbf{X}(t; s, \mathbf{x})) \ , \quad t \in (0, T) \\[2mm] \mathbf{X}(s; s, \mathbf{x}) = \mathbf{x} \ . \end{cases}$$

The equations (13.1) are then discretized along this characteristic flow direction. In the simplest one-step case we go from t_n to t_{n+1} by facing the Stokes problem:

(13.6.2)
$$\begin{cases} \mathbf{u}^{n+1} - \nu \, \Delta t \, \Delta \mathbf{u}^{n+1} + \Delta t \, \nabla p^{n+1} = \mathbf{u}^n \circ \mathbf{X}^n + \Delta t \, \mathbf{f}^{n+1} & \text{in } \Omega \\[2mm] \operatorname{div} \mathbf{u}^{n+1} = 0 & \text{in } \Omega \ , \end{cases}$$

where $\mathbf{X}^n$ is a suitable approximation of $\mathbf{X}(t_n; t_{n+1}, \mathbf{x})$. The first term on the right hand side must be traced back to the preceding time level t_n through solving, for each gridpoint $\mathbf{x}_k$, system (13.6.1) starting at time $t = t_{n+1}$ (see Section 12.5).

If used in the framework of piecewise linear finite elements in space, the error behaviour of the characteristic scheme is

$$\|\mathbf{u}^n - \mathbf{u}(t_n)\|_{L^2(\Omega)} = O\left(h + \Delta t + \frac{h^2}{\Delta t} \right) \ ,$$

uniformly with respect to ν. For references on this approach see Pironneau (1982), Süli (1988). A high order method of this type has been proposed by Ho, Maday, Patera and Rønquist (1990), Maday, Patera and Rønquist (1990) and Süli and Ware (1992) in the framework of spectral methods.

Recently, for the approximation of turbulent flows, Marion and Temam (1989) have introduced the so-called *nonlinear Galerkin method* with the purpose of approximating the inertial manifold. The latter is an analytical tool for the description of the mechanism through which the energy is shifted from low into high frequency modes.

This method is quite natural to implement in a spectral context, taking as basis functions the eigenfunctions of the Stokes operator. The momentum equation is projected first onto the subspace of low modes, then onto its orthogonal. The nonlinear interaction between the high frequency component ($\mathbf{u}_{\text{high}}$) and the low frequency component ($\mathbf{u}_{\text{low}}$) of the solution is driven by the convective term (see Foias, Manley and Temam (1988)). Assuming the existence of an analytic relation $\mathbf{u}_{\text{high}} = \mathcal{M}(\mathbf{u}_{\text{low}})$ (the "inertial manifold") and eliminating $\mathbf{u}_{\text{high}}$ from the low frequency equation one obtains a nonlinear momentum equation for the low frequency solution $\mathbf{u}_{\text{low}}$. In the latter equation, after replacing $\mathcal{M}$ by a finite dimensional approximation $\mathcal{M}_N$, one gets a finite dimensional problem (the *nonlinear Galerkin* problem).

An analysis of the stability and convergence of the nonlinear spectral Galerkin method for the Navier-Stokes equations has been given by Jauberteau, Rosier and Temam (1990) and Devulder, Marion and Titi (1993) (see also the references therein). Other related results have been provided by Shen (1990), Titi (1991), Foias, Jolly, Kevrekidis and Titi (1991) and Temam (1991b). The use of the finite element method in this context has been advocated in Marion and Temam (1990).

13.7 Complements

Stabilization methods like those introduced in Sections 8.3 and 9.4 have been used in flow computations, sometimes combined with suitable fractional-step methods (see, e.g., Hughes and Brooks (1982), Hansbo and Szepessy (1990), Franca and Frey (1992), Tezduyar, Mittal and Shih (1991), Tezduyar, Mittal, Ray and Shih (1992)).

An ad-hoc approach for the simulation of vorticity is the one based on the so-called *vortex method*. Among the first contributions to this subject, let us quote the papers by Chorin (1973, 1980), Hald (1979), Leonard (1980) and Beale and Majda (1982). For more recent results, see Anderson and Greengard (1991) and the references therein.

14. Hyperbolic Problems

The numerical approximation of hyperbolic equations is a very active area of research. The main distinguishing feature of these initial-boundary value problems is the fact that perturbations propagate with finite speed. Another characterizing aspect is that the boundary treatment is not as simple as that for elliptic or parabolic equations. According to the sign of the equation coefficients, the inflow and outflow boundary regions determine, from case to case, where boundary conditions have to be prescribed. The situation becomes more complex for systems of hyperbolic equations, where the boundary treatment must undergo a local characteristic analysis. If not implemented conveniently, the numerical realization of boundary conditions is a potential source of spurious instabilities.

Hyperbolic problems also feature the presence of discontinuous solutions, arising in nonlinear equations, as well as in linear problems with discontinuous initial data. In order to account for unsmooth solutions, the problem is not set in differential form but rather in a weak form in which spatial derivatives are no longer acting on the solution but only on smooth test functions. Roughly speaking, both finite difference and collocation approximations are derived directly from the differential form of the equation. Galerkin methods (including finite elements and spectral methods) stem from the weak formulation. A third way, the finite volume method, is very popular especially for hyperbolic systems of conservation laws. It is derived from the integral form of the equations, by restricting the integration to subregions of the computational domain, called control volumes. This ensures conservation within each control volume.

Most part of this Chapter is devoted to finite difference schemes for both linear and nonlinear equations. This is covered in Section 14.2, after addressing some remarkable examples of hyperbolic equations and their principal mathematical properties in Section 14.1. A less detailed description of both finite element and spectral approximations is presented in Sections 14.3 and 14.4, respectively. A short presentation of the wave equation and its finite element approximation is included in Section 14.5. Finally, Section 14.6 is concerned with the finite volume method.

14.1 Some Instances of Hyperbolic Equations

To begin with, let us consider some examples of hyperbolic initial value and initial-boundary value problems.

14.1.1 Linear Scalar Advection Equations

The simplest example of hyperbolic equation is provided by the linear advection equation

$$(14.1.1) \qquad \frac{\partial u}{\partial t} + a \frac{\partial u}{\partial x} = 0 \quad , \quad t > 0 \, , \, x \in \mathbb{R} \ ,$$

where $a \in \mathbb{R} \setminus \{0\}$.

The solution to the Cauchy problem defined by (14.1.1) and the initial condition

$$(14.1.2) \qquad u(0, x) = u_0(x) \quad , \quad x \in \mathbb{R} \ ,$$

is simply the wave travelling with speed a:

$$u(t, x) = u_0(x - at) \quad , \quad t \geq 0 \ .$$

The solution is constant along the *characteristics*, i.e., the curves $X(t)$ in the (t, x)-plane satisfying the ordinary differential equations

$$\begin{cases} X'(t) = a \, , \quad t > 0 \\ X(0) = x_0 \ . \end{cases}$$

In the more general case

$$(14.1.3) \qquad \frac{\partial u}{\partial t} + a \frac{\partial u}{\partial x} + a_0 u = f \quad , \quad t > 0 \, , \, x \in \mathbb{R} \ ,$$

where a, a_0, f are given functions of (t, x), defining again the characteristic curves $X(t)$ as the solution to

$$(14.1.4) \qquad \begin{cases} X'(t) = a(t, X(t)) \, , \quad t > 0 \\ X(0) = x_0 \ , \end{cases}$$

it turns out that the solution to (14.1.3) satisfies

$$\frac{d}{dt} u(t, X(t)) = f(t, X(t)) - a_0(t, X(t)) \, u(t, X(t)) \ .$$

The solution along the characteristic curves is no longer constant, and it cannot be determined in a simple form any longer; however, it can be obtained solving two sets of ordinary differential equations. Thus, a smooth solution exists for all time $t > 0$ provided a, a_0, f and u_0 are smooth.

Let us return to equation (14.1.1). Although we have assumed smoothness on u, so that first derivatives make sense in the previous process, the case of non-smooth initial data can be covered as well. Indeed, it is easily speculated from the previous derivation that, if u_0 has a discontinuity across a point x_0, then it is propagated along the characteristic issuing from x_0. Of course, this process can be given a mathematical sense after defining the concept of weak solution to the hyperbolic equation.

The non-smooth case can be switched to a smooth one if, e.g., the initial data $u_0(x)$ is approximated by a sequence of smooth functions $u_0^\epsilon(x)$, $\epsilon > 0$. The solution $u^\epsilon(t, x)$ of problem (14.1.1) corresponding to the smooth data becomes $u^\epsilon(t, x) = u_0^\epsilon(x - at)$, $t \geq 0$, $x \in \mathbb{R}$. If $||u_0 - u_0^\epsilon||_{L^1(\mathbb{R})} \leq \epsilon$ where

$$(14.1.5) \qquad ||v||_{L^1(\mathbb{R})} := \int_{-\infty}^{\infty} |v(x)| dx$$

is the L^1-norm, then

$$\lim_{\epsilon \to 0} u^\epsilon(t, x) = \lim_{\epsilon \to 0} u_0^\epsilon(x - at) = u_0(x - at) \ ,$$

the limit being taken with respect to the L^1-norm, and for each $t \geq 0$.

This approach would be uneffective for nonlinear hyperbolic equations, as the latter can develop singularities even for smooth initial data. An alternative approach is to regularize the equation (14.1.1) rather than the initial datum u^0, replacing it by the *parabolic* advection-diffusion equation

$$\frac{\partial u^\epsilon}{\partial t} + a \frac{\partial u^\epsilon}{\partial x} = \varepsilon \frac{\partial^2 u^\epsilon}{\partial x^2} \ , \quad t > 0 \ , \ x \in \mathbb{R} \ ,$$

for small $\varepsilon > 0$. The function $\lim_{\varepsilon \to 0} u^\epsilon(t, x)$ is called the *vanishing viscosity solution*, and is equal to $u_0(x - at)$. This approach can be generalized to nonlinear equations as well. However, we warn the reader that when (14.1.1) is studied on a finite space interval, the approximating advection-diffusion equation needs to be supplemented by an extra boundary condition, that has to be chosen conveniently in order to avoid the onset of a spurious boundary layer (see Chapters 8 and 12).

14.1.2 Linear Hyperbolic Systems

Now consider the case of linear systems of the form

$$(14.1.6) \qquad \frac{\partial \mathbf{U}}{\partial t} + A \frac{\partial \mathbf{U}}{\partial x} = \mathbf{0} \ , \quad t > 0 \ , \ x \in \mathbb{R} \ ,$$

where $\mathbf{U} : [0, \infty) \times \mathbb{R} \to \mathbb{R}^p$, and $A \in \mathbb{R}^{p \times p}$ is a constant matrix. This system is said to be *hyperbolic* if A is diagonalizable with real eigenvalues, so that one can consider the decomposition

$$(14.1.7) \qquad A = T \Lambda T^{-1} \ ,$$

where $\Lambda := \operatorname{diag}(\lambda_1, ..., \lambda_p)$ is a diagonal matrix of eigenvalues, and $T := (\omega^1|\omega^2|...|\omega^p)$ is the matrix of right eigenvectors, i.e.,

$$A\omega^k = \lambda_k \omega^k \ , \quad k = 1, ..., p \ .$$

We assume that $\lambda_1, ..., \lambda_{p_0}$ are non-negative, while $\lambda_{p_0+1}, ..., \lambda_p$ are negative. *Strict* hyperbolicity holds if the eigenvalues are distinct.

Introducing the *characteristic variables* $\mathbf{W} := T^{-1}\mathbf{U}$ in (14.1.6) gives

$$(14.1.8) \qquad \frac{\partial \mathbf{W}}{\partial t} + \Lambda \frac{\partial \mathbf{W}}{\partial x} = \mathbf{0} \ .$$

This decouples into p independent scalar advection equations

$$(14.1.9) \qquad \frac{\partial W_k}{\partial t} + \lambda_k \frac{\partial W_k}{\partial x} = 0 \ , \quad k = 1, ..., p \ ,$$

whose solutions are $W_k(t, x) = W_k(0, x - \lambda_k t)$. The solution $\mathbf{U} = T\mathbf{W}$ to the initial problem (14.1.6) can therefore be expressed as

$$(14.1.10) \qquad \mathbf{U}(t, x) = \sum_{k=1}^{p} W_k(0, x - \lambda_k t)\omega^k \ .$$

The curves $X(t) = x_0 + \lambda_k t$ of the (t, x)-plane satisfying $X'(t) = \lambda_k$ are the k-characteristics. Any characteristic variable W_k is constant along each k-characteristic. For a strictly hyperbolic system, p distinct characteristic curves pass through each point in the (t, x)-plane. As a consequence, for any fixed $\overline{x}$ and $\overline{t}$, $u(\overline{t}, \overline{x})$ depends only on the initial data at the p points $\overline{x} - \lambda_k \overline{t}$. For this reason the set of p points (feet of the characteristics issuing from the point $(\overline{t}, \overline{x})$)

$$(14.1.11) \qquad D(\overline{t}, \overline{x}) := \{x \in \mathbb{R} \,|\, x = \overline{x} - \lambda_k \overline{t} \ , \quad k = 1, ..., p\} \ ,$$

is called the *domain of dependence* of the solution $\mathbf{U}$ at the point $(\overline{t}, \overline{x})$.

A classic example of a hyperbolic equation is the second order *wave equation*

$$(14.1.12) \qquad \frac{\partial^2 u}{\partial t^2} = \gamma^2 \frac{\partial^2 u}{\partial x^2} \ , \quad t > 0 \ , \ x \in \mathbb{R} \ ,$$

with initial data

$$(14.1.13) \qquad u(0, x) = u_0(x) \ \text{ and } \ \frac{\partial u}{\partial t}(0, x) = v_0(x) \ .$$

By the change of variables

$$(14.1.14) \qquad U_1 := \frac{\partial u}{\partial x} \ , \quad U_2 := \frac{\partial u}{\partial t} \ ,$$

we obtain from (14.1.12) a first order system of the form (14.1.6) with

$$(14.1.15) \qquad \mathbf{U} = (U_1, U_2) \ , \quad A = \begin{pmatrix} 0 & -1 \\ -\gamma^2 & 0 \end{pmatrix} \ ,$$

and initial conditions $U_1(0, x) = u_0'(x)$ and $U_2(0, x) = v_0(x)$.

Hyperbolicity comes from the fact that the eigenvalues of A are the two distinct real numbers $\pm\gamma$ (the wave speeds). The solution components U_1 and U_2 can be obtained by the diagonalization procedure illustrated on (14.1.6), after computing the eigenvectors of A. The result is

$$U_1(t, x) = \frac{1}{2} \left[u_0'(\xi) + \frac{1}{\gamma} v_0(\xi) + u_0'(\eta) - \frac{1}{\gamma} v_0(\eta) \right]$$

$$U_2(t, x) = \frac{1}{2} \left[\gamma u_0'(\xi) + v_0(\xi) - \gamma u_0'(\eta) + v_0(\eta) \right] \ ,$$

with $\xi := x + \gamma t$ and $\eta := x - \gamma t$. Then $u(t, x)$ can be obtained through (14.1.14).

As before, smooth initial data give smooth solutions, whereas any singularities in the initial data can propagate only along characteristics.

14.1.3 Initial-Boundary Value Problems

Let us now consider the case of initial-boundary value problems, and assume, therefore, that all hyperbolic equations considered so far are set in a finite interval, say $x \in I = (0, 1)$, rather than on the real line. Then, we need to prescribe boundary data in a way which is compatible with the nature of the problem at hand.

For the advection equation (14.1.1) the boundary condition takes the form

$$(14.1.16) \qquad \begin{cases} u(t, 0) = \varphi_-(t) & \text{if } a > 0 \\ u(t, 1) = \varphi_+(t) & \text{if } a < 0 \ , \end{cases}$$

where $\varphi_\pm(t)$ are given functions. If $a > 0$, the boundary point $x = 0$ is the *inflow* point while $x = 1$ is the *outflow* point. The role is reversed in the case of a negative a. Similarly, for the more general advection equation (14.1.3), the solution u needs to be prescribed at any inflow boundary point. According to the sign of $a(t, x)$, the number of inflow points may be one, two or none.

The solution to the corresponding initial-boundary value problem satisfies an a-priori bound in the L^2-norm. Indeed, multiplying (14.1.3) by u, integrating on x from 0 to 1, and using integration by parts on the advective term we obtain for all $t > 0$

$$\frac{1}{2} \frac{d}{dt} \|u(t)\|_{0,I}^2 + \int_0^1 \mu(t, x) u^2(t, x) dx$$

$$= \int_0^1 f(t, x) u(t, x) dx + \frac{1}{2} [-a(t, 1) u^2(t, 1) + a(t, 0) u^2(t, 0)] \ ,$$

where the usual notation for the L^2-norm has been adopted (see Section 1.2), and we have set

$$\mu(t,x) := a_0(t,x) - \frac{1}{2}\frac{\partial a}{\partial x}(t,x) \ .$$

If we define

$$\Phi(t) := -\sigma_+(t)a(t,1)\varphi_+^2(t) + \sigma_-(t)a(t,0)\varphi_-^2(t) \ ,$$

where

$$\sigma_+(t) := \begin{cases} 1 & \text{if } a(t,1) < 0 \\ 0 & \text{otherwise} \end{cases} \ , \quad \sigma_-(t) := \begin{cases} 1 & \text{if } a(t,0) > 0 \\ 0 & \text{otherwise} \end{cases} \ ,$$

we immediately have $\Phi(t) \geq 0$ and moreover

$$-a(t,1)u^2(t,1) + a(t,0)u^2(t,0) \leq \Phi(t) \quad \forall \, t > 0 \ .$$

Therefore, using Gronwall lemma (see Lemma 1.4.1) for each $t \in [0,T]$ it follows

$$||u(t)||_{0,I}^2 \leq \left(||u_0||_{0,I}^2 + \int_0^t ||f(s)||_0^2 \, ds + \int_0^t \Phi(s) \, ds \right)$$
$$\times \exp\left(\int_0^t [1 + 2\mu^*(s)] \, ds \right) \ ,$$

with $\mu^*(t) := \max_x |\mu(t,x)|$.

We now consider the hyperbolic system (14.1.6) for $x \in I = (0,1)$. We can easily deduce from the diagonal form (14.1.9) that one should prescribe as many conditions at $x = 0$ as the number of positive eigenvalues of A. Simmetrically, at $x = 1$ the number of prescribed data should equate the number of negative eigenvalues of A. Precisely, at each boundary point one has to prescribe the *incoming* variables.

Let us split Λ and $\mathbf{W}$ as follows

$$(14.1.17) \qquad \Lambda = \begin{pmatrix} \Lambda^+ & 0 \\ 0 & \Lambda^- \end{pmatrix} \quad \text{and} \quad \mathbf{W} = (\mathbf{W}^+, \mathbf{W}^-) \ ,$$

where

$$\Lambda^+ := \text{diag}(\lambda_1, ..., \lambda_{p_0}) \ , \quad \Lambda^- := \text{diag}(\lambda_{p_0+1}, ..., \lambda_p)$$
$$\mathbf{W}^+ := (W_1, ..., W_{p_0}) \ , \quad \mathbf{W}^- := (W_{p_0+1}, ..., W_p) \ .$$

Furthermore, for simplicity, assume that Λ is non-singular, so that $\lambda_1, ..., \lambda_{p_0}$ are positive.

The boundary conditions to be prescribed can take the following form:

$$(14.1.18) \qquad \begin{aligned} \mathbf{W}^+(t,0) &= S_0(t)\mathbf{W}^-(t,0) + \mathbf{g}_0(t) & \text{at } x = 0 \ , \\ \mathbf{W}^-(t,1) &= S_1(t)\mathbf{W}^+(t,1) + \mathbf{g}_1(t) & \text{at } x = 1 \ , \end{aligned}$$

where $\mathbf{g}_0$, $\mathbf{g}_1$ are smooth vectors and S_0, S_1 are rectangular reflection matrices. Assuming that the boundary conditions satisfy the dissipativity properties

$$S_0^T(t)\Lambda^+ S_0(t) + \Lambda^- \leq 0$$
$$S_1^T(t)\Lambda^- S_1(t) + \Lambda^+ \geq 0 \ ,$$

then the solution is stable in time with respect to the spatial L^2-norm, i.e., it can be bounded in terms of the initial and boundary data. This can easily be derived from (14.1.8), multiplying by $\mathbf{W}$, integrating in space and then proceeding as done previously for the scalar case.

Now, consider the wave equation (14.1.12) in the finite space interval $(0, 1)$. Several kind of boundary conditions can be prescribed. An instance is provided by the homogeneous Dirichlet conditions

(14.1.19) $u(t, 0) = 0 \ , \quad u(t, 1) = 0 \ , \quad t > 0 \ .$

In this case, u represents the vertical displacement of a vibrating string of length equal to 1 which is at rest at the endpoints. (Neumann conditions prescribing $\frac{\partial u}{\partial x}$ at one (or both) endpoint are also admissible.)

A conservation property can easily be derived for the first derivatives of the solution. Indeed, multiplying both sides of (14.1.12) by $v = u_t$, integrating with respect to x and using (14.1.19) we obtain

$$\frac{d}{dt} \int_0^1 \left[\left(\frac{\partial u}{\partial t}(t, x) \right)^2 + \gamma^2 \left(\frac{\partial u}{\partial x}(t, x) \right)^2 \right] dx = 0$$

and therefore the conservation property

$$\int_0^1 \left[\left(\frac{\partial u}{\partial t}(t, x) \right)^2 + \gamma^2 \left(\frac{\partial u}{\partial x}(t, x) \right)^2 \right] dx$$
$$= \int_0^1 \left[\left(\frac{\partial u}{\partial t}(0, x) \right)^2 + \gamma^2 \left(\frac{\partial u}{\partial x}(0, x) \right)^2 \right] dx \quad \forall\, t \geq 0 \ .$$

14.1.4 Nonlinear Scalar Equations

We now consider the scalar equation

(14.1.20) $\dfrac{\partial u}{\partial t} + \dfrac{\partial}{\partial x} F(u) = 0 \ , \quad t > 0 \, , \ x \in \mathbb{R} \ ,$

where $F(u)$ is a nonlinear function of u, and $u(0, x) = u_0(x)$. Very often, F is a strictly convex function, i.e., $F''(\xi) > 0$ for all $\xi \in \mathbb{R}$.

A classical example of this class is provided by Burgers equation in which $F(u) = u^2/2$, so (14.1.20) becomes

$$(14.1.21) \qquad \frac{\partial u}{\partial t} + u \frac{\partial u}{\partial x} = 0 \ .$$

This is a nonlinear transport equation whose advection velocity is given by the solution u itself. Following (14.1.4), the characteristics satisfy

$$(14.1.22) \qquad X'(t) = u(t, X(t)) \ .$$

Taking the total derivative of u along each characteristic gives

$$\frac{d}{dt}[u(t, X(t))] = \frac{\partial u}{\partial t}(t, X(t)) + X'(t) \frac{\partial u}{\partial x}(t, X(t))$$

$$= \left(\frac{\partial u}{\partial t} + u \frac{\partial u}{\partial x} \right)(t, X(t)) = 0 \ .$$

Hence, u is constant on each characteristic. In turn, owing to (14.1.22) the slope $X'(t)$ is constant, and therefore the characteristics are straight lines whose slope is determined by the initial data. When the latter is smooth, this property entails that the solution $u(t,x)$ has the form

$$u(t, x) = u_0(s) \ ,$$

where $s = s(t, x)$ satisfies

$$(14.1.23) \qquad x = s + u_0(s)t \ .$$

Equation (14.1.23) has a unique solution for t small enough (as long as characteristics don't cross). Whenever u_0' is negative at some point, say s_0, there is a critical time t_c when characteristics first cross, the function $u(t_c, x)$ has an infinite slope at a point x_0 and a discontinuity forms. For Burgers equation x_0 is obtained from (14.1.23) taking $s = s_0$.

The value of the critical time t_c is obtained as follows. From the negativeness assumption on u_0', we can find two values s_1 and s_2 with $s_1 < s_2$ and $u_0(s_1) > u_0(s_2)$. The two characteristics $x = s_1 + u_0(s_1)t$ and $x = s_2 + u_0(s_2)t$ cross at the time $t(s_1, s_2) = -(s_2 - s_1)/[u_0(s_2) - u_0(s_1)]$. Passing to the limit as $s_2 \to s_1$ and taking the minimum over s_1, we find the critical time $t_c = -1/\min u_0'(x)$.

For the general equation (14.1.20) with a strictly convex F, the same argument shows that the critical time is $t_c = -1/\min[F''(u_0(x))u_0'(x)]$.

Beyond the time t_c, a classical solution to (14.1.20) fails to exist, while a *weak* solution (in the sense that we are going to state) at that time becomes discontinuous. The basic goal in defining a weak solution for (14.1.20) is to avoid differentiating $u(t,x)$: definitely, this would require less smoothness.

This process is accomplished in the following way. Let us denote by $C_0^1([0, \infty) \times \mathbb{R})$ the space of functions $\varphi(t, x)$ that are continuously differentiable and with compact support in $[0, \infty) \times \mathbb{R}$, i.e., $\varphi(t, x) = 0$ outside of some bounded set. Multiplying (14.1.20) by $\varphi(t, x)$, integrating over space and time, and using the Green formula we obtain

$$\text{(14.1.24)} \qquad \int_0^\infty \int_{-\infty}^\infty \left[u(t,x) \frac{\partial \varphi}{\partial t}(t,x) + F(u(t,x)) \frac{\partial \varphi}{\partial x}(t,x) \right] dx\, dt$$

$$= - \int_{-\infty}^\infty u_0(x) \varphi(0,x) dx \ .$$

A function $u(t,x)$ is called a *weak solution* of (14.1.20) provided that it is measurable and (14.1.24) holds for all functions $\varphi \in C_0^1([0,\infty) \times \mathbb{R})$.

Weak solutions are not necessarily unique: among them, the physically correct one is determined by requiring that other conditions (the so-called *entropy conditions*) are satisfied. One therefore gets the unique *entropy solution*. In the case of equation (14.1.21) such a solution is the limit, as ε goes to 0, of the solution $u^\varepsilon(t,x)$ of the perturbed viscous equation

$$\text{(14.1.25)} \qquad \frac{\partial u^\varepsilon}{\partial t} + \frac{\partial}{\partial x} F(u^\varepsilon) = \varepsilon \frac{\partial^2 u^\varepsilon}{\partial x^2} \ , \quad t > 0 \ , \ x \in \mathbb{R} \ ,$$

with $u^\varepsilon(0,x) = u_0(x)$. Thus, the entropy solution is a vanishing viscosity solution.

In order to find out the form of the entropy condition, let us start from the case of a conservation law (14.1.20) with piecewise-constant initial data having a single discontinuity. This is known as the *Riemann problem*.

As an example, consider (14.1.21) with piecewise-constant initial data

$$\text{(14.1.26)} \qquad u_0(x) = \begin{cases} u_l & , \ x < 0 \\ u_r & , \ x > 0 \end{cases} \ .$$

If $u_l > u_r$ there is a unique weak solution (which is indeed the entropy solution)

$$\text{(14.1.27)} \qquad u(t,x) = \begin{cases} u_l & , \ x < \sigma t \\ u_r & , \ x > \sigma t \end{cases} \ ,$$

where

$$\text{(14.1.28)} \qquad \sigma = \frac{u_l + u_r}{2} \ ,$$

is the speed at which the discontinuity propagates. The solution has, at each $t \geq 0$, two constant states, and characteristics go *into* the discontinuity from both regions (see Fig. 14.1.1). This kind of discontinuity is called a *shock*.

The shock speed is given in (14.1.28). In order to obtain it we integrate in space (14.1.20) from $\sigma t - 1$ to $\sigma t + 1$ so that

$$\text{(14.1.29)} \qquad \int_{\sigma t - 1}^{\sigma t + 1} \frac{\partial u}{\partial t}(t,x) dx = -F(u_r) + F(u_l)$$

for each $t > 0$. We then obtain

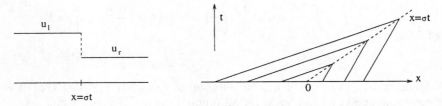

Fig. 14.1.1. Shock-wave (entropic) solution to the Riemann problem

$$0 = \frac{d}{dt}(u_l + u_r) = \frac{d}{dt}\left[\int_{\sigma t-1}^{\sigma t} u(t,x)dx + \int_{\sigma t}^{\sigma t+1} u(t,x)dx\right]$$

$$= \int_{\sigma t-1}^{\sigma t+1} \frac{\partial u}{\partial t}(t,x)dx - \sigma(u_l - u_r) \ ,$$

and therefore

(14.1.30) $$\sigma(u_l - u_r) = F(u_l) - F(u_r) \ .$$

This relation holds for any flux function F. In the case of the Burgers equation, $F(u) = u^2/2$, hence σ takes the form (14.1.28).

Relationship (14.1.30) is called the *Rankine-Hugoniot jump condition*. For more general cases, Rankine-Hugoniot conditions hold across any shock. In this case u_l and u_r denote respectively the left-hand and right-hand values of the solution across the discontinuity, while σ is the corresponding instantaneous speed.

In the case $u_l < u_r$ there are infinitely many weak solutions. One is again (14.1.27) (see Fig. 14.1.2), but now characteristics are going *out* of the discontinuity. It can be shown that this solution is unstable: under small perturbations on the data the solution changes dramatically.

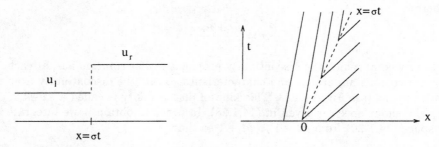

Fig. 14.1.2. Entropy-violating solution to the Riemann problem

Another weak solution which is however stable to perturbations is

(14.1.31)
$$u(t,x) = \begin{cases} u_l & \text{if } x < u_l\,t \\ \dfrac{x}{t} & \text{if } u_l\,t \le x \le u_r\,t \\ u_r & \text{if } x > u_r\,t \ . \end{cases}$$

This is the vanishing viscosity (or entropic) solution to the Burgers equation (see Fig. 14.1.3).

The previous considerations indicate how to devise the form of the entropy condition. As we have seen, a shock should have characteristics going *into* the discontinuity, as time advances. When characteristics are coming *out* of a discontinuity, small perturbations on the initial data or on the equation itself (e.g., adding some viscosity) may result in a strong changement of the solution (typically, a rarefaction of characteristics shows up).

We can therefore derive the following statement:

A discontinuity propagating with speed σ given by (14.1.30) *satisfies the entropy condition if*

(14.1.32)
$$F'(u_l) > \sigma > F'(u_r) \ .$$

For a convex F, the Rankine-Hugoniot condition (14.1.30) entails that σ must lie between $F'(u_l)$ and $F'(u_r)$, so (14.1.32) simply reduces to require $F'(u_l) > F'(u_r)$, or equivalently (again by convexity) that $u_l > u_r$.

In the case of a non-convex F, a more general statement is the following:

A discontinuity propagating with speed σ given by (14.1.30) *satisfies the entropy condition if*

(14.1.33)
$$\frac{F(u) - F(u_l)}{u - u_l} \ge \sigma \ge \frac{F(u) - F(u_r)}{u - u_r}$$

for all u between u_l and u_r.

Another form of the entropy condition entails the use of entropy functions and entropy fluxes. Let us start from the conservation law (14.1.20), and suppose that some functions $\eta(u)$ and $\mu(u)$ satisfy the following equation

(14.1.34)
$$\frac{\partial}{\partial t}\eta(u) + \frac{\partial}{\partial x}\mu(u) = 0 \ .$$

The functions $\eta(u)$ and $\mu(u)$ are called *entropy function* and *entropy flux*, respectively. The functions η, μ and F are related by

(14.1.35)
$$\mu'(u) = \eta'(u)F'(u) \ .$$

Indeed, assuming that u is smooth, from (14.1.34) we have

$$\eta'(u)\frac{\partial u}{\partial t} + \mu'(u)\frac{\partial u}{\partial x} = 0 \ ,$$

while from (14.1.20)

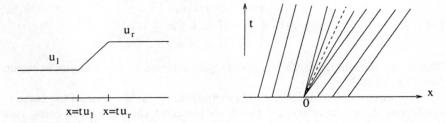

Fig. 14.1.3. Rarefaction-wave (entropic) solution to the Riemann problem

$$0 = \eta'(u)\left[\frac{\partial u}{\partial t} + F'(u)\frac{\partial u}{\partial x}\right] \ .$$

By comparison we obtain (14.1.35).

The third statement for the entropy condition is given as follows:

The function $u(t,x)$ is the entropy solution of (14.1.20) if, for all convex entropy functions $\eta(\cdot)$ and corresponding entropy fluxes $\mu(\cdot)$, the entropy inequality

$$(14.1.36) \qquad \frac{\partial}{\partial t}\eta(u) + \frac{\partial}{\partial x}\mu(u) \le 0$$

is satisfied in the weak sense.

A heuristic explanation is as follows. Using (14.1.25) and (14.1.35) gives:

$$\frac{\partial}{\partial t}\eta(u^\varepsilon) + \frac{\partial}{\partial x}\mu(u^\varepsilon) = \eta'(u^\varepsilon)\left[\frac{\partial u^\varepsilon}{\partial t} + F'(u^\varepsilon)\frac{\partial u^\varepsilon}{\partial x}\right] = \varepsilon\eta'(u^\varepsilon)\frac{\partial^2 u^\varepsilon}{\partial x^2}$$

$$= \varepsilon\left[\frac{\partial}{\partial x}\left(\eta'(u^\varepsilon)\frac{\partial u^\varepsilon}{\partial x}\right) - \eta''(u^\varepsilon)\left(\frac{\partial u^\varepsilon}{\partial x}\right)^2\right] \le \varepsilon\frac{\partial}{\partial x}\left(\eta'(u^\varepsilon)\frac{\partial u^\varepsilon}{\partial x}\right) \ .$$

Letting $\varepsilon \to 0$ and setting $u := \lim_{\varepsilon \to 0} u^\varepsilon$ one formally obtains (14.1.36).

The weak form of (14.1.36) is

$$(14.1.37) \qquad -\int_0^\infty \int_{-\infty}^\infty \left[\eta(u(t,x))\frac{\partial\varphi}{\partial t}(t,x) + \mu(u(t,x))\frac{\partial\varphi}{\partial x}(t,x)\right]dxdt$$
$$\le \int_{-\infty}^\infty \varphi(0,x)\eta(u(0,x))dx$$

for all $\varphi \in C_0^1([0,\infty) \times \mathbb{R})$, with $\varphi(t,x) \ge 0$ for all $(t,x) \in [0,\infty) \times \mathbb{R}$.

In the case of Burgers equation in which $F(u) = u^2/2$, taking for instance $\eta(u) = u^2$ from (14.1.35) we get $\mu'(u) = 2u^2$ and hence $\mu(u) = 2u^3/3$. The entropy inequality (14.1.36) becomes

$$(14.1.38) \qquad \frac{\partial}{\partial t}u^2 + \frac{\partial}{\partial x}\left(\frac{2}{3}u^3\right) \le 0 \ .$$

For smooth solutions this holds with equality. When a discontinuity is present, it is possible to show that (14.1.38) is satisfied if, and only if, $(u_l - u_r)^3 > 0$, i.e., $u_l > u_r$ as expected. For a given numerical method, a discrete form of (14.1.36) is useful to check whether the numerical solution is converging to the entropy solution.

To summarize, we can say that the physical solution $u = \lim_{\varepsilon \to 0} u^\varepsilon$ is a special kind of weak solution of (14.1.20) which is subject to the entropy condition (14.1.36). Kruzkov (1970) proved that the entropy solution exists and is uniquely determined by the initial data $u_0(x)$.

For an in-depth analysis of nonlinear conservation laws we refer the reader to, e.g., Courant and Friedrichs (1948), Oleinik (1957), Lax (1973), Smoller (1983).

14.2 Approximation by Finite Differences

For a space-time approximation based on finite differences, the (t, x)-plane is discretized by choosing a time-step Δt and a mesh-width Δx, and defining the grid-points (t_n, x_j) by

$$t_n := n\Delta t \ , \quad n \in \mathrm{N} \ ; \quad x_j := j\Delta x \ , \quad j \in \mathbb{Z} \ .$$

We define the mesh ratio $\lambda := \Delta t / \Delta x$ and also $x_{j+1/2} := x_j + \Delta x/2$, then we look for discrete solutions u_j^n that approximate $u(t_n, x_j)$ for all j, n. Sometimes, u_j^n will approximate the so-called "cell average"

$$\overline{u}_j^n := \frac{1}{\Delta x} \int_{x_j - 1/2}^{x_j + 1/2} u(t_n, x)dx \ .$$

This is often the case of finite volume approaches.

14.2.1 Linear Scalar Advection Equations and Hyperbolic Systems

Hyperbolic initial value problems are, in general, advanced in time by explicit methods, although this yields a stability restriction on λ that in general implicit methods don't have. Some examples for the linear advection equation (14.1.1) with constant a are listed in Table 14.2.1. Except the leap-frog scheme, all of the methods reported in Table 14.2.1 are two-level ones, as they only involve the time levels t_n and t_{n+1}.

Some of the methods use centered finite differences, i.e., both u_{j-1} and u_{j+1} are used in the approximation of the x-derivative at the node x_j. The *centered forward Euler* method (see Table 14.1.1) is the simplest example in this class. Unfortunately, it is not stable (the definition of stability will be given in the next Section). The *Lax-Friedrichs* method is a slight modification of the Euler one (it merely replaces u_j^n by the mean value between u_{j+1}^n and

Table 14.2.1. Explicit finite difference schemes for problem (14.1.1)

	Difference equations	Numerical flux function $(h_{j+1/2})$
Centered forward Euler	$u_j^{n+1} = u_j^n - \dfrac{\lambda}{2}a(u_{j+1}^n - u_{j-1}^n)$	$\dfrac{1}{2}a(u_{j+1} + u_j)$
Upwind	$u_j^{n+1} = u_j^n - \dfrac{\lambda}{2}a(u_{j+1}^n - u_{j-1}^n)$ $+\dfrac{\lambda}{2}|a|(u_{j+1}^n - 2u_j^n + u_{j-1}^n)$	$\dfrac{1}{2}[a(u_{j+1} + u_j)$ $-|a|(u_{j+1} - u_j)]$
Lax-Friedrichs	$u_j^{n+1} = \dfrac{1}{2}(u_{j+1}^n + u_{j-1}^n)$ $-\dfrac{\lambda}{2}a(u_{j+1}^n - u_{j-1}^n)$	$\dfrac{1}{2}[a(u_{j+1} + u_j)$ $-\dfrac{1}{\lambda}(u_{j+1} - u_j)]$
Lax-Wendroff	$u_j^{n+1} = u_j^n - \dfrac{\lambda}{2}a(u_{j+1}^n - u_{j-1}^n)$ $+\dfrac{\lambda^2}{2}a^2(u_{j+1}^n - 2u_j^n + u_{j-1}^n)$	$\dfrac{1}{2}[a(u_{j+1} + u_j)$ $-\lambda a^2(u_{j+1} - u_j)]$
Warming-Beam $(a > 0)$	$u_j^{n+1} = u_j^n - \dfrac{\lambda}{2}a(3u_j^n - 4u_{j-1}^n + u_{j-2}^n)$ $+\dfrac{\lambda^2}{2}a^2(u_j^n - 2u_{j-1}^n + u_{j-2}^n)$	$\dfrac{1}{2}[a(3u_j - u_{j-1})$ $-\lambda a^2(u_j - u_{j-1})]$
Leap-frog	$u_j^{n+1} = u_j^{n-1} - \lambda a(u_{j+1}^n - u_{j-1}^n)$	

u_{j-1}^n) which enjoyes better stability properties (see Section 14.2.2). Other methods are based on one-sided finite differences. Among the latters are the upwind and Warming-Beam methods.

The *upwind* method is derived as follows:

$$(14.2.1) \qquad \frac{u_j^{n+1} - u_j^n}{\Delta t} = \begin{cases} -\dfrac{a}{\Delta x}(u_{j+1}^n - u_j^n) \quad , \ a < 0 \\[2mm] -\dfrac{a}{\Delta x}(u_j^n - u_{j-1}^n) \quad , \ a > 0 \end{cases} .$$

This can be written equivalently as

$$(14.2.2) \qquad \frac{u_j^{n+1} - u_j^n}{\Delta t} + a\frac{u_{j+1}^n - u_{j-1}^n}{2\Delta x} = \frac{1}{2}|a|\Delta x\frac{u_{j+1}^n - 2u_j^n + u_{j-1}^n}{(\Delta x)^2} .$$

It differs from the centered forward Euler method due to the presence at the right hand side of (14.2.2) of the extra term

$$(14.2.3) \qquad \varepsilon\frac{u_{j+1}^n - 2u_j^n + u_{j-1}^n}{(\Delta x)^2}$$

with $\varepsilon = \frac{1}{2}|a|\Delta x$. Since $\frac{u_{j+1}-2u_j+u_{j-1}}{(\Delta x)^2}$ is an approximation of $\frac{\partial^2 u}{\partial x^2}$ at the point x_j, it follows that (14.2.3) is nothing but a term of *artificial viscosity* (or *numerical dissipation*) that is added to the scheme for stabilization purposes.

The *Lax-Wendroff* method is not directly based on a finite difference approximation to the space and time derivatives in (14.1.1), but rather on Taylor expansion about $t = t_n$ truncated at the second order:

$$u(t_{n+1}, x_j) = u(t_n, x_j) + \Delta t\frac{\partial u}{\partial t}(t_n, x_j) + \frac{1}{2}(\Delta t)^2\frac{\partial^2 u}{\partial t^2}(t_n, x_j) + O((\Delta t)^3) .$$

Using (14.1.1) we have

$$\frac{\partial u}{\partial t} = -a\frac{\partial u}{\partial x} \ , \quad \frac{\partial^2 u}{\partial t^2} = a^2\frac{\partial^2 u}{\partial x^2} .$$

Replacing these terms, retaining only the first three terms of the right hand side of Taylor development, and approximating space derivatives by centered finite differences gives the Lax-Wendroff method.

The *Warming-Beam* method is obtained in a similar manner, but now space derivatives are approximated by one-sided differences. In Table 14.2.1 we report the formula when $a > 0$.

We can easily see that the numerical dissipation introduced by the Lax-Friedrichs method has still the form (14.2.3), with $\varepsilon = \frac{1}{2}\frac{(\Delta x)^2}{\Delta t}$, while for the Lax-Wendroff method we have $\varepsilon = \frac{1}{2}a^2\Delta t$.

The Lax-Wendroff and Warming-Beam schemes are the only two-level schemes of second order accuracy on the supports $(j - 1, j, j + 1)$ and $(j - 2, j-1, j)$, respectively (see, e.g., Hirsch (1988), pp. 357–360). These methods are in fact third order accurate when regarded as a direct approximation of the modified *dispersive* equation

$$\frac{\partial u}{\partial t} + a\frac{\partial u}{\partial x} = \varepsilon\frac{\partial^3 u}{\partial x^3} .$$

Precisely, the dispersion coefficient is $\varepsilon = \frac{\Delta x^2}{6}a(\lambda^2 a^2 - 1)$ for the Lax-Wendroff scheme, while $\varepsilon = \frac{\Delta x^2}{6}a(2 - 3\lambda a + \lambda^2 a^2)$ in the case of the Warming-Beam scheme. The theory of dispersive waves is extensively developed in Whitham (1974).

All two-level explicit finite difference methods can be written in the unified form

$$(14.2.4) \qquad u_j^{n+1} = u_j^n - \lambda(h_{j+1/2}^n - h_{j-1/2}^n) \ ,$$

where $h_{j+1/2} = h(u_j, u_{j+1})$ and the function $h(\cdot, \cdot)$ is called *numerical flux function*. Its expression is reported aside each method in Table 14.1.1.

The advantage of the compact reformulation (14.2.4) is that it can easily be extended to the case of more general hyperbolic problems (remarkably, linear systems and nonlinear equations). When we consider system (14.1.6), the numerical approximation by finite differences can still be accomplished as indicated in Table 14.2.1. Only a little care needs to be paid to the notations. First of all, a should be replaced by the matrix A, a^2 by A^2 and u_j^n by $\mathbf{U}_j^n$, the vector approximating $\mathbf{U}(t_n, x_j)$. Moreover, in the upwind method $|a|$ should be replaced by

$$(14.2.5) \qquad |A| := T|\Lambda|T^{-1} \ ,$$

where $|\Lambda| := \operatorname{diag}(|\lambda_1|, ..., |\lambda_p|)$ (see (14.1.7)).

The rationale behind this formula is easily understood, provided we transform first the system (14.1.6) into the p independent advection equations (14.1.9), and then approximate each of them by the upwind scheme for scalar equations. This yields

$$(W_k)_j^{n+1} = (W_k)_j^n - \frac{\lambda}{2}\lambda_k[(W_k)_{j+1}^n - (W_k)_{j-1}^n]$$
$$+ \frac{\lambda}{2}|\lambda_k|[(W_k)_{j+1}^n - 2(W_k)_j^n + (W_k)_{j-1}^n]$$

for each $k = 1, ..., p$ or, with compact notation,

$$\mathbf{W}_j^{n+1} = \mathbf{W}_j^n - \frac{\lambda}{2}\Lambda(\mathbf{W}_{j+1}^n - \mathbf{W}_{j-1}^n) + \frac{\lambda}{2}|\Lambda|(\mathbf{W}_{j+1}^n - 2\mathbf{W}_j^n + \mathbf{W}_{j-1}^n) \ .$$

Transforming back to the physical variables $\mathbf{U}_j^n$ by means of the transformation $T\mathbf{W}_j^n = \mathbf{U}_j^n$ gives the vector upwind method

$$(14.2.6) \quad \mathbf{U}_j^{n+1} = \mathbf{U}_j^n - \frac{\lambda}{2}A(\mathbf{U}_{j+1}^n - \mathbf{U}_{j-1}^n) + \frac{\lambda}{2}|A|(\mathbf{U}_{j+1}^n - 2\mathbf{U}_j^n + \mathbf{U}_{j-1}^n) \ .$$

Also in the vector case all two-level time-advancing schemes can be reformulated as

$$(14.2.7) \qquad \mathbf{U}_j^{n+1} = \mathbf{U}_j^n - \lambda(\mathbf{H}_{j+1/2}^n - \mathbf{H}_{j-1/2}^n) \ ,$$

where $\mathbf{H}_{j+1/2}$ is the *numerical flux vector function*. Its expression can easily be derived by generalizing the scalar case, by replacing again a, a^2, $|a|$ by A, A^2, $|A|$, respectively.

14.2.2 Stability, Consistency, Convergence

Any two-level method can be written in the compact form

$$\text{(14.2.8)} \qquad \mathbf{u}^{n+1} = E_{\Delta t}(\mathbf{u}^n) \ ,$$

where $\mathbf{u}^n$ represents the vector of approximations $\{u_j^n \mid j \in \mathbb{Z}\}$ at time t_n. Componentwise we have

$$\text{(14.2.9)} \qquad u_j^{n+1} = E_{\Delta t}(\mathbf{u}^n; j) \ .$$

For instance, for the forward Euler method the operator $E_{\Delta t}$ takes the form

$$\text{(14.2.10)} \qquad E_{\Delta t}(\mathbf{u}^n; j) = u_j^n - \frac{\lambda a}{2}(u_{j+1}^n - u_{j-1}^n) \ .$$

A method is said to be *stable* (in the sense of Lax-Richtmyer) if for each time T there is a constant $C_T > 0$ (possibly depending on T) and a value $\delta_0 > 0$ such that

$$\text{(14.2.11)} \qquad ||\mathbf{u}^n||_\Delta \leq C_T ||\mathbf{u}^0||_\Delta$$

for each $n\Delta t \leq T$ and $0 < \Delta t \leq \delta_0$, $0 < \Delta x \leq \delta_0$. Here,

$$\text{(14.2.12)} \qquad ||\mathbf{v}||_\Delta := \Delta x \sum_{j=-\infty}^{\infty} |v_j|$$

is an approximation of the norm of $L^1(\mathbb{R})$ (see (14.1.5)). Since $\mathbf{u}^n = E_{\Delta t}^n(\mathbf{u}^0)$, stability holds if there exists $\beta \geq 0$ such that for each $0 < \Delta t \leq \delta_0$ and $0 < \Delta x \leq \delta_0$

$$||E_{\Delta t}\mathbf{v}||_\Delta \leq (1 + \beta \Delta t)||\mathbf{v}||_\Delta \quad \forall \mathbf{v} \ .$$

As a matter of fact,

$$||\mathbf{u}^n||_\Delta \leq (1 + \beta \Delta t)^n ||\mathbf{u}^0||_\Delta \leq e^{\beta T}||\mathbf{u}^0||_\Delta$$

for all Δt and n such that $n\Delta t \leq T$, hence (14.2.11) would follow.

We now come to the issue of consistency. Let us define the *local truncation error* by

$$(\mathcal{T}_{\Delta t}(t))_j := \frac{1}{\Delta t}[u(t + \Delta t, x_j) - E_{\Delta t}(\mathbf{u}(t); j)] \ ,$$

where $E_{\Delta t}(\mathbf{u}(t); j)$, depending on the two-level method at hand, is defined as in (14.2.9) with u_j^n substituted by $u(t, x_j)$. A two-level method is said to be *consistent* if

(14.2.13) $$\lim_{\Delta t \to 0} ||\mathcal{T}_{\Delta t}(t)||_{\Delta} = 0 \ .$$

for all $t > 0$ and $0 < \Delta x \leq \delta_0$. In addition, the method is of order q_1 in time and q_2 in space if for all sufficiently smooth initial data there is some constant $C > 0$ such that

(14.2.14) $$||\mathcal{T}_{\Delta t}(t)||_{\Delta} \leq C\left[(\Delta t)^{q_1} + (\Delta x)^{q_2}\right] \ , \quad t > 0 \ ,$$

for all $0 < \Delta t \leq \delta_0$ and $0 < \Delta x \leq \delta_0$. Quite often, in the discretization process Δt and Δx are proportional to one another, i.e., $\Delta t = \kappa \Delta x$ for some positive constant κ. In that case, the method is said to be of *order q*, where $q = \min(q_1, q_2)$. (Let us also point out that whenever method (14.2.8) is used to approximate a steady-state solution, then what matters is the order in space, i.e., the exponent q_2 of Δx.)

Let us finally recall that a scheme is said to be *convergent* if

$$\max_{0 \leq n \leq T/\Delta t} ||u(t_n, \cdot) - \mathbf{u}^n||_{\Delta} \to 0$$

as $\Delta t, \Delta x \to 0$.

The fundamental result of the theory of linear hyperbolic initial value problems is the *Lax-Richtmyer equivalence theorem*, which states that for a consistent method, stability is necessary and sufficient for convergence. A proof of the Lax-Richtmyer equivalence theorem may be found in Richtmyer and Morton (1967) (see also Strikwerda (1989)).

We illustrate some results of stability here below. If we consider, e.g., the Lax-Friedrichs method we obtain

$$||\mathbf{u}^{n+1}||_{\Delta} = \Delta x \sum_j |u_j^{n+1}| \leq \frac{\Delta x}{2}\left[\sum_j |(1 - \lambda a)u_{j+1}^n| + \sum_j |(1 + \lambda a)u_{j-1}^n|\right] \ .$$

Assuming that

(14.2.15) $$|a\lambda| = \left|a\frac{\Delta t}{\Delta x}\right| \leq 1 \ ,$$

then both $1 - \lambda a$ and $1 + \lambda a$ are non-negative, and

$$||\mathbf{u}^{n+1}||_{\Delta} \leq \frac{\Delta x}{2}\left[(1 - a\lambda)\sum_j |u_{j+1}^n| + (1 + a\lambda)\sum_j |u_{j-1}^n|\right]$$

$$\leq \frac{1}{2}(1 - a\lambda)||\mathbf{u}^n||_{\Delta} + \frac{1}{2}(1 + a\lambda)||\mathbf{u}^n||_{\Delta} = ||\mathbf{u}^n||_{\Delta} \ .$$

Thus, (14.2.11) holds with $C_T = 1$ if the stability restriction (14.2.15) is satisfied.

With a similar procedure it can be shown that the upwind method is stable provided Δt and Δx satisfy condition (14.2.15). As a matter of fact, when $a > 0$ the upwind scheme (14.2.1) reads

$$u_j^{n+1} = u_j^n - \lambda a(u_j^n - u_{j-1}^n) \ .$$

Then

$$||\mathbf{u}^{n+1}||_\Delta \le \Delta x \sum_j |(1 - \lambda a)u_j^n| + \Delta x \sum_j |\lambda a u_{j-1}^n| \ .$$

If (14.2.15) holds, then the coefficients of u_j^n and u_{j-1}^n are both nonnegative, and therefore $||\mathbf{u}^{n+1}||_\Delta \le ||\mathbf{u}^n||_\Delta$. The same occurs when $a < 0$.

On the contrary, the centered forward Euler method is unstable under a restriction like (14.2.15). However, it becomes stable if $\Delta t = O((\Delta x)^2)$ (see Strikwerda (1989), p. 49).

The same restriction (14.2.15) (but with the strict inequality) is sufficient to guarantee stability for the leap-frog scheme as well. Indeed, if we multiply the leap-frog equation by $u_j^{n+1} + u_j^{n-1}$ we obtain:

$$\sum_j [(u_j^{n+1})^2 + \lambda a(u_j^{n+1}u_{j+1}^n - u_{j-1}^n u_j^{n+1})]$$

$$= \sum_j [(u_j^{n-1})^2 + \lambda a(u_{j-1}^n u_j^{n-1} - u_j^{n-1}u_{j+1}^n)] \ .$$

Adding in both terms $(u_j^n)^2$, and noticing that for all $k \ge 0$

$$\sum_i u_{i-1}^k u_i^{k+1} = \sum_i [u_i^k u_{i+1}^{k+1} - (u_i^k u_{i+1}^{k+1} - u_{i-1}^k u_i^{k+1})] = \sum_i u_i^k u_{i+1}^{k+1} \ ,$$

we obtain from the previous relation

$$\sum_j G_j^{n+1} = \sum_j G_j^n$$

with

$$G_j^n := (u_j^n)^2 + (u_j^{n-1})^2 + a\lambda(u_j^n u_{j+1}^{n-1} - u_{j+1}^n u_j^{n-1}) \ .$$

Therefore $\sum_j G_j^{n+1} = \sum_j G_j^1$, i.e.,

$$\sum_j [(u_j^{n+1})^2 + (u_j^n)^2 + \lambda a(u_j^{n+1}u_{j+1}^n - u_{j+1}^{n+1}u_j^n)]$$

$$= \sum_j [(u_j^1)^2 + (u_j^0)^2 + \lambda a(u_j^1 u_{j+1}^0 - u_{j+1}^1 u_j^0)] \ .$$

Using the inequality $-x^2 - y^2 \le 2xy \le x^2 + y^2$ we can deduce

$$(1 - |\lambda a|) \sum_j [(u_j^{n+1})^2 + (u_j^n)^2] \le (1 + |\lambda a|) \sum_j [(u_j^1)^2 + (u_j^0)^2] \ .$$

If we assume that $|\lambda a| < 1$, we therefore conclude that for each time T there exists a constant $C_T > 0$ such that for all $n\Delta t \le T$

$$||\mathbf{u}^n||_\Delta \leq C_T(||\mathbf{u}^0||_\Delta^2 + ||\mathbf{u}^1||_\Delta^2)^{1/2} \ .$$

This is a stability inequality, that generalizes (14.2.11) to the case of three-level finite difference schemes.

By a similar procedure it can be shown that the Lax-Wendroff method is stable under the same restriction (14.2.15).

Another suitable approach for proving stability is provided by the von Neumann analysis that is based on Fourier transforming the finite difference equation. This approach is extensively pursued, e.g., in Ritchmyer and Morton (1967) and Strikwerda (1989); see also Vichnevetsky and Bowles (1982). A proof of the stability of the Warming-Beam scheme (based on a von Neumann analysis) is given, e.g., in Hirsch (1988), Sect. 9.3. The stability condition is $|a\lambda| \leq 2$. The restriction on λ is weaker than in (14.2.15). The reason is the following: reminding that the *stencil* of a given method is a graph connecting all grid-points that are involved in the computation of u_j^{n+1}, the stencil of the Warming-Beam scheme uses two rather than one point on one side of the point x_j.

To be more precise, a geometrical interpretation of the stability condition can be given as follows. For a finite difference scheme having, for example, a stencil like the one of Lax-Wendroff, the value u_j^{n+1} depends on values of u^n at the three points x_{j+i}, $i = -1, 0, 1$. Continuing back to $t = 0$ we can see that the solution u_j^{n+1} depends only on initial data at the points x_{j+i}, for $i = -(n+1), ..., (n+1)$. Thus the *numerical domain of dependence* $D_{\Delta t}(t_n, x_j)$ of u_j^n satisfies

$$D_{\Delta t}(t_n, x_j) \subset \left\{ x \in \mathbb{R} \mid |x - x_j| \leq n\Delta x = \frac{t_n}{\lambda} \right\} \ .$$

For any fixed point $(\bar{t}, \bar{x})$ we therefore have

$$D_{\Delta t}(\bar{t}, \bar{x}) \subset \left\{ x \in \mathbb{R} \mid |x - \bar{x}| \leq \frac{\bar{t}}{\lambda} \right\} \ .$$

In particular when going to the limit $\Delta t \to 0$, keeping λ fixed, the numerical domain of dependence becomes

$$D_0(\bar{t}, \bar{x}) = \left\{ x \in \mathbb{R} \mid |x - \bar{x}| \leq \frac{\bar{t}}{\lambda} \right\} \ .$$

Condition (14.2.15) is equivalent to

(14.2.16) $D(\bar{t}, \bar{x}) \subset D_0(\bar{t}, \bar{x}) \ ,$

where $D(\bar{t}, \bar{x})$ is the domain of dependence defined in (14.1.11) (where $p = 1$ and $\lambda_1 = a$, as we are now considering the scalar case).

This condition is necessary for stability. Indeed, if it wasn't satisfied, there would be points y^* in the domain of dependence that wouldn't belong to the numerical domain of dependence. Changing the initial data at y^* would change the exact solution but not the numerical one. This would prevent obtaining convergence, and henceforth stability in view of the Lax-Richtmyer equivalence theorem.

The inequality (14.2.15) is called *CFL condition*, after that Courant, Friedrichs and Lewy (1928) showed that it is necessary in order for the numerical solution to be stable. In particular, this entails that there are no explicit, unconditionally stable, consistent finite difference schemes for hyperbolic initial value problems.

In some of the examples above we have shown that the CFL condition can not only be necessary, but also sufficient for the stability of the numerical scheme. However, there exist methods for which the CFL condition is not sufficient for stability.

Similar conditions of stability hold for the vector case as well. For example, any scheme having a stencil like Lax-Wendroff one when applied to system (14.1.6) needs the following restriction on the time-step:

$$(14.2.17) \qquad \left| \lambda_k \frac{\Delta t}{\Delta x} \right| \le 1 \ , \quad k = 1, ..., p \ ,$$

where $\{\lambda_k \mid k = 1, ..., p\}$ are the eigenvalues of A. These conditions can also be written in the form (14.2.16). In other words, each straight line $x = \overline{x} - \lambda_k(\overline{t} - t)$, $k = 1, ..., p$, must intersect the time-line $t = \overline{t} - \Delta t$ at points $x^{(k)}$ falling within the basis of the stencil (see Fig. 14.2.1).

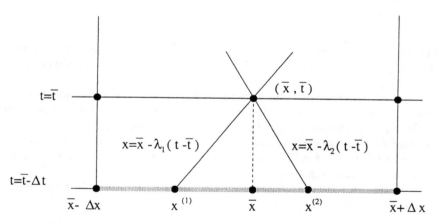

Fig. 14.2.1. Geometrical interpretation of the CFL condition (14.2.17)

Remark 14.2.1 *(Boundary treatment)*. Let us consider the initial-boundary value problem (for $a > 0$)

$$\begin{cases} \dfrac{\partial u}{\partial t} + a\dfrac{\partial u}{\partial x} = 0 \ , \quad t > 0 \ , \ 0 < x < 1 \\[2mm] u(t,0) = \varphi_-(t) \ , \quad t > 0 \\[2mm] u(0,x) = u_0(x) \ , \quad 0 < x < 1 \ . \end{cases}$$

Having chosen a positive integer M, we set $\Delta x := 1/M$ and $x_j := j\Delta x$, $j = 0, ..., M$.

Any finite difference scheme provides a discretization of the differential equation at each internal point x_j, $j = 1, ..., M - 1$. Besides, it must enforce the prescribed boundary condition at the inflow point. Many finite difference schemes, however, also require additional boundary conditions in order to determine a unique solution. These extra conditions are usually referred to as *numerical boundary conditions*. For instance, when using the Lax-Wendroff (or the Lax-Friedrichs) method, the scheme cannot be applied at the outflow point x_M, as it would require the value u_{M+1}^n at the point x_{M+1} which is outside of the computational interval $[0, 1]$. Typically, the value u_M^{n+1} is recovered not from the scheme itself, but through an extrapolation formula that may take several different forms. For instance one can take $u_M^{n+1} = u_{M-1}^{n+1}$ (constant extrapolation) or else $u_M^{n+1} = 2u_{M-1}^{n+1} - u_{M-2}^{n+1}$ (linear extrapolation), but also $u_M^{n+1} = 2u_{M-1}^n - u_{M-2}^{n-1}$ (approximate linear extrapolation on characteristics).

Another possibility consists in deriving numerical boundary conditions using one-sided difference schemes. For instance, for the Lax-Wendroff method we might use

$$u_M^{n+1} = u_M^n - \lambda a(u_M^n - u_{M-1}^n) \ .$$

This formula is obtained by applying the Lax-Wendroff scheme at the outflow point x_M^{n+1} and evaluating u_{M+1}^n by the linear extrapolation formula $u_{M+1}^n = 2u_M^n - u_{M-1}^n$.

We need to emphasize the fact that numerical boundary conditions may sometimes induce instabilities in finite difference schemes that would be stable on the pure initial-value problem. For instance, for the leap-frog scheme both constant and linear extrapolations are unstable, whereas the extrapolation on characteristics is stable. The above conclusions are reversed for the Crank-Nicolson (implicit) scheme.

A thorough analysis of numerical boundary conditions is presented in Strikwerda (1989), Chap. 11.   □

14.2.3 Nonlinear Scalar Equations

We consider the nonlinear, scalar conservation law

$$(14.2.18) \qquad \begin{cases} \dfrac{\partial u}{\partial t} + \dfrac{\partial}{\partial x}F(u) = 0 , & t > 0 , \; x \in \mathbb{R} \\[2mm] u(0,x) = u_0(x) , & x \in \mathbb{R} \end{cases}$$

and denote the characteristic speed by $a(u) := F'(u)$.

We focus on the case of discontinuous solutions. For smooth solutions, the methods we have seen in Section 14.2 generally apply successfully to the linearized equation. However, we know from Section 14.1.4 that discontinuous solutions may easily arise even when data are smooth. Their numerical approximation is a challenging task, and represents a research area which is still very active nowadays. In this Section we simply aim at illustrating a few basic concepts and some examples of shock capturing schemes. For a systematic presentation we refer the reader to Yee (1989), LeVeque (1990), Godlewski and Raviart (1991), and the references therein.

As for the linear case, the general design principle for three-point, explicit finite difference schemes *in conservation form* is

$$(14.2.19) \qquad u_j^{n+1} = u_j^n - \lambda(h_{j+1/2}^n - h_{j-1/2}^n) ,$$

where, as before, $h_{j+1/2}^n := h(u_j^n, u_{j+1}^n)$. $h(\cdot, \cdot)$ is the *numerical flux function*. In the most general case it would be

$$u_j^{n+1} = u_j^n - \lambda[H(u_{j-p}^n, ..., u_{j+q}^n) - H(u_{j-p-1}^n, ..., u_{j+q-1}^n)]$$

for some function H of $p + q + 1$ arguments, $p \geq 0$, $q \geq 0$.

A heuristic explanation of (14.2.19) is provided from the following argument. From (14.2.18) we deduce

$$\int_{x_{j-1/2}}^{x_{j+1/2}} u(t^{n+1}, x)dx = \int_{x_{j-1/2}}^{x_{j+1/2}} u(t^n, x)dx$$

$$- \left[\int_{t^n}^{t^{n+1}} F(u(t, x_{j+1/2}))dt - \int_{t^n}^{t^{n+1}} F(u(t, x_{j-1/2}))dt \right]$$

Dividing by Δx and using the cell average $\bar{u}_j^n$ defined in Section 14.2.3 the above relation gives

$$\bar{u}_j^{n+1} = \bar{u}_j^n - \frac{1}{\Delta x} \left[\int_{t^n}^{t^{n+1}} F(u(t, x_{j+1/2}))dt - \int_{t^n}^{t^{n+1}} F(u(t, x_{j-1/2}))dt \right] .$$

Comparing this to (14.2.19) shows that

$$h_{j+1/2} \simeq \frac{1}{\Delta t} \int_{t^n}^{t^{n+1}} F(u(t, x_{j+1/2}))dt ,$$

i.e., the numerical flux function plays the role of an average flux through $x_{j+1/2}$ over the time interval $[t_n, t_{n+1}]$.

For *consistency* with the conservation law, the numerical flux function is asked to reduce to the true flux F for the case of constant solutions, i.e.,

$$(14.2.20) \qquad h(\overline{u}_0, \overline{u}_0) = F(\overline{u}_0) \quad , \quad \text{if} \quad u(t, x) \equiv \overline{u}_0 \quad .$$

Besides, we must require that h is a Lipschitz continuous function of each variable.

Stability analysis is well established in the linear case, as seen in Section 14.2. However, local stability of linearized equation is neither necessary nor sufficient for the nonlinear case, especially when strong discontinuities are present. A numerical method can be *nonlinearly unstable*, i.e., unstable on the nonlinear problem, even though linearized versions are stable.

Stabilization through adding "linear" numerical dissipation (or "artificial viscosity") has been a common practice (especially in numerical fluid dynamics). This approach alone doesn't guarantee convergence to a physically correct (entropic) solution in the nonlinear case, and not even convergence to a weak solution. However, according to the Lax-Wendroff theorem, *the limit solution (provided it does exist) of any finite difference scheme in conservation form which is consistent with the given conservation law is a weak solution of the conservation law* (Lax and Wendroff (1960)). Unfortunately, it is not assured that the weak solutions obtained in this manner satisfy the entropy condition. As a matter of fact, there are examples of conservative and consistent numerical methods that converge to weak solutions that violate the entropy condition.

As we have seen in Section 14.1, weak solutions are not uniquely determined by their initial values if they don't satisfy the entropy condition. Also, at the finite dimensional level, a discrete form of the entropy condition is needed in order to pick out the physically relevant solutions. After (14.1.36), the discrete form of the entropy inequality can read as follows:

$$\eta(u_j^{n+1}) \leq \eta(u_j^n) - \lambda[\mu_\Delta(\mathbf{u}^n; j) - \mu_\Delta(\mathbf{u}^n; j-1)] \quad .$$

In the above inequality the function μ_Δ is some discrete entropy flux function. It must be consistent with μ in the same way that the numerical flux function $h_{j+1/2}$ is required to be consistent with F.

Let us now recall some definitions. First of all, rewrite (14.2.19) as

$$(14.2.21) \qquad u_j^{n+1} = G(u_{j-1}^n, u_j^n, u_{j+1}^n) \quad .$$

This numerical scheme is said to be *monotone* if G is a monotonic increasing function of each of its arguments. A numerical solution for a single conservation law obtained by a monotone scheme, with $\Delta t / \Delta x$ fixed, always converge to the physically relevant solution as $\Delta t \to 0$. For the proof see Harten, Hyman and Lax (1976), or Crandall and Majda (1980).

The numerical scheme (14.2.21) is said to be *bounded* if there exists $C > 0$ such that

$$\sup_{j,n} |u_j^n| \leq C \ .$$

The scheme is said to be *stable* if the finite difference solutions $\mathbf{u}^n$ and $\mathbf{v}^n$, that are obtained starting from two different initial data $\mathbf{u}^0$ and $\mathbf{v}^0$, satisfy:

$$||\mathbf{u}^n - \mathbf{v}^n||_\Delta \leq C_T ||\mathbf{u}^0 - \mathbf{v}^0||_\Delta$$

for all $n \geq 0$, where $C_T > 0$ and $|| \cdot ||_\Delta$ is the norm introduced in (14.2.12). (For linear problems this definition is equivalent to (14.2.11).)

According to a result of Kuznetsov, monotone schemes are bounded, stable, convergent to the entropy solution and at most first order accurate (Kuznetsov (1976), Harten, Hyman and Lax (1976)). They produce smooth behaviour near discontinuities.

Hereby, we report several instances of numerical flux functions that generate, throughout (14.2.19), some well known finite difference schemes.

Setting for simplicity $F_j := F(u_j)$, the *Lax-Friedrichs* scheme corresponds to taking

$$(14.2.22) \qquad h_{j+1/2} = \frac{1}{2}\left[F_{j+1} + F_j - \frac{1}{\lambda}(u_{j+1} - u_j)\right]$$

and reads therefore

$$(14.2.23) \qquad u_j^{n+1} = \frac{1}{2}(u_{j+1}^n + u_{j-1}^n) - \frac{\lambda}{2}(F(u_{j+1}^n) - F(u_{j-1}^n)) \ .$$

This method can be shown to be consistent and first order accurate. It is monotone provided the CFL condition

$$\left| F'(u_j^n)\frac{\Delta t}{\Delta x} \right| \leq 1 \quad \text{for all } j \in \mathbb{Z} \text{ and all } n \in \mathbb{N}$$

is satisfied.

The *Lax-Wendroff* method is associated to the numerical flux

$$(14.2.24) \qquad h_{j+1/2} = \frac{1}{2}[F_{j+1} + F_j - \lambda a_{j+1/2}(F_{j+1} - F_j)]$$

and therefore has the form

$$(14.2.25) \qquad \begin{aligned} u_j^{n+1} = \ &u_j^n - \frac{\lambda}{2}(F(u_{j+1}^n) - F(u_{j-1}^n)) \\ &+ \frac{\lambda^2}{2}[a_{j+1/2}^n(F(u_{j+1}^n) - F(u_j^n)) - a_{j-1/2}^n(F(u_j^n) - F(u_{j-1}^n))] \end{aligned}$$

Here, we have set

$$a_{j+1/2} := F'\left(\frac{u_{j+1} + u_j}{2}\right) \ .$$

This method is second order accurate (on smooth solutions), not monotone. Unfortunately, it requires the use of the Jacobian $F'(\cdot)$, which may not be an easy task when facing systems of equations. An alternative, still second order accurate and conservative, that avoids using $F'(\cdot)$ is furnished by the following modification of the Lax-Wendroff scheme:

$$(14.2.26) \qquad \begin{aligned} u_{j+1/2}^{n+1/2} &= \frac{1}{2}(u_{j+1}^n + u_j^n) - \frac{\lambda}{2}[F(u_{j+1}^n) - F(u_j^n)] \\ u_j^{n+1} &= u_j^n - \lambda[F(u_{j+1/2}^{n+1/2}) - F(u_{j-1/2}^{n+1/2})] \ . \end{aligned}$$

The numerical flux is given by

$$(14.2.27) \qquad h_{j+1/2} = F\left(\frac{1}{2}(u_{j+1} + u_j) - \frac{\lambda}{2}(F_{j+1} - F_j)\right) \ .$$

Another alternative is the *MacCormack* method, that uses first forward and then backward differencing:

$$(14.2.28) \qquad \begin{aligned} u_j^* &= u_j^n - \lambda[F(u_{j+1}^n) - F(u_j^n)] \\ u_j^{n+1} &= \frac{1}{2}(u_j^n + u_j^*) - \frac{\lambda}{2}[F(u_j^*) - F(u_{j-1}^*)] \ . \end{aligned}$$

The scheme is second order accurate and conservative, and therefore guarantees correct shock capturing, but is not dissipative. It reduces to the Lax-Wendroff method for the linear scalar advection equation. The numerical flux is given by

$$(14.2.29) \qquad h_{j+1/2} = \frac{1}{2}F_{j+1} + \frac{1}{2}F\big(u_j - \lambda(F_{j+1} - F_j)\big)$$

The *Courant-Isaacson-Rees* method reduces instead to the upwind method for the linear scalar advection equation. For the nonlinear problem (14.2.18), in which u is constant along characteristics, it consists in computing u_j^{n+1} by an approximation to u at the point $x_j - F'(u_j^n)\Delta t$ obtained by linear interpolation of the data $\mathbf{u}^n$. The scheme reads (if $F'(u_j^n) > 0$):

$$u_j^{n+1} = u_j^n - \lambda F'(u_j^n)(u_j^n - u_{j-1}^n) \ .$$

In general, it can be written as:

$$(14.2.30) \qquad \begin{aligned} u_j^{n+1} = u_j^n - \lambda &\left[\frac{|F'(u_j^n)| + F'(u_j^n)}{2}(u_j^n - u_{j-1}^n) \right. \\ &\left. + \frac{|F'(u_j^n)| - F'(u_j^n)}{2}(u_{j+1}^n - u_j^n)\right] \ . \end{aligned}$$

Since it is not in conservation form, this method is unsuitable for approximating shock waves.

The *Godunov* method makes use of characteristic information without compromising the conservation form. The breakthrough is that Riemann problems are solved forward in time, rather than following characteristics backward in time. For a given n, a discrete solution $\mathbf{u}^n$ is used to define a piecewise-constant function $v^n(x)$ having the value u_j^n on the whole interval $x_{j-1/2} < x < x_{j+1/2}$. Then, $v^n(x)$ is used as initial datum at time t_n for the conservation law considered on the time interval $(t_n, t_{n+1}]$. If Δt is small enough, the equation can in fact be solved exactly through facing a sequence of Riemann problems. Indeed, Δt has to satisfy a CFL-like condition in order to prevent interaction between neighbouring Riemann problems. The solution $u_*^n(t, x)$ defined on $[t_n, t_{n+1}] \times \mathbb{R}$ is obtained by simply collecting all these Riemann solutions. Finally, the discrete solution $\mathbf{u}^{n+1}$ at the new time-level t_{n+1} is defined by averaging u_*^n as follows:

$$u_j^{n+1} := \frac{1}{\Delta x} \int_{x_{j-1/2}}^{x_{j+1/2}} u_*^n(t_{n+1}, x) dx \quad .$$

Based on these values, a new piecewise-constant function $v^{n+1}(x)$ is constructed and the process is repeated.

The resulting scheme can be written under the conservation form (14.2.19), where the numerical flux function is

$$(14.2.31) \qquad h_{j+1/2}^n = \frac{1}{\Delta t} \int_{t^n}^{t^{n+1}} F(u_*^n(t, x_{j+1/2})) dt \quad ,$$

and it turns out to be monotone and upwind, and satisfies the entropy condition. It can also be shown that the numerical flux function is given by

$$h_{j+1/2} = \begin{cases} \min\{F(u) \mid u_j \leq u \leq u_{j+1}\} & , \quad \text{if } u_j < u_{j+1} \\ \max\{F(u) \mid u_{j+1} \leq u \leq u_j\} & , \quad \text{if } u_j > u_{j+1} \end{cases} \quad .$$

On a linear problem the Godunov scheme reduces to the upwind scheme.

We also want to recall another monotone and upwind method in conservation form, the *Engquist-Osher* scheme, which corresponds to the following numerical flux function

$$(14.2.32) \qquad h_{j+1/2} = F_j + F_{j+1}^- - F_j^- = F_{j+1} - F_{j+1}^+ + F_j^+ \quad ,$$

where

$$F_j^+ := F(\max(u_j, \overline{u})) \quad , \quad F_j^- := F(\min(u_j, \overline{u}))$$

and $\overline{u}$ is the sonic point of $F(u)$, i.e., the value $\overline{u}$ such that $F(\overline{u}) = 0$.

14.2.4 High Order Shock Capturing Schemes

Monotone and first order upwind schemes are too dissipative, therefore they cannot produce accurate solutions for complex flow fields without using a very fine grid. For this purpose, higher order shock capturing schemes ought to be used, with the purpose of including minimal numerical dissipation. (Clearly, some dissipation needs to be used in order to give non-oscillatory shocks and ensure convergence to the entropy solution.)

The basic idea underlying high order methods (say, second order at least), is to move from a potentially high order method and modify it by adding numerical dissipation in the neighbourhood of a discontinuity. The extra term can model a diffusive term proportional to $\frac{\partial^2 u}{\partial x^2}$ (sometimes, $\frac{\partial^4 u}{\partial x^4}$). The coefficient, say ε, must vanish quickly enough as Δt, Δx go to 0, so that the scheme remains consistent with the hyperbolic equation without affecting its high order of accuracy on smooth solutions.

There are two classes of such schemes. Classical methods make use of *linear* numerical dissipation (i.e., the same for any point). Instances are given by the methods of Lax-Wendroff, MacCormack and Warming-Beam.

Modern methods use *nonlinear* numerical dissipation. This means that the diffusion coefficient ε depends on behaviour of the solution itself, being larger near discontinuities than in smooth regions. Second and third (or higher) order space accuracy is achieved by these approaches. Their design principle is:

a. use the conservation form (this ensures the shock capturing property);
b. no spurious overshoots or wiggles are created near discontinuities yet providing sharp shocks;
c. high accuracy is achieved in smooth regions of the flow;
d. entropy condition is satisfied for limit solutions.

Conventional methods (e.g., Lax-Wendroff type schemes) typically don't match requisites b. and d.

Schemes are developed, at first, for a scalar, nonlinear equation in one-space dimension. Extensions to systems come through Riemann solvers (either exact or approximate), or flux vector splitting techniques.

An in-depth analysis of these methods is beyond the scope of this book. Instead, we confine ourselves to illustrate the main concepts and present a few examples falling into this cathegory.

We recall that the total variation of a smooth function $u : \mathbb{R} \to \mathbb{R}$ can be defined as the (finite) value

$$TV(u) := \int_{-\infty}^{\infty} |u'(x)| dx \ .$$

If $\mathbf{u}^n$ is a grid-function, its total variation can be defined as follows

$$(14.2.33) \qquad TV(\mathbf{u}^n) := \sum_j |u_{j+1}^n - u_j^n| \ .$$

A difference method is said to be *Total Variation Diminishing* (briefly TVD) if

$$(14.2.34) \qquad TV(\mathbf{u}^{n+1}) \leq TV(\mathbf{u}^n) \qquad \forall \, n \geq 0 \ , \quad \Delta t > 0 \ .$$

Even if conservative and consistent with the conservation law, a TVD scheme is not necessarily consistent with an entropy inequality.

Monotone methods for scalar conservation laws are TVD (for the proof see Crandall and Majda (1980)).

A family of *five-point* difference schemes in conservation form can be written as

$$(14.2.35) \quad u_j^{n+1} + \theta\lambda(h_{j+1/2}^{n+1} - h_{j-1/2}^{n+1}) = u_j^n - (1-\theta)\lambda(h_{j+1/2}^n - h_{j-1/2}^n) \; ,$$

where $h_{j+1/2}$ denotes the numerical flux function, and in this case it reads

$$h_{j+1/2} = h(u_{j-1}, u_j, u_{j+1}, u_{j+2}) \; .$$

As usual, $\lambda := \Delta t/\Delta x$, while $0 \le \theta \le 1$ is a parameter. If $\theta = 0$ the scheme is explicit, otherwise it is implicit.

We rewrite (14.2.35) as

$$L(\mathbf{u}^{n+1}) = R(\mathbf{u}^n) \; ,$$

where L and R are the finite difference operators:

$$L(\mathbf{u})_j := u_j + \theta\lambda(h_{j+1/2} - h_{j-1/2})$$
$$R(\mathbf{u})_j := u_j - (1-\theta)\lambda(h_{j+1/2} - h_{j-1/2}) \; .$$

As noticed by Harten (1984), sufficient conditions for (14.2.35) to be TVD are

$$TV(R(\mathbf{u}^n)) \le TV(\mathbf{u}^n) \quad \text{and} \quad TV(L(\mathbf{u}^{n+1})) \ge TV(\mathbf{u}^{n+1}) \; .$$

Let us define the new flux

$$(14.2.36) \qquad \overline{h}_{j+1/2}^n := (1-\theta)h_{j+1/2}^n + \theta h_{j+1/2}^{n+1} \; .$$

Thus $\overline{h}_{j+1/2}^n = \overline{h}(u_{j-1}^n, u_j^n, u_{j+1}^n, u_{j+2}^n, u_{j-1}^{n+1}, u_j^{n+1}, u_{j+1}^{n+1}, u_{j+2}^{n+1})$ for a suitable flux function $\overline{h}$. Consistency with (14.1.20) is achieved provided the function $\overline{h}$ satisfies

$$(14.2.37) \qquad \overline{h}(\underbrace{u, u, ..., u}_{8 \text{ times}}) = F(u) \; .$$

Then (14.2.35) becomes

$$(14.2.38) \qquad u_j^{n+1} = u_j^n - \lambda(\overline{h}_{j+1/2}^n - \overline{h}_{j-1/2}^n)$$

(notice the analogy with (14.2.4) for the linear scalar equation).

To give an example of a TVD scheme, let us define

$$\tilde{a}_{j+1/2} := \frac{F_{j+1} - F_j}{u_{j+1} - u_j} \; .$$

For a numerical flux function of the form

(14.2.39) $$h_{j+1/2} = \frac{1}{2}[F_{j+1} + F_j - |\tilde{a}_{j+1/2}|(u_{j+1} - u_j)]$$

(first order Roe upwind flux) a sufficient condition for being TVD is the (CFL-like) condition:

$$|\lambda \tilde{a}_{j+1/2}| \leq \frac{1}{1-\theta}$$

(see Harten (1984)).

We now comment on the design principles for higher order TVD schemes. Most high order TVD schemes can be viewed as centered-difference schemes with an appropriate numerical dissipation or smoothing mechanism, i.e., an automatic feed-back mechanism to control the amount of numerical dissipation (unlike the numerical dissipation used in linear theory).

Generally speaking, given any high order flux function $h(\mathbf{u}; j)$ associated with the scheme (14.2.19), the addition of artificial dissipation amounts to replace this by the modified flux function

(14.2.40) $$h^{art}(\mathbf{u}; j) := h(\mathbf{u}; j) - \Delta x\, \varepsilon(\mathbf{u}; j)\, (u_{j+1} - u_j) \ .$$

Sometimes, instead of (14.2.40), one considers the more general form:

$$h^{art}(\mathbf{u}; j) := h(\mathbf{u}; j) + \Phi(\mathbf{u}; j) \ ,$$

so that the numerical flux function becomes

$$h_{j+1/2} = \frac{1}{2}(F_{j+1} + F_j + \Phi_{j+1/2}) \ .$$

In other words, $h_{j+1/2}$ is the sum of a second order centered scheme and an artificial dissipation (this is the TVD correction).

The critical issue is how to determine an appropriate form of $\varepsilon(\cdot; \cdot)$ (or $\Phi(\cdot; \cdot)$) that achieves both goals of introducing enough dissipation to preserve monotonicity without affecting the level of accuracy.

The fulfillment of this task is not easy to achieve. For this reason, in practice, a more direct approach is often used. Precisely, the main mechanisms for satisfying higher order TVD sufficient conditions are some kind of limiting procedures called *limiters*. They impose constraints on the gradient of the dependent variable u (*slope limiters*) or the flux function F (*flux limiters*). For constant coefficients, the two types of limiters are equivalent.

Let us notice that any high order flux, say h^{high}, can generally be expressed as a low order flux h^{low} plus a correction, i.e.,

$$h^{high}(\mathbf{u}; j) = h^{low}(\mathbf{u}; j) + [h^{high}(\mathbf{u}; j) - h^{low}(\mathbf{u}; j)] \ .$$

A flux limiter method is typically associated with a flux of the form

(14.2.41) $$h^{lim}(\mathbf{u}; j) := h^{low}(\mathbf{u}; j) + \Phi(\mathbf{u}; j)[h^{high}(\mathbf{u}; j) - h^{low}(\mathbf{u}; j)] \ .$$

The correction term is usually referred to as an *antidiffusive flux*, as it compensates the high numerical diffusion that is introduced by the low order flux.

Based on this idea is the so-called *Flux-Corrected-Transport* (FCT) method of Boris and Book (1973), one of the earliest high resolution methods.

A common approach is to set $\Phi(\mathbf{u}; j) = \varphi(\theta_j)$, with $\theta_j := (u_j - u_{j-1})/(u_{j+1} - u_j)$. Two instances are the so-called "superbee" limiter of Roe (1985), for which $\varphi(\theta) := \max(0, \min(1, 2\theta), \min(\theta, 2))$, and the limiter $\varphi(\theta) := (\theta + |\theta|)/(1 + |\theta|)$ of van Leer (1974).

It can be shown that TVD schemes degrade to first order accuracy at local extrema of the solution. Indeed, the formal order of accuracy of TVD schemes is achieved only where the solution is monotone. For this reason, spatial order of accuracy of TVD schemes refers to the numerical solution that is away from shocks.

There are either *upwind* TVD or *symmetric* (or *centered*) TVD schemes. The term "upwind" and "symmetric" pertains to the scheme *before* the application of limiters. Dissipation terms are upwind weighted in the former but not in the latter. In general, symmetric are slightly more diffusive but requires less operation count than upwind TVD schemes.

A very popular TVD scheme is *MUSCL (Monotonic Upstream Centered Schemes for Conservation Laws)*, introduced by van Leer (1979). The idea is to replace, in the Godunov scheme, piecewise-constant initial data of the Riemann problem with piecewise-*parabolic* initial data. This yields second order accuracy in space.

Clearly, a similar approach can be pursued starting from other kinds of numerical flux function. For instance, consider the second order Roe numerical flux given by:

(14.2.42)
$$
h_{j+1/2} = \frac{1}{2}\left[F(u^R_{j+1/2}) + F(u^L_{j+1/2}) \right.
$$
$$
\left. - \left| \frac{F(u^R_{j+1/2}) - F(u^L_{j+1/2})}{u^R_{j+1/2} - u^L_{j+1/2}} \right| (u^R_{j+1/2} - u^L_{j+1/2}) \right] .
$$

The MUSCL approach applied to this flux involves using $u^R_{j+1/2}$ and $u^L_{j+1/2}$ defined by

$$
u^R_{j+1/2} := u_{j+1} - \frac{1}{4}[(1-\eta)\Delta_{j+3/2}u + (1+\eta)\Delta_{j+1/2}u]
$$
$$
u^L_{j+1/2} := u_j + \frac{1}{4}[(1-\eta)\Delta_{j-1/2}u + (1+\eta)\Delta_{j+1/2}u] ,
$$

where
$$
\Delta_{j+1/2}u := u_{j+1} - u_j .
$$

The value of η determines the spatial order of accuracy. Some instances are: $\eta = -1$ (fully upwind scheme), $\eta = 0$ (Fromm scheme), $\eta = 1/3$ (third-order upwind biased scheme, the bias being due to the existence of a downstream point in the stencil), $\eta = 1$ (three-point centered-difference scheme).

Unwanted oscillations can be removed using various slope limiters. This is accomplished by setting

$$u_{j+1/2}^R := u_{j+1} - \frac{1}{4}[(1-\eta)\Delta_{j+3/2}^* u + (1+\eta)\Delta_{j+1/2}^{**} u]$$

$$u_{j+1/2}^L := u_j + \frac{1}{4}[(1-\eta)\Delta_{j-1/2}^{**} u + (1+\eta)\Delta_{j+1/2}^* u]$$

for suitable Δ^* and Δ^{**} operators. For instance, the *minmod* limiter is defined as follows:

$$\Delta_{j+1/2}^* u := \text{minmod}(\Delta_{j+1/2} u, \omega \Delta_{j-1/2} u)$$

$$\Delta_{j+1/2}^{**} u := \text{minmod}(\Delta_{j+1/2} u, \omega \Delta_{j+3/2} u) \ ,$$

where

$$\text{minmod}(x, \omega y) := \text{sign}(x) \max\{0, \min[|x|, \omega\, y\, \text{sign}(x)]\}$$

and $1 \le \omega \le (3-\eta)/(1-\eta)$, $-1 \le \eta < 1$. With compact form this scheme reads

$$(14.2.43) \qquad u_j^{n+1} = u_j^n - \lambda[h(u_{j+1/2}^{R,n}, u_{j+1/2}^{L,n}) - h(u_{j-1/2}^{R,n}, u_{j-1/2}^{L,n})]$$

(notice the formal analogy between (14.2.4) and (14.2.43), as well as the one between (14.2.39) and (14.2.42)).

Other MUSCL TVD schemes are those of Harten (1984), Roe (1985), Sweby (1984), and Osher and Chakravarthy (1986) (see also Osher (1985)), whereas TVD schemes that are not MUSCL are the methods of Davis (1984), Roe (1984) and Yee (1987).

Among the latest developments we mention the TVB and ENO schemes. The numerical method (14.2.35) for the initial-value problem (14.1.20) is said to be *Total Variation Bounded* (briefly, TVB) in $[0, T]$ if there exists $B = B(\mathbf{u}^0) > 0$ such that

$$(14.2.44) \qquad\qquad TV(\mathbf{u}^n) \le B$$

for all n, Δt such that $n\Delta t \le T$.

Obviously, a TVD scheme is also TVB. Advantages of TVB over TVD can be outlined as follows.

a. TVB schemes can be *uniformly* higher order accurate in space including *extrema* points;
b. it is easier to combine interior and boundary schemes in order to obtain a global TVB scheme.

Shu (1987) has shown how to modify some existing TVD schemes (which can be second order accurate at most) so that the resulting schemes are TVB.

ENO, that stands for *Essentially Non-Oscillatory*, is a class of methods introduced by Harten (1986), Harten and Osher (1987). The idea is the following:

a. at the time level t^n, take u_j^n as the cell average on the interval $I_j = (x_{j-1/2}, x_{j+1/2})$;

b. reconstruct a polynomial approximation to $u(t_n, x)$ within cell I_j in a non-oscillatory fashion, up to a desired accuracy;

c. solve the differential equation using the flux given by this polynomial to advance the solution up to the level t_{n+1};

d. repeat the process at time level t_{n+1}.

When accomplishing the interpolation procedure at step b., the idea is to take information from the smoother part of the flow and to choose the interpolant having smaller magnitude. This is a nonlinear interpolant.

It can be also proven that any TVD scheme is a ENO scheme. For an algorithmic description of this method, the reader is referred to Harten, Osher, Engquist and Chakravarthy (1987). An effective implementation based on reconstruction of fluxes is presented in Shu and Osher (1988, 1989). The latter papers extend to ENO schemes the result by Osher and Chakravarthy (1986) on reconstruction of fluxes for MUSCL schemes.

All ENO schemes with moving stencil may permit oscillations up to the order of the truncation error (see Harten, Osher, Engquist and Chakravarthy (1987)). Furthermore, wider stencils are generally involved (e.g., a seven-point stencil is needed for a second order ENO scheme).

14.3 Approximation by Finite Elements

For simplicity, until now in this Chapter we have only considered problems in one-space dimension. However, when introducing a finite element approximation, there is no formal difference in presenting results for multi-dimensional problems. Therefore, below, Ω will denote a domain in $\mathbb{R}^d$, $d = 2, 3$.

Let us consider the linear scalar advection equation

$$(14.3.1) \qquad \begin{cases} \dfrac{\partial u}{\partial t} + \mathbf{a} \cdot \nabla u + a_0 u = f & \text{in } Q_T := (0, T) \times \Omega \\[2mm] u = \varphi & \text{on } \Sigma_T^{in} := (0, T) \times \partial \Omega^{in} \\[2mm] u_{|t=0} = u_0 & \text{on } \Omega \ , \end{cases}$$

where $\mathbf{a} = \mathbf{a}(\mathbf{x})$, $a_0 = a_0(t, \mathbf{x})$ and $f = f(t, \mathbf{x})$ are given functions, as well as $\varphi = \varphi(t, \mathbf{x})$ and $u_0 = u_0(\mathbf{x})$. Moreover, we have defined the *inflow* boundary $\partial \Omega^{in}$ as

$$(14.3.2) \qquad \partial \Omega^{in} := \{ \mathbf{x} \in \partial \Omega \mid \mathbf{a}(\mathbf{x}) \cdot \mathbf{n}(\mathbf{x}) < 0 \} \ ,$$

$\mathbf{n}$ being the unit outward normal vector on $\partial \Omega$. For the sake of simplicity, we have assumed that $\mathbf{a}$ doesn't depend on t; therefore the inflow boundary $\partial \Omega^{in}$ is the same at any time.

14.3.1 Galerkin Method

A finite element semi-discrete approximation of (14.3.1) can be set as follows. Define the spaces

(14.3.3)
$$V_h = X_h^k := \{v_h \in C^0(\overline{\Omega}) \,|\, v_{h|K} \in \mathbb{P}_k(K) \; \forall \, K \in \mathcal{T}_h\}$$
$$V_h^{in} := \{v_h \in V_h \,|\, v_{h|\partial\Omega^{in}} = 0\} \; ,$$

where, as usual, $\mathcal{T}_h$ is a finite element family of triangulations of Ω and K are its triangles (see Chapter 3). Furthermore, let us denote by $u_{0,h}$ and φ_h two suitable finite element approximations of u_0 and φ, respectively. We consider the problem: for any $t \in [0, T]$ find $u_h(t) \in V_h$ such that

(14.3.4)
$$\begin{cases} \dfrac{d}{dt}(u_h(t), v_h) + (\mathbf{a} \cdot \nabla u_h(t), v_h) + (a_0(t)u_h(t), v_h) \\ \qquad\qquad = (f(t), v_h) \quad \forall \, v_h \in V_h^{in} \; , \quad t \in (0, T) \\[4pt] u_h(t) = \varphi_h(t) \quad \text{on } \partial\Omega^{in} \\[4pt] u_h(0) = u_{0,h} \; . \end{cases}$$

Clearly, the variational formulation of (14.3.1) takes the same form of (14.3.4), just dropping out any index h and defining $V = H^1(\Omega)$ and $V^{in} = H^1_{\partial\Omega^{in}}(\Omega) := \{v \in H^1(\Omega) \,|\, v_{|\partial\Omega^{in}} = 0\}$.

A stability analysis for (14.3.4) is easily performed by the energy method. Assume for simplicity that φ (and therefore φ_h) is equal to 0. Then $u_h(t) \in V_h^{in}$, and we are allowed to take $v_h = u_h(t)$ to get:

(14.3.5)
$$\frac{1}{2}\frac{d}{dt}\|u_h(t)\|_0^2 + a_1(u_h(t), u_h(t)) = (f(t), u_h(t)) \; ,$$

where we have set:

(14.3.6) $a_1(z, v) := (\mathbf{a} \cdot \nabla z, v) + (a_0 z, v) \quad \forall \, z \in H^1(\Omega) \; , \quad v \in L^2(\Omega) \; .$

Integrating by parts we obtain

(14.3.7)
$$a_1(z, z) = \int_\Omega \left(-\frac{1}{2}\operatorname{div}\mathbf{a} + a_0 \right) z^2 d\mathbf{x}$$
$$+ \frac{1}{2}\int_{\partial\Omega\setminus\partial\Omega^{in}} \mathbf{a} \cdot \mathbf{n}\, z^2 d\gamma \quad \forall \, z \in V^{in} \; .$$

Let us set

(14.3.8)
$$\mu(t, \mathbf{x}) := -\frac{1}{2}\operatorname{div}\mathbf{a}(\mathbf{x}) + a_0(t, \mathbf{x}) \; .$$

If $\mu(t, \mathbf{x}) > 0$ for all $(t, \mathbf{x}) \in \overline{Q_T}$, taking $\mu_0(t) := \min_{\overline{\Omega}} \mu(t, \mathbf{x})$ and using the inequality

$$(f(t), u_h(t)) \leq \frac{1}{2}\mu_0(t)||u_h(t)||_0^2 + \frac{1}{2\mu_0(t)}||f(t)||_0^2$$

we easily deduce from (14.3.5) the stability inequality:

$$
\begin{aligned}
(14.3.9) \quad & ||u_h(t)||_0^2 + \int_0^t \mu_0(\tau)||u_h(\tau)||_0^2 d\tau + \int_0^t \int_{\partial\Omega\setminus\partial\Omega^{in}} \mathbf{a}\cdot\mathbf{n}\, u_h^2(\tau) d\gamma\, d\tau \\
& \leq ||u_{0,h}||_0^2 + \int_0^t \frac{1}{\mu_0(\tau)}||f(\tau)||_0^2 d\tau \quad \forall\, t \in [0, T] \ .
\end{aligned}
$$

When $\mu(t, \mathbf{x})$ isn't positive, a straightforward application of Gronwall lemma (see Lemma 1.4.1) entails the estimate

$$
\begin{aligned}
(14.3.10) \quad & ||u_h(t)||_0^2 + \int_0^t \int_{\partial\Omega\setminus\partial\Omega^{in}} \mathbf{a}\cdot\mathbf{n}\, u_h^2(\tau) d\gamma\, d\tau \\
& \leq \left(||u_{0,h}||_0^2 + \int_0^t ||f(\tau)||_0^2 d\tau \right) \exp\left(\int_0^t [1 + 2\mu^*(\tau)] d\tau \right)
\end{aligned}
$$

for each $t \in [0, T]$, where $\mu^*(t) := \max_{\overline{\Omega}} |\mu(t, \mathbf{x})|$.

In the above problem (14.3.4) the boundary condition has been enforced in a *strong* (or essential) way. A *weak* treatment of the boundary terms can also be accomplished, as stated in the following finite element problem: for any $t \in [0, T]$ find $u_h(t) \in V_h$ such that

$$
(14.3.11) \quad
\begin{cases}
\dfrac{d}{dt}(u_h(t), v_h) + a_1(u_h(t), v_h) - \displaystyle\int_{\partial\Omega^{in}} \mathbf{a}\cdot\mathbf{n}\, u_h(t) v_h d\gamma \\
\qquad = (f(t), v_h) - \displaystyle\int_{\partial\Omega^{in}} \mathbf{a}\cdot\mathbf{n}\, \varphi_h(t) v_h d\gamma \quad \forall\, v_h \in V_h \\
u_h(0) = u_{0,h}
\end{cases}
$$

for each $t \in (0, T)$, where $V_h = X_h^k$ as before. Notice that now the test functions don't vanish on $\partial\Omega^{in}$ any longer.

A stability analysis can still be accomplished taking $v_h = u_h(t)$. Using (14.3.7) we obtain from (14.3.11):

$$
\begin{aligned}
\frac{1}{2}\frac{d}{dt}||u_h(t)||_0^2 & + (\mu(t)u_h(t), u_h(t)) \\
& + \frac{1}{2}\int_{\partial\Omega\setminus\partial\Omega^{in}} \mathbf{a}\cdot\mathbf{n}\, u_h^2(t) d\gamma + \frac{1}{2}\int_{\partial\Omega^{in}} |\mathbf{a}\cdot\mathbf{n}| u_h^2(t) d\gamma \\
& = (f(t), u_h(t)) + \int_{\partial\Omega^{in}} |\mathbf{a}\cdot\mathbf{n}|\varphi_h(t) u_h(t) d\gamma \ .
\end{aligned}
$$

If $\mu_0(t) > 0$ for each $t \in [0, T]$ we can easily deduce that

$$\|u_h(t)\|_0^2 + \int_0^t \mu_0(\tau)\|u_h(\tau)\|_0^2 d\tau + \int_0^t \int_{\partial\Omega\setminus\partial\Omega^{in}} \mathbf{a}\cdot\mathbf{n}\, u_h^2(\tau) d\gamma\, d\tau$$

(14.3.12)

$$\leq \|u_{0,h}\|_0^2 + \int_0^t \int_{\partial\Omega^{in}} |\mathbf{a}\cdot\mathbf{n}|\varphi_h^2(\tau) d\gamma\, d\tau + \int_0^t \frac{1}{\mu_0(\tau)}\|f(\tau)\|_0^2 d\tau$$

for each $t \in [0,T]$. In the most general case in which $\mu(t,\mathbf{x})$ may change sign, by using Gronwall lemma we obtain the following inequality:

$$\|u_h(t)\|_0^2 + \int_0^t \int_{\partial\Omega\setminus\partial\Omega^{in}} \mathbf{a}\cdot\mathbf{n}\, u_h^2(\tau) d\gamma\, d\tau$$

(14.3.13)

$$\leq \left(\|u_{0,h}\|_0^2 + \int_0^t \int_{\partial\Omega^{in}} |\mathbf{a}\cdot\mathbf{n}|\varphi_h^2(\tau) d\gamma\, d\tau \right.$$

$$\left. + \int_0^t \|f(\tau)\|_0^2 d\tau \right) \exp\left(\int_0^t [1 + 2\mu^*(\tau)] d\tau \right)$$

for each $t \in [0,T]$.

The convergence analysis can be performed in the following way. Let us consider, for instance, formulation (14.3.4), assuming for simplicity that $\varphi = \varphi_h = 0$ and that $\mu(t,\mathbf{x}) \geq \mu_0 > 0$ for each $(t,\mathbf{x}) \in \overline{Q_T}$. (We noticed in Section 12.1 that this last condition is not restrictive, provided that $\mu(t,\mathbf{x})$ is bounded from below, as we can operate the change of variable $u_\lambda(t,\mathbf{x}) := e^{-\lambda t}u(t,\mathbf{x})$ for λ large enough.) Let us introduce the semi-norm

(14.3.14) $$|v|_{\mathbf{a},\Gamma} := \left(\int_\Gamma |\mathbf{a}\cdot\mathbf{n}|v^2 d\gamma \right)^{1/2} ,$$

where Γ is a (non-empty) subset of $\partial\Omega$. We remind that the interpolant $\pi_h^k(v)$ of v belongs to V_h^{in} and satisfies

(14.3.15) $\|v - \pi_h^k(v)\|_0 + h\|v - \pi_h^k(v)\|_1 + h^{1/2}|v - \pi_h^k(v)|_{\mathbf{a},\partial\Omega} \leq Ch^{k+1}|v|_{k+1}$

(see Theorem 3.4.2; the proof of the estimate in the norm $|\cdot|_{\mathbf{a},\partial\Omega}$ can be obtained by a similar procedure). Proceeding now as in the proof of Proposition 11.2.2, setting this time $w_1(t) := u_h(t) - \pi_h^k(u(t))$ and $w_2(t) := \pi_h^k(u(t)) - u(t)$, it follows that

(14.3.16) $$\max_{t\in[0,T]} \|u(t) - u_h(t)\|_0 + \left(\int_0^T |u(t) - u_h(t)|_{\mathbf{a},\partial\Omega}^2 dt \right)^{1/2}$$

$$= O(\|u_0 - u_{0,h}\|_0 + h^k) ,$$

provided $u_0 \in H^k(\Omega)$, $u \in L^2(0,T;H^{k+1}(\Omega))$ and $\frac{\partial u}{\partial t} \in L^2(0,T;H^k(\Omega))$. This rate of convergence is not the optimal one (namely, $O(h^{k+1})$ for the norm $\|\cdot\|_0$ and $O(h^{k+1/2})$ for the norm $|\cdot|_{\mathbf{a},\partial\Omega}$). Moreover, no control on the first order derivatives of the error is obtained. It should also be pointed out that the solution of (14.3.1) is smooth only if the data are smooth. In fact, discontinuities (or steep gradients) are propagated by the characteristics.

14.3.2 Stabilization of the Galerkin Method

Other methods which have better performances are the stabilization methods of SUPG, GALS or DWG type (see Sections 8.3 and 12.3). Still assuming for simplicity that $\varphi = \varphi_h = 0$ and moreover that $|\mathbf{a}(\mathbf{x})| \geq a_* > 0$ for each $\mathbf{x} \in \Omega$, we can consider the following problem: for any $t \in [0, T]$ find $u_h(t) \in V_h^{in}$ such that

$$(14.3.17) \quad \begin{cases} \dfrac{d}{dt}(u_h(t), v_h) + a_1(u_h(t), v_h) \\ \quad + \displaystyle\sum_{K \in \mathcal{T}_h} \delta \left(L_1 u_h(t), \dfrac{h_K}{|\mathbf{a}|} (L_{1,SS} v_h + \rho L_{1,S} v_h) \right)_K \\ = (f(t), v_h) \\ \quad + \displaystyle\sum_{K \in \mathcal{T}_h} \delta \left(f(t), \dfrac{h_K}{|\mathbf{a}|} (L_{1,SS} v_h + \rho L_{1,S} v_h) \right)_K \quad \forall\, v_h \in V_h^{in} \\ u_h(0) = u_{0,h} \end{cases}$$

for each $t \in (0, T)$, where $L_1 v := \mathbf{a} \cdot \nabla v + a_0 v = L_{1,S} v + L_{1,SS} v$ and

$$(14.3.18) \quad \begin{aligned} L_{1,S} v &:= \left(-\frac{1}{2} \operatorname{div} \mathbf{a} + a_0 \right) v \\ L_{1,SS} v &:= \frac{1}{2} \operatorname{div}(\mathbf{a} v) + \frac{1}{2}\mathbf{a} \cdot \nabla v \ . \end{aligned}$$

As usual, SUPG corresponds to the choice $\rho = 0$, GALS to $\rho = 1$ and DWG to $\rho = -1$.

Assuming that $\mu(t, \mathbf{x}) \geq \mu_0 > 0$ in $\overline{Q_T}$, stability follows as in the preceding cases, by choosing $v_h = u_h(t)$ and the product $\delta\, h$ small enough when considering the SUPG and DWG methods (see Propositions 8.4.1-8.4.3). The final result reads

$$(14.3.19) \quad ||u_h(t)||_0^2 + \int_0^t ||u_h(\tau)||_*^2 \leq C \left(||u_{0,h}||_0^2 + \int_0^t ||f(\tau)||_0^2 \right)$$

for each $t \in [0, T]$, where

$$||v||_*^2 := \begin{cases} ||v||_0^2 + |v|_{\mathbf{a},\partial\Omega}^2 \\ \quad + \displaystyle\sum_{K \in \mathcal{T}_h} \delta \left(\dfrac{h_K}{|\mathbf{a}|} L_1 v, L_1 v \right)_K \quad \text{(GALS)} \\ ||v||_0^2 + |v|_{\mathbf{a},\partial\Omega}^2 \\ \quad + \displaystyle\sum_{K \in \mathcal{T}_h} \delta \left(\dfrac{h_K}{|\mathbf{a}|} L_{1,SS} v, L_{1,SS} v \right)_K \quad \text{(SUPG or DWG)} \end{cases}$$

The proof of convergence can be done following the guidelines presented above for problem (14.3.4) (see also Johnson (1987), pp. 182–185, for similar results in the stationary case). Due to the fact that strong consistency does not hold for (14.3.17) (i.e., the exact solution u does not satisfy (14.3.17)), one finally obtains the error estimate

$$(14.3.20) \quad \max_{t \in [0,T]} ||u(t) - u_h(t)||_0 + \left(\int_0^T ||u(t) - u_h(t)||_*^2 \, dt \right)^{1/2}$$
$$= O(||u_0 - u_{0,h}||_0 + h^{1/2}) \ ,$$

independently of the regularity of the solution u. Though the method turns out to be less accurate than (14.3.4), it must be noticed that the norm $||v||_*$ gives also a control of $L_1 v$ (or $L_{1,SS} v$), hence of suitable first derivatives.

To construct more accurate schemes, one has to modify (14.3.17) in order to obtain a strongly consistent method, for instance by seeking the solution of

$$(14.3.21) \quad \begin{cases} \dfrac{d}{dt}(u_h(t), v_h) + a_1(u_h(t), v_h) \\[2mm] \qquad + \displaystyle\sum_{K \in \mathcal{T}_h} \delta \left(\dfrac{\partial u_h}{\partial t}(t) + L_1 u_h(t), \dfrac{h_K}{|\mathbf{a}|} (L_{1,SS} v_h + \rho L_{1,S} v_h) \right)_K \\[2mm] = (f(t), v_h) \\[2mm] \qquad + \displaystyle\sum_{K \in \mathcal{T}_h} \delta \left(f(t), \dfrac{h_K}{|\mathbf{a}|} (L_{1,SS} v_h + \rho L_{1,S} v_h) \right)_K \quad \forall \, v_h \in V_h^{in} \\[2mm] u_h(0) = u_{0,h} \end{cases}$$

for each $t \in (0, T)$. An equivalent approach is as follows: discretize at first in time (e.g., using the backward Euler scheme) and then apply a stabilization method to the resulting stationary advection equation. In this way one is led to consider the totally discretized problem

$$(14.3.22) \quad \begin{aligned} &(u_h^{n+1}, v_h) + \Delta t \, a_1(u_h^{n+1}, v_h) \\ &\quad + \sum_{K \in \mathcal{T}_h} \delta \left(u_h^{n+1} + \Delta t \, L_1 u_h^{n+1}, \frac{h_K}{|\mathbf{a}|} (L_{1,SS} v_h + \rho L_{1,S} v_h) \right)_K \\ &= (u_h^n, v_h) + \Delta t \, (f(t_{n+1}), v_h) \\ &\quad + \sum_{K \in \mathcal{T}_h} \delta \left(u_h^n + \Delta t \, f(t_{n+1}), \frac{h_K}{|\mathbf{a}|} (L_{1,SS} v_h + \rho L_{1,S} v_h) \right)_K \end{aligned}$$

for each $v_h \in V_h^{in}$, $n = 0, 1, ..., \mathcal{N} - 1$. Clearly, the backward Euler scheme applied to (14.3.21), substituting also in the stabilizing term the time derivative of u_h with the backward incremental ratio, leads to the same scheme (14.3.22).

As we already noticed for the case of parabolic problems (see Section 12.3), this scheme turns out to be unstable for $\Delta t / h$ too close to 0, therefore we cannot achieve optimal accuracy in time; a more consistent approach is presented in the next Section.

Remark 14.3.1 If in the stabilizing terms we substitute h_K with h, the SUPG, GALS and DWG methods applied to the stationary hyperbolic equation $L_1 u = f$ in Ω, $u = 0$ on $\partial\Omega^{in}$ can be viewed as examples of Petrov-Galerkin methods (see Section 5.3). Precisely, one finds the formulation there presented by setting $L_h := \delta h \, |\mathbf{a}|^{-1}(L_{1,SS} + \rho L_{1,S})$. $\qquad\qquad\square$

14.3.3 Space-Discontinuous Galerkin Method

An alternative method is based on discontinuous (in space) finite elements and it has been proposed by Lesaint and Raviart (1974) (for the stationary case). This time the finite element space is

$$W_h = Y_h^k := \{v_h \in L^2(\Omega) \,|\, v_{h|K} \in \mathbb{P}_k \;\; \forall K \in T_h\} \ ,$$

i.e., the space of piecewise-polynomials of degree less than or equal to k, $k \geq 0$, with no continuity requirements across interelement boundaries.

The space-discontinuous Galerkin method reads: for any $t \in [0,T]$ find $u_h(t) \in W_h$ such that

(14.3.23)
$$\begin{cases} \dfrac{d}{dt}(u_h(t), v_h) + \displaystyle\sum_{K \in T_h} a_{1,K}(u_h(t), v_h) \\[2mm] \qquad = (f(t), v_h) - \displaystyle\int_{\partial\Omega^{in}} \mathbf{a} \cdot \mathbf{n} \, \varphi_h(t) \, (v_h)_+ \, d\gamma \quad \forall \, v_h \in W_h \\[3mm] u_h(0) = u_{0,h} \end{cases}$$

for each $t \in (0,T)$. Here, we have set

$$a_{1,K}(z,v) := \begin{cases} \displaystyle\int_K (\mathbf{a} \cdot \nabla z + a_0 z) v \, d\mathbf{x} \\[1mm] \qquad - \displaystyle\int_{\partial K^{in}} \mathbf{a} \cdot \mathbf{n}_K \, [z] \, v_+ \, d\gamma \quad \text{if } K \cap \partial\Omega^{in} = \emptyset \\[3mm] \displaystyle\int_K (\mathbf{a} \cdot \nabla z + a_0 z) v \, d\mathbf{x} \\[1mm] \qquad - \displaystyle\int_{\partial K^{in}} \mathbf{a} \cdot \mathbf{n}_K \, z_+ \, v_+ \, d\gamma \quad \text{if } K \cap \partial\Omega^{in} \neq \emptyset \end{cases} ,$$

where $\mathbf{n}_K$ denotes the unit outward normal vector on ∂K, $\partial K^{in} := \{\mathbf{x} \in \partial K \,|\, \mathbf{a}(\mathbf{x}) \cdot \mathbf{n}_K(\mathbf{x}) < 0\}$ and

$$[z](\mathbf{x}) := z_+(\mathbf{x}) - z_-(\mathbf{x}) \ , \quad z_\pm(\mathbf{x}) := \lim_{s \to 0^\pm} z(\mathbf{x} + s\mathbf{a}) \ , \quad \mathbf{x} \in \partial K \ .$$

For the analysis of stability and convergence we refer to Johnson (1987), pp. 189–196 (see also Johnson and Pitkäranta (1986), Falk and Richter (1992)). Still assuming that $\mu(t, \mathbf{x}) \geq \mu_0 > 0$ in $\overline{Q_T}$, the stability result reads:

$$\|u_h(t)\|_0^2 + \int_0^t \|u_h(\tau)\|_{DG}^2$$

(14.3.24)

$$\leq C \left[\|u_{0,h}\|_0^2 + \int_0^t \left(\|f(\tau)\|_0^2 + |\varphi_h(\tau)|_{\mathbf{a}, \partial\Omega^{in}}^2 \right) \right]$$

for each $t \in [0, T]$, where

$$\|v\|_{DG}^2 := \|v\|_*^2 + \sum_{K \in \mathcal{T}_h} \int_{\partial K^{in}} |\mathbf{a} \cdot \mathbf{n}_K| \, [v]^2 \ .$$

On the other hand, the convergence estimate is:

(14.3.25) $$\max_{t \in [0,T]} \|u(t) - u_h(t)\|_0 + \left(\int_0^T \|u(t) - u_h(t)\|_{DG}^2 \, dt \right)^{1/2}$$

$$= O(\|u_0 - u_{0,h}\|_0 + h^{k+1/2}) \ .$$

14.3.4 Schemes for Time-Discretization

Let us turn now to the fully-discrete approximation of (14.3.1). Anyone of the finite difference schemes described in Section 11.3 can be used to advance in time the systems of ordinary differential equations stemming from the semi-discrete approximations introduced above. As usual, both the backward Euler and Crank-Nicolson methods turn out to be unconditionally stable. On the other hand, the forward Euler method is subject to the stability condition $\Delta t = O(h)$. In fact, the spectral radius of all the bilinear forms $\mathcal{A}_1(\cdot, \cdot)$ (appearing at the left hand side of (14.3.4), (14.3.11), (14.3.21) or (14.3.23)) behaves like $O(h^{-1})$. The time-step limitation is not as severe as in the parabolic case, and this explains why explicit schemes are often used in the approximation of hyperbolic problems.

Another approach for performing time-discretization is based on the remark that for hyperbolic problems space and time variables, from the mathematical point of view, play essentially the same role. Therefore it seems very natural to use space-time finite elements as done for the discontinuous (in time) Galerkin method described in Sections 11.3.1 and 12.3. All the methods introduced in this Section can be reformulated in a fully-discrete form by following the guidelines presented there. We leave the reader with the task of deducing the corresponding variational formulations. The resulting schemes are implicit, and are not subject to stability restrictions.

Finally, one could also use the characteristic Galerkin method introduced in Section 12.5. For some remarks on this approach, let us refer to Pironneau (1988), pp. 90–91, for finite element approximation and to Süli and Ware (1991) for the spectral case.

A different approach, which is known as the *Taylor-Galerkin* method and was proposed by Donea (1984), is often used in practical computations. The idea is expand the solution at the point (t_n, x) by means of the Taylor formula in the time increment truncated at the third order, use the hyperbolic equation to replace time derivatives by space derivatives, and finally for space discretization use the Galerkin method based on piecewise-linear finite elements.

For the sake of simplicity, let us consider equation (14.1.1). By proceeding as in the derivation of the Lax-Wendroff scheme, the Taylor expansion truncated at the third order gives

$$u(t_{n+1}, x) = u(t_n, x) + \Delta t \frac{\partial u}{\partial t}(t_n, x) + \frac{1}{2}(\Delta t)^2 \frac{\partial^2 u}{\partial t^2}(t_n, x)$$

$$+ \frac{1}{6}(\Delta t)^3 \frac{\partial^3 u}{\partial t^3}(t_n, x) + O((\Delta t)^4) \ .$$

Using equation (14.1.1) we have

$$\frac{\partial u}{\partial t} = -a\frac{\partial u}{\partial x} \ , \quad \frac{\partial^2 u}{\partial t^2} = a^2\frac{\partial^2 u}{\partial x^2} \ , \quad \frac{\partial^3 u}{\partial t^3} = a^2\frac{\partial^2}{\partial x^2}\left(\frac{\partial u}{\partial t}\right) \ .$$

Approximating in the last term the time derivative by the forward Euler formula

$$\frac{\partial u}{\partial t}(t_n, x) = \frac{u(t_{n+1}, x) - u(t_n, x)}{\Delta t} + O(\Delta t)$$

gives at last the implicit scheme

$$(14.3.26) \qquad \left[1 - \frac{1}{6}(\Delta t)^2 a^2 \frac{\partial^2}{\partial x^2}\right]\frac{u^{n+1} - u^n}{\Delta t} = -a\frac{\partial u^n}{\partial x} + \frac{1}{2}\Delta t\, a^2\frac{\partial^2 u^n}{\partial x^2} \ .$$

Notice that third-order accuracy has been obtained by a simple modification of the operator acting on $(u^{n+1} - u^n)/\Delta t$. Equation (14.3.26) is then discretized in space by the Galerkin method, choosing as finite dimensional space V_h the set of piecewise-linear finite elements on a uniform mesh given by $x_j := j\Delta x$, $j \in \mathbb{Z}$. Denoting by φ_j the basis function related to the node x_j, the mass matrix $M_{ij} := (\varphi_i, \varphi_j)$ takes the form

$$M = \frac{\Delta x}{6}\, \mathrm{tridiag}\,(1, 4, 1) = \Delta x\left[I + \frac{1}{6}\,\mathrm{tridiag}\,(1, -2, 1)\right] \ .$$

On the other hand, the discretization of the second order derivative is given by the matrix $(D_2)_{ij} := -(\varphi_i', \varphi_j')$, which reads

$$D_2 = \frac{1}{\Delta x}\, \mathrm{tridiag}\,(1, -2, 1) \ ,$$

and similarly the discretization of the first order derivative is given by $(D_1)_{ij} := (\varphi_j', \varphi_i)$, i.e.,

$$D_1 = \frac{1}{2}\, \mathrm{tridiag}\,(-1, 0, 1) \ .$$

Setting, as usual, $\lambda = \Delta t / \Delta x$, the resulting scheme therefore reads

$$(1 - \lambda^2 a^2)u_{j+1}^{n+1} + 2(2 + \lambda^2 a^2)u_j^{n+1} + (1 - \lambda^2 a^2)u_{j-1}^{n+1}$$

(14.3.27)
$$= (1 - 3\lambda a + 2\lambda^2 a^2)u_{j+1}^n + 4(1 - \lambda^2 a^2)u_j^n$$
$$+ (1 + 3\lambda a + 2\lambda^2 a^2)u_{j-1}^n \ .$$

The linear system to be solved at each time-step is tridiagonal and symmetric, similarly to the one obtained by applying the procedure we have just described to the classical Lax-Wendroff scheme (in that case, in fact, the system is the one associated to the mass matrix M). Therefore, the advantage here is that, with similar computational cost, we obtain third order accuracy, whereas the Lax-Wendroff scheme is only second order accurate.

The Taylor-Galerkin scheme is proven to be stable under the usual CFL condition $|a\lambda| \leq 1$. The analysis of the method (including the issue of imposing the boundary conditions) and a discussion of its relations with Petrov-Galerkin and characteristic Galerkin methods are given in Donea (1984) and Donea, Quartapelle and Selmin (1987), where numerical results for several linear test problems can also be found. The extension of the method to two-dimensional problems can be easily performed, only requiring some more complicated manipulations.

14.4 Approximation by Spectral Methods

For simplicity, in this Section we return to the one-dimensional case, though a similar approach could be followed in higher spatial dimensions.

Spectral approximations to hyperbolic equations can take the form of Galerkin methods or collocation methods. The Galerkin case falls under the general framework that has been discussed in Section 14.3. Precisely, the spectral Galerkin method can be cast in the form (14.3.4), provided V_h denotes now a space of global (rather than piecewise-) polynomials of degree less than or equal to N. We are not entering further comments here. Instead, we present, in detail, the case of the spectral collocation method.

To begin with, we consider the initial-boundary value problem for the simple one-dimensional linear scalar advection equation:

(14.4.1) $$\frac{\partial u}{\partial t} + a \frac{\partial u}{\partial x} = f \ , \quad t > 0 \ , \ x \in I := (-1, 1)$$

(14.4.2) $$u(t, -1) = \varphi_-(t) \ , \quad t > 0$$

(14.4.3) $$u(0, x) = u_0(x) \ , \quad x \in I \ ,$$

where $f = f(t, x)$ and $u_0 = u_0(x)$ are continuous functions, $\varphi_- = \varphi_-(t)$ is a prescribed boundary datum and $a > 0$ is a given coefficient, which, for simplicity, is assumed to be constant.

14.4.1 Spectral Collocation Method: the Scalar Case

Let $\{x_j \mid 0 \leq j \leq N\}$ denote the $N + 1$ nodes of the Gauss-Lobatto formula in $[-1, 1]$ (see Chapter 4). They can be referred to as Legendre or Chebyshev nodes, according to the kind of Gaussian formula that we are considering. In any case, we order them in a decreasing way, so that $x_0 = 1$, $x_N = -1$ and $x_j > x_{j+1}$ for $j = 0, ..., N - 1$.

Correspondingly, we denote by $\{w_j \mid 0 \leq j \leq N\}$ the associated weights (see Sections 4.3.2 and 4.4.2). Introducing the discrete scalar product

$$(14.4.4) \qquad (z, v)_N := \sum_{j=0}^{N} z(x_j)v(x_j)w_j \quad , \quad z, v \in C^0([-1, 1]) \quad ,$$

we recall that from (4.2.6) it follows

$$(14.4.5) \qquad (z, v)_N = \int_{-1}^{1} z(x)v(x)(1 - x^2)^\alpha dx$$

provided $zv \in \mathbb{P}_{2N-1}$. Here, $\alpha = 0$ when we are using Legendre nodes while $\alpha = -1/2$ for Chebyshev nodes.

The collocation approximation to (14.4.1)-(14.4.3) reads: for each $t \geq 0$ find $u_N = u_N(t) \in \mathbb{P}_N$ such that

$$(14.4.6) \quad \frac{\partial u_N}{\partial t}(t, x_j) + a\frac{\partial u_N}{\partial x}(t, x_j) = f(t, x_j) \quad , \quad 0 \leq j \leq N - 1 \, , \, t > 0$$

$$(14.4.7) \quad u_N(t, x_N) = \varphi_-(t) \quad , \quad t > 0$$

$$(14.4.8) \quad u_N(0, x_j) = u_{0,N}(x_j) \quad , \quad 0 \leq j \leq N \quad ,$$

where $u_{0,N} \in \mathbb{P}_N$ is a suitable approximation of the initial datum u_0.

The differential equation has been enforced at all collocation nodes unless the inflow point $x_N = -1$ where the boundary condition (14.4.2) is prescribed. In particular, it is satisfied also at the outflow boundary point $x_0 = 1$. Similarly, at the time $t = 0$ the approximate solution u_N is asked to match the initial data at all Gauss-Lobatto nodes.

If a is negative and therefore the boundary condition is prescribed at the point $x_0 = 1$, the collocation method is stated similarly, by simply reversing the role of x_0 and x_N.

In the case of Legendre collocation, stability can be easily derived for u_N. As a matter of fact take, for simplicity, $\varphi_- = 0$, $f = 0$ and multiply (14.4.6) by $u_N(t, x_j)w_j$, then sum up on j and use (14.4.5). We obtain for each $t > 0$

$$0 = \frac{1}{2}\frac{d}{dt}(u_N(t), u_N(t))_N + \left(a\frac{\partial u_N}{\partial x}(t), u_N(t)\right)$$

$$= \frac{1}{2}\frac{d}{dt}\|u_N(t)\|_N^2 + \frac{1}{2}au_N^2(t, 1) \quad ,$$

where $||v||_N^2 = (v, v)_N$ is a discrete norm, equivalent to the norm of $L^2(-1, 1)$ on $\mathbb{P}_N$ (see (4.4.16)). Then

$$||u_N(t)||_N^2 + \int_0^t a u_N^2(\tau, 1) \, d\tau = ||u_{0,N}||_N^2 \quad \forall t > 0 .$$

Taking $u_{0,N} = I_N u_0$, the interpolant of the initial datum u_0 assumed to belong to $C^0([-1, 1])$, the $L^2(-1, 1)$-stability follows at once (see also (6.2.27)). The case of non-homogeneous data φ_- and f can be handled similarly, using additionally the Gronwall lemma (see Lemma 1.4.1). A convergence result can be proven in a straightforward manner.

The case of Chebyshev nodes is less trivial. Indeed, as pointed out in Gottlieb and Orszag (1977), p. 89, stability doesn't hold in the norm $||u_N||_N$ introduced in (4.3.14). Instead, it is possible to prove stability in the weaker norm

$$|||u_N|||_N := ||u_N \sqrt{(1 - x)/(1 + x)}||_N .$$

Still in the Legendre case, a slightly different collocation approach is the one based on the weak form of the collocation method, that has already been considered for both elliptic and parabolic equations. In this case, for each $t \geq 0$ we look for $u_N = u_N(t) \in \mathbb{P}_N$ such that

$$(14.4.9) \quad \frac{d}{dt}(u_N(t), v_N)_N - \left(a u_N(t), \frac{\partial v_N}{\partial x}\right)_N + a u_N(t, 1) v_N(1)$$
$$= (f(t), v_N)_N + a \varphi_-(t) v_N(-1) \quad \forall \, v_N \in \mathbb{P}_N , \quad t > 0 .$$

Moreover, u_N is asked to satisfy the initial condition (14.4.8).

The above scheme differs from the one previously introduced in what concerns the boundary condition. Indeed, taking $v_N = \psi_j$ in (14.4.9) for $j = 0, ..., N - 1$, where $\psi_j \in \mathbb{P}_N$ is such that $\psi_j(x_k) = \delta_{kj}$ for $k, j = 0, ..., N$, we can easily show that we obtain the collocation statements (14.4.6). Taking then $v_N = \psi_N$ we obtain

$$(14.4.10) \quad \begin{aligned} a(u_N(t, x_N) &- \varphi_-(t)) \\ &= w_N \left(f - \frac{\partial u_N}{\partial t} - a \frac{\partial u_N}{\partial x}\right)(t, x_N) , \quad t > 0 . \end{aligned}$$

This is the relaxed form of the inflow boundary condition (14.4.8). Since $w_N = 2/[N(N + 1)]$, (14.4.10) states that the boundary data is matched up to a small coefficient times the residual of the differential equations.

This treatment of boundary data for hyperbolic equations has been introduced by Funaro and Gottlieb (1988, 1989) and can be also extended to the Chebyshev collocation method. (See also Funaro (1992), p. 231.)

Stability analysis for (14.4.9) is straightforward. Taking $v_N = u_N(t)$ we can easily deduce a result like the one we have obtained for (14.4.6).

The proof of convergence can be accomplished as follows. The exact solution u of (14.4.1) satisfies

(14.4.11) $\dfrac{d}{dt}(u(t),v) - \left(au(t),\dfrac{\partial v}{\partial x}\right) + au(t,1)v(1)$

$$= (f(t),v) + a\varphi_-(t)v(-1) \quad \forall\, v \in H^1(I) \ , \ t > 0 \ .$$

The interpolant of $u(t)$ at the nodes $\{x_j \mid j = 0, ..., N\}$, say $\tilde{u}(t) := I_N u(t) \in \mathbb{P}_N$ (see Chapter 4), satisfies

$$\dfrac{d}{dt}(\tilde{u}(t),v) - \left(a\tilde{u}(t),\dfrac{\partial v}{\partial x}\right) + a\tilde{u}(t,1)v(1)$$

$$= (f(t),v) + a\varphi_-(t)v(-1)$$

$$+ \left(\dfrac{\partial(\tilde{u}-u)}{\partial t}(t),v\right) - \left(a(\tilde{u}-u)(t),\dfrac{\partial v}{\partial x}\right)$$

for all $v \in H^1(I)$, $t > 0$. In particular, taking $v = v_N \in \mathbb{P}_N$, we obtain

$$\dfrac{d}{dt}(\tilde{u}(t),v_N)_N - \left(a\tilde{u}(t),\dfrac{\partial v_N}{\partial x}\right)_N + a\tilde{u}(t,1)v_N(1)$$

$$= (f(t),v_N) + a\varphi_-(t)v_N(-1)$$

(14.4.12) ,
$$+ \left(\dfrac{\partial(\tilde{u}-u)}{\partial t}(t),v_N\right) - \left(a(\tilde{u}-u)(t),\dfrac{\partial v_N}{\partial x}\right)$$

$$+ \left(\dfrac{\partial\tilde{u}}{\partial t}(t),v_N\right)_N - \left(\dfrac{\partial\tilde{u}}{\partial t}(t),v_N\right) \ .$$

Comparing (14.4.12) and (14.4.9), it follows that $e_N(t) := \tilde{u}(t) - u_N(t) \in \mathbb{P}_N$ satisfies

$$\dfrac{d}{dt}(e_N(t),v_N)_N - \left(ae_N(t),\dfrac{\partial v_N}{\partial x}\right)_N + ae_N(t,1)v_N(1)$$

(14.4.13)
$$= E(f(t),v_N) - E\left(\dfrac{\partial\tilde{u}}{\partial t}(t),v_N\right)$$

$$+ \left(\dfrac{\partial(\tilde{u}-u)}{\partial t}(t),v_N\right) - \left(a(\tilde{u}-u)(t),\dfrac{\partial v_N}{\partial x}\right)$$

for all $v_N \in \mathbb{P}_N$, $t > 0$, where, for each $\varphi \in C^0([-1,1])$, we have set $E(\varphi,v_N) := (\varphi,v_N) - (\varphi,v_N)_N$. Notice that, by proceeding as in Section 4.4.2, it follows

$$E(f(t),v_N) \le C\big(\|f(t) - I_N f(t)\|_0 + \|f(t) - I_{N-1}f(t)\|_0\big)\|v_N\|_0$$

and

$$-E\left(\dfrac{\partial\tilde{u}}{\partial t}(t),v_N\right) \le C\Big(\Big\|\dfrac{\partial u}{\partial t}(t) - I_N\dfrac{\partial u}{\partial t}(t)\Big\|_0$$

$$+ \Big\|\dfrac{\partial u}{\partial t}(t) - I_{N-1}\dfrac{\partial u}{\partial t}(t)\Big\|_0\Big)\|v_N\|_0 \ ,$$

where $\|\cdot\|_k$, $k \ge 0$, denotes the norm in the Sobolev space $H^k(I)$ (see Section 1.2). Hence, employing the interpolation results proven in Section 4.4.2, we find

$$(14.4.14) \quad \begin{aligned} & E(f(t), v_N) - E\left(\frac{\partial \tilde{u}}{\partial t}(t), v_N\right) \\ & \qquad \leq CN^{1-s}\left(\|f(t)\|_{s-1} + \left\|\frac{\partial u}{\partial t}(t)\right\|_{s-1}\right)\|v_N\|_0 \ . \end{aligned}$$

Taking, for each fixed t, $v_N = e_N(t)$ in (14.4.13) yields

$$(14.4.15) \quad \begin{aligned} & \frac{1}{2}\frac{d}{dt}\|e_N(t)\|_0^2 + \frac{1}{2}ae_N^2(t, 1) + \frac{1}{2}ae_N^2(t, -1) \\ & \leq \left[CN^{1-s}\left(\|f(t)\|_{s-1} + \left\|\frac{\partial u}{\partial t}(t)\right\|_{s-1}\right)\right. \\ & \quad + \left.\left\|\frac{\partial(\tilde{u}-u)}{\partial t}(t)\right\|_0 + a\|(\tilde{u}-u)(t)\|_1\right]\|e_N(t)\|_0 \ . \end{aligned}$$

Here, we have integrated by parts the term $(a(\tilde{u}-u)(t), \frac{\partial v_N}{\partial x})$, taking into account that $(\tilde{u}-u)(t, \pm 1) = 0$. Now using Gronwall lemma (see Lemma 1.4.1) furnishes

$$(14.4.16) \quad \begin{aligned} & \|e_N(t)\|_0^2 + a\int_0^t [e_N^2(\tau, 1) + e_N^2(\tau, -1)]d\tau \\ & \leq \left[\|I_N u_0 - u_{0,N}\|_0^2 + CN^{2-2s}\int_0^t \left(\|f(\tau)\|_{s-1}^2 \right.\right. \\ & \quad + \left.\left.\left\|\frac{\partial u}{\partial t}(\tau)\right\|_{s-1}^2 + \|u(\tau)\|_s^2\right)d\tau\right]\exp(Ct) \ , \end{aligned}$$

provided that there exists a suitable $s \geq 2$ for which all the norms at the right-hand side make sense. The error estimates follows from the relation $u - u_N = u - I_N u + e_N$, and reads

$$(14.4.17) \quad \begin{aligned} & \|u(t) - u_N(t)\|_0^2 + a\int_0^t [(u-u_N)^2(\tau, 1) + (u-u_N)^2(\tau, -1)]d\tau \\ & \leq C\left[\|u_0 - u_{0,N}\|_0^2 + N^{2-2s}\int_0^t \left(\|f(\tau)\|_{s-1}^2 \right.\right. \\ & \quad + \left.\left.\left\|\frac{\partial u}{\partial t}(\tau)\right\|_{s-1}^2 + \|u(\tau)\|_s^2\right)d\tau\right]\exp(Ct) \ . \end{aligned}$$

14.4.2 Spectral Collocation Method: the Vector Case

Let us now consider the spectral approximation to the linear *hyperbolic system*

$$(14.4.18) \quad \frac{\partial \mathbf{U}}{\partial t} + A\frac{\partial \mathbf{U}}{\partial x} = \mathbf{F} \ , \quad t > 0 \ , \ x \in I \ ,$$

with initial data

(14.4.19) $$\mathbf{U}(0,x) = \mathbf{U}_0(x) \ , \quad x \in I \ ,$$

where $\mathbf{U} : [0,+\infty) \times I \rightarrow \mathbb{R}^p$ and $\mathbf{U}_0 : I \rightarrow \mathbb{R}^p$. Moreover, $A \in \mathbb{R}^{p \times p}$ is a constant matrix, which, for simplicity, is assumed to be non-singular, so that anyone of its eigenvalues λ_i, $i = 1, ..., p$, is different from 0. Precisely, we indicate by $\lambda_1, ..., \lambda_{p_0}$ the positive eigenvalues, and by $\lambda_{p_0+1}, ..., \lambda_p$ the negative ones.

We diagonalize A as done in (14.1.7), and introduce the characteristic variables $\mathbf{W} = T^{-1}\mathbf{U}$. The latter satisfy the equation

$$\frac{\partial \mathbf{W}}{\partial t} + \Lambda \frac{\partial \mathbf{W}}{\partial x} = T^{-1}\mathbf{F} \ , \quad t > 0 \, , \, x \in I \ ,$$

where $\Lambda = \text{diag} \, (\lambda_1, ..., \lambda_p)$. We split Λ and $\mathbf{W}$ as done in (14.1.17), and the matrix T^{-1} accordingly, setting

$$T^{-1} = \begin{pmatrix} (T^{-1})^+ \\ \\ (T^{-1})^- \end{pmatrix} \ .$$

Here, $(T^{-1})^+$ is the $p_0 \times p$ matrix given by the first p_0 rows of T^{-1}, and $(T^{-1})^-$ is the remaining $(p - p_0) \times p$ matrix. Notice that, if A is symmetric, then the matrix T turns out to be unitary, i.e., $T^{-1} = T^T$. In this case, $(T^{-1})^+$ is the $p_0 \times p$ matrix having the k-th row given by the eigenvector ω^k, $k = 1, ..., p_0$, and analogously for $(T^{-1})^-$.

As discussed in Section 14.1, the system (14.4.18) must be supplemented by p_0 boundary conditions at the left-hand point $x_N = -1$, and $p - p_0$ at the right-hand point $x_0 = 1$, for all $t > 0$. For instance, one assignes $G^-\mathbf{U}$ at $x_0 = 1$, G^- being a $(p - p_0) \times p$ matrix, and $G^+\mathbf{U}$ at $x_N = -1$, with G^+ a $p_0 \times p$ matrix. These matrices must satisfy the following additional condition:

$$\begin{cases} \text{the submatrix given by the first } p_0 \text{ columns of } G^+T \text{ is non-singular} \\ \text{the submatrix given by the last } p - p_0 \text{ columns of } G^-T \text{ is non-singular} \end{cases}$$

(see, e.g., Gastaldi and Quarteroni (1989)).

The spectral collocation approximation to (14.4.18) can be defined as follows. For any $t \geq 0$ we look for $\mathbf{U}_N(t) \in (\mathbb{P}_N)^p$ satisfying the differential equations at each internal point, i.e.,

(14.4.20) $$\left(\frac{\partial \mathbf{U}_N}{\partial t} + A \frac{\partial \mathbf{U}_N}{\partial x} - \mathbf{F} \right)(t, x_j) = 0 \ , \quad 1 \leq j \leq N - 1 \ , \quad t > 0 \ ,$$

and moreover the initial condition $\mathbf{U}_N(0) = \mathbf{U}_{0,N} \in (\mathbb{P}_N)^p$, a suitable approximation of $\mathbf{U}_0$. Besides, at each boundary point we have to enforce the differential equations associated with the outgoing characteristic variables. Precisely

(14.4.21) $$(T^{-1})^- \left(\frac{\partial \mathbf{U}_N}{\partial t} + A \frac{\partial \mathbf{U}_N}{\partial x} - \mathbf{F} \right)(t, x_N) = 0 \ , \quad t > 0$$

(p_0 equations) and

$$(14.4.22) \qquad (T^{-1})^+ \left(\frac{\partial \mathbf{U}_N}{\partial t} + A \frac{\partial \mathbf{U}_N}{\partial x} - \mathbf{F} \right)(t, x_0) = 0 \ , \quad t > 0$$

($p - p_0$ equations).

Introducing the discrete characteristic variables $\mathbf{W}_N := T^{-1}\mathbf{U}_N$, and splitting $\mathbf{W}_N$ into $\mathbf{W}_N^+$ (its first p_0 components) and $\mathbf{W}_N^-$ (the last $p - p_0$ components) (the same notation is adopted for the vector $T^{-1}\mathbf{F}$), we can easily see that (14.4.21) and (14.4.22) are respectively equivalent to:

$$(14.4.23) \qquad \left[\frac{\partial \mathbf{W}_N^-}{\partial t} + \Lambda^- \frac{\partial \mathbf{W}_N^-}{\partial x} - (T^{-1})^- \mathbf{F} \right](t, x_N) = 0 \ , \quad t > 0$$

$$(14.4.24) \qquad \left[\frac{\partial \mathbf{W}_N^+}{\partial t} + \Lambda^+ \frac{\partial \mathbf{W}_N^+}{\partial x} - (T^{-1})^+ \mathbf{F} \right](t, x_0) = 0 \ , \quad t > 0 \ .$$

The system has to be closed enforcing p_0 boundary conditions at $x_N = -1$, and $p - p_0$ at $x_0 = 1$ (for instance, by assigning $G^+\mathbf{U}_N$ at $x_N = -1$ and $G^-\mathbf{U}_N$ at $x_0 = 1$).

This boundary treatment for hyperbolic systems was proposed in Gottlieb, Gunzburger and Turkel (1982) and later generalized in Canuto and Quarteroni (1987). It is stable, and needs to be implemented consistently in both explicit and implicit temporal discretization. For its analysis also see Canuto, Hussaini, Quarteroni and Zang (1988), pp. 428–430.

14.4.3 Time-Advancing and Smoothing Procedures

Before concluding, we briefly address the issue of time-discretization. First of all, let us consider the eigenvalue problem: find $\omega_N \in \mathbb{P}_N$, $\omega_N \neq 0$, and $\lambda \in \mathbb{C}$ such that

$$(14.4.25) \qquad \begin{cases} \dfrac{d\omega_N}{dx}(x_j) = \lambda \omega_N(x_j) \ , & 0 \leq j \leq N - 1 \ , \\[2mm] \omega_N(-1) = 0 \ , \end{cases}$$

which is associated with the scalar hyperbolic problem (14.4.1)-(14.4.3).

For both cases of Legendre and Chebyshev nodes, the real parts of λ are strictly negative, while the modulus satisfies a bound of the form $|\lambda| = O(N^2)$ (see, e.g., Canuto, Hussaini, Quarteroni and Zang (1988), Sect. 11.4.3). This has the consequence that explicit temporal discretization for problems like (14.4.1) or (14.4.18) will undergo a stability limit on the time-step of the form $\Delta t = O(N^{-2})$. In particular, under this restriction all Adams-Bashforth methods are asymptotically stable.

On the other hand, implicit methods like backward Euler or Crank-Nicolson are unconditionally stable. They generate however a linear system to be solved at each time-level, whose matrix has a condition number that behaves like $O(N^2)$. Finite difference based preconditioners can still be used. An account is given in Funaro (1992), pp. 172–174.

When spectral methods are used naively to approximate hyperbolic equations (or systems) with non-smooth data, oscillations grow uncontrolled due to the onset of the Gibbs phenomenon.

Since oscillations are due to the growth of the higher order modes, the cure relies in the use of a filtering mechanism. This requires transformation of the collocation solution from the physical space to the frequency one, i.e., finding its Legendre (or Chebyshev) coefficients and applying to them a suitable cut-off function (e.g., Vandeven (1991) and references therein). By this global smoothing, the accuracy of the solution may be dramatically improved. However, in practical applications only a finite-order decay of the error is usually observed in the region near the discontinuities.

An alternative consists in post-processing the collocation solution by a local smoothing, that is one based on a convolution procedure. The idea has been developed both theoretically and computationally, by Gottlieb and coworkers (see Gottlieb (1985), Gottlieb and Tadmor (1985), Abarbanel, Gottlieb and Tadmor (1986)).

For nonlinear scalar conservation laws like (14.1.20) (with $x \in (-1, 1)$), the so-called *spectral viscosity* method consists in adding to the standard spectral method (Galerkin or collocation) an extra term that has the meaning of nonlinear viscosity. Besides, the resulting solution needs to be post-processed by a local smoothing procedure. This approach, that was first introduced for spectral Fourier approximation of scalar conservation laws with periodic solutions (Tadmor (1989), Maday and Tadmor (1989), Chen, Du and Tadmor (1992)), has been recently extended to the case of Legendre collocation approximation (Maday, Ould Kaber and Tadmor (1993)).

14.5 Second Order Linear Hyperbolic Problems

Let $\Omega \subset \mathbb{R}^d$, $d = 2, 3$, be a given domain, and consider the second order hyperbolic equation

$$\begin{cases} \dfrac{\partial^2 u}{\partial t^2} + Lu = f & \text{in } Q_T := (0,T) \times \Omega \\[2mm] u = 0 & \text{on } \Sigma_{D,T} := (0,T) \times \Gamma_D \\[2mm] \dfrac{\partial u}{\partial n_L} = 0 & \text{on } \Sigma_{N,T} := (0,T) \times \Gamma_N \\[2mm] u_{|t=0} = u_0 & \text{on } \Omega \\[2mm] \dfrac{\partial u}{\partial t}\Big|_{t=0} = u_1 & \text{on } \Omega \ . \end{cases}$$

(14.5.1)

Here $\partial\Omega = \overline{\Gamma_D} \cup \overline{\Gamma_N}$, $\Gamma_D \cap \Gamma_N = \emptyset$, $Lz = -\sum_{i,j=1}^{d} D_i(a_{ij} D_j z)$ and the coefficients a_{ij} satisfy Definition 6.1.1, so that L is an elliptic operator. Moreover, the conormal derivative has been denoted by $\frac{\partial z}{\partial n_L} := \sum_{i,j=1}^{d} a_{ij} D_j z\, n_i$, where $\mathbf{n} = (n_1, ..., n_d)$ is the unit outward normal vector on $\partial\Omega$.

The variational formulation of problem (14.5.1) is as follows: find $u \in C^0([0,T]; V) \cap C^1([0,T]; L^2(\Omega))$ satisfying

$$\begin{cases} \dfrac{d^2}{dt^2}(u(t), v) + a(u(t), v) = (f(t), v) & \forall\, v \in V \\[2mm] u(0) = u_0 \\[2mm] \dfrac{du}{dt}(0) = u_1 \end{cases}$$

(14.5.2)

in the sense of distributions in $(0,T)$, where $V := H^1_{\Gamma_D}(\Omega)$ (see (6.1.6)) and

$$a(z,v) := \sum_{i,j=1}^{d} \int_{\Omega} a_{ij} D_j z D_i v \quad , \quad (z,v) := \int_{\Omega} zv \ .$$

Assuming $u_0 \in V$, $u_1 \in L^2(\Omega)$ and $f \in L^2(Q_T)$, the above problem has a unique solution (for a proof, see, e.g., Lions and Magenes (1968b); also see Raviart and Thomas (1983)).

A semi-discrete approximation of (14.5.2) derived from a Galerkin method can be defined as follows. Let V_h be a finite dimensional subspace of V (whose dimension is N_h), and assume that $u_{0,h}, u_{1,h} \in V_h$ are suitable approximations to u_0 and u_1, respectively. Then for each $t \in [0,T]$ we look for a function $u_h(t) \in V_h$ that satisfies

$$\begin{cases} \dfrac{d^2}{dt^2}(u_h(t), v_h) + a(u_h(t), v_h) \\[2mm] \qquad\qquad = (f(t), v_h) & \forall\, v_h \in V_h \ , \quad t \in (0,T) \\[2mm] u_h(0) = u_{0,h} \\[2mm] \dfrac{du_h}{dt}(0) = u_{1,h} \ . \end{cases}$$

(14.5.3)

Proceeding as in the case of the differential problem (14.5.2) one can show that problem (14.5.3) has a unique solution. Stability is easily proven by taking at each $t > 0$ $v_h = \frac{\partial u_h}{\partial t}(t)$ and integrating in time the resulting equation.

If the bilinear form $a(\cdot, \cdot)$ is coercive in V and symmetric, i.e., $a_{ij} = a_{ji}$ for each $i, j = 1, ..., d$, it can be proven that there exists a sequence of eigenvalues $0 < \mu_{1,h} \leq ... \leq \mu_{N_h,h}$ and corresponding eigenfunctions $\{w_{i,h}\}$ verifying

$$(14.5.4) \qquad a(w_{i,h}, v_h) = \mu_{i,h}(w_{i,h}, v_h) \quad \forall \, v_h \in V_h \ .$$

These eigenfunctions are mutually orthogonal in $L^2(\Omega)$, and provide a complete orthonormal basis of V_h. Setting $\mu_{i,h} = \sqrt{\mu_{i,h}}$, the solution of (14.5.3) takes the form

$$u_h(t) = \sum_{i=1}^{N_h} \left\{ (u_{0,h}, w_{i,h}) \cos(\mu_{i,h}t) + \frac{1}{\mu_{i,h}}(u_{1,h}, w_{i,h}) \sin(\mu_{i,h}t) \right.$$

$$\left. + \frac{1}{\mu_{i,h}} \int_0^t (f(s), w_{i,h}) \sin(\mu_{i,h}(t-s))ds \right\} w_{i,h} \ .$$

For practical purposes the representation above is not useful. With notations similar to those of Section 5.6.1, (14.5.3) can be reformulated in an algebraic form as follows

$$(14.5.5) \qquad \begin{cases} M\dfrac{d^2 \boldsymbol{\xi}}{dt^2}(t) + A\boldsymbol{\xi}(t) = \mathbf{F}(t) \ , \quad t \in (0,T) \\[2mm] \boldsymbol{\xi}(0) = \boldsymbol{\xi}_0 \\[2mm] \dfrac{d\boldsymbol{\xi}}{dt}(0) = \boldsymbol{\xi}_1 \ . \end{cases}$$

Precisely, having denoted with $\{\varphi_j\}_{j=1,...,N_h}$ a basis of V_h, and

$$u_h(t) = \sum_{j=1}^{N_h} \xi_j(t)\varphi_j \ , \quad u_{0,h} = \sum_{j=1}^{N_h} \xi_{0,j}\varphi_j \ , \quad u_{1,h} = \sum_{j=1}^{N_h} \xi_{1,j}\varphi_j \ ,$$

it is

$$M_{ij} := (\varphi_i, \varphi_j) \ , \quad A_{ij} := a(\varphi_j, \varphi_i) \ , \quad F_i(t) := (f(t), \varphi_i) \ .$$

For the time discretization, let us consider the simple ordinary differential equation of second order in time

$$(14.5.6) \qquad \begin{cases} y''(t) = \psi(t, y(t), y'(t)) \ , \quad t \in (0,T) \\[2mm] y(0) = y_0 \\[2mm] y'(0) = z_0 \ , \end{cases}$$

where $\psi : [0,T] \times \mathbb{R} \times \mathbb{R} \rightarrow \mathbb{R}$ is a continuous function. Let us divide $[0,T]$ into subinternals $[t_n, t_{n+1}]$, with $t_0 = 0$, $t_{\mathcal{N}} = T$, $t_{n+1} = t_n + \Delta t$ for $n = 0, ..., \mathcal{N} - 1$. A popular discretization of second order ordinary differential equations like (14.5.6) is the one based on the Newmark method. Denoting y^n the approximation of $y(t_n)$, the Newmark method generates the following sequence: set $y^0 = y_0$, $z^0 = z_0$ and then solve

(14.5.7)
$$\begin{cases} y^{n+1} = y^n + \Delta t\, z^n + (\Delta t)^2 \left[\zeta\psi_{n+1} + \left(\tfrac{1}{2} - \zeta\right)\psi_n \right] \\ z^{n+1} = z^n + \Delta t[\theta\psi_{n+1} + (1 - \theta)\psi_n] \end{cases}$$

for $n = 0, 1, ..., \mathcal{N} - 1$, where ζ and θ are some non-negative parameters, and $\psi_n := \psi(t_n, y^n, z^n)$ (here z^n is an approximation of $y'(t^n)$). The scheme is explicit if $\zeta = \theta = 0$, otherwise for each n a system (which is nonlinear when ψ depends nonlinearly on y and/or y') needs to be solved in order to obtain y^{n+1}, z^{n+1} from y^n and z^n.

When ψ doesn't depend on y', then z^{n+1} can be eliminated from the above equations, and we obtain for $n = 0, 1, ..., \mathcal{N} - 2$

(14.5.8)
$$\begin{aligned} y^{n+2} &- 2y^{n+1} + y^n \\ &= (\Delta t)^2 \left[\zeta\psi_{n+2} + \left(\frac{1}{2} - 2\zeta + \theta\right)\psi_{n+1} + \left(\frac{1}{2} + \zeta - \theta\right)\psi_n \right] . \end{aligned}$$

If $\zeta = 0$ this scheme is explicit (when $\theta = 1/2$ it reduces to the well known leap-frog scheme). By formal Taylor expansion it can be found that the Newmark method (14.5.7) is a second order one if $\theta = 1/2$, whereas it is first order accurate if $\theta \neq 1/2$.

The condition $\theta \geq 1/2$ is necessary for stability. Moreover, if $\theta = 1/2$ and $\zeta = 1/4$, the Newmark method is unconditionally stable. This popular choice is, however, unsuitable for long time integration, as the discrete solution may be affected by parasitic oscillations that are not damped as far as t increases. For long time integration it is therefore preferable to use $\zeta \geq (\theta + 1/2)^2/4$ for a suitable $\theta > 1/2$ (although in such a case the method downgrades to a first order one).

When applied to system (14.5.5), the Newmark method becomes:

(14.5.9)
$$\begin{cases} \dfrac{1}{(\Delta t)^2} M(\xi^{n+1} - \xi^n - \Delta t\sigma^n) + A\left[\zeta\xi^{n+1} + \left(\dfrac{1}{2} - \zeta\right)\xi^n\right] \\ \qquad = \zeta\mathbf{F}(t_{n+1}) + \left(\dfrac{1}{2} - \zeta\right)\mathbf{F}(t_n) \\ \dfrac{1}{\Delta t} M(\sigma^{n+1} - \sigma^n) + A[\theta\xi^{n+1} + (1 - \theta)\xi^n] \\ \qquad = \theta\mathbf{F}(t_{n+1}) + (1 - \theta)\mathbf{F}(t_n) \end{cases}$$

for $n = 0, 1, ..., \mathcal{N} - 1$, with $\xi^0 = \xi_0$ and $\sigma^0 = \xi_1$.

At each step (14.5.9) entails the solution of a linear system

(14.5.10) $$[M + \zeta (\Delta t)^2 A]\, \xi^{n+1} = \eta^n \ ,$$

whose matrix is symmetric and positive definite. If $\zeta = 0$ and M is diagonal (e.g., after a lumping procedure), (14.5.10) provides ξ^{n+1} explicitly.

It can be shown that the solution is unconditionally stable for $\theta = 1/2$ and $\zeta = 1/4$, whereas for the other values of $\theta \geq 1/2$ and $\zeta \geq 0$ the time step-restriction

$$\Delta t \sqrt{\mu_{N_h,h}} \leq C_{\theta,\zeta}$$

has to be assumed. Here $\mu_{N_h,h}$ is the maximum eigenvalue of the bilinear form $a(\cdot,\cdot)$ (see (14.5.4)), and behaves like h^{-2} for finite element subspaces V_h (see (6.3.24)). When $\theta \geq 1/2$ and $\zeta \geq (\theta + 1/2)^2/4$ the constant $C_{\theta,\zeta} > 0$ can be arbitrarily large; otherwise, it is strictly less than $[(\theta + 1/2)^2/4 - \zeta]^{-1}$.

Proofs of these stability results are given in Raviart and Thomas (1983), where the convergence theory is also provided.

14.6 The Finite Volume Method

The finite volume method is especially designed for the approximation of conservation laws. In a very general framework, a conservation law can be written in the vector form as

(14.6.1) $$\frac{d}{dt} \int_D \sigma(\mathbf{w})\, d\mathbf{x} + \int_{\partial D} \varphi(\mathbf{w})\, d\gamma = \mathbf{0} \ ,$$

where $\mathbf{x} \in \Omega \subset \mathbb{R}^d$, $d = 2,3$, D is any open subset of Ω, $\mathbf{w}$ is the unknown function and $\sigma(\mathbf{w})$ is the "matter" that is going to be conserved. Relation (14.6.1) states the property that the variation in time of $\sigma(\mathbf{w})$ in the bounded domain D is equal to the outward normal flux of matter $\varphi(\mathbf{w})$ throughout the boundary ∂D. Equation (14.6.1) can take a scalar form as well, in the case in which both $\sigma(\mathbf{w})$ and $\varphi(\mathbf{w})$ are scalar functions.

When considering an infinitesimal domain, the pointwise expression of the conservation law (14.6.1) reads

(14.6.2) $$\frac{\partial \sigma(\mathbf{w})}{\partial t}(t,\mathbf{x}) + \mathrm{div}\,\mathbf{H}(\mathbf{w})(t,\mathbf{x}) = 0 \ , \quad t > 0 \,, \ \mathbf{x} \in \Omega \ ,$$

where $\mathbf{H}$ is a given tensor, depending on $\mathbf{w}$, such that

$$\int_{\partial D} \mathbf{H}(\mathbf{w})(t,\mathbf{x}) \cdot \mathbf{n}(t,\mathbf{x})\, d\gamma = \int_{\partial D} \varphi(\mathbf{w})(t,\mathbf{x})\, d\gamma \ .$$

Here, we have denoted by $\mathbf{n}(t,\mathbf{x})$ the unit outward normal to the boundary ∂D at the time t and the point $\mathbf{x}$, by $(\mathbf{div}\,\mathbf{H})_i := \sum_j D_j H_{ij}$ and by $(\mathbf{H}\cdot\mathbf{n})_i := \sum_j H_{ij} n_j$.

The perhaps simplest example of conservation law is provided by the one-dimensional scalar problem (14.2.18). Another remarkable instance arises from fluid dynamics, as the momentum, continuity and energy equations for compressible flows, given in (13.1.1)-(13.1.3), are written in the conservation form (14.6.2). In particular, considering, for simplicity, the two-dimensional case, and denoting the unknown vector $\mathbf{w}$ and the function $\sigma(\mathbf{w})$ by

$$(14.6.3) \qquad \mathbf{w} := \begin{pmatrix} \rho \\ u_1 \\ u_2 \\ \vartheta \end{pmatrix} \quad , \quad \sigma(\mathbf{w}) := \begin{pmatrix} \rho \\ \rho u_1 \\ \rho u_2 \\ \rho \eta \end{pmatrix} \quad ,$$

these equations can be written as

$$(14.6.4) \quad \frac{\partial \sigma(\mathbf{w})}{\partial t} + \mathbf{div}[\mathbf{H}_{inv}(\mathbf{w})] = \mathbf{div}[\mathbf{H}_{vis}(\mathbf{w})] \quad \text{in } Q_T := (0,T) \times \Omega \ .$$

Here $\mathbf{H}_{inv}$ and $\mathbf{H}_{vis}$ are the inviscid and viscous fluxes, respectively, defined as follows:

$$
\mathbf{H}_{inv}(\mathbf{w}) := \begin{pmatrix} \rho u_1 & \rho u_2 \\ \rho u_1^2 + \hat{p} & \rho u_1 u_2 \\ \rho u_1 u_2 & \rho u_2^2 + \hat{p} \\ \rho \eta u_1 & \rho \eta u_2 \end{pmatrix}
$$

$$(14.6.5)$$

$$
\mathbf{H}_{vis}(\mathbf{w}) := \begin{pmatrix} 0 & 0 \\ S_{11} & S_{12} \\ S_{21} & S_{22} \\ (\mathbf{S} \cdot \mathbf{u})_1 - q_1 & (\mathbf{S} \cdot \mathbf{u})_2 - q_2 \end{pmatrix} \quad ,
$$

where

$$(14.6.6) \quad \eta := e + \frac{1}{2}|\mathbf{u}|^2 + \frac{\hat{p}}{\rho} \ , \quad S_{ij} := (\zeta - \mu) \operatorname{div} \mathbf{u}\, \delta_{ij} + 2\mu D_{ij} = T_{ij} + \hat{p}\delta_{ij}$$

and the other quantities have been introduced in Section 13.1.

Clearly, these equations can also be written in the integral way (14.6.1), taking the form

$$(14.6.7) \qquad \frac{d}{dt} \int_D \sigma(\mathbf{w})\, d\mathbf{x} + \int_{\partial D} [\mathbf{H}_{inv}(\mathbf{w}) - \mathbf{H}_{vis}(\mathbf{w})] \cdot \mathbf{n}\, d\gamma = 0 \ .$$

System (14.6.4) is the *Navier-Stokes* system for compressible viscous flows written in conservative form, in the absence of external force field and heat supply. The *Euler* system for compressible inviscid flows is obtained by disregarding the right-hand side.

The finite volume method has been introduced for two-dimensional fluid dynamics simulation by McDonald (1971) and MacCormack and Paullay (1972), then extended by Rizzi and Inouye (1973) to three-dimensional flows. It moves from the integral form of conservation laws (14.6.1). Equivalently, it makes use of the Green formula to replace volume averages with surface fluxes when starting from the pointwise form (14.6.2).

A first step toward setting up a finite volume approximation of (14.6.7) consists in subdividing the computational domain into quadrilaterals V which are usually called "volumes" (also for a two-dimensional problem). Subdivisions into triangles are also used, especially in an unstructured grid context (see Fig. 14.6.3 below). At this stage, the same rules of finite element partitions need to be followed. In particular, the volumes should not overlap each other, and the intersection of two of them can be either empty, or a full side, or a vertex. We focus below on this type of space discretization. The way of discretizing the time derivative will be addressed later.

For each *control volume* $V_J \subset \Omega$, equations (14.6.7) yield

$$(14.6.8) \qquad \frac{d}{dt} \int_{V_J} \sigma(\mathbf{w})\, dx + \int_{\partial V_J} [\mathbf{H}_{inv}(\mathbf{w}) - \mathbf{H}_{vis}(\mathbf{w})] \cdot \mathbf{n}\, d\gamma = 0 \ .$$

A control volume can be either a volume or the union of subsets of volumes merging at a certain vertex. See Fig. 14.6.1 for an example referring to both quadrilateral (a) and triangular (b) decompositions.

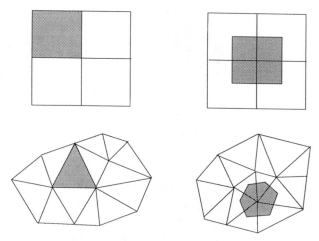

Fig. 14.6.1. Examples of control volumes

Equation (14.6.8) can be given a different meaning according to

a. the assumed polynomial form of the function $\mathbf{w}$ into V_J;
b. the choice of the degrees of freedom for $\mathbf{w}$;
c. the way of computing the boundary fluxes (i.e., the exchange mechanism between adjacent control volumes).

Typically, the volume integral in (14.6.8) is replaced by

$$(14.6.9) \qquad \int_{V_J} \sigma(\mathbf{w})dx \simeq (\sigma(\mathbf{w}))_J\, \mathrm{meas}(V_J) \ .$$

The value $(\sigma(\mathbf{w}))_J$ represents an *average value* of the flow variable $\sigma(\mathbf{w})$ over the control volume V_J. The same approximation is used for source terms, if any.

Concerning the choice of degrees of freedom, we may either have *cell-centered*, or *cell-vertex*, or else *node-centered* approximations (see Fig. 14.6.2 for an illustration in the case of rectangular volumes). In cell-centered finite volumes, the variables are associated with a cell. They therefore represent averaged values over the cell, and can be considered as representative of some point inside the cell (for instance, they provide exactly the center of gravity of the cell for finite volume schemes that are second order in space). In both cell-vertex and node-centered cases, the variable are "attached" to the mesh points.

When considering systems of conservation laws such as the Euler or Navier-Stokes equations for compressible flows, staggered grids may be used for the finite volume discretization. In such a case, different degrees of freedom are associated to each physical unknown (see, e.g., Patankar (1980)).

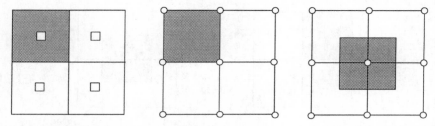

Fig. 14.6.2. Cell-centered (a), cell-vertex (b), and node-centered (c) volumes. The shaded region represents the control volume

With respect to the grid, two types of meshes can be considered:
1. A "finite difference" like mesh, where all mesh points lie on the intersection of two families of lines (e.g., those parallel to the coordinate axes). This is usually designated as a *structured* mesh. Examples are shown in Fig. 14.6.2 (a) and Fig. 14.6.3 (a).
2. A "finite element" like mesh formed by arbitrary combinations of triangular or quadrilateral cells, with mesh points that cannot be identified by a couple of integers but need to be numbered individually, and the connections among the different vertices must be provided explicitly. This type of mesh is designated as *unstructured*. Examples are shown in Fig. 14.6.2 (b) and Fig. 14.6.3 (b)

Unstructured meshes can offer greater flexibility for complicated geometrical configurations.

On the other hand, structured meshes allow an easy mapping of the computational domain into a rectangular one. Moreover, one-dimensional algo-

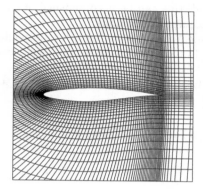 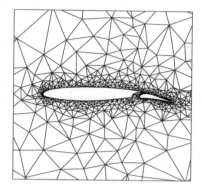

Fig. 14.6.3. Structured (a) and unstructured (b) grid around airfoils

rithms can easily be extended to two-dimensional situations by means of splitting the derivatives along coordinate lines. Tackling complicated geometries by structured meshes is made possible by partitioning the domain into multiblocks.

Concerning the way of choosing the variables, cell-centered discretization are often used in connection with structured meshes. In this framework, the issue is how to enforce boundary conditions, since no mesh point lies on the domain boundary $\partial\Omega$. Often, this problem is solved by resorting to extrapolation procedures. The node-centered approach is typically used in unstructured meshes.

A final issue that characterizes a finite volume method is the strategy adopted for computing the boundary fluxes. The evaluation of the flux components along the edges of ∂V_J (see (14.6.8)) depends on the selected scheme as well as on the location of the flow variables (i.e., the choice of the degrees of freedom). Basically, we can distinguish between centered and upwind discretization schemes.

The former are based on estimating the fluxes symmetrically, taking contributions from both sides. In such cases, a suitable amount of numerical dissipation needs to be artificially added to the scheme in order to prevent the rise of oscillations (see, e.g., Jameson, Schmidt and Turkel (1981)).

The latter determine the fluxes on each side according to the propagation direction of the wave components. We briefly discuss some possible criteria. Let us forget for a while the viscous flux. A first possibility is offered by the replacement of the normal flux $\mathbf{H}_{inv}(\mathbf{w}) \cdot \mathbf{n} =: \mathbf{H}_\Sigma(\mathbf{w})$ on a side Σ by a numerical flux function $\mathbf{H}^{num}(\mathbf{w}^l, \mathbf{w}^r)$ that depends on two states (left and right or upper and lower). The numerical flux function, that must be consistent with the exact one, can be chosen as done in Section 14.2.3 for one dimensional scalar hyperbolic equations. One possibility is to compute $\mathbf{H}_\Sigma(\mathbf{w})$ after applying a Riemann solver to the two states $\mathbf{w}^l$, $\mathbf{w}^r$ at the previous time-level. For a structured grid, this can be accomplished, e.g., as done in Section 14.2.3 for finite difference approximations.

More direct approaches can be pursued also. Let us refer, for simplicity, to the structured grid depicted in Fig. 14.6.4, which is associated to a cell-centered finite volume method.

If V_J is the rectangle $ABCD$, and $\mathbf{H}_{inv} = (\mathbf{F} \,|\, \mathbf{G})$, where $\mathbf{F}$ and $\mathbf{G}$ are four-dimensional column vectors, then

$$(14.6.10) \quad \int_{\partial V_J} \mathbf{H}_{inv}(\mathbf{w}) \cdot \mathbf{n} \, d\gamma = \int_A^B \mathbf{F}(\mathbf{w}) \, dx_2 + \int_B^C \mathbf{G}(\mathbf{w}) \, dx_1$$
$$- \int_C^D \mathbf{F}(\mathbf{w}) \, dx_2 - \int_D^A \mathbf{G}(\mathbf{w}) \, dx_1 \ .$$

In the discretization, the right hand side can be approximated by the *centered* scheme

$$(14.6.11) \quad \mathbf{F}_{AB}|x_{B,2} - x_{A,2}| + \mathbf{G}_{BC}|x_{B,1} - x_{C,1}|$$
$$- \mathbf{F}_{CD}|x_{C,2} - x_{D,2}| - \mathbf{G}_{DA}|x_{A,1} - x_{D,1}| \ .$$

Let us focus, e.g., on $\mathbf{F}_{AB}$. Several alternatives can be considered:

a.   left and right average of fluxes

$$(14.6.12) \qquad \mathbf{F}_{AB} = \frac{1}{2}(\mathbf{F}_{i+1,j} + \mathbf{F}_{i,j}) \text{ with } \mathbf{F}_{i,j} := \mathbf{F}(\mathbf{w}_{i,j}) \ ,$$

where $\mathbf{w}_{i,j}$ is the value of $\mathbf{w}$ at the node (i,j);

b.   vertex average of fluxes

$$(14.6.13) \qquad \mathbf{F}_{AB} = \frac{1}{2}(\mathbf{F}_A + \mathbf{F}_B) \text{ with } \mathbf{F}_A := \mathbf{F}(\mathbf{w}_A), \mathbf{F}_B := \mathbf{F}(\mathbf{w}_B) \ ,$$

where $\mathbf{w}_A$ (respectively, $\mathbf{w}_B$) is the sum of the values of $\mathbf{w}$ on the four cells merging at the point A (respectively, B) divided by 4;

c.   flux of averages of variables

$$(14.6.14) \qquad \mathbf{F}_{AB} = \mathbf{F}\left(\frac{1}{2}(\mathbf{w}_{i+1,j} + \mathbf{w}_{i,j})\right) \ .$$

The computation of the other fluxes is carried out similarly.

For cell-vertex finite volume methods, the choice (14.6.13) looks straightforward, but now $\mathbf{w}_A$ (respectively, $\mathbf{w}_B$) denotes the value of $\mathbf{w}$ at the vertex A (respectively, B).

It is not difficult to see that on a uniform grid the above finite volume methods coincide, indeed, with finite difference schemes. For instance, if the size of each cell is $\Delta x_1 \times \Delta x_2$, the choice (14.6.12) leads to the second-order centered scheme:

$$\frac{\partial \mathbf{w}_{ij}}{\partial t} + \frac{\mathbf{F}_{i+1,j} - \mathbf{F}_{i-1,j}}{2\Delta x_1} + \frac{\mathbf{G}_{i,j+1} - \mathbf{G}_{i,j-1}}{2\Delta x_2} = \mathbf{0} \ .$$

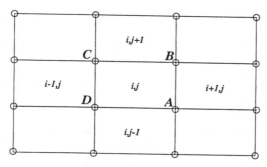

Fig. 14.6.4. Cell-centered finite volumes

As seen in Section 14.2 we may therefore expect oscillations, and this suggests using *upwind* schemes also in this context. This amounts to evaluating a convective flux as a function of the propagation direction of the associated convection speed, which is given by the Jacobian

$$A(\mathbf{w}) = \frac{\partial \mathbf{H}_{inv}(\mathbf{w})}{\partial \mathbf{w}} \quad .$$

For cell-centered finite volumes, the simplest upwind scheme takes the cell-side flux equal to the flux generated in the upstream cell. This yields standard finite difference upwind schemes on uniform meshes.

For cell-vertex finite volume methods, the flux $\mathbf{F}_{AB}$ is replaced by either $\mathbf{F}_{CD}$ or $\mathbf{F}_{EF}$, according to the sign of the advective velocity. The "stencil" of the resulting scheme can therefore be quite large.

As for viscous fluxes, i.e., $\int_{\partial V_J} \mathbf{H}_{vis}(\mathbf{w}) \cdot \mathbf{n} \, d\gamma$ in (14.6.8), these are typically evaluated by centered schemes. There are, however, cases in which these terms have been dealt with according to a finite element strategy (see, e.g., Rostand and Stoufflet (1988), Tidriri (1992), Arminjon and Dervieux (1993)). For other finite volume schemes on structured grids for the compressible Navier-Stokes equations see Crumpton, Mackenzie and Morton (1993) and the references therein.

Time-advancing schemes can be either explicit or implicit. For steady solutions, explicit schemes become competitive provided they are accelerated by suitable techniques: by local time-stepping, by multi-grid technique for the evaluation of residuals, or by a suitable smoothing of residuals (for an account, see, e.g., Johnson (1983), Jameson (1985), Koren (1988), Lallemand and Koren (1989), Lallemand, Steve and Dervieux (1992), Hemker and Desideri (1993)).

Implicit methods are generally more efficient for unsteady problems, especially for viscous flows with high Reynolds numbers. These schemes are usually implemented in the so-called "delta-form", solving at each step a problem for the unknown $\Delta \mathbf{w}^{(n)} := \mathbf{w}^{n+1} - \mathbf{w}^n$. A way for obtaining a problem in this form is outlined below. Starting from (14.6.1) and advancing in time by the backward Euler scheme, we obtain for each control volume V_J

$$\int_{V_J} \frac{\sigma(\mathbf{w}^{n+1}) - \sigma(\mathbf{w}^n)}{\Delta t} \, d\mathbf{x} + \int_{\partial V_J} \varphi(\mathbf{w}^{n+1}) \, d\gamma = 0 \ .$$

Using a first order Taylor expansion for the unknown terms produces the approximate equation:

$$\frac{1}{\Delta t} \int_{V_J} \frac{\partial \sigma}{\partial \mathbf{w}}(\mathbf{w}^n) \, \Delta \mathbf{w}^{(n)} \, d\mathbf{x} + \int_{\partial V_J} \frac{\partial \varphi}{\partial \mathbf{w}}(\mathbf{w}^n) \, \Delta \mathbf{w}^{(n)} \, d\gamma = -\int_{\partial V_J} \varphi(\mathbf{w}^n) \, d\gamma \ ,$$

which is now assumed as a starting point for implementing one of the finite volume schemes illustrated above.

14.7 Complements

A review article on multi-dimensional upwind finite difference methods is the one by van Leer (1992). A general convergence theory is presented in Coquel and Le Floch (1993).

Concerning finite elements, nonlinear hyperbolic equations have been extensively investigated in recent years. For the theoretical analysis we refer to the contributions of Johnson and his coworkers (see, e.g., Johnson and Szepessy (1987), Johnson, Szepessy and Hansbo (1990), Szepessy (1989, 1991)). For the numerical applications to the field of compressible inviscid flows we refer to Hughes and Tezduyar (1984), Shakib (1988), Le Beau and Tezduyar (1991), Le Beau, Ray, Aliabadi and Tezduyar (1993).

An approach that aims at exploiting the geometric generality of finite element tringulations and a finite difference-like treatment of interface fluxes (à la ENO) is the basis of the so-called *discontinuous Galerkin Runge-Kutta* method, that has been proposed and analyzed by Cockburn and Shu (1989, 1991), Cockburn, Lin and Shu (1989), Cockburn, Hou and Shu (1990). The schemes based on this approach are TVB and kth-order accurate (when polynomials of degree less than or equal to $k \geq 1$ are used for defining the finite element subspaces).

For a review paper on the Taylor-Galerkin method for nonlinear hyperbolic equations see Donea and Quartapelle (1992).

References

S. Abarbanel, D. Gottlieb and E. Tadmor (1986), Spectral methods for discontinuous problems, in *Numerical Methods for Fluid Dynamics II*, K.W. Morton and M.J. Baines eds., Oxford University Press, Oxford, pp. 129–153.

M. Abramowitz and I.A. Stegun, eds. (1966), *Handbook of Mathematical Functions with Formulas, Graphs, and Mathematical Tables*, Dover, New York.

R.A. Adams (1975), *Sobolev Spaces*, Academic Press, New York.

S. Agmon (1965), *Lectures on Elliptic Boundary Value Problems*, Van Nostrand, Princeton.

V.I. Agoshkov (1988), Poincaré-Steklov's operators and domain decomposition methods in finite dimensional spaces, in *First International Symposium on Domain Decomposition Methods for Partial Differential Equations*, R. Glowinski, G.H. Golub, G.A. Meurant and J. Périaux eds., SIAM, Philadelphia, pp. 73–112.

V.I. Agoshkov and V.I. Lebedev (1985), Poincaré-Steklov operators and the methods of partition of the domain in variational problems, in *Computational Processes and Systems, No. 2*, G.I. Marchuk ed., Nauka, Moscow, pp. 173–227 [Russian].

M. Ainsworth and J.T. Oden (1993), A unified approach to a posteriori error estimation using element residual methods, Numer. Math. 65, 23–50.

M. Ainsworth, J.Z. Zhu, A.W. Craig and O.C. Zienkiewicz (1989), Analysis of the Zienkiewicz-Zhu a posteriori error estimator in the finite element method, Internat. J. Numer. Methods Engrg. 28, 2161–2174.

E.L. Allgower and K. Georg (1990), *Numerical Continuation Methods: an Introduction*, Springer-Verlag, Berlin.

E.L. Allgower and K. Georg (1993), Continuation and path following, Acta Numer., 1–64.

C.R. Anderson and C. Greengard, eds. (1991), *Vortex Dynamics and Vortex Methods*, American Mathematical Society, Providence.

P. Arminjon and A. Dervieux (1993), Construction of TVD-like artificial viscosities on two-dimensional arbitrary FEM grids, J. Comput. Phys. 106, 176–198.

D.N. Arnold and F. Brezzi (1985), Mixed and non-conforming finite element methods: implementation, post-processing and error estimates, M^2AN 19, 7–35.

D.N. Arnold, F. Brezzi and M. Fortin (1984), A stable finite element for the Stokes equations, Calcolo 21, 337–344.

W.E. Arnoldi (1951), The principle of minimized iterations in the solution of matrix eigenvalue problem, Quart. Appl. Math. 9, 17–29.

K.E. Atkinson (1978), *An Introduction to Numerical Analysis*, John Wiley & Sons, New York.

O. Axelsson and V.A. Barker (1984), *Finite Element Solution of Boundary Value Problems. Theory and Computation*, Academic Press, Orlando.

I. Babuška (1971), Error bounds for the finite element method, Numer. Math. 16, 322–333.

I. Babuška (1973a), The finite element method with Lagrangian multipliers, Numer. Math. 20, 179–192.

I. Babuška (1973b), The finite element method with penalty, Math. Comput. 27, 221–228.

I. Babuška (1976), The selfadaptive approach in the finite element method, in *The Mathematics of Finite Elements and Applications II*, J.R. Whiteman ed., Academic Press, New York, pp. 125–142.

I. Babuška and A.K. Aziz (1972), Survey lectures on the mathematical foundations of the finite element method, in *The Mathematical Foundations of the Finite Element Method with Applications to Partial Differential Equations*, A.K. Aziz ed., Academic Press, New York, pp. 3–359.

I. Babuška, J Chandra and J.E. Flaherty, eds. (1983), *Adaptive Computational Methods for Partial Differential Equations*, SIAM, Philadelphia.

I. Babuška and M.R. Dorr (1981), Error estimates for the combined h and p-version of the finite element method, Numer. Math. 37, 257–277.

I. Babuška and A. Miller (1987), A feedback finite element method with a-posteriori error estimates. Part I: the finite element method and some basic properties of the a-posteriori estimator, Comput. Meth. Appl. Mech. Engrg. 61, 1–40.

I. Babuška, J.T. Oden and J.K. Lee (1977), Mixed-hybrid finite element approximations of second order elliptic boundary-value problems, Comput. Meth. Appl. Mech. Engrg. 11, 175–206.

I. Babuška and J. Osborn (1980), Analysis of finite element methods for second order boundary value problems using mesh dependent norms, Numer. Math. 34, 41–62.

I. Babuška and W.C. Rheinboldt (1978a), Error estimates for adaptive finite element computations, SIAM J. Numer. Anal. 12, 1597–1615.

I. Babuška and W.C. Rheinboldt (1978b), A posteriori error estimates for the finite element method, Internat. J. Numer. Methods Engrg. 15, 736–754.

I. Babuška and M. Suri (1990), The p- and h-p versions of the finite element method, an overview, in *Spectral and High Order Methods for Partial Differential Equations*, C. Canuto and A. Quarteroni eds., North-Holland, Amsterdam, pp. 5–26.

I. Babuška, B.A. Szabó and I.N. Katz (1981), The p-version of the finite element method, SIAM J. Numer. Anal. 18, 512–545.

I. Babuška, O.C. Zienkiewicz, J. Gago and E.R. de Arantes e Oliveira, eds. (1986), *Accuracy Estimates and Adaptive Refinements in Finite Element Computations*, John Wiley & Sons, Chichester.

C. Baiocchi and F. Brezzi (1983), Optimal error estimates for linear parabolic problems under minimal regularity assumptions, Calcolo 20, 143–176.

C. Baiocchi and F. Brezzi (1992), Stabilization of unstable numerical methods, in *Problemi Attuali dell'Analisi e della Fisica Matematica*, P.E. Ricci ed., Taormina, pp. 59–64.

C. Baiocchi, F. Brezzi and L.P. Franca (1993), Virtual bubbles and Galerkin-least-squares type methods (Ga.L.S.), Comput. Meth. Appl. Mech. Engrg. 105, 125–141.

G.A. Baker, J.H. Bramble and V. Thomée (1977), Single step Galerkin approximations for parabolic problems, Math. Comput. 31, 818–847.

G.A. Baker, V.A. Dougalis and O.A. Karakashian (1982), On a higher order accurate fully discrete Galerkin approximation to the Navier-Stokes equations, Math. Comput. 39, 339–375.

R.E. Bank and T. Dupont (1981), An optimal order process for solving elliptic finite element equations, Math. Comput. 36, 35–51.

R.E. Bank and B.D. Welfert (1990), A comparison between the mini-element and the Petrov-Galerkin formulations for the generalized Stokes problem, Comput. Meth. Appl. Mech. Engrg. 83, 61–68.

R.E. Bank and A. Wieser (1985), Some a posteriori error estimators for elliptic partial differential equations, Math. Comput. 44, 283–301.

V. Barwell and J.A. George (1976), A comparison of algorithms for solving symmetric indefinite systems of linear equations, ACM Trans. Math. Soft. 2, 242–251.

J.T. Beale and A. Majda (1982), Vortex methods. I: convergence in three dimensions, Math. Comput. 39, 1–27.

M.A. Behr, L.P. Franca and T.E. Tezduyar (1993), Stabilized finite element methods for the velocity-pressure-stress formulation of incompressible flows, Comput. Meth. Appl. Mech. Engrg. 104, 31–48.

J. Bemelmans (1981), Gleichgewichtsfiguren zäher Flüssigkeiten mit Oberflächenspannung, Analysis 1, 241–282.

C. Bègue, C. Conca, F. Murat and O. Pironneau (1988), Les équations de Stokes et de Navier-Stokes avec des conditions aux limites sur la pression, in *Nonlinear Partial Differential Equations and Their Applications, Collège de France Seminar, Vol. IX*, H. Brezis and J.-L. Lions eds., Longman, Harlow, pp. 179–264.

J.P. Benqué, B. Ibler, A. Keramsi and G. Labadie (1980), A finite element method for Navier-Stokes equations, in *Proceedings of the Third International Conference on Finite Elements in Flow Problems, vol. 1*, D.H. Norrie ed., Banff, pp. 110–120.

M. Bercovier (1978), Perturbation of a mixed variational problem, applications to mixed finite element methods, R.A.I.R.O. Anal. Numér. 12, 211–236.

M. Bercovier and O. Pironneau (1979), Error estimates for finite element solution of the Stokes problem in the primitive variables, Numer. Math. 33, 211–224.

J. Bergh and J. Löfström (1976), *Interpolation Spaces: an Introduction*, Springer-Verlag, Berlin.

C. Bernardi, C. Canuto and Y. Maday (1988), Generalized inf-sup condition for Chebyshev approximation of the Stokes problem, SIAM J. Numer. Anal. 25, 1237–1271.

C. Bernardi and Y. Maday (1989), Properties of some weighted Sobolev spaces, and application to spectral approximations, SIAM J. Numer. Anal. 26, 769–829.

C. Bernardi and Y. Maday (1992), *Approximations Spectrales de Problèmes aux Limites Elliptiques*, Springer-Verlag, Paris.

C. Bernardi, Y. Maday and B. Metivet (1987), Calcul de la pression dans la résolution spectrale du problème de Stokes, Rech. Aérospat., 1–21.

C. Bernardi and G. Raugel (1985), A conforming finite element method for the time-dependent Navier-Stokes equations, SIAM J. Numer. Anal. 22, 455–473.

Å. Björck, R.J. Plemmons and H. Schneider (1981), *Large Scale Matrix Problems*, North-Holland, New York.

P.E. Bjørstad (1989), Multiplicative and additive Schwarz methods: convergence in the two-domain case, in *Domain Decomposition Methods*, T.F. Chan, R. Glowinski, J. Périaux and O.B. Widlund eds., SIAM, Philadelphia, pp. 147–159.

P.E. Bjørstad and O.B. Widlund (1986), Iterative methods for the solution of elliptic problems on regions partitioned into substructures, SIAM J. Numer. Anal. 23, 1097–1120.

J. Boland and R.A. Nicolaides (1983), Stability of finite elements under divergence constraints, SIAM J. Numer. Anal. 20, 722–731.

J.P. Boris and D.L. Book (1973), Flux corrected transport, I. SHASTA, a fluid transport algorithm that works, J. Comput. Phys. 11, 38–69.

J.F. Bourgat, R. Glowinski, P. Le Tallec and M. Vidrascu (1989), Variational formulation and algorithm for trace operator in domain decomposition calculations, in

Domain Decomposition Methods, T.F. Chan, R. Glowinski, J. Périaux and O.B. Widlund eds., SIAM, Philadelphia, pp. 3–16.

J.P. Boyd (1989), *Chebyshev and Fourier Spectral Methods*, Springer-Verlag, Berlin.

J.H. Bramble (1986), The construction of preconditioners for elliptic problems by substructuring, in *Numerical Analysis*, J.P. Hennart ed., Springer-Verlag, Berlin, pp. 158–166.

J.H. Bramble (1993), *Multigrid Methods*, Longman, Harlow.

J.H. Bramble and J.E. Pasciak (1988), A preconditioning technique for indefinite systems resulting from mixed approximations of elliptic problems, Math. Comput. 50, 1–17.

J.H. Bramble, J.E. Pasciak and A.H. Schatz (1986a), An iterative method for elliptic problems on regions partitioned into substructures, Math. Comput. 46, 361–369.

J.H. Bramble, J.E. Pasciak and A.H. Schatz (1986b), The construction of preconditioners for elliptic problems by substructuring. I, Math. Comput. 47, 103–134.

J.H. Bramble, J.E. Pasciak and A.H. Schatz (1987), The construction of preconditioners for elliptic problems by substructuring. II, Math. Comput. 49, 1–16.

J.H. Bramble, J.E. Pasciak and A.H. Schatz (1988), The construction of preconditioners for elliptic problems by substructuring, III, Math. Comput. 51, 415–430.

J.H. Bramble, J.E. Pasciak and A.H. Schatz (1989), The construction of preconditioners for elliptic problems by substructuring, IV, Math. Comput. 53, 1–24.

J.H. Bramble, J.E. Pasciak and J. Xu (1990), Parallel multilevel preconditioners, in *Third International Symposium on Domain Decomposition Methods for Partial Differential Equations*, T.F. Chan, R. Glowinski, J. Périaux and O.B. Widlund eds., SIAM, Philadelphia, pp. 341–357.

J.H. Bramble, A.H. Schatz, V. Thomée and L. Wahlbin (1977), Some convergence estimates for semi-discrete Galerkin type approximations for parabolic equations, SIAM J. Numer. Anal. 14, 218–241.

J.H. Bramble and V. Thomée (1974), Discrete time Galerkin methods for a parabolic boundary value problem, Ann. Mat. Pura Appl. 101, 115–152.

A. Brandt (1977), Multi-level adaptive solution to boundary-value problems, Math. Comput. 31, 333–390.

C.A. Brebbia, J.C.F. Telles and L.C. Wrobel (1984), *Boundary Element Techniques. Theory and Applications in Engineering*, Springer-Verlag, Berlin.

K.E. Brenan, S.L. Campbell and L.R. Petzold (1989), *Numerical Solution of Initial-Value Problems in Differential-Algebraic Equations*, North-Holland, New York.

N. Bressan and A. Quarteroni (1986), Analysis of Chebyshev collocation methods for parabolic equations, SIAM J. Numer. Anal. 23, 1138–1154.

H. Brezis (1983), *Analyse Fonctionnelle*, Masson, Paris.

F. Brezzi (1974), On the existence, uniqueness and approximation of saddle-point problems arising from Lagrange multipliers, R.A.I.R.O. Anal. Numér. 8, 129–151.

F. Brezzi, M.-O. Bristeau, L.P. Franca, M. Mallet and G. Rogé (1992), A relationship between stabilized finite element methods and the Galerkin method with bubble functions, Comput. Meth. Appl. Mech. Engrg. 96, 117–129.

F. Brezzi, C. Canuto and A. Russo (1989), A self-adaptive formulation for the Euler/Navier-Stokes coupling, Comput. Meth. Appl. Mech. Engrg. 73, 317–330.

F. Brezzi, J. Douglas, Jr. (1988), Stabilized mixed methods for the Stokes problem, Numer. Math. 53, 225–235.

F. Brezzi, J. Douglas, Jr., R. Duran and M. Fortin (1987), Mixed finite elements for second order elliptic problems in three variables, Numer. Math. 51, 237–250.

F. Brezzi, J. Douglas, Jr. and L.D. Marini (1985), Two families of mixed finite elements for second order elliptic problems, Numer. Math. 47, 217–235.

F. Brezzi and R.S. Falk (1991), Stability of higher-order Hood-Taylor methods, SIAM J. Numer. Anal. 28, 581–590.

F. Brezzi and M. Fortin (1991), *Mixed and Hybrid Finite Element Methods*, Springer-Verlag, New York.

F. Brezzi and G. Gilardi (1987), Functional analysis & Functional spaces, in *Finite Element Handbook*, Chap. 1 & 2, H. Kardestuncer ed., McGraw-Hill, New York.

F. Brezzi and J. Pitkäranta (1984), On the stabilization of finite element approximations of the Stokes equations, in *Efficient Solutions of Elliptic Systems*, W. Hackbusch ed., Friedr. Vieweg & Sohn, Braunschweig, pp. 11–19.

F. Brezzi, J. Rappaz and P.-A. Raviart (1980), Finite-dimensional approximation of nonlinear problems. Part I: branches of nonsingular solutions, Numer. Math. 36, 1–25.

F. Brezzi, J. Rappaz and P.-A. Raviart (1981a), Finite-dimensional approximation of nonlinear problems. Part II: limit points, Numer. Math. 37, 1–28.

F. Brezzi, J. Rappaz and P.-A. Raviart (1981b), Finite-dimensional approximation of nonlinear problems. Part III: simple bifurcation points, Numer. Math. 38, 1–30.

M.O. Bristeau, R. Glowinski and J. Périaux (1987), Numerical methods for the Navier-Stokes equations. Application to the simulation of compressible and incompressible viscous flows, Comput. Phys. Rep. 6, 73–187.

A.N. Brooks and T.J.R. Hughes (1982), Streamline Upwind/Petrov-Galerkin formulations for convection dominated flows with particular emphasis on the incompressible Navier-Stokes equations, Comput. Meth. Appl. Mech. Engrg. 32, 199–259.

P.N. Brown and Y. Saad (1990), Hybrid Krylov methods for nonlinear systems of equations, SIAM J. Sci. Stat. Comput. 11, 450–481.

J.R. Bunch and D.J. Rose, eds. (1976), *Sparse Matrix Computations*, Academic Press, New York.

J.C. Butcher (1987), *The Numerical Analysis of Ordinary Differential Equations: Runge-Kutta and General Linear Methods*, John Wiley & Sons, Chichester.

B.L. Buzbee, F.W. Dorr, J.A. George and G.H. Golub (1971), The direct solution of the discrete Poisson equation on irregular regions, SIAM J. Numer. Anal. 8, 722–736.

B.L. Buzbee, G.H. Golub and C.V. Nielson (1970), On direct methods for solving Poisson's equation, SIAM J. Numer. Anal. 7, 627–656.

J. Cahouet and J.P. Chabard (1988), Some fast 3-D finite element solvers for the generalized Stokes problem, Internat. J. Numer. Methods Fluids 8, 869–895.

X.-C. Cai (1990), An additive Schwarz algorithm for nonselfadjoint elliptic equations, in *Third International Symposium on Domain Decomposition Methods for Partial Differential Equations*, T.F. Chan, R. Glowinski, J. Périaux and O.B. Widlund eds., SIAM, Philadelphia, pp. 232–244.

X.-C. Cai (1991), Additive Schwarz algorithms for parabolic convection-diffusion equations, Numer. Math. 60, 41–61.

C. Canuto (1988), Spectral methods and a maximum principle, Math. Comput. 51, 615–629.

C. Canuto (1994), Stabilization of spectral methods by finite elements bubble functions, Comput. Meth. Appl. Mech. Engrg. 116, 13–26.

C. Canuto, M.Y. Hussaini, A. Quarteroni and T.A. Zang (1988), *Spectral Methods in Fluid Dynamics*, Springer-Verlag, New York.

C. Canuto and P. Pietra (1991), Boundary and interface conditions within a finite element preconditioner for spectral methods, J. Comput. Phys. 91, 310–343.

C. Canuto and A. Quarteroni (1982), Approximation results for orthogonal polynomials in Sobolev spaces, Math. Comput. 38, 67–86.

C. Canuto and A. Quarteroni (1984), Variational methods in the theoretical analysis of spectral approximations, in *Spectral Methods for Partial Differential Equations*, R.G. Voigt, D. Gottlieb and M.Y. Hussaini eds., SIAM, Philadelphia, pp. 55–78.

C. Canuto and A. Quarteroni (1985), Preconditioned minimal residual methods for Chebyshev spectral calculations, J. Comput. Phys 60, 315–337.

C. Canuto and A. Quarteroni (1987), On the boundary treatment in spectral methods for hyperbolic systems, J. Comput. Phys. 71, 100–110.

C. Canuto and A. Russo (1993), On the elliptic-hyperbolic coupling, I: the advection-diffusion equation via the χ-formulation, M^3AS 3, 145–170.

C. Canuto and G. Sacchi Landriani (1986), Analysis of the Kleiser-Schumann method, Numer. Math. 50, 217–243.

C. Carlenzoli and P. Gervasio (1992), Effective numerical algorithms for the solution of algebraic systems arising in spectral methods, Appl. Numer. Math. 10, 87–113.

A.L. Cauchy (1847), Méthode générale pour résolution des systèmes d'équations simultanées, C.R.H.A.S. Paris 25, 536–538.

T.F. Chan (1987), Analysis of preconditioners for domain decomposition, SIAM J. Numer. Anal. 24, 382–390.

T.F. Chan and T.P. Mathew (1991), An application of the probing techinique to the vertex space method in domain decomposition, in *Fourth International Symposium on Domain Decomposition Methods for Partial Differential Equations*, R. Glowinski, Yu.A. Kuznetsov, G. Meurant, J. Périaux and O.B. Widlund eds., SIAM, Philadelphia, pp. 101–111.

G.-Q. Chen, Q. Du and E. Tadmor (1992), Spectral viscosity approximations to multidimensional scalar conservation laws, Math. Comput. 61, 629–643.

G. Chen and J. Zhou (1992), *Boundary Element Methods*, Academic Press, London.

C. Chinosi and M.I. Comodi (1991), A new numerical approach to the Stokes problem, Numer. Meth. Partial Differ. Eq. 7, 363–374.

A.J. Chorin (1967), The numerical solution of the Navier-Stokes equations for an incompressible fluid, Bull. Amer. Math. Soc. 73, 928–931.

A.J. Chorin (1968), Numerical solution of the Navier-Stokes equations, Math. Comput. 22, 745–762.

A.J. Chorin (1973), Numerical study of slightly viscous flow, J. Fluid Mech. 57, 785–796.

A.J. Chorin (1980), Vortex models and boundary layer instability, SIAM J. Sci. Stat. Comput. 1, 1–21.

Ph.G. Ciarlet (1968), An $O(h^2)$ method for a non-smooth boundary value problem, Aequationes Math. 2, 39–49.

Ph.G. Ciarlet (1978), *The Finite Element Method for Elliptic Problems*, North-Holland, Amsterdam.

Ph.G. Ciarlet (1991), Basic error estimates for elliptic problems, in *Handbook of Numerical Analysis, II*, Ph.G. Ciarlet and J.-L. Lions eds., North-Holland, Amsterdam, pp. 16–351.

Ph.G. Ciarlet and P. Destuynder (1979), A justification of the two-dimensional linear plate model, J. Mécanique 18, 315–344.

A. Cividini, A. Quarteroni and E. Zampieri (1993), Numerical solution of linear elastic problems by spectral collocation methods, Comput. Meth. Appl. Mech. Engrg. 104, 49–76.

Ph. Clément (1975), Approximation by finite element functions using local regularization, R.A.I.R.O. Anal. Numér. 9, 77–84.

B. Cockburn and C.-W. Shu (1989), TVB Runge-Kutta local projection discontinuous Galerkin finite element method for conservation laws II: general framework, Math. Comput. 52, 411–435.

B. Cockburn and C.-W. Shu (1991), The Runge-Kutta local projection $\mathbb{P}^1$ discontinuous Galerkin finite element method for conservation laws, M²AN 25, 337–361.

B. Cockburn, S. Hou and C.-W. Shu (1990), The Runge-Kutta local projection discontinuous Galerkin finite element method for conservation laws IV: the multidimensional case, Math. Comput.54, 545–581.

B. Cockburn, S.-Y. Lin and C.-W. Shu (1989), TVB Runge-Kutta local projection discontinuous Galerkin finite element method for conservation laws III: one-dimensional systems, J. Comput. Phys. 84, 90–113.

C. Conca (1984), Approximation de quelques problèmes de type Stokes par une méthode d'éléments finis mixtes. Math. 45, 75–91.

P. Concus and G.H. Golub (1976), A generalized conjugate gradient method for non-symmetric systems of linear equations, in *Computing Methods in Applied Sciences and Engineering*, R. Glowinski and J.-L. Lions eds., Springer-Verlag, New York, pp. 56–65.

P. Concus, G.H. Golub and D.P. O'Leary (1976), A generalized conjugate gradient method for the numerical solution of elliptic partial differential equations, in *Sparse Matrix Computations*, J.R. Bunch and D.J. Rose eds., Academic Press, New York, pp. 309–332.

F. Coquel and Ph. Le Floch (1993), Convergence of finite difference schemes for conservation laws in several space dimension: a general theory, SIAM J. Numer. Anal. 30, 675–700.

R. Courant and K.O. Friedrichs (1948), *Supersonic Flow and Shock Waves*, Interscience, New York.

R. Courant, K.O. Friedrichs and H. Lewy (1928), Über die partiellen Differenzengleichungen der mathematischen Physik, Math. Ann. 100, 32–74.

R. Courant and D. Hilbert (1953), *Methods of Mathematical Physics, I*, Interscience, New York.

M.G. Crandall and A. Majda (1980), Monotone difference approximations for scalar conservations laws, Math. Comput. 34, 1–21.

M. Crouzeix and A.L. Mignot (1984), *Analyse Numérique des Équations Différentielles*, Masson, Paris.

M. Crouzeix and J. Rappaz (1990), *On Numerical Approximation in Bifurcation Theory*, Masson, Paris.

M. Crouzeix and P.-A. Raviart (1973), Conforming and nonconforming finite element methods for solving the stationary Stokes equations, I, R.A.I.R.O. Anal. Numér. 7, 33–75.

P.I. Crumpton, J.A. Mackenzie and K.W. Morton (1993), Cell vertex algorithms for the compressible Navier-Stokes equations, J. Comput. Phys. 109, 1–15.

P.J. Davis and P. Rabinowitz (1984), *Methods of Numerical Integration*, 2nd ed., Academic Press, Orlando.

S.F. Davis (1984), TVD finite difference schemes and artificial viscosity, ICASE Report n. 84–20, Hampton.

C.N. Dawson and Q. Du (1991), A domain decomposition method for paraboic equations based on finite elements, in *Fourth International Symposium on Domain Decomposition Methods for Partial Differential Equations*, R. Glowinski, Yu.A. Kuznetsov, G. Meurant, J. Périaux and O.B. Widlund eds., SIAM, Philadelphia, pp. 255–263.

C.N. Dawson and T.F. Dupont (1992), Explicit-implicit conservative Galerkin domain decomposition procedures for parabolic problems, Math. Comput. 58, 21–34.

L. Demkowicz, J.T. Oden and I. Babuška, eds. (1992), *Reliability in Computational Mechanics*, North-Holland, Amsterdam.

J.W. Demmel, M.T. Heath and H.A. van der Vorst (1993), Parallel numerical linear algebra, Acta Numer., 111–197.

Y.-H. De Roeck and P. LeTallec (1991), Analysis and test of a local domain-decomposition preconditioner, in *Fourth International Symposium on Domain Decomposition Methods for Partial Differential Equations*, R. Glowinski, Yu.A. Kuznetsov, G. Meurant, J. Périaux and O.B. Widlund eds., SIAM, Philadelphia, pp. 112–128.

M.O. Deville and E.H. Mund (1985), Chebyshev pseudospectral solution of second-order elliptic equations with finite element preconditioning, J. Comput. Phys. 60, 517–533.

M.O. Deville and E.H. Mund (1990), Finite element preconditioning for pseudospectral solutions of elliptic problems, SIAM J. Sci. Stat. Comput. 11, 311–342.

C. Devulder, M. Marion and E.S. Titi (1993), On the rate of convergence of nonlinear Galerkin methods, Math. Comput. 60, 495–514.

J. Donea (1984), A Taylor-Galerkin method for convective transport problems, Internat. J. Numer. Methods Engrg. 20, 101–120.

J. Donea and L. Quartapelle (1992), An introduction to finite element methods for transient advection problems, Comput. Meth. Appl. Mech. Engrg. 95, 169–203.

J. Donea, L. Quartapelle and V. Selmin (1987), An analysis of time discretization in the finite element solution of hyperbolic problems, J. Comput. Phys. 70, 463–499.

J.J. Dongarra, I.S. Duff, D.C. Sorensen and H.A. van der Vorst (1991), *Solving Linear Systems on Vector and Shared Memory Computers*, SIAM, Philadelphia.

F.W. Dorr (1970), The direct solution of the discrete Poisson equation on a rectangle, SIAM Rev. 12, 248–263.

M.R. Dorr (1986), The approximation of solutions of elliptic boundary value problems via the p-version of the finite element method, SIAM J. Numer. Anal. 23, 58–77.

C.C. Douglas and J. Douglas, Jr. (1993), A unified convergence theory for abstract multigrid or multilevel algorithms, serial and parallel, SIAM J. Numer. Anal. 30, 136–158.

J. Douglas, Jr. (1955), On the numerical integration of $u_{xx} + u_{yy} = u_t$ by implicit methods, J. SIAM 3, 42–65.

J. Douglas, Jr. and T. Dupont (1970), Galerkin methods for parabolic equations, SIAM J. Numer. Anal. 7, 575–626.

J. Douglas, Jr. and H.H. Rachford, Jr. (1956), On the numerical solution of heat conduction problems in two and three space variables, Trans. Amer. Math. Soc. 82, 421–439.

J. Douglas, Jr. and J.E. Roberts (1985), Global estimates for mixed methods for second order elliptic problems, Math. Comput. 44, 39–52.

J. Douglas, Jr. and T.F. Russell (1982), Numerical mathods for convection-dominated diffusion problems based on combining the method of characteristics with finite element or finite difference procedures, SIAM J. Numer. Anal. 19, 871–885.

J. Douglas, Jr. and J. Wang (1989), An absolutely stabilized finite element method for the Stokes problem, Math. Comput. 52, 495–508.

M. Dryja (1982), A capacitance matrix method for Dirichlet problem on polygon region, Numer. Math. 39, 51–64.

M. Dryja (1983), A finite element-capacitance matrix method for the elliptic problem, SIAM J. Numer. Anal. 20, 671–680.

M. Dryja (1984), A finite element-capacitance method for elliptic problems on regions partitioned into subregions, Numer. Math. 44, 153–168.

M. Dryja (1989), An additive Schwarz algorithm for two- and three-dimensional finite element elliptic problems, in *Domain Decomposition Methods*, T.F. Chan, R. Glowinski, J. Périaux and O.B. Widlund eds., SIAM, Philadelphia, pp. 168–172.

M. Dryja (1991), Substructuring methods for parabolic problmes, in *Fourth International Symposium on Domain Decomposition Methods for Partial Differential Equations*, R. Glowinski, Yu.A. Kuznetsov, G. Meurant, J. Périaux and O.B. Widlund eds., SIAM, Philadelphia, pp. 264–271.

M. Dryja and O.B. Widlund (1987), An additive variant of the Schwarz alternating method for the case of many subregions, Technical Report 339, Department of Computer Science, Courant Institute of Mathematical Sciences, New York University.

M. Dryja and O.B. Widlund (1990), Towards a unified theory of domain decomposition algorithms for elliptic problems, in *Third International Symposium on Domain Decomposition Methods for Partial Differential Equations*, T.F. Chan, R. Glowinski, J. Périaux and O.B. Widlund eds., SIAM, Philadelphia, pp. 3–21.

M. Dryja and O.B. Widlund (1992), Additive Schwarz methods for elliptic finite element problems in three dimensions, in *Fifth International Symposium on Domain Decomposition Methods for Partial Differential Equations*, D.E. Keyes, T.F. Chan, G. Meurant, J.S. Scroggs and R.G. Voigt eds., SIAM, Philadelphia, pp. 3–18.

I.S. Duff, A.M. Erisman and J.K. Reid (1986), *Direct Methods for Sparse Matrices*, Oxford University Press, Oxford.

T. Dupont (1982), Mesh modification for evolution equations, Math. Comput. 39, 85–107.

T. Dupont and R. Scott (1980), Polynomial approximation of functions in Sobolev spaces, Math. Comput. 34, 441–463.

S.C. Eisenstat, H.C. Elman and M.H. Schultz (1983), Variational iterative methods for non-symmetric systems of linear equations, SIAM J. Numer. Anal. 20, 354–361.

H.C. Elman, Y. Saad and P.E. Saylor (1986), A hybrid Chebyshev Krylov subspace algorithm for solving non-symmetric systems of linear equations, SIAM J. Sci. Stat. Comput. 7, 840–855.

P. Erdös (1961), Problems and results on the theory of interpolation, II, Acta Math. Acad. Sci. Hungar. 12, 235–244.

K. Eriksson, D. Estep, P. Hansbo and C. Johnson (1995), Introduction to adaptive methods for differential equations, Acta Numer., 105–158.

K. Eriksson, C. Johnson and V. Thomée (1985), Time discretization of parabolic problems by the discontinuous Galerkin method, M²AN 19, 611–643.

O. Ernst and G.H. Golub (1994), A domain decomposition approach to solving the Helmholtz equation via a radiation boundary condition, in *Domain Decomposition Methods in Science and Engineering*, A. Quarteroni, J. Périaux, Yu.A. Kuznetsov and O.B. Widlund eds., American Mathematical Society, Providence, pp. 177–192.

D.J. Evans, ed. (1983), *Preconditioning Methods: Analysis and Applications*, Gordon & Breach, New York.

R.E. Ewing, T.F. Russell and M.F. Wheeler (1984), Simulation of miscible displacements using mixed methods and a modified method of characteristics, Comput. Meth. Appl. Mech. Engrg. 47, 73–92.

R.S. Falk and J.E. Osborn (1980), Error estimates for mixed methods, R.A.I.R.O. Anal. Numér. 14, 249–277.

R.S. Falk and G.R. Richter (1992), Local error estimates for a finite element method for hyperbolic and convection-diffusion equations, SIAM J. Numer. Anal. 29, 730–754.

C.A.J. Fletcher (1988a), *Computational Techniques for Fluid Dynamics. 1. Fundamental and General Techniques*, Springer-Verlag, Berlin.

C.A.J. Fletcher (1988b), *Computational Techniques for Fluid Dynamics. 2. Specific Techniques for Different Flow Categories*, Springer-Verlag, Berlin.

R. Fletcher (1976), Conjugate gradient methods for indefinite systems, in *Numerical Analysis*, G.A. Watson ed., Springer-Verlag, Berlin, pp. 73–89.

C. Foias, M. Jolly, I. Kevrekidis and E.S. Titi (1991), Dissipativity of numerical schemes, Nonlinearity 4, 591–631.

C. Foias, O. Manley and R. Temam (1988), Modelling of the interaction of small and large eddies in two dimensional turbulent flows, M^2AN 22, 93–114.

G.E. Forsythe and W.R. Wasow (1960), *Finite Difference Methods for Partial Differential Equations*, John Wiley & Sons, New York.

M. Fortin (1972), *Calcul Numérique des Écoulements des Fluides de Bingham et des Fluides Newtonians Incompressibles par le Méthode des Éléments Finis*, Thèse, Université de Paris VI, Paris.

M. Fortin (1977), An analysis of the convergence of mixed finite element methods, R.A.I.R.O. Anal. Numér. 11, 341–354.

M. Fortin (1989), Some iterative methods for incompressible flow problems, Comput. Phys. Comm. 53, 393–399.

M. Fortin (1993), Finite element solution of the Navier-Stokes equations, Acta Numer., 239–284.

M. Fortin and R. Glowinski (1983), *Augmented Lagrangian Methods: Applications to the Numerical Solution of Boundary-Value Problems*, North-Holland, Amsterdam.

M. Fortin and R. Pierre (1992), Stability analysis of discrete generalized Stokes problems, Numer. Meth. Partial Differ. Eq. 8, 303–323.

B. Fraeijs de Veubeke (1965), Displacement and equilibrium models in the finite element method, in *Stress Analysis*, O.C. Zienkiewicz and G.S. Holister eds., John Wiley & Sons, New York, pp. 145–197.

L.P. Franca and S.L. Frey (1992), Stabilized finite element methods: II. The incompressible Navier-Stokes equations, Comput. Meth. Appl. Mech. Engrg. 99, 209–233.

L.P. Franca, S.L. Frey and T.J.R. Hughes (1992), Stabilized finite element methods, I. Application to the advective-diffusive model, Comput. Meth. Appl. Mech. Engrg. 95, 253–276.

L.P. Franca and R. Stenberg (1991), Error analysis of some Galerkin least squares methods for the elasticity equations, SIAM J. Numer. Anal. 28, 1680–1697.

A. Frati, F. Pasquarelli and A. Quarteroni (1993), Spectral approximation to advection-diffusion problems by the fictitious interface method, J. Comput. Phys. 107, 201–212.

H. Fujita and T. Suzuki (1991), Evolution problems, in *Handbook of Numerical Analysis, II*, Ph.G. Ciarlet and J.-L. Lions eds., North-Holland, Amsterdam, pp. 789–928.

D. Funaro (1981), *Approssimazione Numerica di Problem Parabolici e Iperbolici con Metodi Spettrali*, Tesi, Università di Pavia, Pavia.

D. Funaro (1986), A multidomain spectral approximation of elliptic equations, Numer. Meth. Partial Differ. Eq. 2, 187–205.

D. Funaro (1988), Domain decomposition methods for pseudospectral approximations. Part I: second order equations in one dimension, Numer. Math. 52, 325–344.

D. Funaro (1992), *Polynomial Approximation of Differential Equations*, Springer-Verlag, Berlin.

D. Funaro and D. Gottlieb (1988), A new method of imposing boundary conditions in pseudospectral approximation of hyperbolic equations, Math. Comput. 51, 599–613.

D. Funaro and D. Gottlieb (1989), Convergence results for pseudospectral approximations of hyperbolic systems, ICASE Report n. 89–59, Hampton.

D. Funaro, A. Quarteroni and P. Zanolli (1988), An iterative procedure with interface relaxation for domain decomposition methods, SIAM J. Numer. Anal. 25, 1213–1236.

A.C. Galeão and E.G. Dutra do Carmo (1988), A consistent approximate upwind Petrov-Galerkin method for convection-dominated problems, Comput. Meth. Appl. Mech. Engrg. 68, 83–95.

F. Gastaldi and A. Quarteroni (1989), On the coupling of hyperbolic and parabolic systems: analytical and numerical approach, Appl. Numer. Math. 6, 3–31.

F. Gastaldi, A. Quarteroni and G. Sacchi Landriani (1990), On the coupling of two dimensional hyperbolic and elliptic equations: analytical and numerical approach, in *Third International Symposium on Domain Decomposition Methods for Partial Differential Equations*, T.F. Chan, R. Glowinski, J. Périaux and O.B. Widlund eds., SIAM, Philadelphia, pp. 22–63.

L. Gastaldi and R. Nochetto (1987), Optimal L^∞-estimates for nonconforming and mixed finite element methods of lowest order, Numer. Math. 50, 587–611.

C.W. Gear (1971), *Numerical Initial Value Problems in Ordinary Differential Equations*, Prentice-Hall, Englewood Cliffs.

J.A. George and J.W.H. Liu (1981), *Computer Solution of Large Sparse Positive Definite Systems*, Prentice-Hall, Englewood Cliffs.

V. Girault (1988), Incompressible finite element methods for Navier-Stokes equations with nonstandard boundary conditions in $\mathbb{R}^3$, Math. Comput. 51, 55–74.

V. Girault (1990), Curl-conforming finite element methods for Navier-Stokes equations with non-standard boundary conditions in $\mathbb{R}^3$, in *The Navier-Stokes Equations. Theory and Numerical Methods*, J.G. Heywood, K. Masuda, R. Rautmann and V.A. Solonnikov eds., Springer-Verlag, Berlin, pp. 201–218.

V. Girault and P.-A. Raviart (1986), *Finite Element Methods for Navier-Stokes Equations*, Springer-Verlag, Berlin.

R. Glowinski (1984), *Numerical Methods for Nonlinear Variational Problems*, Springer-Verlag, New York.

R. Glowinski, Q.V. Dinh and J. Périaux (1983), Domain decomposition methods for nonlinear problems in fluid dynamics, Comput. Meth. Appl. Mech. Engrg. 40, 27–109.

R. Glowinski and P. Le Tallec (1989), *Augmented Lagrangian and Operator-Splitting Methods in Nonlinear Mechanics*, SIAM, Philadelphia.

R. Glowinski and P. Le Tallec (1990), Augmented Lagrangian interpretation of the nonoverlapping Schwarz alternating method, in *Third International Symposium on Domain Decomposition Methods for Partial Differential Equations*, T.F. Chan, R. Glowinski, J. Périaux and O.B. Widlund eds., SIAM, Philadelphia, pp. 224–231.

R. Glowinski, B. Mantel, J. Périaux, P. Perrier and O. Pironneau (1982), On an efficient new preconditioned conjugate gradient method. Application to the in-core solution of the Navier-Stokes equations via nonlinear least-squares and finite element methods, in *Finite Elements in Fluids, Vol. 4*, R.H. Gallagher, D.H. Norrie, J.T. Oden and O.C. Zienkiewicz eds., John Wiley & Sons, Chichester, pp. 365–401.

R. Glowinski, T.W. Pan and J. Périaux (1993), Fictitious domain method for the Dirichlet problem and its generalization to some flow problems, in *Finite Elements in Fluids. New Trends and Applications*, K. Morgan, E. Oñate, J. Périaux and J. Peraire eds., Pineridge Press, Swansea, pp. 347–368.

R. Glowinski, T.W. Pan and J. Périaux (1994), A fictitious domain method for unsteady incompressible viscous flow modeled by Navier-Stokes equations, in *Domain Decomposition Methods in Science and Engineering*, A. Quarteroni, J. Périaux, Yu.A. Kuznetsov and O.B. Widlund eds., American Mathematical Society, Providence, pp. 421–430.

R. Glowinski and O. Pironneau (1979), On mixed finite element approximation of the Stokes problem (I) Numer. Math. 33, 397–424.

E. Godlewski and P.-A. Raviart (1991), *Hyperbolic Systems of Conservations Laws*, Ellipse, Paris.

S.K. Godunov and V.S. Ryabenkiĭ (1987), *Theory of Difference Schemes. An Introduction to the Underlying Theory*, North-Holland, Amsterdam. Russian ed.: Gosudarstv. Izdat. Fiz.-Mat. Lit., Moscow, 1962.

G.H. Golub and D. Mayers (1984), The use of pre-conditioning over irregular regions, in *Computing Methods in Applied Sciences and Engineering, VI*, R. Glowinski and J.-L. Lions eds., North-Holland, Amsterdam, pp. 3–14.

G.H. Golub and D.P. O'Leary (1989), Some history of the conjugate gradient and Lanczos algorithms: 1948–1976, SIAM Rev. 31, 50–102.

G.H. Golub and J.M. Ortega (1993), *Scientific Computing. An Introduction with Parallel Computing*, Academic Press, San Diego.

G.H. Golub and C.F. Van Loan (1989), *Matrix Computations*, 2nd ed., The Johns Hopkins University Press, Baltimore.

D. Gottlieb (1985), Spectral methods for compressible flow problems, in *Ninth International Conference on Numerical Methods in Fluid Dynamics*, Soubbarameyer and J.-P. Boujot eds., Springer-Verlag, Berlin, pp. 48–61.

D. Gottlieb, M.D. Gunzburger and E. Turkel (1982), On numerical boundary treatment for hyperbolic systems, SIAM J. Numer. Anal. 19, 671–697.

D. Gottlieb and L. Lustman (1983), The spectrum of the Chebyshev collocation operator for the heat equation, SIAM J. Numer. Anal. 20, 908–921.

D. Gottlieb and S.A. Orszag (1977), *Numerical Analysis of Spectral Methods: Theory and Applications*, SIAM, Philadelphia.

D. Gottlieb and E. Tadmor (1985), Recovering pointwise values of discontinuous data within spectral accuracy, in *Progress and Supercomputing in Computational Fluid Dynamics*, E.M. Murman and S.S. Abarbanel eds., Birkhäuser Verlag, Boston, pp. 357–375.

P.M. Gresho (1990), On the theory of semi-implicit projection methods for viscous incompressible flow and its implementation via a finite element method that also introduces a nearly consistent mass matrix. Part 1: Theory, Internat. J. Numer. Methods Fluids 11, 587–620.

P.M. Gresho (1991), Some current CFD issues relevant to the incompressible Navier-Stokes equations, Comput. Meth. Appl. Mech. Engrg. 87, 201–252.

P.M. Gresho and S.T. Chan (1990), On the theory of semi-implicit projection methods for viscous incompressible flow and its implementation via a finite element method that also introduces a nearly consistent mass matrix. Part. 2: Implementation, Internat. J. Numer. Methods Fluids 11, 621–659.

P. Grisvard (1976), Behaviour of the solutions of an elliptic boundary value problem in a polygonal or polyhedral domain, in *Numerical Solution of Partial Differential Equations, III*, B. Hubbard ed., Academic Press, New York, pp. 207–274.

P. Grisvard (1985), *Elliptic Problems in Non-Smooth Domains*, Pitman, London.

W.D. Gropp (1992), Parallel computing and domain decomposition, in *Fifth International Symposium on Domain Decomposition Methods for Partial Differential Equations*, D.E. Keyes, T.F. Chan, G. Meurant, J.S. Scroggs and R.G. Voigt eds., SIAM, Philadelphia, pp. 349–361.

W.D. Gropp and D.E.Keyes (1988), Complexity of parallel implementation of domain decomposition techniques for elliptic partial differential equations, SIAM J. Sci. Stat. Comput. 9, 312–326.

H. Guillard and J.-A. Desideri (1990), Iterative methods with spectral preconditioning for elliptic equations, Comput. Meth. Appl. Mech. Engrg. 80, 305–315.

M.D. Gunzburger (1989), *Finite Element Methods for Viscous Incompressible Flows. A Guide to Theory, Practice and Algorithms*, Academic Press, San Diego.

M.D. Gunzburger and J.S. Peterson (1988), Finite-element methods for the stream-function-vorticity equations: boundary-condition treatments and multiply connected domains, SIAM J. Sci. Stat. Comput. 9, 650–668.

B. Guo and I. Babuška (1986a), The h-p version of the finite element method. Part 1: The basic approximation results, Comput. Mech. 1, 21–41.

B. Guo and I. Babuška (1986b), The h-p version of the finite element method. Part 2: General results and applications, Comput. Mech. 1, 203–226.

W. Hackbusch (1981a), Multi-grid convergence theory, in *Multi-Grid Methods*, W. Hackbusch and U. Trottenberg eds., Springer-Verlag, Berlin, pp. 177–219.

W. Hackbusch (1981b), Optimal $H^{p,p/2}$ error estimates for a parabolic Galerkin method, SIAM J. Numer. Anal. 18, 681–692.

W. Hackbusch (1985), *Multi-Grid Methods and Applications*, Springer-Verlag, Berlin.

L.A. Hageman and D.M. Young (1981), *Applied Iterative Methods*, Academic Press, New York.

D.B. Haidvogel and T.A. Zang (1979), The accurate solution of Poisson's equation by expansion in Chebyshev polynomials, J. Comput. Phys. 30, 167–180.

O.H. Hald (1979), Convergence of vortex methods for Euler's equations, II, SIAM J. Numer. Anal. 16, 726–755.

P. Haldenwang, G. Labrosse, S. Abboudi and M.O. Deville (1984), Chebyshev 3-D spectral and 2-D pseudospectral solvers for the Helmholtz equation, J. Comput. Phys. 55, 115–128.

P. Hansbo and A. Szepessy (1990), A velocity-pressure streamline diffusion finite element method for the incompressible Navier-Stokes equations, Comput. Meth. Appl. Mech. Engrg. 84, 175–192.

A. Harten (1984), On a class of high-resolution total-variation-stable finite difference schemes, SIAM J. Numer. Anal. 21, 1–23.

A. Harten (1986), On high-order accurate interpolation for non-oscillatory shock capturing schemes, in *Oscillation Theory, Computations and Methods of Compensated Compactness*, C. Dafermos, J.L. Eriksen, D. Kinderlehrer and M. Slemrod eds., Springer-Verlag, New York, pp. 71–105.

A. Harten, J.M. Hyman and P.D. Lax (1976), On finite-difference approximations and entropy conditions for shocks, Comm. Pure Appl. Math. 29, 297–322.

A. Harten and S. Osher (1987), Uniformly high-order accurate essentially non-oscillatory schemes, I, SIAM J. Numer. Anal. 24, 279–309.

A. Harten, S. Osher, B. Engquist and S. Chakravarthy (1987), Some results on uniformly high-order accurate essentially non-oscillatory schemes, Appl. Numer. Math. 2, 347–377.

P. Hartman (1973), *Ordinary Differential Equations*, John Wiley & Sons, Baltimore.

R. Haverkamp (1984), Eine Aussage zur L^∞-Stabilität und zur genauen Konvergenzordnung der H_0^1-Projektionen, Numer. Math. 44, 393–405.

F.-K. Hecht (1981), Construction d'une base de fonctions $\mathbb{P}_1$ non conforme à divergence nulle dans $\mathbb{R}^3$, R.A.I.R.O. Anal. Numér. 15, 119–150.

B. Heinrich (1987), *Finite Difference Methods on Irregular Networks: a Generalized Approach to Second Order Elliptic Problems*, Birkhäuser Verlag, Basel.

W. Heinrichs (1992), A spectral multigrid method for the Stokes problem in streamfunction formulation, J. Comput. Phys. 102, 310–318.

W. Heinrichs (1993), Spectral multigrid techniques for the Navier-Stokes equations, Comput. Meth. Appl. Mech. Engrg. 106, 297–314.

P.W. Hemker and J.-A. Desideri (1993), Convergence behaviour of defect correction for hyperbolic equations, J. Comput. Appl. Math. 45, 357–365.

M.R. Hestenes (1980), *Conjugate Direction Methods in Optimization*, Springer-Verlag, New York.

M.R. Hestenes and E. Stiefel (1952), Methods of conjugate gradients for solving linear systems, J. Res. Nat. Bur. Stand. 49, 409–436.

J.G. Heywood (1980), The Navier-Stokes equations: on the existence, regularity and decay of solutions, Indiana Univ. Math. J. 29, 639–681.

J.G. Heywood and R. Rannacher (1982), Finite element approximation of the nonstationary Navier-Stokes problem. I. Regularity of solutions and second-order error estimates for spatial discretization, SIAM J. Numer. Anal. 19, 275–311.

J.G. Heywood and R. Rannacher (1986a), Finite element approximation of the nonstationary Navier-Stokes problem. Part II: stability of solutions and error estimates uniform in time, SIAM J. Numer. Anal. 23, 750–777.

J.G. Heywood and R. Rannacher (1986b), An analysis of stability concepts for the Navier-Stokes equations, J. Reine Angew. Math. 372, 1–33.

J.G. Heywood and R. Rannacher (1988), Finite element approximation of the nonstationary Navier-Stokes problem. Part III. Smoothing property and higher order error estimates for spatial discretization, SIAM J. Numer. Anal. 25, 489–512.

J.G. Heywood and R. Rannacher (1990), Finite element approximation of the nonstationary Navier-Stokes problem. Part IV: error analysis for second-order time discretization, SIAM J. Numer. Anal. 27, 353–384.

C. Hirsch (1988), *Numerical Computation of Internal and External Flows. Volume I: Fundamentals of Numerical Discretization*, John Wiley & Sons, Chichester.

L.-W. Ho, Y. Maday, A.T. Patera and E.H. Rønquist (1990), A high-order Lagrangian-decoupling method for the incompressible Navier-Stokes equations, in *Spectral and High Order Methods for Partial Differential Equations*, C. Canuto and A. Quarteroni eds., North-Holland, Amsterdam, pp. 65–90.

E. Hopf (1951), Über die Anfangswertaufgabe für die hydrodynamischen Grundgleichungen, Math. Nachr. 4, 213–231.

M. Huang and V. Thomée (1981), Some convergence estimates for semi-discrete type schemes for time-dependent non-selfadjoint parabolic equations, Math. Comput. 37, 327–346.

M. Huang and V. Thomée (1982), On the backward Euler method for parabolic equations with rough initial data, SIAM J. Numer. Anal. 19, 599–603.

T.J.R. Hughes (1987), Recent progress in the development and understanding of SUPG methods with special reference to the compressible Euler and Navier-Stokes equations, Internat. J. Numer. Methods Fluids 7, 1261–1275.

T.J.R. Hughes and A.N. Brooks (1979), A multidimensional upwind scheme with no crosswind diffusion, in *Finite Elements Methods for Convection Dominated Flows*, T.J.R. Hughes ed., The American Society of Mechanical Engineers, New York, pp. 19–35.

T.J.R. Hughes and A.N. Brooks (1982), A theoretical framework for Petrov-Galerkin methods with discontinuous weighting functions: application to the streamline-upwind procedure, in *Finite Elements in Fluids, Vol. 4*, R.H. Gallagher, D.H. Norrie, J.T. Oden and O.C. Zienkiewicz eds., John Wiley & Sons, Chichester, pp. 47–65.

T.J.R. Hughes and L.P. Franca (1987), A new finite element formulation for computational fluid dynamics: VII. The Stokes problem with various well-posed boundary conditions: symmetric formulations that converge for all velocity/pressure spaces, Comput. Meth. Appl. Mech. Engrg. 65, 85–96.

T.J.R. Hughes, L.P. Franca and M. Balestra (1986), A new finite element formulation for computational fluid dynamics: V. Circumventing the Babuška-Brezzi condition: a stable Petrov-Galerkin formulation of the Stokes problem accomodating equal-order interpolations, Comput. Meth. Appl. Mech. Engrg. 59, 85–99.

T.J.R. Hughes, L.P. Franca and G.M. Hulbert (1989), A new finite element formulation for computational fluid dynamics: VIII. The Galerkin/Least-Squares method for advective-diffusive equations, Comput. Meth. Appl. Mech. Engrg. 73, 173–189.

T.J.R. Hughes, L.P. Franca and M. Mallet (1987), A new finite element formulation for computational fluid dynamics: VI. Convergence analysis of the generalized SUPG formulation for linear time dependent multi-dimensional advective-diffusive systems, Comput. Meth. Appl. Mech. Engrg. 63, 97–112.

T.J.R. Hughes, M. Mallet and A. Mizukami (1986), A new finite element formulation for computational fluid dynamics: II. Beyond SUPG, Comput. Meth. Appl. Mech. Engrg. 54, 341–355.

T.J.R. Hughes and T.E. Tezduyar (1984), Finite element methods for first-order hyperbolic systems with particular emphasis on the compressible Euler equations, Comput. Meth. Appl. Mech. Engrg. 45, 217–284.

G. Ierley, B. Spencer and R. Worthing (1992), Spectral methods in time for a class of parabolic partial differential equations, J. Comput. Phys. 102, 88–97.

T. Ikeda (1983), *Maximum Principle in Finite Element Models for Convection-Diffusion Phenomena*, Kinokuniya/North-Holland, Tokyo/Amsterdam.

V.P. Il'in (1966), Application of alternating direction methods to the solution of quasi-linear equations of parabolic and elliptic types, in *Certain Problems of Numerical and Applied Mathematics*, Nauka, Novosibirsk, pp. 101–114 [Russian].

E. Isaacson and H.B. Keller (1966), *Analysis of Numerical Methods*, John Wiley & Sons, New York.

A. Jameson (1985), Numerical solution of the Euler equations for compressible inviscid fluids, in *Numerical Methods for the Euler Equations of Fluid Dynamics*, F. Angrand, A. Dervieux, J.A. Desideri and R. Glowinski eds., SIAM, Philadelphia, pp. 199–245.

A. Jameson, W. Schmidt and E. Turkel (1981), Numerical solution of the Euler equations by finite volume methods using Runge-Kutta time stepping schemes, AIAA Paper n. 81-1259, San Diego.

P. Jamet (1978), Galerkin-type approximations which are discontinuous in time for parabolic equations in a variable domain, SIAM J. Numer. Anal. 15, 912–928.

F. Jauberteau, C. Rosier and R. Temam (1990), A nonlinear Galerkin method for the Navier-Stokes equations, in *Spectral and High Order Methods for Partial Differential Equations*, C. Canuto and A. Quarteroni eds., North-Holland, Amsterdam, pp. 245–260.

G.M. Johnson (1983), Convergence acceleration of viscous flow computations, NASA Technical Memorandum n. TM 83039.

C. Johnson (1987), *Numerical Solution of Partial Differential Equations by the Finite Element Method*, Cambridge University Press, Cambridge.

C. Johnson, U. Nävert and J. Pitkäranta (1984), Finite element methods for linear hyperbolic problems, Comput. Meth. Appl. Mech. Engrg. 45, 285–312.

C. Johnson and J. Pitkäranta (1986), An analysis of the discontinuous Galerkin method for a scalar hyperbolic equation, Math. Comput. 46, 1–26.

C. Johnson and J. Saranen (1986), Streamline diffusion methods for the incompressible Euler and Navier-Stokes equations, Math. Comput. 47, 1–18.

C. Johnson and A. Szepessy (1987), On the convergence of a finite element method for a nonlinear hyperbolic conservation law, Math. Comput. 49, 427–444.

C. Johnson, A. Szepessy and P. Hansbo (1990), On the convergence of shock-capturing streamline diffusion finite element methods for hyperbolic conservation laws, Math. Comput. 54, 107–129.

C. Johnson and V. Thomée (1981), Error estimates for some mixed finite element methods for parabolic type problems, R.A.I.R.O. Anal. Numér. 15, 41–78.

L.V. Kantorovich (1948), Functional analysis and applied mathematics, Uspekhi Mat. Nauk (N.S.) 3, 89–185 [Russian]. English transl.: Nat. Bur. Stand. Rep. 1509 (1952).

N. Kechkar and D. Silvester (1992), Analysis of locally stabilised mixed finite element methods for the Stokes problem, Math. Comput. 58, 1–10.

R.B. Kellogg (1963), Another alternating direction implicit method, J. SIAM 11, 976–979.

J. Kim and P. Moin (1985), Application of a fractional-step method to incompressible Navier-Stokes equations, J. Comput. Phys. 59, 308–323.

L. Kleiser and U. Schumann (1980), Treatment of incompressibility and boundary conditions 3D numerical spectral simulations of plane channel flows, in *Proceedings of the Third GAMM Conference on Numerical Methods in Fluid Mechanics*, E.H. Hirschel ed., Friedr. Vieweg & Sohn, Braunschweig, pp. 165–173.

P. Klouček and F.S. Rys (1994), Stability of the fractional step ϑ- scheme for the nonstationary Navier-Stokes equations, SIAM J. Numer. Anal. 31, 1312–1335.

B. Koren (1988), Defect correction and multigrid for an efficient and accurate computation of airfoil flows, J. Comput. Phys. 77, 183–206.

M.A. Krasnosel'skiĭ and S.G. Kreĭn (1952), An iteration process with minimal residuals, Mat. Sb. 31 (73), 315–334 [Russian].

S.G. Kreĭn and G.I. Laptev (1968), On the problem of the motion of a viscous fluid in an open vessel, Funkcional. Anal. i Priložen. 2, 40–50 [Russian]. English transl.: Functional Anal. Appl. 2 (1968), 38–47.

H.O. Kreiss and J. Lorenz (1989), *Initial-Boundary Value Problems and the Navier-Stokes Equations*, Academic Press, San Diego.

H.O. Kreiss and J. Oliger (1979), Stability of the Fourier method, SIAM J. Numer. Anal. 16, 421–433.

S.N. Kružkov (1970), First order quasilinear equations in several independent variables, Mat. Sb. 81 (123), 228–255 [Russian]. English transl.: Math. USSR Sb. 10 (1970), 217–243.

N.N. Kuznetsov (1976), Accuracy of some approximate methods for computing the weak solutions of a first order quasi-linear equation, Ž. Vyčisl. Mat i Mat. Fiz. 16, 1489–1502 [Russian]. English transl.: U.S.S.R. Comput. Math. and Math. Phys. 16 (1976), 105–119.

Yu.A. Kuznetsov (1969), On the theory of iteration processes, Dokl. Akad. Nauk S.S.S.R. 184, 274–277 [Russian]. English transl.: Soviet Math. Dokl. 10 (1969), 59–62.

Yu.A. Kuznetsov (1991), Overlapping domain decomposition methods for FE-problems with elliptic singular perturbed operators, in *Fourth International Symposium on Domain Decomposition Methods for Partial Differential Equations*, R. Glowinski, Yu.A. Kuznetsov, G. Meurant, J. Périaux and O.B. Widlund eds. (1991), SIAM, Philadelphia, pp. 223–241.

Yu.A. Kuznetsov (1994), Overlapping domain decomposition methods for parabolic problems, in *Domain Decomposition Methods in Science and Engineering*, A. Quarteroni, J. Périaux, Yu.A. Kuznetsov and O.B. Widlund eds., American Mathematical Society, Providence, pp. 63–69.

O.A. Ladyzhenskaya (1958), Solution "in the large" to the boundary-value problem for the Navier-Stokes system in two space variables, Dokl. Akad. Nauk S.S.S.R. 123, 427–429 [Russian]. English transl.: Soviet Phys. Dokl. 123 (1958), 1128–1131.

O.A. Ladyzhenskaya (1959), Solution "in the large" of the nonstationary boundary value problem for the Navier-Stokes system with two space variables, Comm. Pure Appl. Math. 12, 427–433.

O.A. Ladyzhenskaya (1969), *The Mathematical Theory of Viscous Incompressible Flow*, 2nd ed., Gordon and Breach, New York. Russian ed.: Gosudarstv. Izdat. Fiz.-Mat. Lit., Moscow, 1961.

V. Lakshmikantham and D. Trigiante (1988), *Theory of Difference Equations: Numerical Methods and Applications*, Academic Press, San Diego.

M.-H. Lallemand and B. Koren (1989), Iterative defect correction and multigrid accelerated explicit time-stepping schemes for the steady Euler equations, CWI Report n. NM-R8908, Amsterdam.

M.-H. Lallemand, H. Steve and A. Dervieux (1992), Unstructured multigridding by volume agglomeration: current status, Comput. & Fluids 21, 397–433.

J.D. Lambert (1991), *Numerical Methods for Ordinary Differential Systems. The Initial Value Problem*, John Wiley & Sons, Chichester.

L.D. Landau and E.M. Lifshitz (1959), *Fluid Mechanics*, Pergamon Press, Oxford. Russian 2nd. ed.: Gosudarstv. Izdat. Tehn.-Teor. Lit., Moscow, 1953.

I. Lasiecka (1984), Convergence estimates for semi-discrete approximation of non-selfadjoint parabolic equations, SIAM J. Numer. Anal. 21, 894–909.

P.D. Lax (1973), *Hyperbolic Systems of Conservation Laws and the Mathematical Theory of Shock Waves*, SIAM, Philadelphia.

P.D. Lax and B. Wendroff (1960), Systems of conservation laws, Comm. Pure Appl. Math. 13, 217–237.

G.J. Le Beau and T.E. Tezduyar (1991), Finite element computation of compressible flows with the SUPG formulation, in *Advances in Finite Element Analysis in Fluid Dynamics*, M.N. Dhaubhadel, M.S. Engelman and J.N. Reddy eds., The American Society of Mechanical Engineers, New York, pp. 21–27.

G.J. Le Beau, S.E. Ray, S.K. Aliabadi and T.E. Tezduyar (1993), SUPG finite element computation of compressible flows with the entropy and conservation variables formulations, Comput. Meth. Appl. Mech. Engrg. 104, 397–422.

A. Leonard (1980), Vortex methods for flow simulations, J. Comput. Phys. 37, 289–335.

J. Leray (1934), Essai sur les mouvements plans d'un liquide visqueux que limitent des parois, J. Math. Pures Appl. 13, 331–418.

P. Lesaint and P.-A. Raviart (1974), On a finite element method for solving the neutron transport equation, in *Mathematical Aspects of Finite Elements in Partial Differential Equations*, C. de Boor ed., Academic Press, New York, pp. 89–123.

P. Le Tallec (1992), Neumann-Neumann domain decomposition algorithms for solving 2D elliptic problems with nonmatching grids, East-West J. Numer. Math. 1, 129–146.

R.J. LeVeque (1990), *Numerical Methods for Conservation Laws*, Birkhäuser Verlag, Basel.

J.-L. Lions (1969), *Quelques Méthodes de Résolution des Problèmes aux Limites non Linéaires*, Dunod, Paris.

J.-L. Lions and E. Magenes (1968a), *Problèmes aux Limites non Homogènes et Applications, 1*, Dunod, Paris.

J.-L. Lions and E. Magenes (1968b), *Problèmes aux Limites non Homogènes et Applications, 2*, Dunod, Paris.

J.-L. Lions and G. Prodi (1959), Un théorème d'existence et unicité dans les équations de Navier-Stokes en dimension 2, C.R.A.S. Paris 248, 3519–3521.

P.-L. Lions (1988), On the Schwarz alternating method I, in *First International Symposium on Domain Decomposition Methods for Partial Differential Equations*, R. Glowinski, G.H. Golub, G.A. Meurant and J. Périaux eds., SIAM, Philadelphia, pp. 1–42.

P.-L. Lions (1989), On the Schwarz alternating method II: stochastic interpretation and order properties, in *Domain Decomposition Methods*, T.F. Chan, R. Glowinski, J. Périaux and O.B. Widlund eds., SIAM, Philadelphia, pp. 47–70.

P.-L. Lions (1990), On the Schwarz alternating method III: a variant for nonoverlapping subdomains, in *Third International Symposium on Domain Decomposition Methods for Partial Differential Equations*, T.F. Chan, R. Glowinski, J. Périaux and O.B. Widlund eds., SIAM, Philadelphia, pp. 202–231.

P.-L. Lions and B. Mercier (1979), Splitting algorithms for the sum of two nonlinear operators, SIAM J. Numer. Anal. 16, 964–979.

J.W.H. Liu (1992), The multifrontal method for sparse matrix solution: theory and practice, SIAM Rev. 34, 82–109.

D.G. Luenberger (1984), *Linear and Nonlinear Programming*, 2nd ed., Addison-Wesley, Reading.

M. Luskin and R. Rannacher (1982a), On the smoothing property of the Galerkin method for parabolic equations, SIAM J. Numer. Anal. 19, 93–113.

M. Luskin and R. Rannacher (1982b), On the smoothing property of the Crank-Nicolson scheme, Appl. Analysis 14, 117–135.

R.W. MacCormack and A.J. Paullay (1972), Computational efficiency achieved by time splitting of finite difference operators, AIAA Paper n. 72–154, San Diego.

Y. Maday, D. Meiron, A.T. Patera and E.H. Rønquist (1993), Analysis of iterative methods for the steady and unsteady Stokes problem: application to spectral element discretizations, SIAM J. Sci. Comput. 14, 310–337.

Y. Maday, S.M. Ould Kaber and E. Tadmor (1993), Legendre pseudospectral viscosity method for nonlinear conservation laws, SIAM J. Numer. Anal. 30, 321–342.

Y. Maday and A.T. Patera (1989), Spectral element methods for the incompressible Navier-Stokes equations, in *State of the Art Surveys on Computational Mechanics*, A.K. Noor and J.T. Oden eds., The American Society of Mechanical Engineers, New York, pp. 71–142.

Y. Maday, A.T. Patera and E.H. Rønquist (1990), An operator-integration-factor splitting method for time-dependent problems: application to incompressible fluid flow, J. Sci. Comput. 5, 263–292.

Y. Maday and A. Quarteroni (1981), Legendre and Chebyshev spectral approximations of Burgers' equation, Numer. Math. 37, 321–332.

Y. Maday and A. Quarteroni (1982a), Spectral and pseudo-spectral approximations of the Navier-Stokes equations, SIAM J. Numer. Anal. 19, 761–780.

Y. Maday and A. Quarteroni (1982b), Approximation of Burgers' equation by pseudo-spectral methods, R.A.I.R.O. Anal. Numér. 16, 375–404.

Y. Maday and E. Tadmor (1989), Analysis of spectral vanishing viscosity method for periodic conservation laws, SIAM J. Numer. Anal. 26, 854–870.

J. Mandel, S.F. McCormick and R.E. Bank (1987), Variational multi-grid theory, in *Multi-Grid Methods*, S.F. McCormick ed., SIAM, Philadelphia, pp. 131–177.

T.A. Manteuffel (1979), Shifted incomplete Cholesky factorization, in *Sparse Matrix Proceedings 1978*, I.S. Duff and G.W. Stewart eds., SIAM, Philadelphia, pp. 41–61.

C. Marchioro and M. Pulvirenti (1993), *Mathematical Theory of Incompressible Nonviscous Fluids*, Springer-Verlag, New York.

G.I. Marchuk (1990), Splitting and alternating direction methods, in *Handbook of Numerical Analysis, I*, Ph.G. Ciarlet and J.-L. Lions eds., North-Holland, Amsterdam, pp. 197–462.

G.I. Marchuk and Yu.A. Kuznetsov (1974), Méthodes itératives et fonctionnelles quadratiques, in *Sur les Méthodes Numériques en Sciences Physiques et Économiques*, J.-L. Lions and G.I. Marchuk eds., Dunod, Paris, pp. 1–132.

G.I. Marchuk, Yu.A. Kuznetsov and A.M. Matsokin (1986), Fictitious domain and domain decomposition methods, Soviet J. Numer. Anal. Math. Modelling 1, 3–35.

L.D. Marini and A. Quarteroni (1989), A relaxation procedure for domain decomposition methods using finite elements, Numer. Math. 55, 575–598.

M. Marion and R. Temam (1989), Nonlinear Galerkin methods, SIAM J. Numer. Anal. 26, 1139–1157.

M. Marion and R. Temam (1990), Nonlinear Galerkin methods: the finite element case, Numer. Math. 57, 205–226.

A.M. Matsokin (1988), Norm-preserving prolongations of mesh functions, Soviet J. Numer. Anal. Math. Modelling 3, 137–149.

A.M. Matsokin and S.V. Nepomnyashchikh (1985), A Schwarz alternating method in a subspace, Izv. VUZ Mat. 29 (10), 61–66 [Russian]. English transl.: Soviet Math. (Iz. VUZ) 29 (10) (1985), 78–84.

P.W. McDonald (1971), The computation of transonic flow through two-dimensional gas turbine cascades, ASME Paper n. 71-GT-89, New York.

J.A. Meijerink and H.A. van der Vorst (1981), Guidelines for the usage of incomplete decompositions in solving sets of linear equations as they occur in practical problems, J. Comput. Phys. 44, 134–155.

B. Mercier (1989), *An Introduction to the Numerical Analysis of Spectral Methods*, Springer-Verlag, Berlin.

G. Meurant (1988), Domain decomposition versus block preconditioning, in *First International Symposium on Domain Decomposition Methods for Partial Differential Equations*, R. Glowinski, G.H. Golub, G.A. Meurant and J. Périaux eds., SIAM, Philadelphia, pp. 231–249.

G. Meurant (1991), Numerical experiments with a domain decomposition method for parabolic problems on parallel computers, in *Fourth International Symposium on Domain Decomposition Methods for Partial Differential Equations*, R. Glowinski, Yu.A. Kuznetsov, G. Meurant, J. Périaux and O.B. Widlund eds., SIAM, Philadelphia, pp. 394–408.

A.R. Mitchell and D.F. Griffiths (1980), *The Finite Difference Method in Partial Differential Equations*, John Wiley & Sons, Chichester.

Y. Morchoisne (1979), Resolution of Navier-Stokes equations by a space-time pseudospectral method, Rech. Aérospat., 293–306.

K.W. Morton, A. Priestley and E. Süli (1988), Stability of the Lagrange-Galerkin method with non-exact integration, M^2AN 22, 123–151.

R.D. Moser, P. Moin and A. Leonard (1983), A spectral numerical method for the Navier-Stokes equations with applications to Taylor-Couette flow, J. Comput. Phys. 52, 524–544.

I.P. Natanson (1965), *Constructive Function Theory*, Ungar, New York.

J. Nečas (1962), Sur une méthode pour resoudre les equations aux derivées partielles du type elliptique, voisine de la variationnelle, Ann. Scuola Norm. Sup. Pisa 16, 305–326.

J.-C. Nédélec (1980), Mixed finite elements in $\mathbb{R}^3$, Numer. Math. 35, 315–341.

S.V. Nepomnyaschikh (1992), Decomposition and fictitious domains methods for elliptic boundary value problems, in *Fifth International Symposium on Domain Decomposition Methods for Partial Differential Equations*, D.E. Keyes, T.F. Chan, G. Meurant, J.S. Scroggs and R.G. Voigt eds., SIAM, Philadelphia, pp. 62–72.

R.A. Nicolaides (1982), Existence, uniqueness and approximation for generalized saddle-point problems, SIAM J. Numer. Anal. 19, 349–357.

S.M. Nikol'skiĭ (1975), *Approximation of Functions of Several Variables and Imbedding Theorems*, Springer-Verlag, Berlin. Russian ed.: Nauka, Moscow, 1969.

J.A. Nitsche (1977), L^∞-convergence of finite element approximations, in *Mathematical Aspects of Finite Element Methods*, I. Galligani and E. Magenes eds., Springer-Verlag, Berlin, pp. 261–274.

J.T. Oden (1972), *Finite Elements of Nonlinear Continua*, McGraw-Hill, New York.

J.T. Oden, ed. (1990), *Reliability in Computational Mechanics*, North-Holland, Amsterdam.

J.T. Oden (1991), Finite elements: an introduction, in *Handbook of Numerical Analysis, II*, Ph.G. Ciarlet and J.-L. Lions eds., North-Holland, Amsterdam, pp. 3–12.

J.T. Oden and L. Demkowicz (1991), h-p adaptive finite element methods in computational fluid dynamics, Comput. Meth. Appl. Mech. Engrg. 89, 11–40.

J.T. Oden and J.N. Reddy (1976), *The Mathematical Theory of Finite Elements*, Interscience, New York.

H. Okamoto (1982), On the semi-discrete finite element approximation of the non-stationary Navier-Stokes equation, J. Fac. Sci. Univ. Tokyo Sect. IA 29, 613–652.

D.P. O'Leary (1976), *Hybrid Conjugate Gradient Algorithms*, Ph.D. Thesis, Stanford University, Stanford.

D.P. O'Leary and O. Widlund (1979), Capacitance matrix methods for the Helmholtz equation on general three-dimensional regions, Math. Comput. 33, 849–879.

O.A. Oleĭnik (1957), Discontinuous solutions of non-linear differential equations, Uspekhi Mat. Nauk (N.S.) 12, 3–73 [Russian]. English transl.: Amer. Math. Soc. Transl. (Ser. 2) 26 (1963), 95–172.

S.A. Orszag (1980), Spectral methods for problems in complex geometry, J. Comput. Phys. 37, 70–92.

S.A. Orszag, M. Israeli and M.O. Deville (1986), Boundary conditions for incompressible flows, J. Sci. Comput. 1, 75–111.

J.M. Ortega (1988), *Matrix Theory*, Plenum Press, New York.

S. Osher (1985), Convergence of generalized MUSCL schemes, SIAM J. Numer. Anal. 22, 947–961.

S. Osher and S. Chakravarthy (1986), Very high order accurate TVD schemes, in *Oscillation Theory, Computations and Methods of Compensated Compactness*, C. Dafermos, J.L. Eriksen, D. Kinderlehrer and M. Slemrod eds., Springer-Verlag, New York, pp. 229–274.

J.E. Pasciak (1980), Spectral and pseudospectral methods for advection equations, Math. Comput. 35, 1081–1092.

F. Pasquarelli (1991), Domain decomposition for spectral approximation to Stokes equations via divergence-free functions, Appl. Numer. Math. 8, 493–514.

F. Pasquarelli and A. Quarteroni (1994), Effective spectral approximations to convection-diffusion equations, Comput. Meth. Appl. Mech. Engrg. 116, 39–51.

F. Pasquarelli, A. Quarteroni and G. Sacchi Landriani (1987), Spectral approximation of the Stokes problem by divergence-free functions, J. Sci. Comput. 2, 195–226.

S.V. Patankar (1980), *Numerical Heat Transfer and Fluid Flow*, Hemisphere, Washington.

A.T. Patera (1984), A spectral element method for fluid dynamics: laminar flow in a channel expansion, J. Comput. Phys. 54, 468–488.

A. Pazy (1983), *Semigroups of Linear Operators and Applications to Partial Differential Equations*, Springer-Verlag, New York.

D.W. Peaceman and H.H. Rachford, Jr. (1955), The numerical solution of parabolic and elliptic differential equations, J. SIAM 3, 28–41.

J.B. Perot (1993), An analysis of the fractional step method, J. Comput. Phys. 108, 51–58.

R. Peyret and T.D. Taylor (1983), *Computational Methods for Fluid Flow*, Springer-Verlag, Berlin.

R. Pierre (1988), Simple C^0 approximations for the computations of incompressible flows, Comput. Meth. Appl. Mech. Engrg. 68, 205–227.

R. Pierre (1989), Regularization procedures of mixed finite element approximation of the Stokes problem, Numer. Meth. Partial Differ. Eq. 5, 241–258.

O. Pironneau (1982), On the transport-diffusion algorithm and its applications to the Navier-Stokes equations, Numer. Math. 38, 309–332.

O. Pironneau (1986), Conditions aux limites sur la pression pour les équations de Stokes et de Navier-Stokes, C.R.A.S. Paris Série I 303, 403–406.

O. Pironneau (1988), *Méthodes des Éléments Finis pour les Fluides*, Masson, Paris.

A. Priestley (1993), Evaluating the spatial integrals in the Lagrange-Galerkin method, in *Finite Elements in Fluids. New Trends and Applications*, K. Morgan, E. Oñate, J. Périaux and J. Peraire eds., Pineridge Press, Swansea, pp. 146–155.

G. Prodi (1962), Teoremi di tipo locale per il sistema di Navier-Stokes e stabilità delle soluzioni stazionarie, Rend. Sem. Mat. Univ. Padova 32, 374–397.

W. Proskurowski and O. Widlund (1976), On the numerical solution of Helmholtz's equation by the capacitance matrix method, Math. Comput. 30, 433–468.

L. Quartapelle (1993), *Numerical Solution of the Incompressible Navier-Stokes Equations*, Birkhäuser Verlag, Basel.

L. Quartapelle and M. Napolitano (1986), Integral conditions for the pressure in the computation of incompressible viscous flows, J. Comput. Phys. 62, 340–348.

A. Quarteroni (1991), Domain decomposition and parallel processing for the numerical solution of partial differential equations, Surv. Math. Ind. 1, 75–118.

A. Quarteroni, F. Pasquarelli and A. Valli (1992), Heterogeneous domain decomposition: principles, algorithms, applications, in *Fifth International Symposium on Domain Decomposition Methods for Partial Differential Equations*, D.E. Keyes, T.F. Chan, G. Meurant, J.S. Scroggs and R.G. Voigt eds., SIAM, Philadelphia, pp. 129–150.

A. Quarteroni and G. Sacchi Landriani (1989), Domain decomposition preconditioners for the spectral collocation method, J. Sci. Comput. 3, 45–75.

A. Quarteroni, G. Sacchi Landriani and A. Valli (1991), Coupling of viscous and inviscid Stokes equations via a domain decomposition method for finite elements, Numer. Math. 59, 831–859.

A. Quarteroni and A. Valli (1991a), Theory and applications of Steklov-Poincaré operators for boundary-value problems, in *Applied and Industrial Mathematics*, R. Spigler ed., Kluwer, Dordrecht, pp. 179–203.

A. Quarteroni and A. Valli (1991b), Theory and applications of Steklov-Poincaré operators for boundary-value problems: the heterogeneous operator case, in *Fourth International Symposium on Domain Decomposition Methods for Partial Differential Equations*, R. Glowinski, Yu.A. Kuznetsov, G. Meurant, J. Périaux and O.B. Widlund eds., SIAM, Philadelphia, pp. 58–81.

A. Quarteroni and E. Zampieri (1992), Finite element preconditioning for Legendre spectral collocation approximations to elliptic equations and systems, SIAM J. Numer. Anal. 29, 917–936.

G.D. Raithby (1976), A critical evaluation of upstream differencing applied to problems involving fluid flow, Comput. Meth. Appl. Mech. Engrg. 9, 75–103.

R. Rannacher (1982), Stable finite element solutions to nonlinear parabolic problems of Navier-Stokes type, in *Computing Methods in Applied Sciences and Engineering, V*, R. Glowinski and J.-L. Lions eds., North-Holland, Amsterdam, pp. 301–309.

R. Rannacher (1992), On Chorin's projection method for the incompressible Navier-Stokes equations, in *The Navier-Stokes Equations II: Theory and Numerical Methods*, J.G. Heywood, K. Masuda, R. Rautmann and S.A. Solonnikov eds., Springer-Verlag, Berlin, pp. 167–183.

P.-A. Raviart (1973), The use of numerical integration in finite element methods for solving parabolic equations, in *Topics in Numerical Analysis*, J.J.H. Miller ed., Academic Press, New York, pp. 233–264.

P.-A. Raviart and J.-M. Thomas (1977), A mixed finite element method for second order elliptic problems, in *Mathematical Aspects of Finite Element Methods*, I. Galligani and E. Magenes eds., Springer-Verlag, Berlin, pp. 292–315.

P.-A. Raviart and J.-M. Thomas (1979), Dual finite element models for second order elliptic problems, in *Energy Methods in Finite Element Analysis*, R. Glowinski, E.Y. Rodin and O.C. Zienkiewicz eds., John Wiley & Sons, Chichester, pp. 175–191.

P.-A. Raviart and J.-M. Thomas (1983), *Introduction à l' Analyse Numérique des Équations aux Dérivées Partielles*, Masson, Paris.

S.C. Reddy and L.N. Trefethen (1990), Lax-stability of fully discrete spectral methods via stability regions and pseudo-eigenvalues, in *Spectral and High Order Methods for Partial Differential Equations*, C. Canuto and A. Quarteroni eds., North-Holland, Amsterdam, pp. 147–164.

L. Reinhart (1980), *Sur la Résolution Numerique de Problèmes aux Limites non Linéaires par des Methodes de Continuation*, Thèse de 3ème Cycle, Université Pierre et Marie Curie, Paris.

R.D. Richtmyer and K.W. Morton (1967), *Difference Methods for Initial Value Problems*, 2nd ed., Interscience, New York.

T.J. Rivlin (1974), *The Chebyshev Polynomials*, John Wiley & Sons, New York.

A.W. Rizzi and M.Inouye (1973), Time split finite volume method for three dimensional blunt-body flows, AIAA Journal 11, 1478–1485.

J.E. Roberts and J.-M. Thomas (1991), Mixed and hybrid methods, in *Handbook of Numerical Analysis, II*, Ph.G. Ciarlet and J.-L. Lions eds., North-Holland, Amsterdam, pp. 523–639.

R.T. Rockafellar (1970), *Convex Analysis*, Princeton University Press, Princeton.

P.L. Roe (1984), Generalized formulation of TVD Lax-Wendroff schemes, ICASE Report n. 84–53, Hampton.

P.L. Roe (1985), Some contributions to the modelling of discontinuous flows, in *Large-Scale Computations in Fluid Mechanics. Part 2*, B. E. Engquist, S. Osher and R.C.T. Somerville eds., American Mathematical Society, Providence, pp. 163–193.

E.H. Rønquist (1988), *Optimal Spectral Element Methods for the Unsteady Three Dimensional Incompressible Navier-Stokes Equations*, Ph.D. Thesis, Massachussetts Institute of Technology, Cambridge.

P. Rostand and B. Stoufflet (1988), Finite volume Galerkin methods for viscous gas dynamics, Rapport de Recherche INRIA n. 863, Le Chesnay.

F.-X. Roux (1992), Spectral analysis of the interface operators associated with the preconditioned saddle-point principle domain decomposition method, in *Fifth International Symposium on Domain Decomposition Methods for Partial Differ-*

ential Equations, D.E. Keyes, T.F. Chan, G. Meurant, J.S. Scroggs and R.G. Voigt eds., SIAM, Philadelphia, pp. 73–90.

Y. Saad (1981), Krylov subspace methods for solving large non-symmetric linear systems, Math. Comput. 37, 105–126.

Y. Saad (1989), Krylov subspace methods on supercomputers. Sparse matrix algorithms on supercomputers, SIAM J. Sci. Stat. Comput. 10, 1200–1232.

Y. Saad and M.H. Schultz (1986), GMRES: A generalized minimal residual algorithm for solving non-symmetric linear systems, SIAM J. Sci. Statist. Comput. 7, 856–869.

G. Sacchi Landriani (1987), Convergence of the Kleiser-Schumann method for the Navier-Stokes equations, Calcolo 23, 383–406.

G. Sacchi Landriani and H. Vandeven (1989), Polynomial approximation of divergence-free functions, Math. Comput. 52, 103–130.

A.A. Samarskiĭ and V.B. Andreev (1978), *Méthodes aux Différences pour Équations Elliptiques*, Mir, Moscow. Russian ed.: Nauka, Moscow, 1976.

P.H. Sammon (1982), Convergence estimates for semidiscrete parabolic equation approximations, SIAM J. Numer. Anal. 19, 68–92.

R.L. Sani, P.M. Gresho, R.L. Lee and D.F. Griffiths (1981), The cause and cure (?) of the spurious pressures generated by certain FEM solutions of the incompressible Navier-Stokes equations, I, Internat. J. Numer. Methods Fluids 1, 17–43.

R.L. Sani, P.M. Gresho, R.L. Lee, D.F. Griffiths and M. Engelman (1981), The cause and cure (!) of the spurious pressures generated by certain FEM solutions of the incompressible Navier-Stokes equations, II, Internat. J. Numer. Methods Fluids 1, 171–204.

A.H. Schatz (1974), An observation concerning Ritz-Galerkin methods with indefinite bilinear forms, Math. Comput. 28, 959–962.

A.H. Schatz, V. Thomée and L.B. Wahlbin (1980), Maximum norm stability and error estimates in parabolic finite element equations, Comm. Pure Appl. Math. 33, 265–304.

R. Scholz (1977), L^∞-convergence of saddle-point approximations for second order problems, R.A.I.R.O. Anal. Numér. 11, 209–216.

H.A. Schwarz (1869), Über einige Abbildungsdufgaben, J. Reine Angew. Math. 70, 105–120.

R. Scott (1976), Optimal L^∞ estimates for the finite element method on irregular meshes, Math. Comput. 30, 681–697.

F. Shakib (1988), *Finite Element Analysis of the Compressible Euler and Navier-Stokes Equations*, Ph.D. Thesis, Stanford University, Stanford.

J. Shen (1990), Long time stability and convergence for fully discrete nonlinear Galerkin methods, Appl. Anal. 38, 201–229.

J. Shen (1992), On error estimates of projection methods for Navier-Stokes equations: first order schemes, SIAM J. Numer. Anal. 29, 57–77.

C.-W. Shu (1987), TVB uniformly high-order schemes for conservation laws, Math. Comput. 49, 105–121.

C.-W. Shu and S. Osher (1988), Efficient implementation of essentially non-oscillatory shock capturing schemes, J. Comput. Phys. 77, 439–471.

C.-W. Shu and S. Osher (1989), Efficient implementation of essentially non-oscillatory shock capturing schemes, II, J. Comput. Phys. 83, 32–78.

D. Silvester and A. Wathen (1994), Fast iterative solution of stabilised Stokes systems. Part II: using general block preconditioners, SIAM J. Numer. Anal. 31, 1352–1367.

B.F. Smith (1990), *Domain Decomposition Algorithms for the Partial Differential Equations of Linear Elasticity*, Ph.D. Thesis, New York University, New York.

B.F. Smith (1991), A domain decomposition algorithm for elliptic problems in three dimensions, Numer. Math. 60, 219–234.

J. Smoller (1983), *Shock Waves and Reaction-Diffusion Equations*, Springer-Verlag, New York.

A. Solomonoff and E. Turkel (1986), Global collocation methods for approximation and the solution of partial differential equations, ICASE Report n. 86–60, Hampton.

V.A. Solonnikov and V.E. Ščadilov (1973), On a boundary value problem for a stationary system of Navier-Stokes equations, Trudy Mat. Inst. Steklov. 125, 196–210 [Russian]. English transl.: Proc. Steklov Inst. Math. 125 (1973), 186–199.

P. Sonneveld (1989), CGS, a fast Lanczos-type solver for non-symmetric linear systems, SIAM J. Sci. Stat. Comput. 10, 36–52.

R. Stenberg (1984), Analysis of mixed finite element methods for the Stokes problem: a unified approach, Math. Comput. 42, 9–23.

G. Strang (1968), On the construction and comparison of difference schemes, SIAM J. Numer. Anal. 5, 506–517.

G. Strang and G.J. Fix (1973), *An Analysis of the Finite Element Method*, Prentice-Hall, Englewood Cliffs.

J.C. Strikwerda (1989), *Finite Difference Schemes and Partial Differential Equations*, Wadsworth & Brooks/Cole, Pacific Grove.

K. Stüben and U. Trottenberg (1981), Multi-grid methods: fundamental algorithms, model problem analysis and applications, in *Multi-Grid Methods*, W. Hackbusch and U. Trottenberg eds., Springer-Verlag, Berlin, pp. 1–176.

E. Süli (1988), Convergence and nonlinear stability of the Lagrange-Galerkin method for the Navier-Stokes equations, Numer. Math. 53, 459–483.

E. Süli and A.F. Ware (1991), A spectral method of characteristics for first-order hyperbolic equations, SIAM J. Numer. Anal. 28, 423–445.

E. Süli and A.F. Ware (1992), Analysis of the spectral Lagrange-Galerkin method for the Navier-Stokes equations, in *The Navier-Stokes Equations II: Theory and Numerical Methods*, J.G. Heywood, K. Masuda, R. Rautmann and S.A. Solonnikov eds., Springer-Verlag, Berlin, pp. 184–195.

P.K. Sweby (1984), High resolution schemes using flux limiters for hyperbolic conservation laws, SIAM J. Numer. Anal. 21, 995–1011.

B.A. Szabó (1990), The p- and h-p versions of the finite element method in solid mechanics, in *Spectral and High Order Methods for Partial Differential Equations*, C. Canuto and A. Quarteroni eds., North-Holland, Amsterdam, pp. 185–195.

B.A. Szabó and I. Babuška (1991), *Finite Element Analysis*, John Wiley & Sons, New York.

G. Szegö (1959), *Orthogonal Polynomials*, revised ed., American Mathematical Society, New York.

A. Szepessy (1989), Convergence of a shock-capturing streamline diffusion finite element method for a scalar conservation law in two space dimensions, Math. Comput. 53, 527–545.

A. Szepessy (1991), Convergence of a streamline diffusion finite element method for scalar conservation laws with boundary conditions, M^2AN 25, 749–782.

M. Tabata (1986), A theoretical and computational study of upwind-type finite element methods, in *Pattern and Waves. Qualitative Analysis of Nonlinear Differential Equations*, T. Nishida, M. Mimura and H. Fujii eds., Kinokuniya/North-Holland, Tokyo/Amsterdam, pp. 319–356.

E. Tadmor (1989), The convergence of spectral methods for nonlinear conservation laws, SIAM J. Numer. Anal. 26, 30–44.

H. Tal-Ezer (1986), Spectral methods in time for hyperbolic equations, SIAM J. Numer. Anal. 23, 11–26.

H. Tal-Ezer (1989), Spectral methods in time for parabolic problems, SIAM J. Numer. Anal. 26, 1–11.

C. Taylor and P. Hood (1973), A numerical solution of the Navier-Stokes equations using the finite element technique, Comput. & Fluids 1, 73–100.

R. Temam (1969), Sur l'approximation de la solution des équations de Navier-Stokes par la méthode de pas fractionnaires (II), Arch. Rat. Mech. Anal. 33, 377–385.

R. Temam (1982), Behaviour at time $t = 0$ of the solution of semi-linear evolution equations, J. Diff. Eq. 43, 73–92.

R. Temam (1983), *Navier-Stokes Equations and Nonlinear Functional Analysis*, SIAM, Philadelphia.

R. Temam (1984), *Navier-Stokes Equations. Theory and Numerical Analysis*, 3rd ed., North-Holland, Amsterdam.

R. Temam (1991a), Remark on the pressure boundary condition for the projection method, Theoret. Comput. Fluid Dynamics 3, 181–184.

R. Temam (1991b), Stability analysis of the nonlinear Galerkin method, Math. Comput. 57, 477–505.

T.E. Tezduyar (1992), Stabilized finite element formulations for incompressible flow computations, Adv. Appl. Mech. 28, 1–44.

T.E. Tezduyar, M. Behr, S.K. Aliabadi, S. Mittal and S.E. Ray (1994), A new mixed preconditioning method based on the clustered element-by-element preconditioners, in *Domain Decomposition Methods in Science and Engineering*, A. Quarteroni, J. Périaux, Yu.A. Kuznetsov and O.B. Widlund eds., American Mathematical Society, Providence, pp. 215–222.

T.E. Tezduyar, R. Glowinski and J. Liou (1988), Petrov-Galerkin methods on multiply-connected domains for the vorticity-stream function formulation of the incompressible Navier-Stokes equations, Internat. J. Numer. Methods Fluids 8, 1269–1290.

T.E. Tezduyar and J. Liou (1988), Element-by-element and implicit-explicit finite element formulations for computational fluid dynamics, in *First International Symposium on Domain Decomposition Methods for Partial Differential Equations*, R. Glowinski, G.H. Golub, G.A. Meurant and J. Périaux eds., SIAM, Philadelphia, pp. 281–300.

T.E. Tezduyar, J. Liou and D.K. Ganjoo (1990), Incompressible flow computations based on the vorticity-stream function and velocity-pressure formulations, Comput. & Structures 35, 445–472.

T.E. Tezduyar, S. Mittal, S.E. Ray and R. Shih (1992), Incompressible flow computations with stabilized bilinear and linear equal-order-interpolation velocity-pressure elements, Comput. Meth. Appl. Mech. Engrg. 95, 221–242.

T.E. Tezduyar, S. Mittal and R. Shih (1991), Time accurate incompressible flow computations with quadrilateral velocity-pressure elements, Comput. Meth. Appl. Mech. Engrg. 87, 363–384.

J.-M. Thomas (1977), *Sur l'Analyse Numérique des Méthodes d'Éléments Finis Hybrides et Mixtes*, Thèse d'Etat, Université Pierre et Marie Curie, Paris.

F. Thomasset (1981), *Implementation of Finite Element Methods for Navier-Stokes Equations*, Springer-Verlag, Berlin.

V. Thomée (1984), *Galerkin Finite Element Methods for Parabolic Problems*, Springer-Verlag, Berlin.

M.D. Tidriri (1992), *Couplage d'Approximations et de Modèles de Types Différents dans le Calcul d'Écoulements Externes*, Thèse, Université de Paris-Dauphine, Paris.

E.S. Titi (1990), On approximate inertial manifolds to the Navier-Stokes equations, J. Math. Anal. Appl. 149, 540–557.

J. Todd (1977), *Basic Numerical Mathematics, Vol.2*, Birkhäuser Verlag, Basel.

F. Tomarelli (1984), Regularity theorems and optimal error estimates for linear parabolic Cauchy problems, Numer. Math. 45, 23–50.

M. Vajteršic (1993), *Algorithms for Elliptic Problems. Efficient Sequential and Parallel Solvers*, Kluwer, Dordrecht.

H.A. van der Vorst (1989), High performance preconditioning, SIAM J. Sci. Stat. Comput. 10, 1174–1185.

H.A. van der Vorst (1992), Bi-CGSTAB: a fast and smoothly converging variant of Bi-CG for the solution of non-symmetric linear systems, SIAM J. Sci. Stat. Comput. 13, 631–644.

H. Vandeven (1991), Family of spectral filters for discontinuous problems, J. Sci. Comput. 6, 159–192.

B. van Leer (1974), Towards the ultimate conservative difference scheme II. Monotonicity and conservation combined in a second order scheme, J. Comput. Phys. 14, 361–370.

B. van Leer (1979), Towards the ultimate conservative difference scheme V. A second order sequel to Godunov's method, J. Comput. Phys. 32, 101–136.

B. van Leer (1992), Progress in multi-dimensional upwind differencing, ICASE Report n. 92–43, Hampton.

R.S. Varga (1962), *Matrix Iterative Analysis*, Prentice-Hall, Englewood Cliffs.

R. Verfürth (1984a), A combined conjugate gradient-multigrid algorithm for the numerical solution of the Stokes problem, IMA J. Numer. Anal. 4, 441–455.

R. Verfürth (1984b), Error estimates for a mixed finite elements approximation of the Stokes equations, R.A.I.R.O. Anal. Numér. 18, 175–182.

R. Verfürth (1987), Finite element approximation of incompressible Navier-Stokes equations with slip boundary condition, Numer. Math. 50, 697–721.

R. Verfürth (1989), A posteriori error estimators for the Stokes equations, Numer. Math. 55, 309–325.

R. Vichnevetsky and J.B. Bowles (1982), *Fourier Analysis of Numerical Approximation of Hyperbolic Equations*, SIAM, Philadelphia.

P.K.W. Vinsome (1976), Orthomin, an iterative method for solving sparse sets of simultaneous linear equations, in *Proceedings of the Fourth Symposium on Reservoir Simulation, Society of Petroleum Engineers of AIME, 1976*, pp. 149–159.

R.G. Voigt, D. Gottlieb and M.Y. Hussaini, eds. (1984), *Spectral Methods for Partial Differential Equations*, SIAM, Philadelphia.

E.L. Wachpress (1966), *Iterative Solution of Elliptic Systems*, Prentice-Hall, Englewood Cliffs.

A. Wathen and D. Silvester (1993), Fast iterative solution of stabilised Stokes systems. Part I: using simple diagonal preconditioners, SIAM J. Numer. Anal. 30, 630–649.

W.L. Wendland (1979), *Elliptic Systems in the Plane*, Pitman, London.

M.F. Wheeler (1973), A priori L_2 error estimates for Galerkin approximations to parabolic partial differential equations, SIAM J. Numer. Anal. 10, 723–759.

G.B. Whitham (1974), *Linear and Nonlinear Waves*, John Wiley & Sons, New York.

O.B. Widlund (1978), A Lanczos method for a class of non-symmetric systems of linear equations, SIAM J. Numer Anal. 15, 801–812.

O.B. Widlund (1987), An extension theorem for finite element spaces with three applications, in *Numerical Techniques in Continuum Mechanics*, W. Hackbusch and K. Witsch eds., Friedr. Vieweg & Sohn, Braunschweig, pp. 110–122.

O.B. Widlund (1992), Some Schwarz methods for symmetric and nonsymmetric elliptic problems, in *Fifth International Symposium on Domain Decomposition Methods for Partial Differential Equations*, D.E. Keyes, T.F. Chan, G. Meurant, J.S. Scroggs and R.G. Voigt eds., SIAM, Philadelphia, pp. 19–36.

G. Wittum (1987), Mehrgitterverfahren für die Stokes'sche Gleichung, Z. Angew. Math. Mech. 67, 499–501.

G. Wittum (1988), Multigrid methods for Stokes and Navier-Stokes equations. Transforming smoothers: algorithms and numerical results, Numer. Math. 54, 543–563.

N.N. Yanenko (1971), *The Method of Fractional Steps*, Springer-Verlag, Berlin. Russian ed.: Nauka, Novosibirsk, 1967.

H.C. Yee (1987), Construction of explicit and implicit symmetric TVD schemes and their applications, J. Comput. Phys. 68, 151–179.

H.C. Yee (1989), *A Class of High-Resolution Explicit and Implicit Shock-Capturing Methods*, von Kármán Institute for Fluid Dynamics, Rhode-Saint-Genèse.

K. Yosida (1974), *Functional Analysis*, 4th ed., Springer-Verlag, Berlin.

D.M. Young (1971), *Iterative Solution of Large Linear Systems*, Academic Press, New York.

D.M. Young and K.C. Jea (1980), Generalized conjugate gradient acceleration of non-symmetrizable iterative methods, Linear Algebra Appl. 34 (1980), 159–194.

H. Yserentant (1993), Old and new convergence proofs for multigrid methods, Acta Numer., 285–326.

A. Ženíšek (1990), *Nonlinear Elliptic and Evolution Problems and Their Finite Element Approximations*, Academic Press, London.

T.-X. Zhou and M.-F. Feng (1993), A least squares Petrov-Galerkin finite element method for the stationary Navier-Stokes equations, Math. Comput. 60, 531–543.

O.C. Zienkiewicz (1973), Finite elements: the background story, in *The Mathematics of Finite Elements and Applications*, J.R. Whiteman ed., Academic Press, London, pp. 1–35.

O.C. Zienkiewicz (1977), *The Finite Element Method*, 3rd ed., McGraw-Hill, New York.

O.C. Zienkiewicz and J.Z. Zhu (1987), A simple error estimator and adaptive procedures for practical engineering analysis, Internat. J. Numer. Methods Engrg. 24, 337–357.

M. Zlámal (1974), Finite element methods for parabolic equations, Math. Comput. 28, 393–404.

Subject Index

Druck: STRAUSS OFFSETDRUCK, MÖRLENBACH
Verarbeitung: SCHÄFFER, GRÜNSTADT